MATRICES AND DETERMINOIDS

University of Calcutta
Readership Lectures

MATRICES
AND
DETERMINOIDS

BY

C. E. CULLIS, M.A., Ph.D.

HARDINGE PROFESSOR OF HIGHER MATHEMATICS IN THE UNIVERSITY OF CALCUTTA;
LATE PROFESSOR OF MATHEMATICS IN THE PRESIDENCY COLLEGE, CALCUTTA;
FORMERLY FELLOW OF GONVILLE AND CAIUS COLLEGE, CAMBRIDGE

VOLUME II

Cambridge:
at the University Press
1918

CAMBRIDGE UNIVERSITY PRESS
Cambridge, New York, Melbourne, Madrid, Cape Town,
Singapore, São Paulo, Delhi, Mexico City

Cambridge University Press
The Edinburgh Building, Cambridge CB2 8RU, UK

Published in the United States of America by Cambridge University Press, New York

www.cambridge.org
Information on this title: www.cambridge.org/9781107620834

First published 1918
First paperback edition 2013

A catalogue record for this publication is available from the British Library

ISBN 978-1-107-62083-4 Paperback

PREFACE

THE author's chief aim in writing this book was to give a systematic account of certain applications of matrices, particularly of rectangular matrices as distinguished from square matrices, and thereby to illustrate the very great advantages gained by using them in almost all branches of Mathematics. It originated in a habit of using matrices freely in the solution of problems in Algebra, Geometry and Applied Mathematics, and is based on the very extensive manuscript acquired in doing so. To give a satisfactory answer to the frequently propounded question 'What is a matrix ?', it seemed advisable to commence with some account of the theory. Accordingly the course of Readership Lectures in which this work was first made public was divided into two halves, the first half dealing with the theory, and the second half with the applications. The theoretical portion has been constantly increased, in the first place by abstractions from the applications, and in the second place by incorporating the work of other writers. As a consequence the applications have been driven further back, though they still remain the ultimate object of the book.

The first volume contained the foundations of a Calculus of Matrices in which the operations are addition, subtraction and multiplication, and the result of performing any number of these operations with any rectangular matrices whatever is always a completely determinate matrix. It also contained:

> an account of the properties of the determinoid of a matrix, which becomes the determinant of the matrix in the particular case when the matrix is square;
>
> an account of the solution of matrix equations of the first degree, including as a special case the solution of systems of linear algebraic equations;
>
> a precise statement of the Law of Cancellation of matrix factors in a matrix equation.

If we have an equation $ab=0$ in which a and b are scalar numbers, then, by the Law of Cancellation in Algebra, if either one of the two factors a and

b is not 0, it can be cancelled. The formulation of the corresponding Law of Cancellation when a and b are matrices is very necessary for the applications, and was essential to any advance in the general theory of rectangular matrices. This Law of Cancellation is included as a particular case in the important generalisation of Chapter XV regarding the possible ranks of a matrix product.

The continuation of the work has been greatly hindered by untoward circumstances, above all by the difficulty of obtaining sufficient leisure for the final preparation of the manuscript for the press. On this account, and because of the growth of the manuscript, it has been decided with reluctance to publish as a second volume the first half of that portion of the complete work which deals in greater detail with the Theory of Matrices. Accordingly this second volume contains those parts of the theory which naturally precede any investigation of the special properties of functional matrices, i.e. matrices whose elements are rational integral functions of a finite number of variables. It deals almost exclusively with matrices whose elements are constants, which may be arbitrary parameters, and with those transformations of such matrices which are classed as equigradent. It does not however contain all the properties of such matrices. There remain many properties which it will be more convenient to consider after a preliminary study of functional matrices.

The language used in this volume is frequently geometrical, especially in the later chapters. Any set of matrices which are vertically equivalent to one another are regarded as defining a 'spacelet' which is completely represented by any one of them, usually by one which is undegenerate, the spacelet being a 'point' when the common rank of the matrices is 1. Any property of a matrix which remains unaltered when the matrix is replaced by any matrix vertically equivalent to it, is a property of a spacelet, and conversely all properties of spacelets are properties of matrices. Although these definitions are not geometrical, their geometrical interpretations, which will be fully discussed in a later volume, are quite evident. We therefore speak of 'spacelets' and 'points' from the outset as if they were geometrical concepts, and the chapters dealing with them will serve to lighten subsequent chapters on the geometrical applications of matrices. In the chapter on equigradent transformations it is shown that every such transformation of a matrix whose elements are constants corresponds to a linear transformation of the variables in a bilinear or quadratic algebraic form; and therefore everything in that chapter has an immediate algebraic application. Thus although the present volume is avowedly restricted to the *theory* of matrices, it actually contains a large number of geometrical applications, and it also implicitly contains a large number of algebraic applications to which attention will subsequently be directed.

The geometrical applications which occur can be divided into two classes. The first class contains properties which depend only on the notion of connection, and are invariant in every equigradent (or projective) transformation of the points of space. These include the *rank* of a spacelet, the *paratomy* of two spacelets, and the relations between spacelets represented by the terms *intersection* and *join* and by the terms *incident* and *connected*. The second class contains properties which depend on the notion of orthogonality, and are invariant in every semi-unit transformation of the points of space, i.e. in every equigradent transformation which leaves the absolute quadric unaltered. These include the *extravagance* of a spacelet, the *orthotomy* and *cross rank* of two spacelets, and the relations between spacelets represented by the terms *core* and *plenum* and by the terms *orthogonal* and *normal*.

It is often desirable to know what a general theorem concerning matrices becomes in the special case in which the matrices are real. When the theorem is one involving only rational operations on the elements of the matrices, all reference to the special case is rendered unnecessary by enunciating the general theorem for matrices whose elements lie in any domain of rationality Ω; for the special case is then obtained by simply taking Ω to be the domain of all real numbers. Since however the real domain has special properties not possessed by other domains, there are special properties of real matrices which cannot be obtained in this way.

In the figures which are given in some of the articles, spacelets and their intersections are represented by areas. Such representations, though necessarily very imperfect, can be used in simple cases as aids to the imagination. It should particularly be noted that a shaded area always represents a completely extravagant spacelet, i.e. one which, being orthogonal with itself, is a generating spacelet of the absolute quadric.

The references to Vol. II contained in Vol. I have been vitiated by the alterations and re-arrangements of the text which have been made since Vol. I was published; but a use of the Index will probably remove any inconvenience caused by this.

As this work has been built up on an independent plan and based chiefly on applications, it is for the most part not easy to ascribe definite sources to the various articles or the suggestions for them. The following is a list of the books which have had most influence on the work as a whole:

Bôcher's *Introduction to Higher Algebra*,

Heffter and Koehler's *Lehrbuch der Analytischen Geometrie*,

Muth's *Elementarteiler*,

Netto's *Vorlesungen über Algebra*,

Veronese's *Fondamenti di geometria a piu dimensioni*,

Whitehead's *Universal Algebra*.

My indebtedness to these and other writers will be more easily recognised in those portions of the work, occurring chiefly in Vol. III, which are interpolations in the original scheme. Amongst the few articles in the present volume which admit of more detailed references may be mentioned § 120 which was written after reading the appendix in Heffter and Koehler's *Analytische Geometrie*, and § 159 which was written after reading a paper by Schläfli in *Crelle's Journal* for 1866 to which a reference is given in Muth's *Elementarteiler*. The addition of Appendix B was suggested by reading a paper by Mr Haripada Datta in Vol. XXXIV of the *Proceedings of the Edinburgh Mathematical Society*.

A few remarks are added concerning the contents of the individual chapters.

Chapter XII (the first chapter of the present volume) contains the notations for compound and compartite matrices, definitions of the primaries of a minor determinant and of primary superdeterminants and primary subdeterminants, a description of the elimination of a variable from a system of inequalities, and the determination of the possible ranks of a matrix containing a given minor matrix. The notations for compound matrices and their determinoids are the complete generalisations of the notations used in Vol. I.

Chapter XIII deals with relations between the elements and minor determinants of a matrix. Starting with the determination of the connections between the short rows of an undegenerate matrix, we are led to all the most useful relations, and these are finally seen to be all particular cases of, or immediately deducible from, the fundamental identity of § 116. A brief review of the relations and the reasons for their utility is given in Appendix A; and other summaries will be found in the Index under the headings Relations, Standard identities, and Standard equations. Those of the relations which are identities in the elements are of course applicable to functional matrices.

Chapter XIV gives an account of some special properties of square matrices. The earlier articles deal with the properties of two co-joint complete matrices of the minor determinants of a square matrix, and the later articles with symmetric and skew-symmetric matrices. Appendix B should be read in conjunction with the articles on skew-symmetric matrices.

Chapter XV deals with the possible ranks of the product matrix and the factor matrices in any matrix product. The theorem of § 133 and the final results of §§ 135 and 137 constitute the complete generalisation of the Law of Cancellation for matrices. In proving the latter results an indication is given of methods of determining all solutions of any matrix equation of the form $X_1X_2 \dots X_n = C$. The concluding articles deal with the equivalences of matrices, on which the definition of a spacelet is based, and with the joins, intersections and connections of matrices and spacelets.

Chapter XVI deals with equigradent transformations of matrices, and with the reduction of a matrix of given rank r whose elements are constants to standard forms by equigradent transformations. The simplest standard form is a similar matrix which is conventionally equal to the unit matrix $[1]_r^r$; and every matrix of rank r with constant elements can be derived from the unit matrix $[1]_r^r$ by an equigradent transformation. The reductions of symmetric and skew-symmetric matrices by symmetric equigradent transformations receive special attention.

Chapter XVII deals with the solution of matrix equations of the second degree. One of the most important results obtained in this chapter is the general formula for all solutions of any assigned rank ρ of the symmetric equation $\overline{x}_m^{\,s} [x]_s^m = \overline{a}_m^{\,r} [a]_r^m$ in which $[a]_r^m$ is a given matrix of rank r. The general theory of extravagant matrices is largely based on this result, which leads at once to the reductions of the next chapter.

Chapter XVIII deals in the first place with the extravagances of any matrix whose elements are constants, and with certain special kinds of equigradent transformations, a review of which is given in Appendix C. The result of which most use is made is the reduction of a matrix whose elements are constants to a standard form by a unilaterally semi-unit equigradent transformation. This reduction re-appears in the reduction of a matrix to an equivalent undegenerate matrix whose long rows are mutually orthogonal, or to one which is the join of a core and a semi-unit matrix; in the corresponding representations of a spacelet as a join of mutually orthogonal unconnected points; and in the discussion of the properties of mutually normal undegenerate matrices. Further it enables us to complete the discussion of the unconnected mutually orthogonal solutions of any system of homogeneous linear algebraic equations, which was left unfinished in Chapter XI.

Chapter XVIII deals in the second place with the extravagances of spacelets and with semi-unit transformations of the points of space. The extravagance of a spacelet (or the degree of its orthogonality with itself) is that property of it which is next in importance to its rank. It is invariant in every semi-unit transformation of the points of space, and can be interpreted as being the rank of contact of the spacelet with the absolute quadric. A spacelet which has the greatest extravagance consistent with its rank is either completely extravagant or plenarily extravagant. A completely extravagant spacelet is orthogonal with itself, and is therefore a generating spacelet of the absolute quadric; a plenarily extravagant spacelet contains all points orthogonal with itself. With every spacelet is associated a completely extravagant spacelet, called its *core*, which is the locus of all points which lie in the given spacelet and are orthogonal with it, i.e. the locus of the points in which the given spacelet touches the absolute quadric; and a plenarily extravagant spacelet, called its *plenum*, which is the smallest spacelet containing the given spacelet and all points orthogonal with it. A sharp distinction is drawn between mutually orthogonal spacelets and mutually normal spacelets. A given spacelet has one and only one normal, whereas an indefinite number of spacelets are orthogonal with it.

Chapter XIX deals chiefly with the mutual orthotomy of two spacelets, or the degree of their mutual orthogonality. The most interesting results in it are those relating to the greatest possible orthotomy of two spacelets. It is shown that the mutual orthotomy of two arbitrary spacelets of given ranks is greatest when each of them is incident with the normal to the other, i.e. in that one of the two following mutually exclusive cases which is possible:

(1) when the two spacelets are mutually orthogonal; this being the case when their complete intersection is a completely extravagant spacelet ω_π, and the spacelets are the joins of ω_π with two mutually orthogonal non-intersecting spacelets lying in the plenum of ω_π, i.e. orthogonal with ω_π;

(2) when the normals to the two spacelets are mutually orthogonal; this being the case when the complete intersection of their normals is a completely extravagant spacelet ω_π, and the normals are the joins of ω_π with two mutually orthogonal non-intersecting spacelets lying in the plenum of ω_π, i.e. orthogonal with ω_π.

Further it is shown that the mutual orthotomy of two spacelets of given ranks which have a given complete intersection ω_p with core ω_π is greatest in that one of the two following mutually exclusive cases which is possible:

(1) when the two spacelets lie in the plenum of ω_p and are the joins of ω_p with two mutually orthogonal non-intersecting spacelets orthogonal with ω_p;

(2) when the normals to the two spacelets are mutually orthogonal; this being the case when the complete intersection ω_κ of their normals (whose rank κ is known) lies in ω_π, and the normals are the joins of ω_κ with two mutually orthogonal non-intersecting spacelets orthogonal with ω_p (whose join is necessarily complementary to ω_κ in the normal to ω_p).

The corresponding simpler results for real spacelets are also given. Another noteworthy result is the independence of the extravagances of two spacelets of given ranks which have a given complete intersection. All the theorems of Chapter XIX can be applied to common metrical space Ω_{n+1} of n dimensions when we define the paratomy and orthotomy of two spacelets of Ω_{n+1} to be those of their infinite sub-spaces, i.e. those of their intersections with the (homogeneous) infinite sub-space ω_n of Ω_{n+1}.

I owe many thanks to the authorities of the University of Calcutta who have generously undertaken the publishing of this volume, and have now with the sanction of the Governments of Bengal and India selected me as Hardinge Professor of Mathematics in the University. In consequence of the additional leisure thus secured to me from this time it is hoped that there will be no long interval before the appearance of the third volume, completing the theory of matrices and clearing the way for the applications. My special gratitude is due to Sir Asutosh Mukhopadhyay for his stimulating interest and encouragement.

Finally I desire to acknowledge my indebtedness to the officials and staff of the Cambridge University Press for the very great care bestowed on the printing.

C. E. CULLIS.

CALCUTTA,
February, 1918.

CONTENTS

CHAPTER XII

COMPOUND MATRICES

§§		PAGES
98–99.	Compound matrices; constituent matrices. Special notations for scalar, quasi-scalar, zero and one-rowed matrices; matrix one of whose orders is zero. Multiplication of compound matrices	1–6
100	Compartite matrices: parts; compartite matrix in standard form; successive parts; the rank of a compartite matrix is the sum of the ranks of its parts; special notations for a compartite matrix whose parts are quasi-scalar matrices	6–8
101.	The conjugate reciprocals and inverses of certain matrices. Some special square matrices	8–13
102.	The primaries of a minor determinant; horizontal and vertical primaries; number of primaries; notation for primaries. Primary subdeterminants and primary superdeterminants	13–18
103.	Elimination of a variable from a system of inequalities	18–19
104–107.	Possible ranks of a matrix containing a given minor matrix; of a minor of a matrix whose rank is given; rank of a matrix which contains an undegenerate simple minor formed by the addition of 0's to an undegenerate square matrix. Diagonal minors. Possible ranks of a symmetric matrix containing a given diagonal minor; of a diagonal minor of a symmetric matrix whose rank is given	19–36

CHAPTER XIII

RELATIONS BETWEEN THE ELEMENTS AND MINOR DETERMINANTS OF A MATRIX

108.	Connections between the short rows of an undegenerate matrix: identities which serve to express every element of the matrix as a rational function of degree 1 of any one regular simple minor determinant Δ and the elements and primaries of Δ; the identities for a matrix whose orders differ by 1. Connections between the rows of a degenerate matrix	37–46

§§ PAGES

109. Relations between simple minor determinants: identities which serve to express every simple minor determinant D of an undegenerate matrix as a rational function of degree 1 of any one regular simple minor determinant Δ and the primaries of Δ; the standard identities; ascription of arbitrary values to the elements and primaries of any regular simple minor determinant; utility of these identities; determinants of the primaries of a simple minor determinant. Corresponding identities for superior simple minor determinoids 46–66

110. *Sylvester's* identities: identities which serve to express every superdeterminant D of a regular minor determinant Δ as a rational function of degree 1 of Δ and the primary superdeterminants of Δ; determinants of primary superdeterminants . . . 66–70

111. Identities which serve to express every subdeterminant D of a regular minor determinant Δ as a rational function of degree 1 of Δ and the primary subdeterminants of Δ; determinants of primary subdeterminants 70–73

112. Relations between simple minor determinants: identities which give the sums of terms formed from a product ΔD of two simple minor determinants by replacing s rows of Δ by rows of D, and the s rows taken from D by s rows of Δ 73–75

113. Equivalence of two similar undegenerate matrices: two such matrices are mutually equivalent when and only when their correspondingly formed simple minor determinants are proportional; sign of equivalence 75–78

Spacelets of homogeneous space represented by undegenerate matrices 78–80

114. Criteria for the equivalence of two systems of linear algebraic equations. Mutually orthogonal and mutually normal undegenerate matrices and spacelets 80–85

115. Relations between the elements of a matrix of rank r: equations which serve to express every element as a rational function of degree 1 of any one regular minor determinant Δ of order r and the elements and primaries of Δ 85–89

116. Identical relations between the elements of any matrix: the fundamental identity which serves (*see* Appendix A) to express every element of the matrix as a rational function of any one regular minor determinant Δ and the elements, primaries and primary superdeterminants of Δ; deduction of all previously obtained relations from the fundamental identity; factorisation of a matrix. Necessary and sufficient conditions that a matrix may have rank r. Regular subdeterminants and superdeterminants of a regular minor determinant; standard arrangement of the rows of a matrix. Rank and other properties of a matrix of primary superdeterminants 90–98

117. Relations between the minor determinants of order r of a matrix whose rank is r: equations which serve to express every minor determinant of order r as a rational function of degree 1 of any one regular minor determinant Δ of order r and the primaries of Δ 98–106

§§ PAGES

117 *a*. APPENDIX A. Utility of the relations obtained in Chapter XIII: ascription of arbitrary values to the elements, primaries and primary superdeterminants of any regular minor determinant Δ; if the order of Δ is r, and if the matrix of its primary superdeterminants has rank x, then the matrix has rank $r+x$. Summary of the relations of Chapter XIII 515–520

CHAPTER XIV

SOME PROPERTIES OF SQUARE MATRICES

118. Properties of a product of square matrices: recapitulation . . 107–108

119. Co-joint complete matrices of the minor determinants of a square matrix: definition; fundamental property; a square matrix and its reciprocal are two co-joint matrices. Co-joint matrix of a complete matrix of the minor determinants of a rectangular matrix 108–112

120. Determinant of a complete matrix of the minor determinants of a square matrix. Reciprocals of two co-joint matrices . . . 112–114

121–122. Relations between any two anti-correspondent minor determinants, any two anti-correspondent matrices of the minor determinants, and any two corresponding complete matrices of the minor determinants of two co-joint complete matrices of the minor determinants of a square matrix. Rank of a complete matrix of the minor determinants of a rectangular matrix . . . 114–123

123. Expansions of certain bordered determinants in terms of the simple minor determinants of the bordering rows 123–127

124. Properties of the reciprocal of a square matrix: determinant and reciprocal of the reciprocal matrix; rank of the reciprocal matrix; relations between any two anti-correspondent minor determinants, any two anti-correspondent matrices of the minor determinants, and any two corresponding complete matrices of the minor determinants of a square matrix and its reciprocal; reciprocal of any derangement of a square matrix . . . 127–133

125. Properties of a symmetric matrix of rank 1: the diagonal elements do not all vanish; if the matrix is real, the non-vanishing diagonal elements all have the same sign; expression of the matrix as a product of two mutually conjugate one-rowed matrices 133–135

126. Some general properties of symmetric matrices: properties of diagonal minor determinants; rank determined by diagonal minor determinants; if the matrix has rank r, it has a regular diagonal minor determinant of order r; standard arrangement of the rows 135–141

127. Some general properties of skew-symmetric matrices: diagonal elements are all 0's; properties of diagonal minor determinants; rank determined by diagonal minor determinants, is always even; if the matrix has rank r, it has a regular diagonal minor determinant of order r; standard arrangement of the rows . 141–145

§§ PAGES

128 The symmetric matrix of order 3: notation for minor determinants 145–146

129. The symmetric matrix of order 4: notation for minor determinants; reciprocals of square minors of order 3 of the matrix and its reciprocal 147–156

130. Identical relations between the elements of a square matrix: the fundamental identity; properties of a matrix of the super-determinants of a minor determinant; applications of the fundamental identity to symmetric matrices of orders m, 3 and 4 157–164

130 *a.* APPENDIX B. The Pfaffian of a skew-symmetric matrix of even order: definition; Pfaffian co-factors; expansions of a Pfaffian; a skew-symmetric matrix of even order is the square of its Pfaffian; bordered skew-symmetric determinants expressed as products of two Pfaffians; symmetrically bordered skew-symmetric determinants; reciprocals of skew-symmetric matrices . 521–530

CHAPTER XV

RANKS OF MATRIX PRODUCTS AND MATRIX FACTORS

131. Rank of a matrix product in which one of the extreme factor matrices has rank equal to its passivity; applicability of the results to a product of functional matrices 165–167

132. Possible ranks of the solutions of a given matrix equation of the first degree: the equations $AX=C$, $XB=C$, $AXB=C$; the symmetric equation $A'XA=C$. Formulae for the general solutions of these equations 168–177

133. Restrictions on the rank of any matrix product: the rank of the product matrix cannot exceed the rank of any factor matrix; the sum of the rank and the passivities of the product cannot be less than the sum of the ranks of the factor matrices . . . 177–179

134–135. Possible ranks of the product matrix and the factor matrices in a matrix product: products of two matrices; products of three matrices; any matrix product 179–194

136–137. Possible ranks of the product matrix and the factor matrices in a symmetric matrix product; symmetric products of two matrices; symmetric products of three matrices; any symmetric product 194–205

138. Equivalences of two matrices, and equivalent systems of linear algebraic equations: conditions for the horizontal or vertical equivalence of two matrices; equivalences of two similar square matrices; sign of equivalence 205–208

Spacelets represented by matrices which are not necessarily undegenerate 208–209

139. Joins and intersections of spacelets in homogeneous space of $n-1$ dimensions or rank n: join and complete intersection of two spacelets; formulae for two spacelets, their join and their intersection; relation between the ranks of their join and intersection; possible ranks of the join and intersection of two spacelets of given ranks which are otherwise arbitrary, which both lie in or both contain a given spacelet, which both lie in one given spacelet and both contain another given spacelet;

§§ PAGES

spacelets having a given complete intersection with a given spacelet; mutually complementary spacelets; mutually incident spacelets 209–219

Join and complete intersection of any number of spacelets; spacelets which do not intersect either of two given spacelets; spacelets restricted to be real 219–222

Joins and intersections of matrices 222–223

140. Connections between matrices; connections between spacelets; unconnected spacelets, no one of them intersects the join of the others; spacelets which do not intersect one given spacelet and have a given complete intersection with another . . . 223–227

CHAPTER XVI

EQUIGRADENT TRANSFORMATIONS OF A MATRIX WHOSE ELEMENTS ARE CONSTANTS

EQUIGRADENT TRANSFORMATIONS OF ANY MATRIX.

141. Definitions of an equigradent transformation and its inverse; equigradent matrices, have the same rank; symmetric equigradent transformations; symmetrically equigradent matrices; equigradent transformations between two similar matrices; composition of equigradent transformations, resultant equigradent transformation; elementary equigradent transformations; derangements, unitary transformations, quasi-scalar and scalar transformations; equigradent transformations of a compartite matrix and its parts; correspondences between equigradent transformations of a matrix with constant elements and linear transformations of bilinear and quadratic algebraic forms; completion of any two mutually inverse matrices; conversion of any equigradent transformation and its inverse into two mutually inverse equigradent transformations of similar matrices; some general properties of equigradent transformations . . 228–241

SOME SPECIAL EQUIGRADENT TRANSFORMATIONS IN Ω OF ANY MATRIX IN Ω WHOSE ELEMENTS ARE CONSTANTS.

142. Unitary transformations converting the matrix into a bipartite matrix one of whose parts is a given non-zero element or a given undegenerate square minor; special cases . . . 241–252

143. A more general unitary transformation converting the matrix into a compartite matrix of standard form whose non-zero parts are all undegenerate square matrices 252–260

144. A corresponding non-unitary equigradent transformation . . 260–264

145. The corresponding transformations of any symmetric or skew-symmetric matrix in Ω whose elements are constants: conversion into a bipartite or compartite matrix by symmetric unitary (or corresponding non-unitary) equigradent transformations in Ω; special cases 264–271

§§ PAGES

REDUCTION OF ANY MATRIX WHOSE ELEMENTS ARE CONSTANTS TO STANDARD FORMS BY EQUIGRADENT TRANSFORMATIONS.

146. Reduction of any matrix in Ω by transformations in Ω; reduction to a standard form by unrestricted equigradent transformations in Ω; reduction to a standard form by derangements and unitary (or corresponding non-unitary) equigradent transformations in Ω; two matrices are equigradent when and only when they have the same rank 272–278

Equigradent transformations between two similar undegenerate quasi-scalar or square matrices whose determinants are equal in value; reduction of a square matrix whose determinant is ± 1 278–280

147. Reduction of any symmetric matrix in Ω by symmetric transformations: reduction to a standard form by derangements and unitary (or corresponding non-unitary) transformations in Ω; reduction to a standard form by unrestricted symmetric equigradent transformations (not in Ω); two symmetric matrices are symmetrically equigradent when and only when they have the same rank 280–295

148. The signants and signature of a real symmetric matrix: defined; the positive and negative signants remain invariant in all real symmetric equigradent transformations of the matrix; reduction of a real symmetric matrix to a standard form by such transformations; there exists a real symmetric equigradent transformation between two real symmetric matrices when and only when they are equisignant; the positive and negative signants of a real symmetric compartite matrix are respectively the sums of the positive and negative signants of its parts . . . 295–300

149. Definite and indefinite real symmetric matrices: defined; general formula for all definite matrices; all non-vanishing diagonal minor determinants of order s of a definite matrix have the same sign; every symmetrically formed complete matrix of the minor determinants of a definite matrix is definite; roots of a real and definite quadratic form; essentially positive and essentially negative real quadratic forms; semi-definite matrices 300–303

150. Reduction of any skew-symmetric matrix in Ω by symmetric transformations in Ω: reduction to a standard form by derangements and unitary (or corresponding non-unitary) equigradent transformations in Ω; reduction to a standard form by unrestricted symmetric equigradent transformations in Ω; two skew-symmetric matrices are symmetrically equigradent when and only when they have the same rank 303–308

CHAPTER XVII

SOME MATRIX EQUATIONS OF THE SECOND DEGREE

§§ PAGES

GENERAL EQUATIONS.

151. Matrix equations of the second degree : definitions and descriptions 309–310

152–153. $XY=AB$: Determination of all solutions of the equations

$$\overline{\underline{x}}_m [y]_n = \overline{\underline{a}}_m [b]_n, \quad [x]_m^r [y]_r^n = [a]_m^r [b]_r^n, \quad [x]_m^s [y]_s^n = [a]_m^r [b]_r^n,$$

the given factors on the right in the last two equations having rank r; general solutions of the first two equations . . . 310–315

154. $XY=C$: Determination of all solutions of the equations

$$\overline{\underline{x}}_m [y]_n = [c]_m^n, \quad [x]_m^r [y]_r^n = [c]_m^n, \quad [x]_m^s [y]_s^n = [c]_m^n,$$

the given matrix $[c]_m^n$ having rank 0 or 1 in the first equation, and rank r in the second and third equations; general solutions of the first two equations 316–320

SYMMETRIC EQUATIONS.

155. $X'X=I$, where I is a unit matrix: Semi-unit matrices; determination of all semi-unit matrices, of all real semi-unit matrices; enlargement of a given semi-unit matrix (or real semi-unit matrix) by the addition of long rows (or real long rows); number of arbitrary parameters in a general semi-unit matrix; every derangement and every long minor of a semi-unit matrix is a semi-unit matrix. Square semi-unit matrices; derangements of a unit matrix; compound square semi-unit matrix with four real and purely imaginary constituents, all constituents are square; most general square semi-unit matrices of orders 2 and 3 320–330

Rotations of a rigid body represented by square semi-unit matrices of order 3; theorems on rotations about a fixed point . . 330–332

156–157. $X'X=A'A$: Determination of all solutions of the symmetric equations

$$\overline{\underline{x}}_m [x]_m = \overline{\underline{a}}_m [a]_m, \quad \overline{\underline{x}}_m^r [x]_r^m = \overline{\underline{a}}_m^r [a]_r^m, \quad \overline{\underline{x}}_m^s [x]_s^m = \overline{\underline{a}}_m^r [a]_r^m,$$

the given factors on the right in the last two equations having rank r; general solutions of the first two equations; formulae giving all solutions of the third equation of any possible rank ρ; formulae giving all solutions of rank ρ of the equation

$$\overline{\underline{x}}_m^s [x]_s^m = \begin{bmatrix} 1, & 0 \\ 0, & 0 \end{bmatrix}_{r,\, m-r}^{r,\, m-r} \quad . \quad . \quad . \quad . \quad . \quad .$$ 333–344

158–159. $X'X=0$: General formulae for all solutions of rank ρ of the equation $\overline{\underline{x}}_m^s [x]_s^m = 0$; standard formula for all undegenerate solutions of rank r of the equation $[x]_r^n \overline{\underline{x}}_n^r = 0$. 344–352

§§ PAGES

160. $X'X=C$: Determination of all solutions of the symmetric equations

$$\underset{m}{\overline{\underline{x}}}\,[x]_m=[c]_m^m,\quad \underset{m}{\overline{\underline{x}}}^{\,r}[x]_r^m=[c]_m^m,\quad \underset{m}{\overline{\underline{x}}}^{\,s}[x]_s^m=[c]_m^m,$$

the given symmetric matrix $[c]_m^m$ having rank 0 or 1 in the first equation, and rank r in the second and third equations; general solution of the first equation; particular solutions and general solution of the second equation; semi-real solutions of the second equation when $[c]_m^m$ is real. Special symmetric equations of the forms $X'X=A'BA$, $X'BX=A'A$, $X'BX=A'CA$ . . 352–363

161. $X'AX=C$: Determination of all solutions of any symmetric equation of this form 363–365

SPECIAL EQUATIONS

162. Expressions for a symmetric matrix of order 2 as a product of two square factors 365–369

163. Some special equations of the form $[x]_m^2[y]_2^m=[a]_m^2[b]_2^m$ . . 369–373

164. Expressions for a symmetric matrix whose rank does not exceed 2 as a product of two 2-rowed factors; applications to symmetric matrices of orders 3 and 4 373–377

CHAPTER XVIII

THE EXTRAVAGANCES OF MATRICES AND OF SPACELETS IN HOMOGENEOUS SPACE

165. The degeneracy of any matrix: defined 378

The extravagance of an undegenerate matrix: possible values; general formulae for an undegenerate matrix whose orders and extravagance are given; condition that two similar undegenerate matrices may have the same extravagance; non-extravagant, completely extravagant, and plenarily extravagant undegenerate matrices; general formulae for these; possible ranks and extravagances of undegenerate matrices connected with a given undegenerate matrix 378–385

The horizontal and vertical extravagances of any matrix: of a matrix of rank r expressed as a product of passivity r; possible values; horizontally (or vertically) equivalent matrices have the same horizontal (or vertical) extravagance; general formulae for a matrix of which the rank and one or both of the extravagances are given; condition that two matrices may have the same rank and the same horizontal (or vertical) extravagance; matrices whose extravagances are both zero 385–393

The extravagance of a symmetric matrix: possible values; general formulae for a symmetric matrix whose rank and extravagance are given; ranks of the powers of a non-extravagant symmetric matrix 393–395

§§ PAGES

166. Minor determinants of the matrices $[1, a]_m^{m,n}$ and $[a, 1]_m^{n,m}$: simple minor determinants; minor determinants of order s . . . 395–397

167. Properties of two mutually normal undegenerate matrices: they have the same extravagance; the corranged simple minor determinants of either one are proportional to the anti-correspondent affected simple minor determinants of the other; general formulae for two mutually normal undegenerate matrices; possible values of their ranks and common extravagance; one of them is completely extravagant when and only when the other is plenarily extravagant. Possible ranks and extravagances of two mutually orthogonal undegenerate matrices. Definitions of horizontally (or vertically) orthogonal and normal matrices; all matrices horizontally (or vertically) normal to a given matrix are horizontally (or vertically) equivalent; possible ranks and extravagances of any two mutually orthogonal matrices . 398–411

168. Reduction of an undegenerate matrix (of extravagance ρ) to an equivalent similar undegenerate matrix whose long rows are mutually orthogonal; number of extravagant long rows in the equivalent matrix (is equal to ρ); non-extravagant rows at unit intensity; reduction of a non-extravagant (or real) matrix to an equivalent semi-unit (or real semi-unit) matrix 412–415

169. The cores of an undegenerate matrix (of extravagance ρ): defined; are mutually equivalent completely extravagant matrices (of rank ρ); two mutually normal undegenerate matrices have the same cores, these being their complete intersections; determination of the cores 415–422

170. The extravagance, core and plenum of a spacelet in homogeneous space of $n-1$ dimensions or rank n.

Preliminary remarks: Spacelets, their joins, intersections and incidences; mutually complementary spacelets; unconnected spacelets; orthogonal and normal spacelets; general formula for a spacelet which lies in the join of two given spacelets and has a given complete intersection with one of them . . . 422–425

The extravagance of a spacelet: defined, is the degree of its orthogonality with itself; possible values; non-extravagant, completely extravagant and plenarily extravagant spacelets; mutually orthogonal non-extravagant spacelets are unconnected; two spacelets have the same rank and the same extravagance when and only when each is convertible into the other by a semi-unit transformation 425–427

The core of a spacelet (of extravagance ρ): is a completely extravagant spacelet (of rank ρ) which is the locus of all points which lie in the spacelet and are orthogonal with it; general formulae for a spacelet and its core; the core of the join of unconnected mutually orthogonal spacelets is the join of their cores; possible ranks and extravagances of a spacelet which lies in a given spacelet and has a given complete intersection with its core . 427–430

The plenum of a spacelet: is the plenarily extravagant spacelet which is the normal to its core; is the smallest spacelet which contains the given spacelet and all points orthogonal with it . 431–432

§§ PAGES

Properties of normal spacelets: two mutually normal spacelets have the same extravagance, the same core and the same plenum, their common core being their complete intersection, and their common plenum being their join; general formulae for two mutually normal spacelets, their common core and their common plenum; the join (or complete intersection) of the normals to any number of given spacelets is the normal to the complete intersection (or join) of the given spacelets; ranks of the intersection and join of the normals to two given spacelets; two spacelets are non-intersecting when and only when their normals are mutually complementary. Complementary theorems . . 432–434

Interpretations of the terms 'extravagant', 'core', 'orthogonal', 'normal' as denoting relations to the absolute quadric . . 434–435

171. Standard representations of completely and plenarily extravagant spacelets; a plenarily extravagant spacelet is the join of its core and a real spacelet; anti-cores of a spacelet . . . 435–440

172. Unconnected mutually orthogonal solutions of any system of homogeneous linear algebraic equations: methods of determining a complete set of such solutions; number of extravagant solutions in each set 440–446

173. Possible extravagances of two non-intersecting spacelets: their extravagances are independent, i.e. each spacelet can have independently of the other spacelet any extravagance consistent with its rank; the extravagances of unconnected spacelets are independent 446–450

174. Possible values of the rank and extravagance of a spacelet which lies in a given spacelet of homogeneous space ω_n. A spacelet which contains a plenarily extravagant spacelet must itself be plenarily extravagant 450–456

175. Possible values of the rank and extravagance of a spacelet of homogeneous space ω_n which contains a given spacelet. A spacelet which lies in a completely extravagant spacelet must itself be completely extravagant 456–462

175*a*. Appendix C. Equigradent transformations in which one of the transforming factors is a semi-unit matrix: unilaterally semi-unit equigradent transformations of a matrix; semi-unit transformations of a matrix; equigradent transformations of a spacelet; semi-unit transformations of a spacelet . . . 531–534

CHAPTER XIX

THE PARATOMY AND ORTHOTOMY OF TWO MATRICES AND OF TWO SPACELETS OF HOMOGENEOUS SPACE

176. The paratomy, orthotomy and cross rank of any two spacelets in homogeneous space of $n-1$ dimensions or rank n.

The paratomy, orthotomy and cross rank of two matrices which represent spacelets: are those of the two spacelets which they represent. The mutual paratomy (or degree of intersection) of two spacelets: is the rank of their complete intersection; possible values for two arbitrary spacelets of given ranks. The

§§ PAGES

mutual orthotomy (or degree of orthogonality) and the cross rank of two spacelets : defined ; relations between them ; interpretations ; the extravagance of a spacelet is its orthotomy with itself. The two cases in which each of two spacelets is incident with the normal to the other.

Possible values of the cross rank and mutual orthotomy of two arbitrary spacelets of given ranks ; greatest and least values of the orthotomy 463–468

177. Possible values of the cross rank and mutual orthotomy of two spacelets which both lie in a given spacelet; which both contain a given spacelet.

Spacelets which lie in or contain a given non-extravagant spacelet, or which lie in one given non-extravagant spacelet and contain another : correspondences between such spacelets and spacelets of a complete homogeneous space ; correspondences of ranks and cross ranks ; of extravagances and orthotomies . . . 468–474

178. Possible values of the cross rank and mutual orthotomy of two non-intersecting spacelets of given ranks ; greatest and least values of the orthotomy ; possible values for two non-intersecting spacelets of given ranks which both lie in a given non-extravagant spacelet. Corresponding results when one of the two spacelets is given 474–478

179. Possible values of the cross rank and mutual orthotomy of two mutually complementary spacelets of given ranks ; greatest and least values of the orthotomy ; possible values for two spacelets of given ranks which both lie in a given non-extravagant spacelet and are mutually complementary in it 478–480

180. Properties of two spacelets of given ranks which have a given complete intersection ω_p, and which both lie in the plenum of their intersection ; representation of each spacelet as the join of two mutually orthogonal non-intersecting spacelets, one of which is ω_p ; cross rank and mutual orthotomy of these two joins ; extravagance and core of each join.

Possible values of the cross rank and mutual orthotomy of the two spacelets ; greatest and least values of the orthotomy ; possible extravagances of the two spacelets, the extravagance of each is independent of the other.

Possible values of the mutual orthotomy of two spacelets whose complete intersection is a given non-extravagant spacelet ; of two real spacelets whose complete intersection is given . . 480–489

181. Properties of any two spacelets of given ranks which have a given complete intersection ω_p : representation of each spacelet as the join of three unconnected spacelets of which the first is ω_p, the second is orthogonal with ω_p, and the third lies outside the plenum of ω_p ; cross rank and mutual orthotomy of these two joins ; extravagance and core of each join.

Possible values of the cross rank and mutual orthotomy of the two spacelets ; greatest and least values of the orthotomy ; possible extravagances of the two spacelets, the extravagance of each is independent of the other.

Orthotomy and extravagances of two spacelets having a given join . 490–506

§§ PAGES

182. Possible simultaneous values of the paratomy, cross rank and orthotomy of two entirely arbitrary spacelets; of two entirely arbitrary real spacelets; possible simultaneous values of any two of these quantities 506–509

183. Properties of mutually orthogonal spacelets: they have the same complete intersection as their cores; the core of their join is the join of their cores; their cores are their intersections with the core of their join 509–514

117 *a*. APPENDIX A. Utility of the relations obtained in Chapter XIII . 515–520

130 *a*. APPENDIX B. The Pfaffian of a skew-symmetric matrix of even order 521–530

175 *a*. APPENDIX C. Equigradent transformations in which one of the transforming factors is a semi-unit matrix 531–534

INDEX 535–555

CORRIGENDA

Page	4,	line	8 :	*For* "component",	*read* "constituent".
,,	6,	,,	10 :	*For* "component",	*read* "constituent".
,,	89,	,,	2 :	Interchange the two extreme factors on the right of the equation.	
,,	94,	,,	7 :	*For* (A) ,	*read* (A′).
,,	152,	,,	9 :	*For* "reciprocal" ,	*read* "conjugate reciprocal".
,,	152,	,,	10 :	*For* "reciprocal" ,	*read* "conjugate reciprocal".
,,	176,	,,	1 :	*For* "matrix" ,	*read* "matrix equation".
,,	299,	,,	21 :	*For* "equigradent",	*read* "equigradent in the real domain".

N.B. The space occupied by a matrix is counted as one line.

CHAPTER XII

COMPOUND MATRICES

[This Chapter, which is largely introductory to those which follow, contains definitions of compound and compartite matrices and a number of results to which reference may subsequently be made. In § 102 the primaries of any minor determinant of a matrix are defined, and §§ 104—107 deal with the possible ranks of a matrix which contains a given minor matrix.]

§ 98. Compound matrices.

We will use the notations

$$\phi = \begin{bmatrix} a, & b, & \dots c \\ a', & b', & \dots c' \\ \dots & \dots & \dots \\ a'', & b'', & \dots c'' \end{bmatrix}_{u, v, \dots w}^{\alpha, \beta, \dots \gamma}, \quad \phi' = \overline{\underline{\begin{matrix} a, & a', & \dots a'' \\ b, & b', & \dots b'' \\ \dots & \dots & \dots \\ c, & c', & \dots c'' \end{matrix}}}{}^{u, v, \dots w}_{\alpha, \beta, \dots \gamma}$$

for the matrices derived respectively from the schemes

$$\begin{bmatrix} [a]_u^\alpha & [b]_u^\beta & \dots [c]_u^\gamma \\ [a']_v^\alpha & [b']_v^\beta & \dots [c']_v^\gamma \\ \dots & \dots & \dots \\ [a'']_w^\alpha & [b'']_w^\beta & \dots [c'']_w^\gamma \end{bmatrix}, \quad \overline{\underline{\begin{matrix} \overline{\underline{a}}{}^u_\alpha & \overline{\underline{a'}}{}^v_\alpha & \dots \overline{\underline{a''}}{}^w_\alpha \\ \overline{\underline{b}}{}^u_\beta & \overline{\underline{b'}}{}^v_\beta & \dots \overline{\underline{b''}}{}^w_\beta \\ \dots & \dots & \dots \\ \overline{\underline{c}}{}^u_\gamma & \overline{\underline{c'}}{}^v_\gamma & \dots \overline{\underline{c''}}{}^w_\gamma \end{matrix}}}$$

by writing out the inner matrices in full and then discarding all the inner brackets; and for the determinoid of ϕ, which has the same value as the determinoid of ϕ', we will use the notation

$$\det \phi = \begin{pmatrix} a, & b, & \dots c \\ a', & b', & \dots c' \\ \dots & \dots & \dots \\ a'', & b'', & \dots c'' \end{pmatrix}_{u, v, \dots w}^{\alpha, \beta, \dots \gamma}.$$

Thus ϕ and ϕ' are the two mutually conjugate matrices

$$\phi = \begin{bmatrix}
a_{11} & a_{12} & \dots & a_{1\alpha} & b_{11} & b_{12} & \dots & b_{1\beta} & \dots\dots & c_{11} & c_{12} & \dots & c_{1\gamma} \\
a_{21} & a_{22} & \dots & a_{2\alpha} & b_{21} & b_{22} & \dots & b_{2\beta} & \dots\dots & c_{21} & c_{22} & \dots & c_{2\gamma} \\
\hdotsfor{13} \\
a_{u1} & a_{u2} & \dots & a_{u\alpha} & b_{u1} & b_{u2} & \dots & b_{u\beta} & \dots\dots & c_{u1} & c_{u2} & \dots & c_{u\gamma} \\
a'_{11} & a'_{12} & \dots & a'_{1\alpha} & b'_{11} & b'_{12} & \dots & b'_{1\beta} & \dots\dots & c'_{11} & c'_{12} & \dots & c'_{1\gamma} \\
a'_{21} & a'_{22} & \dots & a'_{2\alpha} & b'_{21} & b'_{22} & \dots & b'_{2\beta} & \dots\dots & c'_{21} & c'_{22} & \dots & c'_{2\gamma} \\
\hdotsfor{13} \\
a'_{v1} & a'_{v2} & \dots & a'_{v\alpha} & b'_{v1} & b'_{v2} & \dots & b'_{v\beta} & \dots\dots & c'_{v1} & c'_{v2} & \dots & c'_{v\gamma} \\
\hdotsfor{13} \\
\hdotsfor{13} \\
a''_{11} & a''_{12} & \dots & a''_{1\alpha} & b''_{11} & b''_{12} & \dots & b''_{1\beta} & \dots\dots & c''_{11} & c''_{12} & \dots & c''_{1\gamma} \\
a''_{21} & a''_{22} & \dots & a''_{2\alpha} & b''_{21} & b''_{22} & \dots & b''_{2\beta} & \dots\dots & c''_{21} & c''_{22} & \dots & c''_{2\gamma} \\
\hdotsfor{13} \\
a''_{w1} & a''_{w2} & \dots & a''_{w\alpha} & b''_{w1} & b''_{w2} & \dots & b''_{w\beta} & \dots\dots & c''_{w1} & c''_{w2} & \dots & c''_{w\gamma}
\end{bmatrix},$$

$$\phi' = \begin{bmatrix}
a_{11} & a_{21} & \dots & a_{u1} & a'_{11} & a'_{21} & \dots & a'_{v1} & \dots\dots & a''_{11} & a''_{21} & \dots & a''_{w1} \\
a_{12} & a_{22} & \dots & a_{u2} & a'_{12} & a'_{22} & \dots & a'_{v2} & \dots\dots & a''_{12} & a''_{22} & \dots & a''_{w2} \\
\hdotsfor{13} \\
a_{1\alpha} & a_{2\alpha} & \dots & a_{u\alpha} & a'_{1\alpha} & a'_{2\alpha} & \dots & a'_{v\alpha} & \dots\dots & a''_{1\alpha} & a''_{2\alpha} & \dots & a''_{w\alpha} \\
b_{11} & b_{21} & \dots & b_{u1} & b'_{11} & b'_{21} & \dots & b'_{v1} & \dots\dots & b''_{11} & b''_{21} & \dots & b''_{w1} \\
b_{12} & b_{22} & \dots & b_{u2} & b'_{12} & b'_{22} & \dots & b'_{v2} & \dots\dots & b''_{12} & b''_{22} & \dots & b''_{w2} \\
\hdotsfor{13} \\
b_{1\beta} & b_{2\beta} & \dots & b_{u\beta} & b'_{1\beta} & b'_{2\beta} & \dots & b'_{v\beta} & \dots\dots & b''_{1\beta} & b''_{2\beta} & \dots & b''_{w\beta} \\
\hdotsfor{13} \\
\hdotsfor{13} \\
c_{11} & c_{21} & \dots & c_{u1} & c'_{11} & c'_{21} & \dots & c'_{v1} & \dots\dots & c''_{11} & c''_{21} & \dots & c''_{w1} \\
c_{12} & c_{22} & \dots & c_{u2} & c'_{12} & c'_{22} & \dots & c'_{v2} & \dots\dots & c''_{12} & c''_{22} & \dots & c''_{w2} \\
\hdotsfor{13} \\
c_{1\gamma} & c_{2\gamma} & \dots & c_{u\gamma} & c'_{1\gamma} & c'_{2\gamma} & \dots & c'_{v\gamma} & \dots\dots & c''_{1\gamma} & c''_{2\gamma} & \dots & c''_{w\gamma}
\end{bmatrix}$$

More generally for the two mutually conjugate matrices derived in the same way from the schemes

$$\begin{bmatrix}
[a_{l\lambda}]^{\alpha}_{u} & [b_{m\mu}]^{\beta}_{u} & \dots & [c_{n\nu}]^{\gamma}_{u} \\
[a'_{r\rho}]^{\alpha}_{v} & [b'_{s\sigma}]^{\beta}_{v} & \dots & [c'_{t\tau}]^{\gamma}_{v} \\
\hdotsfor{4} \\
[a''_{x\xi}]^{\alpha}_{w} & [b''_{y\eta}]^{\beta}_{w} & \dots & [c''_{z\zeta}]^{\gamma}_{w}
\end{bmatrix} \quad \text{and} \quad \begin{bmatrix}
\ulcorner a_{l\lambda} \urcorner^{u}_{\alpha} & \ulcorner a'_{r\rho} \urcorner^{v}_{\alpha} & \dots & \ulcorner a''_{x\xi} \urcorner^{w}_{\alpha} \\
\ulcorner b_{m\mu} \urcorner^{u}_{\beta} & \ulcorner b'_{s\sigma} \urcorner^{v}_{\beta} & \dots & \ulcorner b''_{y\eta} \urcorner^{w}_{\beta} \\
\hdotsfor{4} \\
\ulcorner c_{n\nu} \urcorner^{u}_{\gamma} & \ulcorner c'_{t\tau} \urcorner^{v}_{\gamma} & \dots & \ulcorner c''_{z\zeta} \urcorner^{w}_{\gamma}
\end{bmatrix}$$

we will use the notations

$$\begin{bmatrix} a_{l\lambda}, & b_{m\mu}, & \dots & c_{n\nu} \\ a'_{r\rho}, & b'_{s\sigma}, & \dots & c'_{t\tau} \\ \dots & \dots & \dots & \dots \\ a''_{x\xi}, & b''_{y\eta}, & \dots & c''_{z\zeta} \end{bmatrix}^{\alpha,\beta,\dots\gamma}_{u,v,\dots w} \quad \text{and} \quad \overline{\underline{\begin{matrix} a_{l\lambda}, & a'_{r\rho}, & \dots & a''_{x\xi} \\ b_{m\mu}, & b'_{s\sigma}, & \dots & b''_{y\eta} \\ \dots & \dots & \dots & \dots \\ c_{n\nu}, & c'_{t\tau}, & \dots & c''_{z\zeta} \end{matrix}}}^{u,v,\dots w}_{\alpha,\beta,\dots\gamma} \;;$$

and the determinoid of the first of these matrices, which has the same value as the determinoid of the second matrix, will be denoted by

$$\begin{pmatrix} a_{l\lambda}, & b_{m\mu}, & \dots & c_{n\nu} \\ a'_{r\rho}, & b'_{s\sigma}, & \dots & c'_{t\tau} \\ \dots & \dots & \dots & \dots \\ a''_{x\xi}, & b''_{y\eta}, & \dots & c''_{z\zeta} \end{pmatrix}^{\alpha,\beta,\dots\gamma}_{u,v,\dots w}.$$

These notations are generalisations of those used for augmented matrices in the first volume.

Ex. i.

$$\begin{bmatrix} a, & x \\ b, & y \\ c, & z \end{bmatrix}^{2,3}_{3,1,2} \quad \text{and} \quad \overline{\underline{\begin{matrix} a, & b, & c \\ x, & y, & z \end{matrix}}}^{3,1,2}_{2,3}$$

are by definition the respective matrices

$$\begin{bmatrix} a_{11} & a_{12} & x_{11} & x_{12} & x_{13} \\ a_{21} & a_{22} & x_{21} & x_{22} & x_{23} \\ a_{31} & a_{32} & x_{31} & x_{32} & x_{33} \\ b_{11} & b_{12} & y_{11} & y_{12} & y_{13} \\ c_{11} & c_{12} & z_{11} & z_{12} & z_{13} \\ c_{21} & c_{22} & z_{21} & z_{22} & z_{23} \end{bmatrix} \quad \text{and} \quad \begin{bmatrix} a_{11} & a_{21} & a_{31} & b_{11} & c_{11} & c_{21} \\ a_{12} & a_{22} & a_{32} & b_{12} & c_{12} & c_{22} \\ x_{11} & x_{21} & x_{31} & y_{11} & z_{11} & z_{21} \\ x_{12} & x_{22} & x_{32} & y_{12} & z_{12} & z_{22} \\ x_{13} & x_{23} & x_{33} & y_{13} & z_{13} & z_{23} \end{bmatrix}.$$

Ex. ii.

$$\begin{bmatrix} a_{pq}, & x_{\kappa\lambda} \\ b_{uv}, & y_{\mu\nu} \\ c_{rs}, & z_{\sigma\tau} \end{bmatrix}^{2,3}_{3,1,2} \quad \text{and} \quad \overline{\underline{\begin{matrix} a_{pq}, & b_{uv}, & c_{rs} \\ x_{\kappa\lambda}, & y_{\mu\nu}, & z_{\sigma\tau} \end{matrix}}}^{3\ 1,2}_{2,3}$$

are by definition the respective matrices

$$\begin{bmatrix} a_{p_1q_1} & a_{p_1q_2} & x_{\kappa_1\lambda_1} & x_{\kappa_1\lambda_2} & x_{\kappa_1\lambda_3} \\ a_{p_2q_1} & a_{p_2q_2} & x_{\kappa_2\lambda_1} & x_{\kappa_2\lambda_2} & x_{\kappa_2\lambda_3} \\ a_{p_3q_1} & a_{p_3q_2} & x_{\kappa_3\lambda_1} & x_{\kappa_3\lambda_2} & x_{\kappa_3\lambda_3} \\ b_{u_1v_1} & b_{u_1v_2} & y_{\mu_1\nu_1} & y_{\mu_1\nu_2} & y_{\mu_1\nu_3} \\ c_{r_1s_1} & c_{r_1s_2} & z_{\sigma_1\tau_1} & z_{\sigma_1\tau_2} & z_{\sigma_1\tau_3} \\ c_{r_2s_1} & c_{r_2s_2} & z_{\sigma_2\tau_1} & z_{\sigma_2\tau_2} & z_{\sigma_2\tau_3} \end{bmatrix} \quad \text{and} \quad \begin{bmatrix} a_{p_1q_1} & a_{p_2q_1} & a_{p_3q_1} & b_{u_1v_1} & c_{r_1s_1} & c_{r_2s_1} \\ a_{p_1q_2} & a_{p_2q_2} & a_{p_3q_2} & b_{u_1v_2} & c_{r_1s_2} & c_{r_2s_2} \\ x_{\kappa_1\lambda_1} & x_{\kappa_2\lambda_1} & x_{\kappa_3\lambda_1} & y_{\mu_1\nu_1} & z_{\sigma_1\tau_1} & z_{\sigma_2\tau_1} \\ x_{\kappa_1\lambda_2} & x_{\kappa_2\lambda_2} & x_{\kappa_3\lambda_2} & y_{\mu_1\nu_2} & z_{\sigma_1\tau_2} & z_{\sigma_2\tau_2} \\ x_{\kappa_1\lambda_3} & x_{\kappa_2\lambda_3} & x_{\kappa_3\lambda_3} & y_{\mu_1\nu_3} & z_{\sigma_1\tau_3} & z_{\sigma_2\tau_3} \end{bmatrix}.$$

NOTE 1. *Scalar matrices.*

For a scalar matrix of order m in which the elements of the leading diagonal all have the value k, all other elements having the value 0, we have used the notation $k[1]^m_m$. Whenever k is a scalar quantity, and not a symbol which has no meaning until suffixes are attached, we will understand that

$$[k]^m_m = k[1]^m_m = \begin{bmatrix} k & 0 & \dots & 0 \\ 0 & k & \dots & 0 \\ \dots & \dots & \dots & \dots \\ 0 & 0 & \dots & k \end{bmatrix}.$$

It is convenient to use this alternative notation for a scalar matrix when the scalar matrix is a component of a compound matrix. In cases where any doubt is possible it will be expressly stated when this meaning is to be attached to $[k]^m_m$.

Further whenever k is a scalar quantity and $[a]^n_m$ is a matrix, it will be understood that

$$[ka]^n_m = k[a]^n_m.$$

NOTE 2. *Quasi-scalar matrices.*

In the case of quasi-scalar matrices we shall sometimes use such notations as the following:

$${}^{1}[k]_{m} = {}^{1}\ulcorner k \lrcorner_{m} = \begin{bmatrix} k_1 & 0 & \dots & 0 \\ 0 & k_2 & \dots & 0 \\ \dots & \dots & \dots & \dots \\ 0 & 0 & \dots & k_m \end{bmatrix}, \qquad {}^{1}[\sqrt{k}]_{m} = {}^{1}\ulcorner \sqrt{k} \lrcorner_{m} = \begin{bmatrix} \sqrt{k_1} & 0 & \dots & 0 \\ 0 & \sqrt{k_2} & \dots & 0 \\ \dots & \dots & \dots & \dots \\ 0 & 0 & \dots & \sqrt{k_m} \end{bmatrix},$$

$${}^{1}[k^{-1}]_{m} = {}^{1}\ulcorner k^{-1} \lrcorner_{m} = \begin{bmatrix} k_1^{-1} & 0 & \dots & 0 \\ 0 & k_2^{-1} & \dots & 0 \\ \dots & \dots & \dots & \dots \\ 0 & 0 & \dots & k_m^{-1} \end{bmatrix}, \qquad {}^{1}[k^{p}]_{m} = {}^{1}\ulcorner k^{p} \lrcorner_{m} = \begin{bmatrix} k_1^{p} & 0 & \dots & 0 \\ 0 & k_2^{p} & \dots & 0 \\ \dots & \dots & \dots & \dots \\ 0 & 0 & \dots & k_m^{p} \end{bmatrix}.$$

NOTE 3. *Zero matrices.*

The symbol $[0]^n_m$ will be used to denote a zero matrix whose horizontal and vertical orders are m and n, i.e. to denote the matrix $[a]^n_m$ all of whose elements are 0's. This notation will usually only be required when $[0]^n_m$ is a component of a compound matrix.

NOTE 4. *One-rowed matrices.*

For the one-rowed matrix $[x_1 x_2 \dots x_p \; y_1 y_2 \dots y_q \dots z_1 z_2 \dots z_r]$ and its conjugate we shall sometimes use the notations

$$[x, y, \dots z]_{p,\, q,\, \dots\, r} \quad \text{and} \quad \ulcorner \begin{matrix} x \\ y \\ \vdots \\ z \end{matrix} \lrcorner_{p,\, q,\, \dots\, r}$$

respectively. The commas separating the suffixes $p, q, \dots r$ will distinguish these from the single suffix notations for a matrix described in § 2.

Since $[1]^n_m$ is undefined except when $m=n$, we may define $[1]^m_1$ and $\overline{\underline{1}}{}^1_m$ by the equations

$$[1]^1_m=[1,\ 1,\ \ldots\ 1]^{1,1,\ldots 1}_1=[1\ 1\ \ldots\ 1],\quad \overline{\underline{1}}{}^1_m=\begin{bmatrix}1\\1\\ \vdots\\1\end{bmatrix}^1_{1,1,\ldots 1}=\begin{bmatrix}1\\1\\ \vdots\\1\end{bmatrix},$$

where in each case there are m 1's.

NOTE 5. *Matrices one of whose orders is zero.*

Neither of the orders m and n of a matrix $[a]^n_m$ can be less than 1; but for the sake of generality we shall often proceed as if the value 0 were admissible for m and n. Then if either of the orders of a matrix is 0, the matrix is non-existent; it can be regarded as having rank 0; it can be replaced as a factor by the scalar number 0; and its conjugate reciprocal is also a matrix one of whose orders is 0.

Ex. iii. $$[9]^3_3=9[1]^3_3=\begin{bmatrix}9&0&0\\0&9&0\\0&0&9\end{bmatrix};\quad [-\sqrt{2}]^3_3=-\sqrt{2}\,[1]^3_3=\begin{bmatrix}-\sqrt{2}, & 0, & 0\\ 0, & -\sqrt{2}, & 0\\ 0, & 0, & -\sqrt{2}\end{bmatrix}.$$

Ex. iv. $$\begin{bmatrix}a, & -5\\ 0, & 2b\end{bmatrix}^{4,3}_{3,2}=\begin{bmatrix}a_{11}, & a_{12}, & a_{13}, & a_{14}, & -5, & 0, & 0\\ a_{21}, & a_{22}, & a_{23}, & a_{24}, & 0, & -5, & 0\\ a_{31}, & a_{32}, & a_{33}, & a_{34}, & 0, & 0, & -5\\ 0, & 0, & 0, & 0, & 2b_{11}, & 2b_{12}, & 2b_{13}\\ 0, & 0, & 0, & 0, & 2b_{21}, & 2b_{22}, & 2b_{23}\end{bmatrix}.$$

Ex. v. $${}^1[k]_m\,{}^1[k]_m={}^1[k^2]_m;\quad {}^1[\sqrt{k}]_m\,{}^1[\sqrt{k}]_m={}^1[k]_m;\quad {}^1[k]_m\,{}^1[k^{-1}]_m=[1]^m_m.$$

Ex. vi. $$\overline{\underline{\begin{matrix}x\\1\end{matrix}}}_{m,1}=\begin{bmatrix}x_1\\x_2\\ \vdots\\x_m\\1\end{bmatrix};\quad [x,\ 1]^1_m=[x_1\ x_2\ \ldots\ x_m\ 1].$$

Ex. vii. $$\overline{\underline{\begin{matrix}x\\1\end{matrix}}}{}^r_{m,1}=\begin{bmatrix}x_{11} & x_{21} & \ldots & x_{r1}\\ x_{12} & x_{22} & \ldots & x_{r2}\\ \cdots & \cdots & \cdots & \cdots\\ x_{1m} & x_{2m} & \ldots & x_{rm}\\ 1 & 1 & \ldots & 1\end{bmatrix};\quad [x,\ 1]^{m,1}_r=\begin{bmatrix}x_{11} & x_{12} & \ldots & x_{1m} & 1\\ x_{21} & x_{22} & \ldots & x_{2m} & 1\\ \cdots & \cdots & \cdots & \cdots & \cdots\\ x_{r1} & x_{r2} & \ldots & x_{rm} & 1\end{bmatrix}.$$

§ 99. Multiplication of compound matrices.

If we call $u, v, \ldots w$ and $\alpha, \beta, \ldots \gamma$ the partial orders of the compound matrices ϕ and ϕ' in § 98, then the product of two compound matrices whose partial passivities are the same and arranged in the same order can be evaluated in a manner analogous to that in which we evaluate the product of two simple matrices which have the same passivities. For we have

$$\begin{bmatrix} a, & b, & \dots & c \\ a', & b', & \dots & c' \\ \dots & \dots & \dots & \dots \\ a'', & b'', & \dots & c'' \end{bmatrix}^{r,s,\dots t}_{m,n,\dots p} \begin{bmatrix} \alpha, & \alpha', & \dots & \alpha'' \\ \beta, & \beta', & \dots & \beta'' \\ \dots & \dots & \dots & \dots \\ \gamma, & \gamma', & \dots & \gamma'' \end{bmatrix}^{u,v,\dots w}_{r,s,\dots t} = \begin{bmatrix} x, & y, & \dots & z \\ x', & y', & \dots & z' \\ \dots & \dots & \dots & \dots \\ x'', & y'', & \dots & z'' \end{bmatrix}^{u,v,\dots w}_{m,n,\dots p},$$

where

$$[x]^u_m = [a]^r_m [\alpha]^u_r + [b]^s_m [\beta]^u_s + \dots + [c]^t_m [\gamma]^u_t,$$

$$[y]^v_m = [a]^r_m [\alpha']^v_r + [b]^s_m [\beta']^v_s + \dots + [c]^t_m [\gamma']^v_t,$$

$$[y'']^v_p = [a'']^r_p [\alpha']^v_r + [b'']^s_p [\beta']^v_s + \dots + [c'']^t_p [\gamma']^v_t,$$

and so on.

This method of evaluating the product is particularly advantageous when some of the component matrices are zero matrices or scalar matrices.

Ex. i. Since $[a]^n_m [1]^n_n = [1]^m_m [a]^n_m = [a]^n_m$, we see at once that

$$\begin{bmatrix} a, & b, & c \\ 0, & 1, & 0 \\ 0, & 0, & 1 \end{bmatrix}^{4,3,2}_{2,3,2} \begin{bmatrix} \alpha, & 0, & 0 \\ \beta, & 1, & 0 \\ \gamma, & 0, & 1 \end{bmatrix}^{2,3,2}_{4,3,2} = \begin{bmatrix} x, & b, & c \\ \beta, & 1, & 0 \\ \gamma, & 0, & 1 \end{bmatrix}^{2,3,2}_{2,3,2},$$

where $$[x]^2_2 = [a]^4_2 [\alpha]^2_4 + [b]^3_2 [\beta]^2_3 + [c]^2_2 [\gamma]^2_2.$$

Ex. ii. If $i=\sqrt{-1}$, then

$$\begin{bmatrix} 1, & 0, & 0, & 0 \\ 0, & 1, & i, & 0 \end{bmatrix}^{r-\rho,\,\rho,\,\rho,\,s}_{r-\rho,\,\rho} \begin{bmatrix} 0, & 0 \\ 0, & 1 \\ 0, & i \\ 1, & 0 \end{bmatrix}^{s,\,\rho}_{r-\rho,\,\rho,\,\rho,\,s} = 0.$$

Ex. iii. If k is a scalar quantity, we have

$$[k,\ x]^{m,n}_m \begin{bmatrix} -x \\ k \end{bmatrix}^n_{m,n} = -[k]^m_m [x]^n_m + [x]^n_m [k]^n_n = -k[x]^n_m + k[x]^n_m = 0.$$

§ 100. Compartite matrices.

A matrix whose elements all vanish except those lying in a number of mutually complementary minors will be called a compartite matrix, and those mutually complementary minors will be called the *parts* of the matrix.

Every such matrix can be converted by derangements of its horizontal and vertical rows into a matrix of the form

$$\phi = [e]_m^n = \begin{bmatrix} a, & 0, & \dots & 0 \\ 0, & b, & \dots & 0 \\ \dots & \dots & \dots & \dots \\ 0, & 0, & \dots & c \end{bmatrix}_{p, q, \dots r}^{u, v, \dots w}.$$

A matrix ϕ having this special form will be called a *compartite matrix in standard form.* The parts of ϕ are the matrices

$$A = [a]_p^u, \quad B = [b]_q^v, \quad \dots \quad C = [c]_r^w$$

or any derangements of them. We shall speak of $A, B, \dots C$ themselves, taken in this order, as the *successive parts* of ϕ.

Ex. i. *There exists a connection between the horizontal (or vertical) rows of a compartite matrix when and only when there exists a connection between the horizontal (or vertical) rows of one of its parts.*

Since there will be no loss of generality in supposing the compartite matrix to be in standard form, it will be sufficient to prove this theorem for the matrix ϕ shown above.

An equation of the form

$$[\lambda, \mu, \dots \nu]_{p, q, \dots r} [e]_m^n = 0$$

is satisfied when and only when all the equations

$$[\lambda]_p [a]_p^u = 0, \quad [\mu]_q [b]_q^v = 0, \dots [\nu]_r [c]_r^w = 0$$

are satisfied; and the one-rowed matrix $[\lambda, \mu, \dots \nu]_{p, q, \dots r}$ contains a non-vanishing element when and only when one of its minors $[\lambda]_p, [\mu]_q, \dots [\nu]_r$ contains a non-vanishing element. Consequently there exists a connection between the horizontal rows of ϕ when and only when there exists a connection between the horizontal rows of one of its parts.

The truth of the theorem for vertical rows can be proved in a similar way.

Ex. ii. If the parts of ϕ are all square matrices, so that $u=p, v=q, \dots w=r$, then the determinant of ϕ is equal to the product of the determinants of its successive parts. Also the determinant of any matrix of which ϕ is a derangement can only differ in sign from this product.

Ex. iii. Every minor of ϕ is a compartite matrix whose parts are minors of the parts of ϕ.

Theorem. *The rank of a compartite matrix is the sum of the ranks of its parts.*

It will be sufficient to prove this theorem for the matrix ϕ shown above.

Let the parts $A, B, \dots C$ of ϕ have ranks $\alpha, \beta, \dots \gamma$ respectively, and let $R = \alpha + \beta + \dots + \gamma$. Further let $(a_{\kappa\lambda})_\alpha^\alpha, (b_{\mu\nu})_\beta^\beta, \dots (c_{\sigma\tau})_\gamma^\gamma$ be non-vanishing minor determinants of orders $\alpha, \beta, \ldots \gamma$ of $A, B, \ldots C$. Then the compartite matrix

$$\phi' = \begin{bmatrix} a_{\kappa\lambda}, & 0, & \ldots & 0 \\ 0, & b_{\mu\nu}, & \ldots & 0 \\ \cdots & \cdots & \cdots & \cdots \\ 0, & 0, & \ldots & c_{\sigma\tau} \end{bmatrix}^{\alpha, \beta, \ldots \gamma}_{\alpha, \beta, \ldots \gamma}$$

is a square minor of ϕ of order R whose determinant, which is equal to the product $(a_{\kappa\lambda})^{\alpha}_{\alpha}\,(b_{\mu\nu})^{\beta}_{\beta} \ldots (c_{\sigma\tau})^{\gamma}_{\gamma}$, does not vanish. Therefore the rank of ϕ cannot be less than R.

Again if ϕ'' is any square minor of ϕ of order greater than R, then ϕ'' must contain either more than α horizontal rows of A, or more than β horizontal rows of $B, \ldots$ or more than γ horizontal rows of C. Consequently there must be a connection between the horizontal rows of one of the parts of the compartite matrix ϕ'', and therefore a connection between the horizontal rows of ϕ'' itself, i.e. ϕ'' must be degenerate. Thus ϕ cannot contain a non-vanishing minor determinant of order greater than R.

It follows that ϕ must have rank R, where R is the sum of the ranks of the parts of ϕ.

NOTE. *Compartite matrices whose parts are quasi-scalar matrices.*

For a compartite matrix in standard form whose successive parts are the quasi-scalar matrices ${}^{1}[a]_{p}$, ${}^{1}[b]_{q}$, ... ${}^{1}[c]_{r}$ we shall sometimes use one of the notations

$${}^{1}[a, b, \ldots c]_{p, q, \ldots r}, \qquad {}^{1}\begin{bmatrix} a, & 0, & \ldots & 0 \\ 0, & b, & \ldots & 0 \\ \cdots & \cdots & \cdots & \cdots \\ 0, & 0, & \ldots & c \end{bmatrix}_{p, q, \ldots r}.$$

§ 101. The conjugate reciprocals and inverses of certain matrices.

The following examples can be regarded as exercises in the multiplication of compound matrices.

Ex. i. *The conjugate reciprocals of the respective matrices*

$$\begin{bmatrix} a, & b \\ 0, & 1 \end{bmatrix}^{r, s}_{r, s}, \quad \begin{bmatrix} b, & a \\ 1, & 0 \end{bmatrix}^{s, r}_{r, s}, \quad \begin{bmatrix} 0, & 1 \\ a, & b \end{bmatrix}^{r, s}_{s, r}, \quad \begin{bmatrix} 1, & 0 \\ b, & a \end{bmatrix}^{s, r}_{s, r}$$

are

$$\overline{\underline{\begin{matrix} A, & B \\ 0, & \Delta \end{matrix}}}^{r, s}_{r, s}, \quad \overline{\underline{\begin{matrix} 0, & \Delta \\ A, & B \end{matrix}}}^{r, s}_{s, r}, \quad \overline{\underline{\begin{matrix} B, & A \\ \Delta, & 0 \end{matrix}}}^{s, r}_{r, s}, \quad \overline{\underline{\begin{matrix} \Delta, & 0 \\ B, & A \end{matrix}}}^{s, r}_{s, r},$$

where $\overline{\underline{A}}^{r}_{r}$ *is the conjugate reciprocal of* $[a]^{r}_{r}$; $\Delta = (a)^{r}_{r}$; $[\Delta]^{s}_{s} = \Delta\,[1]^{s}_{s}$; *and*

$$\overline{\underline{B}}^{s}_{r} = -\overline{\underline{A}}^{r}_{r}\,[b]^{s}_{r}.$$

It will be observed that $-B_{ij}$ is the determinant formed when the jth vertical row of Δ is replaced by the ith vertical row of $[b]_r^s$.

If these results are true when the matrices are undegenerate, i.e. when $\Delta \neq 0$, they must be true in general. Hence we can at once verify them by multiplication; for we have

$$\begin{bmatrix} a, & b \\ 0, & 1 \end{bmatrix}_{r,s}^{r,s} \overline{\underline{\begin{matrix} A, & B \\ 0, & \Delta \end{matrix}}}_{r,s}^{r,s} = \begin{bmatrix} \Delta, & 0 \\ 0, & \Delta \end{bmatrix}_{r,s}^{r,s} = \Delta[1]_{r+s}^{r+s},$$

and similarly in the other cases.

To prove the first result directly, assume that $\Delta \neq 0$, and let the conjugate reciprocal of $\begin{bmatrix} a, & b \\ 0, & 1 \end{bmatrix}_{r,s}^{r,s}$ be $\begin{bmatrix} x, & y \\ z, & w \end{bmatrix}_{r,s}^{r,s}$, so that

$$\begin{bmatrix} a, & b \\ 0, & 1 \end{bmatrix}_{r,s}^{r,s} \begin{bmatrix} x, & y \\ z, & w \end{bmatrix}_{r,s}^{r,s} = \Delta \begin{bmatrix} 1, & 0 \\ 0, & 1 \end{bmatrix}_{r,s}^{r,s}$$

Then $$[a]_r^r [x]_r^r + [b]_r^s [z]_s^r = \Delta [1]_r^r, \quad [a]_r^r [y]_r^s + [b]_r^s [w]_s^s = 0,$$

$$[z]_s^r = 0, \quad [w]_s^s = \Delta [1]_s^s.$$

Using the last two equations the first two reduce to

$$[a]_r^r [x]_r^r = \Delta [1]_r^r, \quad [a]_r^r [y]_r^s = -\Delta [b]_r^s.$$

Prefixing $\overline{\underline{A}}_r^r$, the conjugate reciprocal of $[a]_r^r$, on both sides, we obtain

$$[x]_r^r = \overline{\underline{A}}_r^r, \quad [y]_r^s = -\overline{\underline{A}}_r^r [b]_r^s.$$

The remaining three results can be deduced, or proved in similar ways.

Ex. ii. The inverses of the same matrices, when they are undegenerate, are

$$\overline{\underline{\begin{matrix} A, & B \\ 0, & 1 \end{matrix}}}_{r,s}^{r,s}, \quad \overline{\underline{\begin{matrix} 0, & 1 \\ A, & B \end{matrix}}}_{s,r}^{r,s}, \quad \overline{\underline{\begin{matrix} B, & A \\ 1, & 0 \end{matrix}}}_{r,s}^{s,r}, \quad \overline{\underline{\begin{matrix} 1, & 0 \\ B, & A \end{matrix}}}_{s,r}^{s,r},$$

where $\overline{\underline{A}}_r^r$ is the inverse of $[a]_r^r$, and $\overline{\underline{B}}_r^s = -\overline{\underline{A}}_r^r [b]_r^s$.

Ex. iii. *The conjugate reciprocals of the respective matrices*

$$\begin{bmatrix} a, & 0 \\ b, & 1 \end{bmatrix}_{r,s}^{r,s}, \quad \begin{bmatrix} b, & 1 \\ a, & 0 \end{bmatrix}_{s,r}^{r,s}, \quad \begin{bmatrix} 0, & a \\ 1, & b \end{bmatrix}_{r,s}^{s,r}, \quad \begin{bmatrix} 1, & b \\ 0, & a \end{bmatrix}_{s,r}^{s,r}$$

are

$$\overline{\underline{\begin{matrix} A, & 0 \\ B, & \Delta \end{matrix}}}_{r,s}^{r,s}, \quad \overline{\underline{\begin{matrix} 0, & A \\ \Delta, & B \end{matrix}}}_{r,s}^{s,r}, \quad \overline{\underline{\begin{matrix} B, & \Delta \\ A, & 0 \end{matrix}}}_{s,r}^{r,s}, \quad \overline{\underline{\begin{matrix} \Delta, & B \\ 0, & A \end{matrix}}}_{s,r}^{s,r},$$

where $\overline{\underline{A}}_r^r$ *is the conjugate reciprocal of* $[a]_r^r$; $\Delta = (a)_r^r$; $[\Delta]_s^s = \Delta[1]_s^s$; *and*

$$\overline{\underline{B}}_s^r = -[b]_s^r \overline{\underline{A}}_r^r.$$

These results can be verified by multiplication, or proved as in Ex. i.

Ex. iv. The conjugate reciprocal of $\begin{bmatrix} 1, & a \\ 0, & 1 \end{bmatrix}_{r,s}^{r,s}$ is $\begin{bmatrix} 1, & -a \\ 0, & 1 \end{bmatrix}_{r,s}^{r,s}$.

Ex. v. The conjugate reciprocal of $\begin{bmatrix} a, & 0 \\ 0, & 1 \end{bmatrix}_{r,s}^{r,s}$ is $\underline{\overline{\begin{matrix} A, & 0 \\ 0, & \Delta \end{matrix}}}_{r,s}^{r,s}$,

where $\underline{\overline{A}}_r^r$ is the conjugate reciprocal of $[a]_r^r$, and $\Delta=(a)_r^r$.

Ex. vi. The conjugate reciprocal of the matrix

$$M=\begin{bmatrix} a_{11} & a_{12} & \dots & a_{1m} & a_{1,m+1} \\ a_{21} & a_{22} & \dots & a_{2m} & a_{2,m+1} \\ \dots & \dots & \dots & \dots & \dots \\ a_{m1} & a_{m2} & \dots & a_{mm} & a_{m,m+1} \\ 0 & 0 & \dots & 0 & 1 \end{bmatrix} \quad \text{is} \quad M'=\begin{bmatrix} A_{11} & A_{21} & \dots & A_{m1} & A_{m+1,1} \\ A_{12} & A_{22} & \dots & A_{m2} & A_{m+1,2} \\ \dots & \dots & \dots & \dots & \dots \\ A_{1m} & A_{2m} & \dots & A_{mm} & A_{m+1,m} \\ 0 & 0 & \dots & 0 & \Delta \end{bmatrix},$$

where $\Delta=(a)_m^m$; $\underline{\overline{A}}_m^m$ is the conjugate reciprocal of $[a]_m^m$;

and
$$\begin{bmatrix} A_{m+1,1} \\ A_{m+1,2} \\ \vdots \\ A_{m+1,m} \end{bmatrix} = -\underline{\overline{A}}_m^m \begin{bmatrix} a_{1,m+1} \\ a_{2,m+1} \\ \vdots \\ a_{m,m+1} \end{bmatrix}.$$

In particular $-A_{m+1,u}$ is the determinant formed when the uth vertical row of $(a)_m^m$ or Δ is replaced by the last vertical row of $[a]_m^{m+1}$.

The matrix M is included in the matrices considered in Ex. i.

Ex. vii. *The conjugate reciprocal of* $\begin{bmatrix} a, & 0, & b \\ c, & 1, & d \\ 0, & 0, & 1 \end{bmatrix}_{r,s,t}^{r,s,t}$ *is* $\underline{\overline{\begin{matrix} A, & 0, & B \\ C, & \Delta, & D \\ 0, & 0, & \Delta \end{matrix}}}_{r,s,t}^{r,s,t}$,

where $\Delta=(a)_r^r$; $\underline{\overline{A}}_r^r$ *is the conjugate reciprocal of* $[a]_r^r$; *and*

$$\underline{\overline{B}}_r^t=-\underline{\overline{A}}_r^r\,[b]_r^t, \quad \underline{\overline{C}}_s^r=-[c]_s^r\,\underline{\overline{A}}_r^r, \quad \underline{\overline{D}}_s^t=[c]_s^r\,\underline{\overline{A}}_r^r\,[b]_r^t-\Delta\,[d]_s^t.$$

We can prove this by putting

$$\begin{bmatrix} a, & 0, & b \\ c, & 1, & d \\ 0, & 0, & 1 \end{bmatrix}_{r,s,t}^{r,s,t} = \begin{bmatrix} a, & \beta \\ 0, & 1 \end{bmatrix}_{r+s,t}^{r+s,t},$$

where
$$[a]_{r+s}^{r+s}=\begin{bmatrix} a, & 0 \\ c, & 1 \end{bmatrix}_{r,s}^{r,s}, \quad [\beta]_{r+s}^{t}=\begin{bmatrix} b \\ d \end{bmatrix}_{r,s}^{t},$$

and applying Exs. i and iii; or we can verify it by multiplication.

Ex. viii. *The conjugate reciprocal of* $\left[\begin{matrix} a, & c, & 0 \\ 0, & 1, & 0 \\ b, & d, & 1 \end{matrix}\right]_{r,\,s,\,t}^{r,\,s,\,t}$ *is* $\overline{\underline{\begin{matrix} A, & C, & 0 \\ 0, & \Delta, & 0 \\ B, & D, & \Delta \end{matrix}}}_{r,\,s,\,t}^{r,\,s,\,t}$,

where $\Delta=(a)_r^r$; $\overline{\underline{A}}_r^r$ *is the conjugate reciprocal of* $[a]_r^r$; *and*

$$\overline{\underline{C}}_r^s=-\overline{\underline{A}}_r^r\,[c]_r^s,\quad \overline{\underline{B}}_t^r=-[b]_t^r\,\overline{\underline{A}}_r^r,\quad \overline{\underline{D}}_t^s=[b]_t^r\,\overline{\underline{A}}_r^r\,[c]_r^s-\Delta\,[d]_t^s.$$

Ex. ix. If $[a]_u^p, [\beta]_v^q, \dots [\gamma]_w^r$ are inverse prefactors of $[a]_p^u, [b]_q^v, \dots [c]_r^w$, (which can only be the case when the latter matrices have ranks $u, v, \dots w$), then

$$\left[\begin{matrix} a, & 0, & \dots & 0 \\ 0, & \beta, & \dots & 0 \\ \dots & \dots & \dots & \dots \\ 0, & 0, & \dots & \gamma \end{matrix}\right]_{u,\,v,\,\dots\,w}^{p,\,q,\,\dots\,r} \text{ is an inverse prefactor of } \left[\begin{matrix} a, & 0, & \dots & 0 \\ 0, & b, & \dots & 0 \\ \dots & \dots & \dots & \dots \\ 0, & 0, & \dots & c \end{matrix}\right]_{p,\,q,\,\dots\,r}^{u,\,v,\,\dots\,w}.$$

Ex. x. If $[a]_u^p, [\beta]_v^q, \dots [\gamma]_w^r$ are inverse postfactors of $[a]_p^u, [b]_q^v, \dots [c]_r^w$, (which can only be the case when the latter matrices have ranks $p, q, \dots r$), then

$$\left[\begin{matrix} a, & 0, & \dots & 0 \\ 0, & \beta, & \dots & 0 \\ \dots & \dots & \dots & \dots \\ 0, & 0, & \dots & \gamma \end{matrix}\right]_{u,\,v,\,\dots\,w}^{p,\,q,\,\dots\,r} \text{ is an inverse postfactor of } \left[\begin{matrix} a, & 0, & \dots & 0 \\ 0, & b, & \dots & 0 \\ \dots & \dots & \dots & \dots \\ 0, & 0, & \dots & c \end{matrix}\right]_{p,\,q,\,\dots\,r}^{u,\,v,\,\dots\,w}.$$

Ex. xi. If $[a]_p^p, [b]_q^q, \dots [c]_r^r$ are undegenerate, and $\overline{\underline{A}}_p^p, \overline{\underline{B}}_q^q, \dots \overline{\underline{C}}_r^r$ are their inverses (or conjugate reciprocals), then

$$\overline{\underline{\begin{matrix} A, & 0, & \dots & 0 \\ 0, & B, & \dots & 0 \\ \dots & \dots & \dots & \dots \\ 0, & 0, & \dots & C \end{matrix}}}_{p,\,q,\,\dots\,r}^{p,\,q,\,\dots\,r} \text{ is the inverse (or conjugate reciprocal) of } \left[\begin{matrix} a, & 0, & \dots & 0 \\ 0, & b, & \dots & 0 \\ \dots & \dots & \dots & \dots \\ 0, & 0, & \dots & c \end{matrix}\right]_{p,\,q,\,\dots\,r}^{p,\,q,\,\dots\,r}.$$

Ex. xii. The conjugate reciprocals (and inverses) of

$$\left[\begin{matrix} 1, & 0 \\ a, & 1 \end{matrix}\right]_{r,\,s}^{r,\,s},\quad \left[\begin{matrix} 1, & a \\ 0, & 1 \end{matrix}\right]_{r,\,s}^{r,\,s} \text{ are } \left[\begin{matrix} 1, & 0 \\ -a, & 1 \end{matrix}\right]_{r,\,s}^{r,\,s},\quad \left[\begin{matrix} 1, & -a \\ 0, & 1 \end{matrix}\right]_{r,\,s}^{r,\,s}.$$

Ex. xiii. The conjugate reciprocal and inverse of the matrix

$$\phi=\left[\begin{matrix} 1 & 0 & 0 & 0 & 0 \\ a_1 & 1 & 0 & 0 & 0 \\ a_2 & b_1 & 1 & 0 & 0 \\ a_3 & b_2 & c_1 & 1 & 0 \\ a_4 & b_3 & c_2 & d_1 & 1 \end{matrix}\right] \text{ is } \Phi=\left[\begin{matrix} 1 & 0 & 0 & 0 & 0 \\ A_1 & 1 & 0 & 0 & 0 \\ A_2 & B_1 & 1 & 0 & 0 \\ A_3 & B_2 & C_1 & 1 & 0 \\ A_4 & B_3 & C_2 & D_1 & 1 \end{matrix}\right],$$

where the elements of Φ are determined in succession from the equations

$$A_1+a_1=0,\quad B_1+b_1=0,\quad C_1+c_1=0,\quad D_1+d_1=0,$$
$$A_2+B_1a_1+a_2=0,\quad B_2+C_1b_1+b_2=0,\quad C_2+D_1c_1+c_2=0,$$
$$A_3+B_2a_1+C_1a_3+a_4=0,\quad B_3+C_2b_1+D_1b_2+b_3=0,$$
$$A_4+B_3a_1+C_2a_2+D_1a_3+a_4=0,$$

which are the conditions that $\Phi\phi=[1]_5^5$.

We may observe that

$$\phi=\begin{bmatrix}1&0&0&0&0\\a_1&1&0&0&0\\a_2&0&1&0&0\\a_3&0&0&1&0\\a_4&0&0&0&1\end{bmatrix}\begin{bmatrix}1&0&0&0&0\\0&1&0&0&0\\0&b_1&1&0&0\\0&b_2&0&1&0\\0&b_3&0&0&1\end{bmatrix}\begin{bmatrix}1&0&0&0&0\\0&1&0&0&0\\0&0&1&0&0\\0&0&c_1&1&0\\0&0&c_2&0&1\end{bmatrix}\begin{bmatrix}1&0&0&0&0\\0&1&0&0&0\\0&0&1&0&0\\0&0&0&1&0\\0&0&0&d_1&1\end{bmatrix},$$

$$\phi=\begin{bmatrix}1&0&0&0&0\\a_1&1&0&0&0\\0&0&1&0&0\\0&0&0&1&0\\0&0&0&0&1\end{bmatrix}\begin{bmatrix}1&0&0&0&0\\0&1&0&0&0\\a_2&b_1&1&0&0\\0&0&0&1&0\\0&0&0&0&1\end{bmatrix}\begin{bmatrix}1&0&0&0&0\\0&1&0&0&0\\0&0&1&0&0\\a_3&b_2&c_1&1&0\\0&0&0&0&1\end{bmatrix}\begin{bmatrix}1&0&0&0&0\\0&1&0&0&0\\0&0&1&0&0\\0&0&0&1&0\\a_4&b_3&c_2&d_1&1\end{bmatrix}$$

Here the orders of the factors on the right cannot be inverted, though the new products thus obtained are the same in form as ϕ. Utilising Ex. xii we can deduce from either of the above equations the value of the inverse of ϕ.

Ex. xiv. The conjugate reciprocal and inverse of the matrix

$$\phi=\begin{bmatrix}1,&0,&0,&0,&0\\a,&1,&0,&0,&0\\a',&b,&1,&0,&0\\a'',&b',&c,&1,&0\\a''',&b'',&c',&d,&1\end{bmatrix}_{p,q,r,s,t}^{p,q,r,s,t}\quad\text{is}\quad\Phi=\begin{bmatrix}1,&0,&0,&0,&0\\A,&1,&0,&0,&0\\A',&B,&1,&0,&0\\A'',&B',&C,&1,&0\\A''',&B'',&C',&D,&1\end{bmatrix}_{p,q,r,s,t}^{p,q,r,s,t},$$

where the various constituent matrices of Φ are determined in succession from the same equations as before when we write

$$A_1=[A]_q^p,\quad A_2=[A']_r^p,\quad A_3=[A'']_s^p,\quad A_4=[A''']_t^p;\quad a_1=[a]_q^p,$$
$$B_1=[B]_r^q,\quad B_2=[B']_s^q,\quad B_3=[B'']_t^q;\quad b_1=[b]_r^q,\quad a_2=[a']_r^p,$$
$$C_1=[C]_s^r,\quad C_2=[C']_t^r;\quad c_1=[c]_s^r,\quad b_2=[b']_s^q,\quad a_3=[a'']_s^p,$$
$$D_1=[D]_t^s;\quad d_1=[d]_t^s,\quad c_2=[c']_t^r,\quad b_3=[b'']_t^q,\quad a_4=[a''']_t^p;$$

these being the conditions that $\Phi\phi=[1]_u^u$ where $u=p+q+r+s+t$.

Ex. xv. Similar remarks apply to the matrices

$$\phi=\begin{bmatrix}1&a_1&a_2&a_3&a_4\\0&1&b_1&b_2&b_3\\0&0&1&c_1&c_2\\0&0&0&1&d_1\\0&0&0&0&1\end{bmatrix},\quad\phi=\begin{bmatrix}1,&a,&a',&a'',&a'''\\0,&1,&b,&b',&b''\\0,&0,&1,&c,&c'\\0,&0,&0,&1,&d\\0,&0,&0,&0,&1\end{bmatrix}_{p,q,r,s,t}^{p,q,r,s,t}$$

In each case the conjugate reciprocal or inverse of ϕ is a matrix of the same form. For the first matrix we now have

$$\phi = \begin{bmatrix} 1&0&0&0&0\\ 0&1&0&0&0\\ 0&0&1&0&0\\ 0&0&0&1&d_1\\ 0&0&0&0&1 \end{bmatrix} \begin{bmatrix} 1&0&0&0&0\\ 0&1&0&0&0\\ 0&0&1&c_1&c_2\\ 0&0&0&1&0\\ 0&0&0&0&1 \end{bmatrix} \begin{bmatrix} 1&0&0&0&0\\ 0&1&b_1&b_2&b_3\\ 0&0&1&0&0\\ 0&0&0&1&0\\ 0&0&0&0&1 \end{bmatrix} \begin{bmatrix} 1&a_1&a_2&a_3&a_4\\ 0&1&0&0&0\\ 0&0&1&0&0\\ 0&0&0&1&0\\ 0&0&0&0&1 \end{bmatrix},$$

$$\phi = \begin{bmatrix} 1&0&0&0&a_4\\ 0&1&0&0&b_3\\ 0&0&1&0&c_2\\ 0&0&0&1&d_1\\ 0&0&0&0&1 \end{bmatrix} \begin{bmatrix} 1&0&0&a_3&0\\ 0&1&0&b_2&0\\ 0&0&1&c_1&0\\ 0&0&0&1&0\\ 0&0&0&0&1 \end{bmatrix} \begin{bmatrix} 1&0&a_2&0&0\\ 0&1&b_1&0&0\\ 0&0&1&0&0\\ 0&0&0&1&0\\ 0&0&0&0&1 \end{bmatrix} \begin{bmatrix} 1&a_1&0&0&0\\ 0&1&0&0&0\\ 0&0&1&0&0\\ 0&0&0&1&0\\ 0&0&0&0&1 \end{bmatrix},$$

and there are similar equations for the second matrix.

§ 102. The primaries of any minor determinant of a matrix.

If Δ is a minor determinant of any order r of a matrix $A = [a]_m^n$, those minor determinants of A of the same order r which differ from Δ in only one horizontal row will be called the *horizontal primaries of* Δ, and those minor determinants of A of the same order r which differ from Δ in only one vertical row will be called the *vertical primaries of* Δ. Any two primaries of Δ will be regarded as distinct from one another when and only when each is not merely a derangement of the other.

Let H be the vertical minor of A formed by striking out those of the vertical rows of A which do not occur in Δ, and let K be the horizontal minor of A formed by striking out those of the horizontal rows of A which do not occur in Δ. These definitions of H and K are independent of the notations employed for A and Δ. When the standard double-suffix notation is employed and $A = [a]_m^n$, $\Delta = (a_{pq})_r^r$, we have $H = [a_{1q}]_m^r$ and $K = [a_{p1}]_r^n$. Then we form a horizontal primary of Δ when we replace any one of the r horizontal rows of Δ by any one of those $m - r$ horizontal rows of H which do not occur in Δ; and we form a vertical primary of Δ when we replace any one of the r vertical rows of Δ by any one of those $n - r$ vertical rows of K which do not occur in Δ.

Consequently any minor determinant of A of order r such as $\Delta = (a_{pq})_r^r$ has exactly $r(m - r)$ distinct horizontal primaries and exactly $r(n - r)$ distinct vertical primaries.

When $n \nless m$, a *simple* minor determinant of A such as $\Delta = (a_{pq})_m^m$ has exactly $m(n - m)$ distinct vertical primaries and no horizontal primaries.

When $m \nless n$, a *simple* minor determinant of A such as $\Delta = (a_{pq})_n^n$ has exactly $n(m-n)$ distinct horizontal primaries and no vertical primaries.

In this article and the next chapter we shall use symbols $\alpha_{x\lambda}$ and $\beta_{\mu y}$ to denote certain determinants derived from any minor determinant Δ of a matrix A.

We define $\alpha_{x\lambda}$ to be the determinant formed from Δ when we replace the λth horizontal row of H occurring in Δ by the xth horizontal row of H (occurring or not occurring in Δ); and we define $\beta_{\mu y}$ to be the determinant formed from Δ when we replace the μth vertical row of K occurring in Δ by the yth vertical row of K (occurring or not occurring in Δ). These definitions of $\alpha_{x\lambda}$ and $\beta_{\mu y}$ are independent of the notations employed for A and Δ.

When $A = [a]_m^n$ and $\Delta = (a_{pq})_r^r = \begin{pmatrix} q_1 q_2 \dots q_r \\ a \\ p_1 p_2 \dots p_r \end{pmatrix}$, then λ must have one of the values $p_1, p_2, \dots p_r$, and to form $\alpha_{x\lambda}$ from Δ we replace the horizontal suffix λ (occurring in Δ) by x; also μ must have one of the values $q_1, q_2, \dots q_r$, and to form $\beta_{\mu y}$ from Δ we replace the vertical suffix μ (occurring in Δ) by y.

Thus $\alpha_{\lambda\lambda} = \Delta$; if the xth horizontal row of H occurs in Δ and $x \neq \lambda$, then $\alpha_{x\lambda} = 0$; if the xth horizontal row of H does not occur in Δ, then $\alpha_{x\lambda}$ is a horizontal primary of Δ.

Also $\beta_{\mu\mu} = \Delta$; if the yth vertical row of K occurs in Δ and $y \neq \mu$, then $\beta_{\mu y} = 0$; if the yth vertical row of K does not occur in Δ, then $\beta_{\mu y}$ is a vertical primary of Δ.

When the standard double-suffix notation is used and $A = [a]_m^n$, we notice the following cases:

CASE I. If $\Delta = (a_{pq})_r^r$, and if $[A_{pq}]_r^r$ is the reciprocal of $[a_{pq}]_r^r$, we have

$$[\alpha_{1p}]_m^r = [a_{1q}]_m^r \overline{\underline{A_{pq}}}_r^r, \quad [\alpha_{pp}]_r^r = [a_{pq}]_r^r \overline{\underline{A_{pq}}}_r^r = \Delta [1]_r^r, \quad \text{......(A)}$$

$$[\beta_{q1}]_r^n = \overline{\underline{A_{pq}}}_r^r [a_{p1}]_r^n, \quad [\beta_{qq}]_r^r = \overline{\underline{A_{pq}}}_r^r [a_{pq}]_r^r = \Delta [1]_r^r. \quad \text{......(B)}$$

Those $r(m-r)$ elements of $[\alpha_{1p}]_m^r$ which do not lie in $[\alpha_{pp}]_r^r$ are the horizontal primaries of Δ; and those $r(n-r)$ elements of $[\beta_{q1}]_r^n$ which do not lie in $[\beta_{qq}]_r^r$ are the vertical primaries of Δ.

CASE II. If $\Delta = (a)_r^r$, and if $[A]_r^r$ is the reciprocal of $[a]_r^r$, then

$$[\alpha]_m^r = [a]_m^r \overline{\underline{A}}_r^r, \quad [\alpha]_r^r = [a]_r^r \overline{\underline{A}}_r^r = \Delta [1]_r^r, \quad \text{............(A')}$$

$$[\beta]_r^n = \overline{\underline{A}}_r^r [a]_r^n, \quad [\beta]_r^r = \overline{\underline{A}}_r^r [a]_r^r = \Delta [1]_r^r. \quad \text{............(B')}$$

The $r(m-r)$ elements of the last $m-r$ horizontal rows of $[\alpha]_m^r$ are the horizontal primaries of Δ; and the $r(n-r)$ elements of the last $n-r$ vertical rows of $[\beta]_r^n$ are the vertical primaries of Δ.

In Case II we can write

$$[a]_m^r \,\underline{\overline{A}}_r^r = \begin{bmatrix} \Delta \\ p \end{bmatrix}_{r,\,m-r}^r, \quad \underline{\overline{A}}_r^r\,[a]_r^n = [\Delta,\, q]_r^{r\;\;n-r}, \quad \text{............(C)}$$

where $[\Delta]_r^r = \Delta\,[1]_r^r$. Then the elements of $[p]_{m-r}^r$ are the horizontal primaries of Δ; and the elements of $[q]_r^{n-r}$ are the vertical primaries of Δ.

Ex. i. *Proof of formulae* (A) *and* (B).

Let $[a_{1p}]_m^r$ and $[\beta_{q1}]_r^n$ be defined by those equations, and let $H=[a_{1q}]_m^r$ and $K=[a_{p1}]_r^n$. Then by the properties of active rows we have

$$a_{xp_i} = \det\,[a_{xq_1} \ldots a_{xq_r}] \begin{bmatrix} A_{p_iq_1} \\ \vdots \\ A_{p_iq_r} \end{bmatrix}, \quad \beta_{q_jy} = \det\,[A_{p_1q_j} \ldots A_{p_rq_j}] \begin{bmatrix} a_{p_1y} \\ \vdots \\ a_{p_ry} \end{bmatrix},$$

or

$$a_{xp_i} = A_{p_iq_1}a_{xq_1} + A_{p_iq_2}a_{xq_2} + \ldots + A_{p_iq_r}a_{xq_r},$$

$$\beta_{q_jy} = A_{p_1q_j}a_{p_1y} + A_{p_2q_j}a_{p_2y} + \ldots + A_{p_rq_j}a_{p_ry}.$$

This shows that a_{xp_i} is the determinant formed from Δ when we replace the p_ith horizontal row of H (occurring in Δ) by the xth horizontal row of H, and that β_{q_jy} is the determinant formed from Δ when we replace the q_jth vertical row of K (occurring in Δ) by the yth vertical row of K.

Ex. ii. Let $\Delta = \begin{pmatrix} {\scriptstyle 214} \\ a \\ {\scriptstyle 325} \end{pmatrix}$ be a minor determinant of $[a]_5^7$,

so that

$$H = \begin{bmatrix} {\scriptstyle 214} \\ a \\ {\scriptstyle 12345} \end{bmatrix}, \quad K = \begin{bmatrix} {\scriptstyle 1234567} \\ a \\ {\scriptstyle 325} \end{bmatrix}.$$

Then

$$a_{x3} = \begin{pmatrix} {\scriptstyle 214} \\ a \\ {\scriptstyle x25} \end{pmatrix}, \quad a_{x2} = \begin{pmatrix} {\scriptstyle 214} \\ a \\ {\scriptstyle 3x5} \end{pmatrix}, \quad a_{x5} = \begin{pmatrix} {\scriptstyle 214} \\ a \\ {\scriptstyle 32x} \end{pmatrix},$$

$$\beta_{2y} = \begin{pmatrix} {\scriptstyle y14} \\ a \\ {\scriptstyle 325} \end{pmatrix}, \quad \beta_{1y} = \begin{pmatrix} {\scriptstyle 2y4} \\ a \\ {\scriptstyle 325} \end{pmatrix}, \quad \beta_{4y} = \begin{pmatrix} {\scriptstyle 21y} \\ a \\ {\scriptstyle 325} \end{pmatrix},$$

where x receives the values 1, 2, 3, 4, 5, and y the values 1, 2, 3, 4, 5, 6, 7. In particular

$$a_{33} = a_{22} = a_{55} = \Delta\,; \quad \beta_{22} = \beta_{11} = \beta_{44} = \Delta\,;$$

$$\begin{bmatrix} {\scriptstyle 325} \\ a \\ {\scriptstyle 325} \end{bmatrix} = \Delta\,[1]_3^3, \quad \begin{bmatrix} {\scriptstyle 214} \\ \beta \\ {\scriptstyle 214} \end{bmatrix} = \Delta\,[1]_3^3.$$

If $\begin{bmatrix} {\scriptstyle 325} \\ A \\ {\scriptstyle 214} \end{bmatrix}$ is the conjugate reciprocal of $\begin{bmatrix} {\scriptstyle 214} \\ a \\ {\scriptstyle 325} \end{bmatrix}$, we have

$$\begin{bmatrix} {\scriptstyle 214} \\ a \\ {\scriptstyle 12345} \end{bmatrix} \begin{bmatrix} {\scriptstyle 325} \\ A \\ {\scriptstyle 214} \end{bmatrix} = \begin{bmatrix} {\scriptstyle 325} \\ a \\ {\scriptstyle 12345} \end{bmatrix}, \quad \begin{bmatrix} {\scriptstyle 325} \\ A \\ {\scriptstyle 214} \end{bmatrix} \begin{bmatrix} {\scriptstyle 1234567} \\ a \\ {\scriptstyle 325} \end{bmatrix} = \begin{bmatrix} {\scriptstyle 1234567} \\ \beta \\ {\scriptstyle 214} \end{bmatrix}.$$

Ex. iii. Let $\Delta = \begin{vmatrix} b_3 & a_3 & d_3 \\ b_2 & a_2 & d_2 \\ b_5 & a_5 & d_5 \end{vmatrix}$ be a minor determinant of $[a\,b\,c\,d\,e\,f\,g]_{12345}$,

so that
$$H = \begin{bmatrix} b_1 & a_1 & d_1 \\ b_2 & a_2 & d_2 \\ b_3 & a_3 & d_3 \\ b_4 & a_4 & d_4 \\ b_5 & a_5 & d_5 \end{bmatrix}, \quad K = \begin{bmatrix} a_3 & b_3 & c_3 & d_3 & e_3 & f_3 & g_3 \\ a_2 & b_2 & c_2 & d_2 & e_2 & f_2 & g_2 \\ a_5 & b_5 & c_5 & d_5 & e_5 & f_5 & g_5 \end{bmatrix},$$

or
$$H = [b\,a\,d]_{12345}, \qquad K = [a\,b\,c\,d\,e\,f\,g]_{325}.$$

Then
$$\alpha_{43} = (b\,a\,d)_{425}, \quad \alpha_{13} = (b\,a\,d)_{125}, \quad \alpha_{15} = (b\,a\,d)_{321}, \ldots,$$
$$\beta_{23} = (c\,a\,d)_{325}, \quad \beta_{27} = (g\,a\,d)_{325}, \quad \beta_{47} = (b\,a\,g)_{325}, \ldots.$$

In particular
$$\alpha_{33} = \alpha_{22} = \alpha_{55} = \Delta, \quad \beta_{22} = \beta_{11} = \beta_{44} = \Delta,$$
$$\begin{bmatrix} {}^{325} \\ \alpha \\ {}_{325} \end{bmatrix} = \Delta[1]_3^3, \quad \begin{bmatrix} {}^{214} \\ \beta \\ {}_{214} \end{bmatrix} = \Delta[1]_3^3.$$

If $\begin{bmatrix} B_3 & B_2 & B_5 \\ A_3 & A_2 & A_5 \\ D_3 & D_2 & D_5 \end{bmatrix}$ is the conjugate reciprocal of $\begin{bmatrix} b_3 & a_3 & d_3 \\ b_2 & a_2 & d_2 \\ b_5 & a_5 & d_5 \end{bmatrix}$, we have

$$H \begin{bmatrix} B_3 & B_2 & B_5 \\ A_3 & A_2 & A_5 \\ D_3 & D_2 & D_5 \end{bmatrix} = \begin{bmatrix} {}^{325} \\ \alpha \\ {}_{12345} \end{bmatrix}, \quad \begin{bmatrix} B_3 & B_2 & B_5 \\ A_3 & A_2 & A_5 \\ D_3 & D_2 & D_5 \end{bmatrix} K = \begin{bmatrix} {}^{1234567} \\ \beta \\ {}_{214} \end{bmatrix}.$$

Ex. iv. Let $\Delta = (a_{pq})_r^r$ be a minor determinant of $[a]_m^n$, and let $\boxed{A_{pq}}_r^r$ be the conjugate reciprocal of $[a_{pq}]_r^r$.

Then if $[u]_t$ and $[v]_t$ are any minors of $[1\,2\,\ldots\,m]$ and $[1\,2\,\ldots\,n]$ we have

$$[a_{uq}]_t^r \, \boxed{A_{pq}}_r^r = [\alpha_{up}]_m^r, \quad \boxed{A_{pq}}_r^r \, [a_{pv}]_r^t = [\beta_{qv}]_r^t. \quad \ldots\ldots\ldots\ldots\ldots\ldots\text{(D)}$$

In each case the suffixes retained in the product matrix are the active suffixes of the two factor matrices on the left.

These results follow at once from formulae (A) and (B).

Ex. v. Let $\Delta = \begin{pmatrix} a_{pq}, & a_{p\mu} \\ a_{\lambda q}, & a_{\lambda\mu} \end{pmatrix}_{h,\rho}^{k,\sigma}$ be a minor determinant of $[a]_m^n$,

so that
$$k + \sigma = h + \rho,$$

and let $\begin{bmatrix} A_{pq}, & A_{p\mu} \\ A_{\lambda q}, & A_{\lambda\mu} \end{bmatrix}_{h,\rho}^{k,\sigma}$ be the reciprocal of $\begin{bmatrix} a_{pq}, & a_{p\mu} \\ a_{\lambda q}, & a_{\lambda\mu} \end{bmatrix}_{h,\rho}^{k,\sigma}$.

Then
$$[a_{1q}, a_{1\mu}]_m^{k,\sigma} \boxed{\begin{matrix} A_{pq}, & A_{\lambda q} \\ A_{p\mu}, & A_{\lambda\mu} \end{matrix}}_{k,\sigma}^{h,\rho} = [\alpha_{1p}, \alpha_{1\lambda}]_m^{h,\rho}, \quad \ldots\ldots\ldots\ldots\ldots\text{(A}_1\text{)}$$

and
$$\boxed{\begin{matrix} A_{pq}, & A_{\lambda q} \\ A_{p\mu}, & A_{\lambda\mu} \end{matrix}}_{k,\sigma}^{h,\rho} \begin{bmatrix} a_{p1} \\ a_{\lambda 1} \end{bmatrix}_{h,\rho}^n = \begin{bmatrix} \beta_{q1} \\ \beta_{\mu 1} \end{bmatrix}_{k,\sigma}^n. \quad \ldots\ldots\ldots\ldots\ldots\ldots\text{(B}_1\text{)}$$

These formulae are equivalent to formulae (A) and (B). We may if we please prove them directly by the method followed in Ex. i.

We observe that $\begin{bmatrix} a_{pp}, & a_{p\lambda} \\ a_{\lambda p}, & a_{\lambda\lambda} \end{bmatrix}_{h,\rho}^{h,\rho} = \Delta \begin{bmatrix} 1, & 0 \\ 0, & 1 \end{bmatrix}_{h,\rho}^{h,\rho} = \Delta\,[1]_{h+\rho}^{h+\rho},$

and that the remaining $(h+\rho)(m-h-\rho)$ elements of $[a_{1p}, a_{1\lambda}]_m^{h,\rho}$ are the horizontal primaries of Δ.

Also $\begin{bmatrix} \beta_{qq}, & \beta_{q\mu} \\ \beta_{\mu q}, & \beta_{\mu\mu} \end{bmatrix}_{k,\sigma}^{k,\sigma} = \Delta \begin{bmatrix} 1, & 0 \\ 0, & 1 \end{bmatrix}_{k,\sigma}^{k,\sigma} = \Delta\,[1]_{k+\sigma}^{k+\sigma},$

and the remaining $(k+\sigma)(n-k-\sigma)$ elements of $\begin{bmatrix} \beta_{q1} \\ \beta_{\mu 1} \end{bmatrix}_{k,\sigma}^{n}$ are the vertical primaries of Δ.

More generally we have with the notation of Ex. v

$$\begin{bmatrix} a_{uq}, & a_{u\mu} \\ a_{xq}, & a_{x\mu} \end{bmatrix}_{r,s}^{k,\sigma} \underline{\overline{\begin{matrix} A_{pq}, & A_{\lambda q} \\ A_{p\mu}, & A_{\lambda\mu} \end{matrix}}}_{k,\sigma}^{h,\rho} = \begin{bmatrix} a_{up}, & a_{u\lambda} \\ a_{xp}, & a_{x\lambda} \end{bmatrix}_{r,s}^{h,\rho}, \quad \ldots\ldots\ldots\ldots(A_2)$$

$$\underline{\overline{\begin{matrix} A_{pq}, & A_{\lambda q} \\ A_{p\mu}, & A_{\lambda\mu} \end{matrix}}}_{k,\sigma}^{h,\rho} \begin{bmatrix} a_{pv}, & a_{py} \\ a_{\lambda v}, & a_{\lambda y} \end{bmatrix}_{h,\rho}^{r,s} = \begin{bmatrix} \beta_{qv}, & \beta_{qy} \\ \beta_{\mu v}, & \beta_{\mu y} \end{bmatrix}_{k,\sigma}^{r,s}, \quad \ldots\ldots\ldots\ldots(B_2)$$

where $[u]_r$ and $[x]_s$ are any minors of $[1\,2\,\ldots\,m]$ in formula (A_2),
and $[v]_r$ and $[y]_s$ are any minors of $[1\,2\,\ldots\,n]$ in formula (B_2).

Formulae (A_2) and (B_2) follow from formulae (A_1) and (B_1) by the properties of active rows.

In these formulae $a_{u_i p_j}$ is the determinant formed from Δ when in it we replace the horizontal suffix p_j by u_i, $\beta_{q_i y_j}$ is the determinant formed from Δ when in it we replace the vertical suffix q_i by y_j, and so on.

It will be observed that in each of these formulae the suffixes retained in the product matrix on the right are the active suffixes of the two factor matrices on the left.

NOTE 1. *Primary superdeterminants of any minor determinant of a matrix.*

If $\Delta = (a_{pq})_r^r$ is any minor determinant of order r of a matrix $A = [a]_m^n$, those minor determinants of A of order $r+1$ which contain Δ as a minor may be called the primary superdeterminants of Δ in the matrix A.

If we write

$$P_{u_i v_j} = \left(a \begin{matrix} q_1 q_2 \ldots q_r v_j \\ \\ p_1 p_2 \ldots p_r u_i \end{matrix} \right) = \begin{vmatrix} a_{p_1 q_1} & a_{p_1 q_2} & \cdots & a_{p_1 q_r} & a_{p_1 v_j} \\ a_{p_2 q_1} & a_{p_2 q_2} & \cdots & a_{p_2 q_r} & a_{p_2 v_j} \\ \cdots & \cdots & \cdots & \cdots & \cdots \\ a_{p_r q_1} & a_{p_r q_2} & \cdots & a_{p_r q_r} & a_{p_r v_j} \\ a_{u_i q_1} & a_{u_i q_2} & \cdots & a_{u_i q_r} & a_{u_i v_j} \end{vmatrix},$$

then $P_{u_i v_j}$ is a primary superdeterminant of Δ when u_i is an element of the sequence $[1\,2\,\ldots\,m]$ other than $p_1, p_2, \ldots p_r$, and v_j is an element of the sequence $[1\,2\,\ldots\,n]$ other than $q_1, q_2, \ldots q_r$; and $P_{u_i v_j} = 0$ when u_i is an element of the minor sequence $[p_1 p_2 \ldots p_r]$ or v_j is an element of the minor sequence $[q_1 q_2 \ldots q_r]$.

Further if we consider that two primary superdeterminants of Δ in A are distinct from one another only when one is not merely a derangement of the other, then every primary superdeterminant of Δ in A can be expressed in the above form, and there are $(m-r)(n-r)$ distinct primary superdeterminants of Δ, viz. the elements of the matrix $[P_{uv}]_{m-r}^{n-r}$, where $[u]_{m-r}$ is a minor sequence of $[1\ 2 \dots m]$ complementary to $[p]_r$, and $[v]_{n-r}$ is a minor sequence of $[1\ 2 \dots n]$ complementary to $[q]_r$.

More generally a derived determinant of A of any order which contains Δ as a minor may be called a superdeterminant of Δ in A.

NOTE 2. *Primary subdeterminants of any determinant.*

If $\Delta = (a)_r^r$ is any determinant of order r, the minor determinants of $[a]_r^r$ of order $r-1$ may be called the primary subdeterminants of Δ; and if we regard two such primary subdeterminants as being distinct from one another only when one is not merely a derangement of the other, we can identify the primary subdeterminants of Δ with the elements of the matrix $[A]_r^r$ reciprocal to $[a]_r^r$.

§ 103. Elimination of a variable from a system of inequalities.

We can determine real quantities (or integers) $x, u_1, u_2, \dots u_m, v_1, v_2, \dots v_n$ so that the $m+n$ inequalities

$$x \not< u_1,\quad x \not< u_2, \dots x \not< u_m;\quad x \not> v_1,\quad x \not> v_2, \dots x \not> v_n \quad\text{..............(A)}$$

are satisfied when and only when we can determine the real quantities (or integers) $u_1, u_2, \dots u_m, v_1, v_2, \dots v_n$ so that the mn inequalities

$$u_i \not> v_j, \quad (i=1, 2, \dots m;\ j=1, 2, \dots n) \quad\text{..........................(B)}$$

are satisfied.

Thus if x is a real quantity (or an integer) which is variable, and the u's and v's are real quantities (or integers) which are variable or constant, the inequalities (A) are possible when and only when the inequalities (B) are possible.

The inequalities (B) will be said to be obtained by the elimination of x from the inequalities (A).

When the u's and v's are not all independent, in particular when they are all constants, some of the inequalities (B) will be necessary consequences of the rest. These are then superfluous, and we can reduce (B) to a simpler system of inequalities by omitting them.

Again when x is variable and all letters denote real quantities (or integers) the inequalities and equation

$$x \not< u_1,\quad x \not< u_2, \dots x \not< u_m;\quad x \not> v_1,\quad x \not> v_2, \dots x \not> v_n;\quad x = w \quad\text{...........(A')}$$

can be satisfied when and only when the inequalities

$$w \not< u_1,\quad w \not< u_2, \dots w \not< u_m;\quad w \not> v_1,\quad w \not> v_2, \dots w \not> v_n \quad\text{.................(B')}$$

can be satisfied, and the inequalities (B′) will be said to be obtained from (A′) by the elimination of x.

If then we are given any system of linear inequalities and equations in which all constants and variables are real quantities (or integers), we can eliminate all or any number of the variables by successive steps of the above kind. We can thus determine whether it is or is not possible to assign such values to the variables that the given inequalities and equations are satisfied; and we can determine the limitations which the given inequalities and equations impose on the values of any one of the variables or on the values of any number of the variables.

Ex. i. If n and r are given integers, we can determine an integer ρ satisfying the conditions

$$\rho \nless 0, \quad \rho \ngtr r, \quad \rho \ngtr n-r \quad \text{.....................(1)}$$

when and only when n and r satisfy the conditions

$$r \nless 0, \quad r \ngtr n. \quad \text{.....................(2)}$$

For when we eliminate ρ from (1) we obtain $r \nless 0$, $n-r \nless 0$.

We see that n and r must be *positive* integers, and that n must not be less than r.

Ex. ii. If n, r and ρ are given positive integers, we can determine integers s and σ satisfying the conditions

$$s+\sigma \nless r+\rho, \quad s-\sigma \nless r-\rho, \quad \sigma \nless 0, \quad s+\sigma \ngtr n \quad \text{.....................(3)}$$

when and only when n, r and ρ satisfy the conditions

$$r \ngtr n, \quad \rho \ngtr n-r. \quad \text{.....................(4)}$$

Writing the conditions (3) in the form

$$\sigma \nless r+\rho-s, \quad \sigma \ngtr s-r+\rho, \quad \sigma \nless 0, \quad \sigma \ngtr n-s,$$

and eliminating σ, we obtain

$$s-r+\rho \nless r+\rho-s, \quad s-r+\rho \nless 0, \quad n-s \nless r+\rho-s, \quad n-s \nless 0,$$

or

$$s \nless r, \quad s \nless r-\rho, \quad r+\rho \ngtr n, \quad s \ngtr n.$$

Since the second of these last conditions is superfluous, we see that the conditions (3) can be satisfied when and only when we can determine s so as to satisfy the conditions

$$s \nless r, \quad s \ngtr n, \quad r+\rho \ngtr n. \quad \text{.....................(5)}$$

Eliminating s from (5) we obtain the conditions (4).

Ex. iii. If m and r are given integers and u, s and ρ are arbitrary integers satisfying the conditions

$$u \nless 0, \quad s \nless 0, \quad \rho \nless 0, \quad u \ngtr r, \quad u+s+\rho = m-r, \quad \text{.....................(6)}$$

the possible values of u are those consistent with the conditions

$$u \nless 0, \quad u \ngtr r, \quad u \ngtr m-r. \quad \text{.....................(7)}$$

Eliminating ρ from (6) we see that the conditions to be satisfied by m, r, u and s are

$$u \nless 0, \quad s \nless 0, \quad s \ngtr m-r-u, \quad u \ngtr r.$$

Again eliminating s, we obtain (7) as the conditions to be satisfied by u, r and m.

§ 104. Possible ranks of a matrix containing a given minor matrix.

1. *Matrices of the form* $[c, a]_m^{r,p}$ *where* $[c]_m^r$ *is given.*

We will prove three lemmas leading up to Theorem I a.

LEMMA 1. *If* $[c]_{r+1}^r$ *is a given undegenerate matrix of rank* r, *we can always construct a matrix* $[a]_{r+1}^1$ *so that* $[c, a]_{r+1}^{r,1}$ *is undegenerate.*

Denoting the affected simple minor determinants of $[c]_{r+1}^r$ formed by striking out its first, second, ... last horizontal rows by $C_1, C_2, \dots C_{r+1}$ respectively, it is sufficient to choose the elements of $[a]_{r+1}^1$ so that

$$a_{11}C_1 + a_{21}C_2 + \dots + a_{r+1,1}C_{r+1} \neq 0.$$

Since $C_1, C_2, \dots C_{r+1}$ do not all vanish, this is always possible.

LEMMA 2. *If* $[c]_m^r$ *is a given undegenerate matrix of rank* r, *and if* $r<m$, *we can always construct* $[a]_m^1$ *so that* $[c, a]_m^{r,1}$ *has rank* $r+1$.

Let $[c_{u1}]_{r+1}^r$ be any simple minor of $[c]_m^r$ of rank r formed with $r+1$ of its horizontal rows, so that $[u]_{r+1}$ is a minor of the sequence $[1\ 2 \ldots m]$. Then it is sufficient to choose the elements of $[a]_m^1$ so that $[c_{u1}, a_{u1}]_{r+1}^{r,1}$ is undegenerate, and by Lemma 1 this is always possible.

LEMMA 3. *If* $[c]_m^r$ *is a given matrix whose vertical rows are unconnected, and if* $r<m$, *we can construct a matrix* $[a]_m^x$ *so that the vertical rows of* $[c, a]_m^{r,x}$ *are unconnected when and only when*

$$x+r \ngtr m.$$

For by Lemma 2 we can form in succession matrices

$$[c, a]_m^{r,1}, \quad [c, a]_m^{r,2}, \ldots [c, a]_m^{r,x-1}, \quad [c, a]_m^{r,x}$$

of ranks $r+1, r+2, \ldots r+x-1, r+x$ provided that

$$r+1<m, \quad r+2<m, \ldots r+x-1<m,$$

i.e. provided that $x<m-r+1$, or $x \ngtr m-r$.

Theorem Ia. *If* $[c]_m^r$ *is a given matrix of rank* ρ (*where* $\rho \nless 0$, $\rho \ngtr r$, $\rho \ngtr m$), *and if* p *is a given positive integer, we can construct a matrix* $[a]_m^p$ *such that* $[c, a]_m^{r,p}$ *has rank* R *when and only when*

$$R \nless \rho, \quad R \ngtr \rho+p, \quad R \ngtr m. \quad \ldots\ldots\ldots\ldots\ldots\ldots\ldots\ldots\ldots\ldots\text{(A)}$$

Let $[\gamma]_m^\rho = [c_{1v}]_m^\rho$ be a vertical minor of $[c]_m^r$ of rank ρ. Then $[c, a]_m^{r,p}$ has the same rank as $[\gamma, a]_m^{\rho,p}$, and if this rank is R, it is clear that the conditions (A) must be satisfied.

Now let R be any integer consistent with the conditions (A). Then by Lemma 3 we can construct a matrix $[a]_m^{R-\rho}$ such that $[\gamma, a]_m^{\rho, R-\rho}$ has rank R; and if we form $[a]_m^p$ by inserting in $[a]_m^{R-\rho}$ additional vertical rows, $p+\rho-R$ in number, which are either rows of 0's or are connected with the vertical rows of $[\gamma, a]_m^{\rho, R-\rho}$, the matrices $[\gamma, a]_m^{\rho,p}$ and $[c, a]_m^{r,p}$ have rank R.

Ex. i. *If* $[a]_m^p$ *and* $[b]_m^q$ *are arbitrary subject to the conditions that they are undegenerate and have ranks* p *and* q *respectively, the possible values of the rank* R *of the matrix* $\phi=[a, b]_m^{p,q}$ *are those consistent with the conditions*

$$R \nless p, \quad R \nless q, \quad R \ngtr m, \quad R \ngtr p+q. \quad \ldots\ldots\ldots\ldots\ldots\ldots\ldots\ldots\text{(1)}$$

Clearly R must satisfy these conditions.

Again if we put

$$\phi=[a, b]_m^{p,q} = \begin{bmatrix} 1, & 0, & 1, & 0 \\ 0, & 1, & 0, & 0 \\ 0, & 0, & 0, & 1 \\ 0, & 0, & 0, & 0 \end{bmatrix}_{x,\ p-x,\ q-x,\ m-p-q+x}^{x,\ p-x,\ x,\ q-x},$$

where $\qquad x \nless 0, \quad x \ngtr p, \quad x \ngtr q, \quad x \nless p+q-m, \quad \ldots\ldots\ldots\ldots\ldots\text{(2)}$

we have $\qquad R=p+q-x. \quad \ldots\ldots\ldots\ldots\ldots\ldots\ldots\ldots\ldots\ldots\ldots\ldots\text{(3)}$

Giving to x all integral values consistent with the conditions (2), and eliminating x from (2) and (3), we see that R assumes all integral values consistent with the conditions (1). Thus we can construct ϕ so as to have any rank R which is consistent with the conditions (1).

Ex. ii. *If* $[a]_m^p$ *and* $[b]_m^q$ *are arbitrary subject to the conditions that they have ranks* α *and* β *respectively, and if* R *is the rank of the matrix* $\phi = [a, b]_m^{p, q}$, *the possible values of* R *are those consistent with the conditions*

$$R \not< \alpha, \quad R \not< \beta, \quad R \not> m, \quad R \not> \alpha + \beta. \qquad \text{(A')}$$

For if $[a_{1u}]_m^{\alpha}$ and $[b_{1v}]_m^{\beta}$ are vertical minors of $[a]_m^p$ and $[b]_m^q$ of ranks α and β, the matrix ϕ has the same rank as the matrix $\phi' = [a_{1u}, b_{1v}]_m^{\alpha, \beta}$, and the possible ranks of ϕ' are given by Ex. i.

2. *Matrices of the form* $\begin{bmatrix} c \\ b \end{bmatrix}_{r, p}^{n}$ *where* $[c]_r^n$ *is given.*

The following lemmas and theorem correspond to those of sub-article 1, and can be proved in succession in similar ways.

LEMMA 1. *If* $[c]_r^{r+1}$ *is a given undegenerate matrix of rank* r, *we can always construct a matrix* $[b]_1^{r+1}$ *so that* $\begin{bmatrix} c \\ b \end{bmatrix}_{r, 1}^{r+1}$ *is undegenerate.*

LEMMA 2. *If* $[c]_r^n$ *is a given undegenerate matrix of rank* r, *we can always construct* $[b]_1^n$ *so that* $\begin{bmatrix} c \\ b \end{bmatrix}_{r, 1}^{n}$ *has rank* $r+1$.

LEMMA 3. *If* $[c]_r^n$ *is a given matrix whose horizontal rows are unconnected, and if* $r<n$, *we can construct* $[b]_x^n$ *so that the horizontal rows of* $\begin{bmatrix} c \\ b \end{bmatrix}_{r, x}^{n}$ *are unconnected when and only when*

$$x + r \not> n.$$

Theorem I b. *If* $[c]_r^n$ *is a given matrix of rank* ρ (*where* $\rho \not< 0$, $\rho \not> r$, $\rho \not> n$), *and if* p *is a given positive integer, we can construct a matrix* $[b]_p^n$ *such that* $\begin{bmatrix} c \\ b \end{bmatrix}_{r, p}^{n}$ *has rank* R *when and only when*

$$R \not< \rho, \quad R \not> \rho + p, \quad R \not> n. \qquad \text{(B)}$$

Ex. iii. *If* $[a]_p^n$ *and* $[b]_q^n$ *are arbitrary subject to the conditions that they have ranks* α *and* β *respectively, and if* R *is the rank of the matrix* $\phi = \begin{bmatrix} a \\ b \end{bmatrix}_{p, q}^{n}$, *the possible values of* R *are those consistent with the conditions*

$$R \not< \alpha, \quad R \not< \beta, \quad R \not> n, \quad R \not> \alpha + \beta. \qquad \text{(B')}$$

This is proved in the same way as Ex. ii.

3. *Possible ranks of* $\phi = \begin{bmatrix} c, & a \\ b, & d \end{bmatrix}_{p,\, r}^{q,\, s} = [e]_m^n$ *when* $[c]_p^q$ *is given.*

Theorem II. *If* $[c]_p^q$ *is a given matrix of rank* ρ *(where* $\rho \nless 0$, $\rho \ngtr p$, $\rho \ngtr q$*), and if* R *is the rank of the matrix* ϕ*, the possible values of* R *are those consistent with the conditions*

$$R \nless \rho, \quad R \ngtr p+r, \quad R \ngtr q+s, \quad R \ngtr \rho+r+s, \qquad \text{(C)}$$

or

$$R \nless \rho, \quad R \ngtr m, \quad R \ngtr n, \quad R \ngtr \rho+(m-p)+(n-q). \qquad \text{(C')}$$

For if σ is the rank of $[c, a]_p^{q,\, s}$ it follows from Theorem I a that the possible values of σ are given by

$$\sigma \nless \rho, \quad \sigma \ngtr \rho+s, \quad \rho \ngtr p; \qquad (4)$$

and when σ has any assigned value consistent with these conditions it follows from Theorem I b that the possible values of R are given by

$$R \nless \sigma, \quad R \ngtr \sigma+r, \quad R \ngtr q+s. \qquad (5)$$

Eliminating σ from (4) and (5) we see that the possible values of R are those consistent with the conditions (C).

Corollary 1. *A square matrix* $[a]_m^m$ *which contains the zero minor matrix* $[0]_p^p$ *can be undegenerate when and only when*

$$p+q \ngtr m.$$

This is Ex. v of § 83.

Corollary 2. *If* $[a]_m^n$ *is a matrix of assigned rank* R *whose elements are arbitrary (where* $R \nless 0$, $R \ngtr m$, $R \ngtr n$*), and if* ρ *is the rank of a minor matrix* $[a_{uv}]_p^q$*, the possible values of* ρ *are those consistent with the conditions*

$$\rho \nless 0, \quad \rho \ngtr R+p+q-m-n, \quad \rho \ngtr p, \quad \rho \ngtr q, \quad \rho \ngtr R. \qquad \text{(D)}$$

By the theorem these conditions must be satisfied. Further when they are satisfied, then all the conditions (C') are satisfied, and therefore we can construct a matrix $[a]_m^n$ of rank R containing a given minor matrix $[a_{uv}]_p^q$ of rank R.

Corollary 3. *If a matrix* $[a]_m^n$ *and a minor matrix* $[a_{uv}]_p^q$ *have ranks* R *and* ρ *respectively, we must have*

$$\rho+m+n \nless R+p+q.$$

This is the only one of the inequalities (C') and (D) the truth of which is not immediately obvious.

Corollary 4. *If in the matrix* $\phi = \begin{bmatrix} c, & a \\ b, & d \end{bmatrix}_{p,\, r}^{q,\, s}$ *the mutually complementary matrices* $[c]_p^q$ *and* $[d]_r^s$ *have ranks* γ *and* δ *respectively, then the rank* R *of* ϕ *must satisfy the conditions*

$$R \nless \gamma, \quad R \nless \delta, \quad R \ngtr p+r, \quad R \ngtr q+s,$$

and

$$R-\gamma \ngtr r+s, \quad R-\delta \ngtr p+q.$$

§ 105. Possible ranks of a matrix containing a given zero minor matrix.

We will suppose the matrix to be

$$\phi = \begin{bmatrix} c, & a \\ b, & 0 \end{bmatrix}_{r,\, p}^{s,\, q} = [e]_m^n$$

containing the given zero minor matrix $[0]_p^q$; and we will denote the rank of ϕ by R.

1. **Theorem I.** *When the elements of ϕ not contained in the given zero matrix $[0]_p^q$ are arbitrary, the possible values of R are those consistent with the conditions*

$$R \nless 0, \quad R \ngtr r+p, \quad R \ngtr s+q, \quad R \ngtr r+s, \quad \text{.........................(A)}$$

or $$R \nless 0, \quad R \ngtr m, \quad R \ngtr n, \quad R \ngtr (m-p)+(n-q). \quad \text{.....................(A')}$$

This theorem is a particular case of that proved in sub-article 3 of § 104.

NOTE 1. *Alternative proof of Theorem I.*

We can deduce the theorem from Ex. v of § 83 or Corollary 1 of § 104.3. It is evident that R must satisfy the first three of the conditions (A). Now let Δ be any minor determinant of ϕ of order $r+s+x$, where $x \nless 1$. Then Δ must contain at least $s+x$ of the last p horizontal rows and at least $r+x$ of the last q vertical rows of ϕ, and consequently must contain the zero minor matrix $[0]_{s+x}^{r+x}$. Since $(r+x)+(s+x)>r+s+x$, it follows that $\Delta=0$, i.e. every minor determinant of ϕ of order greater than $r+s$ vanishes. Thus R must also satisfy the last of the conditions (A).

It remains to show that every value of R consistent with the conditions (A) is a possible rank of ϕ.

Case I. Let $r+s$ be the least of the three numbers $r+p$, $s+q$, $r+s$.

In this case we have $$p \nless s, \quad q \nless r.$$

We can put all elements of ϕ equal to zero except those occurring in the minor

$$\phi_1 = \begin{bmatrix} c, & a \\ b, & 0 \end{bmatrix}_{r,\, s}^{s,\, r},$$

so that ϕ has the same rank as ϕ_1. By Corollary 1 of § 104.3 we can choose the elements of ϕ_1 so that it is undegenerate and has rank $r+s$. Then by making rows of ϕ_1 to be rows of 0's we can make ϕ_1 have any smaller rank down to 0. Thus we can so construct ϕ that it has any rank from $r+s$ down to 0.

Case II. Let $r+p$ be the least of the three numbers $r+p$, $s+q$, $r+s$.

In this case we have $$s+q \nless r+p, \quad s \nless p.$$

If $r+p \nless s$, we can in the same way make ϕ have the same rank as its minor

$$\phi_2 = \begin{bmatrix} c, & a \\ b, & 0 \end{bmatrix}_{r,\, p}^{s,\, r+p-s},$$

and we can make ϕ_2 have any rank from $r+p$ down to 0.

If $r+p \ngtr s$, we can make ϕ have the same rank as its minor

$$\phi_2' = \begin{bmatrix} c \\ b \end{bmatrix}_{r,\, p}^{r+p},$$

and make ϕ_2' have any rank from $r+p$ down to 0.

Case III. Let $s+q$ be the least of the three numbers $r+p$, $s+q$, $r+s$.

In this case we have $$r+p \nless s+q, \quad r \nless q.$$

If $s+q \nless r$, we can make ϕ have the same rank as its minor

$$\phi_3 = \begin{bmatrix} c, & a \\ b, & 0 \end{bmatrix}_{r,\, s+q-r}^{s,\, q},$$

and we can make ϕ_3 have any rank from $s+q$ down to 0.

If $s+q \not> r$, we can make ϕ have the same rank as its minor

$$\phi_3' = [c,\ a]_{s+q}^{s,\,q},$$

and we can make ϕ_3' have any rank from $s+q$ down to 0.

Thus in every case we can make ϕ have any rank R consistent with the conditions (A).

NOTE 2. *Correction to Ex. vi of* § 83.

The investigation of the possible ranks of ϕ given in Ex. vi of § 83 is incorrect, it having been there implicitly assumed that $\sigma \not> \rho + s$.

2. **Theorem II a.** *When the elements of* $[c]_r^s$, $[a]_r^q$, $[b]_p^s$ *are arbitrary subject to the condition that* $\begin{bmatrix} c \\ b \end{bmatrix}_{r,\,p}^{s}$ *is undegenerate and has rank* s, *the possible values of* R *are those consistent with the conditions*

$$R \not< s, \quad R \not> r+p, \quad R \not> s+q, \quad R \not> r+s. \quad \text{.........................(B)}$$

The conditions (B) are clearly necessary, and they are consistent with one another, since s cannot be greater than any one of the three numbers $r+p$, $s+q$, $r+s$.

To show that every value of R consistent with the conditions (B) is possible, we may proceed as in Note 1 of sub-article 1.

In Case I we can make ϕ_1 have rank $r+s$, in which case $\begin{bmatrix} c \\ b \end{bmatrix}_{r,\,s}^{s}$ has rank s. Then by replacing vertical rows of $[a]_r^r$ by rows of 0's we can make ϕ_1 have any smaller rank which is not less than s.

In Case II we must have $r+p \not< s$, and we can make ϕ_2 have any rank from $r+p$ down to s whilst the matrix $\begin{bmatrix} c \\ b \end{bmatrix}_{r,\,p}^{s}$ always retains the rank s.

In Case III when $s+q \not< r$, we can make ϕ_3 have any rank from $s+q$ down to s whilst $\begin{bmatrix} c \\ b \end{bmatrix}_{r,\,s+q-r}^{s}$ always retains the rank s; and when $s+q \not> r$, we can make ϕ_3' have any rank from $s+q$ down to s whilst $[c]_{s+q}^{s}$ always has rank s.

COROLLARY. *If the only restriction on the elements of* $[c]_r^s$, $[a]_r^q$, $[b]_p^s$ *is that* $\begin{bmatrix} c \\ b \end{bmatrix}_{r,\,p}^{s}$ *has rank* σ, *the possible values of* R *are those consistent with the conditions*

$$R \not< \sigma, \quad R \not> r+p, \quad R \not> \sigma+q, \quad R \not> r+\sigma. \quad \text{.......................}(B_1)$$

For ϕ has the same rank as $\begin{bmatrix} c_{1v}, & a \\ b_{1v}, & 0 \end{bmatrix}_{r,\,p}^{\sigma,\,q}$, where $\begin{bmatrix} c_{1v} \\ b_{1v} \end{bmatrix}_{r,\,p}^{\sigma}$ is a minor of $\begin{bmatrix} c \\ b \end{bmatrix}_{r,\,p}^{s}$ of rank σ.

In a similar way we can prove the following theorem and corollary:

Theorem II b. *When the elements of* $[c]_r^s$, $[a]_r^q$, $[b]_p^s$ *are arbitrary subject to the condition that* $[c,\ a]_r^{s,\,q}$ *is undegenerate and has rank* r, *the possible values of* R *are those consistent with the conditions*

$$R \not< r, \quad R \not> r+p, \quad R \not> s+q, \quad R \not> r+s. \quad \text{.......................(B')}$$

COROLLARY. *If the only restriction on the elements of* $[c]_r^s$, $[a]_r^q$, $[b]_p^s$ *is that* $[c, a]_r^{s, q}$ *has rank* ρ, *the possible values of* R *are those consistent with the conditions*

$$R \not< \rho, \quad R \not> \rho + p, \quad R \not> s + q, \quad R \not> \rho + s. \quad \text{.........................}(\mathrm{B}_1')$$

3. **Theorem III.** *When the elements of* $[c]_r^s$, $[a]_r^q$, $[b]_p^s$ *are arbitrary subject to the conditions that* $\begin{bmatrix} c \\ b \end{bmatrix}_{r, p}^{s}$ *and* $[c, a]_r^{s, q}$ *are undegenerate and have ranks* s *and* r *respectively, the possible values of* R *are those consistent with the conditions*

$$R \not< r, \quad R \not< s, \quad R \not> r + p, \quad R \not> s + q, \quad R \not> r + s. \quad \text{..................}(\mathrm{C})$$

The conditions (C) are clearly necessary and consistent; and they involve the necessary conditions

$$r + p \not< s, \quad s + q \not< r.$$

It remains to show that we can always determine the elements of ϕ so that ϕ has any rank R consistent with these conditions.

We may assume without loss of generality that $s \not< r$, and consider the same cases as in Note 1 of sub-article 1.

In Case I we can in succession put all elements of ϕ equal to zero except those occurring in the minors

$$\begin{bmatrix} c, & a \\ b, & 0 \end{bmatrix}_{r, s}^{s, r}, \quad \begin{bmatrix} c, & a \\ b, & 0 \end{bmatrix}_{r, s-1}^{s, r-1}, \quad \begin{bmatrix} c, & a \\ b, & 0 \end{bmatrix}_{r, s-2}^{s, r-2}, \ldots \begin{bmatrix} c \\ b \end{bmatrix}_{r, s-r}^{s}.$$

Then choosing the elements of these minors (see Corollary 1 in § 104. 3) so that they are undegenerate, we shall have determined values of ϕ having ranks $r+s$, $r+s-1$, $r+s-2$, ... s in which the conditions of the problem are satisfied.

In Case II we have $r + p \not< s$, and we proceed in a similar way making the minors

$$\begin{bmatrix} c, & a \\ b, & 0 \end{bmatrix}_{r, p}^{s, r+p-s}, \quad \begin{bmatrix} c, & a \\ b, & 0 \end{bmatrix}_{r, p-1}^{s, r+p-s-1}, \ldots \begin{bmatrix} c \\ b \end{bmatrix}_{r, s-r}^{s}$$

undegenerate, and putting all other elements of ϕ equal to zero.

In Case III we have $s + q \not< r$, and we make the minors

$$\begin{bmatrix} c, & a \\ b, & 0 \end{bmatrix}_{r, s+q-r}^{s, q}, \quad \begin{bmatrix} c, & a \\ b, & 0 \end{bmatrix}_{r, s+q-r-1}^{s, q-1}, \ldots \begin{bmatrix} c \\ b \end{bmatrix}_{r, s-r}^{s}$$

undegenerate, putting all other elements of ϕ equal to zero.

4. *If* $[b]_p^s$ *is a given matrix of rank* β, *and if the elements of* $[c, a]_r^{s, q}$ *are arbitrary, the possible values of* R *are those consistent with the conditions*

$$R \not< \beta, \quad R \not> r + \beta, \quad R \not> s + q. \quad \text{..............................}(\mathrm{D})$$

If $[a]_r^q$ *is a given matrix of rank* a, *and if the elements of* $\begin{bmatrix} c \\ b \end{bmatrix}_{r, p}^{s}$ *are arbitrary, the possible values of* R *are those consistent with the conditions*

$$R \not< a, \quad R \not> r + p, \quad R \not> s + a. \quad \text{..............................}(\mathrm{D}')$$

These results are particular cases of Theorems I a and I b of § 104. For in the first case $[b, 0]_p^{s, q}$ is a given matrix of rank β, and in the second case $\begin{bmatrix} a \\ 0 \end{bmatrix}_{r, p}^{s}$ is a given matrix of rank a.

5. **Theorem IV.** *If the matrix* $[a]_r^r$ *is undegenerate, then the matrices*

$$\psi = \begin{bmatrix} a, & c \\ 0, & b \end{bmatrix}_{r,\,m}^{r,\,n}, \qquad \psi' = \begin{bmatrix} a, & 0 \\ c, & b \end{bmatrix}_{r,\,n}^{r,\,m}$$

have rank $r+\rho$, *where* ρ *is the rank of* $[b]_m^n$ *or* $[b]_n^m$.

Let $[b_{uv}]_\rho^\rho$ be an undegenerate square minor of $[b]_m^n$ of order ρ. Then by Theorem VI of § 71 the matrix ψ has the same rank as the matrix

$$\psi_1 = \begin{bmatrix} a, & c \\ 0, & b_{u1} \end{bmatrix}_{r,\,\rho}^{r,\,n},$$

and therefore its rank cannot exceed $r+\rho$.

Moreover ψ_1 and ψ contain the non-vanishing minor determinant

$$\begin{pmatrix} a, & c_{1v} \\ 0, & b_{uv} \end{pmatrix}_{r,\,\rho}^{r,\,\rho} = (a)_r^r\,(b_{uv})_\rho^\rho \neq 0$$

of order $r+\rho$, and therefore they both have rank $r+\rho$.

In Theorem IV, which will often be used, our notation has been altered. In the theorems which follow we resume our original notation, and ϕ and R will have the same meanings as in the preceding sub-articles.

Theorem IV a. *If* $[a]_r^q$ *and* $[b]_p^s$ *are given undegenerate matrices of ranks* q *and* p, *and if* $[c]_r^s$ *is arbitrary, the possible values of* R *are those consistent with the conditions*

$$R \nless p+q, \quad R \ngtr r+p, \quad R \ngtr s+q \quad \text{..............................(E)}$$

or

$$R \nless p+q, \quad R \ngtr m, \quad R \ngtr n. \quad \text{..............................(E')}$$

Let $[a_{u1}]_q^q$ be an undegenerate square minor of $[a]_r^q$ of order q. Then by Theorem IV the minor matrix of ϕ

$$\begin{bmatrix} c_{u1}, & a_{u1} \\ b\ \ , & 0 \end{bmatrix}_{q,\,p}^{s,\,q}$$

has rank $p+q$ and is undegenerate. Thus R must satisfy the first of the conditions (E), and by sub-article 1 it must satisfy the other two conditions. It remains to show that ϕ can have any rank R consistent with the conditions (E).

Let $[v]_{r-q}$ be a minor sequence of $[1\,2\ldots r]$ complementary to $[u]_q$. Then if we put $[c_{u1}]_q^s = 0$, the matrix ϕ is a derangement of, and has the same rank as, the matrix

$$\phi' = \begin{bmatrix} 0\ \ , & a_{u1} \\ c_{v1}, & a_{v1} \\ b\ \ , & 0 \end{bmatrix}_{q,\,r-q,\,p}^{s,\,q}.$$

By Theorem IV the matrix ϕ' has rank $q+x$, where x is the rank of

$$\phi'' = \begin{bmatrix} c_{v1} \\ b \end{bmatrix}_{r-q,\,p}^{s}.$$

By Theorem Ib of § 104 we can choose the elements of $[c_{v1}]_{r-q}^s$ so that ϕ'' has rank x when and only when

$$x \nless p, \quad x \ngtr r-q+p, \quad x \ngtr s$$

or

$$q+x \nless p+q, \quad q+x \ngtr r+p, \quad q+x \ngtr s-q.$$

This shows that we can choose the elements of $[c]_r^s$ so that ϕ has any rank R consistent with the conditions (E).

Theorem IV b. *If* $[a]_r^q$ *and* $[b]_p^s$ *are given matrices of ranks* a *and* β, *and if* $[c]_r^s$ *is arbitrary, the possible values of* R *are given by*

$$R \not< a+\beta, \quad R \not> r+\beta, \quad R \not> s+a.$$

For if $[a_{1v}]_r^a$, $[b_{u1}]_\beta^s$ are vertical and horizontal minors of $[a]_r^q$, $[b]_p^s$ of ranks a, β, then ϕ has the same rank as the matrix

$$\phi' = \begin{bmatrix} c, & a_{1v} \\ b_{u1}, & 0 \end{bmatrix}_{r,\beta}^{s,a},$$

and the possible ranks of ϕ' are given by Theorem IV a.

NOTE 3. *Alternative proof of Theorem IV a.*

By Ex. xii of § 71 or by § 146 there exist undegenerate square matrices

$$[h]_r^r, \quad [k]_s^s, \quad [u]_p^p, \quad [v]_q^q$$

such that

$$[h]_r^r [a]_r^q [v]_q^q = \begin{bmatrix} 1 \\ 0 \end{bmatrix}_{q,r-q}^q, \quad [u]_p^p [b]_p^s [k]_s^s = [1, \ 0]_p^{p,s-p}.$$

If we write

$$[a]_r^q = \begin{bmatrix} 1 \\ 0 \end{bmatrix}_{q,r-q}^q, \quad [\beta]_p^s = [1, \ 0]_p^{p,s-p}, \quad [h]_r^r [c]_r^s [k]_s^s = [\gamma]_r^s,$$

we have

$$\begin{bmatrix} h, & 0 \\ 0, & u \end{bmatrix}_{r,p}^{r,p} \begin{bmatrix} c, & a \\ b, & 0 \end{bmatrix}_{r,p}^{r,p} \begin{bmatrix} k, & 0 \\ 0, & v \end{bmatrix}_{s,q}^{s,q} = \begin{bmatrix} \gamma, & a \\ \beta, & 0 \end{bmatrix}_{r,p}^{s,q},$$

where the prefactor and postfactor on the left are undegenerate square matrices. This shows that the matrices

$$\phi = \begin{bmatrix} c, & a \\ b, & 0 \end{bmatrix}_{r,p}^{s,q}, \quad \psi = \begin{bmatrix} \gamma, & a \\ \beta, & 0 \end{bmatrix}_{r,p}^{s,q}$$

have the same rank. Moreover each of the matrices $[c]_r^s$, $[\gamma]_r^s$ receives all possible values when the other receives all possible values. Therefore the possible values of ϕ when $[c]_r^s$ is arbitrary are the same as the possible values of ψ when $[\gamma]_r^s$ is arbitrary.

Now by Theorem IV the rank R of ψ is given by

$$R = p+q+x$$

where x is the rank of the matrix formed from ψ by striking out its first q horizontal rows and its first p vertical rows, i.e. where x is the rank of the matrix $[c]_{r-q}^{s-p}$ whose elements are arbitrary. Since the possible values of x are those consistent with the conditions

$$x \not< 0, \quad x \not> r-q, \quad x \not> s-p,$$

it follows that the possible values of R are those consistent with the conditions (E).

NOTE 4. We can deduce Theorem IV from Theorem IV b as a particular case.

6. **Theorem V.** *If* $[c]_r^s$ *is a given matrix of rank* ρ (*where* $\rho \not< 0$, $\rho \not> r$, $\rho \not> s$), *and if the elements of* $[a]_r^q$ *and* $[b]_p^s$ *are arbitrary, the possible values of* R *are those consistent with the conditions*

$$R \not< \rho, \quad R \not> r+p, \quad R \not> s+q, \quad R \not> r+s, \quad R \not> \rho+p+q. \ldots\ldots\ldots\ldots\ldots\text{(F)}$$

By Theorem II of § 104 the integer R must satisfy the conditions (F). It remains to show that we can construct ϕ so as to have any rank R consistent with these conditions.

By Ex. xii of § 71 or § 146 there exist undegenerate square matrices $[h]_r^r$, $[k]_s^s$ such that

$$[h]_r^r [c]_r^s [k]_s^s = \begin{bmatrix} 1, & 0 \\ 0, & 0 \end{bmatrix}_{\rho,\, r-\rho}^{\rho,\, s-\rho}.$$

If we write

$$[\gamma]_r^s = \begin{bmatrix} 1, & 0 \\ 0, & 0 \end{bmatrix}_{\rho,\, r-\rho}^{\rho,\, s-\rho}, \quad [a]_r^q = [h]_r^r [a]_r^q, \quad [\beta]_p^s = [b]_p^s [k]_s^s,$$

we have

$$\begin{bmatrix} h, & 0 \\ 0, & 1 \end{bmatrix}_{r,\, p}^{r,\, p} \begin{bmatrix} c, & a \\ b, & 0 \end{bmatrix}_{r,\, p}^{s,\, q} \begin{bmatrix} k, & 0 \\ 0, & 1 \end{bmatrix}_{s,\, q}^{s,\, q} = \begin{bmatrix} \gamma, & a \\ \beta, & 0 \end{bmatrix}_{r,\, p}^{s,\, q}.$$

Since the prefactor and postfactor on the left are undegenerate, we see that the matrices

$$\phi = \begin{bmatrix} c, & a \\ b, & 0 \end{bmatrix}_{r,\, p}^{s,\, q}, \quad \psi = \begin{bmatrix} \gamma, & a \\ \beta, & 0 \end{bmatrix}_{r,\, p}^{s,\, q}$$

have the same rank. Moreover $[a]_r^q$, $[b]_p^s$ are arbitrary, i.e. can assume all possible values, when and only when $[a]_r^q$, $[\beta]_p^s$ are arbitrary. Consequently the possible ranks of ϕ when $[c]_r^s$ is a given matrix of rank ρ and $[a]_r^q$ and $[b]_p^s$ are arbitrary are the same as the possible ranks of ψ when $[a]_r^q$ and $[\beta]_p^s$ are arbitrary.

Now consider the special matrix

$$\psi' = \begin{bmatrix} 1, & 0, & 0, & 0, & 1, & 0, & 0 \\ 0, & 1, & 0, & 0, & 0, & 0, & 0 \\ 0, & 0, & 0, & 0, & 0, & 1, & 0 \\ 0, & 0, & 0, & 0, & 0, & 0, & 0 \\ 1, & 0, & 0, & 0, & 0, & 0, & 0 \\ 0, & 0, & 1, & 0, & 0, & 0, & 0 \\ 0, & 0, & 0, & 0, & 0, & 0, & 0 \end{bmatrix}_{u,\, \rho-u,\, x,\, r-\rho-x,\, u,\, y,\, p-u-y}^{u,\, \rho-u,\, y,\, s-\rho-y,\, u,\, x,\, q-u-x}$$

in which u, x and y have any particular values consistent with the conditions

$$u \nless 0, \quad u \ngtr \rho; \quad x \nless 0, \quad x \ngtr r-\rho, \quad x \ngtr q-u; \quad y \nless 0, \quad y \ngtr s-\rho, \quad y \ngtr p-u. \quad \text{......(1)}$$

The matrix ψ' has the form of ψ, and by repeated applications of Theorem IV it will be seen that ψ' has rank R given by

$$R = \rho + u + x + y. \quad \text{..(2)}$$

When we eliminate u, x and y from (1) and (2) we obtain for R the inequalities (F). Thus when u, x and y receive all possible values consistent with the conditions (1), the various values assumed by R are those consistent with the conditions (F); and it follows that we can construct ψ', and therefore also ψ, so as to have any rank R consistent with the conditions (F).

COROLLARY. *If the matrix* $\phi = \begin{bmatrix} c, & a \\ b, & 0 \end{bmatrix}_{r,\, p}^{s,\, q}$ *has the assigned rank* R (*where* $R \nless 0$, $R \ngtr r+p$, $R \ngtr s+q$, $R \ngtr r+s$), *and if* $[c]_r^s$ *has rank* ρ, *the possible values of* ρ *are those consistent with the conditions*

$$\rho \nless 0, \quad \rho \nless R-p-q, \quad \rho \ngtr r, \quad \rho \ngtr s, \quad \rho \ngtr R. \quad \text{........................(F')}$$

For by Theorem V the conditions (F′) must be satisfied, and whenever they are satisfied we can determine ϕ so as to have rank R.

§ 106. Possible ranks of a symmetric matrix containing a given diagonal minor.

1. *Preliminary observations.*

A *diagonal minor* of a square matrix $[a]_m^m$ is one of the form $[a_{uu}]_r^r$, where $[u]_r$ is a minor of the sequence $[1\,2\,...\,m]$, or it is a square minor in which the elements of the leading diagonal are elements of the leading diagonal of the fundamental matrix. Accordingly the diagonal minors of a square matrix are the square minors obtained by striking out corresponding horizontal and vertical rows and then applying corresponding derangements to the retained horizontal and vertical rows.

Every symmetric or self-conjugate matrix ϕ of order $p+r$ can be expressed in the form

$$\phi=\begin{bmatrix} a, & b \\ b', & c \end{bmatrix}_{p,\,r}^{p,\,r},$$

where $a_{ij}=a_{ji}$, $c_{ij}=c_{ji}$, $b'_{ij}=b_{ji}$ for all values of the suffixes i and j which are attached to a, b, b', c in the matrix, so that

$$[b']_r^p = \overline{\underline{b}}_r^p.$$

2. *Possible ranks of the symmetric matrix* $\phi=\begin{bmatrix} a, & b \\ b', & c \end{bmatrix}_{m,\,1}^{m,\,1}$ *when* $[a]_m^m$ *is a given symmetric matrix.*

Theorem I a. *If* $[a]_m^m$ *is a given undegenerate symmetric matrix, the only possible ranks of the symmetric matrix* ϕ *are* m *and* $m+1$. *Whatever the value of* $[b]_m^1$ *may be, we can make* ϕ *have either one of these ranks by suitably choosing the value of* c_{11}.

It is to be understood that $\quad a_{ij}=a_{ji},\quad b'_{1i}=b_{i1}.$

Let $$\Delta=\det\phi=\begin{pmatrix} a, & b \\ b', & c \end{pmatrix}_{m,\,1}^{m,\,1},\quad \Delta_0=\begin{pmatrix} a, & b \\ b', & 0 \end{pmatrix}_{m,\,1}^{m,\,1}.$$

Then $$\Delta=(a)_m^m\, c_{11}+\Delta_0.$$

Since $(a)_m^m\neq 0$, we can choose c_{11}, whatever the value of $[b]_m^1$ may be, so that Δ has any assigned value. If we choose it so that $\Delta=0$, then ϕ has rank m; and if we choose it so that $\Delta\neq 0$, then ϕ has rank $m+1$.

In particular we can make ϕ have rank m by putting $[b]_m^1=0$, $c_{11}=0$; and we can make it have rank $m+1$ by putting $[b]_m^1=0$, $c_{11}=1$.

Theorem I b. *If* $[a]_m^m$ *is a given symmetric matrix of rank* ρ, *where* $\rho<m$, *the possible values of the rank* R *of the symmetric matrix* ϕ *when* c_{11} *and the elements of* $[b]_m^1$ *are arbitrary are* ρ, $\rho+1$ *and* $\rho+2$.

Since the rank of $[a,\ b]_m^{m,\,1}$ cannot exceed the rank of $[a]_m^m$ by more than 1, and the rank of ϕ cannot exceed the rank of $[a,\ b]_m^{m,\,1}$ by more than 1, it follows that R cannot have any value other than ρ, $\rho+1$ or $\rho+2$. It remains to show that R can have any one of these three values.

By § 126 the matrix $[a]_m^m$ must have some undegenerate diagonal minor $[a_{uu}]_\rho^\rho$ of order ρ. We can now distinguish two principal cases.

Case I. *When* $[a,\ b]_m^{m,\,1}$ *and its conjugate* $\begin{bmatrix} a \\ b' \end{bmatrix}_{m,\,1}^m$ *have rank* ρ.

This case occurs when $[b]_m^1$ is connected with the vertical rows of $[a]_m^m$ and therefore $[b']_1^m$ with the horizontal rows of $[a]_m^m$.

It follows from Theorem VI of § 71 that in this case ϕ has the same rank as the matrix

$$\phi' = \begin{bmatrix} a_{uu}, & b_{u1} \\ b'_{1u}, & c \end{bmatrix}_{\rho,\,1}^{\rho,\,1}.$$

By sub-article 1 the only possible values of ϕ' are ρ and $\rho+1$, and we can make ϕ' have either one of these ranks by suitably choosing the value of c_{11}. Thus in this case the possible values of R are ρ and $\rho+1$.

Case II. *When* $[a,\ b]_m^{m,\,1}$ *and its conjugate* $\begin{bmatrix} a \\ b' \end{bmatrix}_{m,\,1}^m$ *have rank* $\rho+1$.

This case occurs when $[b]_m^1$ is not connected with the vertical rows of $[a]_m^m$, and therefore $[b']_1^m$ is not connected with the horizontal rows of $[a]_m^m$. In this case the last horizontal row of ϕ cannot be connected with the preceding horizontal rows, and therefore ϕ has rank $\rho+2$, i.e. the only possible value of R is $\rho+2$.

We see then that it is always possible to construct ϕ so that its rank R has any one of the three values $\rho,\ \rho+1,\ \rho+2$.

In the particular case when $\rho=m$ Case II cannot occur.

Theorem I c. *If* $[a]_m^m$ *is a given symmetric matrix of rank* ρ, *the possible values of the rank* R *of the symmetric matrix* ϕ *when* c_{11} *and the elements of* $[b]_m^1$ *are arbitrary are those consistent with the conditions*

$$R \nless \rho, \quad R \ngtr \rho+2, \quad R \ngtr m. \qquad \text{(A)}$$

This statement summarises the results obtained in Theorems I *a* and I *b*.

3. *Possible ranks of the symmetric matrix* $\phi = \begin{bmatrix} a, & b \\ b', & c \end{bmatrix}_{p,\,r}^{p,\,r} = [a]_m^m$ *when* $[a]_p^p$ *is a given symmetric matrix.*

Theorem II. *If* $[a]_p^p$ *is a given symmetric matrix of rank* ρ (*where* $\rho \nless 0,\ \rho \ngtr p$), *and if the elements of* $[b]_p^r$ *and* $[c]_r^r$ *are arbitrary subject to the restriction that* ϕ *is symmetric, then the possible values of the rank* R *of the symmetric matrix* ϕ *are those consistent with the conditions*

$$R \nless \rho, \quad R \ngtr p+r, \quad R \ngtr \rho+2r \qquad \text{(B)}$$

or

$$R \nless \rho, \quad R \ngtr m, \quad R \ngtr \rho+2(m-p). \qquad \text{(B}')$$

From Theorem II of § 104 we see that R must satisfy the conditions (B). It remains to show that R can have any value consistent with these conditions.

Let the symmetric matrices

$$\begin{bmatrix} a, & b \\ b', & c \end{bmatrix}_{p,1}^{p,1}, \quad \begin{bmatrix} a, & b \\ b', & c \end{bmatrix}_{p,2}^{p,2}, \quad \ldots \begin{bmatrix} a, & b \\ b', & c \end{bmatrix}_{p,r-1}^{p,r-1}, \quad \begin{bmatrix} a, & b \\ b', & c \end{bmatrix}_{p,r}^{p,r}$$

be formed in succession, and let their ranks be denoted by $\rho_1, \rho_2, \ldots \rho_{r-1}, R$. Then by Theorem I*c* the possible values of these numbers are those consistent with the conditions

$$\begin{array}{llll} (1) & \rho_1 \not< \rho, & \rho_1 \not> \rho+2, & \rho_1 \not> p+1; \\ (2) & \rho_2 \not< \rho_1, & \rho_2 \not> \rho_1+2, & \rho_2 \not> p+2; \\ (3) & \rho_3 \not< \rho_2, & \rho_3 \not> \rho_2+2, & \rho_3 \not> p+3; \\ \multicolumn{4}{c}{\cdots\cdots\cdots\cdots\cdots\cdots\cdots\cdots\cdots\cdots} \\ (r) & R \not> \rho_{r-1}, & R \not> \rho_{r+1}+2, & R \not> p+r. \end{array}$$

The result of eliminating ρ_1 from (1) and (2) is

$$\rho_2 \not< \rho, \quad \rho_2 \not> \rho+4, \quad \rho_2 \not> p+2.$$

The result of eliminating ρ_2 from these conditions and (3) is

$$\rho_3 \not< \rho, \quad \rho_3 \not> \rho+6, \quad \rho_3 \not> p+3.$$

Generally we can show by induction that the result of eliminating $\rho_1, \rho_2, \ldots \rho_{i-1}$ from (1), (2), ... (i) is

$$\rho_i \not< \rho, \quad \rho_i \not> \rho+2i, \quad \rho_i \not> p+i.$$

Thus when we eliminate $\rho_1, \rho_2 \ldots \rho_{r-1}$ from (1), (2), ... (r), we obtain the inequalities (B). This shows that we can determine integers $\rho_1, \rho_2, \ldots \rho_{r-1}$ such that the conditions (1), (2), ... (r) are satisfied when and only when the integer R satisfies the conditions (B).

COROLLARY. *If the elements of the symmetric matrix $\phi = [a]_m^m$ are arbitrary subject to the restriction that the matrix is symmetric and has the assigned rank R (where $R \not< 0$, $R \not> m$), and if the diagonal minor $[a_{uu}]_p^p$ has rank ρ, then the possible values of ρ are those consistent with the conditions*

$$\rho \not< 0, \quad \rho \not< R-2(m-p), \quad \rho \not> R, \quad \rho \not> p. \quad \text{..........(C)}$$

For by Theorem II the conditions (C) must be satisfied, and if ρ is any integer satisfying the conditions (C), then the conditions (B′) are satisfied, and we can determine a symmetric matrix $\phi = [a]_m^m$ which has rank R and has a diagonal minor $[a_{uu}]_p^p$ of rank ρ.

NOTE. *Alternative proof of Theorem II.*

By § 147.2 we can determine an undegenerate matrix $[h]_p^p$ such that

$$\overline{h}_p^p \, [a]_p^p \, [h]_p^p = \begin{bmatrix} 1, & 0 \\ 0, & 0 \end{bmatrix}_{\rho, p-\rho}^{\rho, p-\rho}.$$

Writing

$$[a]_p^p = \begin{bmatrix} 1, & 0 \\ 0, & 0 \end{bmatrix}_{\rho, p-\rho}^{\rho, p-\rho}, \qquad [\beta]_p^r = \overline{h}_p^p \, [b]_p^r, \qquad [\beta']_r^p = [b']_r^p \, [h]_p^p, \qquad [\gamma]_r^r = [c]_r^r,$$

we have

$$\overline{\underline{\begin{matrix} h, & 0 \\ 0, & 1 \end{matrix}}}_{p,r}^{p,r} \begin{bmatrix} a, & b \\ b', & c \end{bmatrix}_{p,r}^{p,r} \begin{bmatrix} h, & 0 \\ 0, & 1 \end{bmatrix}_{p,r}^{p,r} = \begin{bmatrix} a, & \beta \\ \beta', & \gamma \end{bmatrix}_{p,r}^{p,r},$$

where the prefactor and postfactor on the left are undegenerate square matrices. This shows that the symmetric matrices

$$\phi = \begin{bmatrix} a, & b \\ b', & c \end{bmatrix}_{p,r}^{p,r}, \qquad \psi = \begin{bmatrix} a, & \beta \\ \beta', & \gamma \end{bmatrix}_{p,r}^{p,r}$$

have the same rank. Moreover each of the matrices $[b]_p^r$, $[\beta]_p^r$ receives all possible values when the other receives all possible values. It follows that the possible ranks of ϕ when $[b]_p^r$ and $[c]_r^r$ are arbitrary are the same as the possible ranks of ψ when $[\beta]_p^r$ and $[\gamma]_r^r$ are arbitrary, it being understood that ϕ and ψ are always symmetric.

Consider now the symmetric matrix

$$\psi' = \begin{bmatrix} 1, & 0, & 0, & 0, & 0, & 0, & 0, & 0 \\ 0, & 0, & 0, & 0, & 0, & 0, & 0, & 0 \\ 0, & 0, & 0, & 1, & 0, & 0, & 0, & 0 \\ 0, & 0, & 1, & 0, & 0, & 0, & 0, & 0 \\ 0, & 0, & 0, & 0, & 1, & 0, & 0, & 0 \\ 0, & 0, & 0, & 0, & 0, & 0, & 0, & 0 \\ 0, & 0, & 0, & 0, & 0, & 0, & 0, & 1 \\ 0, & 0, & 0, & 0, & 0, & 0, & 1, & 0 \end{bmatrix}_{\rho,\, x,\, y,\, y,\, \sigma,\, u,\, v,\, v}^{\rho,\, x,\, y,\, y,\, \sigma,\, u,\, v,\, v}$$

in which x, y, σ, u, v are any integers satisfying the conditions

$$x+y=p-\rho, \quad y+\sigma+u+2v=r\,; \quad \text{.................................(1)}$$

$$x\not<0, \quad y\not<0, \quad \sigma\not<0, \quad u\not<0, \quad v\not<0. \quad \text{.............................(2)}$$

This matrix has the form of the matrix ψ, and by repeated applications of Theorem IV of § 105 we see that it has rank R given by

$$R=\rho+\sigma+2y+2v. \quad \text{..(3)}$$

When we eliminate x, y, σ, u, v from the equation (3) and the inequalities (1) and (2), we obtain the inequalities (B). Accordingly when x, y, σ, u, v receive all integral values consistent with (1) and (2), R receives all integral values consistent with (B).

Thus we can construct ψ', and therefore also ψ, so as to have any value R consistent with the conditions (B).

§ 107. Possible ranks of a symmetric matrix which contains a given zero diagonal minor.

We will suppose that the symmetric matrix contains the zero diagonal minor $[0]_p^p$, and that it is

$$\phi=\begin{bmatrix} c\,, & a \\ a', & 0 \end{bmatrix}_{r,\,p}^{r,\,p}=[e]_m^m$$

where $a'_{ij}=a_{ji}$, $c_{ij}=c_{ji}$; and we will denote its rank by R.

1. **Theorem I.** *When the elements of $[c]_r^r$ and $[a]_r^p$ are arbitrary subject to the condition that ϕ is symmetric, the possible values of R are those consistent with the conditions*

$$R\not<0, \quad R\not>r+p, \quad R\not>2r \quad \text{...................................(A)}$$

or

$$R\not<0, \quad R\not>m, \quad R\not>2(m-p). \quad \text{..........................(A')}$$

This is a particular case of Theorem II of § 106.

Corollary. *If a symmetric matrix $[a]_m^m$ of order m contains a zero diagonal minor $[0]_p^p$ of order p, then if $2p>m$, the determinant $(a)_m^m$ necessarily vanishes; but if $2p\not>m$, the other elements of the matrix can be so chosen that $(a)_m^m\neq 0$.*

This follows from Theorem I, but can be proved independently.

The first part follows from Ex. v of § 83.

To prove the second part we observe that when $m \not< 2p$, the symmetric matrix

$$[a]_m^m = \begin{bmatrix} 0, & 1, & 0 \\ 1, & 0, & 0 \\ 0, & 0, & 1 \end{bmatrix}_{p,\,p,\,m-2p}^{p,\,p,\,m-2p},$$

which is a derangement of $[1]_m^m$, is undegenerate.

NOTE 1. *Alternative proof of Theorem I.*

By Theorem V of § 105 the number R must satisfy the conditions (A). We have to show further that ϕ can have any rank R consistent with these conditions. We consider two cases.

Case I. $2r \not> r+p$, i.e. $r \not> p$.

We can put in succession all elements of ϕ equal to zero except those occurring in the minors

$$\begin{bmatrix} c\,, & a \\ a', & 0 \end{bmatrix}_{r,\,r}^{r,\,r}, \quad \begin{bmatrix} c\,, & a \\ a', & 0 \end{bmatrix}_{r,\,r-1}^{r,\,r-1}, \quad \begin{bmatrix} c\,, & a \\ a', & 0 \end{bmatrix}_{r,\,r-2}^{r,\,r-2}, \ldots [c]_r^r, \quad [c]_{r-1}^{r-1}, \ldots;$$

and then by the above Corollary (which has been proved independently of the theorem) we can choose the elements of these minors so that they are undegenerate. We can thus obtain values of ϕ having the ranks $2r,\ 2r-1,\ 2r-2, \ldots 2,\ 1,\ 0$.

Case II. $r+p \not> 2r$, i.e. $p \not> r$.

We can put in succession all elements of ϕ equal to zero except those occurring in the minors

$$\begin{bmatrix} c\,, & a \\ a', & 0 \end{bmatrix}_{r,\,p}^{r,\,p}, \quad \begin{bmatrix} c\,, & a \\ a', & 0 \end{bmatrix}_{r,\,p-1}^{r,\,p-1}, \quad \begin{bmatrix} c\,, & a \\ a', & 0 \end{bmatrix}_{r,\,p-2}^{r,\,p-2}, \ldots [c]_r^r, \quad [c]_{r-1}^{r-1}, \ldots;$$

and then by the above Corollary we can choose the elements of these minors so that they are undegenerate. We can thus obtain values of ϕ having the ranks

$$r+p,\ r+p-1,\ r+p-2, \ldots 2,\ 1,\ 0.$$

2. **Theorem II.** *When the elements of* $[c]_r^r$ *and* $[a]_r^p$ *are arbitrary subject to the condition that* $[c, a]_r^{r,\,p}$ *and its conjugate are undegenerate or have rank* r, *the possible values of* R *are those consistent with the conditions*

$$R \not< r, \quad R \not> r+p, \quad R \not> 2r. \quad \ldots\ldots\ldots\ldots\ldots\ldots\ldots\ldots\ldots\ldots\text{(B)}$$

The conditions (B) are clearly necessary. To show that ϕ can have any rank R consistent with them, we consider the special matrix

$$\phi' = \begin{bmatrix} 1, & 0, & 0, & 0 \\ 0, & 0, & 1, & 0 \\ 0, & 1, & 0, & 0 \\ 0, & 0, & 0, & 0 \end{bmatrix}_{r-x,\,x,\,x,\,p-x}^{r-x,\,x,\,x,\,p-x}$$

for which $\qquad R = r+x.$

When $r \not> p$, or $2r \not> r+p$, we can give to x the values 0, 1, 2, ... r, and so obtain values of ϕ having the ranks r, $r+1$, ... $2r$.

When $p \not> r$, or $r+p \not> 2r$, we can give to x the values 0, 1, 2, ... p, and so obtain values of ϕ having the ranks r, $r+1$, ... $r+p$.

3. **Theorem III.** *If $[a]_r^p$ is a given undegenerate matrix of rank p, and if $[c]_r^r$ is arbitrary subject to the condition that it is symmetric, the possible values of R are those consistent with the conditions*

$$R \not< 2p, \quad R \not> r+p. \qquad \text{(C)}$$

From Theorem IVa of § 105 we see that R must satisfy the conditions (C). It remains to show that ϕ can have any rank R consistent with these conditions.

Let $[a_{ul}]_p^p$ be an undegenerate square minor of $[a]_r^p$ of order p, and let $[v]_{r-p}$ be a minor sequence of $[1\ 2\ \dots\ r]$ complementary to $[u]_p$. Then if we put $[c_{ul}]_p^r=0$, ϕ is a symmetric derangement of, and has the same rank as, the symmetric matrix

$$\phi' = \begin{bmatrix} 0, & 0, & a_{ul} \\ 0, & c_{vv}, & a_{vl} \\ a'_{1u}, & a'_{1v}, & 0 \end{bmatrix}_{p,\, r-p,\, p}^{p,\, r-p,\, p}.$$

By Theorem IV of § 105 the matrix ϕ' has rank $2p+x$, where x is the rank of the symmetric matrix $[c_{vv}]_{r-p}^{r-p}$; and since the possible values of x are those consistent with the conditions $x \not< 0$, $x \not> r-p$, it follows that the possible ranks R of ϕ' are those consistent with the conditions (C).

Thus we can so choose $[c]_r^r$ that ϕ has any rank R consistent with the conditions (C).

COROLLARY. *If $[a]_r^p$ is a given matrix of rank a and if $[c]_r^r$ is arbitrary subject to the condition that it is symmetric, the possible values of R are those consistent with the conditions*

$$R \not< 2a, \quad R \not> r+a. \qquad \text{(C')}$$

For if $[a_{1v}]_r^a$ is an undegenerate vertical minor of $[a]_r^p$ of rank a, the matrix ϕ has the same rank as the matrix

$$\phi'' = \begin{bmatrix} c, & a_{1v} \\ a'_{v1}, & 0 \end{bmatrix}_{r,\, a}^{r,\, a},$$

and the possible ranks of ϕ'' are given by the theorem.

NOTE 2. *Alternative proof of Theorem III.*

By § 146 it is possible to determine undegenerate square matrices $[h]_r^r$ and $[k]_p^p$

such that $\quad \overline{\underline{h}}_r^r\,[a]_r^p\,[k]_p^p = \begin{bmatrix} 1 \\ 0 \end{bmatrix}_{p,\, r-p}^{p}, \qquad \overline{\underline{k}}_p^p\,[a']_p^r\,[h]_r^r = [1,\ 0]_p^{p,\, r-p}.$

If we write

$$[a]_r^p = \begin{bmatrix} 1 \\ 0 \end{bmatrix}_{p,\, r-p}^{p}, \qquad [a']_p^r = [1,\ 0]_p^{p,\, r-p}, \qquad [\gamma]_r^r = \overline{\underline{h}}_r^r\,[c]_r^r\,[h]_r^r,$$

we have

$$\overline{\underline{\begin{matrix} h, & 0 \\ 0, & k \end{matrix}}}_{r,\, p}^{r,\, p} \begin{bmatrix} c, & a \\ a', & 0 \end{bmatrix}_{r,\, p}^{r,\, p} \begin{bmatrix} h, & 0 \\ 0, & k \end{bmatrix}_{r,\, p}^{r,\, p} = \begin{bmatrix} \gamma, & a \\ a', & 0 \end{bmatrix}_{r,\, p}^{r,\, p},$$

where the prefactor and postfactor on the left are undegenerate square matrices. This shows that the symmetric matrices

$$\phi = \begin{bmatrix} c, & a \\ a', & 0 \end{bmatrix}_{r,\,p}^{r,\,p}, \qquad \psi = \begin{bmatrix} \gamma, & a \\ a', & 0 \end{bmatrix}_{r,\,p}^{r,\,p}$$

have the same rank. Moreover each of the matrices $[c]_r^r$ and $[\gamma]_r^r$ assumes all possible values when the other assumes all possible values. Consequently the possible ranks of ϕ when $[c]_r^r$ is arbitrary but symmetric are the same as the possible ranks of ψ when $[\gamma]_r^r$ is arbitrary but symmetric.

We now consider the special matrix

$$\psi' = \begin{bmatrix} 0, & 0, & 0, & 1 \\ 0, & 0, & 0, & 0 \\ 0, & 0, & 1, & 0 \\ 1, & 0, & 0, & 0 \end{bmatrix}_{p,\, x,\, r-p-x,\, p}^{p,\, x,\, r-p-x,\, p}$$

which has the form of ψ, and whose rank R is given by

$$R = r + p - x.$$

When we give to x the values $r-p$, $r-p-1$, ... 1, 0, we obtain values of ψ having ranks $2p$, $2p+1$, ... $r+p-1$, $r+p$.

Thus we can construct ψ, and therefore also ϕ, so as to have any rank R consistent with the conditions (C).

4. **Theorem IV.** *If $[c]_r^r$ is a given symmetric matrix of rank ρ (where $\rho \nless 0$, $\rho \ngtr r$), and if the elements of $[a]_r^p$ are arbitrary, the possible values of R are those consistent with the conditions*

$$R \nless \rho, \quad R \ngtr r+p, \quad R \ngtr 2r, \quad R \ngtr \rho + 2p, \quad \text{..........................(D)}$$

or

$$R \nless \rho, \quad R \ngtr m, \quad R \ngtr 2r, \quad R \ngtr \rho + 2(m-r). \quad \text{..................(D')}$$

By Theorem V of § 105 the integer R must satisfy the conditions (D). It remains to show that ϕ can have any rank R consistent with those conditions.

By § 147.2 we can determine an undegenerate square matrix $[h]_r^r$ such that

$$\overline{h}_r^r\,[c]_r^r\,[h]_r^r = \begin{bmatrix} 1, & 0 \\ 0, & 0 \end{bmatrix}_{\rho,\, r-\rho}^{\rho,\, r-\rho}.$$

Then if we write

$$[a]_r^p = \overline{h}_r^r\,[a]_r^p, \qquad [a']_p^r = [a']_p^r\,[h]_r^r, \qquad [\gamma]_r^r = \begin{bmatrix} 1, & 0 \\ 0, & 0 \end{bmatrix}_{\rho,\, r-\rho}^{\rho,\, r-\rho},$$

we have

$$\overline{\begin{bmatrix} h, & 0 \\ 0, & 1 \end{bmatrix}}_{r,\,p}^{r,\,p} \begin{bmatrix} c, & a \\ a', & 0 \end{bmatrix}_{r,\,p}^{r,\,p} \begin{bmatrix} h, & 0 \\ 0, & 1 \end{bmatrix}_{r,\,p}^{r,\,p} = \begin{bmatrix} \gamma, & a \\ a', & 0 \end{bmatrix}_{r,\,p}^{r,\,p},$$

where the prefactor and postfactor on the left are undegenerate square matrices. It follows that the matrices

$$\phi = \begin{bmatrix} c, & a \\ a', & 0 \end{bmatrix}_{r,\,p}^{r,\,p}, \qquad \psi = \begin{bmatrix} \gamma, & a \\ a', & 0 \end{bmatrix}_{r,\,p}^{r,\,p}$$

have the same rank. Moreover each of the matrices $[a]_r^p$, $[a]_r^p$ assumes all possible values when the other assumes all possible values. Consequently the possible ranks of ϕ when $[c]_r^r$ is a given symmetric matrix of rank ρ and $[a]_r^p$ is arbitrary are the same as the possible ranks of ψ when $[a]_r^p$ is arbitrary.

We now consider the special symmetric matrix

$$\psi' = \begin{bmatrix} 1, & 0, & 0, & 0, & 1, & 0, & 0 \\ 0, & 1, & 0, & 0, & 0, & 0, & 0 \\ 0, & 0, & 0, & 0, & 0, & 1, & 0 \\ 0, & 0, & 0, & 0, & 0, & 0, & 0 \\ 1, & 0, & 0, & 0, & 0, & 0, & 0 \\ 0, & 0, & 1, & 0, & 0, & 0, & 0 \\ 0, & 0, & 0, & 0, & 0, & 0, & 0 \end{bmatrix}^{u,\, \rho-u,\, x,\, r-\rho-x,\, u,\, x,\, p-u-x}_{u,\, \rho-u,\, x,\, r-\rho-x,\, u,\, x,\, p-u-x},$$

where u and x have any particular values consistent with the conditions

$$u \not< 0, \quad u \not> \rho; \quad x \not< 0, \quad x \not> r-\rho, \quad x \not> p-u. \quad \text{.....................(1)}$$

This matrix ψ' has the form of ψ, and by repeated applications of Theorem IV of § 105 we see that it has rank R given by

$$R = \rho + u + 2x. \quad \text{..(2)}$$

When we eliminate u and x from (1) and (2), we obtain the inequalities (D). Hence when we give to u and x in ψ' all integral values consistent with the conditions (1), the various integral values assumed by R are those consistent with the conditions (D). Thus we can construct ψ, and therefore also ϕ, so as to have any rank R consistent with the conditions (D).

COROLLARY. *If the symmetric matrix* $\phi = \begin{bmatrix} c, & a \\ a', & 0 \end{bmatrix}^{r,p}_{r,p}$ *has the assigned rank* R *(where* $R \not< 0$, $R \not> r+p$, $R \not> 2r$*), and if* $[c]^r_r$ *has rank* ρ*, then the possible values of* ρ *are those consistent with the conditions*

$$\rho \not< R - 2p, \quad \rho \not> R. \quad \text{.....................................(E)}$$

For by Theorem IV the conditions (E) must be satisfied, and when they are satisfied, then all the conditions (D) are satisfied and we can determine ϕ so that it has rank R whilst $[c]^r_r$ has rank ρ.

CHAPTER XIII

RELATIONS BETWEEN THE ELEMENTS AND MINOR DETERMINANTS OF A MATRIX

[In §§ 108 and 109 we obtain identities which give connections between the short rows of an undegenerate matrix and relations between its simple minor determinants; in §§ 115 and 117 we obtain corresponding equations satisfied by the elements and the minor determinants of order r of a matrix whose rank is r; and in § 116 we obtain the fundamental identity from which all the preceding results are deducible. *Sylvester's* identities involving the primary superdeterminants of a determinant and the corresponding identities involving the primary subdeterminants of a determinant are given in §§ 110 and 111. In §§ 113 and 114 we determine conditions for the equivalence of two similar undegenerate matrices and the equivalence of two systems of linear algebraic equations; also we regard a system of equivalent similar undegenerate matrices as representing a spacelet, and define mutually orthogonal and mutually normal undegenerate matrices and spacelets.]

§ 108. Connections between the short rows of an undegenerate matrix.

1. *Connections between the short rows when they are vertical.*

Let $A = [a]_m^n$ be an undegenerate matrix with m long rows and n short rows, the short rows being vertical so that $n \not< m$; and let $[a_{1q}]_m^m$ be the minor matrix formed by any m unconnected short rows of A. Then every short row of A has a unique connection with the m short rows which occur in $[a_{1q}]_m^m$. The connection must be unique, because if there were two different connections for the same short row, there would necessarily be a connection between the short rows of $[a_{1q}]_m^m$. We shall proceed to determine all such connections.

Let $\Delta = (a_{1q})_m^m$, let $[A_{1q}]_m^m$ be the reciprocal matrix of $[a_{1q}]_m^m$, and let

$$\overline{\underline{A_{1q}}}{}_m^m \, [a]_m^n = [\beta_{q1}]_m^n, \quad \text{so that} \quad [\beta_{qq}]_m^m = \Delta\,[1]_m^m.$$

Then, as in § 102, $\beta_{q_i j}$ is the determinant formed from Δ when in it we replace the q_ith vertical row of $[a]_m^n$ by the jth vertical row of $[a]_m^n$, i.e. the determinant formed from $\Delta = \begin{pmatrix} q_1 q_2 \dots q_m \\ a \\ 12 \dots m \end{pmatrix}$ when we replace q_i by j; and the $m(n-m)$ elements of $[\beta_{q1}]^n_m$ which do not occur in $[\beta_{qq}]^m_m$ are the primaries of Δ.

Since the vth vertical row of $[a]^n_m$ is connected with the vertical rows of $[a_{1q}]^m_m$, there exists a relation of the form

$$\begin{bmatrix} a_{1v} \\ a_{2v} \\ \vdots \\ a_{mv} \end{bmatrix} = [a_{1q}]^m_m \begin{bmatrix} x_1 \\ x_2 \\ \vdots \\ x_m \end{bmatrix}. \qquad \dots\dots\dots\dots\dots\dots\dots\dots (1)$$

Solving this equation for $\overline{\underline{x}}_m$ as in § 78 by prefixing the matrix $\overline{\underline{A_{1q}}}^m_m$ on both sides, we obtain the unique solution

$$\Delta \begin{bmatrix} x_1 \\ x_2 \\ \vdots \\ x_m \end{bmatrix} = \overline{\underline{A_{1q}}}^m_m \begin{bmatrix} a_{1v} \\ a_{2v} \\ \vdots \\ a_{mv} \end{bmatrix} = \begin{bmatrix} \beta_{q_1 v} \\ \beta_{q_2 v} \\ \vdots \\ \beta_{q_m v} \end{bmatrix}. \qquad \dots\dots\dots\dots\dots\dots (2)$$

Again since every vertical row of $[a]^n_m$ is connected with the vertical rows of $[a_{1q}]^m_m$, there exists a relation of the form

$$[a]^n_m = [a_{1q}]^m_m [x]^n_m. \qquad \dots\dots\dots\dots\dots\dots\dots\dots\dots\dots (3)$$

Solving this equation for $[x]^n_m$ in the same way we obtain the unique solution

$$\Delta [x]^n_m = \overline{\underline{A_{1q}}}^m_m [a]^n_m = [\beta_{q1}]^n_m. \qquad \dots\dots\dots\dots\dots\dots\dots\dots (4)$$

Substituting in (1) and (3) the values of $\overline{\underline{x}}_m$ and $[x]^n_m$ given by (2) and (4), we obtain the following theorem:

Theorem I a. *If $\Delta = (a_{1q})^m_m$ is a non-vanishing simple minor determinant of $[a]^n_m$, the connection of the vth vertical row of $[a]^n_m$ with the vertical rows of $[a_{1q}]^m_m$ is given by the formula*

$$(a_{1q})^m_m \begin{bmatrix} a_{1v} \\ a_{2v} \\ \vdots \\ a_{mv} \end{bmatrix} = [a_{1q}]^m_m \overline{\underline{A_{1q}}}^m_m \begin{bmatrix} a_{1v} \\ a_{2v} \\ \vdots \\ a_{mv} \end{bmatrix} = [a_{1q}]^m_m \begin{bmatrix} \beta_{q_1 v} \\ \beta_{q_2 v} \\ \vdots \\ \beta_{q_m v} \end{bmatrix}; \qquad \dots\dots\dots (a)$$

and the connections of all the vertical rows of $[a]^n_m$ with the vertical rows of $[a_{1q}]^m_m$ are given by the formula

$$(a_{1q})^m_m [a]^n_m = [a_{1q}]^m_m \overline{\underline{A_{1q}}}^m_m [a]^n_m = [a_{1q}]^m_m [\beta_{q1}]^n_m. \qquad \dots\dots\dots\dots (A)$$

Clearly (A) is a succinct expression for all such formulae as (a), and we can derive (a) from (A) by equating the vth vertical rows on both sides.

Since
$$[a_{1q}]_m^m \, \overline{\underline{A_{1q}}}_m^m = \Delta\,[1]_m^m = (a_{1q})_m^m\,[1]_m^m,$$

formulae (a) and (A) are simply modes of expressing the respective identities

$$[a_{1q}]_m^m \, \overline{\underline{A_{1q}}}_m^m \cdot \begin{bmatrix} a_{1v} \\ a_{2v} \\ \vdots \\ a_{mv} \end{bmatrix} = [a_{1q}]_m^m \cdot \overline{\underline{A_{1q}}}_m^m \begin{bmatrix} a_{1v} \\ a_{2v} \\ \vdots \\ a_{mv} \end{bmatrix}, \quad \ldots\ldots\ldots\ldots (\mathrm{a}_1)$$

$$[a_{1q}]_m^m \, \overline{\underline{A_{1q}}}_m^m \cdot [a]_m^n = [a_{1q}]_m^n \cdot \overline{\underline{A_{1q}}}_m^m \, [a]_m^n . \ldots\ldots\ldots\ldots (\mathrm{A}_1)$$

Consequently formulae (a) *and* (A) *are identities in the elements of the matrix* $A = [a]_m^n$ *and are always true, even when* $\Delta = (a_{1q})_m^m = 0$, *and even when* A *is degenerate, having rank less than* m.

These identities give connections between every set of $m+1$ short (or vertical) rows of A, and in particular they give the connections of every short row with every set of m unconnected short rows.

When $\Delta \neq 0$ they express every element of A in terms of Δ, the primaries of Δ, and those elements of A which occur in Δ.

NOTE 1. *Special case when* $\Delta = (a)_m^m$.

Let $[A]_m^m$ be the reciprocal matrix of $[a]_m^m$, and let

$$\overline{\underline{A}}_m^m \, [a]_m^n = [\beta]_m^n, \quad \text{so that} \quad [\beta]_m^m = \overline{\underline{A}}_m^m \, [a]_m^m = \Delta\,[1]_m^m.$$

Then β_{ij} is the determinant formed from Δ when in it we replace the ith vertical row of $[a]_m^n$ or $[a]_m^m$ by the jth vertical row of $[a]_m^n$, i.e. the determinant formed from Δ when we replace the vertical suffix i by j; and the $m(n-m)$ elements of the last $n-m$ vertical rows of $[\beta]_m^n$ are the primaries of Δ.

If $\Delta \neq 0$, then to find the unique connection of the vth vertical row of A or the unique connections of all the vertical rows of A with the vertical rows of $[a]_m^m$ we solve the respective equations

$$\begin{bmatrix} a_{1v} \\ a_{2v} \\ \vdots \\ a_{mv} \end{bmatrix} = [a]_m^m \begin{bmatrix} x_1 \\ x_2 \\ \vdots \\ x_m \end{bmatrix}, \quad [a]_m^n = [a]_m^m \, [x]_m^n$$

by prefixing $\overline{\underline{A}}_m^m$ on both sides. Substituting the values of $\overline{\underline{x}}_m$ and $[x]_m^n$ thus found, we obtain the following theorem:

Theorem II a. *If* $\Delta=(a)^m_m\neq 0$, *the connection of the* v*th vertical row of* $[a]^n_m$ *with the vertical rows of* $[a]^m_m$ *is given by the formula*

$$(a)^m_m\begin{bmatrix}a_{1v}\\a_{2v}\\\vdots\\a_{mv}\end{bmatrix}=[a]^m_m\,\overline{\underline{A}}{}^m_m\begin{bmatrix}a_{1v}\\a_{2v}\\\vdots\\a_{mv}\end{bmatrix}=[a]^m_m\begin{bmatrix}\beta_{1v}\\\beta_{2v}\\\vdots\\\beta_{mv}\end{bmatrix};\quad\text{.....................(b)}$$

and the connections of all the vertical rows of $[a]^n_m$ *with the vertical rows of* $[a]^m_m$ *are given by the formula*

$$(a)^m_m[a]^n_m=[a]^m_m\,\overline{\underline{A}}{}^m_m\,[a]^n_m=[a]^n_m\,[\beta]^n_m.\quad\text{..........................(B)}$$

Formulae (b) and (B) are simply modes of expressing the identities

$$[a]^m_m\,\overline{\underline{A}}{}^m_m\cdot\begin{bmatrix}a_{1v}\\a_{2v}\\\vdots\\a_{mv}\end{bmatrix}=[a]^m_m\cdot\overline{\underline{A}}{}^m_m\begin{bmatrix}a_{1v}\\a_{2v}\\\vdots\\a_{mv}\end{bmatrix},\quad\text{.....................}(b_1)$$

$$[a]^m_m\,\overline{\underline{A}}{}^m_m\cdot[a]^n_m=[a]^m_m\cdot\overline{\underline{A}}{}^m_m\,[a]^n_m.\quad\text{....................................}(B_1)$$

Consequently formulae (b) *and* (B) *are identities in the elements of* $[a]^n_m$ *and are always true, even when* $(a)^m_m=0$, *and even when* $[a]^n_m$ *is degenerate.*

Since $[\beta]^m_m=\Delta\,[1]^m_m$, we can in this special case write

$$\overline{\underline{A}}{}^m_m\,[a]^n_m=[\Delta,\,q]^{m,\,n-m}_m,$$

where the q's are the primaries of Δ, and in particular $q_{ij}=\beta_{i,\,m+j}$ is the determinant formed when the ith vertical row of Δ is replaced by the $(m+j)$th vertical row of A. We can then write formula (B) more fully in the form

$$\Delta\begin{bmatrix}a_{11}&a_{12}&\dots&a_{1n}\\a_{21}&a_{22}&\dots&a_{2n}\\\dots&\dots&\dots&\dots\\a_{m1}&a_{m2}&\dots&a_{mn}\end{bmatrix}=\begin{bmatrix}a_{11}&a_{12}&\dots&a_{1m}\\a_{21}&a_{22}&\dots&a_{2m}\\\dots&\dots&\dots&\dots\\a_{m1}&a_{m2}&\dots&a_{mm}\end{bmatrix}\begin{bmatrix}\Delta&0&\dots&0&q_{11}&q_{12}&\dots&q_{1,n-m}\\0&\Delta&\dots&0&q_{21}&q_{22}&\dots&q_{2,n-m}\\\dots&\dots&\dots&\dots&\dots&\dots&\dots&\dots\\0&0&\dots&\Delta&q_{m1}&q_{m2}&\dots&q_{m,n-m}\end{bmatrix}$$

or

$$\Delta\,[a]^n_m=[a]^m_m\,[\Delta,\,q]^{m,\,n-m}_m.\quad\text{....................................(C)}$$

NOTE 2. *General principle underlying formulae* (A) *and* (B).

These formulae are particular cases of that given by the following theorem:

Theorem III a. *Let* $\Delta=(a)^m_m$ *be any determinant of order* m, *and let* $[b]^n_m$ *be any matrix with* m *horizontal rows. Then if* δ_{ij} *is the determinant formed from* Δ *when we replace the* i*th vertical row of* Δ *by the* j*th vertical row of* $[b]^n_m$, *we have*

$$(a)^m_m[b]^n_m=[a]^m_m[\delta]^n_m,\quad\text{....................................(D)}$$

this equation being an identity in the elements of $[a]^m_m$ *and* $[b]^n_m$.

For if $\overline{\underline{A}}{}^m_m$ is the conjugate reciprocal of $[a]^m_m$, we have

$$(a)^m_m[b]^n_m=[a]^m_m\,\overline{\underline{A}}{}^m_m\cdot[b]^n_m=[a]^m_m\cdot\overline{\underline{A}}{}^m_m\,[b]^n_m=[a]^m_m[\delta]^n_m.$$

2. *Connections between the short rows when they are horizontal.*

Let $A=[a]_m^n$ be an undegenerate matrix in which the short rows are horizontal, so that $m \nless n$; and let $[a_{p1}]_n^n$ be the minor matrix formed by any n unconnected short rows of A. Then every short row of A has a unique connection with the n short rows which occur in $[a_{p1}]_n^n$.

Let $\Delta=(a_{p1})_n^n$, let $[A_{p1}]_n^n$ be the reciprocal matrix of $[a_{p1}]_n^n$, and let

$$[a]_m^n \overline{\underline{A_{p1}}}_m^n=[\alpha_{1p}]_m^n, \quad \text{so that} \quad [\alpha_{pp}]_n^n=\Delta\,[1]_n^n.$$

Then, as in § 102, α_{ip_j} is the determinant formed from Δ when in it we replace the p_jth horizontal row of $[a]_m^n$ by the ith horizontal row of $[a]_m^n$, or the determinant formed from $\Delta=\begin{pmatrix} 12 \ldots n \\ a \\ p_1\, p_2 \ldots p_n \end{pmatrix}$ when we replace p_j by i; and the $n(m-n)$ elements of $[\alpha_{1p}]_m^n$ which do not occur in $[\alpha_{pp}]_n^n$ are the primaries of Δ.

To find the connection of the uth horizontal row of A or the connections of all the horizontal rows of A with the horizontal rows of $[a_{p1}]_n^n$, we solve the respective equations

$$[a_{u1}\, a_{u2} \ldots a_{un}]=[x_1\, x_2 \ldots x_n]\,[a_{p1}]_n^n, \quad [a]_m^n=[x]_m^n\,[a_{p1}]_n^n$$

by postfixing $\overline{\underline{A_{p1}}}_n^n$ on both sides. We then obtain the following theorem:

Theorem I b. *If* $\Delta=(a_{p1})_n^n$ *is a non-vanishing simple minor determinant of* $[a]_m^n$, *the connections of all the horizontal rows of* $[a]_m^n$ *with the horizontal rows of* $[a_{p1}]_n^n$ *are given by the formula*

$$(a_{p1})_n^n\,[a]_m^n=[a]_m^n\,\overline{\underline{A_{p1}}}_n^n\,[a_{p1}]_n^n=[\alpha_{1p}]_m^n\,[a_{p1}]_n^n. \quad \text{............(A')}$$

Equating the uth horizontal rows on both sides, we obtain the connection of the uth horizontal row of $[a]_m^n$ with the horizontal rows of $[a_{p1}]_n^n$.

Formula (A′) is simply a mode of expressing the identity

$$[a]_m^n \cdot \overline{\underline{A_{p1}}}_n^n\,[a_{p1}]_n^n=[a]_m^n\,\overline{\underline{A_{p1}}}_n^n \cdot [a_{p1}]_n^n. \quad \text{..............(A}_1'\text{)}$$

Consequently formula (A′) *is an identity in the elements of the matrix* $A=[a]_m^n$ *and is always true, even when* $\Delta=(a_{p1})_n^n=0$, *and even when* A *is degenerate, having rank less than* n.

When $\Delta \neq 0$, it expresses all the elements of A in terms of Δ, the primaries of Δ, and those elements of A which occur in Δ.

NOTE 3. *Special case when* $\Delta=(a)^n_n$.

Let $[A]^n_n$ be the reciprocal matrix of $[a]^n_n$, and let

$$[a]^n_m \overline{A}^{\,n}_{\,n}=[\alpha]^n_m, \text{ so that } [\alpha]^n_n=[a]^n_n \overline{A}^{\,n}_{\,n}=\Delta[1]^n_n.$$

Then α_{ij} is the determinant formed when we replace the jth horizontal row of Δ by the ith horizontal row of A, i.e. the determinant formed from Δ when we replace the horizontal suffix j by i; and the $n(m-n)$ elements of the last $m-n$ horizontal rows of $[\alpha]^n_m$ are the primaries of Δ.

If $\Delta\neq0$, then to find the unique connection of the uth horizontal row of A or the unique connections of all the horizontal rows of A with the horizontal rows of $[a]^n_n$, we solve the respective equations

$$[a_{u1}\ a_{u2}\ \dots\ a_{un}]=[x_1\ x_2\ \dots\ x_n][a]^n_n,\quad [a]^n_m=[x]^n_m[a]^n_n$$

by postfixing $\overline{A}^{\,n}_{\,n}$ on both sides. We then obtain the following theorem:

Theorem II b. *If* $\Delta=(a)^n_n\neq0$, *the connections of all the horizontal rows of* $[a]^n_m$ *with the horizontal rows of* $[a]^n_n$ *are given by the formula*

$$(a)^n_n[a]^n_m=[a]^n_m \overline{A}^{\,n}_{\,n}[a]^n_n=[\alpha]^n_m[a]^n_n. \qquad \text{(B')}$$

Formula (B′) is simply a mode of expressing the identity

$$[a]^n_m.\overline{A}^{\,n}_{\,n}[a]^n_n=[a]^n_m\overline{A}^{\,n}_{\,n}.[a]^n_n \qquad \text{(B}_1'\text{)}$$

Consequently formula (B′) *is an identity in the elements of* $[a]^n_m$ *and is always true, even when* $(a)^n_n=0$, *and even when* $[a]^n_m$ *is degenerate.*

Since $[\alpha]^n_n=\Delta[1]^n_n$, we can in this special case write

$$[a]^n_m\overline{A}^{\,n}_{\,n}=\begin{bmatrix}\Delta\\ p\end{bmatrix}^n_{n,\,m-n},$$

where the p's are the primaries of Δ, and in particular $p_{ij}=\alpha_{n+i,j}$ is the determinant formed when the jth horizontal row of Δ is replaced by the $(n+i)$th horizontal row of A. We can then write formula (B′) more fully in the form

$$\Delta\begin{bmatrix}a_{11} & a_{12} & \dots & a_{1n}\\ a_{21} & a_{22} & \dots & a_{2n}\\ \dots & \dots & \dots & \dots\\ a_{m1} & a_{m2} & \dots & a_{mn}\end{bmatrix}=\begin{bmatrix}\Delta & 0 & \dots & 0\\ 0 & \Delta & \dots & 0\\ \dots & \dots & \dots & \dots\\ 0 & 0 & \dots & \Delta\\ p_{11} & p_{12} & \dots & 0\\ p_{21} & p_{22} & \dots & 0\\ \dots & \dots & \dots & \dots\\ p_{m-n,1} & p_{m-n,2} & \dots & p_{m-n,n}\end{bmatrix}\begin{bmatrix}a_{11} & a_{12} & \dots & a_{1n}\\ a_{21} & a_{22} & \dots & a_{2n}\\ \dots & \dots & \dots & \dots\\ a_{n1} & a_{n2} & \dots & a_{nn}\end{bmatrix}$$

or

$$\Delta[a]^n_m=\begin{bmatrix}\Delta\\ p\end{bmatrix}^n_{n,\,m-n}[a]^n_n. \qquad \text{(C')}$$

NOTE 4. *General principle underlying formulae* (A′) *and* (B′).

These formulae are particular cases of that given by the following theorem:

Theorem III b. *Let* $\Delta=[a]_n^n$ *be any determinant of order n, and let* $[b]_m^n$ *be any matrix with n vertical rows. Then if* δ_{ij} *is the determinant formed from* Δ *when we replace the jth horizontal row of* Δ *by the ith horizontal row of* $[b]_m^n$, *we have*

$$(a)_n^n [b]_m^n = [\delta]_m^n [a]_n^n, \quad \text{.................................(D}')$$

this equation being an identity in the elements of $[a]_n^n$ *and* $[b]_m^n$.

For if $\overline{\underline{A}}_n^n$ is the conjugate reciprocal of $[a]_n^n$, we have

$$(a)_n^n [b]_m^n = [b]_m^n \cdot \overline{\underline{A}}_n^n [a]_n^n = [b]_m^n \overline{\underline{A}}_n^n \cdot [a]_n^n = [\delta]_m^n [a]_n^n.$$

Ex. i. Let $[a]_m^{m+1}$ be a matrix with m long rows and $m+1$ short rows, and let $A_1, A_2, \ldots A_{m+1}$ be the affected simple minor determinants formed by omitting the 1st, 2nd, ... $(m+1)$th short rows. Then by § 27 we have

$$[a]_m^{m+1}\begin{bmatrix} A_1 \\ A_2 \\ \vdots \\ A_{m+1} \end{bmatrix}=0.$$

Similarly for the matrix $[a]_{m+1}^m$ we have the identity

$$[A_1 A_2 \ldots A_{m+1}]\,[a]_{m+1}^n=0.$$

All the connections given by the formulae of the text are included in these two results.

Ex. ii. In the matrix $[a\,b\,c\,d]_{12}$ let $(ab)\neq 0$, and let it be required to connect the last two vertical rows with the first two vertical rows.

We assume that $$\begin{bmatrix} c_1 & d_1 \\ c_2 & d_2 \end{bmatrix}=\begin{bmatrix} a_1 & b_1 \\ a_2 & b_2 \end{bmatrix}\begin{bmatrix} x_1 & y_1 \\ x_2 & y_2 \end{bmatrix}.$$

Prefixing the conjugate reciprocal of $[a\,b]_{12}$ on both sides, we obtain

$$(ab)\begin{bmatrix} x_1 & y_1 \\ x_2 & y_2 \end{bmatrix}=\begin{bmatrix} (cb) & (db) \\ (ac) & (ad) \end{bmatrix},\quad (ab)\begin{bmatrix} c_1 & d_1 \\ c_2 & d_2 \end{bmatrix}=\begin{bmatrix} a_1 & b_1 \\ a_2 & b_2 \end{bmatrix}\begin{bmatrix} (cb) & (db) \\ (ac) & (ad) \end{bmatrix}.$$

Thus $$(ab)\begin{bmatrix} c_1 \\ c_2 \end{bmatrix}=\begin{bmatrix} a_1 & b_1 \\ a_2 & b_2 \end{bmatrix}\begin{bmatrix} (cb) \\ (ac) \end{bmatrix},\quad (ab)\begin{bmatrix} d_1 \\ d_2 \end{bmatrix}=\begin{bmatrix} a_1 & b_1 \\ a_2 & b_2 \end{bmatrix}\begin{bmatrix} (db) \\ (ad) \end{bmatrix},$$

or $$(ab)\,c_1=(cb)\,a_1+(ac)\,b_1,\quad (ab)\,d_1=(db)\,a_1+(ad)\,b_1,$$

$$(ab)\,c_2=(cb)\,a_2+(ac)\,b_2,\quad (ab)\,d_2=(db)\,a_2+(ad)\,b_2.$$

These relations remain true when $(ab)=0$.

Ex. iii. In the matrix $[a\,b\,c\,d\,e]_{123}$ let $(cde)\neq 0$, and let it be required to connect the first short row with the last three short rows.

We assume that $$\begin{bmatrix} a_1 \\ a_2 \\ a_3 \end{bmatrix}=\begin{bmatrix} c_1 & d_1 & e_1 \\ c_2 & d_2 & e_2 \\ c_3 & d_3 & e_3 \end{bmatrix}\begin{bmatrix} \lambda_1 \\ \lambda_2 \\ \lambda_3 \end{bmatrix},$$

and solve for the last factor matrix on the right by prefixing $\begin{bmatrix} C_1 & C_2 & C_3 \\ D_1 & D_2 & D_3 \\ E_1 & E_2 & E_3 \end{bmatrix}$, the conjugate reciprocal of $[cde]_{123}$. We thus obtain

$$(cde)\begin{bmatrix} a_1 \\ a_2 \\ a_3 \end{bmatrix} = \begin{bmatrix} c_1 & d_1 & e_1 \\ c_2 & d_2 & e_2 \\ c_3 & d_3 & e_3 \end{bmatrix} \begin{bmatrix} (ade) \\ (cae) \\ (cda) \end{bmatrix}.$$

If we use the identity

$$\begin{bmatrix} c_1 & d_1 & e_1 \\ c_2 & d_2 & e_2 \\ c_3 & d_3 & e_3 \end{bmatrix} \begin{bmatrix} C_1 & C_2 & C_3 \\ D_1 & D_2 & D_3 \\ E_1 & E_2 & E_3 \end{bmatrix} \cdot \begin{bmatrix} a_1 \\ a_2 \\ a_3 \end{bmatrix} = \begin{bmatrix} c_1 & d_1 & e_1 \\ c_2 & d_2 & e_2 \\ c_3 & d_3 & e_3 \end{bmatrix} \cdot \begin{bmatrix} C_1 & C_2 & C_3 \\ D_1 & D_2 & D_3 \\ E_1 & E_2 & E_3 \end{bmatrix} \begin{bmatrix} a_1 \\ a_2 \\ a_3 \end{bmatrix},$$

we obtain the same result without assuming that $(cde) \neq 0$.

Ex. iv. By connecting $\overline{\underline{x}}_2$, $\overline{\underline{x}}_3$, $\overline{\underline{x}}_4$, ... with the vertical rows of $[ab]_{12}$, $[abc]_{123}$, $[abcd]_{1234}$, ... respectively, we obtain the identities

$$(ab)\begin{bmatrix} x_1 \\ x_2 \end{bmatrix} = \begin{bmatrix} a_1 & b_1 \\ a_2 & b_2 \end{bmatrix} \begin{bmatrix} (xb) \\ (ax) \end{bmatrix},$$

$$(abc)\begin{bmatrix} x_1 \\ x_2 \\ x_3 \end{bmatrix} = \begin{bmatrix} a_1 & b_1 & c_1 \\ a_2 & b_2 & c_2 \\ a_3 & b_3 & c_3 \end{bmatrix} \begin{bmatrix} (xbc) \\ (axc) \\ (abx) \end{bmatrix},$$

$$(abcd)\begin{bmatrix} x_1 \\ x_2 \\ x_3 \\ x_4 \end{bmatrix} = \begin{bmatrix} a_1 & b_1 & c_1 & d_1 \\ a_2 & b_2 & c_2 & d_2 \\ a_3 & b_3 & c_3 & d_3 \\ a_4 & b_4 & c_4 & d_4 \end{bmatrix} \begin{bmatrix} (xbcd) \\ (axcd) \\ (abxd) \\ (abcx) \end{bmatrix},$$

and so on. The non-trivial identities given by formulae (a) and (b) are all reducible to these forms.

The third of the equations given above is equivalent to the identity

$$\begin{bmatrix} a_1 & b_1 & c_1 & d_1 \\ a_2 & b_2 & c_2 & d_2 \\ a_3 & b_3 & c_3 & d_3 \\ a_4 & b_4 & c_4 & d_4 \end{bmatrix} \begin{bmatrix} A_1 & A_2 & A_3 & A_4 \\ B_1 & B_2 & B_3 & B_4 \\ C_1 & C_2 & C_3 & C_4 \\ D_1 & D_2 & D_3 & D_4 \end{bmatrix} \cdot \begin{bmatrix} x_1 \\ x_2 \\ x_3 \\ x_4 \end{bmatrix} = \begin{bmatrix} a_1 & b_1 & c_1 & d_1 \\ a_2 & b_2 & c_2 & d_2 \\ a_3 & b_3 & c_3 & d_3 \\ a_4 & b_4 & c_4 & d_4 \end{bmatrix} \cdot \begin{bmatrix} A_1 & A_2 & A_3 & A_4 \\ B_1 & B_2 & B_3 & B_4 \\ C_1 & C_2 & C_3 & C_4 \\ D_1 & D_2 & D_3 & D_4 \end{bmatrix} \begin{bmatrix} x_1 \\ x_2 \\ x_3 \\ x_4 \end{bmatrix},$$

where $[ABCD]_{1234}$ is the reciprocal of $[abcd]_{1234}$.

Ex. v. Using formula (B) we have

$$(abc)\begin{bmatrix} a_1 & b_1 & c_1 & x_1 & y_1 \\ a_2 & b_2 & c_2 & x_2 & y_2 \\ a_3 & b_3 & c_3 & x_3 & y_3 \end{bmatrix} = \begin{bmatrix} a_1 & b_1 & c_1 \\ a_2 & b_2 & c_2 \\ a_3 & b_3 & c_3 \end{bmatrix} [\beta]_3^5 = \begin{bmatrix} a_1 & b_1 & c_1 \\ a_2 & b_2 & c_2 \\ a_3 & b_3 & c_3 \end{bmatrix} \begin{bmatrix} (abc) & 0 & 0 & (xbc) & (ybc) \\ 0 & (abc) & 0 & (axc) & (ayc) \\ 0 & 0 & (abc) & (abx) & (aby) \end{bmatrix}.$$

Ex. vi. Using formula (A) we have

$$\begin{pmatrix}251\\a\\123\end{pmatrix}[a]_3^5=\begin{bmatrix}251\\a\\123\end{bmatrix}[\beta_{q1}]_3^5=\begin{bmatrix}251\\a\\123\end{bmatrix}\begin{bmatrix}0 & \begin{pmatrix}251\\a\\123\end{pmatrix} & \begin{pmatrix}351\\a\\123\end{pmatrix} & \begin{pmatrix}451\\a\\123\end{pmatrix} & 0\\ 0 & 0 & \begin{pmatrix}231\\a\\123\end{pmatrix} & \begin{pmatrix}241\\a\\123\end{pmatrix} & \begin{pmatrix}251\\a\\123\end{pmatrix}\\ \begin{pmatrix}251\\a\\123\end{pmatrix} & 0 & \begin{pmatrix}253\\a\\123\end{pmatrix} & \begin{pmatrix}254\\a\\123\end{pmatrix} & 0\end{bmatrix}.$$

The corresponding result in the single-suffix notation is

$$(bea)[abcde]_{123}=[bea]_{123}\begin{bmatrix}0 & (bea) & (cea) & (dea) & 0\\ 0 & 0 & (bca) & (bda) & (bea)\\ (bea) & 0 & (bec) & (bed) & 0\end{bmatrix}.$$

Ex. vii. *First extensions of formulae* (A) *and* (A′).

Let $\Delta=(a_{pq})_r^r$ be any minor determinant of $[a]_m^n$, and let $[A_{pq}]_r^r$ be the reciprocal of $[a_{pq}]_r^r$. Also let $[a_{pv}]_r^t$ be any vertical minor of the horizontal minor $[a_{p1}]_r^n$ in which Δ lies, and let $[a_{uq}]_t^r$ be any horizontal minor of the vertical minor $[a_{1q}]_m^r$ in which Δ lies. Then by formulae (D) in Ex. iv of § 102 we have

$$(a_{pq})_r^r[a_{pv}]_r^t=[a_{pq}]_r^r\left\lceil A_{pq}\right\rfloor_r^r[a_{pv}]_r^t=[a_{pq}]_r^r[\beta_{qv}]_r^t, \quad\text{......................(E)}$$

$$(a_{pq})_r^r[a_{uq}]_t^r=[a_{uq}]_t^r\left\lceil A_{pq}\right\rfloor_r^r[a_{pq}]_r^r=[a_{up}]_t^r[a_{pq}]_r^r, \quad\text{......................(E′)}$$

where the a's and β's have the meanings assigned to them in § 102.

These results are simply formulae (A) and (A′) for the matrices $[a_{p1}]_r^n$ and $[a_{1q}]_m^r$.

If $\Delta=(a_{pq},a_{p\mu})_r^{k,\sigma}$, so that $k+\sigma=r$, we can replace (E) by the equivalent formula

$$(a_{pq},a_{p\mu})_r^{k,\sigma}[a_{pv}]_r^t=[a_{pq},a_{p\mu}]_r^{k,\sigma}\begin{bmatrix}\beta_{qv}\\ \beta_{\mu v}\end{bmatrix}_{k,\sigma}^t; \quad\text{......................(F)}$$

and if $\Delta=\begin{pmatrix}a_{pq}\\ a_{\lambda q}\end{pmatrix}_{h,\rho}^r$, so that $h+\rho=r$, we can replace (E′) by the equivalent formula

$$\begin{pmatrix}a_{pq}\\ a_{\lambda q}\end{pmatrix}_{h,\rho}^r[a_{uq}]_t^r=[a_{up},a_{u\lambda}]_t^{h,\rho}\begin{bmatrix}a_{pq}\\ a_{\lambda q}\end{bmatrix}_{h,\rho}^r. \quad\text{.........................(F′)}$$

In both cases the a's and β's have the meanings ascribed to them in § 102, so that we obtain $a_{x\lambda}$ from Δ by replacing the horizontal suffix λ by x, and we obtain $\beta_{\mu y}$ from Δ by replacing the vertical suffix μ by y.

The left-hand sides of these formulae can all be written down immediately by using Theorems III *a* and III *b*.

Ex. viii. *Second extensions of formulae* (A) *and* (A′).

If $k+\sigma=h+\rho$, we see from formulae (B_2) and (A_2) of § 102 that

$$\begin{pmatrix}a_{pq}, & a_{p\mu}\\ a_{\lambda q}, & a_{\lambda\mu}\end{pmatrix}_{h,\rho}^{k,\sigma}\begin{bmatrix}a_{pv}\\ a_{\lambda v}\end{bmatrix}_{h,\rho}^t=\begin{bmatrix}a_{pq}, & a_{p\mu}\\ a_{\lambda q}, & a_{\lambda\mu}\end{bmatrix}\begin{bmatrix}\beta_{qv}\\ \beta_{\mu v}\end{bmatrix}_{k,\sigma}^t, \quad\text{..................(G)}$$

$$\begin{pmatrix}a_{pq}, & a_{p\mu}\\ a_{\lambda q}, & a_{\lambda\mu}\end{pmatrix}_{h,\rho}^{k,\sigma}[a_{uq},\ a_{u\mu}]_t^{k,\sigma}=[a_{up},\ a_{u\lambda}]_t^{h,\rho}\begin{bmatrix}a_{pq}, & a_{p\mu}\\ a_{\lambda q}, & a_{\lambda\mu}\end{bmatrix}_{h,\rho}^{k,\sigma}, \quad\text{...........(G′)}$$

where $\beta_{q_i v_j}$ and $\beta_{\mu_i v_j}$ are the determinants formed from Δ when we replace the vertical suffix q_i by v_j and when we replace the vertical suffix μ_i by v_j, and where $\alpha_{u_i p_j}$ and $\alpha_{u_i \lambda_j}$ are the determinants formed from Δ when we replace the horizontal suffix p_j by u_i and when we replace the horizontal suffix λ_j by u_i.

The left-hand sides of the formulae can be written down immediately when we use Theorems III *a* and III *b*.

NOTE 5. *Connections between the rows of a degenerate matrix.*

For a degenerate matrix the formulae of the text all assume the trivial form $0=0$.

Let $[a]_m^n$ be any degenerate matrix of rank r. Then every row of the matrix has a unique connection with every set of r parallel unconnected rows.

If $(a_{pq})_r^r \neq 0$, then we can by sub-article 1 determine the unique connection of the vth vertical row of $[a_{p1}]_r^n$ with the vertical rows of $[a_{pq}]_r^r$, and this is also the unique connection of the vth vertical row of $[a]_m^n$ with the vertical rows of $[a_{1q}]_m^r$.

Also we can by sub-article 2 determine the unique connection of the uth horizontal row of $[a_{1q}]_m^r$ with the horizontal rows of $[a_{pq}]_r^r$, and this is also the unique connection of the uth horizontal row of $[a]_m^n$ with the horizontal rows of $[a_{p1}]_r^n$.

§ 109. Identities which serve to express all simple minor determinants of an undegenerate matrix in terms of any one non-vanishing simple minor determinant and its primaries.

Throughout the present article Δ will denote a simple minor determinant of the matrix $A=[a]_m^n$, and the α's and β's will denote the determinants formed from Δ in the manner described in § 102.

By an identity is meant an identity in the elements of the matrix A.

1. *The identities when the short rows of the matrix are vertical.*

Let $A=[a]_m^n$ be any matrix, not necessarily undegenerate, in which the short rows are vertical, so that $n \nless m$. Then we will prove the theorems which follow.

Theorem I a. *If $\Delta=(a_{1q})_m^m$ and $D=(a_{1v})_m^m$ are any two simple minor determinants of A, each formed with any m of its vertical rows arranged in any order, then the equation*

$$\Delta^{m-1} D = (\beta_{qv})_m^m \quad \text{.............................(A)}$$

is an identity in the elements of Δ and D, and when $\Delta \neq 0$ serves to express D as a homogeneous rational function of degree 1 of Δ and its primaries.

Theorem II a. *More generally if $[A]_\nu$ and $[Q]_\nu$ are correspondingly formed complete matrices of the simple minor determinants of $[a]_m^n$ and $[\beta_{q1}]_m^n$, then the equation*

$$\Delta^{m-1}[A_1 A_2 \dots A_\nu]=[Q_1 Q_2 \dots Q_\nu] \quad \text{....................(a)}$$

is an identity in the elements of A, and when $\Delta \neq 0$ serves to express all the simple minor determinants of A as homogeneous rational functions of degree 1 of Δ and its primaries.

Here $\beta_{\mu y}$ is the determinant formed from Δ when we replace the μth vertical row of A (occurring in Δ) by the yth vertical row of A, or when in Δ we replace the vertical suffix μ by y.

We denote the reciprocal of $[a_{1q}]_m^m$ by $[A_{1q}]_m^m$, and have

$$[\beta_{q1}]_m^n = \boxed{A_{1q}}_m^m [a]_m^n, \quad [\beta_{qq}]_m^m = \Delta [1]_m^m. \quad \text{..............(1)}$$

Those elements of $[\beta_{q1}]_m^n$ which do not occur in $[\beta_{qq}]_m^m$ are the primaries of Δ.

Proof of Theorem Ia. From formula (A) of § 108 we deduce the obviously true identity

$$(a_{1q})_m^m [a_{1v}]_m^m = [a_{1q}]_m^m \boxed{A_{1q}}_m^m \cdot [a_{1v}]_m^m = [a_{1q}]_m^m \cdot \boxed{A_{1q}}_m^m [a_{1v}]_m^m = [a_{1q}]_m^m [\beta_{qv}]_m^m,$$

and by equating the determinants of both sides we obtain

$$\Delta^m (a_{1v})_m^m = \Delta (\beta_{qv})_m^m.$$

Since this is an identity in the elements of Δ and D, and Δ is a rational integral function of its elements which does not vanish identically, we can cancel the common factor Δ on both sides, and so obtain formula (A).

We can also obtain (A) by equating the determinants of both sides in the identity

$$\boxed{A_{1q}}_m^m [a_{1v}]_m^m = [\beta_{qv}]_m^m,$$

and using the equation $(A_{1q})_m^m = \Delta^{m-1}$ (see Ex. xi of § 68) which is an identity in the elements of Δ.

Every element of $[\beta_{qv}]_m^m$ is either Δ or 0 or a primary of Δ.

Therefore when $\Delta \neq 0$, formula (A) expresses D as a fraction whose denominator is Δ^{m-1}, and whose numerator is a homogeneous rational integral function of degree m of Δ and its primaries.

Proof of Theorem IIa. If we equate correspondingly formed complete matrices of the simple minor determinants of both sides in the equation (A) of § 108, we obtain

$$\Delta^m [A_1 A_2 \dots A_\nu] = \Delta [Q_1 Q_2 \cdots Q_\nu].$$

Since Δ is a rational integral function of the elements of A which does not vanish identically, we can cancel the common factor Δ on both sides, and so obtain formula (a).

We can also obtain (a) by equating the determinants of both sides in the identity

$$\overline{\underline{A_{1q}}}_m^m [a]_m^n = [\beta_{q1}]_m^n,$$

and using the identity $(A_{1q})_m^m = \Delta^{m-1}$.

If in formula (A) of Theorem I a any element of $[\beta_{qv}]_m^m$ has the value Δ, then every other element in the same vertical row has the value 0, and we can reduce formula (A) by cancelling an additional common factor Δ on both sides. This case occurs when and only when Δ and D have a vertical row in common. By such reductions we obtain Theorems III a and IV a which follow.

Theorem III a. *If the simple minor determinants* $\Delta = (a_{1q})_m^m$ *and* $D = (a_{1v})_m^m$ *have* r *vertical rows in common forming the minor matrix* $[a_{1\mu}]_m^r$ *of* $[a]_m^n$, *then we can reduce formula* (A) *of Theorem I a to the identity*

$$\Delta^{m-r-1} D = (-1)^{\omega} (\beta_{xy})_{m-r}^{m-r}, \qquad \text{(B)}$$

where $[x]_{m-r}$, $[y]_{m-r}$ *are the minors of the sequences* $[q]_m$, $[v]_m$ *formed by striking out their common elements* $\mu_1, \mu_2, \ldots \mu_r$, *and where* ω *is the difference of the affects of* $[\mu]_r$ *in* $[q]_m$ *and* $[v]_m$.

If further the determinants Δ *and* D *have no other vertical rows in common, then in formula* (B) *all the elements of* $[\beta_{xy}]_{m-r}^{m-r}$ *are primaries of* Δ.

Theorem IV a. *If* $\Delta = (a_{1\mu}, a_{1x})_m^{r, m-r}$ *and* $D = (a_{1\mu}, a_{1y})_m^{r, m-r}$ *are two simple minor determinants of* A *whose first* r *vertical rows are the same, then the equation*

$$\Delta^{m-r-1} D = (\beta_{xy})_{m-r}^{m-r} \qquad \text{(C)}$$

is an identity in the elements of Δ *and* D.

The identity (B) *can always be reduced to the standard form* (C).

Proof of Theorem III a. Suppose that $[a_{1q}]_m^m$ and $[a_{1v}]_m^m$ have r vertical rows in common which form a minor $[a_{1\mu}]_m^r$ of $[a]_m^n$, and let $[a_{1x}]_m^{m-r}$, $[a_{1y}]_m^{m-r}$ be the minors of $[a_{1q}]_m^m$, $[a_{1v}]_m^m$ formed by striking out these r common vertical rows. Then $[\mu]_r$ is a minor of both the sequences $[q]_m$ and $[v]_m$, and since $[\beta_{qq}]_m^m = \Delta [1]_m^m$, we have $[\beta_{\mu\mu}]_r^r = \Delta [1]_r^r$.

Expanding the determinant $(\beta_{qv})_m^m$ in terms of the simple minor determinants of its minor matrix $[\beta_{q\mu}]_m^r$, and observing that all these determinants vanish except $(\beta_{\mu\mu})_r^r$ which has the value Δ^r, we obtain

$$(\beta_{qv})_m^m = (-1)^{\omega} (\beta_{\mu\mu})_r^r (\beta_{xy})_{m-r}^{m-r} = (-1)^{\omega} \Delta^r (\beta_{xy})_{m-r}^{m-r},$$

where ω is the affect of $[\beta_{\mu\mu}]_r^r$ in $[\beta_{qv}]_m^m$, i.e. the sum (or difference) of the affects of $[\mu]_r$ in $[q]_m$ and $[v]_m$. Substituting this value of $(\beta_{qv})_m^m$ in formula (A), and then cancelling the factor Δ^r which is common to both sides, we obtain formula (B). If $[a_{1q}]_m^m$ and $[a_{1v}]_m^m$ have no more vertical rows in common besides those specified above, then $[x]_{m-r}$ and $[y]_{m-r}$ have no elements in common, and therefore $[q]_r$ and $[y]_{m-r}$ have no elements in common; consequently every element of $[\beta_{xy}]_{m-r}^{m-r}$ is one of the primaries of Δ.

Proof of Theorem IV a. Formula (C) is a particular case of formula (B). To obtain (C) directly we write

$$[h]_m^m = [a_{1\mu}, a_{1x}]_m^{r, m-r}, \quad [k]_m^m = [a_{1\mu}, a_{1y}]_m^{r, m-r}, \quad [H]_m^m = [A_{1\mu}, A_{1x}]_m^{r, m-r},$$

where $[H]_m^m$ is the reciprocal of $[h]_m^m$; so that

$$(h)_m^m = \Delta, \quad (k)_m^m = D, \quad (H)_m^m = \Delta^{m-1}.$$

Then observing that by formula (B_2) of § 102 or Ex. vii of § 108

$$\overline{\underline{\begin{matrix} A_{1\mu} \\ A_{1x} \end{matrix}}}_{r, m-r}^{m} [a_{1\mu}, a_{1y}]_m^{r, m-r} = \begin{bmatrix} \beta_{\mu\mu}, & \beta_{\mu y} \\ \beta_{x\mu}, & \beta_{xy} \end{bmatrix}_{r, m-r}^{r, m-r} = \begin{bmatrix} \Delta, & \beta_{\mu y} \\ 0, & \beta_{xy} \end{bmatrix}_{r, m-r}^{r, m-r},$$

we equate the determinants of both sides in the identity

$$(h)_m^m [k]_m^m = [h]_m^m \overline{\underline{H}}_m^m . [k]_m^m = [h]_m^m . \overline{\underline{H}}_m^m [k]_m^m = [h]_m^m \begin{bmatrix} \Delta, & \beta_{\mu y} \\ 0, & \beta_{xy} \end{bmatrix}_{r, m-r}^{r\ m-r}$$

or in the identity

$$\overline{\underline{H}}_m^m [k]_m^m = \begin{bmatrix} \Delta, & \beta_{\mu y} \\ 0, & \beta_{xy} \end{bmatrix}_{r, m-r}^{r, m-r},$$

cancelling the factor Δ^{r+1} common to both sides in the first case, and the factor Δ^r common to both sides in the second case.

To show that formula (B) can be reduced to formula (C), let ω_1 and ω_2 be the affects of $[\mu]_r$ in $[q]_m$ and $[v]_m$ in the general case in which formula (B) is applicable, so that formula (B) is equivalent to

$$\Delta^{m-r-1} D = (-1)^{\omega_1+\omega_2} (\beta_{xy})_{m-r}^{m-r}. \quad \text{.....................(2)}$$

Let Δ', D', $\beta'_{x_i y_j}$ be the determinants formed from Δ, D, $\beta_{x_i y_j}$ when we bring $[a_{1\mu}]_m^r$, whose vertical rows are common to them all, to the leading position in each by ω_1, ω_2, ω_1 moves. Then we have

$$\Delta = (-1)^{\omega_1} \Delta', \quad D = (-1)^{\omega_2} D', \quad \beta'_{x_i y_j} = (-1)^{\omega_1} \beta_{x_i y_j},$$

and therefore formula (2) is equivalent to

$$\Delta'^{m-r-1}D' = (\beta'_{xy})^{m-r}_{m-r}. \quad\text{..........................(3)}$$

Since $\beta'_{x_i y_j}$ is now the determinant formed from Δ' when in it we replace the x_ith vertical row of A by the y_jth vertical row of A, we see that the identity (B) can always be reduced to the form (C).

NOTE 1. *Special case when* $\Delta = (a)^m_m$.

In this case we have $\Delta = (a)^m_m$, $D = (a_{1v})^m_m$; and if $[A]^m_m$ is the reciprocal of $[a]^m_m$, we can write

$$\overline{\underline{A}}^m_m [a]^n_m = [\beta]^n_m = [\Delta, q]^{m,\, n-m}_m \quad\text{..............................(4)}$$

where β_{ij} is obtained by replacing the ith vertical row of Δ by the jth vertical row of A, and the q's are the primaries of Δ. From the identical equation (B) of § 108 or the identical equation (4) we deduce as in the general case the identities

$$\Delta^{m-1}D = (\beta_{1v})^m_m, \quad\text{....................................}(a)$$

$$\Delta^{m-1}[A_1 A_2 \ldots A_\nu] = [Q_1 Q_2 \ldots Q_\nu], \quad\text{...........................}(a_1)$$

where in the last case $[A]_\nu$ and $[Q]_\nu$ are correspondingly formed complete matrices of the simple minor determinants of $[a]^n_m$ and of $[\beta]^n_m$ or $[\Delta, q]^{m,\, n-m}_m$.

When Δ and D have exactly r vertical rows in common forming the minor $[a_{1\mu}]^r_m$ of $[a]^n_m$, formula (a) can be reduced to

$$\Delta^{m-r-1}D = (-1)^\omega (q_{xv})^{m-r}_{m-r}, \quad\text{..............................}(\beta)$$

where $[x_1 x_2 \ldots x_{m-r}]$ and $[m+y_1,\ m+y_2,\ \ldots m+y_{m-r}]$ are the minors of the sequences $[1\,2 \ldots m]$ and $[v_1 v_2 \ldots v_m]$ formed by striking out their common elements $\mu_1, \mu_2, \ldots \mu_r$, and where ω is the difference of the affects of $[\mu]_r$ in $[1\,2 \ldots m]$ and $[v_1 v_2 \ldots v_m]$.

When $[v]_r = [\mu]_r = [1\,2 \ldots r]$, formula (β) becomes

$$\Delta^{m-r-1}D = (q_{xy})^{m-r}_{m-r}. \quad\text{......................................}(\gamma)$$

NOTE 2. *Number of distinct non-trivial identities included in* (A).

For a given value of r all identities given by (B) are reducible to the same standard form (C).

If $r = m$, we have $D = \Delta$, and (C) has the trivial form $1 = 1$.

If $r = m-1$, the identity (C) has the trivial form $\beta_{xy} = \beta_{xy}$.

If r has one of the values $m-2$, $m-3$, ... 2, 1, 0, the identity (C) is non-trivial, i.e. the expressions occurring on the two sides are different in form.

Thus formula (C), and therefore also formulae (B) and (A), can only give rise to $m-1$ distinct non-trivial identities corresponding respectively to the values $m-2$, $m-3$, ... 2, 1, 0 of r.

If $n \nless 2m$, all these $m-1$ non-trivial identities occur.

If $n < 2m$, then only $n-m-1$ distinct non-trivial identities occur corresponding to the values $m-2$, $m-3$, ... $2m-n$ of r.

The forms of these non-trivial standard identities are shown more fully in sub-article 3.

NOTE 3. *General principle underlying these identities.*

The identities given by formulae (C) and (B) are clearly the same as those given by the following theorems:

Theorem V a. *Let* $\Delta=(a)_m^m$ *and* $D=(b)_m^m$ *be two determinants of the same order* m *whose first* r *vertical rows are the same, and let* $r+s=m$. *Then if* δ_{ij} *is the determinant formed when the* $(r+i)$*th vertical row of* Δ *is replaced by the* $(r+j)$*th vertical row of* D, *we have*

$$\Delta^{s-1}D=(\delta)_s^s. \qquad \text{(C}_1\text{)}$$

Theorem VI a. *Let the two determinants* $\Delta=(a)_m^m$ *and* $D=(b)_m^m$ *have in common at least* r *vertical rows (differently arranged and differently placed in each) forming the matrix* $[c]_m^r$, *and let* $[x]_m^s$, $[y]_m^s$ *be the matrices formed from* $[a]_m^m$, $[b]_m^m$ *by striking out these* r *common vertical rows. Then if* δ_{ij} *is the determinant formed from* Δ *when we replace the* i*th vertical row of* $[x]_m^s$ *(occurring in* Δ*) by the* j*th vertical row of* $[y]_m^s$ *(occurring in* D*), we have*

$$\Delta^{s-1}D=(-1)^{\omega}(\delta)_s^s, \qquad \text{(B}_1\text{)}$$

where ω *is the difference of the affects of* $[c]_m^r$ *in* $[a]_m^m$ *and* $[b]_m^m$.

Proof of Theorem Va. We denote the conjugate reciprocal of $[a]_m^m$ by $\overline{\underline{A}}_m^m$, and use the identity

$$[a]_m^m\,\overline{\underline{A}}_m^m\,.\,[b]_m^m=[a]_m^m\,.\,\overline{\underline{A}}_m^m\,[b]_m^m. \qquad (5)$$

Writing $\quad [b]_m^m=[a,\,y]_m^{r,\,m-r},\quad \overline{\underline{A}}_m^m\,[b]_m^m=[\xi,\,\eta]_m^{r,\,m-r},$

and observing that then

$$\overline{\underline{A}}_m^m\,[a]_m^m=\Delta\,[1]_m^m,\quad [\xi]_m^r=\overline{\underline{A}}_m^m\,[a]_m^r=\begin{bmatrix}\Delta\\0\end{bmatrix}_{r,\,m-r}^r,\quad \eta_{r+i,\,r+j}=\delta_{ij},$$

we see that we can write (5) in the form

$$\Delta\,[b]_m^m=[a]_m^m\begin{bmatrix}\Delta, & \eta\\ 0, & \delta\end{bmatrix}_{r,\,m-r}^{r,\,m-r}.$$

Equating the determinants of both sides, and cancelling the factor Δ^{r+1} common to both sides, we obtain formula (C_1).

We can also obtain (C_1) by equating the determinants of both sides in the identity

$$\overline{\underline{A}}_m^m\,[b]_m^m=\overline{\underline{A}}_m^m\,[a,\,y]_m^{r,\,m-r}=\begin{bmatrix}\Delta, & \eta\\ 0, & \delta\end{bmatrix}_{r,\,m-r}^{r,\,m-r},$$

and using the identity $(A)_m^m=\Delta^{m-1}$.

Proof of Theorem VI a. Let $\Delta'=(c,\,x)_m^{r,\,m-r}$, $D'=(c,\,y)_m^{r,\,m-r}$, and let δ'_{ij} be the determinant formed from Δ' when in it we replace the ith vertical row of $[x]_m^{m-r}$ by the jth vertical row of $[y]_m^{m-r}$. Then by Theorem V a we have

$$\Delta'^{s-1}D'=(\delta')_s^s. \qquad (6)$$

Now let ω_1 and ω_2 be the affects of $[c]_m^r$ in $[a]_m^m$ and $[b]_m^m$.

Then ω_1, ω_2, ω_1 moves of vertical rows applied to Δ, D, δ_{ij} bring $[c]_m^r$ to the leading position in each and convert them into Δ', D', δ'_{ij}; and therefore

$$\Delta'=(-1)^{\omega_1}\Delta, \quad D'=(-1)^{\omega_2}D, \quad \delta'_{ij}=(-1)^{\omega_1}\delta_{ij}.$$

Substituting these values of Δ', D', δ'_{ij} in (6) we obtain formula (B_1).

When m is given, then (C_1) is a non-trivial identity for the values 2, 3, ... m of s, or the values $m-2$, $m-3$, ... 0 of r.

Ex. i. In the case of the matrix $[a\,b\,c\,d\,e\,f]_{123}$, let $\Delta=(daf)_{123}$, and let $[D\,A\,F]_{123}$ be the reciprocal of $[d\,a\,f]_{123}$. Then by formula (A) of § 108 we have the identity

$$(daf)_{123}[a\,b\,c\,d\,e\,f]_{123}=\begin{bmatrix} d_1 & a_1 & f_1 \\ d_2 & a_2 & f_2 \\ d_3 & a_3 & f_3 \end{bmatrix}\begin{bmatrix} D_1 & D_2 & D_3 \\ A_1 & A_2 & A_3 \\ F_1 & F_2 & F_3 \end{bmatrix}\begin{bmatrix} a_1 & b_1 & c_1 & d_1 & e_1 & f_1 \\ a_2 & b_2 & c_2 & d_2 & e_2 & f_2 \\ a_3 & b_3 & c_3 & d_3 & e_3 & f_3 \end{bmatrix}$$

$$=\begin{bmatrix} d_1 & a_1 & f_1 \\ d_2 & a_2 & f_2 \\ d_3 & a_3 & f_3 \end{bmatrix}\begin{bmatrix} 0 & \beta_{42} & \beta_{43} & \Delta & \beta_{45} & 0 \\ \Delta & \beta_{12} & \beta_{13} & 0 & \beta_{15} & 0 \\ 0 & \beta_{62} & \beta_{63} & 0 & \beta_{65} & \Delta \end{bmatrix},$$

where

$$\beta_{42}=(baf), \quad \beta_{43}=(caf), \quad \beta_{45}=(eaf),$$
$$\beta_{12}=(dbf), \quad \beta_{13}=(dcf), \quad \beta_{15}=(def),$$
$$\beta_{62}=(dab), \quad \beta_{63}=(dac), \quad \beta_{65}=(dae).$$

The matrix has 20 distinct simple minor determinants, and 9 of these denoted by the β's above are primary to Δ.

If we equate correspondingly formed matrices of the simple minor determinants of both sides in the above identity, we obtain 20 equations of which 10 are trivial and 10 are non-trivial. The latter 10 non-trivial equations express the remaining 10 simple minor determinants in terms of Δ and its 9 primaries when $\Delta \neq 0$.

By the theorems of the text and Note 2 the 10 non-trivial equations so obtained are reducible to 2 distinct identities corresponding to the values 1 and 0 of r in formula (C).

To obtain the determinant (abe) corresponding to $r=1$, we have

$$(daf)\,[a\,b\,e]_{123}=[d\,a\,f]_{123}\begin{bmatrix} 0 & \beta_{42} & \beta_{45} \\ \Delta & \beta_{12} & \beta_{15} \\ 0 & \beta_{62} & \beta_{65} \end{bmatrix},$$

whence

$$\Delta\,(abe)=-\begin{vmatrix} \beta_{42} & \beta_{45} \\ \beta_{62} & \beta_{65} \end{vmatrix}, \quad \text{i.e. } (adf)\,(abe)=\begin{vmatrix} (abf) & (aef) \\ (adb) & (ade) \end{vmatrix}.$$

To obtain the determinant (cbe) corresponding to $r=0$, we have

$$(daf)\,[c\,b\,e]_{123}=[d\,a\,f]_{123}\begin{bmatrix} \beta_{43} & \beta_{42} & \beta_{45} \\ \beta_{13} & \beta_{12} & \beta_{15} \\ \beta_{63} & \beta_{62} & \beta_{65} \end{bmatrix},$$

whence

$$\Delta\,(cbe)=\begin{vmatrix} \beta_{43} & \beta_{42} & \beta_{45} \\ \beta_{13} & \beta_{12} & \beta_{15} \\ \beta_{63} & \beta_{62} & \beta_{65} \end{vmatrix}, \quad \text{i.e. } (daf)\,(cbe)=\begin{vmatrix} (caf) & (baf) & (eaf) \\ (dcf) & (dbf) & (def) \\ (dac) & (dab) & (dae) \end{vmatrix}.$$

Ex. ii. In the case of the matrix $[a\,b\,c\,d\,e]_{123}$, let $\Delta=(abc)_{123}$. Then by formula (C) of § 108 we have the identity

$$\Delta\begin{bmatrix} a_1 & b_1 & c_1 & d_1 & e_1 \\ a_2 & b_2 & c_2 & d_2 & e_2 \\ a_3 & b_3 & c_3 & d_3 & e_3 \end{bmatrix} = \begin{bmatrix} a_1 & b_1 & c_1 \\ a_2 & b_2 & c_2 \\ a_3 & b_3 & c_3 \end{bmatrix}\begin{bmatrix} \Delta & 0 & 0 & q_{11} & q_{12} \\ 0 & \Delta & 0 & q_{21} & q_{22} \\ 0 & 0 & \Delta & q_{31} & q_{32} \end{bmatrix},$$

where
$$q_{11}=(dbc),\quad q_{21}=(adc),\quad q_{31}=(abd),$$
$$q_{12}=(ebc),\quad q_{22}=(aec),\quad q_{32}=(abe).$$

The last six determinants are the primaries of Δ.

To obtain the identity which expresses (bde) in terms of Δ and its primaries when $\Delta \neq 0$ we use the derived identity

$$\Delta\begin{bmatrix} b_1 & d_1 & e_1 \\ b_2 & d_2 & e_2 \\ b_3 & d_3 & e_3 \end{bmatrix} = \begin{bmatrix} a_1 & b_1 & c_1 \\ a_2 & b_2 & c_2 \\ a_3 & b_3 & c_3 \end{bmatrix}\begin{bmatrix} 0 & q_{11} & q_{12} \\ \Delta & q_{21} & q_{22} \\ 0 & q_{31} & q_{32} \end{bmatrix}.$$

Equating the determinants of both sides and cancelling Δ^2, we have

$$\Delta(bde)=q_{12}q_{31}-q_{11}q_{32},\quad \text{or}\quad (bac)(bde)=\begin{vmatrix} (dbc) & (ebc) \\ (abd) & (abe) \end{vmatrix}.$$

To obtain the identity which expresses (ace) in terms of Δ and its primaries when $\Delta \neq 0$ we use the derived identity

$$\Delta\begin{bmatrix} a_1 & c_1 & e_1 \\ a_2 & c_2 & e_2 \\ a_3 & c_3 & e_3 \end{bmatrix} = \begin{bmatrix} a_1 & b_1 & c_1 \\ a_2 & b_2 & c_2 \\ a_3 & b_3 & c_3 \end{bmatrix}\begin{bmatrix} \Delta & 0 & q_{12} \\ 0 & 0 & q_{22} \\ 0 & \Delta & q_{32} \end{bmatrix}.$$

Equating the determinants of both sides and cancelling Δ^3, we have the trivial identity

$$(ace)=-q_{22},\quad \text{or}\quad (ace)=(ace).$$

Ex. iii. In the case of the matrix $[a\,b\,c\,d\,e]_{123}$ let $\Delta=(abc)_{123}=(abc)$. Then there are 10 simple minor determinants, viz.

$$\Delta,\ q_{11},\ q_{21},\ q_{31},\ q_{12},\ q_{22},\ q_{32},\ p_1,\ p_2,\ p_3,$$

where
$$p_1=(ade),\quad p_2=(dbe),\quad p_3=(dec),$$

and the q's have the values given in Ex. ii. The q's are those which are primary to Δ.

If in the first identity of Ex. ii we equate correspondingly formed matrices of the simple minor determinants of both sides, we obtain after cancelling Δ on both sides

$$\Delta^2[\Delta,\ q_{11},\ q_{21},\ q_{31},\ q_{12},\ q_{22},\ q_{32},\ p_1,\ p_2,\ p_3]$$
$$=\left[\Delta^3,\ \Delta^2q_{11},\ \Delta^2q_{21},\ \Delta^2q_{31},\ \Delta^2q_{12},\ \Delta^2q_{22},\ \Delta^2q_{32},\ \Delta\begin{pmatrix}12\\ q\\ 23\end{pmatrix},\ \Delta\begin{pmatrix}12\\ q\\ 13\end{pmatrix},\ \Delta\begin{pmatrix}12\\ q\\ 12\end{pmatrix}\right].$$

This gives 3 non-trivial identities corresponding to the value 1 of r in formula (B) or (C) and serving to determine p_1, p_2, p_3 in terms of Δ and its primaries when $\Delta \neq 0$. These are

$$\Delta p_1=q_{21}q_{32}-q_{22}q_{31},\quad \Delta p_2=q_{11}q_{32}-q_{12}q_{31},\quad \Delta p_3=q_{11}q_{22}-q_{12}q_{21},$$

or
$$(abc)(ade)=(adc)(abe)-(aec)(abd),$$
$$(abc)(dbe)=(dbc)(abe)-(ebc)(abd),$$
$$(abc)(dec)=(dbc)(aec)-(ebc)(adc).$$

In accordance with Note 2 these three identities are not essentially distinct, being all derivable from the common form

$$(abc)(axy)=(axc)(aby)-(ayc)(abx)=\begin{vmatrix}(axc) & (ayc)\\ (abx) & (aby)\end{vmatrix}.$$

Ex. iv. In the case of the matrix $[a\,b\,c\,d\,e]_{123}$ the equation which serves to express (acd) in terms of (bce) and its primaries when $(bce)\neq 0$ is

$$(bce)(acd)=(ace)(bcd)-(dce)(bca).$$

Using the third of the equations obtained in Ex. iii to reduce this to a relation between (abc) and its primaries, we obtain in accordance with Note 7 the identity

$$(bce)(acd)=(ace)(bcd)+\{(dbc)(aec)-(ebc)(adc)\}.$$

Thus the above equation is deducible from the three equations of Ex. iii.

Ex. v. In the case of the matrix $[a\,b\,c\,d]_{12}$, the equation which serves to express (cd) in terms of (ab) and its primaries when $(ab)\neq 0$ is

$$(bc)(ad)+(ac)(bd)+(ab)(cd)=0.$$

In accordance with Note 7 this is the only relation which can exist between the simple minor determinants of the matrix when its elements are arbitrary.

Ex. vi. *Extensions of formulae* (A), (B) *and* (C).

(1) If $\Delta=(a_{pq})^r_r$ and $D=(a_{pv})^r_r$ are two simple minor determinants of the same horizontal minor $K=[a_{pl}]^n_r$ of $[a]^n_m$, then the equation

$$\Delta^{r-1}D=(\beta_{qv})^r_r \quad \text{......................................}(A_2)$$

is an identity in the elements of Δ and D.

(2) If the two determinants Δ and D have in common exactly k vertical rows forming the minor matrix $[a_{p\mu}]^k_r$ of $[a]^n_m$; if $[y]_{r-k}$, $[\eta]_{r-k}$ are the minors of $[q]_r$, $[v]_r$ formed by striking out their common elements $\mu_1, \mu_2, \dots \mu_k$; and if ω is the difference of the affects of $[\mu]_k$ in $[q]_r$ and $[v]_r$; then the equation

$$\Delta^{r-k-1}D=(-1)^{\omega}(\beta_{y\eta})^{r-k}_{r-k} \quad \text{.................................}(B_2)$$

is an identity in the elements of Δ and D, and all the elements of $[\beta_{y\eta}]^{r-k}_{r-k}$ are primaries of Δ.

(3) If in (B_2) we reduce the determinants Δ and D to the forms

$$\Delta=(a_{p\mu},\, a_{py})^{k,\,r-k}_r, \quad D=(a_{p\mu},\, a_{p\eta})^{k,\,r-k}_r,$$

then the identity (B_2) is reduced to the form

$$\Delta^{r-k-1}D=(\beta_{y\eta})^{r-k}_{r-k}. \quad \text{....................................}(C_2)$$

These results are obtained when we apply Theorems I *a*, III *a* and IV *a* to the matrix $[a_{pl}]^n_r$. We can also deduce (B_2) and (C_2) from Theorems VI *a* and V *a*.

To prove (A_2) directly we use Ex. iv of § 102, and equate the determinants of both sides in the identity

$$(a_{pq})^r_r\,[a_{pv}]^r_r=[a_{pq}]^r_r\,\overline{\underline{A_{pq}}}^r_r\,[a_{pv}]^r_r=[a_{pq}]^r_r\,[\beta_{qv}]^r_r,$$

or in the identity

$$\overline{\underline{A_{pq}}}^r_r\,[a_{pv}]^r_r=[\beta_{qv}]^r_r.$$

To deduce (B_2) we expand $(\beta_{qv})_r^r$ in terms of the simple minor determinants of its minor matrix $[\beta_{q\mu}]_r^k$, observing that all these vanish except $(\beta_{\mu\mu})_k^k$ which has the value Δ^k.

The identities (B_2) and (C_2) are trivial when $k=r$ or $r-1$, and are non-trivial in other cases.

Ex. vii. *Further extension of formula* (C).

If
$$\Delta = \begin{pmatrix} a_{pq}, & a_{p\mu} \\ a_{\lambda q}, & a_{\lambda\mu} \end{pmatrix}_{h,\,\rho}^{k,\,\sigma} \quad \textit{and} \quad D = \begin{pmatrix} a_{pq}, & a_{pv} \\ a_{\lambda q}, & a_{\lambda v} \end{pmatrix}_{h,\,\rho}^{k,\,\sigma}, \quad \textit{where } k+\sigma = h+\rho,$$
are two simple minor determinants of the same horizontal minor K of $[a]_m^n$ *whose first k vertical rows are the same, then the equation*
$$\Delta^{\sigma-1} D = (\beta_{\mu v})_\sigma^\sigma \quad \text{......................................}(C_3)$$
is an identity in the elements of Δ *and D.*

Writing
$$[u]_m^m = \begin{bmatrix} a_{pq}, & a_{p\mu} \\ a_{\lambda q}, & a_{\lambda\mu} \end{bmatrix}_{h,\,\rho}^{k,\,\sigma}, \quad [v]_m^m = \begin{bmatrix} a_{pq}, & a_{pv} \\ a_{\lambda q}, & a_{\lambda v} \end{bmatrix}_{h,\,\rho}^{k,\,\sigma}, \quad [U]_m^m = \begin{bmatrix} A_{pq}, & A_{p\mu} \\ A_{\lambda q}, & A_{\lambda\mu} \end{bmatrix}_{h,\,\rho}^{k,\,\sigma},$$
where $[U]_m^m$ is the reciprocal of $[u]_m^m$, so that
$$(u)_m^m = \Delta, \quad (v)_m^m = D, \quad (U)_m^m = \Delta^{m-1},$$
we have by formula (B_2) of § 102 or Ex. viii of § 108
$$\overline{\underline{U}}_m^m [v]_m^m = \overline{\underline{\begin{matrix} A_{pq}, & A_{\lambda q} \\ A_{p\mu}, & A_{\lambda\mu} \end{matrix}}}_{k,\,\sigma}^{h,\,\rho} \begin{bmatrix} a_{pq}, & a_{pv} \\ a_{\lambda q}, & a_{\lambda v} \end{bmatrix}_{h,\,\rho}^{k,\,\sigma} = \begin{bmatrix} \beta_{qq}, & \beta_{qv} \\ \beta_{\mu q}, & \beta_{\mu v} \end{bmatrix}_{k,\,\sigma}^{k,\,\sigma} = \begin{bmatrix} \Delta, & \beta_{qv} \\ 0, & \beta_{\mu v} \end{bmatrix}_{k,\,\sigma}^{k,\,\sigma}.$$

We obtain (C_3) by equating the determinants of both sides in either one of the identities
$$(u)_m^m [v]_m^m = [u]_m^m \overline{\underline{U}}_m^m \cdot [v]_m^m = [u]_m^m \cdot \overline{\underline{U}}_m^m [v]_m^m = [u]_m^m \begin{bmatrix} \Delta, & \beta_{qv} \\ 0, & \beta_{\mu v} \end{bmatrix}_{k,\,\sigma}^{k,\,\sigma},$$
$$\overline{\underline{U}}_m^m [v]_m^m = \begin{bmatrix} \Delta, & \beta_{qv} \\ 0, & \beta_{\mu v} \end{bmatrix}_{k,\,\sigma}^{k,\,\sigma},$$
cancelling the factor Δ^{k+1} common to both sides in the first case, and the factor Δ^k common to both sides in the second case.

We can also write down formula (C_3) immediately by applying Theorem V a.

2. *The identities when the short rows of the matrix are horizontal.*

Let $A = [a]_m^n$ be any matrix, not necessarily undegenerate, in which the short rows are horizontal, so that $m \nless n$. Then we will prove the theorems which follow.

Theorem I b. *If* $\Delta = (a_{p1})_n^n$ *and* $D = (a_{u1})_n^n$ *are any two simple minor determinants of A, each formed with any n of its horizontal rows arranged in any order, then the equation*
$$\Delta^{n-1} D = (\alpha_{up})_n^n \quad \text{..............................}(A')$$
is an identity in the elements of Δ *and D, and when* $\Delta \neq 0$ *serves to express D as a homogeneous rational function of degree* 1 *of* Δ *and its primaries.*

Theorem II b. *More generally if* $\overline{\underline{A}}_\mu$ *and* $\overline{\underline{P}}_\mu$ *are correspondingly formed complete matrices of the simple minor determinants of* $[a]_m^n$ *and* $[\alpha_{1p}]_m^n$, *then the equation*

$$\Delta^{m-1}\begin{bmatrix} A_1 \\ A_2 \\ \vdots \\ A_\mu \end{bmatrix} = \begin{bmatrix} P_1 \\ P_2 \\ \vdots \\ P_\mu \end{bmatrix} \qquad \text{(a}')$$

is an identity in the elements of A, and when $\Delta \neq 0$ *serves to express all the simple minor determinants of A as homogeneous rational functions of degree* 1 *of* Δ *and its primaries.*

Here $\alpha_{x\lambda}$ is the determinant formed from Δ when we replace the λth horizontal row of A (occurring in Δ) by the xth horizontal row of A, or when in Δ we replace the horizontal suffix λ by x.

We denote the reciprocal of $[a_{p1}]_n^n$ by $[A_{p1}]_n^n$, and have

$$[\alpha_{1p}]_m^n = [a]_m^n \overline{\underline{A_{p1}}}_n^n, \qquad [\alpha_{pp}]_n^n = \Delta[1]_n^n. \qquad (1')$$

Those elements of $[\alpha_{1p}]_m^n$ which do not occur in $[\alpha_{pp}]_n^n$ are the primaries of Δ.

Proof of Theorem I b. From formula (A′) of § 108 we deduce the obviously true identity

$$(a_{p1})_n^n [a_{u1}]_n^n = [a_{u1}]_n^n \cdot \overline{\underline{A_{p1}}}_n^n [a_{p1}]_n^n = [a_{u1}]_n^n \overline{\underline{A_{p1}}}_n^n \cdot [a_{p1}]_n^n = [\alpha_{up}]_n^n [a_{p1}]_n^n,$$

and by equating the determinants of both sides and cancelling the factor Δ common to both sides, we obtain formula (A′), in which every element of $[\alpha_{up}]_n^n$ is either Δ or 0 or a primary of Δ. We can also obtain formula (A′) by equating the determinants of both sides in the identity

$$[a_{u1}]_n^n \overline{\underline{A_{p1}}}_n^n = [\alpha_{up}]_n^n.$$

Proof of Theorem II b. We obtain formula (a′) by equating correspondingly formed complete matrices of the simple minor determinants of both sides in the identity (A′) of § 108 or in the identity

$$[a]_m^n \overline{\underline{A_{p1}}}_n^n = [\alpha_{1p}]_m^n.$$

If in formula (A′) of Theorem I *b* any element of $(\alpha_{up})_n^n$ has the value Δ, then every other element in the same horizontal row has the value 0, and we can reduce formula (A′) by cancelling an additional common factor Δ on both sides. This case occurs when and only when Δ and D have a horizontal row in common. By such reductions we obtain Theorems III *b* and IV *b* which follow.

Theorem III b. *If the simple minor determinants* $\Delta = (a_{p1})_n^n$ *and* $D = (a_{u1})_n^n$ *have r horizontal rows in common forming the minor matrix* $[a_{\lambda 1}]_r^n$ *of* $[a]_m^n$, *then we can reduce formula* (A′) *of Theorem I b to the identity*

$$\Delta^{n-r-1} D = (-1)^{\omega} (a_{yx})_{n-r}^{n-r}, \quad \text{.....................(B′)}$$

where $[x]_{n-r}$, $[y]_{n-r}$ *are the minors of the sequences* $[p]_n$, $[u]_n$ *formed by striking out their common elements* $\lambda_1, \lambda_2, \ldots \lambda_r$, *and where* ω *is the difference of the affects of* $[\lambda]_r$ *in* $[p]_n$ *and* $[u]_n$.

If further the determinants Δ *and* D *have no other horizontal rows in common, then in formula* (B′) *all the elements of* $[\alpha_{yx}]_{n-r}^{n-r}$ *are primaries of* Δ.

Theorem IV b. *If* $\Delta = \begin{bmatrix} a_{\lambda 1} \\ a_{x1} \end{bmatrix}_{r,\, n-r}^{n}$ *and* $D = \begin{bmatrix} a_{\lambda 1} \\ a_{y1} \end{bmatrix}_{r,\, n-r}^{n}$ *are two simple minor determinants of* A *whose first r horizontal rows are the same, then the equation*

$$\Delta^{n-r-1} D = (\alpha_{yx})_{n-r}^{n-r} \quad \text{..........................(C′)}$$

is an identity in the elements of Δ *and* D.

The identity (B′) *can always be reduced to the standard form* (C′).

Proof of Theorem III b. Expanding the determinant $(\alpha_{up})_n^n$ in terms of the simple minor determinants of its minor matrix $[\alpha_{\lambda p}]_r^n$, and observing that all these determinants vanish except $(\alpha_{\lambda\lambda})_r^r$ which has the value Δ^r, we obtain

$$(\alpha_{up})_n^n = (-1)^{\omega} (\alpha_{\lambda\lambda})_r^r (\alpha_{yx})_{n-r}^{n-r} = (-1)^{\omega} \Delta^r (\alpha_{yx})_{n-r}^{n-r}.$$

Substituting this value of $(\alpha_{up})_n^n$ in formula (A′), and then cancelling the factor Δ^r which is common to both sides, we obtain formula (B′).

If Δ and D have no more horizontal rows in common besides those specified in the theorem, then $[x]_{n-r}$ and $[y]_{n-r}$ have no elements in common, and therefore $[p]_r$ and $[y]_{n-r}$ have no elements in common; consequently every element of $[\alpha_{yx}]_{n-r}^{n-r}$ is one of the primaries of Δ.

Proof of Theorem IV b. Formula (C′) is a particular case of formula (B′). To obtain (C′) directly we write

$$[h]_n^n = \begin{bmatrix} a_{\lambda 1} \\ a_{x1} \end{bmatrix}_{r,\, n-r}^{n}, \quad [k]_n^n = \begin{bmatrix} a_{\lambda 1} \\ a_{y1} \end{bmatrix}_{r,\, n-r}^{n}, \quad [H]_n^n = \begin{bmatrix} A_{\lambda 1} \\ A_{x1} \end{bmatrix}_{r,\, n-r}^{n},$$

where $[H]_n^n$ is the reciprocal of $[h]_n^n$; so that

$$(h)_n^n = \Delta, \quad (k)_n^n = D, \quad (H)_n^n = \Delta^{n-1}.$$

Then observing that by formula (A_2) of § 102 or Ex. viii of § 108

$$\begin{bmatrix} a_{\lambda 1} \\ a_{y1} \end{bmatrix}^{n}_{r,\,n-r} \overline{\underline{A_{\lambda 1}\, A_{x1}}}^{\,r,\,n-r}_{n} = \begin{bmatrix} \alpha_{\lambda\lambda}, & \alpha_{\lambda x} \\ \alpha_{y\lambda}, & \alpha_{yx} \end{bmatrix}^{r,\,n-r}_{r,\,n-r} = \begin{bmatrix} \Delta\;, & 0 \\ \alpha_{y\lambda}, & \alpha_{yx} \end{bmatrix}^{r,\,n-r}_{r,\,n-r},$$

we equate the determinants of both sides in the identity

$$(h)^{n}_{n}\,[k]^{n}_{n} = [k]^{n}_{n} \cdot \overline{\underline{H}}^{n}_{n}\,[h]^{n}_{n} = [k]^{n}_{n}\,\overline{\underline{H}}^{n}_{n} \cdot [h]^{n}_{n} = \begin{bmatrix} \Delta\;, & 0 \\ \alpha_{y\lambda}, & \alpha_{yx} \end{bmatrix}^{r,\,n-r}_{r,\,n-r} [h]^{n}_{n},$$

or in the identity

$$[k]^{n}_{n}\,\overline{\underline{H}}^{n}_{n} = \begin{bmatrix} \Delta\;, & 0 \\ \alpha_{y\lambda}, & \alpha_{yx} \end{bmatrix}^{r,\,n-r}_{r,\,n-r},$$

cancelling the factor Δ^{r+1} common to both sides in the first case, and the factor Δ^{r} common to both sides in the second case.

To reduce (B′) to the form (C′), let ω_1 and ω_2 be the affects of $[\lambda]_r$ in $[p]_n$ and $[u]_n$ in Theorem III *b*, and let Δ', D', $\alpha'_{y_i x_j}$ be the determinants formed from Δ, D, $\alpha_{y_i x_j}$ when we bring $[a_{\lambda 1}]^{n}_{r}$ to the leading position in each by ω_1, ω_2, ω_1 moves of horizontal rows.

Then we have

$$\Delta = (-1)^{\omega_1}\Delta', \quad D = (-1)^{\omega_2} D', \quad \alpha_{y_i x_j} = (-1)^{\omega_1}\alpha'_{y_i x_j}.$$

When we substitute these values in (B), we reduce it to

$$\Delta'^{\,n-r-1} D' = (\alpha'_{yx})^{n-r}_{n-r}, \quad \text{.........................(3′)}$$

which is of the form (C′).

NOTE 4. *Special case when* $\Delta = (a)^{n}_{n}$.

In this case we have $\Delta = (a)^{n}_{n}$, $D = (a_{u1})^{n}_{n}$; and if $[A]^{n}_{n}$ is the reciprocal of $[a]^{n}_{n}$, we can write

$$[a]^{n}_{m}\,\overline{\underline{A}}^{n}_{n} = [\alpha]^{n}_{m} = \begin{bmatrix} \Delta \\ p \end{bmatrix}^{n}_{n,\,m-n}, \quad \text{............................(4′)}$$

where α_{ij} is obtained by replacing the jth horizontal row of Δ by the ith horizontal row of A, and the p's are the primaries of Δ.

From the identical equation (B′) of § 108 or the identical equation (4′) we deduce as in the general case the identities

$$\Delta^{n-1} D = (a_{u1})^{n}_{n}, \quad \text{.......................................}(a')$$

$$\Delta^{n-1}\,\overline{\underline{A}}_{\mu} = \overline{\underline{P}}_{\mu}, \quad \text{.......................................}(a_1')$$

where in the last case $\overline{\underline{A}}_{\mu}$ and $\overline{\underline{P}}_{\mu}$ are correspondingly formed complete matrices of the simple minor determinants of $[a]^{n}_{m}$ and of $[\alpha]^{n}_{m}$ or $\begin{bmatrix} \Delta \\ p \end{bmatrix}^{n}_{n,\,m-n}$.

When Δ and D have exactly r horizontal rows in common forming the minor matrix $[a_{\lambda 1}]^{n}_{r}$ of $[a]^{n}_{m}$, formula (a') can be reduced to

$$\Delta^{n-r-1} D = (-1)^{\omega} (p_{yx})^{n-r}_{n-r}, \quad \text{...............................}(\beta')$$

where $[x_1 x_2 \ldots x_{n-r}]$ and $[n+y_1, n+y_2, \ldots n+y_{n-r}]$ are the minors of the sequences $[1\,2 \ldots n]$ and $[u_1 u_2 \ldots u_n]$ formed by striking out their common elements $\lambda_1, \lambda_2, \ldots \lambda_r$, and where ω is the difference of the affects of $[\lambda]_r$ in $[1\,2 \ldots n]$ and $[u_1 u_2 \ldots u_n]$.

When $[u]_r = [\lambda]_r = [1\,2 \ldots r]$, formula ($\beta'$) becomes

$$\Delta^{n-r-1} D = (p_{yx})^{n-r}_{n-r}. \qquad \text{(}\gamma'\text{)}$$

NOTE 5. *Number of distinct non-trivial identities included in* (A′).

For a given value of r all identities given by (A′) are reducible to the same standard form (C′); and the identity (C′) is trivial when $r=n$ or $n-1$, and is non-trivial in other cases.

Thus formula (C′), and therefore also formulae (B′) and (A′), can only give rise to $n-1$ distinct non-trivial identities corresponding respectively to the values $n-2$, $n-3$, ... 2, 1, 0 of r.

If $m \nless 2n$, all these $n-1$ distinct non-trivial identities occur.

If $m < 2n$, then only $m-n-1$ of them occur, corresponding respectively to the values $n-2, n-3, \ldots 2n-m$ of r.

NOTE 6. *General principle underlying these identities.*

The identities given by formulae (C′) and (B′) are clearly the same as those given by the following theorems:

Theorem V b. *Let $\Delta=(a)^n_n$ and $D=(b)^n_n$ be two determinants of the same order n whose first r horizontal rows are the same, and let $r+s=n$. Then if δ_{ij} is the determinant formed when the $(r+j)$th horizontal row of Δ is replaced by the $(r+i)$th horizontal row of D, we have*

$$\Delta^{s-1} D = (\delta)^s_s. \qquad \text{(C}_1'\text{)}$$

Theorem VI b. *Let the two determinants $\Delta=(a)^n_n$ and $D=(b)^n_n$ have in common at least r horizontal rows (differently arranged and differently placed in each) forming the matrix $[c]^n_r$, and let $[x]^n_s$, $[y]^n_s$ be the matrices formed from $[a]^n_n$, $[b]^n_n$ by striking out those r common horizontal rows. Then if δ_{ij} is the determinant formed from Δ when we replace the jth horizontal row $[x]^n_s$ (occurring in Δ) by the ith horizontal row of $[y]^n_s$ (occurring in D), we have*

$$\Delta^{s-1} D = (-1)^{\omega} (\delta)^s_s, \qquad \text{(B}_1'\text{)}$$

where ω is the difference of the affects of $[c]^n_r$ in $[a]^n_n$ and $[b]^n_n$.

Proof of Theorem Vb. We denote the conjugate reciprocal of $[a]^n_n$ by $\overline{\underline{A}}^n_n$, and use the identity

$$[b]^n_n \cdot \overline{\underline{A}}^n_n \, [a]^n_n = [b]^n_n \, \overline{\underline{A}}^n_n \cdot [a]^n_n. \qquad \text{(5}'\text{)}$$

Writing
$$[b]^n_n = \begin{bmatrix} a \\ y \end{bmatrix}^n_{r,\,n-r}, \qquad [b]^n_n \, \overline{\underline{A}}^n_n = \begin{bmatrix} \xi \\ \eta \end{bmatrix}^n_{r,\,n-r},$$

and observing that then

$$[a]^n_n \, \overline{\underline{A}}^n_n = \Delta\,[1]^n_n, \qquad [\xi]^n_r = [a]^n_r \, \overline{\underline{A}}^n_n = [\Delta,\, 0]^{r,\,n-r}_r, \qquad \eta_{r+i,\,r+j} = \delta_{ij},$$

we see that we can write (5′) in the form

$$\Delta\,[b]^n_n = \begin{bmatrix} \Delta, & 0 \\ \eta, & \delta \end{bmatrix}^{r,\,n-r}_{r,\,n-r} [a]^n_n.$$

Equating the determinants of both sides, and cancelling the factor Δ^{r+1} common to both sides, we obtain formula (C_1').

We can also obtain (C_1') by equating the determinants of both sides in the identity

$$[b]^n_n \, \overline{\underline{A}}^n_n = \begin{bmatrix} a \\ y \end{bmatrix}^n_{r,\, n-r} \overline{\underline{A}}^n_n = \begin{bmatrix} \Delta, & 0 \\ \eta, & \delta \end{bmatrix}^{r,\, n-r}_{r,\, n-r},$$

and using the identity $\qquad (A)^n_n = \Delta^{n-1}.$

Proof of Theorem VI b. Let $\Delta' = \begin{bmatrix} c \\ x \end{bmatrix}^n_{r,\, n-r}$, $D' = \begin{bmatrix} c \\ y \end{bmatrix}^n_{r,\, n-r}$, and let δ'_{ij} be the determinant formed from Δ' when in it we replace the jth horizontal row of $[x]^n_{n-r}$ by the ith horizontal row of $[y]^n_{n-r}$. Then by Theorem V b we have

$$\Delta'^{s-1} D' = (\delta')^s_s. \qquad \text{(6')}$$

Now let ω_1 and ω_2 be the affects of $[c]^n_r$ in $[a]^n_n$ and $[b]^n_n$.

Then ω_1, ω_2, ω_1 moves of horizontal rows applied to Δ, D, δ_{ij} bring $[c]^n_r$ to the leading position in each and convert them into Δ', D', δ'_{ij}; and therefore

$$\Delta' = (-1)^{\omega_1}\Delta, \quad D' = (-1)^{\omega_2} D, \quad \delta'_{ij} = (-1)^{\omega_1}\delta_{ij}.$$

Substituting these values in (6') we obtain formula (B_1').

When n is given, then (C_1') is a non-trivial identity for the values 2, 3, ... n of s, or the values $n-2$, $n-3$, ... 0 of r.

Ex. viii. In the case of the matrix $[a]^5_9$ let $\Delta = \begin{pmatrix} 12345 \\ a \\ 12479 \end{pmatrix}$, so that

$$[a]^5_9 \, \overline{\underline{A}}^{12479}_{12345} = \begin{bmatrix} 12479 \\ a \\ 12\ldots9 \end{bmatrix}, \quad \begin{bmatrix} 12479 \\ a \\ 12479 \end{bmatrix} = \Delta\,[1]^5_5.$$

Then to express the simple minor determinant $D = \begin{pmatrix} 12345 \\ a \\ 24578 \end{pmatrix}$ in terms of Δ and its primaries when $\Delta \neq 0$, we use the identities

$$\Delta\,[a]^5_9 = \begin{bmatrix} 12479 \\ a \\ 12\ldots9 \end{bmatrix}\begin{bmatrix} 12345 \\ a \\ 12479 \end{bmatrix}, \quad \Delta \begin{bmatrix} 12345 \\ a \\ 24578 \end{bmatrix} = \begin{bmatrix} 12479 \\ a \\ 24578 \end{bmatrix}\begin{bmatrix} 12345 \\ a \\ 12479 \end{bmatrix}.$$

Equating the determinants of both sides in the last identity, we obtain

$$\Delta^4 \begin{pmatrix} 12345 \\ a \\ 24578 \end{pmatrix} = \begin{pmatrix} 12345 \\ a \\ 24578 \end{pmatrix}, \quad \text{or} \quad \Delta \begin{pmatrix} 12345 \\ a \\ 24578 \end{pmatrix} = \begin{vmatrix} a_{51} & a_{59} \\ a_{81} & a_{89} \end{vmatrix},$$

which has the standard form

$$\begin{pmatrix} 12345 \\ a \\ 24719 \end{pmatrix}\begin{pmatrix} 12345 \\ a \\ 24758 \end{pmatrix} = \begin{pmatrix} 19 \\ a \\ 58 \end{pmatrix} = \begin{vmatrix} \begin{pmatrix} 12345 \\ a \\ 52479 \end{pmatrix} & \begin{pmatrix} 12345 \\ a \\ 12475 \end{pmatrix} \\ \begin{pmatrix} 12345 \\ a \\ 82479 \end{pmatrix} & \begin{pmatrix} 12345 \\ a \\ 12478 \end{pmatrix} \end{vmatrix}.$$

Ex. ix. *Extensions of formulae* (A'), (B') *and* (C').

(1) If $\Delta = (a_{pq})^r_r$ and $D = (a_{uq})^r_r$ are two simple minor determinants of the same vertical minor $H = [a_{1q}]^r_m$ of $[a]^n_m$, then the equation

$$\Delta^{r-1} D = (a_{up})^r_r \qquad (A_2')$$

is an identity in the elements of Δ and D.

(2) If the two determinants Δ and D have in common exactly h horizontal rows forming the minor matrix $[a_{\lambda q}]_h^r$ of $[a]_m^n$; if $[x]_{r-h}$, $[\xi]_{r-h}$ are the minors of $[p]_r$, $[u]_r$ formed by striking out their common elements $\lambda_1, \lambda_2, \ldots \lambda_h$; and if ω is the difference of the affects of $[\lambda]_h$ in $[p]_r$ and $[u]_r$; then the equation

$$\Delta^{r-h-1}D=(-1)^{\omega}\,(a_{\xi x})_{r-h}^{r-h} \quad \ldots\ldots\ldots\ldots\ldots\ldots\ldots\ldots (B_2')$$

is an identity in the elements of Δ and D, and all the elements of $[a_{\xi x}]_{r-h}^{r-h}$ are primaries of Δ.

(3) If in (B_2') we reduce the determinants Δ and D to the forms

$$\Delta=\begin{pmatrix} a_{\lambda q} \\ a_{xq} \end{pmatrix}_{h,\, r-h}^{r}, \quad D=\begin{pmatrix} a_{\lambda q} \\ a_{\xi q} \end{pmatrix}_{h,\, r-h}^{r},$$

then the identity (B_2') is reduced to the form

$$\Delta^{r-h-1}D=(a_{\xi x})_{r-h}^{r-h}. \quad \ldots\ldots\ldots\ldots\ldots\ldots\ldots\ldots (C_2')$$

These results are obtained when we apply Theorems I *b*, III *b* and IV *b* to the matrix $[a_{1q}]_m^r$. We can also deduce (B_2') and (C_2') from Theorems VI *b* and V *b*.

To prove (A_2') directly we use Ex. iv of § 102, and equate the determinants of both sides in the identity

$$(a_{pq})_r^r\,[a_{uq}]_r^r=[a_{uq}]_r^r\,\overline{\underline{A_{pq}}}_r^r\,[a_{pq}]_r^r=[a_{up}]_r^r\,[a_{pq}]_r^r,$$

or in the identity

$$[a_{uq}]_r^r\,\overline{\underline{A_{pq}}}_r^r=[a_{up}]_r^r.$$

To deduce (B_2') we expand $[a_{up}]_r^r$ in terms of the simple minor determinants of its minor matrix $[a_{\lambda p}]_h^r$, observing that all these vanish except $(a_{\lambda\lambda})_h^h$ which has the value Δ^h.

The identities (B_2') and (C_2') are trivial when $h=r$ or $r-1$, and are non-trivial in other cases.

Ex. x. *Further extension of formula* (C′).

If $\Delta=\begin{pmatrix} a_{pq}, & a_{p\mu} \\ a_{\lambda q}, & a_{\lambda\mu} \end{pmatrix}_{h,\,\rho}^{k,\,\sigma}$ *and* $D=\begin{pmatrix} a_{pq}, & a_{p\mu} \\ a_{xq}, & a_{x\mu} \end{pmatrix}_{h,\,\rho}^{k,\,\sigma}$, *where* $k+\sigma=h+\rho$,

are two simple minor determinants of the same vertical minor H of $[a]_m^n$ whose first h horizontal rows are the same, then the equation

$$\Delta^{\rho-1}D=(a_{x\lambda})_\rho^\rho \quad \ldots\ldots\ldots\ldots\ldots\ldots\ldots\ldots (C_3')$$

is an identity in the elements of Δ and D.

Writing

$$[u]_m^m=\begin{bmatrix} a_{pq}, & a_{p\mu} \\ a_{\lambda q}, & a_{\lambda\mu} \end{bmatrix}_{h,\,\rho}^{k,\,\sigma}, \quad [v]_m^m=\begin{bmatrix} a_{pq}, & a_{p\mu} \\ a_{xq}, & a_{x\mu} \end{bmatrix}_{h,\,\rho}^{k,\,\sigma}, \quad [U]_m^m=\begin{bmatrix} A_{pq}, & A_{p\mu} \\ A_{\lambda q}, & A_{\lambda\mu} \end{bmatrix}_{h,\,\rho}^{k,\,\sigma},$$

where $[U]_m^m$ is the reciprocal of $[u]_m^m$, so that

$$(u)_m^m=\Delta, \quad (v)_m^m=D, \quad (U)_m^m=\Delta^{m-1},$$

we have by formula (A_2) of § 102 or Ex. viii of § 108

$$[v]_m^m\,\overline{\underline{U}}_m^m=\begin{bmatrix} a_{pq}, & a_{p\mu} \\ a_{xq}, & a_{x\mu} \end{bmatrix}_{h,\,\rho}^{k,\,\sigma}\overline{\underline{\begin{matrix} A_{pq}, & A_{\lambda q} \\ A_{p\mu}, & A_{\lambda\mu} \end{matrix}}}_{k,\,\sigma}^{h,\,\rho}=\begin{bmatrix} a_{pp}, & a_{p\lambda} \\ a_{xp}, & a_{x\lambda} \end{bmatrix}_{h,\,\rho}^{h,\,\rho}=\begin{bmatrix} \Delta, & 0 \\ a_{xp}, & a_{x\lambda} \end{bmatrix}_{h,\,\rho}^{h,\,\rho}.$$

We obtain (C_3') by equating the determinants of both sides in either one of the identities

$$(u)_m^m [v]_m^m = [v]_m^m \cdot \overline{\underline{U}}_m^m [u]_m^m = [v]_m^m \overline{\underline{U}}_m^m \cdot [u]_m^m = \begin{bmatrix} \Delta, & 0 \\ a_{xp}, & a_{x\lambda} \end{bmatrix}_{h,\rho}^{h,\rho} [u]_m^m,$$

$$[v]_m^m \overline{\underline{U}}_m^m = \begin{bmatrix} \Delta, & 0 \\ a_{xp}, & a_{x\lambda} \end{bmatrix}_{h,\rho}^{h,\rho},$$

cancelling the factor Δ^{h+1} common to both sides in the first case, and the factor Δ^h common to both sides in the second case.

We can also write down formula (C_3') immediately by applying Theorem V *b*.

Ex. xi. *If A is a matrix whose elements are arbitrary, and if Δ is any one of its simple minor determinants, then:*

(1) *We can ascribe an arbitrary non-zero value to Δ, and arbitrary values to the primaries of Δ. The remaining simple minor determinants of A are then completely and uniquely determinate.*

(2) *We can ascribe arbitrary values consistent with the condition $\Delta \neq 0$ to the elements of Δ, and arbitrary values to the primaries of Δ. The remaining elements of A are then completely and uniquely determinate.*

There will be no loss of generality in supposing that the long rows of A are horizontal, and that $A = [a, b]_m^{m, n-m}$, $\Delta = (a)_m^m$.

We can prove (1) by considering the special matrix $A = [\delta, q]_m^{m, n-m}$ in which δ is a scalar quantity and $[\delta]_m^m = \delta [1]_m^m$. For this matrix we have $\Delta = \delta^m$, and the primaries of Δ are the elements of the matrix $\delta^{m-1} [q]_m^{n-m}$. The result (1) follows when we observe that we can ascribe an arbitrary non-zero value to δ and arbitrary values to the elements of $[q]_m^{n-m}$.

To prove (2) we assume that $\Delta = (a)_m^m \neq 0$, denote the reciprocal of $[a]_m^m$ by $[A]_m^m$, and write

$$A = [a, b]_m^{m, n-m}, \quad \overline{\underline{A}}_m^m [a, b]_m^{m, n-m} = [\Delta, q]_m^{m, n-m},$$

so that the q's are the primaries of Δ. We then have

$$\Delta [a, b]_m^{m, n-m} = [a]_m^m [\Delta, q]_m^{m, n-m}, \quad \Delta [b]_m^{n-m} = [a]_m^m [q]_m^{n-m}.$$

If we ascribe arbitrary values to the elements of $[a]_m^m$ consistent with the condition $\Delta = (a)_m^m \neq 0$ and arbitrary values to the elements of $[q]_m^{n-m}$, the last equation determines the values of the elements of $[b]_m^{n-m}$, which are the remaining elements of A, completely and uniquely, and the matrix $A = [a, b]_m^{m, n-m}$ thus determined has the required properties.

Ex. xii. *Let Δ be any one of the simple minor determinants of the matrix $A = [a]_m^n$ whose elements are arbitrary, and let $p_1, p_2, p_3, \ldots$ be the primaries of Δ. Then if $g(\Delta, p_1, p_2, p_3, \ldots)$ is a rational integral function of Δ and its primaries, and if the equation*

$$g(\Delta, p_1, p_2, p_3, \ldots) = 0$$

is true for all values of the elements of A, it must be an identity in $\Delta, p_1, p_2, p_3, \ldots$.

For by Ex. xi the above equation regarded as an equation in the variables $\Delta, p_1, p_2, p_3, \ldots$ is true for all finite values of these quantities for which $\Delta \neq 0$. It is therefore true for all finite values of these quantities, and is an identity in them.

NOTE 7. *Utility of the identities of this article.*

Let Δ be any one simple minor determinant of a matrix $A=[a]_m^n$ whose elements are arbitrary; let $p_1, p_2, \ldots$ be the primaries of Δ; let $P_1, P_2, \ldots$ be the remaining simple minor determinants of A; and let the system of equations which serve to express $P_1, P_2, \ldots$ in terms of Δ and its primaries when $\Delta \neq 0$ be denoted by E.

Then if $f(\Delta, p_1, p_2, \ldots P_1, P_2, \ldots)$ is any rational integral function of the simple minor determinants of A, we can determine by means of the equations E a power Δ^i of Δ and a rational integral function $g(\Delta, p_1, p_2, \ldots)$ of Δ and its primaries such that

$$\Delta^i f = g(\Delta, p_1, p_2, \ldots).$$

Since Δ is a function of the elements of A which does not vanish identically, it follows that $f=0$ is an identity in the elements of A when and only when $g=0$ is an identity in the elements of A, i.e. by Ex. xii when and only when $g=0$ is an identity in $\Delta, p_1, p_2, \ldots$. The relation $f=0$ between the simple minor determinants of A is then deducible from the equations E between the simple minor determinants of A and the equation $g=0$ which is an identity in Δ and its primaries.

Thus the equations E (obtained in this article) and those deducible from them are the only relations between the simple minor determinants of the matrix A which are identities in the elements of A, or are true for all values of the elements of A.

The utility of the equations E rests on this fact.

Illustrations are given in Exs. iv and v.

3. *General formula for the standard identities.*

For a matrix which has m long rows let the simple minor determinant formed by its u_1th, u_2th, ... u_mth short rows be denoted for the sake of brevity by $(u_1 u_2 \ldots u_m)$. Then the results given by Theorems IV a and IV b or by Theorems V a and V b can be summarised in the following theorem:

Theorem VII. *In a matrix with m long rows and an unrestricted number of short rows there exist between the simple minor determinants non-trivial identities of the $m-1$ different forms*

$$\begin{vmatrix} (p_1p_2\ldots p_r\; y_1x_2x_3\ldots x_s), & (p_1p_2\ldots p_r\; y_2x_2x_3\ldots x_s), & \ldots (p_1p_2\ldots p_r\; y_sx_2x_3\ldots x_s) \\ (p_1p_2\ldots p_r\; x_1y_1x_3\ldots x_s), & (p_1p_2\ldots p_r\; x_1y_2x_3\ldots x_s), & \ldots (p_1p_2\ldots p_r\; x_1y_sx_3\ldots x_s) \\ \cdots & \cdots & \cdots \\ (p_1p_2\ldots p_r\; x_1x_2\ldots x_{s-1}y_1), & (p_1p_2\ldots p_r\; x_1x_2\ldots x_{s-1}y_2), & \ldots (p_1p_2\ldots p_r\; x_1x_2\ldots x_{s-1}y_s) \end{vmatrix}$$
$$= (p_1p_2\ldots p_r\; x_1x_2\ldots x_s)^{s-1}(p_1p_2\ldots p_r\; y_1y_2\ldots y_s), \quad\ldots\ldots\ldots\ldots(D)$$

where $r+s=m$, and s receives the values $2, 3, \ldots m$.

The last of these identities is

$$\begin{vmatrix} (y_1x_2x_3\ldots x_m), & (y_2x_2x_3\ldots x_m), & \ldots (y_mx_2x_3\ldots x_m) \\ (x_1y_1x_3\ldots x_m), & (x_1y_2x_3\ldots x_m), & \ldots (x_1y_mx_3\ldots x_m) \\ \cdots & \cdots & \cdots \\ (x_1x_2\ldots x_{m-1}y_1), & (x_1x_2\ldots x_{m-1}y_2), & \ldots (x_1x_2\ldots x_{m-1}y_m) \end{vmatrix}$$
$$= (x_1x_2x_3\ldots x_m)^{m-1}(y_1y_2y_3\ldots y_m), \quad\ldots\ldots\ldots\ldots(E)$$

and from this one all the others can be derived by supposing respectively that 1, 2, ... $m-1$ *of the numbers* $y_1, y_2, \ldots y_m$ *belong to the sequence* $[x_1 x_2 \ldots x_m]$.

Formula (E) is equivalent to formula (A) or formula (A′) according as the long rows are horizontal or vertical; and formula (D) is equivalent to formulae (C) and (C_1) or to formulae (C′) and (C_1') according as the long rows are horizontal or vertical.

If in (D) we put $x_1 = y_1 = p_{r+1}$, we can divide both sides by

$$(p_1 p_2 \ldots p_{r+1}\ x_2 x_3 \ldots x_s)$$

and so obtain the corresponding formula when r is replaced by $r+1$.

Ex. xiii. For $m=2$ there is only one distinct non-trivial identity, viz.

$$(1) \qquad (ab)(xy) = \begin{vmatrix} (xb) & (yb) \\ (ax) & (ay) \end{vmatrix}.$$

For $m=3$ there are two distinct non-trivial identities, viz.

$$(2a) \qquad (abc)(xyz) = \begin{vmatrix} (ayc) & (azc) \\ (aby) & (abz) \end{vmatrix};$$

$$(2b) \qquad (abc)^2(xyz) = \begin{vmatrix} (xbc) & (ybc) & (zbc) \\ (axc) & (ayc) & (azc) \\ (abx) & (aby) & (abz) \end{vmatrix}.$$

For $m=4$ there are three distinct non-trivial identities, viz.

$$(3a) \qquad (abcd)(abzw) = \begin{vmatrix} (abzd) & (abwd) \\ (abcz) & (abcw) \end{vmatrix};$$

$$(3b) \qquad (abcd)^2(ayzw) = \begin{vmatrix} (aycd) & (azcd) & (awcd) \\ (abyd) & (abzd) & (abwd) \\ (abcy) & (abcz) & (abcw) \end{vmatrix};$$

$$(3c) \qquad (abcd)^3(xyzw) = \begin{vmatrix} (xbcd) & (ybcd) & (zbcd) & (wbcd) \\ (axcd) & (aycd) & (azcd) & (awcd) \\ (abxd) & (abyd) & (abzd) & (abwd) \\ (abcx) & (abcy) & (abcz) & (abcw) \end{vmatrix}.$$

Ex. xiv. In Ex. xiii we can regard (ab), (azc), $(abyd)$ as standing for $(ab)_{12}$, $(azc)_{123}$, $(abyd)_{1234}$, and so on. We can then prove (3 c) by equating the determinants of both sides in the identity

$$\begin{bmatrix} A_1 & A_2 & A_3 & A_4 \\ B_1 & B_2 & B_3 & B_4 \\ C_1 & C_2 & C_3 & C_4 \\ D_1 & D_2 & D_3 & D_4 \end{bmatrix} \begin{bmatrix} x_1 & y_1 & z_1 & w_1 \\ x_2 & y_2 & z_2 & w_2 \\ x_3 & y_3 & z_3 & w_3 \\ x_4 & y_4 & z_4 & w_4 \end{bmatrix} = \begin{bmatrix} (xbcd) & (ybcd) & (zbcd) & (wbcd) \\ (axcd) & (aycd) & (azcd) & (awcd) \\ (abxd) & (abyd) & (abzd) & (abwd) \\ (abcx) & (abcy) & (abcz) & (abcw) \end{bmatrix}$$

where $[A\,B\,C\,D]_{1234}$ is the reciprocal of $[a\,b\,c\,d]_{1234}$, making use of the identity

$$(ABCD)_{1234} = \{(abcd)_{1234}\}^3 = (abcd)^3.$$

We can prove (3 b) by putting $[x_1 x_2 x_3 x_4]=[a_1 a_2 a_3 a_4]$ and then equating the determinants of both sides.

We can prove (3 a) by putting $[x_1 x_2 x_3 x_4]=[a_1 a_2 a_3 a_4]$, $[y_1 y_2 y_3 y_4]=[b_1 b_2 b_3 b_4]$, and then equating the determinants of both sides.

Ex. xv. It will sometimes be convenient to use for the minor matrices and minor determinoids of $[a]_m^n$ the notations

$$\begin{bmatrix} v_1 v_2 \dots v_s \\ a \\ u_1 u_2 \dots u_r \end{bmatrix} = \begin{bmatrix} v_1 v_2 \dots v_s \\ u_1 u_2 \dots u_r \end{bmatrix}, \quad \begin{pmatrix} v_1 v_2 \dots v_s \\ a \\ u_1 u_2 \dots u_r \end{pmatrix} = \begin{pmatrix} v_1 v_2 \dots v_s \\ u_1 u_2 \dots u_r \end{pmatrix},$$

the letter a being omitted, and only the horizontal and vertical suffixes being shown.

If then $$\Delta = \begin{pmatrix} q_1 q_2 \dots q_k & \mu_1 \mu_2 \dots \mu_s \\ p_1 p_2 \dots p_h & \lambda_1 \lambda_2 \dots \lambda_r \end{pmatrix} = \begin{pmatrix} a_{pq}, & a_{p\mu} \\ a_{\lambda q}, & a_{\lambda\mu} \end{pmatrix}_{h,\, r}^{k,\, s}, \text{ where } k+s=h+r,$$

Theorem VII is equivalent to the two formulae

$$\begin{pmatrix} q_1 q_2 \dots q_k & \mu_1 \mu_2 \dots \mu_s \\ p_1 p_2 \dots p_h & \lambda_1 \lambda_2 \dots \lambda_r \end{pmatrix}^{r-1} \begin{pmatrix} q_1 q_2 \dots q_k & \mu_1 \mu_2 \dots \mu_s \\ p_1 p_2 \dots p_h & x_1 x_2 \dots x_r \end{pmatrix} = (a_{x\lambda})_r^r, \quad \dots\dots\dots\dots\text{(F)}$$

$$\begin{pmatrix} q_1 q_2 \dots q_k & \mu_1 \mu_2 \dots \mu_s \\ p_1 p_2 \dots p_h & \lambda_1 \lambda_2 \dots \lambda_r \end{pmatrix}^{s-1} \begin{pmatrix} q_1 q_2 \dots q_k & y_1 y_2 \dots y_s \\ p_1 p_2 \dots p_h & \lambda_1 \lambda_2 \dots \lambda_r \end{pmatrix} = (\beta_{\mu y})_s^s, \quad \dots\dots\dots\dots\text{(F}')$$

where $a_{x_i \lambda_j}$ and $\beta_{\mu_i y_j}$ are the determinants formed respectively from Δ when we replace the horizontal suffix λ_j by x_i and when we replace the vertical suffix μ_i by y_j.

These are equivalent to the formulae in Exs. x and vii.

Note 8. *Corresponding identities for minor determinoids.*

All the formulae of this article can be generalised by taking D to be a *superior simple minor determinoid* of the matrix of which Δ is a simple minor determinant.

If $$\Delta=(a_{pq})_r^r, \quad D=(a_{uq})_t^r, \quad D'=(a_{pv})_r^t,$$

we have $$[a_{uq}]_t^r \overline{\underline{A_{pq}}}_r^r = [a_{up}]_t^r, \quad \overline{\underline{A_{pq}}}_r^r [a_{pv}]_r^t = [\beta_{qv}]_r^t;$$

and when $t \nless r$, we see by equating the determinoids of both sides that

$$\Delta^{r-1} D = (a_{up})_t^r, \quad \Delta^{r-1} D' = (\beta_{qv})_r^t. \quad \dots\dots\dots\dots\text{(G)}$$

When Δ and D have h horizontal rows in common, and when Δ and D' have k vertical rows in common, these formulae can be reduced to

$$\Delta^{r-h-1} D = (-1)^{\omega} (a_{\xi x})_{t-h}^{r-h}, \quad \Delta^{r-k-1} D' = (-1)^{\omega'} (\beta_{y\eta})_{r-k}^{t-k}. \quad \dots\dots\dots\text{(H)}$$

In the first formula $[x]_{r-h}$, $[\xi]_{t-h}$ are the minors of $[p]_r$, $[u]_t$ formed by striking out their common elements $\lambda_1, \lambda_2, \dots \lambda_h$, and ω is the difference of the affects of $[\lambda]_h$ in $[p]_r$ and $[u]_t$.

In the second formula $[y]_{r-k}$, $[\eta]_{t-k}$ are the minors of $[q]_r$, $[v]_t$ formed by striking out their common elements $\mu_1, \mu_2, \dots \mu_k$, and ω' is the difference of the affects of $[\mu]_k$ in $[q]_r$ and $[v]_t$.

If $$\Delta=\begin{pmatrix} a_{pq}, & a_{p\mu} \\ a_{\lambda q}, & a_{\lambda\mu} \end{pmatrix}_{h,\rho}^{k,\sigma}, \quad D=\begin{pmatrix} a_{pq}, & a_{p\mu} \\ a_{xq}, & a_{x\mu} \end{pmatrix}_{h,r}^{k,\sigma}, \quad D'=\begin{pmatrix} a_{pq}, & a_{pv} \\ a_{\lambda q}, & a_{\lambda v} \end{pmatrix}_{h,\rho}^{k,s},$$

where $$k+\sigma=h+\rho, \quad r \not< \rho, \quad s \not< \sigma,$$

we have $$\Delta^{\rho-1}D=(a_{x\lambda})_{r}^{\rho}, \quad \Delta^{\sigma-1}D'=(\beta_{\mu v})_{\sigma}^{s}. \qquad \text{(I)}$$

The corresponding generalisation of formula (D) is

$$\begin{vmatrix} (p_1\dots p_r\ y_1x_2x_3\dots x_s), & (p_1\dots p_r\ y_2x_2x_3\dots x_s), & \dots & (p_1\dots p_r\ y_tx_2x_3\dots x_s) \\ (p_1\dots p_r\ x_1y_1x_3\dots x_s), & (p_1\dots p_r\ x_1y_2x_3\dots x_s), & \dots & (p_1\dots p_r\ x_1y_tx_3\dots x_s) \\ \dots & \dots & \dots & \dots \\ (p_1\dots p_r\ x_1x_2\dots x_{s-1}y_1), & (p_1\dots p_r\ x_1x_2\dots x_{s-1}y_2), & \dots & (p_1\dots p_r\ x_1x_2\dots x_{s-1}y_t) \end{vmatrix}$$
$$=(p_1p_2\dots p_r\ x_1x_2\dots x_s)^{s-1}(p_1p_2\dots p_r\ y_1y_2\dots y_t), \qquad \text{(J)}$$

where $t \not< s$, and where $(p_1p_2\dots p_r\ y_1y_2\dots y_t)$ is now the determinoid formed by the p_1th, p_2th, ... p_rth, y_1th, y_2th, ... y_tth short rows of the matrix which as before contains $r+s$ long rows.

§ 110. Sylvester's identities satisfied by those primary super-determinants of one determinant Δ which lie in another determinant D containing Δ.

Using the notation described in Ex. xv of § 109 for the minor determinants of the matrix $[a]_m^n$ we have the following theorem:

Theorem. *In the case of any matrix $A = [a]_m^n$ there exist identities of the form*

$$\begin{vmatrix} \begin{pmatrix} q_1q_2\dots q_r\ v_1 \\ p_1p_2\dots p_r\ u_1 \end{pmatrix}, & \begin{pmatrix} q_1q_2\dots q_r\ v_2 \\ p_1p_2\dots p_r\ u_1 \end{pmatrix}, & \dots & \begin{pmatrix} q_1q_2\dots q_r\ v_s \\ p_1p_2\dots p_r\ u_1 \end{pmatrix} \\ \begin{pmatrix} q_1q_2\dots q_r\ v_1 \\ p_1p_2\dots p_r\ u_2 \end{pmatrix}, & \begin{pmatrix} q_1q_2\dots q_r\ v_2 \\ p_1p_2\dots p_r\ u_2 \end{pmatrix}, & \dots & \begin{pmatrix} q_1q_2\dots q_r\ v_s \\ p_1p_2\dots p_r\ u_2 \end{pmatrix} \\ \dots & \dots & \dots & \dots \\ \begin{pmatrix} q_1q_2\dots q_r\ v_1 \\ p_1p_2\dots p_r\ u_s \end{pmatrix}, & \begin{pmatrix} q_1q_2\dots q_r\ v_2 \\ p_1p_2\dots p_r\ u_s \end{pmatrix}, & \dots & \begin{pmatrix} q_1q_2\dots q_r\ v_s \\ p_1p_2\dots p_r\ u_s \end{pmatrix} \end{vmatrix}$$
$$=\begin{pmatrix} q_1q_2\dots q_r \\ p_1p_2\dots p_r \end{pmatrix}^{s-1}\begin{pmatrix} q_1q_2\dots q_r\ v_1v_2\dots v_s \\ p_1p_2\dots p_r\ u_1u_2\dots u_s \end{pmatrix}, \qquad \text{(A)}$$

where $[p_1p_2\dots p_r\,u_1u_2\dots u_s]$ and $[q_1q_2\dots q_r\,v_1v_2\dots v_s]$ are any minors of order $r+s$ of the respective sequences $[1\,2\dots m]$ and $[1\,2\dots n]$.

The identity (A) in the elements of $[a]_m^n$ will be called *Sylvester's Identity*.

If we put

$$\Delta=\begin{pmatrix} q_1q_2\dots q_r \\ p_1p_2\dots p_r \end{pmatrix}=(a_{pq})_r^r,$$

$$D=\begin{pmatrix} q_1q_2\dots q_r\ v_1v_2\dots v_s \\ p_1p_2\dots p_r\ u_1u_2\dots u_s \end{pmatrix}=\begin{pmatrix} a_{pq}, & a_{pv} \\ a_{uq}, & a_{uv} \end{pmatrix}_{r,s}^{r,s},$$

$$P_{u_i v_j} = \begin{pmatrix} q_1 q_2 \dots q_r & v_j \\ p_1 p_2 \dots p_r & u_i \end{pmatrix} = \begin{vmatrix} a_{p_1 q_1} \dots a_{p_1 q_r} & a_{p_1 v_j} \\ \dots\dots\dots & \dots \\ a_{p_r q_1} \dots a_{p_r q_r} & a_{p_r v_j} \\ a_{u_i q_1} \dots a_{u_i q_r} & a_{u_i v_j} \end{vmatrix},$$

we can express (A) more briefly in the forms

$$\begin{pmatrix} q_1 q_2 \dots q_r \\ p_1 p_2 \dots p_r \end{pmatrix}^{s-1} \begin{pmatrix} q_1 q_2 \dots q_r & v_1 v_2 \dots v_s \\ p_1 p_2 \dots p_r & u_1 u_2 \dots u_s \end{pmatrix} = (P_{uv})_s^s, \quad \text{.........(B)}$$

and

$$\Delta^{s-1} D = (P_{uv})_s^s. \quad \text{...............................(C)}$$

When $\Delta \neq 0$, these formulae express D, which is any superdeterminant of Δ, as a function of Δ and those primary superdeterminants of Δ which are contained in D.

We can deduce (C) from Ex. x or Theorems IV *b* and V *b* of § 109.

If
$$\Delta = \begin{pmatrix} a_{pq}, & a_{pv} \\ a_{\lambda q}, & a_{\lambda v} \end{pmatrix}_{r,s}^{r,s}, \qquad D = \begin{pmatrix} a_{pq}, & a_{pv} \\ a_{uq}, & a_{uv} \end{pmatrix}_{r,s}^{r,s},$$

we have
$$\Delta^{s-1} D = (\alpha_{u\lambda})_s^s, \quad \text{...............................(1)}$$

where $\alpha_{u_i \lambda_j}$ is obtained from Δ by replacing the horizontal suffix λ_j by u_i.

Now let
$$[a_{\lambda q}, \; a_{\lambda v}]_s^{r,s} = [0, \; 1]_s^{r,s}.$$

Then
$$\Delta = (a_{pq})_r^r, \qquad \alpha_{u_i \lambda_j} = \begin{pmatrix} q_1 q_2 \dots q_r v_j \\ a \\ p_1 p_2 \dots p_r u_i \end{pmatrix} = P_{u_i v_j},$$

and the equation (1) becomes (C).

In a similar way we can deduce (C) from Ex. vii or Theorems IV *a* and V *a* of § 109.

Ex. i. *Direct proof of the theorem.*

Let
$$[h]_m^m = \begin{bmatrix} a_{pq}, & a_{pv} \\ 0, & 1 \end{bmatrix}_{r,s}^{r,s}, \qquad [k]_m^m = \begin{bmatrix} a_{pq}, & a_{pv} \\ a_{uq}, & a_{uv} \end{bmatrix}_{r,s}^{r,s},$$

so that
$$\Delta = (h)_m^m = (a_{pq})_r^r, \qquad D = (k)_m^m = \begin{pmatrix} q_1 \dots q_r \; v_1 \dots v_s \\ a \\ p_1 \dots p_r \; u_1 \dots u_s \end{pmatrix}, \qquad m = r + s,$$

and let $[H]_m^m$ be the reciprocal of $[h]_m^m$, so that $(H)_m^m = \Delta^{m-1}$. Then we can obtain formula (C) by equating the determinants of both sides in the identity

$$(h)_m^m [k]_m^m = [k]_m^m . \overline{\underline{H}}_m^m [h]_m^m = [k]_m^m \overline{\underline{H}}_m^m . [h]_m^m = [\delta]_m^m [h]_m^m, \quad \text{...............(2)}$$

or in the identity

$$[k]_m^m \overline{\underline{H}}_m^m = [\delta]_m^m, \quad \text{.......................................(3)}$$

where δ_{ij} is the determinant of the matrix formed when we replace the jth horizontal row of $[h]_m^m$ by the ith horizontal row of $[k]_m^m$.

By § 101 we have
$$\boxed{H}_m^m = \boxed{\begin{matrix} A_{pq}, & A_{uq} \\ 0\ , & \Delta \end{matrix}}_{r,s}^{r,s},$$

where $\boxed{A_{pq}}_r^r$ is the conjugate reciprocal of $[a_{pq}]_r^r$, $\Delta=(a_{pq})_r^r$,

and
$$\boxed{A_{uq}}_r^s = -\boxed{A_{pq}}_r^r \, [a_{pv}]_r^s .$$

Therefore we can write
$$[\delta]_m^m = \begin{bmatrix} a_{pq}, & a_{pv} \\ a_{uq}, & a_{uv} \end{bmatrix}_{r,s}^{r,s} \boxed{\begin{matrix} A_{pq}, & A_{uq} \\ 0\ , & \Delta \end{matrix}}_{r,s}^{r,s} = \begin{bmatrix} \Delta\ , & 0 \\ x_{up}, & x_{uv} \end{bmatrix}_{r,s}^{r,s},$$

where
$$[x_{uv}]_s^s = [a_{uq}]_s^r \boxed{A_{uq}}_r^s + \Delta [a_{uv}]_s^s = \Delta [a_{uv}]_s^s - [a_{uq}]_s^r \boxed{A_{pq}}_r^r \, [a_{pv}]_r^s .$$

As shown in § 116 it follows by Ex. xi of § 62 that
$$x_{u_i v_j} = \begin{pmatrix} q_1 q_2 \ldots q_r v_j \\ a \\ p_1 p_2 \ldots p_r u_i \end{pmatrix} = P_{u_i v_j}, \qquad [x_{uv}]_s^s = [P_{uv}]_s^s .$$

We have therefore
$$[\delta]_m^m = \begin{bmatrix} \Delta\ , & 0 \\ x_{up}, & P_{uv} \end{bmatrix}_{r,s}^{r,s},$$

and we obtain $\quad \Delta^m D = \Delta^{r+1} (P_{uv})_s^s, \quad \Delta^{s-1} D = (P_{uv})_s^s \quad$ from (2),

or $\quad \Delta^{m-1} D = \Delta^r (P_{uv})_s^s, \quad \Delta^{s-1} D = (P_{uv})_s^s \quad$ from (3).

Ex. ii. Particular cases of the identity (A) are

(1)
$$\begin{vmatrix} \begin{pmatrix} uv\xi \\ a \\ pqx \end{pmatrix} & \begin{pmatrix} uv\eta \\ a \\ pqx \end{pmatrix} \\ \begin{pmatrix} uv\xi \\ a \\ pqy \end{pmatrix} & \begin{pmatrix} uv\eta \\ a \\ pqy \end{pmatrix} \end{vmatrix} = \begin{pmatrix} uv \\ a \\ pq \end{pmatrix} \begin{pmatrix} uv\xi\eta \\ a \\ pqxy \end{pmatrix};$$

(2)
$$\begin{vmatrix} \begin{pmatrix} uv\xi \\ a \\ pqx \end{pmatrix} & \begin{pmatrix} uv\eta \\ a \\ pqx \end{pmatrix} & \begin{pmatrix} uv\zeta \\ a \\ pqx \end{pmatrix} \\ \begin{pmatrix} uv\xi \\ a \\ pqy \end{pmatrix} & \begin{pmatrix} uv\eta \\ a \\ pqy \end{pmatrix} & \begin{pmatrix} uv\zeta \\ a \\ pqy \end{pmatrix} \\ \begin{pmatrix} uv\xi \\ a \\ pqz \end{pmatrix} & \begin{pmatrix} uv\eta \\ a \\ pqz \end{pmatrix} & \begin{pmatrix} uv\zeta \\ a \\ pqz \end{pmatrix} \end{vmatrix} = \begin{pmatrix} uv \\ a \\ pq \end{pmatrix}^2 \begin{pmatrix} uv\xi\eta\zeta \\ a \\ pqxyz \end{pmatrix}.$$

The corresponding results when the single-suffix notation is used are

(1)
$$\begin{vmatrix} (abx)_{pq\lambda} & (abx)_{pq\mu} \\ (aby)_{pq\lambda} & (aby)_{pq\mu} \end{vmatrix} = (ab)_{pq} \, (abxy)_{pq\lambda\mu};$$

(2)
$$\begin{vmatrix} (abx)_{pq\lambda} & (abx)_{pq\mu} & (abx)_{pq\nu} \\ (aby)_{pq\lambda} & (aby)_{pq\mu} & (aby)_{pq\nu} \\ (abz)_{pq\lambda} & (abz)_{pq\mu} & (abz)_{pq\nu} \end{vmatrix} = (ab)_{pq}^2 \, (abxyz)_{pq\lambda\mu\nu}.$$

We can obtain the last result by writing

$$[h]_5^5 = \begin{bmatrix} a_p & b_p & 0 & 0 & 0 \\ a_q & b_q & 0 & 0 & 0 \\ a_\lambda & b_\lambda & 1 & 0 & 0 \\ a_\mu & b_\mu & 0 & 1 & 0 \\ a_\nu & b_\nu & 0 & 0 & 1 \end{bmatrix}, \quad [k]_5^5 = \begin{bmatrix} a_p & b_p & x_p & y_p & z_p \\ a_q & b_q & x_q & y_q & z_q \\ a_\lambda & b_\lambda & x_\lambda & y_\lambda & z_\lambda \\ a_\mu & b_\mu & x_\mu & y_\mu & z_\mu \\ a_\nu & b_\nu & x_\nu & y_\nu & z_\nu \end{bmatrix},$$

denoting the reciprocal of $[h]_5^5$ by $[H]_5^5$, so that $(h)_5^5 = (ab)_{pq} = \Delta$, $(H)_5^5 = \Delta^4$, and equating the determinants of both sides in the identity

$$[h]_5^5 \, \overline{\underline{H}}_5^5 \, . \, [k]_5^5 = [h]_5^5 \, . \, \overline{\underline{H}}_5^5 \, [k]_5^5,$$

i.e.

$$(h)_5^5 [k]_5^5 = [h]_5^5 [\delta]_5^5,$$

or in the identity

$$\overline{\underline{H}}_5^5 \, [k]_5^5 = [\delta]_5^5,$$

where δ_{ij} is the determinant of the matrix formed when the ith vertical row of $[h]_5^5$ is replaced by the jth vertical row of $[k]_5^5$.

Ex. iii. If we expand the determinant on the left in the second result of Ex. ii in terms of the minor determinants of the last two vertical rows and simplify each minor determinant by means of the first result, we obtain

$$\begin{pmatrix} u\,v\,\xi \\ a \\ p\,q\,x \end{pmatrix} \begin{pmatrix} u\,v\,\eta\,\zeta \\ a \\ p\,q\,y\,z \end{pmatrix} + \begin{pmatrix} u\,v\,\xi \\ a \\ p\,q\,y \end{pmatrix} \begin{pmatrix} u\,v\,\eta\,\zeta \\ a \\ p\,q\,z\,x \end{pmatrix} + \begin{pmatrix} u\,v\,\xi \\ a \\ p\,q\,z \end{pmatrix} \begin{pmatrix} u\,v\,\eta\,\zeta \\ a \\ p\,q\,x\,y \end{pmatrix} = \begin{pmatrix} u\,v \\ a \\ p\,q \end{pmatrix} \begin{pmatrix} u\,v\,\xi\,\eta\,\zeta \\ a \\ p\,q\,x\,y\,z \end{pmatrix}.$$

This result can be generalised.

Ex. iv. *Generalisation of the theorem of the text.*

Formula (A) is included in the more general formula

$$\begin{vmatrix} \begin{pmatrix} q_1\,q_2 \ldots q_r\,v_1 \\ p_1\,p_2 \ldots p_r\,u_1 \end{pmatrix}, & \begin{pmatrix} q_1\,q_2 \ldots q_r\,v_2 \\ p_1\,p_2 \ldots p_r\,u_1 \end{pmatrix}, & \ldots & \begin{pmatrix} q_1\,q_2 \ldots q_r\,v_t \\ p_1\,p_2 \ldots p_r\,u_1 \end{pmatrix} \\ \begin{pmatrix} q_1\,q_2 \ldots q_r\,v_1 \\ p_1\,p_2 \ldots p_r\,u_2 \end{pmatrix}, & \begin{pmatrix} q_1\,q_2 \ldots q_r\,v_2 \\ p_1\,p_2 \ldots p_r\,u_2 \end{pmatrix}, & \ldots & \begin{pmatrix} q_1\,q_2 \ldots q_r\,v_t \\ p_1\,p_2 \ldots p_r\,u_2 \end{pmatrix} \\ \cdots\cdots & \cdots\cdots & \cdots\cdots & \cdots\cdots \\ \begin{pmatrix} q_1\,q_2 \ldots q_r\,v_1 \\ p_1\,p_2 \ldots p_r\,u_s \end{pmatrix}, & \begin{pmatrix} q_1\,q_2 \ldots q_r\,v_2 \\ p_1\,p_2 \ldots p_r\,u_s \end{pmatrix}, & \ldots & \begin{pmatrix} q_1\,q_2 \ldots q_r\,v_t \\ p_1\,p_2 \ldots p_r\,u_s \end{pmatrix} \end{vmatrix}$$

$$= \begin{pmatrix} q_1\,q_2 \ldots q_r \\ p_1\,p_2 \ldots p_r \end{pmatrix}^{s-1} \begin{pmatrix} q_1\,q_2 \ldots q_r\,v_1\,v_2 \ldots v_t \\ p_1\,p_2 \ldots p_r\,u_1\,u_2 \ldots u_s \end{pmatrix}, \quad \ldots\ldots\ldots\ldots\ldots\ldots(\mathrm{A}')$$

where $t \not< s$, which can be proved in the same way by using Note 8 of § 109.

Writing

$$\Delta = \begin{pmatrix} a_{pq}, & a_{p\mu} \\ a_{uq}, & a_{u\mu} \end{pmatrix}_{r,s}^{r,s}, \quad D = \begin{pmatrix} a_{pq}, & a_{pv} \\ a_{uq}, & a_{uv} \end{pmatrix}_{r,s}^{r,t},$$

we have

$$\Delta^{s-1} D = (\beta_{\mu v})_s^t.$$

Now let

$$\begin{bmatrix} a_{p\mu} \\ a_{u\mu} \end{bmatrix}_{r,s}^{s} = \begin{bmatrix} 0 \\ 1 \end{bmatrix}_{r,s}^{s}.$$

Then
$$\Delta=(a_{pq})^r_r,\quad \beta_{\mu_i v_j}=\begin{pmatrix} q_1 q_2 \dots q_r\, v_j \\ a \\ p_1 p_2 \dots p_r\, u_i\end{pmatrix}=P_{u_i v_j},$$

and we have
$$\Delta^{s-1}D=(P_{uv})^t_s. \quad \dots\dots\dots\dots(C')$$

Ex. v.
$$\begin{vmatrix} \begin{pmatrix}u v \xi\\ a\\ p q x\end{pmatrix} & \begin{pmatrix}u v \eta\\ a\\ p q x\end{pmatrix} & \begin{pmatrix}u v \zeta\\ a\\ p q x\end{pmatrix} & \begin{pmatrix}u v \omega\\ a\\ p q x\end{pmatrix} \\ \begin{pmatrix}u v \xi\\ a\\ p q y\end{pmatrix} & \begin{pmatrix}u v \eta\\ a\\ p q y\end{pmatrix} & \begin{pmatrix}u v \zeta\\ a\\ p q y\end{pmatrix} & \begin{pmatrix}u v \omega\\ a\\ p q y\end{pmatrix} \\ \begin{pmatrix}u v \xi\\ a\\ p q z\end{pmatrix} & \begin{pmatrix}u v \eta\\ a\\ p q z\end{pmatrix} & \begin{pmatrix}u v \zeta\\ a\\ p q z\end{pmatrix} & \begin{pmatrix}u v \omega\\ a\\ p q z\end{pmatrix}\end{vmatrix} = \begin{pmatrix}u v\\ a\\ p q\end{pmatrix}^2 \begin{pmatrix}u v \xi \eta \zeta \omega\\ a\\ p q x y z\end{pmatrix}.$$

Ex. vi.
$$\begin{vmatrix} (abx)_{pq\lambda}, & (abx)_{pq\mu}, & (abx)_{pq\nu}, & (abx)_{pq\pi} \\ (aby)_{pq\lambda}, & (aby)_{pq\mu}, & (aby)_{pq\nu}, & (aby)_{pq\pi} \\ (abz)_{pq\lambda}, & (abz)_{pq\mu}, & (abz)_{pq\nu}, & (abz)_{pq\pi}\end{vmatrix} = (ab)^2_{pq}(abxyz)_{pq\lambda\mu\nu\pi}.$$

§ 111. Identities satisfied by those primary subdeterminants of one determinant Δ which contain another determinant D lying in Δ.

We again use the notation described in Ex. xv of § 109.

Theorem. *For any matrix $A=[a]^n_m$ there exist identities of the form*

$$\begin{vmatrix} +\begin{pmatrix} q_1q_2\dots q_r\ \mu_2\mu_3\dots\mu_s \\ p_1p_2\dots p_r\ \lambda_2\lambda_3\dots\lambda_s\end{pmatrix}, & -\begin{pmatrix} q_1q_2\dots q_r\ \mu_1\mu_3\dots\mu_s \\ p_1p_2\dots p_r\ \lambda_2\lambda_3\dots\lambda_s\end{pmatrix}, & \dots & \pm\begin{pmatrix} q_1q_2\dots q_r\ \mu_1\mu_2\dots\mu_{s-1} \\ p_1p_2\dots p_r\ \lambda_2\lambda_3\dots\lambda_s\end{pmatrix} \\ -\begin{pmatrix} q_1q_2\dots q_r\ \mu_2\mu_3\dots\mu_s \\ p_1p_2\dots p_r\ \lambda_1\lambda_3\dots\lambda_s\end{pmatrix}, & +\begin{pmatrix} q_1q_2\dots q_r\ \mu_1\mu_3\dots\mu_s \\ p_1p_2\dots p_r\ \lambda_1\lambda_3\dots\lambda_s\end{pmatrix}, & \dots & \mp\begin{pmatrix} q_1q_2\dots q_r\ \mu_1\mu_2\dots\mu_{s-1} \\ p_1p_2\dots p_r\ \lambda_1\lambda_3\dots\lambda_s\end{pmatrix} \\ \dots\dots & \dots\dots & \dots\dots & \dots\dots \\ \pm\begin{pmatrix} q_1q_2\dots q_r\ \mu_2\mu_3\dots\mu_s \\ p_1p_2\dots p_r\ \lambda_1\lambda_2\dots\lambda_{s-1}\end{pmatrix}, & \mp\begin{pmatrix} q_1q_2\dots q_r\ \mu_1\mu_3\dots\mu_s \\ p_1p_2\dots p_r\ \lambda_1\lambda_2\dots\lambda_{s-1}\end{pmatrix}, & \dots & \pm\begin{pmatrix} q_1q_2\dots q_r\ \mu_1\mu_2\dots\mu_{s-1} \\ p_1p_2\dots p_r\ \lambda_1\lambda_2\dots\lambda_{s-1}\end{pmatrix}\end{vmatrix}$$

$$=\begin{vmatrix} \begin{pmatrix} q_1q_2\dots q_r\ \mu_2\mu_3\dots\mu_s \\ p_1p_2\dots p_r\ \lambda_2\lambda_3\dots\lambda_s\end{pmatrix}, & \begin{pmatrix} q_1q_2\dots q_r\ \mu_1\mu_3\dots\mu_s \\ p_1p_2\dots p_r\ \lambda_2\lambda_3\dots\lambda_s\end{pmatrix}, & \dots & \begin{pmatrix} q_1q_2\dots q_r\ \mu_1\mu_2\dots\mu_{s-1} \\ p_1p_2\dots p_r\ \lambda_2\lambda_3\dots\lambda_s\end{pmatrix} \\ \begin{pmatrix} q_1q_2\dots q_r\ \mu_2\mu_3\dots\mu_s \\ p_1p_2\dots p_r\ \lambda_1\lambda_3\dots\lambda_s\end{pmatrix}, & \begin{pmatrix} q_1q_2\dots q_r\ \mu_1\mu_3\dots\mu_s \\ p_1p_2\dots p_r\ \lambda_1\lambda_3\dots\lambda_s\end{pmatrix}, & \dots & \begin{pmatrix} q_1q_2\dots q_r\ \mu_1\mu_2\dots\mu_{s-1} \\ p_1p_2\dots p_r\ \lambda_1\lambda_3\dots\lambda_s\end{pmatrix} \\ \dots\dots & \dots\dots & \dots\dots & \dots\dots \\ \begin{pmatrix} q_1q_2\dots q_r\ \mu_2\mu_3\dots\mu_s \\ p_1p_2\dots p_r\ \lambda_1\lambda_2\dots\lambda_{s-1}\end{pmatrix}, & \begin{pmatrix} q_1q_2\dots q_r\ \mu_1\mu_3\dots\mu_s \\ p_1p_2\dots p_r\ \lambda_1\lambda_2\dots\lambda_{s-1}\end{pmatrix}, & \dots & \begin{pmatrix} q_1q_2\dots q_r\ \mu_1\mu_2\dots\mu_{s-1} \\ p_1p_2\dots p_r\ \lambda_1\lambda_2\dots\lambda_{s-1}\end{pmatrix}\end{vmatrix}$$

$$=\begin{pmatrix} q_1q_2\dots q_r\ \mu_1\mu_2\dots\mu_s \\ p_1p_2\dots p_r\ \lambda_1\lambda_2\dots\lambda_s\end{pmatrix}^{s-1}\begin{pmatrix} q_1q_2\dots q_r \\ p_1p_2\dots p_r\end{pmatrix}, \quad \dots\dots\dots\dots(A)$$

where $[p_1p_2\dots p_r\ \lambda_1\lambda_2\dots\lambda_s]$ and $[q_1q_2\dots q_r\ \mu_1\mu_2\dots\mu_s]$ are any minors of the respective sequences $[1\,2\dots m]$ and $[1\,2\dots n]$ of the same order $r+s$.

The sign prefixed to each minor determinant occurring as an element in the first large determinant on the left is that determined by its affect in the determinant
$$\Delta=\begin{pmatrix} q_1q_2\dots q_r\ \mu_1\mu_2\dots\mu_s \\ p_1p_2\dots p_r\ \lambda_1\lambda_2\dots\lambda_s\end{pmatrix},$$
and therefore the sign prefixed to the minor determinant forming the element common to the uth horizontal row and the vth vertical row is the same as the sign of $(-1)^{u+v}$. The signs occurring in each horizontal and vertical row are alternately positive and negative.

The second large determinant on the left is obtained from the first by changing the signs of all elements in alternate horizontal rows and alternate vertical rows.

If we put $$\Delta=\begin{pmatrix} q_1 q_2 \dots q_r & \mu_1 \mu_2 \dots \mu_s \\ p_1 p_2 \dots p_r & \lambda_1 \lambda_2 \dots \lambda_s \end{pmatrix}=\begin{pmatrix} a_{pq}, & a_{p\mu} \\ a_{\lambda q}, & a_{\lambda\mu} \end{pmatrix}_{r,\,s}^{r,\,s},$$

$$D=\begin{pmatrix} q_1 q_2 \dots q_r \\ p_1 p_2 \dots p_r \end{pmatrix}=(a_{pq})_r^r,$$

and if $A_{\lambda_i \mu_j}$ is the co-factor of $a_{\lambda_i \mu_j}$ in Δ, or if $A_{\lambda_i \mu_j}$ is the primary subdeterminant of Δ formed by striking out the horizontal and vertical rows whose suffixes are λ_i and μ_j, we can express (A) more briefly in the forms

$$\begin{pmatrix} q_1 q_2 \dots q_r & \mu_1 \mu_2 \dots \mu_s \\ p_1 p_2 \dots p_r & \lambda_1 \lambda_2 \dots \lambda_s \end{pmatrix}^{s-1}\begin{pmatrix} q_1 q_2 \dots q_r \\ p_1 p_2 \dots p_r \end{pmatrix}=(A_{\lambda\mu})_s^s, \quad \text{.................(B)}$$

and $$\Delta^{s-1} D=(A_{\lambda\mu})_s^s. \quad \text{....................................(C)}$$

It is clearly immaterial which definition we adopt for $A_{\lambda_i \mu_j}$.

When $\Delta \neq 0$, these formulae express D, which is a minor determinant of Δ, as a function of Δ and those primary subdeterminants of Δ which contain D.

We can deduce (C) from Ex. x or Theorems IV *b* and V *b* of § 109.

If $$\Delta=\begin{pmatrix} a_{pq}, & a_{p\mu} \\ a_{\lambda q}, & a_{\lambda\mu} \end{pmatrix}_{r,\,s}^{r,\,s}, \quad D=\begin{pmatrix} a_{pq}, & a_{p\mu} \\ a_{xq}, & a_{x\mu} \end{pmatrix}_{r,\,s}^{r,\,s},$$

we have $$\Delta^{s-1} D=(a_{x\lambda})_s^s, \quad \text{....................................(1)}$$

where $a_{x_i \lambda_j}$ is obtained from Δ by replacing the horizontal suffix λ_j by x_i.

Now let $$[a_{xq}, \ a_{x\mu}]_s^{r,\,s}=[0, \ 1]_s^{r,\,s},$$

and let the second definition of $A_{\lambda_i \mu_j}$ be adopted.

Then $$D=(a_{pq})_r^r, \quad a_{x_i \lambda_j}=A_{\lambda_j \mu_i},$$

and the equation (1) becomes (C).

In a similar way we can deduce (C) from Ex. vii or Theorems IV *a* and V *a* of § 109.

NOTE. The identity (B) is equivalent to the well-known relation existing between any minor determinant of the reciprocal of a square matrix and the anti-correspondent minor determinant of the matrix itself which is given in formulae (A′) and (E) of § 121 and again in formula (D) of § 124.

Ex. i. *Direct proof of the theorem.*

Let $$[h]_m^m=\begin{bmatrix} a_{pq}, & a_{p\mu} \\ a_{\lambda q}, & a_{\lambda\mu} \end{bmatrix}_{r,\,s}^{r,\,s}, \quad [k]_m^m=\begin{bmatrix} a_{pq}, & 0 \\ a_{\lambda q}, & 1 \end{bmatrix}_{r,\,s}^{r,\,s},$$

so that $$\Delta=(h)_m^m=\begin{pmatrix} q_1 \dots q_r \ \mu_1 \dots \mu_s \\ a \\ p_1 \dots p_r \ \lambda_1 \dots \lambda_s \end{pmatrix}, \quad D=(k)_m^m=(a_{pq})_r^r, \quad m=r+s,$$

and let $[H]_m^m=\begin{bmatrix} A_{pq}, & A_{p\mu} \\ A_{\lambda q}, & A_{\lambda\mu} \end{bmatrix}_{r,\,s}^{r,\,s}$ be the reciprocal of $[h]_m^m$, so that $A_{\lambda_i \mu_j}$ is the co-factor of $a_{\lambda_i \mu_j}$ in Δ, and $(H)_m^m=\Delta^{m-1}$.

Then we can obtain formula (C) by equating the determinants of both sides in the identity

$$(h)_m^m [k]_m^m = [h]_m^m \overline{\underline{H}}_m^m \cdot [k]_m^m = [h]_m^m \cdot \overline{\underline{H}}_m^m [k]_m^m = [h]_m^m [\delta]_m^m, \quad\ldots\ldots\ldots\ldots(2)$$

or in the identity

$$\overline{\underline{H}}_m^m [k]_m^m = [\delta]_m^m, \quad\ldots\ldots\ldots\ldots(3)$$

where

$$[\delta]_m^m = \left\lceil\begin{matrix} A_{pq}, & A_{\lambda q} \\ A_{p\mu}, & A_{\lambda\mu} \end{matrix}\right\rfloor_{r,s}^{r,s} \begin{bmatrix} a_{pq}, & 0 \\ a_{\lambda q}, & 1 \end{bmatrix}_{r,s}^{r,s} = \left\lceil\begin{matrix} \Delta, & A_{\lambda q} \\ 0, & A_{\lambda\mu} \end{matrix}\right\rfloor_{r,s}^{r,s}.$$

After cancellation of the factor Δ^{r+1} or Δ^r common to both sides, we obtain

$$\Delta^{s-1} D = (A_{\lambda\mu})_s^s.$$

Here δ_{ij} is the determinant formed from Δ when we replace the ith vertical row of $[h]_m^m$ by the jth vertical row of $[k]_m^m$.

Ex. ii. Particular cases of the identity (A) are

$$(1)\qquad \begin{vmatrix} \begin{pmatrix} u\eta\zeta \\ a \\ pyz \end{pmatrix} & \begin{pmatrix} u\xi\zeta \\ a \\ pyz \end{pmatrix} & \begin{pmatrix} u\xi\eta \\ a \\ pyz \end{pmatrix} \\ \begin{pmatrix} u\eta\zeta \\ a \\ pxz \end{pmatrix} & \begin{pmatrix} u\xi\zeta \\ a \\ pxz \end{pmatrix} & \begin{pmatrix} u\xi\eta \\ a \\ pxz \end{pmatrix} \\ \begin{pmatrix} u\eta\zeta \\ a \\ pxy \end{pmatrix} & \begin{pmatrix} u\xi\zeta \\ a \\ pxy \end{pmatrix} & \begin{pmatrix} u\xi\eta \\ a \\ pxy \end{pmatrix} \end{vmatrix} = a_{pu} \begin{pmatrix} u\xi\eta\zeta \\ a \\ pxyz \end{pmatrix}^2;$$

$$(2)\qquad \begin{vmatrix} \begin{pmatrix} uv\eta\zeta \\ a \\ pqyz \end{pmatrix} & \begin{pmatrix} uv\xi\zeta \\ a \\ pqyz \end{pmatrix} & \begin{pmatrix} uv\xi\eta \\ a \\ pqyz \end{pmatrix} \\ \begin{pmatrix} uv\eta\zeta \\ a \\ pqxz \end{pmatrix} & \begin{pmatrix} uv\xi\zeta \\ a \\ pqxz \end{pmatrix} & \begin{pmatrix} uv\xi\eta \\ a \\ pqxz \end{pmatrix} \\ \begin{pmatrix} uv\eta\zeta \\ a \\ pqxy \end{pmatrix} & \begin{pmatrix} uv\xi\zeta \\ a \\ pqxy \end{pmatrix} & \begin{pmatrix} uv\xi\eta \\ a \\ pqxy \end{pmatrix} \end{vmatrix} = \begin{pmatrix} uv \\ a \\ pq \end{pmatrix} \begin{pmatrix} uv\xi\eta\zeta \\ a \\ pqxyz \end{pmatrix}^2.$$

The corresponding identities when the single-suffix notation is used are

$$(1)\qquad \begin{vmatrix} (ayz)_{p\mu\nu} & (ayz)_{p\lambda\nu} & (ayz)_{p\lambda\mu} \\ (axz)_{p\mu\nu} & (axz)_{p\lambda\nu} & (axz)_{p\lambda\mu} \\ (axy)_{p\mu\nu} & (axy)_{p\lambda\nu} & (axy)_{p\lambda\mu} \end{vmatrix} = a_p (axyz)^2_{p\lambda\mu\nu};$$

$$(2)\qquad \begin{vmatrix} (abyz)_{pq\mu\nu} & (abyz)_{pq\lambda\nu} & (abyz)_{pq\lambda\mu} \\ (abxz)_{pq\mu\nu} & (abxz)_{pq\lambda\nu} & (abxz)_{pq\lambda\mu} \\ (abxy)_{pq\mu\nu} & (abxy)_{pq\lambda\nu} & (abxy)_{pq\lambda\mu} \end{vmatrix} = (ab)_{pq} (abxyz)^2_{pq\lambda\mu\nu}.$$

The last result can be obtained by equating the determinants of both sides in the identity

$$\begin{bmatrix} A_p & A_q & A_\lambda & A_\mu & A_\nu \\ B_p & B_q & B_\lambda & B_\mu & B_\nu \\ X_p & X_q & X_\lambda & X_\mu & X_\nu \\ Y_p & Y_q & Y_\lambda & Y_\mu & Y_\nu \\ Z_p & Z_q & Z_\lambda & Z_\mu & Z_\nu \end{bmatrix} \begin{bmatrix} a_p & b_p & 0 & 0 & 0 \\ a_q & b_q & 0 & 0 & 0 \\ a_\lambda & b_\lambda & 1 & 0 & 0 \\ a_\mu & b_\mu & 0 & 1 & 0 \\ a_\nu & b_\nu & 0 & 0 & 1 \end{bmatrix} = \begin{bmatrix} \Delta & 0 & A_\lambda & A_\mu & A_\nu \\ 0 & \Delta & B_\lambda & B_\mu & B_\nu \\ 0 & 0 & X_\lambda & X_\mu & X_\nu \\ 0 & 0 & Y_\lambda & Y_\mu & Y_\nu \\ 0 & 0 & Z_\lambda & Z_\mu & Z_\nu \end{bmatrix},$$

where $[A\,B\,X\,Y\,Z]_{pq\lambda\mu\nu}$ is the reciprocal of $[a\,b\,x\,y\,z]_{pq\lambda\mu\nu}$, $\Delta=(abxyz)_{pq\lambda\mu\nu}$, and

$$(ABXYZ)_{pq\lambda\mu\nu}=(abxyz)^4_{pq\lambda\mu\nu}.$$

§ 112. Some other identities satisfied by the simple minor determinants of any matrix.

If in the general formula (D) of Theorem VII in § 109.3 we write

$$\Delta=(p_1p_2\dots p_r\ x_1x_2\dots x_s),$$

all the minor determinants of orders $s-1$, $s-2$, ... 2 of the large matrix on the left have respectively Δ^{s-2}, Δ^{s-3}, ... Δ as factors, as can be seen by applying the same formula for smaller values of s. Hence by expanding the large determinant on the left in terms of products of minor determinants of orders u and $s-u$ belonging to two complementary simple minor matrices of reduced orders u and $s-u$, where $u \not> s-1$, and cancelling the factor Δ^{s-2} which is then common to both sides of (D) we obtain a number of expressions for the product $(p_1p_2\dots p_r\ x_1x_2\dots x_s)\,(p_1p_2\dots p_r\ y_1y_2\dots y_s)$. The corresponding expressions obtained from two simple minor matrices of reduced orders u and $s-u$ which have rows in common must vanish identically.

Ex. i. The following are identities which can be found in this way:

(1) $$(xycd)\,(abzw)+(xzcd)\,(aybw)+(xwcd)\,(ayzb)$$
$$+(yzcd)\,(xabw)+(ywcd)\,(xazb)+(zwcd)\,(xyab)$$
$$=(abcd)\,(xyzw).$$

(2) $$(xycd)\,(bczw)+(xzcd)\,(bycw)+(xwcd)\,(byzc)$$
$$+(yzcd)\,(xbcw)+(ywcd)\,(xbzc)+(zwcd)\,(xybc)=0.$$

(3) $$(\lambda\mu xycd)\,(\lambda\mu abzw)+(\lambda\mu xzcd)\,(\lambda\mu aybw)+(\lambda\mu xwcd)\,(\lambda\mu ayzb)$$
$$+(\lambda\mu yzcd)\,(\lambda\mu xabw)+(\lambda\mu ywcd)\,(\lambda\mu xazb)+(\lambda\mu zwcd)\,(\lambda\mu xyab)$$
$$=(\lambda\mu abcd)\,(\lambda\mu xyzw).$$

(4) $$(\lambda\mu xycd)\,(\lambda\mu bczw)+(\lambda\mu xzcd)\,(\lambda\mu bycw)+(\lambda\mu xwcd)\,(\lambda\mu byzc)$$
$$+(\lambda\mu yzcd)\,(\lambda\mu xbcw)+(\lambda\mu ywcd)\,(\lambda\mu xbzc)+(\lambda\mu zwcd)\,(\lambda\mu xybc)=0.$$

In (4) we can interpret $(\lambda\mu bycw)$ to mean the simple minor determinant formed by the λth, μth, bth, yth, cth and wth short rows of a matrix which contains 6 long rows; or we can interpret it to mean the simple minor determinant $(\lambda\mu bycw)_{123456}$ of the matrix $[\lambda\,\mu\,a\,b\,c\,d\,x\,y\,z\,w\dots]_{123456}$.

Ex. ii. A simpler set of similar identities is:

(1) $$(xbcd)\,(ayzw)+(ybcd)\,(xazw)+(zbcd)\,(xyaw)+(wbcd)\,(xyza)$$
$$=(abcd)\,(xyzw).$$

(2) $$(xbcd)\,(byzw)+(ybcd)\,(xbzw)+(zbcd)\,(xybw)+(wbcd)\,(xyzb)=0.$$

(3) $$(\lambda\mu xbcd)\,(\lambda\mu ayzw)+(\lambda\mu ybcd)\,(\lambda\mu xazw)$$
$$+(\lambda\mu zbcd)\,(\lambda\mu xyaw)+(\lambda\mu wbcd)\,(\lambda\mu xyza)$$
$$=(\lambda\mu abcd)\,(\lambda\mu xyzw).$$

(4) $$(\lambda\mu xbcd)\,(\lambda\mu byzw)+(\lambda\mu ybcd)\,(\lambda\mu xbzw)$$
$$+(\lambda\mu zbcd)\,(\lambda\mu xybw)+(\lambda\mu wbcd)\,(\lambda\mu xyzb)=0.$$

The general theorem giving these identities is as follows:

Theorem I. *Let* $[a]^m_m$ *and* $[x]^m_m$ *be two square matrices of the same order* m, *which may or may not have vertical rows in common; let all possible distinct sets of* s *corresponding* *vertical rows of* $[a]^m_m$ *and* $[x]^m_m$ *be formed and arranged in the same manner; and let a selection be made of any two fixed sets, the ith and jth, belonging to* $[a]^m_m$. *Further let* Δ_{ik} *be the determinant formed when the ith set of vertical rows of* $(a)^m_m$ *is replaced by the kth set of vertical rows of* $(x)^m_m$, *and let* Δ'_{kj} *be the determinant formed when the kth set of vertical rows of* $(x)^m_m$ *is replaced by the jth set of vertical rows of* $(a)^m_m$. *Then when all the possible* $\binom{m}{s}$ *values are given to k, we have the identity*

$$\Sigma\, \Delta_{ik}\, \Delta'_{kj} = (a)^m_m\, (x)^m_m \text{ or } 0 \qquad \text{.....(A)}$$

according as $j = i$ *or* $j \neq i$.

To prove this theorem let $[b]^\mu_\mu$ be a complete matrix of the minor determinants of $[a]^m_m$ of order s, so that $\mu = \binom{m}{s}$, and let $[y]^\mu_\mu$ be the similarly formed complete matrix of the minor determinants of $[x]^m_m$ of order s. Further let B_{uv} be the co-factor of b_{uv} in $[a]^m_m$, and let Y_{uv} be the co-factor of y_{uv} in $[x]^m_m$. The determinants which form the ith vertical row of $[b]^\mu_\mu$ or $[y]^\mu_\mu$ will be considered to belong to the ith set of vertical rows of $[a]^m_m$ or $[x]^m_m$.

Writing $(a)^m_m = a$, $(x)^m_m = \xi$, it follows from Theorem II of § 32 or Theorem II of § 119 that

$$\overline{\underline{B}}^\mu_\mu\, [b]^\mu_\mu = a[1]^\mu_\mu, \quad [y]^\mu_\mu\, \overline{\underline{Y}}^\mu_\mu = \xi[1]^\mu_\mu,$$

and therefore

$$\overline{\underline{B}}^\mu_\mu\, [y]^\mu_\mu\, \overline{\underline{Y}}^\mu_\mu\, [b]^\mu_\mu = a\xi\, [1]^\mu_\mu. \qquad \text{.....(1)}$$

Now

$$B_{1i}y_{1k} + B_{2i}y_{2k} + \ldots + B_{\mu i}y_{\mu k} = \Delta_{ik},$$
$$Y_{1k}b_{1j} + Y_{2k}b_{2j} + \ldots + Y_{\mu k}b_{\mu j} = \Delta'_{kj}.$$

Therefore equation (1) can be written

$$[\Delta]^\mu_\mu\, [\Delta']^\mu_\mu = a\xi\, [1]^\mu_\mu, \qquad \text{.....(2)}$$

and when we equate corresponding elements on both sides, we obtain the equations (A).

In the particular case when the first r vertical rows of $[a]^m_m$ and $[x]^m_m$ are the same, we can write

$$[a]^m_m = [\lambda, c]^{r,\rho}_m, \quad [x]^m_m = [\lambda, z]^{r,\rho}_m,$$

where $r + \rho = m$, and deduce from Theorem I the following theorem:

Theorem II. *Let all possible distinct sets of* σ *corresponding vertical rows of* $[c]^\rho_m$ *and* $[z]^\rho_m$ *be formed and arranged in the same manner; and let a selection be made of any two fixed sets, the ith and jth, belonging to* $[c]^\rho_m$. *Further let* δ_{ik} *be the determinant formed when in* $(\lambda, c)^{r,\rho}_m$ *the ith set of vertical rows of* $[c]^\rho_m$ *is replaced by the kth set of vertical rows of* $[z]^\rho_m$, *and let* δ'_{kj} *be the determinant formed when in* $(\lambda, z)^{r,\rho}_m$ *the kth set of vertical rows of* $[z]^\rho_m$ *is replaced by the jth set of vertical rows of* $[c]^\rho_m$. *Then when all the possible*

$\binom{\rho}{\sigma}$ *values are given to k, we have*

$$\Sigma\, \delta_{ik}\, \delta'_{kj} = (\lambda,\, c)^{\rho}_{m}\, (\lambda,\, z)^{\rho}_{m} \text{ or } 0, \quad \ldots\ldots\ldots\ldots\ldots\ldots\ldots\ldots(B)$$

according as $j=i$ or $j \neq i$.

To prove Theorem II let $s=r+\sigma$, and in applying Theorem I in this case let the ith and jth sets of s vertical rows of $[a]^{m}_{m}$, i.e. of $[\lambda,\, c]^{\rho}_{m}$, both contain all the vertical rows of $[\lambda]^{r}_{m}$. Then if the kth set of vertical rows of $[x]^{m}_{m}$, i.e. of $[\lambda,\, z]^{r,\,\rho}_{m}$, does not contain all the vertical rows of $[\lambda]^{r}_{m}$, but only contains $r-u$ of them, Δ'_{kj} contains $r+u$ vertical rows belonging to $[\lambda]^{r}_{m}$, and therefore vanishes.

Consequently the equations (A) are reduced to the equations (B).

Ex. iii. The simplest cases of these identities are

(1) $(xb)\,(ay)+(yb)\,(xa)=(ab)\,(xy)$.

(2) $(xb)\,(by)+(yb)\,(xb)=0$.

(3) $(xbc)\,(ayz)+(ybc)\,(xaz)+(zbc)\,(xya)=(abc)\,(xyz)$.

(4) $(xbc)\,(byz)+(ybc)\,(xbz)+(zbc)\,(xyb)=0$.

We prove (3) and (4) by equating corresponding elements of both sides in the equation

$$\begin{bmatrix} A_1 & A_2 & A_3 \\ B_1 & B_2 & B_3 \\ C_1 & C_2 & C_3 \end{bmatrix} \begin{bmatrix} x_1 & y_1 & z_1 \\ x_2 & y_2 & z_2 \\ x_3 & y_3 & z_3 \end{bmatrix} \begin{bmatrix} X_1 & X_2 & X_3 \\ Y_1 & Y_2 & Y_3 \\ Z_1 & Z_2 & Z_3 \end{bmatrix} \begin{bmatrix} a_1 & b_1 & c_1 \\ a_2 & b_2 & c_2 \\ a_3 & b_3 & c_3 \end{bmatrix} = (abc)\,(xyz)\,[1]^{3}_{3},$$

or

$$\begin{bmatrix} (xbc) & (ybc) & (zbc) \\ (axc) & (ayc) & (azc) \\ (abx) & (aby) & (abz) \end{bmatrix} \begin{bmatrix} (ayz) & (byz) & (cyz) \\ (xaz) & (xbz) & (xcz) \\ (xya) & (xyb) & (xyc) \end{bmatrix} = (abc)\,(xyz) \begin{bmatrix} 1 & 0 & 0 \\ 0 & 1 & 0 \\ 0 & 0 & 1 \end{bmatrix},$$

where $[A\,B\,C]_{123}$, $[X\,Y\,Z]_{123}$ are the reciprocals of $[a\,b\,c]_{123}$, $[x\,y\,z]_{123}$.

113. Equivalence of similar undegenerate matrices.

Two similar undegenerate matrices $A=[a]^{n}_{m}$ and $B=[b]^{n}_{m}$ will be said to be *mutually equivalent* when the long rows of one (and therefore of either) are connected with the long rows of the other.

Thus if $m \ngtr n$, so that both matrices have rank m, the two matrices are mutually equivalent when and only when there exists a square matrix $[h]^{m}_{m}$, necessarily undegenerate, such that

$$[b]^{n}_{m} = [h]^{m}_{m}\,[a]^{n}_{m}, \quad \text{or} \quad [a]^{n}_{m} = \overline{\underline{H}}^{m}_{m}\,[b]^{n}_{m}, \quad \ldots\ldots\ldots\ldots(1)$$

where $\overline{\underline{H}}^{m}_{m}$ is the inverse of $[h]^{m}_{m}$; or when and only when $\begin{bmatrix} a \\ b \end{bmatrix}^{n}_{m,\,m}$ also has rank m.

And if $n \not> m$, so that both matrices have rank n, the two matrices are mutually equivalent when and only when there exists a square matrix $[k]_n^n$, necessarily undegenerate, such that

$$[b]_m^n = [a]_m^n [k]_n^n, \quad \text{or} \quad [a]_m^n = [b]_m^n \overline{\underline{K}}_n^n, \quad \text{..............(2)}$$

where $\overline{\underline{K}}_n^n$ is the inverse of $[k]_n^n$; or when and only when $[a, b]_m^{n, n}$ also has rank n.

Two undegenerate *square* matrices of the same order are always mutually equivalent.

Ex. i. If either one of the matrices A and B is undegenerate and has rank m, and if the conditions (1) are satisfied, where $[h]_m^m$ is undegenerate, then the other matrix also has rank m, and the two matrices are mutually equivalent.

Ex. ii. If either one of the matrices A and B is undegenerate and has rank n, and if the conditions (2) are satisfied, where $[k]_n^n$ is undegenerate, then the other matrix also has rank n, and the two matrices are mutually equivalent.

Ex. iii. *Let $[a]_r^n$ be an undegenerate matrix of rank r. Then if*

$$[w_1 w_2 \ldots w_n] = [\lambda_1 \lambda_2 \ldots \lambda_r][a]_r^n$$

is any non-vanishing one-rowed matrix connected with the horizontal rows of $[a]_r^n$, so that $\lambda_1, \lambda_2, \ldots \lambda_r$ are any quantities which do not all vanish, and if $\lambda_k \neq 0$, the matrix $[b]_r^n$ obtained when we replace the kth horizontal row of $[a]_r^n$ by $[w_1 w_2 \ldots w_n]$ is an undegenerate matrix equivalent to $[a]_r^n$.

We will now prove two theorems regarding the equivalence of two similar undegenerate matrices whose long rows are horizontal. There are clearly corresponding theorems when the long rows are vertical.

Theorem I. *Two similar undegenerate matrices $A = [a]_m^n$ and $B = [b]_m^n$ whose long rows are horizontal are mutually equivalent when and only when*

$$[B_1 B_2 \ldots B_\nu] = \sigma [A_1 A_2 \ldots A_\nu], \quad \text{or} \quad \frac{B_1}{A_1} = \frac{B_2}{A_2} = \ldots = \frac{B_\nu}{A_\nu} = \sigma, \text{......(A)}$$

where $[A]_\nu$ and $[B]_\nu$ are any two similarly formed complete matrices of the simple minor determinants of A and B, and σ is an unspecified scalar quantity which is neither zero nor infinite.

Theorem II. *Two similar undegenerate matrices $A = [a]_m^n$ and $B = [b]_m^n$ whose long rows are horizontal are mutually equivalent when and only when*

$$[\beta q_1 q_2 \ldots q_N] = \sigma [\alpha p_1 p_2 \ldots p_N], \quad \text{or} \quad \frac{\beta}{\alpha} = \frac{q_1}{p_1} = \frac{q_2}{p_2} = \ldots = \frac{q_N}{p_N} = \sigma, \text{...(B)}$$

where $\alpha, p_1, p_2, \ldots p_N$ *are any non-vanishing simple minor determinants of* A *and its primaries;* $\beta, q_1, q_2, \ldots q_N$ *are the similarly formed simple minor determinants of* B*; and* σ *is an unspecified scalar quantity which is neither zero nor infinite.*

In proving these theorems there will be no loss of generality in supposing that in the formation of the simple minor determinants the long rows of A and B are underanged.

First suppose that the matrices A *and* B *are mutually equivalent.*

Then we have $[b]^n_m = [h]^m_m [a]^n_m$, where $[h]^m_m$ is undegenerate.

If $\alpha = (a_{1v})^m_m$ and $\beta = (b_{1v})^m_m$ are any two correspondingly formed simple minor determinants of A and B, it follows that

$$[b_{1v}]^m_m = [h]^m_m [a_{1v}]^m_m, \quad (b_{1v})^m_m = (h)^m_m (a_{1v})^m_m.$$

Thus $\beta = \sigma\alpha$ where $\sigma = (h)^m_m$ and is neither zero nor infinite; and α and β are either both zero or both not zero.

Since corresponding elements of $[A_1 A_2 \ldots A_\nu]$ and $[B_1 B_2 \ldots B_\nu]$ and also corresponding elements of $[\alpha\, p_1 p_2 \ldots p_N]$ and $[\beta q_1 q_2 \ldots q_N]$ are correspondingly formed simple minor determinants of A and B, we see that the equations (A) and (B) are necessarily satisfied, where $\sigma = (h)^m_m$.

Next suppose that the equations (B) *are true where* $\alpha = (a_{1v})^m_m \neq 0$, *and therefore* $\beta = (b_{1v})^m_m \neq 0$.

Let $\overline{\underline{A_{1v}}}{}^m_m$ and $\overline{\underline{B_{1v}}}{}^m_m$ be the conjugate reciprocals of $[a_{1v}]^m_m$ and $[b_{1v}]^m_m$, and let

$$\overline{\underline{A_{1v}}}{}^m_m [a]^n_m = [\alpha]^n_m, \quad \overline{\underline{B_{1v}}}{}^m_m [b]^n_m = [\beta]^n_m.$$

Then two corresponding elements α_{ij}, β_{ij} of $[\alpha]^n_m$ and $[\beta]^n_m$ are either both zero or are corresponding elements of the sequences $[\alpha\, p_1 p_2 \ldots p_N]$, $[\beta q_1 q_2 \ldots q_N]$, and are therefore connected by the relation $\beta_{ij} = \sigma\alpha_{ij}$. It follows that

$$[\beta]^n_m = \sigma[\alpha]^n_m, \quad \text{i.e.} \quad \overline{\underline{B_{1v}}}{}^m_m [b]^n_m = \sigma\, \overline{\underline{A_{1v}}}{}^m_m [a]^n_m.$$

Prefixing $[b_{1v}]^m_m$ on both sides of the last equation, we obtain

$$[b]^n_m = [h]^m_m [a]^m_m, \quad \text{where} \quad \alpha[h]^m_m = [b_{1v}]^m_m \overline{\underline{A_{1v}}}{}^m_m.$$

Since the square matrix $[h]^m_m$ thus defined is undegenerate, it follows that A and B are mutually equivalent.

Finally suppose that the equations (A) *are true.*

Then we may suppose that $A_i = \alpha = (a_{1v})^m_m \neq 0$, $B_i = \beta = (b_{1v})^m_m \neq 0$; and when $p_1, p_2, \ldots p_N$ and $q_1, q_2, \ldots q_N$ are defined as in Theorem II, it follows from (A) that the equations (B) are true. Therefore, as just shown, the matrices A and B are mutually equivalent.

It may be observed that the two sets of equations (A) and (B) are each necessary consequences of the other. For we can regard $[\alpha p_1 p_2 \ldots p_N]$ and $[\beta q_1 q_2 \ldots q_N]$ as corresponding minors of the sequences $[A]_\nu$ and $[B]_\nu$, and any two corresponding elements A_i and B_i of the latter sequences are by § 109 the same homogeneous rational functions of degree 1 of the elements of $[\alpha\, p_1 p_2 \ldots p_N]$ and $[\beta q_1 q_2 \ldots q_N]$ respectively.

NOTE 1. The equivalences of matrices which are not necessarily undegenerate will be defined in § 138.

NOTE 2. *The sign of equivalence.*

We shall use the sign $\equiv$ to denote the equivalence of two matrices. Thus if two similar undegenerate matrices $[a]^n_m$ and $[b]^n_m$ are mutually equivalent, we write

$$[b]^n_m \equiv [a]^n_m.$$

If both matrices have rank m, this notation indicates the existence of an undegenerate square matrix $[h]^m_m$ such that

$$[b]^n_m = [h]^m_m [a]^n_m;$$

if both matrices have rank n, it indicates the existence of an undegenerate square matrix $[k]^n_n$ such that

$$[b]^n_m = [a]^n_m [k]^n_n.$$

NOTE 3. *Undegenerate matrices regarded as spacelets.*

As will be shown in a later chapter an undegenerate matrix $\lceil a \rfloor^r_n$ of rank r, where $r < n$, represents a flat locus of $r-1$ dimensions in homogeneous space of $n-1$ dimensions, i.e. it represents a certain $(r-1)$-way sub-space or *spacelet* ω_r of the complete $(n-1)$-way homogeneous space ω_n. We shall call r the rank of the spacelet ω_r, and in particular n is the rank of the complete space ω_n. The elements of the successive r vertical rows of $\lceil a \rfloor^r_n$ are the (projective) co-ordinates of r unconnected points of ω_n which lie in ω_r and completely determine ω_r. Every other similar undegenerate matrix $\lceil b \rfloor^r_n$ which is equivalent to $\lceil a \rfloor^r_n$ (so that $\lceil b \rfloor^r_n = \lceil a \rfloor^r_n [k]^r_r$, where $[k]^r_r$ is undegenerate) represents the same spacelet ω_r.

Accordingly when n is given and $\lceil a \rfloor^r_n$ is an undegenerate matrix of rank r, we will write

$$\omega_r \equiv \lceil a \rfloor^r_n$$

and call ω_r, which is completely determined by $\boxed{a}_n^{\,r}$ or by any equivalent similar undegenerate matrix, an $(r-1)$-way spacelet or a spacelet of rank r. In particular $\omega_1 \equiv \boxed{a}_n^{\,1}$ is a 0-way spacelet of rank 1 or a point, $\omega_{n-1} \equiv \boxed{a}_n^{\,n-1}$ is an $(n-2)$-way spacelet of rank $n-1$ or a plane, and $\omega_n \equiv \boxed{a}_n^{\,n}$ is the complete $(n-1)$-way space of rank n, the matrix in each case being undegenerate.

When we speak of a spacelet $\omega_r \equiv \boxed{a}_n^{\,r}$, it will be understood that the matrix $\boxed{a}_n^{\,r}$ has rank r, and that it is to be regarded as replaceable by, and not distinct from, any equivalent similar undegenerate matrix.

When we speak of a spacelet $\omega_r = \boxed{a}_n^{\,r}$, using the ordinary sign of equality, we shall mean that the spacelet is to be represented by the particular matrix $\boxed{a}_n^{\,r}$.

When we speak of a spacelet $\omega_r \equiv \boxed{a}_n^{\,p}$, where $p \neq r$, it will be understood (see Note 2 of § 138) that $\boxed{a}_n^{\,p}$ has rank r, and that $\omega_r \equiv \boxed{a}_n^{\,r}$, where $\boxed{a}_n^{\,r}$ is an undegenerate vertical minor of $\boxed{a}_n^{\,p}$.

We shall speak of spacelets of rank 0 with the understanding that every such spacelet is non-existent.

When it is desired to indicate the complete space in which a spacelet lies, we may use the fuller notations

$$\omega_n^r \equiv \boxed{a}_n^{\,r}, \quad \text{or} \quad \omega_n^r \equiv \boxed{a}_n^{\,p}$$

to denote a spacelet of rank r lying in the complete homogeneous space ω_n^n of rank n, it being understood that $\boxed{a}_n^{\,r}$ has rank r in the former case, and that $\boxed{a}_n^{\,p}$ has rank r in the latter case.

For the present spacelets will always be sub-spaces of homogeneous space. Sub-spaces or spacelets of affine space and metrical space will be considered in the appropriate chapters.

NOTE 4. *Sub-spaces of a spacelet.*

The point $\omega_1 \equiv \boxed{x}_n$ will be said to lie in the spacelet $\omega_r \equiv \boxed{a}_n^{\,r}$ when the vertical row of $\boxed{x}_n$ is connected with the vertical rows of $\boxed{a}_n^{\,r}$, so that we can write $\boxed{x}_n = \boxed{a}_n^{\,r} \boxed{\lambda}_r$.

If $s < r$, the spacelet $\omega_s \equiv \boxed{b}_n^{\,s}$ will be said to lie in the spacelet $\omega_r \equiv \boxed{a}_n^{\,r}$, or to be a sub-space of ω_r, when all the points of ω_s lie in ω_r, i.e. when every vertical row of $\boxed{b}_n^{\,s}$ is connected with the vertical rows of $\boxed{a}_n^{\,r}$, so that we can write

$$\boxed{b}_n^{\,s} = \boxed{a}_n^{\,r} \boxed{\lambda}_r^{\,s},$$

where the matrix $\boxed{\lambda}_r^{\,s}$ necessarily has rank s.

Ex. iv. *Spacelets in homogeneous 3-dimensional space.*

In homogeneous 3-way space ω_4 of rank 4 or dimensions 3 the possible kinds of spacelets, apart from the complete space itself, are:

(1) Spacelets of rank 1, or 0-way spacelets, such as $\omega_1 \equiv \overline{\underline{x}}{}_4^1$. These are points.

(2) Spacelets of rank 2, or 1-way spacelets, such as $\omega_2 \equiv \overline{\underline{x}}{}_4^2$. These are straight lines.

(3) Spacelets of rank 3, or 2-way spacelets, such as $\omega_3 \equiv \overline{\underline{x}}{}_4^3$. These are planes.

Ex. v. *Spacelets in homogeneous 4-dimensional space.*

In homogeneous 4-way space ω_5 of rank 5 or dimensions 4 the possible kinds of spacelets, apart from the complete space itself, are:

(1) Spacelets of rank 1, or 0-way spacelets, such as $\omega_1 \equiv \overline{\underline{x}}{}_5^1$. These are points.

(2) Spacelets of rank 2, or 1-way spacelets, such as $\omega_2 \equiv \overline{\underline{x}}{}_5^2$. These are straight lines.

(3) Spacelets of rank 3, or 2-way spacelets, such as $\omega_3 \equiv \overline{\underline{x}}{}_5^3$.

(4) Spacelets of rank 4, or 3-way spacelets, such as $\omega_4 \equiv \overline{\underline{x}}{}_5^4$. These are planes.

§ 114. Criteria for the equivalence of two systems of linear algebraic equations.

Two systems of linear algebraic equations in the same variables, both of which admit of solution, will be said to be mutually equivalent when every solution of either system is also a solution of the other system, i.e. when the two systems have identical solutions. If one of two equivalent systems has finite solutions (or only infinite solutions), then the other also has finite solutions (or only infinite solutions).

The system of equations $[a]_p^{n+1} \underline{\overline{\begin{matrix} x \\ 1 \end{matrix}}}_{n,1} = 0$ has $\overline{\underline{x}}_n = 0$ for a solution when and only when the last vertical row of $[a]_p^{n+1}$ is a row of 0's, i.e. when and only when the system is homogeneous. Consequently if two systems of equations in the same variables $x_1, x_2, \ldots x_n$ are equivalent, they must be either both homogeneous or both non-homogeneous; for $\overline{\underline{x}}_n = 0$ is a solution of one system when and only when it is a solution of the other system.

Let $[a]_p^n \overline{\underline{x}}_n = 0$, $[b]_q^n \overline{\underline{x}}_n = 0$ be two homogeneous systems of equations in which $[a]_p^n$ and $[b]_q^n$ have ranks r and s respectively. By § 89 their general solutions contain $n-r$ and $n-s$ arbitrary parameters respectively; and by

§ 90 they have respectively $n-r$ and $n-s$ unconnected finite non-zero solutions. Hence a necessary condition for their mutual equivalence is $r=s$.

Let $[a]_p^{n+1} \begin{bmatrix} x \\ 1 \end{bmatrix}_{n,1} = 0$, $[b]_q^{n+1} \begin{bmatrix} x \\ 1 \end{bmatrix}_{n,1} = 0$ be two solvable non-homogeneous systems of equations in which $[a]_p^{n+1}$ and $[b]_q^{n+1}$ have ranks r and s respectively, where $r \not> n$ and $s \not> n$. If they both have finite solutions, $[a]_p^n$ and $[b]_q^n$ also have ranks r and s. In this case by § 88 their general solutions contain respectively $n-r$ and $n-s$ arbitrary parameters, and by § 92 they have respectively $n-r+1$ and $n-s+1$ unconnected finite solutions. If they both have only infinite solutions, $[a]_p^n$ and $[b]_q^n$ have ranks $r-1$ and $s-1$ respectively. In this case by § 89 their general solutions contain respectively $n-r+1$ and $n-s+1$ finite arbitrary parameters, and by § 90 they have respectively $n-r+1$ and $n-s+1$ unconnected solutions. In both cases a necessary condition for the equivalence of the two systems is $r=s$.

We have therefore the following theorem:

Theorem I. *If two systems of linear algebraic equations in the same variables are mutually equivalent, they must either both have or both not have finite solutions; they must be either both homogeneous or both non-homogeneous; and the number of unconnected equations must be the same in the two systems.*

We can now proceed to prove the following two theorems:

Theorem II. *Two systems of r unconnected linear equations*

$$[a]_r^{n+1} \begin{bmatrix} x \\ 1 \end{bmatrix}_{n,1} = 0 \quad \text{and} \quad [b]_r^{n+1} \begin{bmatrix} x \\ 1 \end{bmatrix}_{n,1} = 0,$$

which both have finite solutions, are mutually equivalent when and only when the two matrices $[a]_r^{n+1}$ and $[b]_r^{n+1}$ are mutually equivalent. This is true both when the two systems are non-homogeneous and when they are homogeneous.

Theorem II a. *Two homogeneous systems of r unconnected equations $[a]_r^n \overline{x}_n = 0$ and $[b]_r^n \overline{x}_n = 0$ are mutually equivalent when and only when the two matrices $[a]_r^n$ and $[b]_r^n$ are mutually equivalent.*

Using these theorems the conditions for mutual equivalence can be expressed in the ways shown in § 113.

If the two systems of Theorem II are homogeneous, the last vertical rows of $[a]_r^{n+1}$ and $[b]_r^{n+1}$ are rows of 0's, and these two matrices are mutually equivalent when and only when the matrices $[a]_r^n$ and $[b]_r^n$ are mutually equivalent. Consequently the proof of Theorem II will include that of

Theorem II a, where the two systems have in common the finite solution $\overline{x}_n = 0$; and it will be sufficient to prove Theorem II.

First suppose that the two systems of Theorem II are equivalent.

Then every common solution of the r scalar equations of the first system must satisfy every one of the r scalar equations of the second system. Therefore by Theorem I of § 96 every one of the r scalar equations of the second system is connected with the r scalar equations of the first system, and therefore there exists a relation of the form

$$[b]_r^{n+1} = [h]_r^r \, [a]_r^{n+1}. \quad \text{..............................(1)}$$

Since $[b]_r^{n+1}$ has rank r, it follows from Theorem IV of § 71 that $[h]_r^r$ has rank r and is undegenerate. Hence the matrices $[a]_r^{n+1}$ and $[b]_r^{n+1}$ are mutually equivalent.

Next suppose that in Theorem II the matrices $[a]_r^{n+1}$ and $[b]_r^{n+1}$ are mutually equivalent.

Then there exists a relation of the form (1) in which $[h]_r^r$ is undegenerate, and we have

$$[b]_r^{n+1} \overline{\begin{matrix} x \\ 1 \end{matrix}}_{n,1} = [h]_r^r \, [a]_r^{n+1} \overline{\begin{matrix} x \\ 1 \end{matrix}}_{n,1} \quad \text{........................(2)}$$

for all values of $x_1, x_2, \dots x_n$. The product matrix on the left vanishes or has rank 0 when and only when the product matrix on the right has rank 0 i.e. when and only when $[a]_r^{n+1} \overline{\begin{matrix} x \\ 1 \end{matrix}}_{n,1}$ has rank 0 or vanishes. Consequently the two systems of equations have identical solutions and are mutually equivalent.

Thus Theorem II is completely proved.

Ex. i. If the two systems of equations in Theorem II have infinite solutions but no finite solutions, which is the case when and only when $[a]_r^r$ and $[b]_r^n$ have rank $r-1$, these infinite solutions are found by solving the equations

$$[a']_{r-1}^n \, \overline{x}_n = 0, \quad [b']_{r-1}^n \, \overline{x}_n = 0,$$

where $[a']_{r-1}^n$ and $[b']_{r-1}^n$ are any horizontal minors of $[a]_r^n$ and $[b]_r^n$ of rank $r-1$, and the two systems are mutually equivalent when and only when the two matrices $[a']_{r-1}^n$ and $[b']_{r-1}^n$ are mutually equivalent.

Ex. ii. If $[b]_m^{n+1} = [h]_m^m [a]_m^{n+1}$, where $[h]_m^m$ is undegenerate, the two systems of equations

$$[a]_m^{n+1} \overline{\begin{matrix} x \\ 1 \end{matrix}}_{m,1} = 0, \qquad [b]_m^{n+1} \overline{\begin{matrix} x \\ 1 \end{matrix}}_{m,1} = 0$$

are mutually equivalent.

Ex. iii. If $[a]_p^{n+1}\overline{\underline{\begin{matrix}x\\1\end{matrix}}}_{n,1}=0$ and $[b]_q^{n+1}\overline{\underline{\begin{matrix}x\\1\end{matrix}}}_{n,1}=0$ are any two systems of equations, the first of which has finite solutions, then a necessary and sufficient condition that all solutions of the first system shall also be solutions of the second system is the existence of a relation of the form

$$[b]_q^{n+1}=[h]_q^p\,[a]_p^{n+1}.$$

If the equations of the second system are unconnected, the matrix $[h]_p^q$ must have rank q; and if the equations of the first system are unconnected, the matrix $[h]_p^q$ must have the same rank as $[b]_q^{n+1}$.

Ex. iv. If all solutions of one of the systems of Theorem II are also solutions of the other system, then the two systems are mutually equivalent.

NOTE 1. *Alternative proof of Theorem II.*

Let $a=(a_{1v})_r^r$ be any non-vanishing simple minor determinant of $[a]_r^n$, let $\beta=(b_{1v})_r^r$ be the corresponding simple minor determinant of $[b]_r^n$, and let $p_1, p_2, \dots p_\nu$ and $q_1, q_2, \dots q_\nu$ be the simple minor determinants of $[a]_r^{n+1}$ and $[b]_r^{n+1}$ primary to a and β respectively. Also let $[u]_{n-r+1}$ be a minor sequence of $[1\,2\dots(n+1)]$ complementary to $[v]_r$. Then if we introduce the convention that $x_{n+1}=1$, the two systems of equations can be written

$$[a_{1v}]_r^r\,\overline{\underline{x_v}}_r=-[a_{1u}]_r^{n-r+1}\,\overline{\underline{x_u}}_{n-r+1}, \quad\text{.............................(3)}$$

$$[b_{1v}]_r^r\,\overline{\underline{x_v}}_r=-[b_{1u}]_r^{n-r+1}\,\overline{\underline{x_u}}_{n-r+1}. \quad\text{.............................(4)}$$

Let the reciprocals of $[a_{1v}]_r^r$, $[b_{1v}]_r^r$ be $[A_{1v}]_r^r$, $[B_{1v}]_r^r$.

First suppose that the two systems of equations are equivalent. Then all solutions of (3) satisfy (4). Solving (3) for $\overline{\underline{x_v}}_r$ by prefixing $\overline{\underline{A_{1v}}}_r^r$, substituting the value thus obtained in (4), and observing that the resulting equation must be satisfied identically, we see that

$$[b_{1v}]_r^r\,\overline{\underline{A_{1v}}}_r^r\,[a_{1u}]_r^{n-r+1}=a\,[b_{1u}]_r^{n-r+1}. \quad\text{.........................(5)}$$

Equation (5) shows that every vertical row of $[b_{1u}]_r^{n-r+1}$ is connected with the vertical rows of $[b_{1v}]_r^r$. Therefore $[b]_r^{n+1}$ has the same rank as $[b_{1v}]_r^r$, i.e. $[b_{1v}]_r^r$ is undegenerate and $\beta\neq0$.

Prefixing $\overline{\underline{B_{1v}}}_r^r$ on both sides of (5) we obtain

$$\beta\,\overline{\underline{A_{1v}}}_r^r\,[a_{1u}]_r^{n-r+1}=a\,\overline{\underline{B_{1v}}}_r^r\,[b_{1u}]_r^{n-r+1},$$

i.e. $$\beta\,[p_1 p_2 \dots p_\nu]=a\,[q_1 q_2 \dots q_\nu],$$

or $$\frac{\beta}{a}=\frac{q_1}{p_1}=\frac{q_2}{p_2}=\dots=\frac{q_\nu}{p_\nu}=\sigma,$$

where σ is neither zero nor infinite.

Hence by § 113 the matrices $[a]_r^{n+1}$ and $[b]_r^{n+1}$ are mutually equivalent.

Next suppose that these two matrices are mutually equivalent. Then we can show as in the text that the two systems of equations are mutually equivalent.

NOTE 2. *The solutions of the irreducible matrix equation* $[a]_r^n \,\overline{\underline{x}}_n = 0$ *form a spacelet of rank* $n-r$ *in homogeneous space of rank* n.

Let $\overline{\underline{b}}{}_n^{n-r}$ be a matrix of rank $n-r$ whose vertical rows are a complete set of $n-r$ unconnected solutions of the above equation.

Then the non-zero matrix $\overline{\underline{x}}_n$ is a solution of the equation when and only when it is connected with the vertical rows of $\overline{\underline{b}}{}_n^{n-r}$, i.e. when and only when the point $\omega_1 \equiv \overline{\underline{x}}_n$ lies in the spacelet

$$\omega_{n-r} = \overline{\underline{b}}{}_n^{n-r}.$$

If $\overline{\underline{c}}{}_n^{n-r}$ is any other matrix having the same properties as $\overline{\underline{b}}{}_n^{n-r}$, we have

$$\omega_{n-r} \equiv \overline{\underline{b}}{}_n^{n-r} \equiv \overline{\underline{c}}{}_n^{n-r}.$$

NOTE 3. *Mutually orthogonal undegenerate matrices and mutually orthogonal spacelets.*

Two undegenerate matrices $[a]_r^n$ and $[b]_s^n$ of ranks r and s with the same number n of vertical rows (or two undegenerate matrices $\overline{\underline{a}}{}_n^r$ and $\overline{\underline{b}}{}_n^s$ of ranks r and s with the same number n of horizontal rows), will be said to be *mutually orthogonal* when every long row of either matrix is orthogonal with every long row of the other matrix, i.e. when

$$[a]_r^n \,\overline{\underline{b}}{}_n^s = 0, \text{ and therefore } [b]_s^n \,\overline{\underline{a}}{}_n^r = 0. \quad \text{......................(6)}$$

This is the case when and only when every one-rowed matrix connected with the long rows of $[a]_r^n$ (or $\overline{\underline{a}}{}_n^r$) is orthogonal with every one-rowed matrix connected with the long rows of $[b]_s^n$ (or $\overline{\underline{b}}{}_n^s$). When the conditions (6) are satisfied and $[\alpha]_r^n \equiv [a]_r^n$, $[\beta]_r^n \equiv [b]_r^n$, we also have

$$[\alpha]_r^n \,\overline{\underline{\beta}}{}_n^s = 0, \text{ and therefore } [\beta]_s^n \,\overline{\underline{\alpha}}{}_n^r = 0.$$

The two points $\overline{\underline{x}}_n$ and $\overline{\underline{y}}_n$ of homogeneous space ω_n are *mutually orthogonal* when

$$[x]_n \,\overline{\underline{y}}_n = 0, \quad [y]_n \,\overline{\underline{x}}_n = 0.$$

The two spacelets $\omega_r \equiv \overline{\underline{a}}{}_n^r$ and $\omega_s \equiv \overline{\underline{b}}{}_n^s$ of homogeneous space ω_n will be said to be *mutually orthogonal* when every point of ω_r is orthogonal with every point of ω_s, i.e. when the conditions (6) are satisfied. This is the case when and only when every sub-space of ω_r is orthogonal with every sub-space of ω_s.

More generally the two spacelets $\omega_r \equiv \overline{\underline{a}}{}_n^p$, $\omega_s \equiv \overline{\underline{b}}{}_n^q$, of ranks r and s, will be said to be *mutually orthogonal* when every point of ω_r is orthogonal with every point of ω_s, i.e. when the conditions

$$[a]^n_p \overline{\underline{b}}{}^{q}_{n} = 0, \quad [b]^n_q \overline{\underline{a}}{}^{p}_{n} = 0$$

are satisfied.

NOTE 4. *Mutually normal undegenerate matrices and mutually normal spacelets.*

Two undegenerate matrices $[a]^n_r$ and $[b]^n_s$ or $\overline{\underline{a}}{}^{r}_{n}$ and $\overline{\underline{b}}{}^{s}_{n}$ of ranks r and s will be said to be *mutually normal* when

$$[a]^n_r \overline{\underline{b}}{}^{s}_{n} = 0, \text{ and } r+s=n, \quad \text{.................................(7)}$$

i.e. when they are mutually orthogonal and the condition $r+s=n$ is satisfied.

This is the case when and only when the vertical rows of $\overline{\underline{b}}{}^{s}_{n}$ are a complete set of $n-r$ unconnected solutions of the equation $[a]^n_r \overline{\underline{x}}_{n} = 0$, and the vertical rows of $\overline{\underline{a}}{}^{r}_{n}$ are a complete set of $n-s$ unconnected solutions of the equation $[b]^n_s \overline{\underline{x}}_{n} = 0$.

We then call $[b]^n_s$ a matrix normal to $[a]^n_r$, and $[a]^n_r$ a matrix normal to $[b]^n_s$. All undegenerate matrices normal to a given undegenerate matrix are similar to one another and mutually equivalent; and all mutually equivalent similar undegenerate matrices have the same normal matrices.

The two spacelets $\omega_r \equiv \overline{\underline{a}}{}^{r}_{n}$, $\omega_s \equiv \overline{\underline{b}}{}^{s}_{n}$ will be said to be *mutually normal* when the conditions (7) are satisfied. Each of them is completely and uniquely determinate when any representation of the other is known. We call ω_s the spacelet normal to ω_r, and we call ω_r the spacelet normal to ω_s.

More generally the two spacelets $\omega_r \equiv \overline{\underline{a}}{}^{p}_{n}$, $\omega_s \equiv \overline{\underline{b}}{}^{q}_{n}$, of ranks r and s, are mutually normal when the conditions

$$[a]^n_p \overline{\underline{b}}{}^{q}_{n} = 0, \text{ and } r+s=n$$

are satisfied.

Ex. v. *If two spacelets $\omega_r \equiv \overline{\underline{a}}{}^{r}_{n}$ and $\omega_s \equiv \overline{\underline{b}}{}^{s}_{n}$ are mutually orthogonal, then ω_s is a sub-space of the spacelet normal to ω_r, and ω_r is a sub-space of the spacelet normal to ω_s.*

For all points orthogonal with ω_r (or ω_s) lie in the spacelet normal to ω_r (or ω_s).

§ 115. Relations between the elements of a matrix of rank *r*.

Theorem. *If $A = [a]^n_m$ is a matrix of rank r, and if $\Delta = (a_{pq})^r_r$ is any one of its non-vanishing minor determinants of order r, then all elements of A can be expressed as homogeneous rational functions of degree 1 of the elements of the simple minor matrices $H = [a_{1q}]^r_m$, $K = [a_{p1}]^n_r$ which contain Δ by means of the equation*

$$(a_{pq})^r_r [a]^n_m = [a_{1q}]^r_m \overline{\underline{A_{pq}}}{}^{r}_{r} [a_{p1}]^n_r, \quad \text{.....................(A)}$$

where $[A_{pq}]^r_r$ is the reciprocal of $[a_{pq}]^r_r$.

Since all horizontal rows and all vertical rows of A are connected respectively with those horizontal and vertical rows of A which occur in $[a_{pq}]_r^r$, there exist relations of the forms

$$[a]_m^n = [h]_m^r\,[a_{p1}]_r^n, \quad [a]_m^n = [a_{1q}]_m^r\,[k]_r^n, \qquad\qquad (1)$$

where by Theorem IV of § 71 both the matrices $[h]_m^r$ and $[k]_r^n$ necessarily have rank r. From the equations (1) we deduce by the properties of active rows that

$$[a_{1q}]_m^r = [h]_m^r\,[a_{pq}]_r^r, \quad [a_{p1}]_r^n = [a_{pq}]_r^r\,[k]_r^n.$$

Solving these latter equations by respectively postfixing and prefixing $\overline{\underline{A_{pq}}}_r^r$ on both sides, we obtain the unique solutions

$$\Delta\,[h]_m^r = [a_{1q}]_m^r\,\overline{\underline{A_{pq}}}_r^r, \qquad \Delta\,[k]_r^n = \overline{\underline{A_{pq}}}_r^r\,[a_{p1}]_r^n. \qquad\qquad (2)$$

Substituting the values of $[h]_m^r$ and $[k]_r^n$ given by (2) in (1), we obtain the formula (A); and since the elements of $[A_{pq}]_r^r$ are homogeneous rational integral functions of the elements of $[a_{pq}]_r^r$ of degree $r-1$, this equation expresses every element of A as a fraction whose denominator is Δ, and whose numerator is a homogeneous rational integral function of degree $r+1$ of the elements of H and K.

If we use the notations of § 102, we have

$$[a_{1q}]_m^r\,\overline{\underline{A_{pq}}}_r^r = [\alpha_{1p}]_m^r, \qquad \overline{\underline{A_{pq}}}_r^r\,[a_{p1}]_r^n = [\beta_{q1}]_r^n,$$

$$[\alpha_{pp}]_r^r = \Delta\,[1]_r^r, \qquad [\beta_{qq}]_r^r = \Delta\,[1]_r^r,$$

$$\Delta\,[h]_m^r = [\alpha_{1p}]_m^r, \qquad \Delta\,[k]_r^n = [\beta_{q1}]_r^n,$$

where α_{ip_j} is the determinant formed from Δ when we replace the p_jth horizontal row of H (occurring in Δ) by the ith horizontal row of H, and β_{q_ij} is the determinant formed from Δ when we replace the q_ith vertical row of K (occurring in Δ) by the jth vertical row of K.

We can then replace (A) by either of the equivalent equations

$$\Delta\,[a]_m^n = [\alpha_{1p}]_m^r\,[a_{p1}]_r^n, \quad \Delta\,[a]_m^n = [a_{1q}]_m^r\,[\beta_{q1}]_r^n. \qquad\qquad (B)$$

Again since $\quad [a_{pq}]_r^r\,\overline{\underline{A_{pq}}}_r^r = \overline{\underline{A_{pq}}}_r^r\,[a_{pq}]_r^r = \Delta\,[1]_r^r,$

we deduce from (A) that

$$\Delta^2[a]_m^n = [a_{1q}]_m^r\,\overline{\underline{A_{pq}}}_r^r\,[a_{pq}]_r^r\,\overline{\underline{A_{pq}}}_r^r\,[a_{p1}]_r^n = [\alpha_{1p}]_m^r\,[a_{pq}]_r^r\,[\beta_{q1}]_r^n; \quad (C)$$

and when $\Delta \neq 0$, the equation (C) is equivalent to the equation (A).

NOTE 1. *Special case when* $\Delta = (a)_r^r \neq 0$.

If we denote the reciprocal of $[a]_r^r$ by $[A]_r^r$, and write $H = [a]_m^r$, $K = [a]_r^n$, formula (A) becomes

$$(a_{pq})_r^r \, [a]_m^n = [a]_m^r \, \underline{\overline{A}}_r^r \, [a]_r^n \, ; \qquad \text{(A}')$$

and this equation expresses all elements of A as homogeneous rational functions of degree 1 of the elements of the simple minor matrices H and K which contain Δ.

In this case there exist relations of the forms

$$[a]_m^n = [h]_m^r \, [a]_r^n, \quad [a]_m^n = [a]_m^r \, [k]_r^n, \qquad (1')$$

which lead to

$$[a]_m^r = [h]_m^r \, [a]_r^r, \quad [a]_r^n = [a]_r^r \, [k]_r^n.$$

Solving the last equations by postfixing and prefixing $\underline{\overline{A}}_r^r$, we obtain the unique solutions

$$\Delta \, [h]_m^r = [a]_m^r \, \underline{\overline{A}}_r^r, \quad \Delta \, [k]_r^n = \underline{\overline{A}}_r^r \, [a]_r^n, \qquad (2')$$

and substituting these values in (1′) we obtain the equation (A′).

Defining the α's and β's and the p's and q's as in § 102, we have

$$[a]_m^r \, \underline{\overline{A}}_r^r = [a]_m^r = \begin{bmatrix} \Delta \\ p \end{bmatrix}_{r,\, m-r}^r, \quad \underline{\overline{A}}_r^r \, [a]_r^n = [\Delta, \, q]_r^{r,\, n-r},$$

and we can replace (A′) by either of the equivalent equations

$$\Delta \, [a]_m^n = \begin{bmatrix} \Delta \\ p \end{bmatrix}_{r,\, m-r}^r [a]_r^n, \quad \Delta \, [a]_m^n = [a]_m^r \, [\Delta, \, q]_r^{r,\, n-r}. \qquad \text{(B}')$$

Also we deduce from (A′) that

$$\Delta^2 [a]_m^n = [a]_m^r \, \underline{\overline{A}}_r^r \, [a]_r^r \, \underline{\overline{A}}_r^r \, [a]_r^n = \begin{bmatrix} \Delta \\ p \end{bmatrix}_{r,\, m-r}^r [a]_r^r \, [\Delta, \, q]_r^{r,\, n-r}; \qquad \text{(C}')$$

and when $\Delta \neq 0$ the equation (C′) is equivalent to the equation (A′).

NOTE 2. *Range of validity of the equation* (A).

It will be shown in § 116 that this equation is true whenever the rank of the matrix A does not exceed r, no matter whether Δ is or is not a non-vanishing determinant.

In the special case when $r = m$, it is the identity (A) of § 108.

In the special case when $r = n$, it is the identity (B) of § 108.

In other cases it is not an identity in the elements of A.

NOTE 3. *Utility of the formula* (A).

If $[a_{pq}]_r^r$ is any arbitrarily given square minor of rank r of the matrix $A = [a]_m^n$, we can always assign arbitrary values to the remaining elements of the minor matrices $H = [a_{1q}]_m^r$, $K = [a_{p1}]_r^n$ and still determine the remaining elements of A so that A shall have rank r. We do this by choosing the remaining elements so that the equation (A) is satisfied.

Consequently every relation between the elements of A which is always true whenever A has rank r and $(a_{pq})_r^r \neq 0$ can be deduced from formula (A). If this were not so, we could not assign arbitrary values to the remaining elements of H and K.

Ex. i. *If* $A=[a]_m^n$ *is a matrix of rank* r, *it can be expressed as a product of two undegenerate matrices of rank* r *in the form*

$$[a]_m^n=[h]_m^r\,[k]_r^n. \qquad \text{(D)}$$

We can prove this in several ways.

(1) If $(a_{pq})_r^r=\Delta\neq 0$, formula (A) assumes this form when we write

$$\frac{1}{\Delta}[a_{1q}]_m^r\,\overline{\underline{A_{pq}}}_r^r=[h]_m^r,\quad [a_{p1}]_r^n=[k]_r^n.$$

(2) If $[a_{p1}]_r^n$ is any horizontal minor of A of rank r, every horizontal row of A is connected with the horizontal rows of $[a_{p1}]_r^n$, and therefore there exists a relation of the form

$$[a]_m^n=[h]_m^r\,[a_{p1}]_r^n.$$

By Theorem III of § 71 the matrix $[h]_m^r$ must have rank r.

(3) If $[h]_m^r$ is any matrix of rank r whose vertical rows are connected with the vertical rows of A, then by § 82 the equation (D) admits of a unique finite solution for $[k]_r^n$, and by Theorem III of § 71 this solution must have rank r.

Similarly if $[k]_r^n$ is any matrix of rank r whose horizontal rows are connected with the horizontal rows of A, then the equation (D) admits of a unique finite solution for $[h]_m^r$ and this solution must have rank r.

Ex. ii. Let the reciprocal of the matrix $\phi=[a\,b\,c]_{123}$ be $[A\,B\,C]_{123}$.

If the matrix ϕ has rank 2 and if $A_3=b_1c_2-b_2c_1\neq 0$, we have by formula (A)

$$(b_1c_2-b_2c_1)\begin{bmatrix} a_1 & b_1 & c_1 \\ a_2 & b_2 & c_2 \\ a_3 & b_3 & c_3 \end{bmatrix}=\begin{bmatrix} b_1 & c_1 \\ b_2 & c_2 \\ b_3 & c_3 \end{bmatrix}\begin{bmatrix} c_2, & -c_1 \\ -b_2, & b_1 \end{bmatrix}\begin{bmatrix} a_1 & b_1 & c_1 \\ a_2 & b_2 & c_2 \end{bmatrix},$$

and this equation expresses a_3 in terms of the remaining elements of ϕ, for it gives

$$(b_1c_2-b_2c_1)\,a_3=\det[b_3,\ c_3]\begin{bmatrix} c_2, & -c_1 \\ -b_2, & b_1 \end{bmatrix}\begin{bmatrix} a_1 \\ a_2 \end{bmatrix}=a_1(b_3c_2-b_2c_3)+a_2(b_1c_3-b_3c_1)$$

or

$$A_3a_3=-(A_1a_1+A_2a_2).$$

Corresponding to formula (B) the above equation can be written in the forms

$$(b_1c_2-b_2c_1)\begin{bmatrix} a_1 & b_1 & c_1 \\ a_2 & b_2 & c_2 \\ a_3 & b_3 & c_3 \end{bmatrix}=\begin{bmatrix} A_3, & 0 \\ 0, & A_3 \\ -A_1, & -A_2 \end{bmatrix}\begin{bmatrix} a_1 & b_1 & c_1 \\ a_2 & b_2 & c_2 \end{bmatrix}=\begin{bmatrix} b_1 & c_1 \\ b_2 & c_2 \\ b_3 & c_3 \end{bmatrix}\begin{bmatrix} -B_3, & A_3, & 0 \\ -C_3, & 0, & A_3 \end{bmatrix}.$$

Formula (C) is in this case

$$(b_1c_2-b_2c_1)^2\begin{bmatrix} a_1 & b_1 & c_1 \\ a_2 & b_2 & c_2 \\ a_3 & b_3 & c_3 \end{bmatrix}=\begin{bmatrix} b_1 & c_1 \\ b_2 & c_2 \\ b_3 & c_3 \end{bmatrix}\begin{bmatrix} c_2, & -c_1 \\ -b_2, & b_1 \end{bmatrix}\begin{bmatrix} b_1 & c_1 \\ b_2 & c_2 \end{bmatrix}\begin{bmatrix} c_2, & -c_1 \\ -b_2, & b_1 \end{bmatrix}\begin{bmatrix} a_1 & b_1 & c_1 \\ a_2 & b_2 & c_2 \end{bmatrix}$$

$$=\begin{bmatrix} A_3, & 0 \\ 0, & A_3 \\ -A_1, & -A_2 \end{bmatrix}\begin{bmatrix} b_1 & c_1 \\ b_2 & c_2 \end{bmatrix}\begin{bmatrix} -B_3, & A_3, & 0 \\ -C_3, & 0, & A_3 \end{bmatrix},$$

and this also expresses a_3 in terms of the remaining elements of ϕ.

Ex. iii. If the matrix $\begin{bmatrix} a & h & g & u \\ h & b & f & v \\ g & f & c & w \\ u & v & w & d \end{bmatrix}$ has rank 2, we have

$$(bc-f^2)\begin{bmatrix} a & h & g & u \\ h & b & f & v \\ g & f & c & w \\ u & v & w & d \end{bmatrix} = \begin{bmatrix} h & b & f & v \\ g & f & c & w \end{bmatrix} \begin{bmatrix} c, & -f \\ -f, & b \end{bmatrix} \begin{bmatrix} h & g \\ b & f \\ f & c \\ v & w \end{bmatrix},$$

and when $bc-f^2 \neq 0$, this equation expresses a, d and u in terms of the remaining elements.

Ex. iv. *The reciprocal and conjugate reciprocal of a matrix* $[a]$ *which has only one element* a *are both equal to the matrix* $[1]$.

For the conjugate reciprocal of $[a]$ is a matrix $[x]$ such that

$$[a][x]=a[1], \quad \text{or} \quad ax=a.$$

Since this equation is an identity in a, it follows that $x=1$ in all cases including the limiting case when $a=0$.

Ex. v. *Formula* (A) *when* $[a]_m^n$ *has rank* 1.

In the special case when $[a]_m^n$ has rank 1 and $a_{ij} \neq 0$, formula (A) becomes

$$a_{ij}[a]_m^n = \begin{bmatrix} a_{1j} \\ a_{2j} \\ \vdots \\ a_{mj} \end{bmatrix} [a_{i1}\, a_{i2} \dots a_{in}]. \quad \text{.............................(E)}$$

If in this special case we have $a_{ij}=0$, then since corresponding elements of any two vertical rows of $[a]_m^n$ are proportional, it follows that if the vertical row containing a_{ij} is not a row of 0's, the horizontal row containing a_{ij} must be a row of 0's.

Thus in accordance with Ex. iv and Note 2 the equation (E) remains true when $a_{ij}=0$.

Ex. vi. *A matrix of rank* 1 *can be expressed as a product of two undegenerate one-rowed matrices.*

This is a special case of Ex. i, and follows from Ex. v.

Ex. vii. *Generalisation of the equation* (A).

If $[a_{uv}]_s^t$ is any minor of $[a]_m^n$, it follows from the equation (A) that

$$(a_{pq})_r^r\,[a_{uv}]_s^t = [a_{uq}]_s^r\,\overline{\underline{A_{pq}}}_r^r\,[a_{pv}]_r^t. \quad \text{..........................(F)}$$

Putting $(a_{pq})_r^r = \Delta$, and using the notations of § 102, we have

$$[a_{uq}]_s^r\,\overline{\underline{A_{pq}}}_r^r = [a_{up}]_s^r, \quad \overline{\underline{A_{pq}}}_r^r\,[a_{pv}]_r^t = [\beta_{qv}]_r^t,$$

and (F) can be expressed in the forms

$$(a_{pq})_r^r\,[a_{uv}]_s^t = [a_{up}]_s^r\,[a_{pv}]_r^t = [a_{uq}]_s^r\,[\beta_{qv}]_r^t. \quad \text{........................(G)}$$

These equations are true whenever $[a_{pq}]_r^r$ and $[a_{uv}]_s^t$ are minors of a matrix $[a]_m^n$ whose rank does not exceed r.

§ 116. Identical relations between the elements of any matrix.

Let $A = [a]_m^n$ be any matrix whatever; let $\Delta = (a_{pq})_r^r$ be any one of the derived determinants of A of order r, r being any integer which does not exceed either m or n; and let $[A_{pq}]_r^r$ be the conjugate reciprocal of $[a_{pq}]_r^r$, so that

$$[a_{pq}]_r^r \boxed{A_{pq}}_r^r = \boxed{A_{pq}}_r^r [a_{pq}]_r^r = \Delta [1]_r^r.$$

Also let

$$P_{uv} = \begin{vmatrix} a_{p_1q_1} & a_{p_1q_2} & \dots & a_{p_1q_r} & a_{p_1v} \\ a_{p_2q_1} & a_{p_2q_2} & \dots & a_{p_2q_r} & a_{p_2v} \\ \dots & \dots & \dots & \dots & \dots \\ a_{p_rq_1} & a_{p_rq_2} & \dots & a_{p_rq_r} & a_{p_rv} \\ a_{uq_1} & a_{uq_2} & \dots & a_{uq_r} & a_{uv} \end{vmatrix}, \quad P'_{uv} = \begin{vmatrix} a_{p_1q_1} & a_{p_1q_2} & \dots & a_{p_1q_r} & a_{p_1v} \\ a_{p_2q_1} & a_{p_2q_2} & \dots & a_{p_2q_r} & a_{p_2v} \\ \dots & \dots & \dots & \dots & \dots \\ a_{p_rq_1} & a_{p_rq_2} & \dots & a_{p_rq_r} & a_{p_rv} \\ a_{uq_1} & a_{uq_2} & \dots & a_{uq_r} & 0 \end{vmatrix}, \quad \dots(1)$$

where u and v are any elements of the sequences $[1\ 2 \dots m]$ and $[1\ 2 \dots n]$ respectively.

Then P_{uv} is either 0 or a primary superdeterminant of A in Δ; P'_{uv} is the determinant obtained from P_{uv} by putting $a_{uv} = 0$; and we can also write

$$\Delta = \begin{pmatrix} q_1\, q_2 \dots q_r \\ a \\ p_1\, p_2 \dots p_r \end{pmatrix}, \quad P_{uv} = \begin{pmatrix} q_1\, q_2 \dots q_r\, v \\ a \\ p_1\, p_2 \dots p_r\, u \end{pmatrix}.$$

Using these notations we will prove the following theorem:

Theorem I. *If* $[P]_m^n$ *and* $[P']_m^n$ *are the matrices whose elements are defined by* (1), *then the equations*

$$(a_{pq})_r^r [a]_m^n - [a_{1q}]_m^r \boxed{A_{pq}}_r^r [a_{p1}]_r^n = [P]_m^n, \quad \dots\dots\dots\dots\dots(\mathrm{A})$$

$$- [a_{1q}]_m^r \boxed{A_{pq}}_r^r [a_{p1}]_r^n = [P']_m^n, \quad \dots\dots\dots\dots(\mathrm{B})$$

$$\Delta [a]_m^n + [P']_m^n = [P]_m^n \quad \dots\dots\dots\dots\dots(\mathrm{C})$$

are identities in the elements of $[a]_m^n$, *and are true whatever the rank of* $[a]_m^n$ *may be.*

First let
$$- [a_{1q}]_m^r \boxed{A_{pq}}_r^r [a_{p1}]_r^n = [x]_m^n.$$

Then by equating corresponding elements on both sides and making use of Ex. xi of § 62, we obtain

$$x_{uv} = -\det [a_{uq_1} a_{uq_2} \dots a_{uq_r}] \boxed{A_{pq}}_r^r \begin{bmatrix} a_{p_1v} \\ a_{p_2v} \\ \vdots \\ a_{p_rv} \end{bmatrix} = \begin{vmatrix} a_{p_1q_1} & \dots & a_{p_1q_r} & a_{p_1v} \\ \dots & \dots & \dots & \dots \\ a_{p_rq_1} & \dots & a_{p_rq_r} & a_{p_rv} \\ a_{uq_1} & \dots & a_{uq_r} & 0 \end{vmatrix} = P'_{uv}.$$

Thus $[x]_m^n = [P']_m^n$; and this proves formula (B).

Next let $$(a_{pq})_r^r [a]_m^n + [P']_m^n = [y]_m^n.$$

Then by equating corresponding elements on both sides, we obtain

$$y_{uv} = (a_{pq})_r^r a_{uv} + P'_{uv}$$

$$= \begin{vmatrix} a_{p_1q_1} \dots a_{p_1q_r} & 0 \\ \dots\dots\dots\dots & \\ a_{p_rq_1} \dots a_{p_rq_r} & 0 \\ 0 \;\dots\; 0 & a_{uv} \end{vmatrix} + \begin{vmatrix} a_{p_1q_1} \dots a_{p_1q_r} & a_{p_1v} \\ \dots\dots\dots\dots & \\ a_{p_rq_1} \dots a_{p_rq_r} & a_{p_rv} \\ a_{uq_1} \dots a_{uq_r} & 0 \end{vmatrix} = \begin{vmatrix} a_{p_1q_1} \dots a_{p_1q_r} & a_{p_1v} \\ \dots\dots\dots\dots & \\ a_{p_rq_1} \dots a_{p_rq_r} & a_{p_rv} \\ a_{uq_1} \dots a_{uq_r} & a_{uv} \end{vmatrix} = P_{uv}.$$

Thus $[y]_m^n = [P]_m^n$; and this proves formulae (A) and (C).

In the formulae of §§ 108 and 115 the expression on the left in formula (A) above has the value 0. We have here generalised those formulae and have obtained the value of that expression in all cases.

It will be observed that formula (C) above is merely a succinct expression for the identities $P_{uv} = \Delta a_{uv} + P'_{uv}$ given by Ex. x of § 62.

As regards the matrix $[P]_m^n$, we observe that if u is an element of the sequence $[p_1 p_2 \dots p_r]$ or if v is an element of the sequence $[q_1 q_2 \dots q_r]$, then $P_{uv} = 0$; i.e. we have

$$[P_{p1}]_r^n = 0, \quad [P_{1q}]_m^r = 0.$$

If $r = m$ or $r = n$, these are all the elements of $[P]_m^n$; but if $r < m$ and $r < n$, the remaining elements of $[P]_m^n$ are those minor determinants of A of order $r+1$ which contain Δ as a minor, i.e. they are the primary superdeterminants of Δ in A.

Let $[\mu]_{m-r}$ be the corranged complement of $[p]_r$ in $[1\,2 \dots m]$, and let $[\nu]_{n-r}$ be the corranged complement of $[q]_r$ in $[1\;2 \dots n]$. Then the elements of $[P_{\mu\nu}]_{m-r}^{n-r}$ are the primary superdeterminants of Δ in A, and $[P]_m^n$ is formed from its minor $[P_{\mu\nu}]_{m-r}^{n-r}$ by inserting r additional horizontal rows of 0's and r additional vertical rows of 0's in such a manner that these added rows are the p_1th, p_2th, ... p_rth horizontal rows and the q_1th, q_2th, ... q_rth vertical rows of $[P]_m^n$.

NOTE 1. *Special case when* $\Delta = (a)_r^r$.

In this case we have

$$P_{uv} = \begin{vmatrix} a_{11} & a_{12} \dots a_{1r} & a_{1v} \\ a_{21} & a_{22} \dots a_{2r} & a_{2v} \\ \dots\dots & \dots\dots & \\ a_{r1} & a_{r2} \dots a_{rr} & a_{rv} \\ a_{u1} & a_{u2} \dots a_{ur} & a_{uv} \end{vmatrix}, \quad P'_{uv} = \begin{vmatrix} a_{11} & a_{12} \dots a_{1r} & a_{1v} \\ a_{21} & a_{22} \dots a_{2r} & a_{2v} \\ \dots\dots & \dots\dots & \\ a_{r1} & a_{r2} \dots a_{rr} & a_{rv} \\ a_{u1} & a_{u2} \dots a_{ur} & 0 \end{vmatrix}, \quad \dots\dots\dots\dots(2)$$

so that
$$\Delta=\begin{pmatrix}12\dots r\\ a\\ 12\dots r\end{pmatrix},\quad P_{uv}=\begin{pmatrix}12\dots rv\\ a\\ 12\dots ru\end{pmatrix};$$

and if $\overline{\underline{A}}^r_r$ is the reciprocal of $[a]^r_r$, Theorem I assumes the following form:

Theorem II. *If $[P]^n_m$ and $[P']^n_m$ are the matrices whose elements are defined by* (2), *then the equations*

$$(a)^r_r[a]^n_m-[a]^r_m\overline{\underline{A}}^r_r[a]^n_r=[P]^n_m, \quad\text{.............................(A')}$$

$$-[a]^r_m\overline{\underline{A}}^r_r[a]^n_r=[P']^n_m, \quad\text{..........................(B')}$$

$$\Delta[a]^n_m+[P']^n_m=[P]^n_m, \quad\text{.............................(C')}$$

are identities in the elements of $[a]^n_m$, and are true whatever the rank of $[a]^n_m$ may be.

Writing
$$[\mu_1\mu_2\dots\mu_{m-r}]=[r+1,\ r+2,\dots m],\quad [\nu_1\nu_2\dots\nu_{n-r}]=[r+1,\ r+2,\dots n],$$

we have
$$[P]^n_m=\begin{bmatrix}0, & 0\\ 0, & P_{\mu\nu}\end{bmatrix}^{r,\,n-r}_{r,\,m-r};$$

and the elements of $[P_{\mu\nu}]^{n-r}_{m-r}$ are the primary superdeterminants of Δ in $[a]^n_m$.

Writing $Q_{i-1,\,j-1}=\begin{pmatrix}12\dots rj\\ a\\ 12\dots ri\end{pmatrix}$, we can replace (A') by

$$(a)^r_r[a]^n_m-[a]^r_m\overline{\underline{A}}^r_r[a]^n_r=\begin{bmatrix}0, & 0\\ 0, & Q\end{bmatrix}^{r,\,n-r}_{r,\,m-r}. \quad\text{........................(a)}$$

Note 2. *Generalisation of Theorem I.*

If $\Delta=(a_{pq})^r_r$ is any derived determinant and $[a_{uv}]^\sigma_\rho$ any derived matrix of $A=[a]^n_m$, and if $[P]^n_m$ and $[P']^n_m$ are the matrices whose elements are defined by (1), then the equations

$$(a_{pq})^r_r[a_{uv}]^\sigma_\rho-[a_{uq}]^r_\rho\overline{\underline{A_{pq}}}^r_r[a_{pv}]^\sigma_r=[P_{uv}]^\sigma_\rho, \quad\text{.....................(A'')}$$

$$-[a_{uq}]^r_\rho\overline{\underline{A_{pq}}}^r_r[a_{pv}]^\sigma_r=[P'_{uv}]^\sigma_\rho, \quad\text{....................(B'')}$$

$$\Delta[a_{uv}]^\sigma_\rho+[P'_{uv}]^\sigma_\rho=[P_{uv}]^\sigma_\rho, \quad\text{....................(C'')}$$

are identities in the elements of A.

We deduce these formulae from (A), (B), (C) by the properties of active rows.

Note 3. *Deduction of the formulae of* §§ 108 *and* 115 *from formulae* (A) *and* (A').

(1) When $m \ngtr n$ and $r=m$, the elements of $[P]^n_m$ all vanish identically, and formulae (A) and (A') become

$$(a_{1q})^m_m[a]^n_m=[a_{1q}]^m_m\overline{\underline{A_{1q}}}^m_m[a]^n_m, \quad\text{..............................(A}_1\text{)}$$

$$(a)^m_m[a]^n_m=[a]^m_m\overline{\underline{A}}^m_m[a]^n_m. \quad\text{.................................(A}_1'\text{)}$$

These are formulae (A) and (B) of § 108. They are identities in the elements of $[a]^n_m$.

(2) When $n \not> m$ and $r=n$, the elements of $[P]^n_m$ all vanish identically, and formulae (A) and (A′) become

$$(a_{p1})^n_n \,[a]^n_m = [a]^n_m \,\boxed{A_{p1}}^n_n \,[a_{p1}]^n_n , \quad\text{..............................}(A_2)$$

$$(a)^n_n \,[a]^n_m = [a]^n_m \,\boxed{A}^n_n \,[a]^n_n . \quad\text{..............................}(A_2')$$

These are formulae (A′) and (B′) of § 108. They are identities in the elements of $[a]^n_m$.

(3) When the rank of $[a]^n_m$ does not exceed r, $[P]^n_m$ has rank 0, and we deduce from (A) and (A′) the equations

$$(a_{pq})^r_r \,[a]^n_m = [a_{1q}]^r_m \,\boxed{A_{pq}}^r_r \,[a_{p1}]^n_r , \quad\text{..............................}(A_3)$$

$$(a)^r_r \,[a]^n_m = [a]^r_m \,\boxed{A}^r_r \,[a]^n_r . \quad\text{..............................}(A_3')$$

These are formulae (A) and (A′) of § 115. They are not identities in the elements of $[a]^n_m$; but the first formula is true both when $(a_{pq})^r_r \neq 0$ and when $(a_{pq})^r_r = 0$, and the second formula is true both when $(a)^r_r \neq 0$ and when $(a)^r_r = 0$, provided only that the rank of $[a]^n_m$ does not exceed r.

NOTE 4. *Necessary and sufficient conditions that a matrix* $A = [a]^n_m$ *shall have rank* r.

Let A have a non-vanishing minor determinant $\Delta = (a_{pq})^r_r$ of order r.

First suppose that $[P]^n_m = 0$ in formula (A) so that

$$(a_{pq})^r_r \,[a]^n_m = [a_{1q}]^r_m \,\boxed{A_{pq}}^r_r \,[a_{p1}]^n_r .$$

Since the rank of the product matrix on the right cannot exceed r, the rank of A cannot exceed r; and since A has a non-vanishing minor determinant of order r, it follows that A has rank r.

Next suppose that A has rank r. Then clearly all the elements of $[P]^n_m$ vanish, these being minor determinants of A of order $r+1$; i.e. we have $[P]^n_m = 0$.

Thus $[P]^n_m = 0$ is a necessary and sufficient condition that the rank of A shall be equal to r.

We have thus proved the following result:

Theorem. *If a matrix A has a non-vanishing minor determinant Δ of order r, then the rank of A is equal to r when and only when all the primary superdeterminants of Δ in A vanish.*

These are the necessary and sufficient conditions obtained in § 71.

NOTE 5. *Factorisation of* $[a]^n_m$.

If $\qquad [a_{1p}]^r_m = [a_{1q}]^r_m \,\boxed{A_{pq}}^r_r , \quad [a_{p\mu}]^{m-r}_r = 0, \quad [a_{\mu\mu}]^{m-r}_{m-r} = [1]^{m-r}_{m-r}$

and $\qquad [\beta_{q1}]^n_r = \boxed{A_{pq}}^r_r \,[a_{p1}]^n_r , \quad [\beta_{\nu q}]^r_{n-r} = 0, \quad [\beta_{\nu\nu}]^{n-r}_{n-r} = [1]^{n-r}_{n-r} ,$

the equation (A) is equivalent to the identity

$$\Delta^2 [a]_m^n = [a_{1p}]_m^r [a_{pq}]_r^r [\beta_{q1}]_r^n + \Delta [a_{1\mu}]_m^{m-r} [P_{\mu\nu}]_{m-r}^{n-r} [\beta_{\nu 1}]_{n-r}^n$$
$$= [a]_m^m [a']_m^n [\beta]_n^n, \quad \text{...}(A_4)$$

where $[a']_m^n$ is the compartite matrix in which

$$[a'_{pq}]_r^r = [a_{pq}]_r^r, \quad [a'_{\mu\nu}]_{m-r}^{n-r} = \Delta [P_{\mu\nu}]_{m-r}^{n-r},$$

and all other elements are 0's.

In particular the equation (A) is equivalent to the identity

$$\Delta^2 [a]_m^n = \begin{bmatrix} \Delta \\ p \end{bmatrix}_{r,\, m-r}^{r} [a]_r^r [\Delta, q]_r^{r,\, n-r} + \Delta \overline{\underline{\begin{matrix} 0 \\ 1 \end{matrix}}}_{r,\, m-r}^{m-r} [Q]_{m-r}^{n-r} [0, 1]_{n-r}^{r,\, n-r}$$
$$= \begin{bmatrix} \Delta, & 0 \\ p, & 1 \end{bmatrix}_{r,\, m-r}^{r,\, m-r} \begin{bmatrix} a, & 0 \\ 0, & \Delta Q \end{bmatrix}_{r,\, m-r}^{r,\, n-r} \begin{bmatrix} \Delta, & q \\ 0, & 1 \end{bmatrix}_{r,\, n-r}^{r,\, n-r}, \quad \text{........................}(A_4')$$

where
$$\begin{bmatrix} \Delta \\ p \end{bmatrix}_{r,\, m-r}^{r} = [a]_m^r \overline{\underline{A}}_r^r, \quad [\Delta, q]_r^{r,\, n-r} = \overline{\underline{A}}_r^r [a]_r^n,$$

and $[Q]_{m-r}^{n-r}$ is defined as in Note 1.

These identities are employed in § 141 and the following articles of Chapter XVI.

Ex. i. *If* $A = [a]_m^n$ *is a matrix of rank* r *and if* $i \not> r$, *then every non-vanishing minor determinant of* A *of order* $i-1$ *has a non-vanishing primary superdeterminant in* A *of order* i, *and every non-vanishing derived determinant of* A *of order* i *has a non-vanishing minor determinant of order* $i-1$.

This follows from Note 4.

Consequently we can convert $[a]_m^n$ by derangements of its horizontal and vertical rows into a matrix $[b]_m^n$ whose first r leading derived determinants

$$\Delta_1 = b_{11}, \quad \Delta_2 = (b)_2^2, \quad \Delta_3 = (b)_3^3, \ \dots \ \Delta_r = (b)_r^r$$

are all different from zero. Further b_{11} can be chosen to be any assigned element of A; or as an alternative $(b)_r^r$ can be chosen to be any assigned non-vanishing derived determinant of A of order r.

We give in Exs. ii—vi some illustrations of formula (A).

Ex. ii. If $A = [a\,b\,c\,d]_{12345}$ and $\Delta = (db)_{41}$, we have

$$(db)_{41} \begin{bmatrix} a_1 & b_1 & c_1 & d_1 \\ a_2 & b_2 & c_2 & d_2 \\ a_3 & b_3 & c_3 & d_3 \\ a_4 & b_4 & c_4 & d_4 \\ a_5 & b_5 & c_5 & d_5 \end{bmatrix} - \begin{bmatrix} d_1 & b_1 \\ d_2 & b_2 \\ d_3 & b_3 \\ d_4 & b_4 \\ d_5 & b_5 \end{bmatrix} \begin{bmatrix} b_1, & -b_4 \\ -d_1, & d_4 \end{bmatrix} \begin{bmatrix} a_4 & b_4 & c_4 & d_4 \\ a_1 & b_1 & c_1 & d_1 \end{bmatrix}$$
$$= \begin{bmatrix} 0 & 0 & 0 & 0 \\ (dba)_{412} & 0 & (dbc)_{412} & 0 \\ (dba)_{413} & 0 & (dbc)_{413} & 0 \\ 0 & 0 & 0 & 0 \\ (dba)_{415} & 0 & (dbc)_{415} & 0 \end{bmatrix}.$$

Ex. iii. If $A=[a\,b\,c\,d\,e]_{1234}$ and $\Delta=(db)_{13}$, we have

$$(db)_{13}[a\,b\,c\,d\,e]_{1234}-[d\,b]_{1234}\begin{bmatrix} b_3\,, & -b_1 \\ -d_3, & d_1 \end{bmatrix}[a\,b\,c\,d\,e]_{13}$$

$$=\begin{bmatrix} 0 & 0 & 0 & 0 & 0 \\ (dba)_{132} & 0 & (dbc)_{132} & 0 & (dbe)_{132} \\ 0 & 0 & 0 & 0 & 0 \\ (dba)_{134} & 0 & (dbc)_{134} & 0 & (dbe)_{134} \end{bmatrix}.$$

Ex. iv. If $\begin{bmatrix} A & H & G & U \\ H & B & F & V \\ G & F & C & W \\ U & V & W & D \end{bmatrix}$ is the reciprocal of $\begin{bmatrix} a & h & g & u \\ h & b & f & v \\ g & f & c & w \\ u & v & w & d \end{bmatrix}$, we have

$$(ca-g^2)\begin{bmatrix} a & h & g & u \\ h & b & f & v \\ g & f & c & w \\ u & v & w & d \end{bmatrix}-\begin{bmatrix} g & a \\ f & h \\ c & g \\ w & u \end{bmatrix}\begin{bmatrix} a\,, & -g \\ -g, & c \end{bmatrix}\begin{bmatrix} g & f & c & w \\ a & h & g & u \end{bmatrix}=\begin{bmatrix} 0 & 0 & 0 & 0 \\ 0 & D & 0 & -V \\ 0 & 0 & 0 & 0 \\ 0 & -V & 0 & B \end{bmatrix}$$

Ex. v. If $A=[a]_m^n$ and $\Delta=a_{ij}$, we have (see Ex. iv of § 115)

$$a_{ij}[a]_m^n=\begin{bmatrix} a_{1j} \\ a_{2j} \\ \vdots \\ a_{mj} \end{bmatrix}[a_{i1}\,a_{i2}\,\dots\,a_{in}]+[P]_m^n,$$

where
$$P_{uv}=\begin{vmatrix} a_{ij} & a_{iv} \\ a_{uj} & a_{uv} \end{vmatrix}=a_{ij}\,a_{uv}-a_{uj}\,a_{iv},$$

and where the ith horizontal row and the jth vertical row of $[P]_m^n$ are rows of 0's.

In the special case when $\Delta=a_{11}$, we have

$$a_{11}[a]_m^n=[a]_m^1[a]_1^n+\begin{bmatrix} 0, & 0 \\ 0, & b \end{bmatrix}_{1,\,m-1}^{1,\,n-1},$$

where
$$b_{uv}=\begin{pmatrix} 1,\,1+v \\ a \\ 1,\,1+u \end{pmatrix}.$$

Ex. vi. If $[a]_m^n$ is a matrix whose rank is less than 2, and if a_{ij} is any one of its elements, we have

$$a_{ij}[a]_m^n=\begin{bmatrix} a_{1j} \\ a_{2j} \\ \vdots \\ a_{mj} \end{bmatrix}[a_{i1}\,a_{i2}\,\dots\,a_{in}].$$

The next examples deal with properties of the matrices $[P]_m^n$ and $[P_{\mu\nu}]_{m-r}^{n-r}$ in formulae (A) and (A′).

Ex. vii. Let $(P_{uv})_s^s$ be any minor determinant of $[P]_m^n$ or $[P_{\mu\nu}]_{m-r}^{n-r}$ in formula (A), in which $\Delta=(a_{pq})_r^r=\begin{pmatrix} q_1\, q_2 \dots q_r \\ a \\ p_1\, p_2 \dots p_r \end{pmatrix}$, and let

$$\Delta_{r+s}=\begin{pmatrix} a_{pq}, & a_{pv} \\ a_{uq}, & a_{uv} \end{pmatrix}_{r,\,s}^{r,\,s}=\begin{pmatrix} q_1\, q_2 \dots q_r\, v_1\, v_2 \dots v_s \\ a \\ p_1\, p_2 \dots p_r\, u_1\, u_2 \dots u_s \end{pmatrix}.$$

Then by Sylvester's identity of § 110 we have

$$(P_{uv})_s^s=\Delta^{s-1}\,\Delta_{r+s}. \qquad\qquad \text{(D)}$$

If $(P_{uv})_s^s$ is not a minor determinant of $[P_{\mu\nu}]_{m-r}^{n-r}$, it vanishes identically.

From (D) we see that if $\Delta \neq 0$, then $(P_{uv})_s^s$ vanishes when and only when Δ_{r+s} vanishes.

Ex. viii. *Rank of* $[P]_m^n$ *and* $[P_{\mu\nu}]_{m-r}^{n-r}$ *when* $\Delta \neq 0$.

Theorem. *If ρ is the rank of the matrix $[a]_m^n$ in formula* (A) *and if $\Delta=(a_{pq})_r^r \neq 0$, then the matrices $[P]_m^n$ and $[P_{\mu\nu}]_{m-r}^{n-r}$ have rank $\rho-r$.*

If $\rho=r$, then by Note 3 the matrix $[P]_m^n$ has rank 0 and the theorem is true. If $\rho>r$, then by Note 3 the rank of $[P]_m^n$ is greater than 0.

In general we see from formula (D) in Ex. vii that if every minor determinant $(P_{uv})_s^s$ of $[P]_m^n$ of order s vanishes, then all such determinants as Δ_{r+s} vanish, and therefore by Theorem II *b* of § 71 the rank of $[a]_m^n$ cannot exceed $r+s-1$, and we have

$$\rho \not> r+s-1, \quad \text{or} \quad s \not< \rho-r+1.$$

Hence if s is the common rank of $[P]_m^n$ and $[P_{\mu\nu}]_{m-r}^{n-r}$, we have

$$s+1 \not< \rho-r+1, \quad \text{or} \quad s \not< \rho-r;$$

and we further see from formula (D) that there must exist some determinant Δ_{r+s} which does not vanish, i.e. we must have

$$\rho \not< r+s, \quad \text{or} \quad s \not> \rho-r.$$

Thus the common rank s must have the value $\rho-r$.

Ex. ix. *If $[a]_m^n$ has rank $r+1$, and if $\Delta=(a_{pq})_r^r$ is one of its non-vanishing minor determinants of order r, then in formula* (A) *the matrices $[P]_m^n$ and $[P_{\mu\nu}]_{m-r}^{n-r}$ must have rank* 1.

This is a particular case of Ex. viii. It can however be proved independently in another way.

Let σ be the rank of $[P_{\mu\nu}]_{m-r}^{n-r}$. By § 73 every complete matrix of the minor determinants of $[a]_m^n$ of order $r+1$ has rank 1. Since $[P_{\mu\nu}]_{m-r}^{n-r}$ is a minor of such a matrix, it follows that $\sigma \not> 1$.

Now if σ were 0, it would follow from formula (A) that the rank of $[a]_m^n$ is not greater than r. Hence σ cannot be 0 and must be 1.

Ex. x. *Rank of* $[P]_m^n$ *and* $[P_{\mu\nu}]_{m-r}^{n-r}$ *when* $\Delta=0$.

Theorem. *If* ρ *is the rank of* $[a]_m^n$ *in formula* (A), *and if* $\Delta=(a_{pq})_r^r=0$, *then the rank of* $[P]_m^n$ *cannot exceed* 1.

For the identity (D) shows that all minor determinants of $[P]_m^n$ of order greater than 1 vanish.

Ex. xi. *If* $[a_{\mu\nu}]_{m-r}^{n-r}$ *is the reciprocal of the matrix* $[a_{\mu\nu}]_{m-r}^{n-r}$ *formed from* $[a]_m^n$ *by striking out the horizontal and vertical rows which occur in* $[a_{pq}]_r^r$, *and if* $[\Pi_{\mu\nu}]_{m-r}^{n-r}$ *is the matrix whose elements are the co-factors in* $[a]_m^n$ *of the corresponding elements of* $[P_{\mu\nu}]_{m-r}^{n-r}$, *then*

$$[\Pi_{\mu\nu}]_{m-r}^{n-r}=(-1)^{\omega}[a_{\mu\nu}]_{m-r}^{n-r}, \quad \text{.................................(E)}$$

where ω *is the affect of* $(a_{pq})_r^r$ *in* $[a]_m^n$.

For if $(a_{xy})_{m-r}^{n-r}$ is the corranged minor determinoid of $[a]_m^n$ formed by striking out the p_1th, p_2th, ... p_rth and μ_ith horizontal rows and the q_1th, q_2th, ... q_rth and ν_jth vertical rows, we have

$$a_{\mu_i\nu_j}=(-1)^{i+j-2}(a_{xy})_{m-r}^{n-r}, \quad \Pi_{\mu_i\nu_j}=(-1)^{\eta}(a_{xy})_{m-r}^{n-r},$$

where
$$\begin{aligned}\eta &= \text{affect of the determinant } P_{\mu_i\nu_j} \text{ in } [a]_m^n\\ &= \text{aff. } [p_1p_2\ldots p_r\,\mu_i] \text{ in } [1\,2\ldots m]+\text{aff. } [q_1q_2\ldots q_r\,\nu_j] \text{ in } [1\,2\ldots n]\\ &= \text{aff. } [p]_r \text{ in } [1\,2\ldots m]+\text{aff. } \mu_i \text{ in } [\mu]_{m-r}\\ &\quad +\text{aff. } [q]_r \text{ in } [1\,2\ldots n]+\text{aff. } \nu_j \text{ in } [\nu]_{n-r}\\ &= \omega+(i-1)+(j-1).\end{aligned}$$

Ex. xii. *If* $[a_{\mu\nu}]_{m-r}^{n-r}$ *is the reciprocal of* $[a_{\mu\nu}]_{m-r}^{n-r}$, *and if* $[a'_{\mu\nu}]_{m-r}^{n-r}$ *is the matrix formed from* $[a_{\mu\nu}]_{m-r}^{n-r}$ *when each element of the latter matrix is replaced by that minor determinant of* $[a]_m^n$ *of which it is the co-factor, then*

$$[P_{\mu\nu}]_{m-r}^{n-r}=(-1)^{\omega}[a'_{\mu\nu}]_{m-r}^{n-r}, \quad \text{.................................(F)}$$

where ω *has the same value as in Ex.* xi.

We deduce (F) from (E) by replacing each element of each matrix by that minor determinant of $[a]_m^n$ of which it is the co-factor.

Ex. xiii. As an illustration of formulae (E) and (F) we will consider the case in which $[a]_m^n=[a\,b\,c\,d\,e]_{1234}$ and $(a_{pq})_r^r=(db)_{13}$. In this case we have (see Ex. iii)

$$m=4,\ n=5,\ r=2,\ \omega=5,\ [\mu]_2=[2\,4],\ [\nu]_3=[2\,3\,5],$$

$$[P_{\mu\nu}]_2^3=\begin{bmatrix}(dba)_{132} & (dbc)_{132} & (dbe)_{132}\\ (dba)_{134} & (dbc)_{134} & (dbe)_{134}\end{bmatrix}, \quad [a_{\mu\nu}]_2^3=\begin{bmatrix}a_2 & c_2 & e_2\\ a_4 & c_4 & e_4\end{bmatrix},$$

$$[\Pi_{\mu\nu}]_2^3=\begin{bmatrix}-|c_4e_4|, & |a_4e_4|, & -|a_4c_4|\\ |c_2e_2|, & -|a_2e_2|, & |a_2c_2|\end{bmatrix}, \quad [a_{\mu\nu}]_2^3=\begin{bmatrix}|c_4e_4|, & -|a_4e_4|, & |a_4c_4|\\ -|c_2e_2|, & |a_2e_2|, & -|a_2c_2|\end{bmatrix}.$$

Thus
$$[\Pi_{\mu\nu}]_2^3=(-1)^{\omega}[a_{\mu\nu}]_2^3.$$

The following additional examples depend on the properties of $[P']_m^n$ in Theorem I.

Ex. xiv. From formula (B) we see that the rank of $[P']_m^n$ cannot exceed r whatever values the elements of $[a]_m^n$ may have. Hence if $\Delta=(a_{pq})_r^r$ is any minor determinant of $[a]_m^n$, all minor determinants of order greater than r of the matrix

$$[P']_m^n=[P]_m^n-\Delta[a]_m^n$$

vanish identically.

Ex. xv. Putting $[a_{uv}]_s^s=0$ in formula (D) of Ex. vii, and writing $\Delta=(a_{pq})_r^r$, we obtain for any minor determinant $(P'_{uv})_s^s$ of $[P']_m^n$ the equation

$$(P'_{uv})_s^s=\Delta^{s-1}\begin{pmatrix}a_{pq}, & a_{pv}\\ a_{uq}, & 0\end{pmatrix}_{r,s}^{r,s}, \qquad \text{(D')}$$

which is an identity in the elements of $[a]_m^n$.

Ex. xvi. If $[a_{pq}]_r^r$ and $[a_{uv}]_s^s$ are any square minors of $[a]_m^n$ and if

$$\Delta=(a_{pq})_r^r, \qquad P_{u_iv_j}=\begin{pmatrix}q_1\, q_2 \ldots q_r\, v_j\\ a\\ p_1\, p_2 \ldots p_r\, u_i\end{pmatrix},$$

the equation

$$\det\left\{[P_{uv}]_s^s-\Delta[a_{uv}]_s^s\right\}=\Delta^{s-1}\begin{pmatrix}a_{pq}, & a_{pv}\\ a_{uq}, & 0\end{pmatrix}_{r,s}^{r,s}$$

is an identity in the elements of $[a]_m^n$.

This follows from formula (B) and Ex. xv.

Ex. xvii. *Kronecker's Identity.* If $s>r$, it follows from Ex. xvi or Ex. xiv that the equation

$$\det\left\{\Delta[a_{uv}]_s^s-[P_{uv}]_s^s\right\}=0 \qquad \text{(G)}$$

is an identity in the elements of $[a]_m^n$.

Ex. xviii. When $s=r$ the identity of Ex. xvi becomes

$$\det\left\{\Delta[a_{uv}]_r^r-[P_{uv}]_r^r\right\}=\Delta^{s-1}(a_{uq})_r^r(a_{pv})_r^r.$$

§ 117. Equations which serve to express all the minor determinants of order r of a matrix of rank r in terms of any one non-vanishing minor determinant of order r and its primaries.

Throughout the present article $\Delta=(a_{pq})_r^r$ will denote a minor determinant of order r of a matrix $A=[a]_m^n$ whose rank is r, and the α's and β's will have the same meanings as in § 102.

Accordingly, $[A_{pq}]_r^r$ being the reciprocal of $[a_{pq}]_r^r$, we shall have

$$[\alpha_{1p}]_m^r = [a_{1q}]_m^r \overline{\underline{|A_{pq}|}}_r^r, \quad [\beta_{q1}]_r^n = \overline{\underline{|A_{pq}|}}_r^r [a_{p1}]_r^n, \quad \text{...........(1)}$$

$$[\alpha_{pp}]_r^r = \Delta [1]_r^r, \qquad [\beta_{qq}]_r^r = \Delta [1]_r^r. \quad \text{.................(2)}$$

We will prove theorems corresponding to those of § 109.

Theorem I. *If* $\Delta = (a_{pq})_r^r$ *and* $D = (a_{uv})_r^r$ *are any two minor determinants of order* r *of the matrix* $A = [a]_m^n$, *then the equations*

$$(a_{pq})_r^r (a_{uv})_r^r = (a_{uq})_r^r (a_{pv})_r^r, \quad \text{.....................(A)}$$

$$\Delta^{2r-1} D = (\alpha_{up})_r^r (\beta_{qv})_r^r \quad \text{.......................(B)}$$

are always true when the rank of A *does not exceed* r; *and when* $\Delta \neq 0$, *the second equation expresses* D *as a homogeneous rational function of degree* 1 *of* Δ *and its primaries.*

Theorem II. *If* $[\mathbf{A}]_\mu^\nu$ *is any complete matrix of the minor determinants of order* r *of the matrix* $A = [a]_m^n$, *so that* $\mu = \binom{m}{r}$ *and* $\nu = \binom{n}{r}$; *if* $\overline{\underline{|H}}_\mu$ *and* $[K]_\nu$ *are the correspondingly formed complete matrices of the simple minor determinants of* $[a_{1q}]_m^r$ *and* $[a_{p1}]_r^n$; *and if* $\overline{\underline{|P}}_\mu$ *and* $[Q]_\nu$ *are the correspondingly formed complete matrices of the simple minor determinants of* $[\alpha_{1p}]_m^r$ *and* $[\beta_{q1}]_r^n$; *then the equations*

$$(a_{pq})_r^r [\mathbf{A}]_\mu^\nu = \begin{bmatrix} H_1 \\ H_2 \\ \vdots \\ H_\mu \end{bmatrix} [K_1 K_2 \dots K_\nu] = \begin{bmatrix} H_1K_1, & H_1K_2, & \dots & H_1K_\nu \\ H_2K_1, & H_2K_2, & \dots & H_2K_\nu \\ \dots & \dots & \dots & \dots \\ H_\mu K_1, & H_\mu K_2, & \dots & H_\mu K_\nu \end{bmatrix}, \quad \text{...(a)}$$

$$\Delta^{2r-1} [\mathbf{A}]_\mu^\nu = \begin{bmatrix} P_1 \\ P_2 \\ \vdots \\ P_\mu \end{bmatrix} [Q_1 Q_2 \dots Q_\nu] = \begin{bmatrix} P_1Q_1, & P_1Q_2, & \dots & P_1Q_\nu \\ P_2Q_1, & P_2Q_2, & \dots & P_2Q_\nu \\ \dots & \dots & \dots & \dots \\ P_\mu Q_1, & P_\mu Q_2, & \dots & P_\mu Q_\nu \end{bmatrix} \quad \text{...(b)}$$

are always true when the rank of A *does not exceed* r; *and when* $\Delta \neq 0$, *the second equation expresses all the minor determinants of* A *of order* r *as homogeneous rational functions of degree* 1 *of* Δ *and its primaries.*

Proof of Theorem I. The expression $(a_{pq})^r_r\,(a_{uv})^r_r-(a_{uq})^r_r\,(a_{pv})^r_r$ in which $[p]_r$ and $[u]_r$ are minors of $[1\,2\ldots m]$, and $[q]_r$ and $[v]_r$ are minors of $[1\,2\ldots n]$, vanishes identically when $[u]_r$ is merely a derangement of $[p]_r$, and when $[v]_r$ is merely a derangement of $[q]_r$. In other cases it is a minor determinant of order 2 of a complete matrix of the minor determinants of A of order r, and therefore by § 73 vanishes whenever the rank of A does not exceed r; for every complete matrix of the minor determinants of A of order r has then rank not exceeding 1. Thus the equation (A) is true whenever the rank of A does not exceed r.

From the identities (1) we deduce as in Exs. vi and ix of § 109 the equations

$$[a_{uq}]^r_r\,\boxed{A_{pq}}^{\,r}_{\,r}=[a_{up}]^r_r,\qquad \boxed{A_{pq}}^{\,r}_{\,r}\,[a_{pv}]^r_r=[\beta_{qv}]^r_r,$$

$$\Delta^{r-1}(a_{uq})^r_r=(a_{up})^r_r,\qquad \Delta^{r-1}(a_{pv})^r_r=(\beta_{qv})^r_r,\qquad \text{.....................(3)}$$

which are identities in the elements of A. Multiplying both sides of (A) by Δ^{2r-2}, and substituting from the equations (3), we obtain the equation (B).

The last part of Theorem I follows from the fact that the elements of $[a_{up}]^r_r$ and $[\beta_{qv}]^r_r$ which are not Δ or 0 are primaries of Δ.

When $\Delta\neq 0$ we can deduce (A) from formula (A) of § 115, from which we obtain in succession

$$\Delta\,[a_{uv}]^r_r=[a_{uq}]^r_r\,\boxed{A_{pq}}^{\,r}_{\,r}\,[a_{pv}]^r_r,\qquad \Delta^r(a_{uv})^r_r=\Delta^{r-1}(a_{uq})^r_r\,(a_{pv})^r_r.$$

Also when $\Delta\neq 0$ we can deduce (B) from formula (C) of § 115, from which we obtain in succession

$$\Delta^2\,[a_{uv}]^r_r=[a_{up}]^r_r\,[a_{pq}]^r_r\,[\beta_{qv}]^r_r,\qquad \Delta^{2r}(a_{uv})^r_r=\Delta\,(a_{up})^r_r\,(\beta_{qv})^r_r.$$

Proof of Theorem II. The equation (a) is a succinct expression for $\mu\nu$ equations of the form (A), and the equation (b) is a succinct expression for $\mu\nu$ equations of the form (B). Consequently the equations (a) and (b) are true whenever the rank of the matrix A does not exceed r. The last part of Theorem II follows from the corresponding property in Theorem I.

We can deduce (b) from (a) by means of the identities

$$\Delta^{r-1}\,\boxed{H}_{\,\mu}=\boxed{P}_{\,\mu},\qquad \Delta^{r-1}[K]_\nu=[Q]_\nu\qquad \text{..........................(4)}$$

obtained from (1) by equating correspondingly formed complete matrices of the simple minor determinants of both sides in each equation.

When $\Delta\neq 0$ we can deduce (a) from formula (A) of § 115 by equating correspondingly formed complete matrices of the minor determinants of order r on both sides.

Also when $\Delta\neq 0$, we can in a similar way deduce (b) from formula (C) of § 115.

When any horizontal row of D is taken from the same horizontal row of the matrix A as one of the horizontal rows of Δ, or when any vertical row of D is taken from the same vertical row of the matrix A as one of the vertical rows of Δ, i.e. when the minor sequences $[p]_r$ and $[u]_r$ have an element in common, or when the minor sequences $[q]_r$ and $[v]_r$ have an element in common, the equation (B) can be reduced in degree by the removal of a factor Δ common to both sides. The reduced forms of the equation are given in the theorems which follow.

Theorem III. *If h horizontal rows of A taken from the minor matrix $[a_{\lambda 1}]_h^n$ and k vertical rows of A taken from the minor matrix $[a_{1\mu}]_m^k$ occur in both the minor determinants $\Delta = (a_{pq})_r^r$ and $D = (a_{uv})_r^r$, then the equation* (B) *can always be reduced to*

$$\Delta^{\rho+\sigma-1} D = (-1)^{\omega+\omega'} (\alpha_{\xi x})_\rho^\rho (\beta_{y\eta})_\sigma^\sigma, \quad \dots\dots\dots\dots\dots\dots(C)$$

where $\rho = r - h$, $\sigma = r - k$; $[x]_\rho$ and $[\xi]_\rho$ are the minors of the sequences $[p]_r$ and $[u]_r$ formed by striking out their common elements $\lambda_1, \lambda_2, \dots \lambda_h$; $[y]_\sigma$ and $[\eta]_\sigma$ are the minors of the sequences $[q]_r$ and $[v]_r$ formed by striking out their common elements $\mu_1, \mu_2, \dots \mu_k$; ω is the difference of the affects of $[\lambda]_h$ in $[p]_r$ and $[u]_r$; and ω' is the difference of the affects of $[\mu]_k$ in $[q]_r$ and $[v]_r$.

If further there are no other rows of A which occur in both Δ and D, then in the equation (C) *all the elements of $[\alpha_{\xi x}]_\rho^\rho$ and $[\beta_{y\eta}]_\sigma^\sigma$ are primaries of Δ.*

Theorem IV. *If $\Delta = \begin{pmatrix} a_{\lambda\mu}, & a_{\lambda y} \\ a_{x\mu}, & a_{xy} \end{pmatrix}_{h,\rho}^{k,\sigma}$ and $D = \begin{pmatrix} a_{\lambda\mu}, & a_{\lambda\eta} \\ a_{\xi\mu}, & a_{\xi\eta} \end{pmatrix}_{h,\rho}^{k,\sigma}$ are two minor determinants of A of order r whose first h horizontal rows and first k vertical rows are taken respectively from the same horizontal rows and the same vertical rows of A, then the equation*

$$\Delta^{\rho+\sigma-1} D = (\alpha_{\xi x})_\rho^\rho (\beta_{y\eta})_\sigma^\sigma \quad \dots\dots\dots\dots\dots\dots\dots\dots(D)$$

is always true when the rank of A does not exceed r.

The equation (C) *can always be reduced to the standard form* (D).

Proof of Theorem III. In the case considered in the theorem we have by Exs. vi and ix of § 109 the identities

$$\Delta^{\rho-1} (a_{uq})_r^r = (-1)^\omega (\alpha_{\xi x})_\rho^\rho, \qquad \Delta^{\sigma-1} (a_{pv})_r^r = (-1)^{\omega'} (\beta_{y\eta})_\sigma^\sigma, \quad \dots\dots\dots(5)$$

which are derived from the identities (1); and when we multiply both sides of the equation (A) by $\Delta^{\rho+\sigma-2}$, we convert it by means of these identities into the equation (C).

If $[\xi]_\rho$ has no element in common with $[p]_r$, then all the elements of $[\alpha_{\xi x}]_\rho^\rho$ are horizontal primaries of Δ; and if $[\eta]_\sigma$ has no element in common with $[q]_r$, then all the elements of $[\beta_{y\eta}]_\sigma^\sigma$ are vertical primaries of Δ.

When $\Delta \neq 0$ we can deduce (C) from the equation (B) by expanding $(\alpha_{up})_r^r$ and $(\beta_{qv})_r^r$ in terms of the minor determinants of their minor matrices $[\alpha_{\lambda p}]_h^r$ and $[\beta_{q\mu}]_r^k$, on doing which we obtain the identities

$$(\alpha_{up})_r^r = (-1)^\omega \Delta^h (\alpha_{\xi x})_\rho^\rho, \qquad (\beta_{qv})_r^r = (-1)^{\omega'} \Delta^k (\beta_{y\eta})_\sigma^\sigma.$$

Proof of Theorem IV. We derive the equation (D) from the equation (C) by putting $\omega=0$, $\omega'=0$. That (C) can always be reduced to the form (D) can be seen as in the proofs of Theorems IV *a* and IV *b* of § 109.

Theorem V. *When the matrix A has rank r, all the equations* (A), (B), *and* (C) *in which Δ and D are any minor determinants whatever of A of order r can be deduced by means of the identities of* § 109 *from those of the equations* (A) *in which Δ is any one particular non-vanishing minor determinant of A of order r, i.e. from those of the equations* (A) *which are necessary and sufficient conditions that A shall have rank r.*

Proof of Theorem V. Since the equations (B) and (C) have been deduced from the equation (A) by means of the identities of § 109, it will be sufficient to show that all equations of the form (A) can be deduced from those in which $\Delta=(a_{pq})_r^r$ is any one particular non-vanishing minor determinant of order r.

Using the notations $(a_{uv})_r^r = A_{uv}$, $(\alpha_{uv})_r^r = \mathrm{A}_{uv}$, $(\beta_{uv})_r^r = \mathrm{B}_{uv}$, so that $\Delta = A_{pq}$, we will obtain the equation

$$A_{hk}\,A_{\lambda\mu}=A_{\lambda k}\,A_{h\mu} \quad \text{or} \quad A_{hk}\,A_{\lambda\mu}-A_{\lambda k}\,A_{h\mu}=0.$$

As shown in the derivation of (B) from (A), we can by means of the identities of § 109 deduce from the equations

$$A_{pq}\,A_{hk}=A_{hq}\,A_{pk},\quad A_{pq}\,A_{\lambda\mu}=A_{\lambda q}\,A_{p\mu},\quad A_{pq}\,A_{\lambda k}=A_{\lambda q}\,A_{pk},\quad A_{pq}\,A_{h\mu}=A_{hq}\,A_{p\mu}$$

respectively the equations

$$\Delta^{2r-1}\,A_{hk}=\mathrm{A}_{hp}\,\mathrm{B}_{qk},\quad \Delta^{2r-1}\,A_{\lambda\mu}=\mathrm{A}_{\lambda p}\,\mathrm{B}_{q\mu},\quad \Delta^{2r-1}\,A_{\lambda k}=\mathrm{A}_{\lambda p}\,\mathrm{B}_{qk},\quad \Delta^{2r-1}\,A_{h\mu}=\mathrm{A}_{hp}\,\mathrm{B}_{q\mu},$$

and from the last four equations we have

$$\Delta^{4r-2}\{A_{hk}\,A_{\lambda\mu}-A_{\lambda k}\,A_{h\mu}\}=0,\quad A_{hk}\,A_{\lambda\mu}-A_{\lambda k}\,A_{h\mu}=0.$$

Thus we have deduced the equation $(a_{hk})_r^r\,(a_{\lambda\mu})_r^r=(a_{\lambda k})_r^r\,(a_{h\mu})_r^r$ from those of the equations (A) in which Δ has the particular non-vanishing value $(a_{pq})_r^r$, and this establishes Theorem V.

Those of the equations (A) in which Δ is any one particular non-vanishing minor determinant of A of order r are necessary and sufficient conditions that $[a]_m^n$ shall have rank r, or that $[\mathbf{A}]_\mu^\nu$ shall have rank 1. Hence when they are satisfied, all minor determinants of $[\mathbf{A}]_\mu^\nu$ of order 1 must vanish, i.e. all such equations as (A) must be satisfied.

NOTE 1. *The standard forms of the equations of this article.*

If
$$\Delta=(a_{pq})_r^r=\begin{pmatrix}a_{\lambda\mu}, & a_{\lambda y}\\ a_{x\mu}, & a_{xy}\end{pmatrix}_{h,\rho}^{k,\sigma},\qquad D=(a_{uv})_r^r=\begin{pmatrix}a_{\lambda\mu}, & a_{\lambda\eta}\\ a_{\xi\mu}, & a_{\xi\eta}\end{pmatrix}_{h,\rho}^{k,\sigma},$$

so that
$$(a_{uq})_r^r=\begin{pmatrix}a_{\lambda\mu}, & a_{\lambda y}\\ a_{\xi\mu}, & a_{\xi y}\end{pmatrix}_{h,\rho}^{k,\sigma},\qquad (a_{pv})_r^r=\begin{pmatrix}a_{\lambda\mu}, & a_{\lambda\eta}\\ a_{x\mu}, & a_{x\eta}\end{pmatrix}_{h,\rho}^{k,\sigma},$$

we have obtained in Theorem IV the equation

$$\Delta^{\rho+\sigma-1}D=(a_{\xi x})_\rho^\rho\,(\beta_{y\eta})_\sigma^\sigma,\ \ldots\ldots\ldots\ldots\text{(D)}$$

which is derived from $(a_{pq})_r^r\,(a_{uv})_r^r=(a_{uq})_r^r\,(a_{pv})_r^r$ by the identities

$$\Delta^{\rho-1}(a_{uq})_r^r=(a_{\xi x})_\rho^\rho,\qquad \Delta^{\sigma-1}(a_{pv})_r^r=(\beta_{y\eta})_\sigma^\sigma.$$

For a given pair of values of h and k all the equations given by formula (C) or (B) can be reduced to the same standard form (D).

If we use the notation of Ex. xv of § 109, and write

$$\Delta=\begin{pmatrix}\mu_1\mu_2\ldots\mu_k & y_1y_2\ldots y_\sigma\\ \lambda_1\lambda_2\ldots\lambda_h & x_1x_2\ldots x_\rho\end{pmatrix},$$

we have as an equivalent of (D) the *standard equation*

$$\begin{pmatrix}\mu_1\mu_2\ldots\mu_k & y_1y_2\ldots y_\sigma\\ \lambda_1\lambda_2\ldots\lambda_h & x_1x_2\ldots x_\rho\end{pmatrix}^{\rho+\sigma-1}\begin{pmatrix}\mu_1\mu_2\ldots\mu_k & \eta_1\eta_2\ldots\eta_\sigma\\ \lambda_1\lambda_2\ldots\lambda_h & \xi_1\xi_2\ldots\xi_\rho\end{pmatrix}=(a_{\xi x})_\rho^\rho(\beta_{y\eta})_\sigma^\sigma, \quad\ldots\ldots(\mathrm{D}')$$

where $$h+\rho=k+\sigma=r;$$

and the equation (D′) is derived from the equation

$$\begin{pmatrix}\mu_1\ldots\mu_k & y_1\ldots y_\sigma\\ \lambda_1\ldots\lambda_h & x_1\ldots x_\rho\end{pmatrix}\begin{pmatrix}\mu_1\ldots\mu_k & \eta_1\ldots\eta_\sigma\\ \lambda_1\ldots\lambda_h & \xi_1\ldots\xi_\rho\end{pmatrix}=\begin{pmatrix}\mu_1\ldots\mu_k & y_1\ldots y_\sigma\\ \lambda_1\ldots\lambda_h & \xi_1\ldots\xi_\rho\end{pmatrix}\begin{pmatrix}\mu_1\ldots\mu_k & \eta_1\ldots\eta_\sigma\\ \lambda_1\ldots\lambda_h & x_1\ldots x_\rho\end{pmatrix}$$

by means of the identities

$$\begin{pmatrix}\mu_1\ldots\mu_k & y_1\ldots y_\sigma\\ \lambda_1\ldots\lambda_h & x_1\ldots x_\rho\end{pmatrix}^{\rho-1}\begin{pmatrix}\mu_1\ldots\mu_k & y_1\ldots y_\sigma\\ \lambda_1\ldots\lambda_h & \xi_1\ldots\xi_\rho\end{pmatrix}=(a_{\xi x})_\rho^\rho,$$

$$\begin{pmatrix}\mu_1\ldots\mu_k & y_1\ldots y_\sigma\\ \lambda_1\ldots\lambda_h & x_1\ldots x_\rho\end{pmatrix}^{\sigma-1}\begin{pmatrix}\mu_1\ldots\mu_k & \eta_1\ldots\eta_\sigma\\ \lambda_1\ldots\lambda_h & x_1\ldots x_\rho\end{pmatrix}=(\beta_{y\eta})_\sigma^\sigma,$$

which are included in the general formula (D) of § 109. 3.

NOTE 2. *Classification of the standard equations.*

The standard equation (D) or (D′) has as many distinct forms as there are possible values of ρ and σ.

The possible values of ρ are 0, 1, 2, ... r, or 0, 1, 2, ... $r-m$ according as $m \not< 2r$ or $m<2r$.

The possible values of σ are 0, 1, 2, ... r, or 0, 1, 2, ... $r-n$ according as $n \not< 2r$ or $n<2r$.

When $\rho=0$, or $h=r$, then (D) or (D′) is one of the standard identities considered in § 109. 1, and is trivial when $\sigma=0$ or 1, and non-trivial in other cases.

When $\sigma=0$. or $k=r$, then (D) or (D′) is one of the standard identities considered in § 109. 2, and is trivial when $\rho=0$ or 1, and non-trivial in other cases.

When $\rho\neq0$ and $\sigma\neq0$, then (D) and (D′) are not identities in the elements of A.

NOTE 3. *General method of expressing all minor determinants of A of order r in terms of Δ and its primaries when $\Delta\neq0$.*

Let $A=[a]_m^n$ be a matrix of rank r, let $\Delta=(a_{pq})_r^r\neq0$, and let $[\mathbf{A}]_\mu^\nu$ be a complete matrix of the minor determinants of A of order r in which $\Delta=\mathbf{A}_{pq}$, so that equation (a) has the form

$$\mathbf{A}_{pq}[\mathbf{A}]_\mu^\nu=\begin{bmatrix}\mathbf{A}_{1q}\\ \mathbf{A}_{2q}\\ \vdots\\ \mathbf{A}_{\mu q}\end{bmatrix}[\mathbf{A}_{p1}\,\mathbf{A}_{p2}\ldots\mathbf{A}_{p\nu}].$$

We write down

(1) The non-trivial equations expressing the fact that the rank of $[\mathbf{A}]_\mu^\nu$ is less than 2. These are the $(\mu-1)(\nu-1)$ equations of the form

$$\mathbf{A}_{pq}\mathbf{A}_{uv}=\mathbf{A}_{uq}\mathbf{A}_{pv} \quad\text{or}\quad \begin{vmatrix} \mathbf{A}_{pq} & \mathbf{A}_{pv} \\ \mathbf{A}_{uq} & \mathbf{A}_{uv} \end{vmatrix}=0,$$

where $u \neq p$ and $v \neq q$; i.e. the equations

$$(a_{pq})_r^r (a_{uv})_r^r = (a_{uq})_r^r (a_{pv})_r^r,$$

where $[u]_r$ is not merely a derangement of $[p]_r$, and $[v]_r$ is not merely a derangement of $[q]_r$.

(2) The non-trivial identities of § 109 which serve to express the simple minor determinants of $[a_{1q}]_m^r$ and $[a_{p1}]_r^n$ in terms of Δ and its primaries.

By means of these two sets of equations we can express all other minor determinants of A of order r in terms of Δ and its primaries in the simplest possible ways.

Ex. i. Let the matrix $A=[a]_5^6$ have rank 3, and let $\Delta=\begin{pmatrix}\lambda\mu\nu\\ a\\ lmn\end{pmatrix}=\begin{pmatrix}\lambda\mu\nu\\ lmn\end{pmatrix}$.

Then the standard equation (D) or (D′) has the following 9 distinct forms when it is non-trivial:

$$\begin{pmatrix}\lambda\mu\nu\\ lmn\end{pmatrix}\begin{pmatrix}\lambda\mu\nu\\ lyz\end{pmatrix}=\begin{pmatrix}mn\\ a\\ yz\end{pmatrix},\qquad \begin{pmatrix}\lambda\mu\nu\\ lmn\end{pmatrix}\begin{pmatrix}\lambda\eta\zeta\\ lmn\end{pmatrix}=\begin{pmatrix}\eta\zeta\\ \beta\\ \mu\nu\end{pmatrix},\qquad \begin{pmatrix}\lambda\mu\nu\\ lmn\end{pmatrix}^2\begin{pmatrix}\xi\eta\zeta\\ lmn\end{pmatrix}=\begin{pmatrix}\xi\eta\zeta\\ \beta\\ \lambda\mu\nu\end{pmatrix},$$

$$\begin{pmatrix}\lambda\mu\nu\\ lmn\end{pmatrix}\begin{pmatrix}\lambda\mu\zeta\\ lmz\end{pmatrix}=a_{zn}\beta_{\nu\zeta},\qquad \begin{pmatrix}\lambda\mu\nu\\ lmn\end{pmatrix}^2\begin{pmatrix}\lambda\mu\zeta\\ lyz\end{pmatrix}=\begin{pmatrix}mn\\ a\\ yz\end{pmatrix}\beta_{\nu\zeta},\qquad \begin{pmatrix}\lambda\mu\nu\\ lmn\end{pmatrix}^2\begin{pmatrix}\lambda\eta\zeta\\ lmz\end{pmatrix}=a_{zn}\begin{pmatrix}\eta\zeta\\ \beta\\ \mu\nu\end{pmatrix},$$

$$\begin{pmatrix}\lambda\mu\nu\\ lmn\end{pmatrix}^3\begin{pmatrix}\lambda\eta\zeta\\ lyz\end{pmatrix}=\begin{pmatrix}mn\\ a\\ yz\end{pmatrix}\begin{pmatrix}\eta\zeta\\ \beta\\ \mu\nu\end{pmatrix},\qquad \begin{pmatrix}\lambda\mu\nu\\ lmn\end{pmatrix}^3\begin{pmatrix}\xi\eta\zeta\\ lmz\end{pmatrix}=a_{zn}\begin{pmatrix}\xi\eta\zeta\\ \beta\\ \lambda\mu\nu\end{pmatrix},$$

$$\begin{pmatrix}\lambda\mu\nu\\ lmn\end{pmatrix}^4\begin{pmatrix}\xi\eta\zeta\\ lyz\end{pmatrix}=\begin{pmatrix}mn\\ a\\ yz\end{pmatrix}\begin{pmatrix}\xi\eta\zeta\\ \beta\\ \lambda\mu\nu\end{pmatrix}.$$

These correspond respectively to the values

$$2,0;\ 0,2;\ 0,3;\ 1,1;\ 2,1;\ 1,2;\ 2,2;\ 1,3;\ 2,3 \quad\text{of}\quad \rho,\sigma.$$

Here $$\begin{pmatrix}mn\\ a\\ yz\end{pmatrix}=\begin{vmatrix}\begin{pmatrix}\lambda\mu\nu\\ lyn\end{pmatrix} & \begin{pmatrix}\lambda\mu\nu\\ lmy\end{pmatrix}\\ \begin{pmatrix}\lambda\mu\nu\\ lzn\end{pmatrix} & \begin{pmatrix}\lambda\mu\nu\\ lmz\end{pmatrix}\end{vmatrix},\qquad \begin{pmatrix}\xi\eta\zeta\\ \beta\\ \lambda\mu\nu\end{pmatrix}=\begin{vmatrix}\begin{pmatrix}\xi\mu\nu\\ lmn\end{pmatrix} & \begin{pmatrix}\eta\mu\nu\\ lmn\end{pmatrix} & \begin{pmatrix}\zeta\mu\nu\\ lmn\end{pmatrix}\\ \begin{pmatrix}\lambda\xi\nu\\ lmn\end{pmatrix} & \begin{pmatrix}\lambda\eta\nu\\ lmn\end{pmatrix} & \begin{pmatrix}\lambda\zeta\nu\\ lmn\end{pmatrix}\\ \begin{pmatrix}\lambda\mu\xi\\ lmn\end{pmatrix} & \begin{pmatrix}\lambda\mu\eta\\ lmn\end{pmatrix} & \begin{pmatrix}\lambda\mu\zeta\\ lmn\end{pmatrix}\end{vmatrix}$$

and so on.

By means of these equations every other minor determinant of A of order 3 can be expressed in terms of Δ and its primaries when $\Delta \neq 0$.

The first three equations are identities in the elements of A, and by means of them we can deduce the other equations from the equation

$$\begin{pmatrix}\lambda\mu\nu\\ lmn\end{pmatrix}\begin{pmatrix}\xi\eta\zeta\\ xyz\end{pmatrix}=\begin{pmatrix}\lambda\mu\nu\\ xyz\end{pmatrix}\begin{pmatrix}\xi\eta\zeta\\ lmn\end{pmatrix}.$$

Ex. ii. Let the matrix $A=[a\,b\,c\,d]_{1234}$ have rank 2; let $\Delta=(ab)_{12}\neq 0$; and let

$$\begin{bmatrix} A_1 & B_1 \\ A_2 & B_2 \end{bmatrix} = \begin{bmatrix} b_2, & -a_2 \\ -b_1, & a_1 \end{bmatrix} \text{ be the reciprocal of } \begin{bmatrix} a_1 & b_1 \\ a_2 & b_2 \end{bmatrix}.$$

Then by formula (A) of § 115 we have

$$\Delta\,[a\,b\,c\,d]_{1234} = \begin{bmatrix} a_1 & b_1 \\ a_2 & b_2 \\ a_3 & b_3 \\ a_4 & b_4 \end{bmatrix} \begin{bmatrix} A_1 & A_2 \\ B_1 & B_2 \end{bmatrix} \begin{bmatrix} a_1 & b_1 & c_1 & d_1 \\ a_2 & b_2 & c_2 & d_2 \end{bmatrix}.$$

Equating corresponding matrices of the minor determinants of order 2 on both sides, we obtain

$$\Delta \begin{bmatrix} (ab)_{12}, & (ac)_{12}, & \dots & (cd)_{12} \\ (ab)_{13}, & (ac)_{13}, & \dots & (cd)_{13} \\ \dots & \dots & \dots & \dots \\ (ab)_{34}, & (ac)_{34}, & \dots & (cd)_{34} \end{bmatrix} = \begin{bmatrix} (ab)_{12} \\ (ab)_{13} \\ \vdots \\ (ab)_{34} \end{bmatrix} [(ab)_{12}, \ (ac)_{12}, \ \dots \ (cd)_{12}].$$

This is the equation corresponding to formula (a), and it is equivalent to 36 equations of the form

$$(ab)_{12}(xy)_{\lambda\mu} = (ab)_{\lambda\mu}(xy)_{12}. \quad \dots\dots(6)$$

Of these equations there are 11 which are trivial, viz. those in which $(xy)_{\lambda\mu}$ is an element of the leading horizontal row or the leading vertical row in the large matrix. The remaining 25 equations are non-trivial.

The non-trivial identical relations between the simple minor determinants of the minor matrices $[a\,b]_{1234}$, $[a\,b\,c\,d]_{12}$ determined as in § 109 are

$$\Delta\,(ab)_{34} = \begin{vmatrix} (ab)_{32}, & (ab)_{42} \\ (ab)_{13}, & (ab)_{14} \end{vmatrix}, \qquad \Delta\,(cd)_{12} = \begin{vmatrix} (cb)_{12}, & (ac)_{12} \\ (db)_{12}, & (ad)_{12} \end{vmatrix}. \quad \dots\dots(7)$$

By means of the 25 non-trivial equations (6) and the two identities (7) we can express every other minor determinant of A of order 2 in terms of Δ and its primaries.

The non-trivial equations expressing these relations when reduced to their standard forms by means of the two identities (7) are of 6 different types. In particular the equations for $(ac)_{13}$ and $(ac)_{34}$ are

$$\Delta\,(ac)_{13} = a_{32}\,\beta_{23} = (ab)_{13}(ac)_{12};$$

$$\Delta^2\,(ac)_{34} = \begin{pmatrix} 12 \\ a \\ 34 \end{pmatrix} \beta_{23} = \begin{vmatrix} (ab)_{32} & (ab)_{13} \\ (ab)_{42} & (ab)_{14} \end{vmatrix} (ac)_{12}.$$

Ex. iii. Let $\Phi = \begin{bmatrix} A & H & G \\ H & B & F \\ G & F & C \end{bmatrix}$ be the reciprocal of $\phi = \begin{bmatrix} a & h & g \\ h & b & f \\ g & f & c \end{bmatrix}$, so that Φ is a complete matrix of the minor determinants of ϕ of order 2.

Suppose that ϕ has rank 2 and that $C=ab-h^2\neq 0$.

Then F and G are the primaries of C, and to express A, B, H in terms of C, F, G we may use:

(1) The equations expressing the fact that the rank of Φ does not exceed 1.

(2) Those non-trivial identical relations between the simple minor determinants of $\begin{bmatrix} a & h & g \\ h & b & f \end{bmatrix}$ which serve to express them all in terms of C and its primaries.

Since the identities (2) are non-existent, we use the equations (1) only, which are

$$CA - G^2 = 0, \quad CB - F^2 = 0, \quad CH - FG = 0. \qquad \text{.......................(8)}$$

In this case the equations (A) and (a) of the text are

$$C \begin{bmatrix} a & h & g \\ h & b & f \\ g & f & c \end{bmatrix} = \begin{bmatrix} a & h \\ h & b \\ g & f \end{bmatrix} \begin{bmatrix} b, & -h \\ -h, & a \end{bmatrix} \begin{bmatrix} a & h & g \\ h & b & f \end{bmatrix}, \quad C \begin{bmatrix} A & H & G \\ H & B & F \\ G & F & C \end{bmatrix} = \begin{bmatrix} G \\ F \\ C \end{bmatrix} [G\, F\, C].$$

The last equation is equivalent to

$$C \begin{bmatrix} A & H \\ H & B \end{bmatrix} = \begin{bmatrix} G \\ F \end{bmatrix} [G\, F]$$

and gives the equations (8).

CHAPTER XIV

SOME PROPERTIES OF SQUARE MATRICES

[This Chapter begins with an account of the properties of two co-joint complete matrices of the minor determinants of a fundamental square matrix; their determinants are evaluated; and the relation between any two of their anti-correspondent minor determinants is found. In § 123 these properties are used in obtaining the expansions of certain bordered determinants; and in § 124 a separate account is given of the properties of the reciprocal of a square matrix. The next three articles deal with properties of symmetric and skew-symmetric matrices; criteria for the determination of their ranks are given; and they are reduced to standard forms by symmetric derangements of their rows. The last article contains an additional property and further illustrations of the fundamental identity of Chapter XIII.]

§ 118. Recapitulation of properties of a product of square matrices.

The following results have been proved in earlier chapters:

(1) A product of a square matrix and its conjugate reciprocal is always commutative, both when the square matrix is undegenerate and when it is degenerate.

(2) Whenever a product of two square matrices of the same order is a non-zero scalar matrix, the product is commutative, and each of the factor matrices is a scalar multiple of the conjugate reciprocal of the other factor matrix.

(3) The reciprocal of a standard product of any number of square matrices taken in any given order is equal to the product of the reciprocals of the factor matrices taken in the same order.

(4) The conjugate reciprocal of a standard product of any number of square matrices taken in any given order is equal to the product of the conjugate reciprocals of the factor matrices taken in the reverse order.

The first property has been proved in § 46.5 and exemplified in § 67; the second property has been proved and illustrated in Exs. i and ii of § 67 and Ex. xi of § 79; and the last two properties have been proved in § 67.

If $A = [a]_m^m$ is any square matrix, a minor of A of the form $[a_{pp}]_r^r$ is called a *diagonal minor*, and a minor determinant of A of the form $(a_{pp})_r^r$ is called a *diagonal minor determinant.* Thus a diagonal minor of A is a square minor the elements of whose leading diagonal are all elements of the leading diagonal of A; and similarly the elements of the leading diagonal of a diagonal minor determinant are all elements of the leading diagonal of A.

§ 119. Co-joint matrices of the minor determinants of a square matrix.

1. *Definition of two co-joint matrices.*

Let $[b]_\mu^\mu$, where $\mu = \binom{m}{r}$, be a complete matrix of the minor determinants of order r of a fundamental square matrix $A = [a]_m^m$, and let B_{ij} be the co-factor of b_{ij} in A, so that B_{ij} is equal to the corranged minor determinant of A complementary to b_{ij} provided with the sign determined by the affect of the minor determinant b_{ij} in A.

Then $[B]_\mu^\mu$ is a complete matrix of the minor determinants of A of order $m-r$, and it will be called the *co-joint matrix* of $[b]_\mu^\mu$ with respect to the fundamental matrix A.

We proceed to prove the following theorem:

Theorem I. *If $[B]_\mu^\mu$ is the co-joint matrix of $[b]_\mu^\mu$ with respect to the square matrix $A = [a]_m^m$, then $[b]_\mu^\mu$ is the co-joint matrix of $[B]_\mu^\mu$ with respect to A.*

Let b'_{ij} and B'_{ij} be the corranged minor determinants of A formed with the same horizontal and vertical rows as b_{ij} and B_{ij} respectively, so that b_{ij} and B_{ij} are respectively derangements of the corranged minor determinants b'_{ij} and B'_{ij}.

Let ω = affect of b'_{ij} in A, ω' = affect of B'_{ij} in A,

η = affect of b_{ij} in A, η' = affect of B_{ij} in A,

σ = affect of b_{ij} in b'_{ij}, σ' = affect of B_{ij} in B'_{ij}.

Then by Theorem II *a* of § 26 and Ex. xv of § 25 we have

$$\eta = \omega + \sigma, \qquad \eta' = \omega' + \sigma', \qquad \text{......(1)}$$

$$b_{ij} = (-1)^\sigma b'_{ij}, \quad B_{ij} = (-1)^{\sigma'} B'_{ij}. \qquad \text{......(2)}$$

Since by definition $B_{ij} = (-1)^\eta B'_{ij}$, it follows from (2) and from § 26 that

$$\sigma' \equiv \eta \pmod{2}, \quad \omega' \equiv \omega \pmod{2}. \qquad \text{......(3)}$$

From (1) and (3) we have $\eta' \equiv \sigma \pmod{2}$ and therefore

$$b_{ij} = (-1)^\sigma b'_{ij} = (-1)^{\eta'} b'_{ij}. \qquad \text{......(4)}$$

Equation (4) shows that b_{ij} is equal to the corranged minor determinant of A complementary to B_{ij} provided with the sign determined by the affect of B_{ij} in A, and this establishes Theorem I.

Accordingly we call $[b]_{\mu}^{\mu}$ and $[B]_{\mu}^{\mu}$ *two co-joint matrices* of the minor determinants of A, and we see that:

In two co-joint matrices $[b]_{\mu}^{\mu}$ *and* $[B]_{\mu}^{\mu}$ *of the minor determinants of orders* r *and* $m-r$ *of the fundamental square matrix* $A=[a]_{m}^{m}$ *every element of either matrix is the co-factor in* A *of the corresponding element of the other matrix, i.e.* b_{ij} *is the co-factor of* B_{ij}, *and* B_{ij} *is the co-factor of* b_{ij}.

2. *Fundamental property of two co-joint matrices.*

The most important property of two co-joint matrices is that given by the following theorem:

Theorem II. *If* $[b]_{\mu}^{\mu}$ *and* $[B]_{\mu}^{\mu}$ *are two co-joint matrices of the minor determinants of orders* r *and* $m-r$ *of the fundamental square matrix* $A=[a]_{m}^{m}$, *then*

$$[b]_{\mu}^{\mu}\,\overline{\underline{B}}_{\mu}^{\mu}=\overline{\underline{b}}_{\mu}^{\mu}\,[B]_{\mu}^{\mu}=[B]_{\mu}^{\mu}\,\overline{\underline{b}}_{\mu}^{\mu}=\overline{\underline{B}}_{\mu}^{\mu}\,[b]_{\mu}^{\mu}=\Delta\,[1]_{\mu}^{\mu}, \quad \text{......(A)}$$

where $\Delta=(a)_{m}^{m}=\det[a]_{m}^{m}$.

These results can easily be deduced from Ex. ix of § 32. We will however deduce them directly from Theorem II of § 32. It will be sufficient to prove that

$$[b]_{\mu}^{\mu}\,\overline{\underline{B}}_{\mu}^{\mu}=\Delta\,[1]_{\mu}^{\mu}, \qquad \overline{\underline{b}}_{\mu}^{\mu}\,[B]_{\mu}^{\mu}=\Delta\,[1]_{\mu}^{\mu}; \quad \text{............(5)}$$

for the other two results follow when we equate the conjugates of both sides.

Let b'_{ij}, B'_{ij} be the corranged minor determinants of A of which the determinants b_{ij}, B_{ij} are respectively derangements, and let B''_{ij} be the co-factor of b'_{ij} in A, these notations being valid for all the values $1, 2, \ldots \mu$ of i and j.

Then by Theorem II of § 32 we have

$$b'_{i1}B''_{j1}+b'_{i2}B''_{j2}+\ldots+b'_{iu}B''_{ju}+\ldots+b'_{i\mu}B''_{j\mu}=0 \text{ or } \Delta, \quad \text{......(6)}$$

$$b'_{1i}B''_{1j}+b'_{2i}B''_{2j}+\ldots+b'_{ui}B''_{uj}+\ldots+b'_{\mu i}B''_{\mu j}=0 \text{ or } \Delta, \quad \text{......(6')}$$

according as $j \neq i$ or $j=i$.

We will prove the two results (5) by deducing from (6) and (6') that

$$b_{i1}B_{j1}+b_{i2}B_{j2}+\ldots+b_{iu}B_{ju}+\ldots+b_{i\mu}B_{j\mu}=0 \text{ or } \Delta, \quad \text{.........(7)}$$

$$b_{1i}B_{1j}+b_{2i}B_{2j}+\ldots+b_{ui}B_{uj}+\ldots+b_{\mu i}B_{\mu j}=0 \text{ or } \Delta, \quad \text{.........(7')}$$

according as $j \neq i$ or $j=i$.

Let $\omega_1 =$ affect of b'_{ju} in A, $\eta_1 =$ affect of b_{ju} in A,

$\sigma_1 =$ affect of b_{ju} in b'_{ju}, $\rho_1 =$ affect of b_{iu} in b'_{iu},

so that

$$\eta_1 = \omega_1 + \sigma_1.$$

Since $B''_{ju} = (-1)^{\omega_1} B'_{ju}$, $B_{ju} = (-1)^{\eta_1} B'_{ju}$, we have $B_{ju} = (-1)^{\sigma_1} B''_{ju}$; and since $b_{iu} = (-1)^{\rho_1} b'_{iu}$, it follows that

$$b_{iu} B_{ju} = (-1)^{\rho_1 + \sigma_1} b'_{iu} B''_{ju}. \quad \text{.........................(8)}$$

If $\sigma_2 =$ affect of b_{uj} in b'_{uj}, $\rho_2 =$ affect of b_{ui} in b'_{ui},

we can show in the same way that

$$b_{ui} B_{uj} = (-1)^{\rho_2 + \sigma_2} b'_{ui} B''_{uj}. \quad \text{.........................(8')}$$

Now let

$$b_{iu} = (a_{iu})^r_r, \quad b_{ju} = (b_{ju})^r_r, \quad b'_{iu} = (a_{IU})^r_r, \quad b'_{ju} = (a_{JU})^r_r,$$

so that $[i]_r$, $[j]_r$, $[u]_r$ are derangements of $[I]_r$, $[J]_r$, $[U]_r$, which are themselves corranged minors of $[1\,2 \ldots m]$.

Then

$\rho_1 =$ affect of $[i_1 i_2 \ldots i_r]$ in $[I_1 I_2 \ldots I_r]$ + affect of $[u_1 u_2 \ldots u_r]$ in $[U_1 U_2 \ldots U_r]$,

$\sigma_1 =$ affect of $[j_1 j_2 \ldots j_r]$ in $[J_1 J_2 \ldots J_r]$ + affect of $[u_1 u_2 \ldots u_r]$ in $[U_1 U_2 \ldots U_r]$,

and therefore

$$\rho_1 + \sigma_1 \equiv \text{affect of } [i]_r \text{ in } [I]_r + \text{affect of } [j]_r \text{ in } [J]_r \pmod{2}.$$

If $j \neq i$, the sequences $[i]_r$ and $[j]_r$ are different, and

$$\rho_1 + \sigma_1 \equiv \omega_{ij} \pmod{2},$$

where ω_{ij} depends only on i and j, and is independent of u.

If $j = i$, the sequences $[i]_r$ and $[j]_r$ are the same, and

$$\rho_1 + \sigma_1 \equiv 0 \pmod{2}.$$

Giving to u the values $1, 2, \ldots \mu$, it follows that

$$\text{if } j \neq i, \quad \sum_u b_{iu} B_{ju} = (-1)^{\omega_{ij}} \sum_u b'_{iu} B''_{ju} = 0;$$

$$\text{if } j = i, \quad \sum_u b_{iu} B_{ju} = \sum_u b'_{iu} B''_{iu} = \Delta.$$

We have thus proved (7) by means of (8); and in a similar way we can prove (7′) by means of (8′).

Thus (7) and (7′) are true, and therefore Theorem II is true.

Ex. i. *If* $[A]_m^m$ *is the reciprocal of the square matrix* $A=[a]_m^m$, *then* $[a]_m^m$ *and* $[A]_m^m$ *are two co-joint matrices of the minor determinants of orders* 1 *and* $m-1$ *of* A.

Ex. ii. If

$$\phi=[a\,b\,c\,d]_{1234},$$

$$\psi=\begin{bmatrix} (ab)_{12}, & (ac)_{12}, & (ad)_{12}, & (bc)_{12}, & (bd)_{12}, & (cd)_{12} \\ (ab)_{13}, & (ac)_{13}, & (ad)_{13}, & (bc)_{13}, & (bd)_{13}, & (cd)_{13} \\ (ab)_{14}, & (ac)_{14}, & (ad)_{14}, & (bc)_{14}, & (bd)_{14}, & (cd)_{14} \\ (ab)_{23}, & (ac)_{23}, & (ad)_{23}, & (bc)_{23}, & (bd)_{23}, & (cd)_{23} \\ (ab)_{24}, & (ac)_{24}, & (ad)_{24}, & (bc)_{24}, & (bd)_{24}, & (cd)_{24} \\ (ab)_{34}, & (ac)_{34}, & (ad)_{34}, & (bc)_{34}, & (bd)_{34}, & (cd)_{34} \end{bmatrix},$$

$$\psi'=\begin{bmatrix} +(cd)_{34}, & -(bd)_{34}, & +(bc)_{34}, & +(ad)_{34}, & -(ac)_{34}, & +(ab)_{34} \\ -(cd)_{24}, & +(bd)_{24}, & -(bc)_{24}, & -(ad)_{24}, & +(ac)_{24}, & -(ab)_{24} \\ +(cd)_{23}, & -(bd)_{23}, & +(bc)_{23}, & +(ad)_{23}, & -(ac)_{23}, & +(ab)_{23} \\ +(cd)_{14}, & -(bd)_{14}, & +(bc)_{14}, & +(ad)_{14}, & -(ac)_{14}, & +(ab)_{14} \\ -(cd)_{13}, & +(bd)_{13}, & -(bc)_{13}, & -(ad)_{13}, & +(ac)_{13}, & -(ab)_{13} \\ +(cd)_{12}, & -(bd)_{12}, & +(bc)_{12}, & +(ad)_{12}, & -(ac)_{12}, & +(ab)_{12} \end{bmatrix},$$

then ψ and ψ' are two co-joint matrices of the minor determinants of orders 2 and 2 of ϕ such that the horizontal and vertical rows of ψ follow the common scheme

$$(12,\ 13,\ 14,\ 23,\ 24,\ 34).$$

Ex. iii. If one of two co-joint matrices is composed of unaffected corranged minor determinants (or affected minor determinants) of the fundamental matrix, then the other is composed of affected minor determinants (or unaffected corranged minor determinants) of the fundamental matrix.

Ex. iv. *Both the products which can be formed with any complete matrix of the minor determinants of any order r of a square matrix and the conjugate of its co-joint matrix are commutative both when the fundamental square matrix is undegenerate and when it is degenerate.*

Ex. v. If $[b]_\mu^\mu$ is a complete matrix of the minor determinants of any order r of an undegenerate (or arbitrary) square matrix $[a]_m^m$, and if $\overline{\underline{b}}{}_\mu^\mu\,[x]_\mu^\mu=\Delta\,[1]_\mu^\mu$, where $\Delta=(a)_m^m$, then $[x]_\mu^\mu$ is the co-joint matrix of $[b]_\mu^\mu$ with respect to $[a]_m^m$.

NOTE. *Co-joint matrix of any complete matrix of the minor determinants of order r of a rectangular matrix.*

Let $A=[a]_m^n$ be any matrix, not necessarily square, and let $\Delta=\det A=(a)_m^n$. Also let $[b]_\mu^\nu$ be any complete matrix of the minor determinants of order r of A, and let $[B]_\mu^\nu$ be a matrix each of whose elements B_{ij} is the co-factor in A of the corresponding determinantal element b_{ij} of $[b]_\mu^\nu$. Then we may call $[B]_\mu^\nu$ the co-joint matrix of $[b]_\mu^\nu$ with respect to A.

From Ex. ix of § 32 we see that:

(1) If $m \ngtr n$, then $[b]_{\mu}^{\nu} \overline{\underline{B}}_{\nu}^{\mu} = [B]_{\mu}^{\nu} \overline{\underline{b}}_{\nu}^{\mu} = \Delta [1]_{\mu}^{\mu}$.

(2) If $n \ngtr m$, then $\overline{\underline{b}}_{\nu}^{\mu} [B]_{\mu}^{\nu} = \overline{\underline{B}}_{\nu}^{\mu} [b]_{\mu}^{\nu} = \Delta [1]_{\nu}^{\nu}$.

Thus that product of $[b]_{\mu}^{\nu}$ and the conjugate co-joint matrix of $[b]_{\mu}^{\nu}$ in which long rows are active is a scalar matrix whose argument is Δ.

This is a generalisation of the theorem (A) in § 46.

§ 120. Determinant of the minor determinants of order r of a given fundamental square matrix.

The following theorem gives the value of the determinant of any complete matrix of the minor determinants of order r of a given square matrix.

Theorem. *If $[b]_{\mu}^{\mu}$ is any complete matrix of the minor determinants of order r of the square matrix $A = [a]_{m}^{m}$, so that $\mu = \binom{m}{r}$, and if $(a)_{m}^{m} = \Delta$, then*

$$(b)_{\mu}^{\mu} = \pm \Delta^{\rho}, \quad \textit{where } \rho = \binom{m-1}{r-1}. \qquad \text{(A)}$$

Here the upper sign is always to be taken when in the formation of $[b]_{\mu}^{\mu}$ the successive horizontal and vertical minors of A follow a common scheme.

Let $[B]_{\mu}^{\mu}$ be the co-joint matrix of $[b]_{\mu}^{\mu}$ with respect to A, so that

$$[b]_{\mu}^{\mu} \overline{\underline{B}}_{\mu}^{\mu} = \overline{\underline{b}}_{\mu}^{\mu} [B]_{\mu}^{\mu} = \Delta [1]_{\mu}^{\mu}. \qquad \text{(1)}$$

Equating the determinants of both sides in (1) we have

$$(b)_{\mu}^{\mu} (B)_{\mu}^{\mu} = \Delta^{\mu}. \qquad \text{(2)}$$

In this equation, which is an identity in the elements of A, the quantities Δ, $(b)_{\mu}^{\mu}$, $(B)_{\mu}^{\mu}$ are homogeneous rational integral functions of the elements of A, and also homogeneous rational integral functions of the elements of each selected row of A. Regarding Δ as a linear function of the elements of any one of its rows, the identity (2) shows that $(b)_{\mu}^{\mu}$ and $(B)_{\mu}^{\mu}$ are rational integral factors of the μth power of that linear function. But an integral power of a linear function has no rational integral factors except integral powers of that same linear function. Hence we conclude from (2) that

$$(b)_{\mu}^{\mu} = k\Delta^{\rho}, \quad (B)_{\mu}^{\mu} = \frac{1}{k}\Delta^{\rho'}, \qquad \text{(3)}$$

where ρ and ρ' are positive integers, and k is independent of the elements of each selected row of A, and is therefore a numerical constant.

We can obtain (3) more simply by observing that Δ is an irresoluble function of its elements.

Equating the degrees of both sides of each of the equations (3) in the elements of A, we have

$$\rho = \binom{m-1}{r-1}, \quad \rho' = \binom{m-1}{r}, \quad \rho + \rho' = \mu. \quad \text{...........(4)}$$

Since the equations (3) are identities in the elements of A, we can determine the value of k by considering the special case in which $[a]_m^m = [1]_m^m$. In this case $\Delta = 1$; also in each row of $(b)_\mu^\mu$ one of the elements is ± 1, and the other elements are 0's, and therefore $(b)_\mu^\mu$ has one of the values ± 1. It follows that in all cases

$$k = \pm 1, \quad \text{..................................(5)}$$

the sign depending on the schemes of formation of $[b]_\mu^\mu$. We see then that

$$(b)_\mu^\mu = \epsilon \Delta^\rho, \quad (B)_\mu^\mu = \epsilon \Delta^{\rho'}, \quad \text{.......................(B)}$$

where ϵ is either $+1$ or -1.

When the same scheme is followed for the vertical rows as for the horizontal rows in the formation of $[b]_\mu^\mu$, b_{ii} has the form $(a_{ii})_r^r$; accordingly for the special matrix $[a]_m^m = [1]_m^m$ the diagonal element b_{ii} is the determinant of a unit matrix and has the value $+1$, and therefore $(b)_\mu^\mu$ is also the determinant of a unit matrix and has the value $+1$. Thus in these particular cases we have $k = +1$, and therefore $\epsilon = +1$.

Ex. i. If the same scheme is followed for the vertical rows as for the horizontal rows in the formation of $[b]_\mu^\mu$ from $[a]_m^m$, then the results $\epsilon = 1$, $(b)_\mu^\mu = \Delta^\rho$ remain true when we prefix to every determinantal element b_{ij} of $[b]_\mu^\mu$ the sign determined by its affect in A.

Ex. ii. *Every complete matrix of the minor determinants of any given order of a square matrix A is undegenerate or degenerate according as A is undegenerate or degenerate.*

For in the equation (A) the determinant $(b)_\mu^\mu$ does or does not vanish according as Δ does or does not vanish.

Ex. iii. *Reciprocals of two co-joint matrices.*

Let $[c]_\mu^\mu$, $[C]_\mu^\mu$ be the reciprocals of the matrices $[b]_\mu^\mu$, $[B]_\mu^\mu$ in the text. Then by the fundamental property of a reciprocal matrix we have

$$[c]_\mu^\mu \overline{\underline{b}}_\mu^\mu = (b)_\mu^\mu [1]_\mu^\mu = \epsilon \Delta^\rho [1]_\mu^\mu.$$

Postfixing $[B]_\mu^\mu$ on both sides, we see that

$$\Delta [c]_\mu^\mu = (b)_\mu^\mu [B]_\mu^\mu = \epsilon \Delta^\rho [B]_\mu^\mu. \quad \text{...........................(C)}$$

Since the last equation is an identity in the elements of A and Δ is a function of those elements which does not vanish identically, it follows that the equations

$$[c]_{\mu}^{\mu}=\epsilon\Delta^{\rho-1}[B]_{\mu}^{\mu},\quad \Delta[c]_{\mu}^{\mu}=(b)_{\mu}^{\mu}[B]_{\mu}^{\mu}$$

are identities in the elements of A, and are true both when $\Delta\neq 0$ and when $\Delta=0$.

We can deduce or prove in a similar manner that

$$[C]_{\mu}^{\mu}=\epsilon\Delta^{\rho-1}[b]_{\mu}^{\mu},\quad \Delta[C]_{\mu}^{\mu}=(B)_{\mu}^{\mu}[b]_{\mu}^{\mu}. \quad \text{.......................(D)}$$

Thus each of two co-joint matrices differs from the reciprocal of the other only by a scalar factor of the form $\pm\Delta^k$, *where* Δ *is the determinant of the fundamental square matrix. The upper sign is always to be taken when horizontal and vertical rows follow a common scheme.*

Ex. iv. Let $\phi=[a\,b\,c\,d]_{1234}$, $\det\phi=(abcd)_{1234}=\Delta$.

Then if ψ and ψ' are defined as in Ex. ii of § 119, formula (A) gives

$$\det\psi=\Delta^3,\quad \det\psi'=\Delta^3.$$

To prove this in the way the theorem of the text was proved, we may denote the conjugate of ψ by $\overline{\psi}$. We then have the equation

$$\overline{\psi}\psi'=\Delta[1]_6^6,$$

from which it follows that

$$\det\overline{\psi}\,.\det\psi'=\Delta^6,\quad\text{or}\quad\det\psi\,.\det\psi'=\Delta^6.$$

The last equation shows that

$$\det\psi=k\Delta^{\rho},\qquad \det\psi'=\frac{1}{k}\Delta^{\rho'},$$

where k and $\frac{1}{k}$ are numerical factors, and ρ and ρ' are positive integers.

Considering first the degrees of both sides of the last two equations, and then the special case in which $[a\,b\,c\,d]_{1234}=[1]_4^4$, we find that

$$\rho=\rho'=3,\quad k=1.$$

Ex. v. *Determinant of the reciprocal of a square matrix.*

Let $[A]_m^m$ be the reciprocal of a square matrix $[a]_m^m$, and let $(a)_m^m=\Delta$.

Then $[A]_m^m$ is a complete matrix of the minor determinants of order $m-1$ of $[a]_m^m$ in the formation of which a common scheme is used for horizontal and vertical rows.

Hence as a particular case of formula (A) we have

$$(A)_m^m=\Delta^{m-1}.$$

§ 121. Relations between anti-correspondent minor determinants of two co-joint matrices.

1. *Definitions of correspondence and anti-correspondence.*

Let $[b]_{\mu}^{\mu}$ and $[B]_{\mu}^{\mu}$, where $\mu=\binom{m}{r}$, be two co-joint matrices of the minor determinants of orders r and $m-r$ of the fundamental square matrix $A=[a]_m^m$, and let $(a)_m^m=\Delta$, so that

$$[b]_{\mu}^{\mu}\,\overline{\underline{B}}_{\mu}^{\mu}=\overline{\underline{b}}_{\mu}^{\mu}\,[B]_{\mu}^{\mu}=\Delta[1]_{\mu}^{\mu}. \quad \text{......................(1)}$$

If $\mathbf{b}$ is any minor determinant of $[b]_\mu^\mu$, $\mathbf{b}'$ the co-factor of $\mathbf{b}$ in $[b]_\mu^\mu$, and if $\mathbf{B}$ and $\mathbf{B}'$ are the minor determinants of $[B]_\mu^\mu$ which are formed from $[B]_\mu^\mu$ in the same way as $\mathbf{b}$ and $\mathbf{b}'$ are formed from $[b]_\mu^\mu$, then we will call

$\mathbf{b}$ and $\mathbf{B}$ corresponding minor determinants,

$\mathbf{b}'$ and $\mathbf{B}'$ corresponding minor determinants,

$\mathbf{b}$ and $\mathbf{B}'$ anti-correspondent minor determinants,

$\mathbf{b}'$ and $\mathbf{B}$ anti-correspondent minor determinants

of the two co-joint matrices $[b]_\mu^\mu$ and $[B]_\mu^\mu$.

Thus the anti-correspondent of any minor determinant $\mathbf{b}$ of $[b]_\mu^\mu$ is that minor determinant of $[B]_\mu^\mu$ which corresponds to the co-factor of $\mathbf{b}$ in $[b]_\mu^\mu$.

If one of two anti-correspondent minor determinants is an unaffected corranged (or affected) minor determinant of $[b]_\mu^\mu$, then the other is an affected (or unaffected corranged) minor determinant of $[B]_\mu^\mu$.

Again let $[\mathbf{b}]_\nu^\nu$ and $[\mathbf{b}']_\nu^\nu$, where $\nu = \binom{\mu}{s}$, be any two co-joint complete matrices of the minor determinants of $[b]_\mu^\mu$ of orders s and $\mu - s$, and let $[\mathbf{B}]_\nu^\nu$ and $[\mathbf{B}']_\nu^\nu$ be the similarly formed co-joint complete matrices of the minor determinants of $[B]_\mu^\mu$ of orders s and $\mu - s$. Then we will call

$[\mathbf{b}]_\nu^\nu$ and $[\mathbf{B}]_\nu^\nu$ corresponding matrices of the minor determinants,

$[\mathbf{b}']_\nu^\nu$ and $[\mathbf{B}']_\nu^\nu$ corresponding matrices of the minor determinants,

$[\mathbf{b}]_\nu^\nu$ and $[\mathbf{B}']_\nu^\nu$ anti-correspondent matrices of the minor determinants,

$[\mathbf{b}']_\nu^\nu$ and $[\mathbf{B}]_\nu^\nu$ anti-correspondent matrices of the minor determinants

of the two co-joint matrices $[b]_\mu^\mu$ and $[B]_\mu^\mu$.

If i and j are any elements of the sequence $[1\ 2 \dots \nu]$, then clearly

$\mathbf{b}_{ij}$ and $\mathbf{B}_{ij}$ are corresponding minor determinants,

$\mathbf{b}_{ij}$ and $\mathbf{B}'_{ij}$ are anti-correspondent minor determinants

of the two co-joint matrices $[b]_\mu^\mu$ and $[B]_\mu^\mu$.

We can define in the same way correspondences and anti-correspondences for any two similar matrices $[p]_\lambda^\mu$, $[P]_\lambda^\mu$.

2. *Relation between any two anti-correspondent matrices of the minor determinants of two co-joint matrices.*

Let $[b]_\mu^\mu$ and $[B]_\mu^\mu$ be two co-joint complete matrices of the minor determinants of orders r and $m-r$ of the fundamental square matrix $A=[a]_m^m$ as in sub-article 1, and let $[\mathbf{b}]_\nu^\nu$, $[\mathbf{b}']_\nu^\nu$, $[\mathbf{B}]_\nu^\nu$, $[\mathbf{B}']_\nu^\nu$ also be defined as in sub-article 1.

Then by formula (B) of § 120 we have

$$(b)_\mu^\mu = \epsilon\Delta^\rho, \quad (B)_\mu^\mu = \epsilon\Delta^{\rho'}, \text{ ....(2)}$$

where $$\rho=\binom{m-1}{r-1}, \quad \rho'=\binom{m-1}{r}, \quad \rho+\rho'=\mu, \quad \mu=\binom{m}{r}, \text{ ......(3)}$$

and where ϵ is either $+1$ or -1.

Again by formula (A) of § 119 we have

$$[\mathbf{b}]_\nu^\nu \ulcorner\mathbf{b}'\lrcorner_\nu^\nu = \ulcorner\mathbf{b}\lrcorner_\nu^\nu [\mathbf{b}']_\nu^\nu = (b)_\mu^\mu [1]_\nu^\nu = \epsilon\Delta^\rho [1]_\nu^\nu, \text{ ....(4)}$$

$$[\mathbf{B}]_\nu^\nu \ulcorner\mathbf{B}'\lrcorner_\nu^\nu = \ulcorner\mathbf{B}\lrcorner_\nu^\nu [\mathbf{B}']_\nu^\nu = (B)_\mu^\mu [1]_\nu^\nu = \epsilon\Delta^{\rho'} [1]_\nu^\nu. \text{ ....(4')}$$

Further by equating correspondingly formed matrices of the minor determinants of order s, and correspondingly formed matrices of the minor determinants of order $\mu-s$ on both sides of the equation (1) we obtain

$$[\mathbf{b}]_\nu^\nu \ulcorner\mathbf{B}\lrcorner_\nu^\nu = \ulcorner\mathbf{b}\lrcorner_\nu^\nu [\mathbf{B}]_\nu^\nu = \Delta^s [1]_\mu^\mu, \text{ ....(5)}$$

$$[\mathbf{b}']_\nu^\nu \ulcorner\mathbf{B}'\lrcorner_\nu^\nu = \ulcorner\mathbf{b}'\lrcorner_\nu^\nu [\mathbf{B}']_\nu^\nu = \Delta^{\mu-s} [1]_\mu^\mu. \text{ ....(5')}$$

Here (5′) is deducible from (4), (4′) and (5) and is superfluous.

The products occurring on the left in equations (4), (4′), (5) and (5′) are commutative. Hence if in the equation

$$[\mathbf{b}]_\nu^\nu \ulcorner\mathbf{B}\lrcorner_\nu^\nu = \Delta^s [1]_\mu^\mu$$

we prefix $\ulcorner\mathbf{b}'\lrcorner_\nu^\nu$, the conjugate co-joint matrix of $[\mathbf{b}]_\nu^\nu$ with respect to $[b]_\mu^\mu$, on both sides, and again in the same equation postfix $[\mathbf{B}']_\nu^\nu$, the conjugate co-joint matrix of $\ulcorner\mathbf{B}\lrcorner_\nu^\nu$ with respect to $[B]_\mu^\mu$, on both sides, we obtain

$$\Delta^s [\mathbf{b}']_\nu^\nu = (b)_\mu^\mu [\mathbf{B}]_\nu^\nu = \epsilon\Delta^\rho [\mathbf{B}]_\nu^\nu, \text{ ....(A)}$$

$$\Delta^s [\mathbf{B}']_\nu^\nu = (B)_\mu^\mu [\mathbf{b}]_\nu^\nu = \epsilon\Delta^{\rho'} [\mathbf{b}]_\nu^\nu. \text{ ....(B)}$$

Here $\Delta = (a)^m_m$; ϵ is $+1$ or -1; and if a common scheme is used for horizontal and vertical rows in the formation of $[b]^\mu_\mu$ from the fundamental square matrix $A = [a]^m_m$, then $\epsilon = +1$.

We can deduce or prove in the same way the formulae

$$\Delta^{\mu - s}[\mathbf{b}]^\nu_\nu = (b)^\mu_\mu [\mathbf{B}']^\nu_\nu = \epsilon \Delta^\rho [\mathbf{B}']^\nu_\nu, \qquad \text{..............(C)}$$

$$\Delta^{\mu - s}[\mathbf{B}]^\nu_\nu = (B)^\mu_\mu [\mathbf{b}']^\nu_\nu = \epsilon \Delta^{\rho'} [\mathbf{b}']^\nu_\nu. \qquad \text{..............(D)}$$

Since (A), (B), (C) and (D) are identities in the elements of A and Δ is a function of those elements which does not vanish identically, we can in each equation cancel the highest power of Δ occurring as a common factor of both sides, and the reduced equations thus obtained are also identities in the elements of A. We have therefore established the following theorem:

Theorem I. *Any two anti-correspondent matrices of the minor determinants of two co-joint complete matrices of the minor determinants of a fundamental square matrix* $[a]^m_m$ *differ only by a scalar factor of the form* $\pm \Delta^k$, *where* $\Delta = (a)^m_m$, *and* k *is a positive integer.*

In particular when a common scheme is used for horizontal and vertical rows in the formation of the two co-joint matrices, the scalar factor has the form $+\Delta^k$.

The value of the index k in any equation given by Theorem I and the side of the equation on which the scalar factor occurs can always be determined from the degrees of both sides in the elements of the fundamental square matrix, and the scalar factor is $\pm \Delta^k$ according as the value assumed by the determinant of either one of the two co-joint matrices is ± 1 in the special case when $[a]^m_m$ is a unit matrix.

Clearly (C) is equivalent to (B), and (D) is equivalent to (A). Also each of the equations (A) and (B) is deducible from the other.

Ex. i. If we prefix $\overline{\underline{\mathbf{b}'}}^\nu_\nu$ and postfix $[\mathbf{B}']^\nu_\nu$ on both sides of the first of equations (5), we obtain

$$\overline{\underline{\mathbf{b}'}}^\nu_\nu [\mathbf{b}]^\nu_\nu \overline{\underline{\mathbf{B}}}^\nu_\nu [\mathbf{B}']^\nu_\nu = \Delta^s \overline{\underline{\mathbf{b}'}}^\nu_\nu [\mathbf{B}']^\nu_\nu.$$

Using (4), (4′) and (3), this is seen to be equivalent to the second of the equations (5′). Similarly the first of the equations (5′) can be deduced from the second of the equations (5) and from (4), (4′) and (3).

Ex. ii. ***Relation between any two anti-correspondent matrices of the minor determinants of a square matrix and its reciprocal.***

Let $[A]^m_m$ be the reciprocal of a square matrix $A = [a]^m_m$.

Then $[a]_m^m$ and $[A]_m^m$ are two co-joint matrices of the minor determinants of orders 1 and $m-1$ of A in the formation of which a common scheme is used for horizontal and vertical rows.

Let $[\mathbf{a}]_\nu^\nu$, $[\mathbf{a}']_\nu^\nu$, where $\nu=\binom{m}{s}$, be two co-joint complete matrices of the minor determinants of orders s and $m-s$ of $[a]_m^m$, and let $[\mathbf{A}]_\nu^\nu$, $[\mathbf{A}']_\nu^\nu$ be the similarly formed co-joint complete matrices of the minor determinants of orders s and $m-s$ of $[A]_m^m$. Then as particular cases of formulae (A) and (B) we have

$$[\mathbf{A}]_\nu^\nu = \Delta^{s-1}[\mathbf{a}']_\nu^\nu, \quad \text{.......................................(A')}$$

$$[\mathbf{A}']_\nu^\nu = \Delta^{m-s-1}[\mathbf{a}]_\nu^\nu, \quad \text{...................................(B')}$$

where $\Delta=(a)_m^m$.

Ex. iii. *The reciprocal of the reciprocal of* $[a]_m^m$ *is* $\Delta^{m-2}[a]_m^m$, *where* $\Delta=(a)_m^m$.

For if $[A]_m^m$ is the reciprocal of $[a]_m^m$, and $[\mathfrak{a}]_m^m$ is the reciprocal of $[A]_m^m$, then $[a]_m^m$ and $[\mathfrak{a}]_m^m$ are anti-correspondent matrices of the minor determinants of orders 1 and $m-1$ of $[a]_m^m$ and $[A]_m^m$. Accordingly putting $s=1$ in formula (B') we have

$$[\mathfrak{a}]_m^m = \Delta^{m-2}[a]_m^m.$$

3. *Relation between any two anti-correspondent minor determinants of two co-joint matrices.*

Let i and j be any two elements, not necessarily different, of the sequence $[1\,2\,\ldots\,\nu]$. Then from equations (A), (B), (C) and (D) we deduce that

$$\Delta^s\,\mathbf{b}'_{ij} = (b)_\mu^\mu\,\mathbf{B}_{ij} = \epsilon\Delta^\rho\,\mathbf{B}_{ij}, \quad \text{.......................(a)}$$

$$\Delta^s\,\mathbf{B}'_{ij} = (B)_\mu^\mu\,\mathbf{b}_{ij} = \epsilon\Delta^{\rho'}\,\mathbf{b}_{ij}, \quad \text{.......................(b)}$$

$$\Delta^{\mu-s}\,\mathbf{b}_{ij} = (b)_\mu^\mu\,\mathbf{B}'_{ij} = \epsilon\Delta^\rho\,\mathbf{B}'_{ij}, \quad \text{.......................(c)}$$

$$\Delta^{\mu-s}\,\mathbf{B}_{ij} = (B)_\mu^\mu\,\mathbf{b}'_{ij} = \epsilon\Delta^{\rho'}\,\mathbf{b}'_{ij}. \quad \text{.......................(d)}$$

Here $\mathbf{b}_{ij}$ and $\mathbf{B}'_{ij}$ are anti-correspondent minor determinants of orders s and $\mu-s$ of $[b]_\mu^\mu$ and $[B]_\mu^\mu$; $\mathbf{b}'_{ij}$ and $\mathbf{B}_{ij}$ are anti-correspondent minor determinants of orders $\mu-s$ and s of $[b]_\mu^\mu$ and $[B]_\mu^\mu$; and ϵ is $+1$ or -1 as in formula (B) of § 120. Formulae (c) and (d) are clearly equivalent to (b) and (a) respectively; and each of the formulae (a) and (b) is deducible from the other. These formulae lead to the following theorem, which is of course included in Theorem I.

Theorem II. *If* $[b]_\mu^\mu$ *and* $[B]_\mu^\mu$ *are any two co-joint complete matrices of the minor determinants of the fundamental square matrix* $A=[a]_m^m$, *then any two anti-correspondent minor determinants of* $[b]_\mu^\mu$ *and* $[B]_\mu^\mu$ *differ only by a scalar factor of the form* $\pm\Delta^k$, *where* $\Delta=(a)_m^m$, *and k is a positive integer.*

When in particular a common scheme is used for horizontal and vertical rows in the formation of $[b]_\mu^\mu$ *and* $[B]_\mu^\mu$, *the scalar factor has the form* $+\Delta^k$.

The value of the index k in any equation given by Theorem II and the side of that equation on which the scalar factor occurs can always be determined from a consideration of the degrees of both sides of the equation in the elements of A; and the scalar factor is $\pm\Delta^k$ according as either one of the determinants $(b)_\mu^\mu$ and $(B)_\mu^\mu$ assumes the value ± 1 in the special case when A is a unit matrix.

Ex. iv. Let $\phi=[a\,b\,c\,d]_{1234}$, $\Delta=\det\phi=(abcd)_{1234}$; and let ψ and ψ' be defined as in Ex. ii of § 119. Then by Theorem II we have

$$\begin{vmatrix} +(cd)_{34}, & +(bc)_{34}, & +(ad)_{34}, & +(ab)_{34} \\ -(cd)_{24}, & -(bc)_{24}, & -(ad)_{24}, & -(ab)_{24} \\ -(cd)_{13}, & -(bc)_{13}, & -(ad)_{13}, & -(ab)_{13} \\ +(cd)_{12}, & +(bc)_{12}, & +(ad)_{12}, & +(ab)_{12} \end{vmatrix} = \Delta \begin{vmatrix} (ac)_{14}, & (bd)_{14} \\ (ac)_{23}, & (bd)_{23} \end{vmatrix}, \qquad \text{............(6)}$$

$$\begin{vmatrix} -(cd)_{24}, & +(ac)_{24}, & -(ab)_{24} \\ +(cd)_{14}, & -(ac)_{14}, & +(ab)_{14} \\ -(cd)_{13}, & +(ac)_{13}, & -(ab)_{13} \end{vmatrix} = - \begin{vmatrix} (ac)_{12}, & (ad)_{12}, & (bc)_{12} \\ (ac)_{14}, & (ad)_{14}, & (bc)_{14} \\ (ac)_{34}, & (ad)_{34}, & (bc)_{34} \end{vmatrix}. \qquad \text{............(7)}$$

For ψ and ψ' are co-joint matrices of the minor determinants of ϕ in the formation of which a common scheme is used for horizontal and vertical rows. In (6) the determinant on the right provided with a positive sign is the anti-correspondent of the determinant on the left. In (7) the determinant on the right provided with a negative sign is the anti-correspondent of the determinant on the left. In these two equations the scalar factor Δ^k is Δ^1 and Δ^0 respectively, the index k being so chosen that the two sides of each equation have the same total degree in the elements of ϕ. We must clearly affix the factor Δ^k to the determinant of lower degree.

Ex. v. *Relation between any two anti-correspondent minor determinants of a square matrix and its reciprocal.*

Let $[A]_m^m$ be the reciprocal of the square matrix $A=[a]_m^m$, and let $\Delta=(a)_m^m$. Then $[a]_m^m$ and $[A]_m^m$ are two co-joint matrices of the minor determinants of orders 1 and $m-1$ of A in the formation of which a common scheme is used for horizontal and vertical rows. By Theorem II any two anti-correspondent minor determinants of $[a]_m^m$ and $[A]_m^m$ differ only by a scalar factor of the form Δ^k.

If $\mathbf{A}$ is any minor determinant of order s of $[A]_m^m$, and if $\mathbf{a}'$ is the anti-correspondent minor determinant of order $m-s$ of $[a]_m^m$, we have

$$\mathbf{A}=\Delta^{s-1}\mathbf{a}'. \qquad \text{..(E)}$$

We can obtain this formula by determining the index of Δ so that both sides have the same total degree in the elements of $[a]_m^m$, or we can deduce it from formula (a) or formula (b) as a particular case. It is equivalent (see Ex. iv of § 124) to the identity (A) of § 111.

Ex. vi. Let $\phi=[a\,b\,c\,d]_{1234}$, $\Phi=[A\,B\,C\,D]_{1234}$, $\psi=[\alpha\,\beta\,\gamma\,\delta]_{1234}$ be three matrices such that the second is the reciprocal of the first, and the third is the reciprocal of the second. Then ϕ and Φ are co-joint matrices of the minor determinants of orders 1 and 3 of ϕ; ϕ and ψ are anti-correspondent matrices of the minor determinants of orders 1 and 3 of the two co-joint matrices ϕ and Φ; and corresponding elements of ϕ and ψ are anti-correspondent minor determinants of orders 1 and 3 of ϕ and Φ. Accordingly if $\Delta=\det\phi=(abcd)_{1234}$, we have (as in Ex. iii)

$$[\alpha\,\beta\,\gamma\,\delta]_{1234}=\Delta^2\,[a\,b\,c\,d]_{1234},$$

i.e. $$(BCD)_{234}=\Delta^2 a_1,\quad (ACD)_{234}=-\Delta^2 b_1,\quad (ABD)_{234}=\Delta^2 c_1,\quad \text{etc.}$$

§ 122. Relations between corresponding complete matrices of the minor determinants of two co-joint matrices.

We will now generalise the result given in Ex. iii of § 120.

Theorem. *If $[b]_\mu^\mu$ and $[B]_\mu^\mu$ are any two co-joint complete matrices of the minor determinants of a fundamental square matrix $A=[a]_m^m$, then each of two similarly formed complete matrices of the minor determinants of $[b]_\mu^\mu$ and $[B]_\mu^\mu$ differs from the reciprocal of the other only by a scalar factor of the form $\pm\Delta^k$, where $\Delta=(a)_m^m$, and k is a positive integer.*

When in particular every complete matrix of minor determinants is formed with a common scheme for horizontal and vertical rows, the scalar factor has the form $+\Delta^k$.

Let $[b]_\mu^\mu$ and $[B]_\mu^\mu$ be two co-joint complete matrices of the minor determinants of orders r and $m-r$ of A; let $[\mathbf{b}]_\nu^\nu$, $[\mathbf{b}']_\nu^\nu$ be two co-joint complete matrices of the minor determinants of orders s and $\mu-s$ of $[b]_\mu^\mu$; let $[\mathbf{B}]_\nu^\nu$, $[\mathbf{B}']_\nu^\nu$ be the two similarly formed co-joint complete matrices of the minor determinants of orders s and $\mu-s$ of $[B]_\mu^\mu$; and let $[\mathbf{c}]_\nu^\nu$, $[\mathbf{c}']_\nu^\nu$, $[\mathbf{C}]_\nu^\nu$, $[\mathbf{C}']_\nu^\nu$ be the reciprocals of $[\mathbf{b}]_\nu^\nu$, $[\mathbf{b}']_\nu^\nu$, $[\mathbf{B}]_\nu^\nu$, $[\mathbf{B}']_\nu^\nu$ respectively.

From Ex. iii of § 120 we see that $[\mathbf{c}]_\nu^\nu$ differs from $[\mathbf{b}']_\nu^\nu$ by a scalar factor which is a power of $[b]_\mu^\mu$ and therefore a power of Δ, and from Theorem I of § 121 we see that $[\mathbf{b}']_\nu^\nu$ differs from $[\mathbf{B}]_\nu^\nu$ by a scalar factor which is a power of Δ. Thus $[\mathbf{c}]_\nu^\nu$ differs from $[\mathbf{B}]_\nu^\nu$ only by a scalar factor which is a power of Δ, and from this the theorem follows. We will however give the proof in greater detail.

By formula (B) of § 120 we have

$$(b)_{\mu}^{\mu} = \epsilon\Delta^{\rho}, \qquad (B)_{\mu}^{\mu} = \epsilon\Delta^{\rho'},$$

$$(\mathbf{b})_{\nu}^{\nu} = \theta\,\{(b)_{\mu}^{\mu}\}^{\sigma} = \theta\epsilon^{\sigma}\,\Delta^{\rho\sigma}, \quad (\mathbf{b}')_{\nu}^{\nu} = \theta\,\{(b)_{\mu}^{\mu}\}^{\sigma'} = \theta\epsilon^{\sigma'}\,\Delta^{\rho\sigma'},$$

$$(\mathbf{B})_{\nu}^{\nu} = \theta\,\{(B)_{\mu}^{\mu}\}^{\sigma} = \theta\epsilon^{\sigma}\,\Delta^{\rho'\sigma}, \quad (\mathbf{B}')_{\nu}^{\nu} = \theta\,\{(B)_{\mu}^{\mu}\}^{\sigma'} = \theta\epsilon^{\sigma'}\,\Delta^{\rho'\sigma'};$$

where $$\mu = \binom{m}{r},\ \nu = \binom{\mu}{s};\quad \rho + \rho' = \mu,\ \sigma + \sigma' = \nu;$$

$$\rho = \binom{m-1}{r-1},\ \rho' = \binom{m-1}{r};\quad \sigma = \binom{\mu-1}{s-1},\ \sigma' = \binom{\mu-1}{s};$$

and where ϵ is either $+1$ or -1, and θ is either $+1$ or -1.

From Ex. iii of § 120 and formulae (A), (B), (C), (D) of § 121 we have

$$(b)_{\mu}^{\mu}\,[\mathbf{c}]_{\nu}^{\nu} = (\mathbf{b})_{\nu}^{\nu}\,[\mathbf{b}']_{\nu}^{\nu}, \quad \Delta^{s}\,[\mathbf{b}']_{\nu}^{\nu} = (b)_{\mu}^{\mu}\,[\mathbf{B}]_{\nu}^{\nu}, \quad \text{..........(1)}$$

$$(B)_{\mu}^{\mu}\,[\mathbf{C}]_{\nu}^{\nu} = (\mathbf{B})_{\nu}^{\nu}\,[\mathbf{B}']_{\nu}^{\nu}, \quad \Delta^{s}\,[\mathbf{B}']_{\nu}^{\nu} = (B)_{\mu}^{\mu}\,[\mathbf{b}]_{\nu}^{\nu}, \quad \text{..........(2)}$$

$$(b)_{\mu}^{\mu}\,[\mathbf{c}']_{\nu}^{\nu} = (\mathbf{b}')_{\nu}^{\nu}\,[\mathbf{b}]_{\nu}^{\nu}, \quad \Delta^{\mu-s}\,[\mathbf{b}]_{\nu}^{\nu} = (b)_{\mu}^{\mu}\,[\mathbf{B}']_{\nu}^{\nu}, \quad \text{..........(3)}$$

$$(B)_{\mu}^{\mu}\,[\mathbf{C}']_{\nu}^{\nu} = (\mathbf{B}')_{\nu}^{\nu}\,[\mathbf{B}]_{\nu}^{\nu}, \quad \Delta^{\mu-s}\,[\mathbf{B}]_{\nu}^{\nu} = (B)_{\mu}^{\mu}\,[\mathbf{b}']_{\nu}^{\nu}; \quad \text{..........(4)}$$

and from these results it follows that

$$\Delta^{s}\,[\mathbf{c}]_{\nu}^{\nu} = (\mathbf{b})_{\nu}^{\nu}\,[\mathbf{B}]_{\nu}^{\nu} = \theta\epsilon^{\sigma}\,\Delta^{\rho\sigma}\,[\mathbf{B}]_{\nu}^{\nu}, \quad \text{...............(A)}$$

$$\Delta^{s}\,[\mathbf{C}]_{\nu}^{\nu} = (\mathbf{B})_{\nu}^{\nu}\,[\mathbf{b}]_{\nu}^{\nu} = \theta\epsilon^{\sigma}\,\Delta^{\rho'\sigma}\,[\mathbf{b}]_{\nu}^{\nu}, \quad \text{...............(B)}$$

$$\Delta^{\mu-s}\,[\mathbf{c}']_{\nu}^{\nu} = (\mathbf{b}')_{\nu}^{\nu}\,[\mathbf{B}']_{\nu}^{\nu} = \theta\epsilon^{\sigma'}\,\Delta^{\rho\sigma'}\,[\mathbf{B}']_{\nu}^{\nu}, \quad \text{...............(C)}$$

$$\Delta^{\mu-s}\,[\mathbf{C}']_{\nu}^{\nu} = (\mathbf{B}')_{\nu}^{\nu}\,[\mathbf{b}']_{\nu}^{\nu} = \theta\epsilon^{\sigma'}\,\Delta^{\rho'\sigma'}\,[\mathbf{b}']_{\nu}^{\nu}. \quad \text{...............(D)}$$

Equations (A), (B), (C), (D), all of which are deducible from any one of them, show that the theorem given above is true in the general case, for $\theta\epsilon^{\sigma}$ is either $+1$ or -1. It is also true in the particular case, for then $\epsilon = +1$ and $\theta = +1$.

The index k of the scalar factor $\pm\Delta^{k}$ in any equation given by the theorem can always be determined by equating the degrees of both sides in the elements of the fundamental matrix.

Ex. i. *Relation between any two corresponding complete matrices of the minor determinants of a square matrix and its reciprocal.*

Let $[A]_{m}^{m}$ be the reciprocal of the square matrix $A = [a]_{m}^{m}$, and let $(a)_{m}^{m} = \Delta$, so that $(A)_{m}^{m} = \Delta^{m-1}$. Then as a particular case of the theorem of the text we see that each of any two similarly formed complete matrices of the minor determinants of $[a]_{m}^{m}$ and $[A]_{m}^{m}$ differs from the reciprocal of the other only by a scalar factor of the form $\pm\Delta^{k}$, and that the upper sign must be taken when in the formation of these matrices a common scheme is used for horizontal and vertical rows.

Let $[\mathbf{a}]_\nu^\nu$, $[\mathbf{a}']_\nu^\nu$ be any two co-joint complete matrices of the minor determinants of orders s and $m-s$ of $[a]_m^m$; let $[\mathbf{A}]_\nu^\nu$, $[\mathbf{A}']_\nu^\nu$ be the similarly formed co-joint complete matrices of the minor determinants of orders s and $m-s$ of $[A]_m^m$; and let $[\mathbf{c}]_\nu^\nu$, $[\mathbf{c}']_\nu^\nu$, $[\mathbf{C}]_\nu^\nu$, $[\mathbf{C}']_\nu^\nu$ be the reciprocals of $[\mathbf{a}]_\nu^\nu$, $[\mathbf{a}']_\nu^\nu$, $[\mathbf{A}]_\nu^\nu$, $[\mathbf{A}']_\nu^\nu$ respectively. Then putting

$$r=1,\quad \mu=m,\quad \rho=1,\quad \rho'=m-1,\quad \epsilon=+1,\quad \nu=\binom{m}{s},$$

we have $(\mathbf{a})_\nu^\nu=\theta\Delta^\sigma$, $(\mathbf{a}')_\nu^\nu=\theta\Delta^{\sigma'}$, $(\mathbf{A})_\nu^\nu=\theta\Delta^{(m-1)\sigma}$, $(\mathbf{A}')_\nu^\nu=\theta\Delta^{(m-1)\sigma'}$,

where $\sigma=\binom{m-1}{s-1}$, $\sigma'=\binom{m-1}{s}$, and θ is either $+1$ or -1; and equations (1), (2), (3), (4) and (A), (B), (C), (D) become

$$\Delta[\mathbf{c}]_\nu^\nu=(\mathbf{a})_\nu^\nu[\mathbf{a}']_\nu^\nu,\qquad \Delta^s[\mathbf{a}']_\nu^\nu=\Delta[\mathbf{A}]_\nu^\nu, \quad \text{.....................(1')}$$

$$\Delta^{m-1}[\mathbf{C}]_\nu^\nu=(\mathbf{A})_\nu^\nu[\mathbf{A}']_\nu^\nu,\qquad \Delta^s[\mathbf{A}']_\nu^\nu=\Delta^{m-1}[\mathbf{a}]_\nu^\nu, \quad \text{.................(2')}$$

$$\Delta[\mathbf{c}']_\nu^\nu=(\mathbf{a}')_\nu^\nu[\mathbf{a}]_\nu^\nu,\qquad \Delta^{m-s}[\mathbf{a}]_\nu^\nu=\Delta[\mathbf{A}']_\nu^\nu, \quad \text{.....................(3')}$$

$$\Delta^{m-1}[\mathbf{C}']_\nu^\nu=(\mathbf{A}')_\nu^\nu[\mathbf{A}]_\nu^\nu,\qquad \Delta^{m-s}[\mathbf{A}]_\nu^\nu=\Delta^{m-1}[\mathbf{a}']_\nu^\nu; \quad \text{..............(4')}$$

$$\Delta^s[\mathbf{c}]_\nu^\nu=(\mathbf{a})_\nu^\nu[\mathbf{A}]_\nu^\nu=\theta\Delta^\sigma[\mathbf{A}]_\nu^\nu, \quad \text{..............................(A')}$$

$$\Delta^s[\mathbf{C}]_\nu^\nu=(\mathbf{A})_\nu^\nu[\mathbf{a}]_\nu^\nu=\theta\Delta^{(m-1)\sigma}[\mathbf{a}]_\nu^\nu, \quad \text{.......................(B')}$$

$$\Delta^{m-s}[\mathbf{c}']_\nu^\nu=(\mathbf{a}')_\nu^\nu[\mathbf{A}']_\nu^\nu=\theta\Delta^{\sigma'}[\mathbf{A}']_\nu^\nu, \quad \text{..........................(C')}$$

$$\Delta^{m-s}[\mathbf{C}']_\nu^\nu=(\mathbf{A}')_\nu^\nu[\mathbf{a}']_\nu^\nu=\theta\Delta^{(m-1)\sigma'}[\mathbf{a}']_\nu^\nu. \quad \text{...................(D')}$$

Ex. ii. *Alternative proof of Theorem II of* § 73.

Let $[a]_m^n$ be a matrix of rank r; let $\Delta=(a_{pq})_r^r$ be one of its non-vanishing derived determinants of order r; and let $[A_{pq}]_r^r$ be the reciprocal of $[a_{pq}]_r^r$. Then by § 115 or § 116 we have

$$(a_{pq})_r^r\,[a]_m^n=[a_{1q}]_m^r\,\overline{\underline{A_{pq}}}_r^r\,[a_{p1}]_r^n. \quad \text{..............................(E)}$$

Now let $[\mathbf{a}]_\mu^\nu$ be any complete matrix of the minor determinants of order s of $[a]_m^n$, where $s \not> r$. Then $[\mathbf{a}]_\mu^\nu$ has a minor matrix $[\mathbf{a}_{uv}]_\rho^\rho$ which is a complete matrix of the minor determinants of order s of $[a_{pq}]_r^r$.

Here $$\mu=\binom{m}{s},\quad \nu=\binom{n}{s},\quad \rho=\binom{r}{s};$$

also $\rho \not> \mu$, $\rho \not> \nu$, because $r \not> m$, $r \not> n$.

If $[\mathbf{A}_{uv}]_\rho^\rho$ is the reciprocal of $[\mathbf{a}_{uv}]_\rho^\rho$, we will deduce from (E) the equation

$$(\mathbf{a}_{uv})_\rho^\rho\,[\mathbf{a}]_\mu^\nu=[\mathbf{a}_{1v}]_\mu^\rho\,\overline{\underline{\mathbf{A}_{uv}}}_\rho^\rho\,[\mathbf{a}_{u1}]_\rho^\nu. \quad \text{..........................(F)}$$

Equating correspondingly formed complete matrices of the minor determinants of order s on both sides of (E), we obtain

$$\Delta^s[\mathbf{a}]_\mu^\nu = [\mathbf{a}_{1v}]_\mu^\rho \boxed{\mathbf{b}_{uv}}_\rho^\rho [\mathbf{a}_{u1}]_\rho^\nu, \quad \text{.................................(5)}$$

where $[\mathbf{b}_{uv}]_\rho^\rho$ is a complete matrix of the minor determinants of order s of $[A_{pq}]_r^r$ formed from $[A_{pq}]_r^r$ in the same way as $[\mathbf{a}_{uv}]_\rho^\rho$ is formed from $[a_{pq}]_r^r$.

Now by the theorem of the text $[\mathbf{b}_{uv}]_\rho^\rho$ and $[\mathbf{A}_{uv}]_\rho^\rho$ differ only by a scalar factor of the form $\pm\Delta^k$, and by formula (A′) or formula (A) we have

$$\Delta^s[\mathbf{A}_{uv}]_\rho^\rho = (\mathbf{a}_{uv})_\rho^\rho [\mathbf{b}_{uv}]_\rho^\rho. \quad \text{.................................(6)}$$

Substituting from (6) in (5) we obtain (F).

By § 120 we have $(\mathbf{a}_{uv})_\rho^\rho = \pm\Delta^\sigma \neq 0$, where $\sigma = \binom{r-1}{s-1}$.

Hence all three factors on the right in (F) have rank ρ, and it follows by Ex. xi of § 71 that $[\mathbf{a}]_\mu^\nu$ has rank ρ or $\binom{r}{s}$.

The method by which (F) has been derived from (E) remains valid when $(a_{pq})_r^r = \Delta = 0$. We can also deduce (F) from the equation (C) of § 115.

§ 123. Expansions of certain bordered determinants in terms of the simple minor determinants of the bordering rows.

Let
$$\Omega = \begin{vmatrix} a_{11} & a_{12} & \cdots & a_{1m} & y_{11} & y_{12} & \cdots & y_{1r} \\ a_{21} & a_{22} & \cdots & a_{2m} & y_{21} & y_{22} & \cdots & y_{2r} \\ \cdots & \cdots & \cdots & \cdots & \cdots & \cdots & \cdots & \cdots \\ a_{m1} & a_{m2} & \cdots & a_{mm} & y_{m1} & y_{m2} & \cdots & y_{mr} \\ x_{11} & x_{12} & \cdots & x_{1m} & 0 & 0 & \cdots & 0 \\ x_{21} & x_{22} & \cdots & x_{2m} & 0 & 0 & \cdots & 0 \\ \cdots & \cdots & \cdots & \cdots & \cdots & \cdots & \cdots & \cdots \\ x_{r1} & x_{r2} & \cdots & x_{rm} & 0 & 0 & \cdots & 0 \end{vmatrix} = \begin{pmatrix} a, & y \\ x, & 0 \end{pmatrix}_{m,\,r}^{m,\,r},$$

so that Ω is a determinant formed from the determinant

$$\Delta = (a)_m^m$$

by bordering it with r additional final horizontal rows and r additional final vertical rows of the character shown; and let $[A]_m^m$ be the reciprocal of $[a]_m^m$.

Proceeding to evaluate the determinant Ω we shall obtain the following results:

Case I. *If $r > m$, then $\Omega = 0$.*

Case II. *If $r = m$, then $\Omega = (-1)^m (x)_m^m (y)_m^m$.*

CASE III. *If $r \not> m$, then*

$$(1)\quad \Omega = (-1)^r \det[X_1\, X_2 \ldots X_\mu]\, \overline{\underline{\mathbf{a}'}}^{\mu}_{\mu} \begin{bmatrix} Y_1 \\ Y_2 \\ \vdots \\ Y_\mu \end{bmatrix} = (-1)^r \det [X]_\mu\, \overline{\underline{\mathbf{a}'}}^{\mu}_{\mu}\, \overline{\underline{Y}}_{\mu}\,.$$

$$(2)\quad \Delta^{r-1}\Omega = (-1)^r \det[x]^m_r\, \overline{\underline{A}}^m_m\, [y]^r_m = (-1)^r \det [X]_\mu\, \overline{\underline{\mathbf{A}}}^{\mu}_{\mu}\, \overline{\underline{Y}}_{\mu}\,.$$

$$(3)\quad \epsilon\Delta^{\rho-1}\Omega = (-1)^{r-1} \begin{vmatrix} \mathbf{a}_{11} & \mathbf{a}_{12} & \ldots & \mathbf{a}_{1\mu} & Y_1 \\ \mathbf{a}_{21} & \mathbf{a}_{22} & \ldots & \mathbf{a}_{2\mu} & Y_2 \\ \cdots & \cdots & \cdots & \cdots & \cdots \\ \mathbf{a}_{\mu 1} & \mathbf{a}_{\mu 2} & \ldots & \mathbf{a}_{\mu\mu} & Y_\mu \\ X_1 & X_2 & \ldots & X_\mu & 0 \end{vmatrix} = (-1)^r \det [X]_\mu\, \overline{\underline{\mathbf{c}}}^{\mu}_{\mu}\, \overline{\underline{Y}}_{\mu}\,.$$

Here $[X]_\mu$ and $\overline{\underline{Y}}_\mu$ are any complete matrices of the simple minor determinants of $[x]^m_r$ and $[y]^r_m$; $[\mathbf{a}]^\mu_\mu$ and $[\mathbf{A}]^\mu_\mu$ are complete matrices of the minor determinants of order r of $[a]^m_m$ and $[A]^m_m$ in the formation of which the successive selections of vertical and horizontal rows follow the same schemes as in the formation of $[X]_\mu$ and $\overline{\underline{Y}}_\mu$ from $[x]^m_r$ and $[y]^r_m$; $[\mathbf{a}']^\mu_\mu$ is the co-joint matrix of $[\mathbf{a}]^\mu_\mu$ with respect to $[a]^m_m$; and $[\mathbf{c}]^\mu_\mu$ is the reciprocal of $[\mathbf{a}]^\mu$. Further $\mu = \binom{m}{r}$, $\rho = \binom{m-1}{r-1}$; and ϵ is either $+1$ or -1, being so chosen that $(\mathbf{a})^\mu_\mu = \epsilon\Delta^\rho$.

The results given in Cases I and II are obtained immediately when we expand Ω in terms of the simple minor determinants of its last r horizontal rows or its last r vertical rows. Accordingly we can confine ourselves to Case III.

In the formation of $[X]_\mu$ from $[x]^m_r$ let the successive selections of vertical rows of $[x]^m_r$ follow the scheme $(T_1, T_2, \ldots T_q, \ldots T_\mu)$, the horizontal rows being underanged; and in the formation of $\overline{\underline{Y}}_\mu$ from $[y]^r_m$ let the successive selections of horizontal rows of $[y]^r_m$ follow the scheme $(S_1, S_2, \ldots S_p, \ldots S_\mu)$, the vertical rows being underanged.

Then $(S_1, S_2, \ldots S_p, \ldots S_\mu)$ and $(T_1, T_2, \ldots T_q, \ldots T_\mu)$ are any two sets of μ distinct minors of order r of the sequence $[1\,2 \ldots m]$.

Let $S_p = [p_1\, p_2 \ldots p_r]$, $T_q = [q_1\, q_2 \ldots q_r]$; let $[u]_{m-r}$, $[v]_{m-r}$ be the corranged complements of $[p]_r$ and $[q]_r$ in $[1\,2\ldots m]$; and let ω_1, ω_2 be the affects of $[p]_r, [q]_r$ in $[1\,2\ldots m]$, so that

$$\omega = \omega_1 + \omega_2 = \text{affect of } (a_{pq})_r^r \text{ in } [a]_m^m.$$

Then

$$X_q = (x_{1q})_r^r, \quad Y_p = (y_{p1})_r^r, \quad \mathbf{a}_{pq} = (a_{pq})_r^r, \quad \mathbf{a}'_{pq} = (-1)^\omega (a_{uv})_{m-r}^{m-r}.$$

Expanding Ω in terms of the minor determinants of order r belonging to its last r vertical rows, we obtain

$$\Omega = \sum_p (-1)^h Y_p \xi_p, \quad \ldots\ldots\ldots\ldots\ldots\ldots(1)$$

where h = affect of Y_p or $(y_{p1})_r^r$ in $\Omega = \omega_1 + rm$,

$$\xi_p = \text{corranged complement of } Y_p \text{ in } \Omega = \begin{pmatrix} a_{u1} \\ x \end{pmatrix}_{m-r,\, r}^{m},$$

and $\sum_p$ means summation for all the values $S_1, S_2, \ldots S_\mu$ of $[p]_r$.

Again expanding ξ_p in terms of the minor determinants of order r belonging to its last r horizontal rows, we obtain

$$\xi_p = \sum_q (-1)^k X_q (a_{uv})_{m-r}^{m-r}, \quad \ldots\ldots\ldots\ldots\ldots\ldots(2)$$

where k = affect of X_q or $(x_{1q})_r^r$ in $\xi_p = \omega_2 + r(m-r)$, and $\sum_q$ means summation for all the values $T_1, T_2, \ldots T_\mu$ of $[q]_r$.

Since $h + k = \omega_1 + \omega_2 - r^2 \equiv \omega + r$ (mod. 2), we see that

$$\begin{aligned}\Omega &= \sum_p \sum_q (-1)^{\omega + r} X_q Y_p (a_{uv})_{m-r}^{m-r} \\ &= (-1)^r \sum_p \sum_q X_q Y_p \mathbf{a}'_{pq} = (-1)^r \det [X]_\mu \, \overline{\underline{\mathbf{a}'}}_\mu^\mu \, \overline{\underline{Y}}_\mu .\end{aligned}$$

This proves the first result in Case III.

We can deduce the second result from the first by using the relations

$$\mathbf{A}_{pq} = \Delta^{r-1} \mathbf{a}'_{pq}, \quad [\mathbf{A}]_\mu^\mu = \Delta^{r-1} [\mathbf{a}']_\mu^\mu$$

given by Ex. ii or Ex. v of § 121, or we can deduce it from the equation

$$\Omega = \sum_p \sum_q (-1)^{\omega+r} (x_{1q})_r^r \, (a_{uv})_{m-r}^{m-r} (y_{p1})_r^r$$

by means of the equation

$$(A_{pq})_r^r = \Delta^{r-1} . (-1)^\omega (a_{uv})_{m-r}^{m-r} .$$

We deduce the third result from the first by using the equation

$$\Delta [\mathbf{c}]_\mu^\mu = (\mathbf{a})_\mu^\mu [\mathbf{a}']_\mu^\mu, \quad \text{or} \quad \epsilon \Delta^{\rho - 1} [\mathbf{a}']_\nu^\nu = [\mathbf{c}]_\mu^\mu$$

given by Ex. iii of § 120, and the equation

$$\begin{vmatrix} \mathbf{a}_{11} & \mathbf{a}_{12} & \dots & \mathbf{a}_{1\mu} & Y_1 \\ \mathbf{a}_{21} & \mathbf{a}_{22} & \dots & \mathbf{a}_{2\mu} & Y_2 \\ \dots & \dots & \dots & \dots & \dots \\ \mathbf{a}_{\mu 1} & \mathbf{a}_{\mu 2} & \dots & \mathbf{a}_{\mu\mu} & Y_\mu \\ X_1 & X_2 & \dots & X_\mu & 0 \end{vmatrix} = -\det[X]_\mu \,\overline{\underline{\mathbf{c}}}^{\mu}_{\mu}\, \overline{\underline{Y}}_{\mu},$$

which is the particular case of the first result proved in Ex. xi of § 62.

The first result gives the expansion of Ω in terms of the determinants $X_1, X_2, \dots X_\mu, Y_1, Y_2, \dots Y_\mu$ in all cases, both when $\Delta \neq 0$ and when $\Delta = 0$. The second and third results give the expansion only when $\Delta \neq 0$ in general, but they give the expansion in all cases when $r = 1$.

Ex. i. In the special case when $r=1$ and therefore $\mu = m$, we have

$$[\mathbf{a}]^m_m = [a]^m_m, \quad [\mathbf{a}']^m_m = [A]^m_m, \quad [\mathbf{c}]^m_m = [A]^m_m,$$

and the three formulae obtained all have the common form

$$\begin{vmatrix} a_{11} & a_{12} & \dots & a_{1m} & y_1 \\ a_{21} & a_{22} & \dots & a_{2m} & y_2 \\ \dots & \dots & \dots & \dots & \dots \\ a_{m1} & a_{m2} & \dots & a_{mm} & y_m \\ x_1 & x_2 & \dots & x_m & 0 \end{vmatrix} = -\det[x_1\, x_2 \dots x_m] \begin{bmatrix} A_{11} & A_{21} & \dots & A_{m1} \\ A_{12} & A_{22} & \dots & A_{m2} \\ \dots & \dots & \dots & \dots \\ A_{1m} & A_{2m} & \dots & A_{mm} \end{bmatrix} \begin{bmatrix} x_1 \\ x_2 \\ \vdots \\ x_m \end{bmatrix}.$$

This is the special case considered in Ex. xi of § 62.

Ex. ii. Let the notation of § 129 be used, and let

$$U = \begin{vmatrix} a & h & g & u & \lambda_1 & \lambda_2 \\ h & b & f & v & \mu_1 & \mu_2 \\ g & f & c & w & \nu_1 & \nu_2 \\ u & v & w & d & \pi_1 & \pi_2 \\ l_1 & m_1 & n_1 & p_1 & 0 & 0 \\ l_2 & m_2 & n_2 & p_2 & 0 & 0 \end{vmatrix}, \quad V = \begin{vmatrix} a & h & g & u & \lambda_1 & \lambda_2 & \lambda_3 \\ h & b & f & v & \mu_1 & \mu_2 & \mu_3 \\ g & f & c & w & \nu_1 & \nu_2 & \nu_3 \\ u & v & w & d & \pi_1 & \pi_2 & \pi_3 \\ l_1 & m_1 & n_1 & p_1 & 0 & 0 & 0 \\ l_2 & m_2 & n_2 & p_2 & 0 & 0 & 0 \\ l_3 & m_3 & n_3 & p_3 & 0 & 0 & 0 \end{vmatrix}.$$

Then if $[\xi]_6$, $[\eta]_6$ are the complete matrices of the unaffected simple minor determinants of $[l\,m\,n\,p]_{12}$ and $[\lambda\,\mu\,\nu\,\pi]_{12}$ formed according to the scheme (23, 31, 12, 14, 24, 34); and if $[X]_4$, $[Y]_4$ are the complete matrices of the affected simple minor determinants of $[l\,m\,n\,p]_{123}$, $[\lambda\,\mu\,\nu\,\pi]_{123}$ formed according to the scheme (234, 134, 124, 123); we have

$$U = \det[\xi]_6 \begin{bmatrix} A_2 & H_2 & G_2 & U_0 & W_2 & V_1 \\ H_2 & B_2 & F_2 & W_1 & V_0 & U_2 \\ G_2 & F_2 & C_2 & V_2 & U_1 & W_0 \\ U_0 & W_1 & V_2 & A_1 & H_1 & G_1 \\ W_2 & V_0 & U_1 & H_1 & B_1 & F_1 \\ V_1 & U_2 & W_0 & G_1 & F_1 & C_1 \end{bmatrix} \overline{\underline{\eta}}_6, \quad \Delta^2 U = -\begin{vmatrix} A_1 & H_1 & G_1 & U_0 & W_1 & V_2 & \eta_1 \\ H_1 & B_1 & F_1 & W_2 & V_0 & U_1 & \eta_2 \\ G_1 & F_1 & C_1 & V_1 & U_2 & W_0 & \eta_3 \\ U_0 & W_2 & V_1 & A_2 & H_2 & G_2 & \eta_4 \\ W_1 & V_0 & U_2 & H_2 & B_2 & F_2 & \eta_5 \\ V_2 & U_1 & W_0 & G_2 & F_2 & C_2 & \eta_6 \\ \xi_1 & \xi_2 & \xi_3 & \xi_4 & \xi_5 & \xi_6 & 0 \end{vmatrix};$$

$$V = -\det [X]_4 \begin{bmatrix} a & h & g & u \\ h & b & f & v \\ g & f & c & w \\ u & v & w & d \end{bmatrix} \overline{\underline{Y}}_4, \qquad \Delta^2 V = \begin{vmatrix} A & H & G & U & Y_1 \\ H & B & F & V & Y_2 \\ G & F & C & W & Y_3 \\ U & V & W & D & Y_4 \\ X_1 & X_2 & X_3 & X_4 & 0 \end{vmatrix}.$$

Ex. iii. Let $[b]_{\mu}^{\mu}$ and $[B]_{\mu}^{\mu}$ be any two co-joint matrices of the minor determinants of orders r and $m-r$ of the square matrix $A=[a]_m^m$, let $\Delta=(a)_m^m$, and let

$$\Omega = \begin{pmatrix} b, & y \\ x, & 0 \end{pmatrix}_{\mu, s}^{\mu, s}.$$

Then if $s \not> \mu$, we have the following results which are deducible from and include those of the text:

$$(1) \qquad \Omega = (-1)^s \det [X]_1^{\nu} \overline{\underline{\mathbf{b}'}}_{\nu}^{\nu} [Y]_{\nu}^1.$$

$$(2) \qquad \Delta^s \Omega = (-1)^s (b)_{\mu}^{\mu} \det [x]_s^{\mu} \overline{\underline{B}}_{\mu}^{\mu} [y]_{\mu}^s = (-1)^s (b)_{\mu}^{\mu} \det [X]_1^{\nu} \overline{\underline{\mathbf{B}}}_{\nu}^{\nu} [Y]_{\nu}^1.$$

$$(3) \quad (\mathbf{b})_{\nu}^{\nu}\, \Omega = (-1)^{s-1} (b)_{\mu}^{\mu} \begin{pmatrix} \mathbf{b}, & Y \\ X, & 0 \end{pmatrix}_{\nu, 1}^{\nu, 1} = (-1)^s (b)_{\mu}^{\mu} \det [X]_1^{\nu} \overline{\underline{\mathbf{c}}}_{\nu}^{\nu} [Y]_{\nu}^1.$$

Here $[X]_1^{\nu}$, $[Y]_{\nu}^1$ are complete matrices of the simple minor determinants of order s of $[x]_s^{\mu}$ and $[y]_{\mu}^s$; $[\mathbf{b}]_{\nu}^{\nu}$ and $[\mathbf{B}]_{\nu}^{\nu}$ are complete matrices of the minor determinants of order s of $[b]_{\mu}^{\mu}$ and $[B]_{\mu}^{\mu}$ having the same schemes of formation for vertical and horizontal rows as $[X]_1^{\nu}$ and $[Y]_{\nu}^1$; $[\mathbf{b}']_{\nu}^{\nu}$ is the co-joint matrix of $[\mathbf{b}]_{\nu}^{\nu}$ with respect to $[b]_{\mu}^{\mu}$; and $[\mathbf{c}]_{\nu}^{\nu}$ is the reciprocal of $[\mathbf{b}]_{\nu}^{\nu}$.

We have $\mu = \binom{m}{r}$, $\nu = \binom{\mu}{s}$; and if $\rho = \binom{m-1}{r-1}$, $\sigma = \binom{\mu-1}{s-1}$, we have

$$(b)_{\mu}^{\mu} = \epsilon \Delta^{\rho}, \quad (\mathbf{b})_{\nu}^{\nu} = \theta \left\{(b)_{\mu}^{\mu}\right\}^{\sigma} = \theta \epsilon^{\sigma} \Delta^{\rho\sigma},$$

where ϵ is $+1$ or -1, and θ is $+1$ or -1.

In the special case when $s=1$, we have

$$\Omega = \begin{pmatrix} b, & y \\ x, & 0 \end{pmatrix}_{\mu, 1}^{\mu, 1} = -\det [x]_1^{\mu} \overline{\underline{c}}_{\mu}^{\mu} [y]_{\mu}^1,$$

$$\Delta\Omega = -(b)_{\mu}^{\mu} \det [x]_1^{\mu} \overline{\underline{B}}_{\mu}^{\mu} [y]_{\mu}^1 = -\epsilon \Delta^{\rho} \det [x]_1^{\mu} \overline{\underline{B}}_{\mu}^{\mu} [y]_{\mu}^1,$$

where $[c]_{\mu}^{\mu}$ is the reciprocal of $[b]_{\mu}^{\mu}$.

§ 124. Properties of the reciprocal of a square matrix.

The present article contains a recapitulation and independent proofs of the properties of the reciprocal of a square matrix. We shall take the square matrix to be $A=[a]_m^m$; we shall denote the reciprocal of $[a]_m^m$ by $[A]_m^m$; and we shall write $(a)_m^m = \Delta$. Since $[a]_m^m$ and $[A]_m^m$ are co-joint complete matrices of the minor determinants of orders 1 and $m-1$ of A, the properties of the reciprocal matrix $[A]_m^m$ are included amongst the properties of co-joint matrices given in the immediately preceding articles.

1. *Fundamental property of the reciprocal matrix.*

The reciprocal matrix is defined in § 28. Its fundamental property is that given by the equations

$$[a]_m^m \,\underline{\overline{A}}_m^m = \underline{\overline{a}}_m^m \,[A]_m^m = \underline{\overline{A}}_m^m \,[a]_m^m = [A]_m^m \,\underline{\overline{a}}_m^m = \Delta\,[1]_m^m. \quad \text{............(A)}$$

These equations are equivalent to those given in Ex. i of § 28.

2. *Value of the determinant of the reciprocal matrix.*

Equating the determinants of both sides in any one of the equations (A), we obtain

$$\Delta\,(A)_m^m = \Delta^m.$$

Since this equation is an identity in the elements of A, and Δ is a function of those elements which does not vanish identically, it follows that

$$(A)_m^m = \Delta^{m-1}. \quad \text{..(B)}$$

This last equation is an identity in the elements of A, and therefore remains true when $\Delta = 0$.

3. *Reciprocal of the reciprocal matrix.*

Let $[a]_m^m$ be the reciprocal matrix of $[a]_m^m$. Equating the reciprocals (or correspondingly formed complete matrices of the minor determinants of order $m-1$) on both sides of the fundamental equation $\underline{\overline{a}}_m^m \,[A]_m^m = \Delta\,[1]_m^m$, we obtain

$$\underline{\overline{A}}_m^m \,[a]_m^m = \Delta^{m-1}\,[1]_m^m.$$

Prefixing $[a]_m^m$ on both sides, and making use of (A), we deduce that

$$\Delta\,[a]_m^m = \Delta^{m-1}\,[a]_m^m.$$

Cancelling the common factor Δ on both sides as in sub-article 2, we obtain the final result

$$[a]_m^m = \Delta^{m-2}\,[a]_m^m. \quad \text{..(C)}$$

This equation is an identity in the elements of A, and therefore remains true when $\Delta = 0$.

4. *Relations between the minor determinants of the reciprocal matrix and the anti-correspondent minor determinants of the fundamental matrix.*

Let $[\mathbf{A}]_\mu^\mu$, where $\mu = \binom{m}{r}$, be any complete matrix of the minor determinants of order r of $[A]_m^m$; let $[\mathbf{a}]_\mu^\mu$ be the similarly formed complete matrix of the minor determinants of order r of $[a]_m^m$; and let $[\mathbf{a}']_\mu^\mu$ be the matrix each of whose elements is the co-factor in $[a]_m^m$ of the corresponding element of $[\mathbf{a}]_\mu^\mu$. Then $[\mathbf{a}']_\mu^\mu$ is the co-joint matrix of $[\mathbf{a}]_\mu^\mu$ with respect to $[a]_m^m$, and is a complete matrix of the minor determinants of order $m-r$ of $[a]_m^m$.

The matrices $[\mathbf{A}]_{\mu}^{\mu}$, $[\mathbf{a}]_{\mu}^{\mu}$ are corresponding complete matrices of the minor determinants of order r of $[A]_{m}^{m}$, $[a]_{m}^{m}$.

The matrices $[\mathbf{A}]_{\mu}^{\mu}$, $[\mathbf{a}']_{\mu}^{\mu}$ are anti-correspondent complete matrices of the minor determinants of orders r, $m-r$ of $[A]_{m}^{m}$, $[a]_{m}^{m}$.

If $\mathbf{A}_{pq}$, a minor determinant of order r of $[A]_{m}^{m}$, is any element of $[\mathbf{A}]_{\mu}^{\mu}$, then $\mathbf{a}_{pq}$ is the corresponding minor determinant of order r of $[a]_{m}^{m}$, and $\mathbf{a}'_{pq}$ is the anti-correspondent minor determinant of order $m-r$ of $[a]_{m}^{m}$.

Equating correspondingly formed complete matrices of the minor determinants of order r on both sides of the fundamental equation $\overline{\underline{a}}\,_{m}^{m}\,[A]_{m}^{m}=\Delta\,[1]_{m}^{m}$, we obtain

$$\overline{\underline{\mathbf{a}}}\,_{\mu}^{\mu}\,[\mathbf{A}]_{\mu}^{\mu}=\Delta^{r}\,[1]_{\mu}^{\mu}.$$

Prefixing $[\mathbf{a}']_{\mu}^{\mu}$ on both sides, and observing that $[\mathbf{a}']_{\mu}^{\mu}\,\overline{\underline{\mathbf{a}}}\,_{\mu}^{\mu}=\Delta\,[1]_{\mu}^{\mu}$, as shown in § 119.2, we deduce that

$$\Delta\,[\mathbf{A}]_{\mu}^{\mu}=\Delta^{r}\,[\mathbf{a}']_{\mu}^{\mu}.$$

Cancelling the factor Δ common to both sides as in sub-article 2, we have

$$[\mathbf{A}]_{\mu}^{\mu}=\Delta^{r-1}[\mathbf{a}']_{\mu}^{\mu},\quad \mathbf{A}_{pq}=\Delta^{r-1}\mathbf{a}'_{pq}. \qquad \text{(D)}$$

These equations are identities in the elements of A and remain true when $\Delta=0$. They show that:

Every complete matrix of the minor determinants of order r of $[A]_{m}^{m}$ is equal to the anti-correspondent complete matrix of the minor determinants of order $m-r$ of $[a]_{m}^{m}$ multiplied by the scalar factor Δ^{r-1}.

Every minor determinant of order r of the reciprocal matrix $[A]_{m}^{m}$ is equal to the anti-correspondent minor determinant of order $m-r$ of the fundamental matrix $[a]_{m}^{m}$ multiplied by the scalar factor Δ^{r-1}.

The relations (D) are, as shown in Ex. iv, equivalent to the identities which were obtained in § 111.

5. *Relations between two corresponding complete matrices of the minor determinants of a square matrix and its reciprocal.*

Let $[\mathbf{a}]_{\mu}^{\mu}$, $[\mathbf{A}]_{\mu}^{\mu}$ be two corresponding complete matrices of the minor determinants of order r of $[a]_{m}^{m}$, $[A]_{m}^{m}$; and let $[\mathbf{c}]_{\mu}^{\mu}$, $[\mathbf{C}]_{\mu}^{\mu}$ be the reciprocals of $[\mathbf{a}]_{\mu}^{\mu}$, $[\mathbf{A}]_{\mu}^{\mu}$.

Equating corresponding matrices of the minor determinants of order r on both sides of the equation $\overline{\underline{a}}\,_{m}^{m}\,[A]_{m}^{m}=\Delta\,[1]_{m}^{m}$, and using the fundamental properties of the reciprocal matrices $[\mathbf{c}]_{\mu}^{\mu}$, $[\mathbf{C}]_{\mu}^{\mu}$, we have

$$\overline{\underline{\mathbf{a}}}\,_{\mu}^{\mu}\,[\mathbf{A}]_{\mu}^{\mu}=\Delta^{r}\,[1]_{\mu}^{\mu},\qquad \overline{\underline{\mathbf{a}}}\,_{\mu}^{\mu}\,[\mathbf{c}]_{\mu}^{\mu}=(\mathbf{a})_{\mu}^{\mu}[1]_{\mu}^{\mu},$$

$$\overline{\underline{\mathbf{A}}}\,_{\mu}^{\mu}\,[\mathbf{a}]_{\mu}^{\mu}=\Delta^{r}\,[1]_{\mu}^{\mu},\qquad \overline{\underline{\mathbf{A}}}\,_{\mu}^{\mu}\,[\mathbf{C}]_{\mu}^{\mu}=(\mathbf{A})_{\mu}^{\mu}\,[1]_{\mu}^{\mu},$$

and it follows that

$$\overline{\underline{\mathbf{a}}}^{\mu}_{\mu}\left\{(\mathbf{a})^{\mu}_{\mu}[\mathbf{A}]^{\mu}_{\mu}-\Delta^{r}[\mathbf{c}]^{\mu}_{\mu}\right\}=0,\quad \overline{\underline{\mathbf{A}}}^{\mu}_{\mu}\left\{(\mathbf{A})^{\mu}_{\mu}[\mathbf{a}]^{\mu}_{\mu}-\Delta^{r}[\mathbf{C}]^{\mu}_{\mu}\right\}=0.$$

If $\Delta \neq 0$, then $[a]^m_m$ and $[A]^m_m$ are undegenerate, and therefore by § 73 the matrices $[\mathbf{a}]^{\mu}_{\mu}$ and $[\mathbf{A}]^{\mu}_{\mu}$ are undegenerate; consequently by § 84 we have

$$(\mathbf{a})^{\mu}_{\mu}[\mathbf{A}]^{\mu}_{\mu}=\Delta^{r}[\mathbf{c}]^{\mu}_{\mu},\quad (\mathbf{A})^{\mu}_{\mu}[\mathbf{a}]^{\mu}_{\mu}=\Delta^{r}[\mathbf{C}]^{\mu}_{\mu}. \qquad \text{......................(E)}$$

Since the equations (E) are true whenever $\Delta \neq 0$, they are identities in the elements of $[a]^m_m$, and remain true when $\Delta = 0$.

By § 120 we have $(\mathbf{a})^{\mu}_{\mu}=\epsilon\Delta^{\rho}$, $(\mathbf{A})^{\mu}_{\mu}=\epsilon\Delta^{(m-1)\rho}$, where $\rho=\binom{m-1}{r-1}$, and where ϵ is either $+1$ or -1, and is $+1$ when a common scheme is used for horizontal and vertical rows in the formation of $[\mathbf{a}]^{\mu}_{\mu}$ and $[\mathbf{A}]^{\mu}_{\mu}$ from $[a]^m_m$ and $[A]^m_m$. We can therefore replace the equations (E) by

$$\epsilon\Delta^{\rho}[\mathbf{A}]^{\mu}_{\mu}=\Delta^{r}[\mathbf{c}]^{\mu}_{\mu},\quad \epsilon\Delta^{(m-1)\rho}[\mathbf{a}]^{\mu}_{\mu}=\Delta^{r}[\mathbf{C}]^{\mu}_{\mu}. \qquad \text{......................(E')}$$

Cancelling the highest power of Δ which is a common factor of both sides in each of the equations (E′) we see that:

Each of the two corresponding matrices $[\mathbf{a}]^{\mu}_{\mu}$, $[\mathbf{A}]^{\mu}_{\mu}$ of the minor determinants of order r of $[a]^m_m$, $[A]^m_m$ differs from the reciprocal of the other only by a scalar factor of the form $\pm\Delta^k$, the upper sign being taken whenever in the formation of them a common scheme is used for horizontal and vertical rows.

This result was obtained in another way in Ex. i of § 122.

6. *Rank of the reciprocal matrix.*

The rank of the reciprocal matrix $[A]^m_m$ depends on the rank of $[a]^m_m$ in the way shown in the scheme which follows.

Rank of $[a]^m_m$	m	$m-1$	$<m-1$
Rank of $[A]^m_m$	m	1	0

Case I. *Let $[a]^m_m$ have rank m.*

In this case $(a)^m_m=\Delta\neq 0$, $(A)^m_m=\Delta^{m-1}\neq 0$, and therefore $[A]^m_m$ is undegenerate. The same result follows from the equation $\overline{\underline{a}}^m_m[A]^m_m=\Delta[1]^m_m$.

Case II. *Let $[a]^m_m$ have rank $m-1$.*

In this case $[a]^m_m$ has a non-vanishing minor determinant of order $m-1$, and therefore $[A]^m_m$ has a non-vanishing element of order 1. Again since $\Delta=0$, it follows from (D), putting $r=2$, that every minor determinant of order 2 of $[A]^m_m$ vanishes. Consequently $[A]^m_m$ must have rank 1.

CASE III. *Let the rank of* $[a]_m^m$ *be less than* $m-1$.

In this case every minor determinant of order $m-1$ of $[a]_m^m$ vanishes, and therefore every element of $[A]_m^m$ vanishes, i.e. $[A]_m^m$ must have rank 0.

The above results form a particular case of the general theorems of § 73.

7. *Reciprocals of the derangements of a square matrix.*

If $[A]_m^m$ is the reciprocal of $[a]_m^m$ and if $[a_{pq}]_m^m$ is a derangement of $[a]_m^m$ having affect ω in $[a]_m^m$, we will show that:

The reciprocal of the derangement $[a_{pq}]_m^m$ *is* $(-1)^\omega [A_{pq}]_m^m$. ....(F)

If a matrix is formed from $[a]_m^m$ by interchanging two consecutive parallel rows, it is clear that the reciprocal of that matrix is -1 times the matrix formed from $[A]_m^m$ by interchanging the two corresponding consecutive parallel rows; and by repeated applications of this result we obtain (F).

To prove (F) directly, let the reciprocal of $[a_{pq}]_m^m$ be $[A'_{pq}]_m^m$, and let the minors of the sequence $[1\ 2 \dots m]$ formed by striking out its elements p_i, q_j be respectively $[\lambda_1 \lambda_2 \dots \lambda_{m-1}]$, $[\mu_1 \mu_2 \dots \mu_{m-1}]$, so that

$$A'_{p_i q_j} = (-1)^{i+j} \begin{pmatrix} q_1 \dots q_{j-1} q_{j+1} \dots q_m \\ a \\ p_1 \dots p_{i-1} p_{i+1} \dots p_m \end{pmatrix}, \quad A_{p_i q_j} = (-1)^{p_i+q_j} \begin{pmatrix} \mu_1 \mu_2 \dots \mu_{m-1} \\ a \\ \lambda_1 \lambda_2 \dots \lambda_{m-1} \end{pmatrix}.$$

Then if

$$\eta_1 = \text{aff.}\ [p_1 \dots p_{i-1} p_{i+1} \dots p_m] \text{ in } [\lambda]_{m-1}, \quad \eta_2 = \text{aff.}\ [q_1 \dots q_{j-1} q_{j+1} \dots q_m] \text{ in } [\mu]_{m-1},$$

we have
$$A'_{p_i q_j} = (-1)^{\omega'} A_{p_i q_j},$$
where
$$\omega' = (p_i - i) + (q_j - j) + \eta_1 + \eta_2.$$

Let
$$\omega_1 = \text{aff.}\ [p]_m \text{ in } [1\ 2 \dots m], \quad \omega_2 = \text{aff.}\ [q]_m \text{ in } [1\ 2 \dots m].$$

Then
$$\omega_1 \equiv (p_i - i) + \eta_1 \pmod{2}, \quad \omega_2 \equiv (q_j - j) + \eta_2 \pmod{2};$$

for, when the modulus is 2, we have by Theorems V *b* and III of § 19,

$$\omega_1 \equiv \text{aff.}\ [p]_m \text{ in } [p_i p_1 \dots p_{i-1} p_{i+1} \dots p_m] + \text{aff.}\ [p_i p_1 \dots p_{i-1} p_{i+1} \dots p_m] \text{ in } [1\ 2 \dots m]$$
$$\equiv (i-1) + (p_i - 1) + \text{aff.}\ [p_1 \dots p_{i-1} p_{i+1} \dots p_m] \text{ in } [1\ 2 \dots m]$$
$$\equiv (i-1) + (p_i - 1) + \eta_1 \equiv (p_i - i) + \eta_1.$$

It follows that
$$\omega = \omega_1 + \omega_2 \equiv \omega' \pmod{2},$$

and therefore
$$A'_{p_i q_j} = (-1)^\omega A_{p_i q_j}, \quad [A'_{pq}]_m^m = (-1)^\omega [A_{pq}]_m^m.$$

Ex. i. Let
$$\phi = [a\, b\, c]_{123}, \quad \Phi = [A\ B\ C]_{123}, \quad \psi = [\alpha\, \beta\, \gamma]_{123}$$
be three matrices such that the second is the reciprocal of the first, and the third is the reciprocal of the second; and let $\Delta = \det \phi = (abc)_{123}$.

Then by formula (B) we have

$$(ABC)_{123} = \Delta^2, \quad (\alpha\beta\gamma)_{123} = (\Delta^2)^2 = \Delta^4.$$

Also by formula (C) or formula (D) we have

$$[\alpha\beta\gamma]_{123} = \Delta[a\,b\,c]_{123},$$

i.e.

$$(BC)_{23} = \Delta a_1, \quad (CA)_{23} = \Delta b_1, \quad (AB)_{23} = \Delta c_1,$$

and so on.

Ex. ii. Let the reciprocal of $\phi = [a\,b\,c\,d]_{1234}$ be $\Phi = [A\,B\,C\,D]_{1234}$, and let

$$\Delta = \det \phi = (abcd)_{1234}.$$

Also let

$$\Psi = \begin{bmatrix} (AB)_{12}, & (AC)_{12}, & (AD)_{12}, & (BC)_{12}, & (BD)_{12}, & (CD)_{12} \\ (AB)_{13}, & (AC)_{13}, & (AD)_{13}, & (BC)_{13}, & (BD)_{13}, & (CD)_{13} \\ (AB)_{14}, & (AC)_{14}, & (AD)_{14}, & (BC)_{14}, & (BD)_{14}, & (CD)_{14} \\ (AB)_{23}, & (AC)_{23}, & (AD)_{23}, & (BC)_{23}, & (BD)_{23}, & (CD)_{23} \\ (AB)_{24}, & (AC)_{24}, & (AD)_{24}, & (BC)_{24}, & (BD)_{24}, & (CD)_{24} \\ (AB)_{34}, & (AC)_{34}, & (AD)_{34}, & (BC)_{34}, & (BD)_{34}, & (CD)_{34} \end{bmatrix},$$

$$\psi' = \begin{bmatrix} +(cd)_{34}, & -(bd)_{34}, & +(bc)_{34}, & +(ad)_{34}, & -(ac)_{34}, & +(ab)_{34} \\ -(cd)_{24}, & +(bd)_{24}, & -(bc)_{24}, & -(ad)_{24}, & +(ac)_{24}, & -(ab)_{24} \\ +(cd)_{23}, & -(bd)_{23}, & +(bc)_{23}, & +(ad)_{23}, & -(ac)_{23}, & +(ab)_{23} \\ +(cd)_{14}, & -(bd)_{14}, & +(bc)_{14}, & +(ad)_{14}, & -(ac)_{14}, & +(ab)_{14} \\ -(cd)_{13}, & +(bd)_{13}, & -(bc)_{13}, & -(ad)_{13}, & +(ac)_{13}, & -(ab)_{13} \\ +(cd)_{12}, & -(bd)_{12}, & +(bc)_{12}, & +(ad)_{12}, & -(ac)_{12}, & +(ab)_{12} \end{bmatrix},$$

so that Ψ is a complete matrix of the minor determinants of order 2 of Φ, and ψ' is the matrix of the anti-correspondent minor determinants of order 2 of ϕ. Then by formula (D) we have

$$\Psi = \Delta\psi'.$$

To prove this let ψ be the matrix formed from ϕ in the same way as Ψ is formed from Φ, so that ψ has the same value as in Ex. ii of § 119, and let the conjugates of ϕ and ψ be denoted by $\bar{\phi}$ and $\bar{\psi}$. Then

$$\Phi\bar{\phi} = \Delta[1]_4^4, \quad \bar{\psi}\psi' = \Delta[1]_6^6.$$

From the first of these equations it follows that

$$\Psi\bar{\psi} = \Delta^2[1]_6^6.$$

Then postfixing ψ' and using the second equation, we obtain

$$\Delta\Psi = \Delta^2\psi', \quad \text{or} \quad \Psi = \Delta\psi'.$$

Ex. iii. If the reciprocal of $[a\,b\,c]_{123}$ is $[A\,B\,C]_{123}$, then by formula (F) the reciprocals of $[a\,b\,c]_{213}$, $[c\,b\,a]_{213}$ are $-[A\,B\,C]_{213}$, $+[C\,B\,A]_{213}$.

Ex. iv. *To deduce the relation between any two anti-correspondent minor determinants of* $[a]_m^m$ *and* $[A]_m^m$ *given by formula* (D) *from the general identity of* § 111.

Let $(A_{pq})_r^r$ be any minor determinant of $[A]_m^m$ of order r; let $[\lambda]_{m-r}$, $[\mu]_{m-r}$ be the corranged complements of $[p]_r$, $[q]_r$ in $[1\,2\ldots m]$; let ω_1=affect of $[p]_r$ in $[1\,2\ldots m]$,

ω_2 = affect of $[q]_r$ in $[1\,2\ldots m]$, $\omega=\omega_1+\omega_2$ = affect of $(A_{pq})_r^r$ in $[A]_m^m$; and let $\Delta=(a_{\lambda\mu})_{m-r}^{m-r}$. Then by formula (D) we have

$$(A_{pq})_r^r=\Delta^{r-1}.(-1)^{\omega}(a_{\lambda\mu})_{m-r}^{m-r}\text{.................................(1)}$$

We have to deduce (1) from the identity of § 111.

Let $\Phi=\begin{bmatrix} A'_{\lambda\mu}, & A'_{\lambda q} \\ A'_{p\mu}, & A'_{pq}\end{bmatrix}_{m-r,\,r}^{m-r,\,r}$ be the reciprocal of $\begin{bmatrix} a_{\lambda\mu}, & a_{\lambda q} \\ a_{p\mu}, & a_{pq}\end{bmatrix}_{m-r,\,r}^{m-r,\,r}$, and let

$$\Delta'=\begin{pmatrix} a_{\lambda\mu}, & a_{\lambda q} \\ a_{p\mu}, & a_{pq}\end{pmatrix}_{m-r,\,r}^{m-r,\,r}.$$

Then by § 111 we have

$$(A'_{pq})_r^r=\Delta'^{r-1}(a_{\lambda\mu})_{m-r}^{m-r}.\text{ ....(2)}$$

Now let

$$\omega_1'=\text{aff. }[\lambda,\ p]_{m-r,\,r}\text{ in }[1\,2\ldots m],$$

$$\omega_2'=\text{aff. }[\mu,\ q]_{m-r,\,r}\text{ in }[1\,2\ldots m],$$

so that $\omega'=\omega_1'+\omega_2'$ = affect of ϕ in $[a]_m^m$.

Then $\omega_1'\equiv r(m-r)+\omega_1,\quad \omega_2'\equiv r(m-r)+\omega_2\quad (\text{mod. } 2)$;

for ω_1 moves change $[1\,2\ldots m]$ into $[p,\lambda]_{r,\,m-r}$, and $r(m-r)$ moves change $[p,\lambda]_{r,\,m-r}$ into $[\lambda,p]_{m-r,\,r}$.

Therefore $\omega_1'+\omega_2'\equiv\omega_1+\omega_2$, or $\omega'\equiv\omega$ (mod. 2).

From Ex. ii it now follows that

$$\begin{bmatrix} A'_{\lambda\mu}, & A'_{\lambda q} \\ A'_{p\mu}, & A'_{pq}\end{bmatrix}_{m-r,\,r}^{m-r,\,r}=(-1)^{\omega}\begin{bmatrix} A_{\lambda\mu}, & A_{\lambda q} \\ A_{p\mu}, & A_{pq}\end{bmatrix}_{m-r,\,r}^{m-r,\,r}.$$

Consequently $[A'_{pq}]_r^r=(-1)^{\omega}[A_{pq}]_r^r,\quad (A'_{pq})_r^r=(-1)^{r\omega}(A_{pq})_r^r.$

Also $\Delta'=(-1)^{\omega}\Delta,\quad \Delta'^{r-1}=(-1)^{(r-1)\omega}\Delta^{r-1}.$

These last two results show that (2) is equivalent to (1).

§ 125. Properties of a symmetric matrix of rank 1.

The square matrix $[a]_m^m$ is *symmetric* or *self-conjugate* when it is equal to its conjugate, i.e. when $\overline{\underline{a}}{}_m^m=[a]_m^m$, or when $a_{ij}=a_{ji}$ for the values 1, 2, ... m of i and j.

The following theorems relate to a symmetric matrix of rank 1.

Theorem I. *If a symmetric matrix has rank* 1, *the elements of its leading diagonal cannot all vanish.*

Theorem II. *All the elements of a symmetric matrix of rank* 1 *can be expressed in terms of the elements of its leading diagonal.*

Theorem III. *If a symmetric matrix whose elements are all real has rank* 1, *the elements of its leading diagonal all have the same sign.*

To prove Theorem I, let $A=[a]_m^m$ be a symmetric matrix in which the elements of the leading diagonal all vanish. If the rank of A is not 0, we may suppose that $a_{ij}\neq 0$, where $j\neq i$. Then the matrix has a non-vanishing minor determinant $a_{ii}a_{jj}-a^2_{ij}=-a^2_{ij}$ of order 2, and its rank cannot be less than 2. Thus the rank of a symmetric matrix in which the elements of the leading diagonal all vanish must be either 0 or greater than 1; it cannot be equal to 1. This establishes Theorem I.

To prove Theorem II, let $A=[a]_m^m$ be a symmetric matrix of rank 1 in which $a_{ii}\neq 0$. Since the rank is 1, there exist finite quantities $k_1, k_2, \dots k_j, \dots k_m$, of which $k_i=1$, such that

$$\begin{bmatrix} a_{1j} \\ a_{2j} \\ \vdots \\ a_{mj} \end{bmatrix} = k_j \begin{bmatrix} a_{1i} \\ a_{2i} \\ \vdots \\ a_{mi} \end{bmatrix}, \quad [a]_m^m = \begin{bmatrix} a_{1i} \\ a_{2i} \\ \vdots \\ a_{mi} \end{bmatrix} [k_1\, k_2 \dots k_m]. \quad \dots\dots(1)$$

Equating the ith horizontal rows on both sides of (1), we have

$$[a_{i1}\, a_{i2} \dots a_{im}] = a_{ii}[k_1\, k_2 \dots k_m]. \quad \dots\dots(2)$$

Making use of (2), and remembering that $a_{ij}=a_{ji}$, equation (1) becomes

$$a_{ii}[a]_m^m = \begin{bmatrix} a_{i1} \\ a_{i2} \\ \vdots \\ a_{im} \end{bmatrix} [a_{i1}\, a_{i2} \dots a_{im}], \quad \dots\dots(3)$$

a result which can be obtained at once from Ex. v of § 115 or Ex. vi of § 116.

If we equate the elements of the leading diagonals on both sides of (3), or equate to zero those diagonal minor determinants of order 2 of A which contain a_{ii}, we obtain such equations as

$$a_{ii}\, a_{jj} = a^2_{ij}, \qquad a_{ij} = \pm\sqrt{a_{ii}}\sqrt{a_{jj}}.$$

Hence the radicals $\sqrt{a_{11}}, \sqrt{a_{22}}, \dots \sqrt{a_{mm}}$ can be so chosen that

$$[a_{i1}\, a_{i2} \dots a_{im}] = \sqrt{a_{ii}}[\sqrt{a_{11}}, \sqrt{a_{22}}, \dots \sqrt{a_{mm}}]. \quad \dots\dots(4)$$

Substituting this value of the matrix on the left in (3), we obtain

$$[a]_m^m = \begin{bmatrix} \sqrt{a_{11}} \\ \sqrt{a_{22}} \\ \vdots \\ \sqrt{a_{mm}} \end{bmatrix} [\sqrt{a_{11}}\sqrt{a_{22}} \dots \sqrt{a_{mm}}]. \quad \dots\dots(A)$$

Equation (A) establishes Theorem II, for it shows that if a_{uv} is any element of A, then

$$a_{uv} = \sqrt{a_{uu}}\sqrt{a_{vv}}.$$

To prove Theorem III, we assume that all elements of the symmetric matrix $A=[a]_m^m$ are real and that $a_{ii}\neq 0$. Then when A has rank 1, the equation $a_{ii}\, a_{jj}=a^2_{ij}$ obtained above shows that when a_{jj} is not zero, it must have the same sign as a_{ii}, i.e. every non-vanishing diagonal element of A has the same sign as a_{ii}.

When $a_{11}, a_{22}, \dots a_{mm}$ are all positive, so that $\sqrt{a_{11}}, \sqrt{a_{22}}, \dots \sqrt{a_{mm}}$ are all real, it is clear from (4) that formula (A) is true when the signs inherent in the radicals are so chosen that any one row (horizontal or vertical) of A, the elements of which are not all zero, is identical with the corresponding row of the product matrix on the right in (A). This may be expressed by saying that the signs of $\sqrt{a_{11}}, \sqrt{a_{22}}, \ldots \sqrt{a_{mm}}$ are either all the same as or all opposite to the signs of the rows of $[a]_m^m$ to which $a_{11}, a_{22}, \ldots a_{mm}$ respectively belong. When $a_{11}, a_{22}, \ldots a_{mm}$ are all negative, the signs of $\sqrt{-a_{11}}, \sqrt{-a_{22}}, \ldots \sqrt{-a_{mm}}$ must be chosen in the same way.

From formula (A) *it is clear that an element a_{ii} of the leading diagonal can only vanish when every element of the horizontal and vertical rows to which it belongs is zero.*

NOTE. Formula (A) is obviously true when $[a]_m^m$ has rank 0. Thus it is true whenever the rank of the symmetric matrix $[a]_m^m$ is less than 2.

Ex. i. The following are symmetric matrices of rank 1, the first two being real:

$$\begin{bmatrix} 4, & -6, & 2, & -4 \\ -6, & 9, & -3, & 6 \\ 2, & -3, & 1, & -2 \\ -4, & 6, & -2, & 4 \end{bmatrix} = \begin{bmatrix} 2 \\ -3 \\ 1 \\ 2 \end{bmatrix} [2, \ -3, \ 1, \ 2],$$

$$\begin{bmatrix} -4, & 6, & -2, & 0 \\ 6, & -9, & 3, & 0 \\ -2, & 3, & -1, & 0 \\ 0, & 0, & 0, & 0 \end{bmatrix} = \begin{bmatrix} 2i \\ -3i \\ i \\ 0 \end{bmatrix} [2i, \ -3i, \ i, \ 0],$$

$$\begin{bmatrix} 2i & , 0, & 2 & , \sqrt{2}(1+i) \\ 0 & , 0, & 0 & , \ 0 \\ 2 & , 0, & -2i & , \sqrt{2}(1-i) \\ \sqrt{2}(1+i), & 0, & \sqrt{2}(1-i), & 2 \end{bmatrix} = \begin{bmatrix} 1+i \\ 0 \\ 1-i \\ \sqrt{2} \end{bmatrix} [1+i, \ 0, \ 1-i, \ \sqrt{2}],$$

where $i = \sqrt{-1}$.

Ex. ii. The radicals $\sqrt{a}, \sqrt{b}, \sqrt{c}, \sqrt{d}$ can be so chosen that

$$\begin{bmatrix} a & h & g & u \\ h & b & f & v \\ g & f & c & w \\ u & v & w & d \end{bmatrix} = \begin{bmatrix} \sqrt{a} \\ \sqrt{b} \\ \sqrt{c} \\ \sqrt{d} \end{bmatrix} [\sqrt{a}, \sqrt{b}, \sqrt{c}, \sqrt{d}],$$

when and only when the symmetric matrix on the left has rank not exceeding 1.

§ 126. Some general properties of symmetric matrices.

Theorem I. *If the rank of a symmetric matrix is r, then the matrix has at least one diagonal minor determinant of order r which does not vanish.*

Let $A = [a]_m^m$ be a symmetric matrix, and let $[A]_\mu^\mu$, where $\mu = \binom{m}{r}$, be a complete matrix of the minor determinants of order r of A in the formation of which the successive selections of r vertical rows follow the same scheme as the successive selections of r horizontal rows. Then $[A]_\mu^\mu$ is also a symmetric matrix, and the elements of its leading diagonal are the diagonal minor determinants of order r of A. If A has rank r, then by § 73 the matrix $[A]_\mu^\mu$ has rank 1, and therefore by Theorem I of § 125, it must have a non-vanishing diagonal element, i.e. A must have a non-vanishing diagonal minor determinant of order r.

Theorem II. *If a symmetric matrix $A = [a]_m^m$ has a diagonal minor determinant Δ_i of order i which does not vanish, and if all those diagonal minor determinants of A of orders $i+1$ and $i+2$ which contain Δ_i vanish, then A has rank i.*

We assume that $i \nless 1$, and $i + 2 \ngtr m$.

Let $[p_1 p_2 \dots p_i u v]$ be any minor of order $i+2$ of the sequence $[1\, 2 \dots m]$, and let

$$\Delta_i = \begin{pmatrix} p_1 p_2 \dots p_i \\ a \\ p_1 p_2 \dots p_i \end{pmatrix} = (a_{pp})_i^i, \quad \Delta_{i+2} = \begin{pmatrix} p_1 p_2 \dots p_i uv \\ a \\ p_1 p_2 \dots p_i uv \end{pmatrix},$$

$$A_{uu} = \begin{pmatrix} p_1 \dots p_i u \\ a \\ p_1 \dots p_i u \end{pmatrix}, \quad A_{vv} = \begin{pmatrix} p_1 \dots p_i v \\ a \\ p_1 \dots p_i v \end{pmatrix}, \quad A_{uv} = \begin{pmatrix} p_1 \dots p_i v \\ a \\ p_1 \dots p_i u \end{pmatrix}, \quad A_{vu} = \begin{pmatrix} p_1 \dots p_i u \\ a \\ p_1 \dots p_i v \end{pmatrix}.$$

If Δ_i is given, Δ_{i+2} can be any diagonal minor determinant of A of order $i+2$ which contains Δ_i, or A_{uv} can be any non-diagonal minor determinant of A of order $i+1$ which contains Δ_i.

If Δ_{i+2} is given, Δ_i can be any diagonal minor determinant of A of order i which is contained in Δ_{i+2}, or A_{uv} can be any non-diagonal minor determinant of A of order $i+1$ which is contained in Δ_{i+2}.

If Δ_i and Δ_{i+2} are given, then A has four and only four minor determinants of order $i+1$ which are contained in Δ_{i+2} and contain Δ_i. Two of these are the diagonal minor determinants A_{uu}, A_{vv}; and the other two are the two mutually conjugate non-diagonal minor determinants A_{uv}, A_{vu}.

Since A is symmetric, we have $A_{vu} = A_{uv}$, and it follows from § 110 that

$$\begin{vmatrix} A_{uu} & A_{uv} \\ A_{vu} & A_{vv} \end{vmatrix} = A_{uu} A_{vv} - A_{uv}^2 = \Delta_i \Delta_{i+2}. \quad \dots\dots\dots\dots (1)$$

To prove Theorem II suppose that Δ_i is a given non-vanishing diagonal minor determinant of A of order i, and that all such minor determinants as A_{uu}, Δ_{i+2} vanish. Then from equation (1) we see that all such determinants as A_{uv} vanish. Thus all minor determinants of A of order $i+1$ which contain Δ_i vanish, and it follows by Note 4 of § 116 that the matrix A has rank i.

Ex. i. *If all diagonal minor determinants of orders* 1 *and* 2 *vanish, the matrix has rank* 0.

For in this case we have

$$a_{uu}=0, \quad a_{vv}=0, \quad \begin{vmatrix} a_{uu} & a_{uv} \\ a_{vu} & a_{vv} \end{vmatrix} = -a^2_{uv}=0, \quad a_{uv}=0,$$

for all values of u and v, i.e. $[a]_m^m=0$.

This is the special form assumed by Theorem II when $i=0$, and $m \nless 2$.

Ex. ii. *If there is a non-vanishing minor determinant of order* $m-1$, *and if* $(a)_m^m=0$, *then the matrix has rank* $m-1$, *and there is a non-vanishing diagonal minor determinant of order* $m-1$.

This is the special form assumed by Theorem II when $i=m-1$.

Theorem III. *If* $A=[a]_m^m$ *is a symmetric matrix, and if all its diagonal minor determinants of orders* $i+1$ *and* $i+2$ *vanish, then the rank of* A *cannot exceed* i.

Here i is any integer satisfying the conditions $i \nless 0$, $i+2 \ngtr m$.

If 0 is a possible value of i, i.e. if $m \nless 2$, we know by Ex. i that the theorem is true when $i=0$. Using this fact we can prove Theorem III by induction.

Let k and $k+1$ be both possible values of i, so that $k \nless 0$ and $k+3 \ngtr m$, and assume that the theorem is true when $i=k$.

This being assumed, suppose that all diagonal minor determinants of A of orders $k+2$ and $k+3$ vanish. Then if all diagonal minor determinants of A of order $k+1$ also vanish, it follows from the assumption that the rank of A cannot exceed k; and if A has a non-vanishing diagonal minor determinant of order $k+1$, it follows from Theorem II that A has rank $k+1$. Consequently the rank of A cannot in any case exceed $k+1$.

Thus if k and $k+1$ are both possible values of i, and if the theorem is true when $i=k$, it is also true when $i=k+1$.

Since the theorem is true for $i=0$, when 0 is a possible value of i, it follows that the theorem is true for $i=1, 2, 3, \ldots$ when these are possible values of i, i.e. it is true for all possible values of i.

In the next two theorems $A=[a]_m^m$ is a symmetric matrix of rank r, and we assume that $i \nless 1$, and $i+2 \ngtr r$. The theorems are however also true (see Ex. iii) when $i=0$ and $r \nless 2$.

Theorem IV a. *If the symmetric matrix A has a non-vanishing diagonal minor determinant Δ_{i+2} of order $i+2$, and if Δ_{i+2} contains no non-vanishing diagonal minor determinant of order $i+1$, then Δ_{i+2} contains a non-vanishing diagonal minor determinant Δ_i of order i.*

Theorem IV b. *If the symmetric matrix A has a non-vanishing diagonal minor determinant Δ_i of order i, and if it has no non-vanishing diagonal minor determinant of order $i+1$ which contains Δ_i, then it has a non-vanishing diagonal minor determinant Δ_{i+2} of order $i+2$ which contains Δ_i.*

In the case of Theorem IV *a* the determinant Δ_{i+2} must contain some non-diagonal minor determinant D_{i+1} of order $i+1$ which does not vanish, and without loss of generality we may suppose that

$$\Delta_{i+2} = \begin{pmatrix} p_1 p_2 \dots p_i uv \\ a \\ p_1 p_2 \dots p_i uv \end{pmatrix}, \qquad D_{i+1} = \begin{pmatrix} p_1 p_2 \dots p_i v \\ a \\ p_1 p_2 \dots p_i u \end{pmatrix}.$$

Then using the same notation as in the proof of Theorem II, so that $A_{uv} = D_{i+1}$, $A_{uu} = 0$, $A_{vv} = 0$, we see from the equation (1) that

$$\Delta_i \Delta_{i+2} = -D_{i+1}^2 \neq 0. \qquad \text{.........................(2)}$$

Therefore $\Delta_i = (a_{pp})_i^i$ is a *non-vanishing* diagonal minor determinant of A of order i which is contained in Δ_{i+2}. It is that one diagonal minor determinant of A of order i which is contained in D_{i+1}.

In the case of Theorem IV *b* the matrix A must contain some non-diagonal minor determinant D_{i+1} of order $i+1$ which contains Δ_i and does not vanish, and we may suppose that

$$\Delta_i = \begin{pmatrix} p_1 p_2 \dots p_i \\ a \\ p_1 p_2 \dots p_i \end{pmatrix}, \qquad D_{i+1} = \begin{pmatrix} p_1 p_2 \dots p_i v \\ a \\ p_1 p_2 \dots p_i u \end{pmatrix}.$$

Then using the same notation as in the proof of Theorem II, so that $A_{uv} = D_{i+1}$, $A_{uu} = 0$, $A_{vv} = 0$, we again have

$$\Delta_i \Delta_{i+2} = -D_{i+1}^2 \neq 0. \qquad \text{.........................(3)}$$

Therefore $\Delta_{i+2} = \begin{pmatrix} p_1 p_2 \dots p_i uv \\ a \\ p_1 p_2 \dots p_i uv \end{pmatrix}$ is a *non-vanishing* diagonal minor determinant of A of order $i+2$ which contains Δ_i. It is that one diagonal minor determinant of A of order $i+2$ which contains D_{i+1}.

NOTE 1. If in either one of Theorems IV *a* and IV *b* we take Δ_{i+1} to be A_{uu} or A_{vv}, then Δ_i, Δ_{i+1}, Δ_{i+2} are three diagonal minor determinants of A of orders i, $i+1$, $i+2$ having the following properties:

(1) Δ_{i+1} contains Δ_i and is contained in Δ_{i+2}.

(2) $\Delta_i \neq 0$, $\Delta_{i+1} = 0$, $\Delta_{i+2} \neq 0$.

(3) If D_{i+1}, D'_{i+1} are those two (mutually conjugate) non-diagonal minor determinants of A of order $i+1$ which are contained in Δ_{i+2} and contain Δ_i, then

$$D'_{i+1} = D_i \neq 0.$$

Ex. iii. *If $r > 1$, and if the symmetric matrix A contains no non-vanishing diagonal element, then it must contain a non-vanishing diagonal minor determinant of order 2.*

Here as in the text $A = [a]_m^m$ is a symmetric matrix of rank r. In this case A must contain some non-diagonal element a_{uv} which does not vanish, and then

$$\Delta_2 = \begin{vmatrix} a_{uu} & a_{uv} \\ a_{vu} & a_{vv} \end{vmatrix} = -a^2_{uv}$$

is a diagonal minor determinant of A of order 2 which does not vanish.

This is the form assumed by Theorem IV *b* when $i = 0$.

Ex. iv. *If Δ_{r-1} is a non-vanishing diagonal minor determinant of order $r-1$ of the symmetric matrix A, then we can determine a diagonal minor determinant Δ_r of A of order r which contains Δ_{r-1} and does not vanish.*

As before $A = [a]_m^m$ is a symmetric matrix of rank r.

If $r = m$, this theorem is obviously true. Accordingly we may assume that $r < m$.

Putting $i = r-1$ in the identity (1) we have in this case

$$A_{uu}A_{vv} - A^2_{uv} = 0, \quad \text{or} \quad A_{uu}A_{vv} = A^2_{uv}.$$

If there is a non-diagonal minor determinant A_{uv} of order r which contains Δ_{r-1} and does not vanish, then A_{uu} and A_{vv} are diagonal minor determinants of order r which contain Δ_{r-1} and do not vanish.

And if every non-diagonal minor determinant of order r containing Δ_{r-1} vanishes, then there must be a diagonal minor determinant of order r containing Δ_{r-1} which does not vanish, otherwise A would only have rank $r-1$.

Thus the above theorem is true in all cases. It is the theorem which replaces Theorem IV *b* when $i = r-1$.

Theorem V. *If $A = [a]_m^m$ is a symmetric matrix of rank r, and if $\Delta_0 = 1$, it is possible to determine a series of diagonal minor determinants*

$$\Delta_0, \Delta_1, \Delta_2, \Delta_3, \dots \Delta_r \quad \dots\dots\dots\dots\dots\dots\dots\dots(4)$$

of orders 0, 1, 2, 3, ... r of A such that:

(1) *Each determinant after the first contains the preceding.*

(2) *The last determinant Δ_r does not vanish.*

(3) *No two consecutive determinants are both zero.*

(4) *If $\Delta_{i+1} = 0$, then all those diagonal minor determinants of A of order $i+1$ which contain Δ_i vanish; but if D_{i+1}, D'_{i+1} are those two (mutually conjugate) non-diagonal minor determinants of A of order $i+1$ which are contained in Δ_{i+2} and contain Δ_i, then*

$$D'_{i+1} = D_{i+1} \neq 0.$$

In forming such a series we can always commence by taking Δ_r to be any non-vanishing diagonal minor determinant of A of order r; when A has at least one non-vanishing diagonal element, we can commence by taking Δ_1 to be any non-vanishing diagonal element of A; and when A has no non-vanishing diagonal element, we can commence by taking Δ_1 to be any non-vanishing non-diagonal element of A.

First let Δ_r be any selected non-vanishing diagonal minor determinant of A of order r. Then Theorem IV a shows that we can determine $\Delta_{r-1}, \Delta_{r-2}, \ldots \Delta_1$ in succession so that the series (4) has the properties mentioned. To see this suppose that $\Delta_r, \Delta_{r-1}, \ldots \Delta_{i+2}$ have been determined and that $\Delta_{i+2} \neq 0$. Then if $i+2>2$, we can extend the series by one or two terms; for if Δ_{i+2} has any non-vanishing diagonal minor determinant of order $i+1$, we can choose that for Δ_{i+1}; and if Δ_{i+2} has no non-vanishing diagonal minor determinant of order $i+1$, we can choose Δ_{i+1} and Δ_i as in Theorem IV a and Note 1, so that $\Delta_{i+1}=0$ and $\Delta_i \neq 0$. And if $i+2=2$, we can clearly complete the series, Δ_1 being 0 when Δ_2 has no non-vanishing diagonal element.

Next let the series be commenced at the other end. If A has at least one non-vanishing diagonal element, we take Δ_1 to be any selected non-vanishing diagonal element, so that $\Delta_1 \neq 0$. If A has no non-vanishing diagonal element, then $\Delta_1=0$, and we choose Δ_2 as in Ex. iii, so that we have $\Delta_1=0, \Delta_2 \neq 0$. Having started the series we can complete it by the use of Theorem IV b and Ex. iv. To see this suppose that $\Delta_1, \Delta_2, \ldots \Delta_i$ have been determined, and that $\Delta_i \neq 0$. Then if $i<r-1$, we can extend the series by one or two terms; for if A has any non-vanishing diagonal minor determinant of order $i+1$ which contains Δ_i, we can choose that for Δ_{i+1}; and if A has no non-vanishing diagonal minor determinant of order $i+1$ containing Δ_i, we have $i+1<r$, and we can choose Δ_{i+1} and Δ_{i+2} as in Theorem IV b and Note 1, so that $\Delta_{i+1}=0, \Delta_{i+2} \neq 0$. And if $i=r-1$, then we can complete the series by Ex. iv.

The determinants of the series (4) have the forms

$$\Delta_0=1,\ \Delta_1=a_{p_1p_1},\ \Delta_2=(a_{pp})_2^2,\ \Delta_3=(a_{pp})_3^3,\ \ldots\ \Delta_r=(a_{pp})_r^r, \ldots\ldots(5)$$

where $[p]_r$ is a minor of order r of the sequence $[1\ 2 \ldots m]$.

NOTE 2. *Reduction of a symmetric matrix to a standard form by symmetric derangements of its rows.*

By like re-arrangements of its horizontal and vertical rows any symmetric matrix of order m and rank r can be converted into a symmetric matrix $[a]_m^m$ which is such that the series of *leading* diagonal minor determinants

$$\Delta_0=1,\ \Delta_1=a_{11},\ \Delta_2=(a)_2^2,\ \Delta_3=(a)_3^3,\ \ldots\ \Delta_r=(a)_r^r \ \ldots\ldots\ldots\ldots\ldots\ldots(6)$$

has the properties mentioned in Theorem V; so that:

(1) The last determinant Δ_r does not vanish.

(2) No two consecutive determinants are both zero.

(3) If $\Delta_{i+1}=0$, then all those diagonal minor determinants of order $i+1$ which contain Δ_i vanish, but

$$\begin{pmatrix}1,2,\ldots i,i+2\\ a\\ 1,2,\ldots i,i+1\end{pmatrix}=\begin{pmatrix}1,2,\ldots i,i+1\\ a\\ 1,2,\ldots i,i+2\end{pmatrix}\neq 0.$$

A symmetric matrix having this character may be said to be in *standard form* as regards symmetric derangements.

To reduce any given symmetric matrix $[a]_m^m$ of rank r to this standard form, we first determine a minor sequence $[p]_r$ of $[1\ 2 \ldots m]$ as in (5), and then bring the p_1th, p_2th, ... p_rth horizontal and vertical rows to the leading positions arranged in this order.

§ 127. Some general properties of skew-symmetric matrices.

A square matrix $[a]_m^m$ is said to be *skew-symmetric* when it is equal to minus its conjugate, i.e. when $\overline{\underline{a}}_m^m = -[a]_m^m$, or when $a_{ij} = -a_{ji}$ for the values 1, 2, ... m of i and j. If we say that two matrices are mutually *skew-conjugate* when each is equal to minus the conjugate of the other, then a skew-symmetric matrix is one which is skew-conjugate with itself.

It follows from the above definition that in the case of a skew-symmetric matrix $[a]_m^m$ we have $a_{ii} = 0$ for the values 1, 2, ... m of i.

Thus the diagonal elements of a skew-symmetric matrix are all 0's.

Ex. i. If $[a]_m^m$ is any square matrix whatever, then $[a]_m^m + \overline{\underline{a}}_m^m$ is a symmetric matrix, and $[a]_m^m - \overline{\underline{a}}_m^m$ is a skew-symmetric matrix. Consequently there exists a symmetric matrix $[u]_m^m$ and a skew-symmetric matrix $[v]_m^m$ such that

$$[u]_m^m = \frac{1}{2}\left\{[a]_m^m + \overline{\underline{a}}_m^m\right\}, \quad [v]_m^m = \frac{1}{2}\left\{[a]_m^m - \overline{\underline{a}}_m^m\right\}, \qquad \ldots\ldots\ldots\ldots(1)$$

$$[a]_m^m = [u]_m^m + [v]_m^m, \qquad \overline{\underline{a}}_m^m = [u]_m^m - [v]_m^m. \qquad \ldots\ldots\ldots\ldots(2)$$

Conversely if $[u]_m^m$ is a symmetric matrix and $[v]_m^m$ a skew-symmetric matrix, then $[u]_m^m + [v]_m^m$ and $[u]_m^m - [v]_m^m$ are two mutually conjugate matrices, and there exists a matrix $[a]_m^m$ such that the equations (2) and (1) are true.

Ex. ii. *No skew-symmetric matrix can have rank* 1.

Let $A = [a]_m^m$ be a skew-symmetric matrix whose rank is less than 2. Then all its diagonal elements vanish, and if a_{ij} is any non-diagonal element, we have

$$\begin{vmatrix} a_{ii} & a_{ij} \\ a_{ji} & a_{jj} \end{vmatrix} = \begin{vmatrix} 0, & a_{ij} \\ -a_{ij}, & 0 \end{vmatrix} = a^2_{ij} = 0, \text{ i.e. } a_{ij} = 0.$$

Thus every element of A vanishes, i.e. the matrix A has rank 0.

Ex. iii. *If $\phi = [a]_m^m$ is a skew-symmetric matrix and if $\Phi = [A]_\mu^\mu$ is any complete matrix of its minor determinants of order s in which a common scheme of formation is used for horizontal and vertical rows, then Φ is a symmetric matrix when s is even and a skew-symmetric matrix when s is odd.*

For if $A_{uv} = \det [a_{pq}]_s^s = (a_{pq})_s^s$, then $A_{vu} = \det [a_{qp}]_s^s = (a_{qp})_s^s$; and the ith vertical row of A_{vu} is the ith horizontal row of A_{uv} with its sign changed. Consequently $A_{vu} = (-1)^s A_{uv}$.

If s is even, we have $A_{vu} = A_{uv}$, and Φ is symmetric.

If s is odd, we have $A_{vu} = -A_{uv}$, and Φ is skew-symmetric.

Making use of Exs. ii and iii, we obtain the following theorems:

Theorem I a. *The rank of a skew-symmetric matrix is always an even integer.*

Theorem I b. *If the rank of a skew-symmetric matrix is r, where r is an even integer greater than* 0, *then the matrix has at least one diagonal minor determinant of order r which does not vanish.*

Theorem I c. *If a matrix is skew-symmetric, every diagonal minor determinant of odd order vanishes.*

Let $\phi = [a]_m^m$ be a skew-symmetric matrix of rank r, where $r > 0$; and let $\Phi = [A]_\mu^\mu$, where $\mu = \binom{m}{r}$, be any symmetrically formed complete matrix of the minor determinants of ϕ of order r. Then by § 73 the matrix Φ has rank 1. If r were odd, then by Ex. iii Φ would be a skew-symmetric matrix of rank 1, which by Ex. ii is impossible. Consequently r must be even. It follows from Ex. iii that Φ is a symmetric matrix of rank 1. Therefore by Theorem I of § 125 the matrix Φ has a diagonal element which does not vanish, i.e. ϕ has a non-vanishing diagonal minor determinant of order r. Thus Theorems I a and I b are proved.

To prove Theorem I c let s be an odd integer not greater than m, and let Δ_s be a diagonal minor determinant of ϕ of order s. Then Δ_s is a diagonal element of a symmetrically formed complete matrix of the minor determinants of ϕ of order s. Therefore by Ex. iii it is a diagonal element of a skew-symmetric matrix and must vanish.

The theorems which follow correspond to the similarly numbered theorems in § 126.

Theorem II. *If a skew-symmetric matrix* $A = [a]_m^m$ *has a diagonal minor determinant* Δ_i *of even order i which does not vanish, and if all those diagonal minor determinants of A of order* $i + 2$ *which contain* Δ_i *vanish, then A has rank i.*

We assume that $i \nless 2$, and $i + 2 \ngtr m$.

Using the same notation as in the proof of Theorem II of § 126, and observing that now $A_{vu} = (-1)^{i+1} A_{uv}$, we have

$$\begin{vmatrix} A_{uu} & A_{uv} \\ A_{vu} & A_{vv} \end{vmatrix} = A_{uu} A_{vv} + (-1)^i A_{uv}^2 = \Delta_i \Delta_{i+2}; \quad \text{..........(3)}$$

and when i is even, we have $A_{uu}=0$, $A_{vv}=0$, and therefore

$$\Delta_i \Delta_{i+2} = A_{uv}^2. \quad\quad (4)$$

To prove Theorem II suppose that Δ_i is a given non-vanishing diagonal minor determinant of A of even order i, and that all such determinants as Δ_{i+2} vanish. Then equation (4) shows that all such determinants as A_{uv} vanish, besides all such determinants as A_{uu}. Thus all those minor determinants of A of order $i+1$ which contain Δ_i vanish, and it follows from Note 4 of § 116 that the matrix A has rank i.

Ex. iv. *If all diagonal minor determinants of A of order* 2 *vanish, then A has rank* 0.

For all diagonal elements of A vanish, and if a_{uv} is any non-diagonal element, we have

$$\begin{vmatrix} a_{uu} & a_{uv} \\ a_{vu} & a_{vv} \end{vmatrix} = \begin{vmatrix} 0, & a_{uv} \\ -a_{uv}, & 0 \end{vmatrix} = a^2_{uv} = 0, \quad \text{i.e.} \quad a_{uv}=0.$$

This is the special form assumed by Theorem II when $i=0$, and $m \nless 2$.

Theorem III. *If $A=[a]_m^m$ is a skew-symmetric matrix, and if all its diagonal minor determinants of even order $i+2$ vanish, then the rank of A cannot exceed i.*

Here i is any even integer satisfying the conditions $i \nless 0$, $i+2 \ngtr m$.

If 0 is a possible value of i, i.e. if $m \nless 2$, we know by Ex. iii that the theorem is true when $i=0$. Using this fact we can prove Theorem III by induction.

Let $2k$ and $2k+2$ be both possible values of i, so that $k \nless 0$ and $2k+4 \ngtr m$, and assume that the theorem is true when $i=2k$.

This being assumed, suppose that all diagonal minor determinants of A of order $2k+4$ vanish. Then if all diagonal minor determinants of A of order $2k+2$ vanish, it follows from the assumption that the rank of A cannot exceed $2k$; and if A has a non-vanishing diagonal minor determinant of order $2k+2$, it follows from Theorem II that A has rank $2k+2$. Consequently the rank of A cannot in any case exceed $2k+2$.

Thus if $2k$ and $2k+2$ are both possible values of i, and if the theorem is true when $i=2k$, it is also true when $i=2k+2$.

Since the theorem is true for $i=0$, when 0 is a possible value of i, it follows that the theorem is true for $i=2, 4, 6, \ldots$ when these are possible values of i, i.e. it is true for all possible values of i.

In the next two theorems $A=[a]_m^m$ is a skew-symmetric matrix of (even) rank r, and we assume that $i \nless 2$, $i+2 \ngtr r$. The theorems are however (see Ex. v) also true when $i=0$ and $r \nless 2$.

Theorem IV a. *If Δ_{i+2} is a non-vanishing diagonal minor determinant of the skew-symmetric matrix A of order $i+2$, where $i+2$ is necessarily even, then Δ_{i+2} contains a non-vanishing diagonal minor determinant Δ_i of order i.*

Theorem IV b. *If Δ_i is a non-vanishing diagonal minor determinant of the skew-symmetric matrix A of order i, where i is necessarily even, then A has a non-vanishing diagonal minor determinant Δ_{i+2} of order $i+2$ which contains Δ_i.*

In the case of Theorem IVa the determinant Δ_{i+2} must contain some non-diagonal minor determinant D_{i+1} of order $i+1$ which does not vanish, and we may suppose that

$$\Delta_{i+2} = \begin{pmatrix} p_1 p_2 \dots p_i u v \\ a \\ p_1 p_2 \dots p_i u v \end{pmatrix}, \quad D_{i+1} = \begin{pmatrix} p_1 p_2 \dots p_i v \\ a \\ p_1 p_2 \dots p_i u \end{pmatrix}.$$

Then, using the same notation as in the proof of Theorem II of § 126, so that $D_{i+1} = A_{uv} \neq 0$, we see from (4) that $\Delta_i = \begin{pmatrix} p_1 p_2 \dots p_i \\ a \\ p_1 p_2 \dots p_i \end{pmatrix}$ is a *non-vanishing* diagonal minor determinant of Δ_{i+2} of order i. It is that one diagonal minor determinant of A of order i which is contained in D_{i+1}.

In the case of Theorem IVb the matrix A must contain some non-diagonal minor determinant D_{i+1} of order $i+1$ which contains Δ_i and does not vanish, and we may suppose that

$$\Delta_i = \begin{pmatrix} p_1 p_2 \dots p_i \\ a \\ p_1 p_2 \dots p_i \end{pmatrix}, \quad D_{i+1} = \begin{pmatrix} p_1 p_2 \dots p_i v \\ a \\ p_1 p_2 \dots p_i u \end{pmatrix}.$$

Then, using the same notation as before, we see from (4) that

$$\Delta_{i+2} = \begin{pmatrix} p_1 p_2 \dots p_i u v \\ a \\ p_1 p_2 \dots p_i u v \end{pmatrix}$$

is a *non-vanishing* diagonal minor determinant of A of order $i+2$ which contains Δ_i. It is that one diagonal minor determinant of A of order $i+2$ which contains D_{i+1}.

NOTE 1. If in either of the last two theorems D_{i+1} and D'_{i+1} are those two (mutually skew-conjugate) non-diagonal minor determinants of A of order $i+1$ which are contained in Δ_{i+2} and contain Δ_i, then

$$D'_{i+1} = -D_{i+1} \neq 0.$$

Ex. v. *If the rank of the skew-symmetric matrix A is not less than 2, then A contains a non-vanishing diagonal minor determinant of order 2.*

For A must contain some non-vanishing non-diagonal element a_{uv}, and then

$$\Delta_2 = \begin{vmatrix} a_{uu} & a_{uv} \\ a_{vu} & a_{vv} \end{vmatrix} = \begin{vmatrix} 0, & a_{uv} \\ -a_{uv}, & 0 \end{vmatrix} = a^2_{uv} \neq 0$$

is a non-vanishing diagonal minor determinant of order 2.

This is the special form assumed by Theorem IVb when $i=0$ and $r \nless 2$.

Ex. vi. *Every undegenerate skew-symmetric matrix of order m, (where m is necessarily even) contains at least one undegenerate diagonal minor of order $m-2$.*

This statement is equivalent to Theorem IVa.

Theorem V. *If $A = [a]_m^m$ is a skew-symmetric matrix of (even) rank r, and if $\Delta_0 = 1$, it is possible to determine a series of diagonal minor determinants*

$$\Delta_0,\ \Delta_1,\ \Delta_2,\ \Delta_3,\ \ldots\ \Delta_r \qquad\ldots\ldots(5)$$

of orders 0, 1, 2, 3, ... r of A such that

(1) *Each determinant after the first contains the preceding.*

(2) *No one of the determinants $\Delta_0, \Delta_2, \Delta_4, \ldots \Delta_r$ vanishes.*

(3) *All the determinants $\Delta_1, \Delta_3, \ldots \Delta_{r-1}$ vanish.*

(4) *If $i+1$ is odd, (so that $\Delta_{i+1} = 0$), and if D_{i+1}, D'_{i+1} are those two (mutually skew-conjugate) non-diagonal minor determinants of A of order $i+1$ which are contained in Δ_{i+2} and contain Δ_i, then*

$$D'_{i+1} = -D_{i+1} \neq 0.$$

If Δ_r is any given non-vanishing diagonal minor determinant of A of order r, we can choose the determinants $\Delta_{r-1}, \Delta_{r-2}, \ldots \Delta_1$ in succession by the use of Theorem IV *a*; and if Δ_2 is any given non-vanishing diagonal minor determinant of A of order 2, we can choose the determinants $\Delta_3, \Delta_4, \ldots \Delta_r$ in succession by the use of Theorem IV *b*.

The determinants (5) have the forms

$$\Delta_0 = 1,\quad \Delta_1 = a_{p_1 p_1},\quad \Delta_2 = (a_{pp})_2^2,\quad \Delta_3 = (a_{pp})_3^3,\ \ldots\ \Delta_r = (a_{pp})_r^r, \ldots(6)$$

where $[p]_r$ is a minor of order r of the sequence $[1\ 2 \ldots m]$.

NOTE 2. *Reduction of a skew-symmetric matrix to a standard form by symmetric derangements of its rows.*

By like derangements of its horizontal and vertical rows we can convert any skew-symmetric matrix of order m and (even) rank r into a skew-symmetric matrix $[a]_m^m$ which is such that the series of *leading* diagonal minor determinants

$$\Delta_0 = 1,\quad \Delta_1 = a_{11},\quad \Delta_2 = (a)_2^2,\quad \Delta_3 = (a)_3^3,\ \ldots\ \Delta_r = (a)_r^r$$

have the properties mentioned in Theorem V.

A skew-symmetric matrix having this character may be said to be in *standard form* as regards symmetric derangements.

§ 128. The symmetric matrix of order 3.

For this matrix and its determinant we shall generally use the notations

$$\phi = \begin{bmatrix} a & h & g \\ h & b & f \\ g & f & c \end{bmatrix},\quad D = \det\phi = \begin{vmatrix} a & h & g \\ h & b & f \\ g & f & c \end{vmatrix}. \qquad\ldots\ldots(1)$$

The reciprocal of ϕ, which is also its conjugate reciprocal, will be denoted by

$$\Phi = \begin{bmatrix} A & H & G \\ H & B & F \\ G & F & C \end{bmatrix} = \begin{bmatrix} bc-f^2, & fg-ch, & hf-bg \\ fg-ch, & ca-g^2, & gh-af \\ hf-bg, & gh-af, & ab-h^2 \end{bmatrix}. \qquad\ldots\ldots(2)$$

The following examples deal with properties of the matrices ϕ and Φ.

Ex. i.
$$\begin{bmatrix} a & h & g \\ h & b & f \\ g & f & c \end{bmatrix}\begin{bmatrix} A & H & G \\ H & B & F \\ G & F & C \end{bmatrix} = \begin{bmatrix} D & 0 & 0 \\ 0 & D & 0 \\ 0 & 0 & D \end{bmatrix}, \text{ or } \phi\Phi = \Phi\phi = D\,[1]_3^3. \quad \text{.........(A)}$$

This is the fundamental property of the reciprocal matrix and is a particular case of formula (A) of § 124.

Ex. ii. Equating the determinants of both sides in (A), we obtain the equation $D \,.\, \det \Phi = D^3$, which is an identity in the elements of ϕ. Since D is a rational integral function of the elements of ϕ which does not vanish identically, we can cancel it on both sides, and so obtain

$$\det \Phi = \begin{vmatrix} A & H & G \\ H & B & F \\ G & F & C \end{vmatrix} = D^2. \quad \text{.................................(B)}$$

This is a particular case of formula (B) of § 124.

Ex. iii. Let the reciprocal of Φ be

$$\begin{bmatrix} a' & h' & g' \\ h' & b' & f' \\ g' & f' & c' \end{bmatrix} = \begin{bmatrix} BC - F^2, & FG - CH, & HF - BG \\ FG - CH, & CA - G^2, & GH - AF \\ HF - BG, & GH - AF, & AB - H^2 \end{bmatrix}. \quad \text{...............(3)}$$

Equating the reciprocals of both sides in (A), we have

$$\begin{bmatrix} A & H & G \\ H & B & F \\ G & F & C \end{bmatrix}\begin{bmatrix} a' & h' & g' \\ h' & b' & f' \\ g' & f' & c' \end{bmatrix} = \begin{bmatrix} D^2 & 0 & 0 \\ 0 & D^2 & 0 \\ 0 & 0 & D^2 \end{bmatrix}$$

Prefixing ϕ on both sides and then using (A), we obtain

$$D\begin{bmatrix} a' & h' & g' \\ h' & b' & f' \\ g' & f' & c' \end{bmatrix} = D^2\begin{bmatrix} a & h & g \\ h & b & f \\ g & f & c \end{bmatrix}, \quad \begin{bmatrix} a' & h' & g' \\ h' & b' & f' \\ g' & f' & c' \end{bmatrix} = D\begin{bmatrix} a & h & g \\ h & b & f \\ g & f & c \end{bmatrix}, \quad \text{.........(C)}$$

where D can be cancelled on both sides because it does not vanish identically.

Thus we have

$$a' = BC - F^2 = Da, \quad h' = FG - CH = Dh, \quad g' = HF - BG = Dg, \text{ etc.};$$

$$(b'c' - f'^2) = D^2(bc - f^2), \quad (f'g' - c'h') = D^2(fg - ch), \quad (h'f' - b'g') = D^2(hf - bg), \text{ etc.}$$

These results are examples of the general theorems of § 124.4 or § 121.

Ex. iv. The rank of Φ is determined by the rank of ϕ as shown in the following table.

Rank of ϕ	3	2	1	0
Rank of Φ	3	1	0	0

This can be deduced from § 73; and can be proved independently with the help of Exs. ii and iii.

Ex. v. If ϕ has rank 1, then a, b, c do not all vanish (see § 125).

If ϕ has rank 2, then Φ has rank 1, and therefore A, B, C do not all vanish.

§ 129. **The symmetric matrix of order 4.**

For this matrix and its determinant we shall generally use the notations

$$\phi=\begin{bmatrix} a & h & g & u \\ h & b & f & v \\ g & f & c & w \\ u & v & w & d \end{bmatrix}, \quad \Delta=\det\phi=\begin{vmatrix} a & h & g & u \\ h & b & f & v \\ g & f & c & w \\ u & v & w & d \end{vmatrix}. \quad \text{.................(1)}$$

The reciprocal or conjugate reciprocal of ϕ will be denoted by

$$\Phi=\begin{bmatrix} A & H & G & U \\ H & B & F & V \\ G & F & C & W \\ U & V & W & D \end{bmatrix}=\begin{bmatrix} (bcd), & -(cdh), & -(bdg), & -(bcu) \\ -(cdh), & (cad), & -(adf), & -(cav) \\ -(bdg), & -(adf), & (abd), & -(abw) \\ -(bcu), & -(cav), & -(abw), & (abc) \end{bmatrix}. \quad \text{......(2)}$$

In the second of the notations (2) the three letters of each bracket can be taken in any order and represent without ambiguity the value of a derived determinant of ϕ in which they are the elements of the leading diagonal. We have

$$A=bcd+2wvf-bw^2-cv^2-df^2, \quad B=cad+2uwg-cu^2-aw^2-dg^2,$$
$$C=abd+2vuh-av^2-bu^2-dh^2, \quad D=abc+2fgh-af^2-bg^2-ch^2,$$
$$-F=adf+(vg+wh)u-avw-dgh-fu^2,$$
$$-G=bdg+(wh+uf)v-bwu-dhf-gv^2,$$
$$-H=cdh+(uf+vg)w-cuv-dfg-hw^2,$$
$$-U=bcu+(vg+wh)f-bgw-chv-uf^2,$$
$$-V=cav+(wh+uf)g-chu-afw-vg^2,$$
$$-W=abw+(uf+vg)h-afv-bgu-wh^2.$$

The matrix of the minor determinants of order 2 of ϕ in which the successive selections of two horizontal rows and the successive selections of two vertical rows follow the common scheme (23, 31, 12, 14, 24, 34) will be denoted by

$$\Psi=\begin{bmatrix} A_1 & H_1 & G_1 & U_0 & W_1 & V_2 \\ H_1 & B_1 & F_1 & W_2 & V_0 & U_1 \\ G_1 & F_1 & C_1 & V_1 & U_2 & W_0 \\ U_0 & W_2 & V_1 & A_2 & H_2 & G_2 \\ W_1 & V_0 & U_2 & H_2 & B_2 & F_2 \\ V_2 & U_1 & W_0 & G_2 & F_2 & C_2 \end{bmatrix}$$

$$=\begin{bmatrix} bc-f^2, & fg-ch, & hf-bg, & hw-gv, & bw-fv, & fw-cv \\ fg-ch, & ca-g^2, & gh-af, & gu-aw, & fu-hw, & cu-gw \\ hf-bg, & gh-af, & ab-h^2, & av-hu, & hv-bu, & gv-fu \\ hw-gv, & gu-aw, & av-hu, & ad-u^2, & dh-uv, & dg-wu \\ bw-fv, & fu-hw, & hv-bu, & dh-uv, & bd-v^2, & df-vw \\ fw-cv, & cu-gw, & gv-fu, & dg-wu, & df-vw, & cd-w^2 \end{bmatrix} \quad \text{.........(3)}$$

The co-joint matrix of Ψ with respect to ϕ is the matrix Ψ' given by

$$\Psi'=\begin{bmatrix} A_2 & H_2 & G_2 & U_0 & W_2 & V_1 \\ H_2 & B_2 & F_2 & W_1 & V_0 & U_2 \\ G_2 & F_2 & C_2 & V_2 & U_1 & W_0 \\ U_0 & W_1 & V_2 & A_1 & H_1 & G_1 \\ W_2 & V_0 & U_1 & H_1 & B_1 & F_1 \\ V_1 & U_2 & W_0 & G_1 & F_1 & C_1 \end{bmatrix} \quad \text{...........................(4)}$$

Corresponding elements of Ψ and Ψ' are co-factors of one another in ϕ. Thus the co-factors in ϕ of

$$A_1, B_1, C_1, F_1, G_1, H_1;\; A_2, B_2, C_2, F_2, G_2, H_2;\; U_0, V_0, W_0;\; U_1, V_1, W_1;\; U_2, V_2, W_2$$

are respectively

$$A_2, B_2, C_2, F_2, G_2, H_2;\; A_1, B_1, C_1, F_1, G_1, H_1;\; U_0, V_0, W_0;\; U_2, V_2, W_2;\; U_1, V_1, W_1.$$

The following examples deal with properties of the matrices ϕ, Φ, Ψ, Ψ'.

Ex. i.
$$\begin{bmatrix} a & h & g & u \\ h & b & f & v \\ g & f & c & w \\ u & v & w & d \end{bmatrix}\begin{bmatrix} A & H & G & U \\ H & B & F & V \\ G & F & C & W \\ U & V & W & D \end{bmatrix}=\begin{bmatrix} \Delta & 0 & 0 & 0 \\ 0 & \Delta & 0 & 0 \\ 0 & 0 & \Delta & 0 \\ 0 & 0 & 0 & \Delta \end{bmatrix}, \quad \ldots\ldots\ldots\ldots\ldots\ldots(\mathrm{A})$$

or
$$\phi\Phi=\Phi\phi=\Delta\,[1]_4^4.$$

This is the fundamental property of the reciprocal matrix proved in Ex. i of § 28.

Ex. ii.
$$\begin{bmatrix} A_1 & H_1 & G_1 & U_0 & W_1 & V_2 \\ H_1 & B_1 & F_1 & W_2 & V_0 & U_1 \\ G_1 & F_1 & C_1 & V_1 & U_2 & W_0 \\ U_0 & W_2 & V_1 & A_2 & H_2 & G_2 \\ W_1 & V_0 & U_2 & H_2 & B_2 & F_2 \\ V_2 & U_1 & W_0 & G_2 & F_2 & C_2 \end{bmatrix}\begin{bmatrix} A_2 & H_2 & G_2 & U_0 & W_2 & V_1 \\ H_2 & B_2 & F_2 & W_1 & V_0 & U_2 \\ G_2 & F_2 & C_2 & V_2 & U_1 & W_0 \\ U_0 & W_1 & V_2 & A_1 & H_1 & G_1 \\ W_2 & V_0 & U_1 & H_1 & B_1 & F_1 \\ V_1 & U_2 & W_0 & G_1 & F_1 & C_1 \end{bmatrix}=\begin{bmatrix} \Delta & 0 & 0 & 0 & 0 & 0 \\ 0 & \Delta & 0 & 0 & 0 & 0 \\ 0 & 0 & \Delta & 0 & 0 & 0 \\ 0 & 0 & 0 & \Delta & 0 & 0 \\ 0 & 0 & 0 & 0 & \Delta & 0 \\ 0 & 0 & 0 & 0 & 0 & \Delta \end{bmatrix},$$
$$\ldots\ldots\ldots(\mathrm{B})$$

or
$$\Psi\Psi'=\Psi'\Psi=\Delta\,[1]_6^6.$$

This is the fundamental property of the two co-joint matrices Ψ and Ψ' proved in § 119 and deducible from Ex. ix of § 32.

Since Ψ and Ψ' are symmetric, each of them is both the co-joint matrix and the conjugate co-joint matrix of the other, and the product on the left in (B) is commutative.

Ex. iii. *Determinant of* Φ. Equating the determinants of both sides in (A) we obtain

$$\Delta \det \Phi=\Delta^4, \quad \text{or} \quad \det \Phi=\Delta^3,$$

where the cancellation of the factor Δ common to both sides is allowable because Δ, regarded as a function of the elements of ϕ, does not vanish identically. Thus we have

$$\det \Phi=\begin{vmatrix} A & H & G & U \\ H & B & F & V \\ G & F & C & W \\ U & V & W & D \end{vmatrix}=\Delta^3. \quad \ldots\ldots\ldots\ldots\ldots\ldots\ldots\ldots\ldots\ldots(\mathrm{C})$$

This is a particular case of formula (B) of § 124 and of formula (A) of § 120.

Ex. iv. *Determinants of* Ψ *and* Ψ'. Clearly the determinants of Ψ and Ψ' are equal in value. Writing $\det\Psi=\det\Psi'=\Omega$, and equating the determinants of both sides in (B), we obtain

$$\Omega^2=\Delta^6, \qquad \Omega=\pm\Delta^3.$$

Since the last equation is an identity in the elements of ϕ, the consideration of the special case in which ϕ is a unit matrix shows that the upper sign must be taken.

Accordingly we have

$$\det\Psi=\det\Psi'=\begin{vmatrix} A_1 & H_1 & G_1 & U_0 & W_1 & V_2 \\ H_1 & B_1 & F_1 & W_2 & V_0 & U_1 \\ G_1 & F_1 & C_1 & V_1 & U_2 & W_0 \\ U_0 & W_2 & V_1 & A_2 & H_2 & G_2 \\ W_1 & V_0 & U_2 & H_2 & B_2 & F_2 \\ V_2 & U_1 & W_0 & G_2 & F_2 & C_2 \end{vmatrix}=\Delta^3. \quad\text{..................(D)}$$

This result is a special case of formula (A) of § 120.

Ex. v. *Minor determinants of* Φ *of order* 3. Let the reciprocal matrix of Φ, which is a matrix of its affected minor determinants of order 3, be

$$\phi'=\begin{bmatrix} a' & h' & g' & u' \\ h' & b' & f' & v' \\ g' & f' & c' & w' \\ u' & v' & w' & d' \end{bmatrix}. \quad\text{.......................................(5)}$$

Equating correspondingly formed matrices of the minor determinants of order 3 on both sides of (A) we obtain

$$\Phi\phi'=\Delta^3[1]_4^4.$$

Prefixing ϕ on both sides and making use of (A), we obtain

$$\Delta\phi'=\Delta^3\phi, \quad\text{or}\quad \phi'=\Delta^2\phi,$$

i.e.

$$\begin{bmatrix} a' & h' & g' & u' \\ h' & b' & f' & v' \\ g' & f' & c' & w' \\ u' & v' & w' & d' \end{bmatrix}=\Delta^2\begin{bmatrix} a & h & g & u \\ h & b & f & v \\ g & f & c & w \\ u & v & w & d \end{bmatrix} \quad\text{............................(E)}$$

We have therefore

$$a'=\begin{vmatrix} B & F & V \\ F & C & W \\ V & W & D \end{vmatrix}=\Delta^2 a, \quad h'=-\begin{vmatrix} H & F & V \\ G & C & W \\ U & W & D \end{vmatrix}=\Delta^2 h, \quad g'=\begin{vmatrix} H & B & V \\ G & F & W \\ U & V & D \end{vmatrix}=\Delta^2 g,$$

and so on. These are examples of the general results obtained in §§ 124 and 121.

Ex. vi. *Minor determinants of* Φ *of order* 2.

Let

$$\psi=\begin{bmatrix} a_1 & h_1 & g_1 & u_0 & w_1 & v_2 \\ h_1 & b_1 & f_1 & w_2 & v_0 & u_1 \\ g_1 & f_1 & c_1 & v_1 & u_2 & w_0 \\ u_0 & w_2 & v_1 & a_2 & h_2 & g_2 \\ w_1 & v_0 & u_2 & h_2 & b_2 & f_2 \\ v_2 & u_1 & w_0 & g_2 & f_2 & c_2 \end{bmatrix}, \quad \psi'=\begin{bmatrix} a_2 & h_2 & g_2 & u_0 & w_2 & v_1 \\ h_2 & b_2 & f_2 & w_1 & v_0 & u_2 \\ g_2 & f_2 & c_2 & v_2 & u_1 & w_0 \\ u_0 & w_1 & v_2 & a_1 & h_1 & g_1 \\ w_2 & v_0 & u_1 & h_1 & b_1 & f_1 \\ v_1 & u_2 & w_0 & g_1 & f_1 & c_1 \end{bmatrix} \quad\text{............(6)}$$

be the two co-joint complete matrices of the minor determinants of order 2 of Φ formed from Φ in the same way as Ψ and Ψ' are formed from ϕ; so that

$$a_1=(BC-F^2), \quad a_2=(AD-U^2), \quad h_1=(FG-CH), \quad h_2=(DH-UV), \quad u_0=(HW-GV),$$

and so on.

Equating correspondingly formed matrices of the minor determinants of order 2 on both sides of (A), we obtain the two equivalent equations

$$\Psi\psi = \Delta^2 [1]_6^6, \quad \Psi'\psi' = \Delta^2 [1]_6^6.$$

Prefixing Ψ' in the first case and Ψ in the second case, making use of (B), and cancelling Δ on both sides, we obtain the two equivalent equations

$$\psi = \Delta\Psi', \quad \psi' = \Delta\Psi,$$

or

$$\begin{bmatrix} a_1 & h_1 & g_1 & u_0 & w_1 & v_2 \\ h_1 & b_1 & f_1 & w_2 & v_0 & u_1 \\ g_1 & f_1 & c_1 & v_1 & u_2 & w_0 \\ u_0 & w_2 & v_1 & a_2 & h_2 & g_2 \\ w_1 & v_0 & u_2 & h_2 & b_2 & f_2 \\ v_2 & u_1 & w_0 & g_2 & f_2 & c_2 \end{bmatrix} = \Delta \begin{bmatrix} A_2 & H_2 & G_2 & U_0 & W_2 & V_1 \\ H_2 & B_2 & F_2 & W_1 & V_0 & U_2 \\ G_2 & F_2 & C_2 & V_2 & U_1 & W_0 \\ U_0 & W_1 & V_2 & A_1 & H_1 & G_1 \\ W_2 & V_0 & U_1 & H_1 & B_1 & F_1 \\ V_1 & U_2 & W_0 & G_1 & F_1 & C_1 \end{bmatrix}. \quad \text{............(F)}$$

This shows that

$$a_1 = (BC - F^2) = \Delta A_2 = \Delta(ad - u^2), \quad u_0 = (HW - GV) = \Delta U_0 = \Delta(hw - gv),$$

$$h_1 = (FG - CH) = \Delta H_2 = \Delta(dh - uv), \quad w_1 = (BW - FV) = \Delta W_2 = \Delta(gu - aw),$$

$$g_1 = (HF - BG) = \Delta G_2 = \Delta(dg - wu), \quad v_2 = (FW - CV) = \Delta V_1 = \Delta(av - hu),$$

and so on.

These results can be obtained at once from the theorems of § 121 or from those of § 124.4.

Ex. vii. *Minor determinants of Ψ of order* 5. Let $[A]_6^6$ be the reciprocal of Ψ, which is a complete matrix of the affected minor determinants of order 5 of Ψ. Since this matrix is symmetric, it is also the conjugate reciprocal of Ψ, and we have therefore

$$[A]_6^6 \Psi = \det \Psi . [1]_6^6 = \Delta^3 [1]_6^6.$$

Postfixing Ψ' on both sides, and making use of (B), we obtain

$$[A]_6^6 = \Delta^2\Psi' = \Delta^2 \begin{bmatrix} A_2 & H_2 & G_2 & U_0 & W_2 & V_1 \\ H_2 & B_2 & F_2 & W_1 & V_0 & U_2 \\ G_2 & F_2 & C_2 & V_2 & U_1 & W_0 \\ U_0 & W_1 & V_2 & A_1 & H_1 & G_1 \\ W_2 & V_0 & U_1 & H_1 & B_1 & F_1 \\ V_1 & U_2 & W_0 & G_1 & F_1 & C_1 \end{bmatrix}. \quad \text{..................(G)}$$

This equation can be deduced at once from § 121.2.

Ex. viii. *Minor determinants of Ψ of order* 4. Let $[\mathbf{A}]_{15}^{15}$ be a complete matrix of the minor determinants of order 4 of Ψ in the formation of which a common scheme is used for horizontal and vertical rows, and let $[\mathbf{B}']_{15}^{15}$ be the anti-correspondent complete matrix of the minor determinants of order 2 of Ψ'. Then by § 121.2 we have

$$[\mathbf{A}]_{15}^{15} = \Delta [\mathbf{B}']_{15}^{15}. \quad \text{.....................................(H)}$$

From equation (H) we see that when ϕ is degenerate, i.e. when $\Delta=0$, the rank of Ψ cannot exceed 3.

Ex. ix. The reciprocals of the minor matrices of Ψ

$$\begin{bmatrix} A_1 & H_1 & G_1 \\ H_1 & B_1 & F_1 \\ G_1 & F_1 & C_1 \end{bmatrix}, \quad \begin{bmatrix} A_2 & H_2 & G_2 \\ H_2 & B_2 & F_2 \\ G_2 & F_2 & C_2 \end{bmatrix}, \quad \begin{bmatrix} U_0, & W_1, & V_2 \\ W_2, & V_0, & U_1 \\ V_1, & U_2, & W_0 \end{bmatrix},$$

$$\begin{bmatrix} C_2, & -F_2, & V_2 \\ -F_2, & B_2, & -W_1 \\ V_2, & -W_1, & A_1 \end{bmatrix}, \quad \begin{bmatrix} C_1, & -F_1, & V_1 \\ -F_1, & B_1, & -W_2 \\ V_1, & -W_2, & A_2 \end{bmatrix}$$

are respectively

$$D\begin{bmatrix} a & h & g \\ h & b & f \\ g & f & c \end{bmatrix}, \quad d\begin{bmatrix} A & H & G \\ H & B & F \\ G & F & C \end{bmatrix}, \quad -\begin{bmatrix} Uu, & Vu, & Wu \\ Uv, & Vv, & Wv \\ Uw, & Vw, & Ww \end{bmatrix}, \quad A\begin{bmatrix} b & f & v \\ f & c & w \\ v & w & d \end{bmatrix}, \quad a\begin{bmatrix} B & F & V \\ F & C & W \\ V & W & D \end{bmatrix},$$

and the values of their determinants are respectively

$$D^2, \quad d^2\Delta, \quad 0, \quad A^2, \quad a^2\Delta.$$

Ex. x. *Ranks of* Φ *and* Ψ.

By Theorem II of § 73 we see that the ranks of Φ and Ψ are determined by the rank of ϕ as shown in the following table.

Rank of ϕ	4	3	2	1	0
Rank of Φ	4	1	0	0	0
Rank of Ψ	6	3	1	0	0

Most of these results can be deduced from the preceding examples.

If the rank of ϕ *is* 4, so that $\Delta \neq 0$, it follows from Exs. iii and iv that Φ has rank 4 and Ψ rank 6.

If the rank of ϕ *is* 3, i.e. if $\Delta=0$ but $\Phi \neq 0$, it follows from (E) and (F) or from § 124.6 that Φ has rank 1. Again by Ex. viii the rank of Ψ cannot exceed 3. Further since Φ has rank 1, we know by § 125 that A, B, C, D do not all vanish, and (using Ex. ix)

$$\begin{vmatrix} A_1 & W_1 & V_2 \\ W_1 & B_2 & F_2 \\ V_2 & F_2 & C_2 \end{vmatrix} = A^2, \quad \begin{vmatrix} B_1 & U_1 & W_2 \\ U_1 & C_2 & G_2 \\ W_2 & G_2 & A_2 \end{vmatrix} = B^2, \quad \begin{vmatrix} C_1 & V_1 & U_2 \\ V_1 & A_2 & H_2 \\ U_2 & H_2 & B_2 \end{vmatrix} = C^2, \quad \begin{vmatrix} A_1 & H_1 & G_1 \\ H_1 & B_1 & F_1 \\ G_1 & F_1 & C_1 \end{vmatrix} = D^2,$$

which do not all vanish, occur amongst the minor determinants of Ψ of order 3. Consequently in this case Ψ has rank 3.

Ex. xi. The minor determinants of ϕ occurring in Ψ satisfy the identities

$$A_1 U_0 + H_1 W_1 + G_1 V_2 = 0, \quad A_2 U_0 + H_2 W_2 + G_2 V_1 = 0,$$
$$H_1 W_2 + B_1 V_0 + F_1 U_1 = 0, \quad H_2 W_1 + B_2 V_0 + F_2 U_2 = 0,$$
$$G_1 V_1 + F_1 U_2 + C_1 W_0 = 0, \quad G_2 V_2 + F_2 U_1 + C_2 W_0 = 0.$$

These are the identical relations which by § 109 exist between the simple minor determinants of each simple minor matrix of ϕ of reduced order 2. They are included in equation (B).

Ex. xii. By multiplying the matrix of each minor determinant of order 3 of ϕ by its reciprocal matrix, and by multiplying the matrix of each minor determinant of order 3 of Φ by its reciprocal matrix, we obtain the equations :

(1)
$$\begin{bmatrix} b & f & v \\ f & c & w \\ v & w & d \end{bmatrix} \begin{bmatrix} C_2, & -F_2, & V_2 \\ -F_2, & B_2, & -W_1 \\ V_2, & -W_1, & A_1 \end{bmatrix} = A \begin{bmatrix} 1 & 0 & 0 \\ 0 & 1 & 0 \\ 0 & 0 & 1 \end{bmatrix},$$

$$\begin{bmatrix} c & g & w \\ g & a & u \\ w & u & d \end{bmatrix} \begin{bmatrix} A_2, & -G_2, & W_2 \\ -G_2, & C_2, & -U_1 \\ W_2, & -U_1, & B_1 \end{bmatrix} = B \begin{bmatrix} 1 & 0 & 0 \\ 0 & 1 & 0 \\ 0 & 0 & 1 \end{bmatrix},$$

$$\begin{bmatrix} a & h & u \\ h & b & v \\ u & v & d \end{bmatrix} \begin{bmatrix} B_2, & -H_2, & U_2 \\ -H_2, & A_2, & -V_1 \\ U_2, & -V_1, & C_1 \end{bmatrix} = C \begin{bmatrix} 1 & 0 & 0 \\ 0 & 1 & 0 \\ 0 & 0 & 1 \end{bmatrix},$$

$$\begin{bmatrix} a & h & g \\ h & b & f \\ g & f & c \end{bmatrix} \begin{bmatrix} A_1 & H_1 & G_1 \\ H_1 & B_1 & F_1 \\ G_1 & F_1 & C_1 \end{bmatrix} = D \begin{bmatrix} 1 & 0 & 0 \\ 0 & 1 & 0 \\ 0 & 0 & 1 \end{bmatrix},$$

$$\begin{bmatrix} f & g & w \\ h & a & u \\ v & u & d \end{bmatrix} \begin{bmatrix} A_2, & -G_2, & W_2 \\ -H_2, & F_2, & -V_0 \\ -V_1, & W_0, & -F_1 \end{bmatrix} = -F \begin{bmatrix} 1 & 0 & 0 \\ 0 & 1 & 0 \\ 0 & 0 & 1 \end{bmatrix},$$

$$\begin{bmatrix} g & h & u \\ f & b & v \\ w & v & d \end{bmatrix} \begin{bmatrix} B_2, & -H_2, & U_2 \\ -F_2, & G_2, & -W_0 \\ -W_1, & U_0, & -G_1 \end{bmatrix} = -G \begin{bmatrix} 1 & 0 & 0 \\ 0 & 1 & 0 \\ 0 & 0 & 1 \end{bmatrix},$$

$$\begin{bmatrix} h & f & v \\ g & c & w \\ u & w & d \end{bmatrix} \begin{bmatrix} C_2, & -F_2, & V_2 \\ -G_2, & H_2, & -U_0 \\ -U_1, & V_0, & -H_1 \end{bmatrix} = -H \begin{bmatrix} 1 & 0 & 0 \\ 0 & 1 & 0 \\ 0 & 0 & 1 \end{bmatrix},$$

$$\begin{bmatrix} u & v & w \\ h & b & f \\ g & f & c \end{bmatrix} \begin{bmatrix} A_1, & V_2, & -W_1 \\ H_1, & U_1, & -V_0 \\ G_1, & W_0, & -U_2 \end{bmatrix} = -U \begin{bmatrix} 1 & 0 & 0 \\ 0 & 1 & 0 \\ 0 & 0 & 1 \end{bmatrix},$$

$$\begin{bmatrix} v & w & u \\ f & c & g \\ h & g & a \end{bmatrix} \begin{bmatrix} B_1, & W_2, & -U_1 \\ F_1, & V_1, & -W_0 \\ H_1, & U_0, & -V_2 \end{bmatrix} = -V \begin{bmatrix} 1 & 0 & 0 \\ 0 & 1 & 0 \\ 0 & 0 & 1 \end{bmatrix},$$

$$\begin{bmatrix} w & u & v \\ g & a & h \\ f & h & b \end{bmatrix} \begin{bmatrix} C_1, & U_2, & -V_1 \\ G_1, & W_1, & -U_0 \\ F_1, & V_0, & -W_2 \end{bmatrix} = -W \begin{bmatrix} 1 & 0 & 0 \\ 0 & 1 & 0 \\ 0 & 0 & 1 \end{bmatrix}.$$

(2)

$$\begin{bmatrix} B & F & V \\ F & C & W \\ V & W & D \end{bmatrix} \begin{bmatrix} C_1, & -F_1, & V_1 \\ -F_1, & B_1, & -W_2 \\ V_1, & -W_2, & A_2 \end{bmatrix} = \Delta a \begin{bmatrix} 1 & 0 & 0 \\ 0 & 1 & 0 \\ 0 & 0 & 1 \end{bmatrix},$$

$$\begin{bmatrix} C & G & W \\ G & A & U \\ W & U & D \end{bmatrix} \begin{bmatrix} A_1, & -G_1, & W_1 \\ -G_1, & C_1, & -U_2 \\ W_1, & -U_2, & B_2 \end{bmatrix} = \Delta b \begin{bmatrix} 1 & 0 & 0 \\ 0 & 1 & 0 \\ 0 & 0 & 1 \end{bmatrix},$$

$$\begin{bmatrix} A & H & U \\ H & B & V \\ U & V & D \end{bmatrix} \begin{bmatrix} B_1, & -H_1, & U_1 \\ -H_1, & A_1, & -V_2 \\ U_1, & -V_2, & C_2 \end{bmatrix} = \Delta c \begin{bmatrix} 1 & 0 & 0 \\ 0 & 1 & 0 \\ 0 & 0 & 1 \end{bmatrix},$$

$$\begin{bmatrix} A & H & G \\ H & B & F \\ G & F & C \end{bmatrix} \begin{bmatrix} A_2 & H_2 & G_2 \\ H_2 & B_2 & F_2 \\ G_2 & F_2 & C_2 \end{bmatrix} = \Delta d \begin{bmatrix} 1 & 0 & 0 \\ 0 & 1 & 0 \\ 0 & 0 & 1 \end{bmatrix},$$

$$\begin{bmatrix} F & G & W \\ H & A & U \\ V & U & D \end{bmatrix} \begin{bmatrix} A_1, & -G_1, & W_1 \\ -H_1, & F_1, & -V_0 \\ -V_2, & W_0, & -F_2 \end{bmatrix} = -\Delta f \begin{bmatrix} 1 & 0 & 0 \\ 0 & 1 & 0 \\ 0 & 0 & 1 \end{bmatrix},$$

$$\begin{bmatrix} G & H & U \\ F & B & V \\ W & V & D \end{bmatrix} \begin{bmatrix} B_1, & -H_1, & U_1 \\ -F_1, & G_1, & -W_0 \\ -W_2, & U_0, & -G_2 \end{bmatrix} = -\Delta g \begin{bmatrix} 1 & 0 & 0 \\ 0 & 1 & 0 \\ 0 & 0 & 1 \end{bmatrix},$$

$$\begin{bmatrix} H & F & V \\ G & C & W \\ U & W & D \end{bmatrix} \begin{bmatrix} C_1, & -F_1, & V_1 \\ -G_1, & H_1, & -U_0 \\ -U_2, & V_0, & -H_2 \end{bmatrix} = -\Delta h \begin{bmatrix} 1 & 0 & 0 \\ 0 & 1 & 0 \\ 0 & 0 & 1 \end{bmatrix},$$

$$\begin{bmatrix} U & V & W \\ H & B & F \\ G & F & C \end{bmatrix} \begin{bmatrix} A_2, & V_1, & -W_2 \\ H_2, & U_2, & -V_0 \\ G_2, & W_0, & -U_1 \end{bmatrix} = -\Delta u \begin{bmatrix} 1 & 0 & 0 \\ 0 & 1 & 0 \\ 0 & 0 & 1 \end{bmatrix},$$

$$\begin{bmatrix} V & W & U \\ F & C & G \\ H & G & A \end{bmatrix} \begin{bmatrix} B_2, & W_1, & -U_2 \\ F_2, & V_2, & -W_0 \\ H_2, & U_0, & -V_1 \end{bmatrix} = -\Delta v \begin{bmatrix} 1 & 0 & 0 \\ 0 & 1 & 0 \\ 0 & 0 & 1 \end{bmatrix},$$

$$\begin{bmatrix} W & U & V \\ G & A & H \\ F & H & B \end{bmatrix} \begin{bmatrix} C_2, & U_1, & -V_2 \\ G_2, & W_2, & -U_0 \\ F_2, & V_0, & -W_1 \end{bmatrix} = -\Delta w \begin{bmatrix} 1 & 0 & 0 \\ 0 & 1 & 0 \\ 0 & 0 & 1 \end{bmatrix}$$

Every one of these equations remains true when we replace both the two factor matrices on the left by their conjugates. This principle adds six other equations of the same character to each of the above two sets.

Ex. xiii. The first set of 144 scalar equations given by the matrix equations of Ex. xii can be expressed in the following more succinct forms:

$$\begin{bmatrix} h & g & u \\ b & f & v \\ f & c & w \\ v & w & d \end{bmatrix} \begin{bmatrix} V_2, & U_1, & W_0, & G_2, & F_2, & C_2 \\ -W_1, & -V_0, & -U_2, & -H_2, & -B_2, & -F_2 \\ A_1, & H_1, & G_1, & U_0, & W_1, & V_2 \end{bmatrix}$$

$$= \begin{bmatrix} -U, & 0, & 0, & 0, & G, & -H \\ 0, & -U, & 0, & -G, & 0, & A \\ 0, & 0, & -U, & H, & -A, & 0 \\ A, & H, & G, & 0, & 0, & 0 \end{bmatrix},$$

$$\begin{bmatrix} a & g & u \\ h & f & v \\ g & c & w \\ u & w & d \end{bmatrix} \begin{bmatrix} -V_2, & -U_1, & -W_0, & -G_2, & -F_2, & -C_2 \\ U_0, & W_2, & V_1, & A_2, & H_2, & G_2 \\ H_1, & B_1, & F_1, & W_2, & V_0, & U_1 \end{bmatrix}$$

$$= \begin{bmatrix} -V, & 0, & 0, & 0, & F, & -B \\ 0, & -V, & 0, & -F, & 0, & H \\ 0, & 0, & -V, & B, & -H, & 0 \\ H, & B, & F, & 0, & 0, & 0 \end{bmatrix},$$

$$\begin{bmatrix} a & h & u \\ h & b & v \\ g & f & w \\ u & v & d \end{bmatrix} \begin{bmatrix} W_1, & V_0, & U_2, & H_2, & B_2, & F_2 \\ -U_0, & -W_2, & -V_1, & -A_2, & -H_2, & -G_2 \\ G_1, & F_1, & C_1, & V_1, & U_2, & W_0 \end{bmatrix}$$

$$= \begin{bmatrix} -W, & 0, & 0, & 0, & C, & -F \\ 0, & -W, & 0, & -C, & 0, & G \\ 0, & 0, & -W, & F, & -G, & 0 \\ G, & F, & C, & 0, & 0, & 0 \end{bmatrix},$$

$$\begin{bmatrix} a & h & g \\ h & b & f \\ g & f & c \\ u & v & w \end{bmatrix} \begin{bmatrix} A_1 & H_1 & G_1 & U_0 & W_1 & V_2 \\ H_1 & B_1 & F_1 & W_2 & V_0 & U_1 \\ G_1 & F_1 & C_1 & V_1 & U_2 & W_0 \end{bmatrix} = \begin{bmatrix} D, & 0, & 0, & 0, & -W, & V \\ 0, & D, & 0, & W, & 0, & -U \\ 0, & 0, & D, & -V, & U, & 0 \\ -U, & -V, & -W, & 0, & 0, & 0 \end{bmatrix}.$$

In each of these results the elements of the second matrix on the left and the elements of the matrix on the right are respectively minor determinants of orders 2 and 3 of the first matrix on the left.

Ex. xiv. The second set of 144 scalar equations given by the matrix equations of Ex. xii can be expressed in the corresponding forms :

$$\begin{bmatrix} H & G & U \\ B & F & V \\ F & C & W \\ V & W & D \end{bmatrix} \begin{bmatrix} V_1, & U_2, & W_0, & G_1, & F_1, & C_1 \\ -W_2, & -V_0, & -U_1, & -H_1, & -B_1, & -F_1 \\ A_2, & H_2, & G_2, & U_0, & W_2, & V_1 \end{bmatrix}$$

$$= \Delta \begin{bmatrix} -u, & 0, & 0, & 0, & g, & -h \\ 0, & -u, & 0, & -g, & 0, & a \\ 0, & 0, & -u, & h, & -a, & 0 \\ a, & h, & g, & 0, & 0, & 0 \end{bmatrix},$$

$$\begin{bmatrix} A & G & U \\ H & F & V \\ G & C & W \\ U & W & D \end{bmatrix} \begin{bmatrix} -V_1, & -U_2, & -W_0, & -G_1, & -F_1, & -C_1 \\ U_0, & W_1, & V_2, & A_1, & H_1, & G_1 \\ H_2, & B_2, & F_2, & W_1, & V_0, & U_2 \end{bmatrix}$$

$$= \Delta \begin{bmatrix} -v, & 0, & 0, & 0, & f, & -b \\ 0, & -v, & 0, & -f, & 0, & h \\ 0, & 0, & -v, & b, & -h, & 0 \\ h, & b, & f, & 0, & 0, & 0 \end{bmatrix},$$

$$\begin{bmatrix} A & H & U \\ H & B & V \\ G & F & W \\ U & V & D \end{bmatrix} \begin{bmatrix} W_2, & V_0, & U_1, & H_1, & B_1, & F_1 \\ -U_0, & -W_1, & -V_2, & -A_1, & -H_1, & -G_1 \\ G_2, & F_2, & C_2, & V_2, & U_1, & W_0 \end{bmatrix}$$

$$= \Delta \begin{bmatrix} -w, & 0, & 0, & 0, & c, & -f \\ 0, & -w, & 0, & -c, & 0, & g \\ 0, & 0, & -w, & f, & -g, & 0 \\ g, & f, & c, & 0, & 0, & 0 \end{bmatrix},$$

$$\begin{bmatrix} A & H & G \\ H & B & F \\ G & F & C \\ U & V & W \end{bmatrix} \begin{bmatrix} A_2 & H_2 & G_2 & U_0 & W_2 & V_1 \\ H_2 & B_2 & F_2 & W_1 & V_0 & U_2 \\ G_2 & F_2 & C_2 & V_2 & U_1 & W_0 \end{bmatrix} = \Delta \begin{bmatrix} d, & 0, & 0, & 0, & -w, & v \\ 0, & d, & 0, & w, & 0, & -u \\ 0, & 0, & d, & -v, & u, & 0 \\ -u, & -v, & -w, & 0, & 0, & 0 \end{bmatrix}.$$

Ex. xv. For the matrix $[a\,b\,c\,d]_{1234}$, we have the identity

$$\begin{bmatrix} b_1 & c_1 & d_1 \\ b_2 & c_2 & d_2 \\ b_3 & c_3 & d_3 \\ b_4 & c_4 & d_4 \end{bmatrix} \begin{bmatrix} (cd)_{23} & (cd)_{31} & (cd)_{12} & (cd)_{14} & (cd)_{24} & (cd)_{34} \\ (db)_{23} & (db)_{31} & (db)_{12} & (db)_{14} & (db)_{24} & (db)_{34} \\ (bc)_{23} & (bc)_{31} & (bc)_{12} & (bc)_{14} & (bc)_{24} & (bc)_{34} \end{bmatrix}$$

$$= \begin{bmatrix} (bcd)_{123} & 0 & 0 & 0 & (bcd)_{124} & (bcd)_{134} \\ 0 & (bcd)_{231} & 0 & (bcd)_{214} & 0 & (bcd)_{234} \\ 0 & 0 & (bcd)_{312} & (bcd)_{314} & (bcd)_{324} & 0 \\ (bcd)_{423} & (bcd)_{431} & (bcd)_{412} & 0 & 0 & 0 \end{bmatrix}.$$

The equations in Exs. xiii and xiv are particular cases of this identity.

Ex. xvi. By means of Exs. xiii and xiv we can express every minor determinant of order 3 of Ψ in which only one of the 1st and 4th horizontal rows, only one of the 2nd and 5th horizontal rows and only one of the 3rd and 6th horizontal rows of Ψ occur in terms of the elements of ϕ and Φ.

Ex. xvii. Replacing the last equation of Ex. xiii by its conjugate and then multiplying it into the last equation of Ex. xiv we obtain

$$\begin{bmatrix} A_1 & H_1 & G_1 \\ H_1 & B_1 & F_1 \\ G_1 & F_1 & C_1 \\ U_0 & W_2 & V_1 \\ W_1 & V_0 & U_2 \\ V_2 & U_1 & W_0 \end{bmatrix} \begin{bmatrix} U_0 & W_2 & V_1 & A_2 & H_2 & G_2 \\ W_1 & V_0 & U_2 & H_2 & B_2 & F_2 \\ V_2 & U_1 & W_0 & G_2 & F_2 & C_2 \end{bmatrix}$$

$$= \begin{bmatrix} D, & 0, & 0, & U \\ 0, & D, & 0, & V \\ 0, & 0, & D, & W \\ 0, & W, & -V, & 0 \\ -W, & 0, & U, & 0 \\ V, & -U, & 0, & 0 \end{bmatrix} \begin{bmatrix} 0, & -w, & v, & d, & 0, & 0 \\ w, & 0, & -u, & 0, & d, & 0 \\ -v, & u, & 0, & 0, & 0, & d \\ 0, & 0, & 0, & u, & v, & w \end{bmatrix}.$$

From the three other pairs of corresponding equations in Exs. xiii and xiv we can deduce three similar equations.

§ 130. Identical relations between the elements of a square matrix.

Let $(a_{pq})_r^r$ be any minor determinant of order r of a square matrix $[a]_m^m$, and let $[\mu]_{m-r}$, $[\nu]_{m-r}$ be the corranged minors of the sequence $[1\,2\,\ldots\,m]$ complementary to $[p]_r$ and $[q]_r$. Then by § 116 we have the identity

$$(a_{pq})_r^r [a]_m^m - [a_{1q}]_m^r \boxed{A_{pq}}_r^r [a_{p1}]_r^m = [P]_m^m, \quad \ldots\ldots\ldots\ldots\ldots(\text{A})$$

where $[A_{pq}]_r^r$ is the reciprocal matrix of $[a_{pq}]_r^r$, and where

$$P_{uv} = \begin{pmatrix} q_1 q_2 \ldots q_r v \\ a \\ p_1 p_2 \ldots p_r u \end{pmatrix}.$$

The matrix $[P]_m^m$ differs by rows of 0's only from its minor matrix $[P_{\mu\nu}]_{m-r}^{m-r}$.

We will now prove the following theorem:

Theorem. *If $[\alpha]_m^m$ is the reciprocal of $[a]_m^m$, and if $[\beta_{\mu\nu}]_{m-r}^{m-r}$ is the reciprocal of $[\alpha_{\mu\nu}]_{m-r}^{m-r}$, then in formula* (A)

$$\Delta^{m-r-2} [P_{\mu\nu}]_{m-r}^{m-r} = (-1)^\omega [\beta_{\mu\nu}]_{m-r}^{m-r}, \quad \ldots\ldots\ldots\ldots\ldots\ldots(\text{B})$$

where $\Delta = (a)_m^m$, and ω is the affect of $[a_{pq}]_r^r$ in $[a]_m^m$.

Let $[c_{\mu\nu}]_{m-r}^{m-r}$ be the reciprocal of $[a_{\mu\nu}]_{m-r}^{m-r}$ and let $C_{\mu_i\nu_j}$ be the co-factor of $c_{\mu_i\nu_j}$ in $[a]_m^m$. Then by equation (F) in Ex. xii of § 116 we have

$$P_{\mu_i\nu_j} = (-1)^\omega C_{\mu_i\nu_j}. \quad \ldots\ldots\ldots\ldots\ldots\ldots\ldots\ldots\ldots(1)$$

Now $\beta_{\mu_i\nu_j}$ is a minor determinant of order $m-r-1$ of $[\alpha]_m^m$,

$c_{\mu_i\nu_j}$ is the corresponding minor determinant of $[a]_m^m$,

and $C_{\mu_i\nu_j}$ is the anti-correspondent minor determinant of $[a]_m^m$.

Therefore by formula (D) of § 124.4 we have

$$\beta_{\mu_i\nu_j} = \Delta^{m-r-2} C_{\mu_i\nu_j}. \quad \ldots\ldots\ldots\ldots\ldots\ldots\ldots\ldots\ldots(2)$$

From (1) and (2) it follows that

$$\Delta^{m-r-2} P_{\mu_i\nu_j} = (-1)^\omega \beta_{\mu_i\nu_j},$$

and this proves the truth of formula (B).

The theorem states that the minor matrix $[P_{\mu\nu}]_{m-r}^{m-r}$ derived from $[P]_m^m$ by striking out the rows of 0's differs only by a scalar factor of the form $\pm\Delta^k$ from the reciprocal of the correspondingly formed minor matrix $[\alpha_{\mu\nu}]_{m-r}^{m-r}$ of $[\alpha]_m^m$, the reciprocal of $[a]_m^m$.

It has been shown in Exs. viii and x of § 116 that if $[a]_m^m$ has rank ρ, then the rank of $[P_{\mu\nu}]_{m-r}^{m-r}$ or $[P]_m^m$ is $\rho - r$ when $(a_{pq})_r^r \neq 0$, either 0 or 1 when $(a_{pq})_r^r = 0$, and 0 when $\rho \ngtr r$.

Ex. i. Let $[a]_m^m = [a\,b\,c\,d\,e]_{12345}$, $(a_{pq})_r^r = (db)_{24}$, so that the identity (A) is

$$(db)_{24}[a\,b\,c\,d\,e]_{12345} - [d\,b]_{12345}\begin{bmatrix} b_4, & -b_2 \\ -d_4, & d_2 \end{bmatrix}[a\,b\,c\,d\,e]_{24}$$

$$= \begin{bmatrix} (dba)_{241}, & 0, & (dbc)_{241}, & 0, & (dbe)_{241} \\ 0 & ,0, & 0 & ,0, & 0 \\ (dba)_{243}, & 0, & (dbc)_{243}, & 0, & (dbe)_{243} \\ 0 & ,0, & 0 & ,0, & 0 \\ (dba)_{245}, & 0, & (dbc)_{245}, & 0, & (dbe)_{245} \end{bmatrix}.$$

In this case we have $m=5$, $r=2$, $\omega=7$, $\Delta^{m-r-2} = \Delta = (abcde)_{12345}$,

and $$[P_{\mu\nu}]_3^3 = \begin{bmatrix} (dba)_{241}, & (dbc)_{241}, & (dbe)_{241} \\ (dba)_{243}, & (dbc)_{243}, & (dbe)_{243} \\ (dba)_{245}, & (dbc)_{245}, & (dbe)_{245} \end{bmatrix}.$$

Denoting the reciprocal of $[a\,b\,c\,d\,e]_{12345}$ by $[A\,B\,C\,D\,E]_{12345}$, we have in this case

$$[\mathrm{a}_{\mu\nu}]_3^3 = \begin{bmatrix} A_1 & C_1 & E_1 \\ A_3 & C_3 & E_3 \\ A_5 & C_5 & E_5 \end{bmatrix}, \quad [\beta_{\mu\nu}]_3^3 = \begin{bmatrix} +(CE)_{35}, & -(AE)_{35}, & +(AC)_{35} \\ -(CE)_{15}, & +(AE)_{15}, & -(AC)_{15} \\ +(CE)_{13}, & -(AE)_{13}, & +(AC)_{13} \end{bmatrix}.$$

Evaluating the elements of the last matrix by means of formula (D) of § 124, we see that in accordance with the theorem

$$[\beta_{\mu\nu}]_3^3 = -\Delta[P_{\mu\nu}]_3^3.$$

Ex. ii. When $[a]_m^m$ is a *symmetric* matrix of rank ρ, and $(a_{pp})_r^r$ is one of its *diagonal* minor determinants of order r, the identity (A) is

$$(a_{pp})_r^r\,[a]_m^m - \boxed{a_{p1}}_m^r\,\boxed{A_{pp}}_r^r\,[a_{p1}]_r^m = [P]_m^m, \quad \text{.......................(A')}$$

where $$P_{uv} = \begin{pmatrix} p_1 p_2 \dots p_r v \\ a \\ p_1 p_2 \dots p_r u \end{pmatrix} = P_{vu}.$$

The matrix $[P]_m^m$ is symmetric, and if $[\mu]_{m-r}$ is the corranged minor of the sequence $[1\,2 \dots m]$ complementary to $[p]_r$, this matrix differs only by rows of 0's from its minor $[P_{\mu\mu}]_{m-r}^{m-r}$.

Further if $\Delta = (a)_m^m$, and if $[\mathrm{a}]_m^m$ is the reciprocal of $[a]_m^m$, then

$$\Delta^{m-r-2}[P_{\mu\mu}]_{m-r}^{m-r} = \text{the reciprocal of } [\mathrm{a}_{\mu\mu}]_{m-r}^{m-r}. \quad \text{..................(B')}$$

The rank of $[P_{\mu\mu}]_{m-r}^{m-r}$ is $\rho - r$ when $(a_{pp})_r^r \neq 0$, either 0 or 1 when $(a_{pp})_r^r = 0$, and 0 when $r \nless \rho$.

Ex. iii. If a_{uu} is any diagonal element of the symmetric matrix $[a]_m^m$ whose rank is ρ, we have

$$a_{uu}[a]_m^m = \begin{bmatrix} a_{u1} \\ a_{u2} \\ \vdots \\ a_{um} \end{bmatrix} [a_{u1}\, a_{u2} \dots a_{um}] + [P]_m^m, \quad \text{.......................(C)}$$

where $$P_{ij} = \begin{vmatrix} a_{uu} & a_{uj} \\ a_{iu} & a_{ij} \end{vmatrix} = a_{uu}a_{ij} - a_{ui}a_{uj} = P_{ji}.$$

The matrix $[P]_m^m$ is symmetric, and its uth horizontal and vertical rows are rows of 0's. If $[\lambda]_{m-1}$ is the minor of the sequence $[1\,2 \dots m]$ formed by striking out the element u, if $[A]_m^m$ is the reciprocal of $[a]_m^m$, and if $(a)_m^m = \Delta$, then

$$\Delta^{m-3}[P_{\lambda\lambda}]_{m-1}^{m-1} = \text{the reciprocal of } [A_{\lambda\lambda}]_{m-1}^{m-1}. \quad \text{.....................(D)}$$

The rank of $[P]_m^m$ is $\rho - 1$ when $a_{uu} \neq 0$, either 0 or 1 when $a_{uu} = 0$, and 0 when $\rho \ngtr 1$.

Ex. iv. If $\begin{vmatrix} a_{uu} & a_{uv} \\ a_{vu} & a_{vv} \end{vmatrix} = a_{uu}a_{vv} - a^2_{uv}$ is any diagonal minor determinant of order 2 of the symmetric matrix $[a]_m^m$ whose rank is ρ, we have

$$(a_{uu}a_{vv} - a^2_{uv})[a]_m^m = \begin{bmatrix} a_{u1} & a_{v1} \\ a_{u2} & a_{v2} \\ \dots & \dots \\ a_{um} & a_{vm} \end{bmatrix} \begin{bmatrix} a_{vv}, & -a_{uv} \\ -a_{vu}, & a_{uu} \end{bmatrix} \begin{bmatrix} a_{u1} & a_{u2} \dots a_{um} \\ a_{v1} & a_{v2} \dots a_{vm} \end{bmatrix} + [P]_m^m, \text{...(E)}$$

where $$P_{ij} = \begin{vmatrix} a_{uu} & a_{uv} & a_{uj} \\ a_{vu} & a_{vv} & a_{vj} \\ a_{iu} & a_{iv} & a_{ij} \end{vmatrix} = P_{ji}.$$

The matrix $[P]_m^m$ is symmetric and its uth and vth horizontal rows are rows of 0's. If $[\lambda]_{m-2}$ is the minor of the sequence $[1\,2 \dots m]$ formed by striking out its elements u and v, if $[A]_m^m$ is the reciprocal of $[a]_m^m$, and if $\Delta = (a)_m^m$, then

$$\Delta^{m-4}[P_{\lambda\lambda}]_{m-2}^{m-2} = \text{the reciprocal of } [A_{\lambda\lambda}]_{m-2}^{m-2}. \quad \text{.....................(F)}$$

The rank of $[P]_m^m$ is $\rho - 2$ when $a_{uu}a_{vv} - a^2_{uv} \neq 0$, either 0 or 1 when $a_{uu}a_{vv} - a^2_{uv} = 0$, and 0 when $\rho \ngtr 2$.

Ex. v. In Ex. iv let $a_{uu} = 0$, $a_{vv} = 0$. Then we have

$$P_{ij} = \begin{vmatrix} 0 & a_{uv} & a_{uj} \\ a_{vu} & 0 & a_{vj} \\ a_{iu} & a_{iv} & a_{ij} \end{vmatrix} = -a_{uv}(a_{uv}a_{ij} - a_{ui}a_{vj} - a_{uj}a_{vi}),$$

and writing $$p_{ij} = a_{uv}a_{ij} - a_{ui}a_{vj} - a_{uj}a_{vi}, \quad [P]_m^m = -a_{uv}[p]_m^m,$$

we obtain from (E) the equations

$$a^2_{uv}\,[a]_m^m = a_{uv}\begin{bmatrix} a_{u1} & a_{v1} \\ a_{u2} & a_{v2} \\ \cdots & \cdots \\ a_{um} & a_{vm}\end{bmatrix}\begin{bmatrix} 0, & 1 \\ 1, & 0\end{bmatrix}\begin{bmatrix} a_{u1} & a_{u2} & \ldots & a_{um} \\ a_{v1} & a_{v2} & \ldots & a_{vm}\end{bmatrix} - [P]_m^m, \qquad \text{(E')}$$

$$a_{uv}\,[a]_m^m = \begin{bmatrix} a_{u1} & a_{v1} \\ a_{u2} & a_{v2} \\ \cdots & \cdots \\ a_{um} & a_{vm}\end{bmatrix}\begin{bmatrix} a_{v1} & a_{v2} & \ldots & a_{vm} \\ a_{u1} & a_{u2} & \ldots & a_{um}\end{bmatrix} + [p]_m^m, \qquad \text{(E'')}$$

which are identities in the remaining elements of $[a]_m^m$.

Here $[P]_m^m$ and $[p]_m^m$ are symmetric matrices in which the uth and vth horizontal and vertical rows are rows of 0's, and when $a_{uv} \neq 0$, they both have rank $\rho - 2$.

Ex. vi. If $(a)_r^r$ is the leading diagonal minor determinant of order r of the symmetric matrix $[a]_m^m$ whose rank is ρ, and if $[A]_r^r$ is the reciprocal of $[a]_r^r$, we have by formula (A′) the identity

$$(a)_r^r\,[a]_m^m = \overline{\underline{a}}{}_m^r\,\overline{\underline{A}}{}_r^r\,[a]_r^m + \begin{bmatrix} 0, & 0 \\ 0, & b\end{bmatrix}_{r,\,m-r}^{r,\,m-r}, \qquad \text{(A'')}$$

where

$$b_{ij} = \begin{pmatrix} 1, 2, \ldots r, r+j \\ a \\ 1, 2, \ldots r, r+i\end{pmatrix} = b_{ji}.$$

The rank of the symmetric matrix $[b]_{m-r}^{m-r}$ is $\rho - r$ when $(a)_r^r \neq 0$, either 0 or 1 when $(a)_r^r = 0$, and 0 when $\rho \not> r$.

Writing $\qquad D = (a)_r^r, \qquad D_{r+s} = \begin{pmatrix} 1, 2, \ldots r, r+v_1, r+v_2, \ldots r+v_s \\ a \\ 1, 2, \ldots r, r+u_1, r+u_2, \ldots r+u_s\end{pmatrix},$

we have by § 110 the identities

$$(b_{uv})_s^s = D^{s-1}D_{r+s}, \quad (b)_s^s = D^{s-1}(a)_{r+s}^{r+s}.$$

These show that when $[a]_m^m$ is in standard form according to Note 2 of § 126 and $D \neq 0$, then $[b]_{m-r}^{m-r}$ is also in standard form; moreover $(b)_s^s = 0$ when and only when $(a)_{r+s}^{r+s} = 0$.

Ex. vii. Using the notation of § 128 we have for the symmetric matrix

$$\phi = \begin{bmatrix} a & h & g \\ h & b & f \\ g & f & c\end{bmatrix}$$

the identities

$$a\phi - \begin{bmatrix} a \\ h \\ g\end{bmatrix}[a\,h\,g] = \begin{bmatrix} 0, & 0\;, & 0 \\ 0, & ab-h^2, & af-gh \\ 0, & af-gh, & ac-g^2\end{bmatrix} = \begin{bmatrix} 0, & 0, & 0 \\ 0, & C, & -F \\ 0, & -F, & B\end{bmatrix},$$

$$b\phi - \begin{bmatrix} h \\ b \\ f\end{bmatrix}[h\,b\,f] = \begin{bmatrix} ba-h^2, & 0, & bg-hf \\ 0\;, & 0, & 0 \\ bg-hf, & 0, & bc-f^2\end{bmatrix} = \begin{bmatrix} C, & 0, & -G \\ 0, & 0, & 0 \\ -G, & 0, & A\end{bmatrix},$$

$$c\phi - \begin{bmatrix} g \\ f \\ c \end{bmatrix} [g\, f\, c] = \begin{bmatrix} ca-g^2, & ch-fg, & 0 \\ ch-fg, & cb-f^2, & 0 \\ 0, & 0, & 0 \end{bmatrix} = \begin{bmatrix} B, & -H, & 0 \\ -H, & A, & 0 \\ 0, & 0, & 0 \end{bmatrix},$$

$$(bc-f^2)\,\phi - \begin{bmatrix} h & g \\ b & f \\ f & c \end{bmatrix} \begin{bmatrix} c, & -f \\ -f, & b \end{bmatrix} \begin{bmatrix} h & b & f \\ g & f & c \end{bmatrix} = \begin{bmatrix} D, & 0, & 0 \\ 0, & 0, & 0 \\ 0, & 0, & 0 \end{bmatrix},$$

$$(ca-g^2)\,\phi - \begin{bmatrix} g & a \\ f & h \\ c & g \end{bmatrix} \begin{bmatrix} a, & -g \\ -g, & c \end{bmatrix} \begin{bmatrix} g & f & c \\ a & h & g \end{bmatrix} = \begin{bmatrix} 0, & 0, & 0 \\ 0, & D, & 0 \\ 0, & 0, & 0 \end{bmatrix},$$

$$(ab-h^2)\,\phi - \begin{bmatrix} a & h \\ h & b \\ g & f \end{bmatrix} \begin{bmatrix} b, & -h \\ -h, & a \end{bmatrix} \begin{bmatrix} a & h & g \\ h & b & f \end{bmatrix} = \begin{bmatrix} 0, & 0, & 0 \\ 0, & 0, & 0 \\ 0, & 0, & D \end{bmatrix}.$$

In accordance with the theorem of the text the matrices

$$\begin{bmatrix} C, & -F \\ -F, & B \end{bmatrix}, \quad \begin{bmatrix} C, & -G \\ -G, & A \end{bmatrix}, \quad \begin{bmatrix} B, & -H \\ -H, & A \end{bmatrix}$$

are the reciprocals of

$$\begin{bmatrix} B & F \\ F & C \end{bmatrix}, \quad \begin{bmatrix} A & G \\ G & C \end{bmatrix}, \quad \begin{bmatrix} A & H \\ H & B \end{bmatrix}.$$

Ex. viii. For the same matrix ϕ we have the following results which are illustrations of formula (E″).

If $b=c=0$, then

$$f\phi - \begin{bmatrix} h & g \\ b & f \\ f & c \end{bmatrix} \begin{bmatrix} g & f & c \\ h & b & f \end{bmatrix} = -\frac{1}{f} \begin{bmatrix} D & 0 & 0 \\ 0 & 0 & 0 \\ 0 & 0 & 0 \end{bmatrix} = \begin{bmatrix} af-2gh, & 0, & 0 \\ 0, & 0, & 0 \\ 0, & 0, & 0 \end{bmatrix}.$$

If $c=a=0$, then

$$g\phi - \begin{bmatrix} g & a \\ f & h \\ c & g \end{bmatrix} \begin{bmatrix} a & h & g \\ g & f & c \end{bmatrix} = -\frac{1}{g} \begin{bmatrix} 0 & 0 & 0 \\ 0 & D & 0 \\ 0 & 0 & 0 \end{bmatrix} = \begin{bmatrix} 0, & 0, & 0 \\ 0, & bg-2hf, & 0 \\ 0, & 0, & 0 \end{bmatrix}.$$

If $a=b=0$, then

$$h\phi - \begin{bmatrix} a & h \\ h & b \\ g & f \end{bmatrix} \begin{bmatrix} h & b & f \\ a & h & g \end{bmatrix} = -\frac{1}{h} \begin{bmatrix} 0 & 0 & 0 \\ 0 & 0 & 0 \\ 0 & 0 & D \end{bmatrix} = \begin{bmatrix} 0, & 0, & 0 \\ 0, & 0, & 0 \\ 0, & 0, & ch-2fg \end{bmatrix}.$$

Ex. ix. Using the notation of § 129 we have for the symmetric matrix

$$\phi = \begin{bmatrix} a & h & g & u \\ h & b & f & v \\ g & f & c & w \\ u & v & w & d \end{bmatrix}$$

the identities

$$a\phi - \begin{bmatrix} a \\ h \\ g \\ u \end{bmatrix} [a\, h\, g\, u] = \begin{bmatrix} 0, & 0\ , & 0\ , & 0 \\ 0, & ab-h^2\ , & af-gh, & av-hu \\ 0, & af-gh\ , & ac-g^2\ , & aw-gu \\ 0, & av-hu\ , & aw-gu, & ad-u^2 \end{bmatrix} = \begin{bmatrix} 0, & 0\ , & 0\ , & 0 \\ 0, & C_1, & -F_1, & V_1 \\ 0, & -F_1, & B_1, & -W_2 \\ 0, & V_1, & -W_2, & A_2 \end{bmatrix},$$

$$b\phi - \begin{bmatrix} h \\ b \\ f \\ v \end{bmatrix} [h\, b\, f\, v] = \begin{bmatrix} ba-h^2\ , & 0, & bg-hf, & bu-hv \\ 0\ , & 0, & 0\ , & 0 \\ bg-hf\ , & 0, & bc-f^2\ , & bw-fv \\ bu-hv\ , & 0, & bw-fv\ , & bd-v^2 \end{bmatrix} = \begin{bmatrix} C_1, & 0, & -G_1, & -U_2 \\ 0\ , & 0, & 0\ , & 0 \\ -G_1, & 0, & A_1, & W_1 \\ -U_2, & 0, & W_1, & B_2 \end{bmatrix},$$

$$c\phi - \begin{bmatrix} g \\ f \\ c \\ w \end{bmatrix} [g\, f\, c\, w] = \begin{bmatrix} ca-g^2\ , & ch-fg\ , & 0, & cu-gw \\ ch-fg\ , & cb-f^2\ , & 0, & cv-fw \\ 0\ , & 0\ , & 0, & 0 \\ cu-gw, & cv-fw, & 0, & cd-w^2 \end{bmatrix} = \begin{bmatrix} B_1, & -H_1, & 0, & U_1 \\ -H_1, & A_1, & 0, & -V_2 \\ 0\ , & 0\ , & 0, & 0 \\ U_1, & -V_2, & 0, & C_2 \end{bmatrix},$$

$$d\phi - \begin{bmatrix} u \\ v \\ w \\ d \end{bmatrix} [u\, v\, w\, d] = \begin{bmatrix} da-u^2\ , & dh-uv, & dg-wu, & 0 \\ dh-uv, & db-v^2\ , & df-vw, & 0 \\ dg-wu, & df-vw, & dc-w^2\ , & 0 \\ 0\ , & 0\ , & 0\ , & 0 \end{bmatrix} = \begin{bmatrix} A_2, & H_2, & G_2, & 0 \\ H_2, & B_2, & F_2, & 0 \\ G_2, & F_2, & C_2, & 0 \\ 0\ , & 0\ , & 0\ , & 0 \end{bmatrix}.$$

It will be seen from Ex. xii of § 129 that in accordance with the theorem of the text, the matrices

$$\Delta \begin{bmatrix} C_1, & -F_1, & V_1 \\ -F_1, & B_1, & -W_2 \\ V_1, & -W_2, & A_2 \end{bmatrix}, \quad \Delta \begin{bmatrix} C_1, & -G_1, & -U_2 \\ -G_1, & A_1, & W_1 \\ -U_2', & W_1, & B_2 \end{bmatrix},$$

$$\Delta \begin{bmatrix} B_1, & -H_1, & U_1 \\ -H_1, & A_1, & -V_2 \\ U_1, & -V_2, & C_2 \end{bmatrix}, \quad \Delta \begin{bmatrix} A_2, & H_2, & G_2 \\ H_2, & B_2, & F_2 \\ G_2, & F_2, & C_2 \end{bmatrix}$$

are respectively the reciprocals of

$$\begin{bmatrix} B & F & V \\ F & C & W \\ V & W & D \end{bmatrix}, \quad \begin{bmatrix} A & G & U \\ G & C & W \\ U & W & D \end{bmatrix}, \quad \begin{bmatrix} A & H & U \\ H & B & V \\ U & V & D \end{bmatrix}, \quad \begin{bmatrix} A & H & G \\ H & B & F \\ G & F & C \end{bmatrix}.$$

Ex. x. For the same matrix ϕ we have such identities as

$$A_1\phi - \begin{bmatrix} h & g \\ b & f \\ f & c \\ v & w \end{bmatrix} \begin{bmatrix} c, & -f \\ -f, & b \end{bmatrix} \begin{bmatrix} h & b & f & v \\ g & f & c & w \end{bmatrix} = \begin{bmatrix} D, & 0, & 0, & -U \\ 0, & 0, & 0, & 0 \\ 0, & 0, & 0, & 0 \\ -U, & 0, & 0, & A \end{bmatrix},$$

$$A_2\phi - \begin{bmatrix} a & u \\ h & v \\ g & w \\ u & d \end{bmatrix} \begin{bmatrix} d, & -u \\ -u, & a \end{bmatrix} \begin{bmatrix} a & h & g & u \\ u & v & w & d \end{bmatrix} = \begin{bmatrix} 0, & 0, & 0, & 0 \\ 0, & C, & -F, & 0 \\ 0, & -F, & B, & 0 \\ 0, & 0, & 0, & 0 \end{bmatrix}.$$

Here $\begin{bmatrix} D, & -U \\ -U, & A \end{bmatrix}$, $\begin{bmatrix} C, & -F \\ -F, & B \end{bmatrix}$ are the reciprocals of $\begin{bmatrix} A & U \\ U & D \end{bmatrix}$, $\begin{bmatrix} B & F \\ F & C \end{bmatrix}$.

When $b=c=0$, the first identity becomes

$$f\phi - \begin{bmatrix} h & g \\ 0 & f \\ f & 0 \\ v & w \end{bmatrix} \begin{bmatrix} g & f & 0 & w \\ h & 0 & f & v \end{bmatrix} = -\frac{1}{f} \begin{bmatrix} D, & 0, & 0, & -U \\ 0, & 0, & 0, & 0 \\ 0, & 0, & 0, & 0 \\ -U, & 0, & 0, & A \end{bmatrix}$$

$$= \begin{bmatrix} af-2gh & , 0, & 0, & uf-vg-wh \\ 0 & , 0, & 0, & 0 \\ 0 & , 0, & 0, & 0 \\ uf-vg-wh, & 0, & 0, & df-2vw \end{bmatrix}.$$

When $a=d=0$, the second identity becomes

$$u\phi - \begin{bmatrix} 0 & u \\ h & v \\ g & w \\ u & 0 \end{bmatrix} \begin{bmatrix} u & v & w & 0 \\ 0 & h & g & u \end{bmatrix} = -\frac{1}{u} \begin{bmatrix} 0, & 0, & 0, & 0 \\ 0, & C, & -F, & 0 \\ 0, & -F, & B, & 0 \\ 0, & 0, & 0, & 0 \end{bmatrix}$$

$$= \begin{bmatrix} 0, & 0 & , & 0 & , 0 \\ 0, & bu-2hv & , & uf-vg-wh, & 0 \\ 0, & uf-vg-wh, & & cu-2gw & , 0 \\ 0, & 0 & , & 0 & , 0 \end{bmatrix}.$$

Ex. xi. For the same matrix ϕ we have the identities

$$A\phi - \begin{bmatrix} h & g & u \\ b & f & v \\ f & c & w \\ v & w & d \end{bmatrix} \begin{bmatrix} C_2, & -F_2, & V_2 \\ -F_2, & B_2, & -W_1 \\ V_2, & -W_1, & A_1 \end{bmatrix} \begin{bmatrix} h & b & f & v \\ g & f & c & w \\ u & v & w & d \end{bmatrix} = \begin{bmatrix} \Delta & 0 & 0 & 0 \\ 0 & 0 & 0 & 0 \\ 0 & 0 & 0 & 0 \\ 0 & 0 & 0 & 0 \end{bmatrix},$$

$$B\phi - \begin{bmatrix} a & g & u \\ h & f & v \\ g & c & w \\ u & w & d \end{bmatrix} \begin{bmatrix} C_2, & -G_2, & -U_1 \\ -G_2, & A_2, & W_2 \\ -U_1, & W_2, & B_1 \end{bmatrix} \begin{bmatrix} a & h & g & u \\ g & f & c & w \\ u & v & w & d \end{bmatrix} = \begin{bmatrix} 0 & 0 & 0 & 0 \\ 0 & \Delta & 0 & 0 \\ 0 & 0 & 0 & 0 \\ 0 & 0 & 0 & 0 \end{bmatrix},$$

$$C\phi - \begin{bmatrix} a & h & u \\ h & b & v \\ g & f & w \\ u & v & d \end{bmatrix} \begin{bmatrix} B_2, & -H_2, & U_2 \\ -H_2, & A_2, & -V_1 \\ U_2, & -V_1, & C_1 \end{bmatrix} \begin{bmatrix} a & h & g & u \\ h & b & f & v \\ u & v & w & d \end{bmatrix} = \begin{bmatrix} 0 & 0 & 0 & 0 \\ 0 & 0 & 0 & 0 \\ 0 & 0 & \Delta & 0 \\ 0 & 0 & 0 & 0 \end{bmatrix},$$

$$D\phi - \begin{bmatrix} a & h & g \\ h & b & f \\ g & f & c \\ u & v & w \end{bmatrix} \begin{bmatrix} A_1, & H_1, & G_1 \\ H_1, & B_1, & F_1 \\ G_1, & F_1, & C_1 \end{bmatrix} \begin{bmatrix} a & h & g & u \\ h & b & f & v \\ g & f & c & w \end{bmatrix} = \begin{bmatrix} 0 & 0 & 0 & 0 \\ 0 & 0 & 0 & 0 \\ 0 & 0 & 0 & 0 \\ 0 & 0 & 0 & \Delta \end{bmatrix}.$$

CHAPTER XV

RANKS OF MATRIX PRODUCTS AND MATRIX FACTORS

[In § 131 it is shown that the rank of a matrix is unaltered when we prefix or postfix another matrix whose rank is equal to its passivity; in § 132 we determine the possible ranks of the solutions of a matrix equation of each of the forms $AX=C$, $XB=C$, $AXB=C$, at the same time giving formulae for the general solutions; and in §§ 133—137 we determine the possible ranks of the product matrix and the factor matrices in any matrix product when some of those matrices are given or have assigned ranks. In the remaining three articles we define horizontally and vertically equivalent matrices, the joins and intersections of spacelets and matrices, and connections between matrices and spacelets.]

§ 131. Rank of a matrix product in which one of the extreme factor matrices has rank equal to its passivity.

We will here prove two theorems which are generalisations of Exs. v—vii of § 71 and Ex. iv of § 73.

Theorem I. *If either of the factor matrices in the standard product*

$$[c]_m^n = [a]_m^s\,[b]_s^n \quad \text{..............................(A)}$$

has rank s equal to its passivity, then the product matrix $[c]_m^n$ *has the same rank as the other factor matrix.*

First let $[a]_m^s$ *have rank s*; and let $[b]_s^n$, $[c]_m^n$ have ranks β, γ. By Ex. v of § 82 the matrix $[a]_m^s$ has an inverse prefactor $\underline{\overline{A}}_s^m$, necessarily of rank s, such that $\underline{\overline{A}}_s^m\,[a]_m^s = [1]_s^s$.

Prefixing this matrix on both sides of (A), we obtain

$$[b]_s^n = \underline{\overline{A}}_s^m\,[c]_m^n. \quad \text{..........................(A')}$$

Applying Theorem III of § 71, it follows from (A) that $\gamma \not> \beta$, and from (A′) that $\beta \not> \gamma$. Accordingly we have $\gamma=\beta$, i.e. $[c]_m^n$ has the same rank as $[b]_s^n$.

Next let $[b]^{n}_{s}$ *have rank* s; and let $[a]^{s}_{m}$, $[c]^{n}_{m}$ have ranks α, γ. By Ex. v of § 81 the matrix $[b]^{n}_{s}$ has an inverse postfactor $\overline{\underline{B}}^{s}_{n}$, necessarily of rank s, such that $[b]^{n}_{s}\,\overline{\underline{B}}^{s}_{n} = [1]^{s}_{s}$.

Postfixing this matrix on both sides of (A), we obtain

$$[a]^{s}_{m} = [c]^{n}_{m}\,\overline{\underline{B}}^{s}_{n}\,. \quad \text{..........................(A}'')$$

Applying Theorem III of § 71, it follows from (A) that $\gamma \not> \alpha$, and from (A″) that $\alpha \not> \gamma$. Accordingly we have $\gamma = \alpha$, i.e. $[c]^{n}_{m}$ has the same rank as $[a]^{s}_{m}$.

Theorem I is equivalent to the following statement:

The rank of a matrix or a matrix product is unaltered when we prefix or postfix a matrix whose rank is equal to its passivity.

Ex. i. *If* $[a]^{s}_{m}$ *has rank* s *in equation* (A), *then all the three matrices* $[c]^{n}_{m}$, $[b]^{n}_{s}$, $\begin{bmatrix} c \\ b \end{bmatrix}^{n}_{m,\,s}$ *have the same rank.*

For since the horizontal rows of $[c]^{n}_{m}$ are connected with the horizontal rows of $[b]^{n}_{s}$ it follows from Theorem VI of § 71 that $\begin{bmatrix} c \\ b \end{bmatrix}^{n}_{m,\,s}$ has the same rank as $[b]^{n}_{s}$.

Ex. ii. *If* $[b]^{n}_{s}$ *has rank* s *in equation* (A), *then all the three matrices* $[c]^{n}_{m}$, $[a]^{s}_{m}$, $[c, a]^{n,\,s}_{m}$ *have the same rank.*

We immediately deduce the following more general theorem in which Theorem I is included:

Theorem II. *If the two extreme factor matrices* $[h]^{r}_{m}$ *and* $[k]^{n}_{s}$ *in the product*

$$[b]^{n}_{m} = [h]^{r}_{m}\,[a]^{s}_{r}\,[k]^{n}_{s} \quad \text{..........................(B)}$$

have ranks r *and* s *respectively, equal to their passivities, then the product matrix* $[b]^{n}_{m}$ *has the same rank as the middle factor matrix* $[a]^{s}_{r}$.

For by Theorem I the matrix $[b]^{n}_{m}$ has the same rank as the product $[a]^{s}_{r}\,[k]^{n}_{s}$, and this product has the same rank as $[a]^{s}_{r}$.

To prove Theorem II directly, let $[h]^{r}_{m}$ have rank r and $[k]^{n}_{s}$ have rank s. Then by Ex. v of § 82 and Ex. v of § 81 we can determine an inverse prefactor $\overline{\underline{H}}{}^{m}_{r}$ of $[h]^{r}_{m}$, necessarily of rank r, and an inverse postfactor $\overline{\underline{K}}{}^{s}_{n}$ of $[k]^{n}_{s}$, necessarily of rank s, such that

$$\overline{\underline{H}}{}^{m}_{r}\,[h]^{r}_{m}=[1]^{r}_{r}, \qquad [k]^{n}_{s}\,\overline{\underline{K}}{}^{s}_{n}=[1]^{s}_{s}.$$

Prefixing $\overline{\underline{H}}{}^{m}_{r}$ and postfixing $\overline{\underline{K}}{}^{s}_{n}$ on both sides of (B), we obtain

$$[a]^{s}_{r}=\overline{\underline{H}}{}^{m}_{r}\,[b]^{n}_{m}\,\overline{\underline{K}}{}^{s}_{n}. \qquad \text{.........................(B')}$$

Now let the ranks of $[a]^{s}_{r}$ and $[b]^{n}_{m}$ be α and β respectively.

Applying Theorem IV of § 71, it follows from (B) that $\beta \not> \alpha$, and from (B′) that $\alpha \not> \beta$. Accordingly we have $\beta=\alpha$, i.e. $[b]^{n}_{m}$ has the same rank as $[a]^{s}_{r}$.

NOTE 1. *Applicability of Theorems I and II to functional matrices.*

It is tacitly assumed in the text that all matrices have constant elements. The theorems however remain true when the elements of the matrices are rational integral functions of certain variables $x, y, z, \ldots$.

For in the case of Theorem II, which includes Theorem I, if r, a, s are the ranks of $[h]^{r}_{m}$, $[a]^{s}_{r}$, $[k]^{n}_{s}$, and if $(h')^{r}_{r}$, $(a')^{a}_{a}$, $(k')^{s}_{s}$ are minor determinants of $[h]^{r}_{m}$, $[a]^{s}_{r}$, $[k]^{n}_{s}$ of orders r, a, s which do not vanish identically, we can assign such constant values to the variables $x, y, z, \ldots$ that the product $(h')^{r}_{r}\,(a')^{a}_{a}\,(k')^{s}_{s}$ does not vanish. Then $[h]^{r}_{m}$, $[a]^{s}_{r}$, $[k]^{n}_{s}$ are matrices of ranks r, a, s with constant elements, and by the theorem as proved $[b]^{n}_{m}$ is a matrix with constant elements having rank a, and therefore having a non-vanishing minor determinant of order a. Thus when $x, y, z, \ldots$ are variables, $[b]^{n}_{m}$ has a minor determinant of order a which does not vanish identically; and by Theorem IV of § 71 every one of its minor determinants of any order greater than a vanishes identically; i.e. the functional matrix $[b]^{n}_{m}$ has rank a.

NOTE 2. *Alternative proof of Theorem II.*

Let the matrices $[h]^{r}_{m}$, $[a]^{s}_{r}$, $[k]^{n}_{s}$ have ranks r, a, s respectively.

By Ex. i of § 115 we can determine matrices $[p]^{a}_{r}$, $[q]^{s}_{a}$ of rank a such that

$$[a]^{s}_{r}=[p]^{a}_{r}\,[q]^{s}_{a},$$

and we then have

$$[b]^{n}_{m}=[u]^{a}_{m}\,[v]^{n}_{a},$$

where

$$[u]^{a}_{m}=[h]^{r}_{m}\,[p]^{a}_{r}, \quad [v]^{n}_{a}=[q]^{s}_{a}\,[k]^{n}_{s}.$$

Now applying Ex. iv of § 73 or Exs. v—vii of § 71, we see in succession that $[u]^{a}_{m}$ has rank a, that $[v]^{n}_{a}$ has rank a, and that $[b]^{n}_{m}$ has rank a.

§ 132. Possible ranks of the solutions of a given matrix equation of the first degree.

1. *The equation $AX = C$, where A and C are given matrices.*

We will first prove the following lemma:

Lemma A. *The general solution of the irreducible equation*

$$[a]^r_a\,[x]^n_r = [c]^n_a \qquad \text{.........(a)}$$

in which $(a)^a_a \neq 0$ can be expressed in the form

$$[x]^n_r = \overline{\underline{A}}^{\,r}_{\,r} \begin{bmatrix} c \\ v \end{bmatrix}^n_{a,\,r-a} \qquad \text{.........(a}_1\text{)}$$

where the elements of $[v]^n_{r-a}$ are arbitrary, and where $\overline{\underline{A}}^{\,r}_{\,r}$ is the inverse of any particular undegenerate square matrix $[a]^r_r$ formed by adding $r-a$ final horizontal rows to $[a]^r_a$.

The equation (a) is irreducible when the matrix $[a]^r_a$ is undegenerate and has rank a. It then always admits of finite solutions.

Defining $[a]^r_r$ as in the enunciation of the lemma, we can replace (a) by

$$[1,\,0]^{a,\,r-a}_a\,[a]^r_r\,[x]^n_r = [c]^n_a.$$

Writing $$[a]^r_r\,[x]^n_r = [y]^n_r,$$

so that $$[x]^n_r = \overline{\underline{A}}^{\,r}_{\,r}\,[y]^n_r,$$

this equivalent equation is satisfied when and only when

$$[1,\,0]^{a,\,r-a}_a\,[y]^n_r = [c]^n_a, \qquad \text{.........(1)}$$

i.e. when $[y]^n_a = [c]^n_a$, the remaining elements of $[y]^n_r$ being arbitrary. Thus the general solution of (1) is

$$[y]^n_r = \begin{bmatrix} c \\ v \end{bmatrix}^n_{a,\,r-a},$$

where the elements of the last $r-a$ horizontal rows of the matrix on the right are arbitrary. It follows that (a$_1$) is the general solution of the equation (a).

Using Lemma A we can prove the following theorem:

Theorem I. *If the given matrices $[a]^r_m$, $[c]^n_m$ have ranks α, γ, and if the equation*

$$[a]^r_m\,[x]^n_r = [c]^n_m \qquad \text{.........(A)}$$

admits of finite solutions, then it has solutions of rank ρ when and only when ρ is an integer satisfying the conditions

$$\rho \nless \gamma, \quad \rho \ngtr n, \quad \rho + \alpha \ngtr \gamma + r. \qquad \text{.........(A}'\text{)}$$

The equation (A) admits of finite solutions when $[a, c]_m^{r, n}$ has the same rank α as $[a]_m^r$, i.e. when the vertical rows of $[c]_m^n$ are connected with the vertical rows of $[a]_m^r$. The conditions (A′) then include $\rho \not> r$ because we must have $\gamma \not> \alpha$.

As in § 81.2 we can reduce the equation (A) to one of the form

$$[a']_\alpha^r [x]_r^n = [c']_\alpha^n,$$

where $[a']_\alpha^r$ and $[c']_\alpha^n$ have ranks α and γ, and by corresponding re-arrangements of corresponding passive rows in the product on the left we can further reduce this equation to one of the form $[a'']_\alpha^r [x']_r^n = [c']_\alpha^n$ where $[a'']_\alpha^r$ and $[x']_r^n$ are derangements of $[a']_\alpha^r$ and $[x]_r^n$, and $(a'')_\alpha^\alpha \neq 0$. Consequently the theorem will be true in general if it is true for the equation (a) in which $[a]_\alpha^r$ and $[c]_\alpha^n$ have ranks α and γ.

Now in (a_1) the matrix $[x]_r^n$ has the same rank as the postfactor on the right in which $[c]_\alpha^n$ is a given matrix of rank γ and the elements of $[v]_{r-\alpha}^n$ are arbitrary; and by Theorem I *b* of § 104 the possible ranks of that postfactor are the values of ρ which satisfy the conditions (A′). Therefore these are also the possible ranks of the solutions $[x]_r^n$ of the equation (a) and the possible ranks of the solutions $[x]_r^n$ of the equation (A).

Ex. i. *If* $[a]_m^r$ *is a given matrix of rank* a, *the possible ranks of the solutions* $[x]_r^n$ *of the equation* $[a]_m^r [x]_r^n = 0$ *are the integral values of* ρ *which satisfy the conditions*

$$\rho \not< 0, \quad \rho \not> n, \quad \rho + a \not> r.$$

This is a particular case of Theorem I, and has been proved before in § 81.6.

Ex. ii. When the equation (A) has infinite solutions but no finite solutions, i.e. when $a < r$ and $[a, c]_m^{r, n}$ has rank greater than a, the solutions are given by $[x]_r^n = k[X]_r^n$, where k is infinite and $[X]_r^n$ is a finite non-zero solution of the equation $[a]_m^r [X]_r^n = 0$.

The possible ranks of $[X]_r^n$ are the integral values of ρ which satisfy the conditions

$$\rho \not< 1, \quad \rho \not> n, \quad \rho + a \not> r.$$

Ex. iii. *A second form of the general solution of the irreducible equation* (a).

If we write $[a]_a^r = [a, p]_a^{a, r-a}$, and take the undegenerate square matrix $[a]_r^r$ to be

$$[a]_r^r = \begin{bmatrix} a, & p \\ 0, & 1 \end{bmatrix}_{a,\ r-a}^{a,\ r-a},$$

then by Ex. ii of § 101 the general solution (a_1) of (a) takes the form

$$[x]^n_r = \begin{bmatrix} \overline{\underline{A}}, & \overline{\underline{P}} \\ 0, & 1 \end{bmatrix}^{a,\, r-a}_{a,\, r-a} \begin{bmatrix} c \\ v \end{bmatrix}^n_{a,\, r-a}, \quad \text{.............................} (a_2)$$

where $\overline{\underline{A}}{}^a_a$ is the inverse of $[a]^a_a$, $\overline{\underline{P}}{}^{r-a}_a = -\overline{\underline{A}}{}^a_a\,[p]^{r-a}_a$, and $[v]^n_{r-a}$ is arbitrary.

Ex. iv. *A third form of the general solution of the irreducible equation* (a).

If the given matrix $[c]^n_a$ in (a) has rank γ, and if we write

$$[c]^n_a = [h]^\gamma_a\,[k]^n_\gamma,$$

where both factor matrices on the right have rank γ, we can replace (a_2) by

$$[x]^n_r = \begin{bmatrix} \overline{\underline{H}}, & \overline{\underline{P}} \\ 0, & 1 \end{bmatrix}^{\gamma,\, r-a}_{a,\, r-a} \begin{bmatrix} k \\ v \end{bmatrix}^n_{\gamma,\, r-a}, \quad \text{.............................} (a_3)$$

where $[v]^n_{r-a}$ is arbitrary, and

$$\overline{\underline{H}}{}^\gamma_a = \overline{\underline{A}}{}^a_a\,[h]^\gamma_a, \quad \overline{\underline{P}}{}^{r-a}_a = -\overline{\underline{A}}{}^a_a\,[p]^{r-a}_a.$$

Since $\overline{\underline{H}}{}^\gamma_a$ has rank γ, it follows from Theorem IV of § 105 that the prefactor on the right in (a_3) has rank $\gamma+r-a$ equal to its passivity. Therefore $[x]^n_r$ has the same rank as the postfactor on the right; and it follows as in the text that the possible ranks ρ of the solutions $[x]^n_r$ are given by (A′).

In formula (a_3) we can regard $\overline{\underline{H}}{}^\gamma_a$ and $[k]^n_\gamma$ as any two particular matrices (both of rank γ) such that

$$\overline{\underline{A}}{}^a_a\,[c]^n_a = \overline{\underline{H}}{}^\gamma_a\,[k]^n_\gamma.$$

Ex. v. The general solution of the equation $[1,\, p]^{a,\, r-a}_a\,[x]^n_r = 0$ can be expressed in the form

$$[x]^n_r = \begin{bmatrix} -p \\ 1 \end{bmatrix}^{r-a}_{a,\, r-a} [v]^n_{r-a},$$

where the postfactor on the right is arbitrary.

2. *The equation $XB = C$, where B and C are given matrices.*

We now have the following lemma:

Lemma B. *The general solution of the irreducible equation*

$$[x]^r_m\,[b]^\beta_r = [c]^\beta_m \quad \text{......................................} (b)$$

in which $(b)^\beta_\beta \neq 0$ *can be expressed in the form*

$$[x]^r_m = [c,\, u]^{\beta,\, r-\beta}_m\,\overline{\underline{B}}{}^r_r, \quad \text{..................................} (b_1)$$

where the elements of $[u]_m^{r-\beta}$ *are arbitrary, and where* $\overline{\underline{B}}_r^{\,r}$ *is the inverse of any particular undegenerate square matrix* $[b]_r^r$ *formed by adding* $r-\beta$ *final vertical rows to* $[b]_r^\beta$.

We replace (b) by

$$[x]_m^r [b]_r^r \begin{bmatrix} 1 \\ 0 \end{bmatrix}_{\beta,\, r-\beta}^{\beta} = [c]_m^\beta,$$

and write

$$[x]_m^r [b]_r^r = [y]_m^r,$$

so that

$$[x]_m^r = [y]_m^r \overline{\underline{B}}_r^{\,r}.$$

Then the equation (b) is satisfied when and only when

$$[y]_m^r \begin{bmatrix} 1 \\ 0 \end{bmatrix}_{\beta,\, r-\beta}^{\beta} = [c]_m^\beta, \quad \text{.................................(2)}$$

i.e. when $[y]_m^\beta = [c]_m^\beta$, the remaining elements of $[y]_m^r$ being arbitrary.

Using Lemma B we can prove the following theorem:

Theorem II. *If the given matrices* $[b]_r^n$, $[c]_m^n$ *have ranks* β, γ, *and if the equation*

$$[x]_m^r [b]_r^n = [c]_m^n \quad \text{............................(B)}$$

admits of finite solutions, then it has solutions of rank ρ *when and only when* ρ *is an integer satisfying the conditions*

$$\rho \not< \gamma, \quad \rho \not> m, \quad \rho + \beta \not> \gamma + r. \quad \text{....................(B')}$$

The equation (B) admits of finite solutions when $\begin{bmatrix} b \\ c \end{bmatrix}_{r,\, m}^{n}$ has the same rank β as $[b]_r^n$, i.e. when the horizontal rows of $[c]_m^n$ are connected with the horizontal rows of $[b]_r^n$. The conditions (B′) then include $\rho \not> r$ because we must have $\gamma \not> \beta$.

Since we can reduce the equation (B) to one of the form (b) in which $[c]_m^\beta$ has rank γ, the theorem will be true in general if it is true when (B) has the special form (b).

In this case the possible ranks of the solutions $[x]_m^r$ are the same as the possible ranks of the prefactor on the right in (b_1) when $[c]_m^\beta$ is a given matrix of rank γ, and by Theorem I a of § 104 these are the values of ρ given by (B′).

Ex. vi. *If* $[b]_r^n$ *is a given matrix of rank* β, *the possible ranks of the solutions* $[x]_m^r$ *of the equation* $[x]_m^r [b]_r^n = 0$ *are the integral values of* ρ *which satisfy the conditions*

$$\rho \not< 0, \quad \rho \not> m, \quad \rho + \beta \not> r.$$

Ex. vii. When the equation (B) has infinite solutions but no finite solutions, i.e. when $\beta < r$ and $\begin{bmatrix} b \\ c \end{bmatrix}_{r,m}^{n}$ has rank greater than β, the solutions are given by $[x]_m^r = k[X]_m^r$, where k is infinite and $[X]_m^r$ is a finite non-zero solution of the equation $[X]_m^r [b]_r^n = 0$.

The possible ranks of $[X]_m^r$ are the integral values of ρ which satisfy the conditions

$$\rho \not< 1, \quad \rho \not> m, \quad \rho + \beta \not> r.$$

Ex. viii. *A second form of the general solution of the irreducible equation* (b).

If we write $[b]_r^\beta = \begin{bmatrix} b \\ q \end{bmatrix}_{\beta,\, r-\beta}^{\beta}$ and take the undegenerate square matrix $[b]_r^r$ to be

$$[b]_r^r = \begin{bmatrix} b, & 0 \\ q, & 1 \end{bmatrix}_{\beta,\, r-\beta}^{\beta,\, r-\beta},$$

then by Ex. iii of § 101 the general solution (b_1) of (b) takes the form

$$[x]_m^r = [c, u]_m^{\beta,\, r-\beta} \overline{\underline{\begin{matrix} B, & 0 \\ Q, & 1 \end{matrix}}}{}_{\beta,\, r-\beta}^{\beta,\, r-\beta}, \qquad \text{..............................}(b_2)$$

where $\overline{\underline{B}}{}_\beta^\beta$ is the inverse of $[b]_\beta^\beta$, $\overline{\underline{Q}}{}_{r-\beta}^\beta = -[q]_{r-\beta}^\beta \overline{\underline{B}}{}_\beta^\beta$, and $[u]_m^{r-\beta}$ is arbitrary.

Ex. ix. *A third form of the general solution of the irreducible equation* (b).

If the given matrix $[c]_m^\beta$ in (b) has rank γ, and if we write $[c]_m^\beta = [h]_m^\gamma [k]_\gamma^\beta$, where both factor matrices on the right have rank γ, we can replace (b_2) by

$$[x]_m^r = [h, u]_m^{\gamma,\, r-\beta} \overline{\underline{\begin{matrix} K, & 0 \\ Q, & 1 \end{matrix}}}{}_{\gamma,\, r-\beta}^{\beta,\, r-\beta} \qquad \text{..............................}(b_3)$$

where $[u]_m^{r-\beta}$ is arbitrary, and

$$\overline{\underline{K}}{}_\gamma^\beta = [k]_\gamma^\beta \overline{\underline{B}}{}_\beta^\beta, \quad \overline{\underline{Q}}{}_{r-\beta}^\beta = -[q]_{r-\beta}^\beta \overline{\underline{B}}{}_\beta^\beta.$$

Since $\overline{\underline{K}}{}_\gamma^\beta$ has rank γ, it follows from Theorem IV of § 105 that the postfactor on the right in (b_3) has rank $\gamma + r - \beta$ equal to its passivity. Therefore $[x]_m^r$ has the same rank as the prefactor on the right, and it follows as in the text that the possible ranks ρ of the solutions $[x]_m^r$ are given by (B′).

In formula (b_3) we can regard $[h]_m^\gamma$ and $\overline{\underline{K}}{}_\gamma^\beta$ as any two particular matrices (both of rank γ) such that

$$[c]_m^\beta \overline{\underline{B}}{}_\beta^\beta = [h]_m^\gamma \overline{\underline{K}}{}_\gamma^\beta.$$

Ex. x. The general solution of the equation $[x]_m^r \begin{bmatrix} 1 \\ q \end{bmatrix}_{\beta,\, r-\beta}^{\beta} = 0$ can be expressed in the form

$$[x]_m^r = [u]_m^{r-\beta} [-q, 1]_{r-\beta}^{\beta,\, r-\beta},$$

where the prefactor on the right is arbitrary.

3. *The equation $AXB = C$, where A, B and C are given matrices.*

We will first prove the following lemma:

Lemma C. *The general solution of the irreducible equation*

$$[a]_\alpha^r\,[x]_r^s\,[b]_s^\beta = [c]_\alpha^\beta \quad\text{.........(c)}$$

in which $(a)_\alpha^\alpha \neq 0$ and $(b)_\beta^\beta \neq 0$ can be expressed in the form

$$[x]_r^s = \underline{\overline{A}}_r^{\,r}\begin{bmatrix} c, & u \\ v, & w \end{bmatrix}_{\alpha,\, r-\alpha}^{\beta,\, s-\beta}\underline{\overline{B}}_s^{\,s}, \quad\text{.........(c}_1\text{)}$$

where the elements of $[u]_\alpha^{s-\beta}$, $[v]_{r-\alpha}^\beta$, $[w]_{r-\alpha}^{s-\beta}$ are arbitrary, and where $\underline{\overline{A}}_r^{\,r}$ and $\underline{\overline{B}}_s^{\,s}$ are the inverses of any two particular undegenerate square matrices $[a]_r^r$ and $[b]_s^s$ formed by adding $r-\alpha$ final horizontal rows to $[a]_\alpha^r$ and $s-\beta$ final vertical rows to $[b]_s^\beta$.

The equation (c) is irreducible when the matrices $[a]_\alpha^r$ and $[b]_s^\beta$ are undegenerate and have ranks α and β. It then always admits of finite solutions.

Defining $[a]_r^r$ and $[b]_s^s$ as in the enunciation of the lemma we can replace (c) by

$$[1,\ 0]_\alpha^{\alpha,\, r-\alpha}[a]_r^r\,[x]_r^s\,[b]_s^s\begin{bmatrix} 1 \\ 0 \end{bmatrix}_{\beta,\, s-\beta}^{\beta} = [c]_\alpha^\beta.$$

Writing $\qquad [a]_r^r\,[x]_r^s\,[b]_s^s = [y]_r^s, \quad [x]_r^s = \underline{\overline{A}}_r^{\,r}\,[y]_r^s\,\underline{\overline{B}}_s^{\,s},$

this equivalent equation is satisfied when and only when

$$[1,\ 0]_\alpha^{\alpha,\, r-\alpha}[y]_r^s\begin{bmatrix} 1 \\ 0 \end{bmatrix}_{\beta,\, s-\beta}^{\beta} = [c]_\alpha^\beta, \quad\text{.........(3)}$$

i.e. when $[y]_\alpha^\beta = [c]_\alpha^\beta$, the remaining elements of $[y]_r^s$ being arbitrary. Thus the general solution of (3) is

$$[y]_r^s = \begin{bmatrix} c, & u \\ v, & w \end{bmatrix}_{\alpha,\, r-\alpha}^{\beta,\, s-\beta},$$

where all elements of the matrix on the right are arbitrary except those occurring in the minor matrix $[c]_\alpha^\beta$.

It follows that (c_1) is the general solution of the equation (c).

Using Lemma C we can prove the following theorem:

Theorem III. *If the given matrices $[a]_m^r$, $[b]_s^n$, $[c]_m^n$ have ranks α, β, γ respectively, and if the equation*

$$[a]_m^r\,[x]_r^s\,[b]_s^n = [c]_m^n \quad\text{.........(C)}$$

admits of finite solutions, then it has solutions of rank ρ when and only when ρ is an integer satisfying the conditions

$$\rho \nless \gamma, \quad \rho \ngtr r, \quad \rho \ngtr s, \quad \rho + \alpha + \beta \ngtr \gamma + r + s. \quad\text{.........(C}'\text{)}$$

The equation (C) admits of finite solutions when and only when $[a, c]_m^{r, n}$ has the same rank α as $[a]_m^r$ and $\begin{bmatrix} b \\ c \end{bmatrix}_{s, m}^n$ has the same rank β as $[b]_s^n$, i.e. when and only when the vertical rows of $[c]_m^n$ are connected with the vertical rows of $[a]_m^r$, and the horizontal rows of $[c]_m^n$ are connected with the horizontal rows of $[b]_s^n$.

SPECIAL CASE. *When the equation* (C) *is irreducible and* $(a)_m^m \neq 0$, $(b)_n^n \neq 0$.

In this case we have $m = \alpha$, $n = \beta$, and (C) is the equation (c) of the lemma. The general solution is given by (c_1), and the possible ranks of the solutions $[x]_r^s$ are the same as the possible ranks of the middle factor matrix on the right in (c_1). Since this is a matrix of the form $[c]_r^s$ which contains a given minor matrix $[c]_\alpha^\beta$ of rank γ and whose remaining elements are arbitrary it follows by Theorem II of § 104 that the possible ranks of this matrix, and therefore the possible ranks of the solutions $[x]_r^s$, are the values of ρ satisfying the conditions (C′).

GENERAL CASE. *When the equation* (C) *is finitely solvable but not necessarily irreducible.*

In this general case the equation (C) can be reduced as in § 83.2 to an irreducible equation of the form (c), and therefore as in the special case the possible values of ρ are given by (C′).

Ex. xi. *If* $[a]_m^r$ *and* $[b]_s^n$ *are given matrices of ranks α and β the possible ranks of the solutions of the equation* $[a]_m^r [x]_r^s [b]_s^n = 0$ *are the integral values of* ρ *which satisfy the conditions*

$$\rho \not< 0, \quad \rho \not> r, \quad \rho \not> s, \quad \rho + \alpha + \beta \not> r + s.$$

Ex. xii. When the equation (C) has infinite solutions but no finite solutions, the solutions are given by $[x]_r^s = k[X]_r^s$, where $[X]_r^s$ is a finite non-zero solution of the equation $[a]_m^r [X]_r^s [b]_s^n = 0$.

The possible ranks of $[X]_r^s$ are the integral values of ρ which satisfy the conditions

$$\rho \not< 1, \quad \rho \not> r, \quad \rho \not> s, \quad \rho + \alpha + \beta \not> r + s.$$

Ex. xiii. *Another form of the general solution of the irreducible equation* (c).

If we write $$[a]_\alpha^r = [\alpha, p]_\alpha^{\alpha, r-\alpha}, \quad [b]_s^\beta = \begin{bmatrix} b \\ q \end{bmatrix}_{\beta, s-\beta}^\beta$$

and take the undegenerate square matrices $[a]_r^r$, $[b]_s^s$ to be

$$[a]_r^r = \begin{bmatrix} \alpha, & p \\ 0, & 1 \end{bmatrix}_{\alpha, r-\alpha}^{\alpha, r-\alpha}, \quad [b]_s^s = \begin{bmatrix} b, & 0 \\ q, & 1 \end{bmatrix}_{\beta, s-\beta}^{\beta, s-\beta},$$

then by Ex. ii of § 101 the equation (c_1) giving the general solution of the equation (c) takes the form

$$[x]_r^s = \overline{\underline{\begin{matrix} A, & P \\ 0, & 1 \end{matrix}}}^{\alpha,\, r-\alpha}_{\alpha,\, r-\alpha} \begin{bmatrix} c, & u \\ v, & w \end{bmatrix}^{\beta,\, s-\beta}_{\alpha,\, r-\alpha} \overline{\underline{\begin{matrix} B, & 0 \\ Q, & 1 \end{matrix}}}^{\beta,\, s-\beta}_{\beta,\, s-\beta}, \qquad \text{.....................}(c_2)$$

where $\overline{\underline{A}}^{\alpha}_{\alpha}$ and $\overline{\underline{B}}^{\beta}_{\beta}$ are the inverses of $[a]^{\alpha}_{\alpha}$ and $[b]^{\beta}_{\beta}$, and

$$\overline{\underline{P}}^{r-\alpha}_{\alpha} = -\overline{\underline{A}}^{\alpha}_{\alpha} [p]^{r-\alpha}_{\alpha}, \qquad \overline{\underline{Q}}^{\beta}_{s-\beta} = -[q]^{\beta}_{s-\beta} \overline{\underline{B}}^{\beta}_{\beta}.$$

Ex. xiv. *General solution of the equation* (C) *when it admits of solution.*

If $[a]^r_m$ and $[b]^n_s$ have ranks α and β, we can by Ex. i of § 115 express these matrices in the forms

$$[a]^r_m = [h]^{\alpha}_m [a']^r_{\alpha}, \quad [b]^n_s = [b']^{\beta}_s [k]^n_{\beta}, \qquad \text{..........................}(4)$$

where the factors on the right are undegenerate matrices having rank α in the first equation and rank β in the second equation.

Let $\overline{\underline{H}}^{m}_{\alpha}$ and $\overline{\underline{K}}^{\beta}_{n}$ be respectively any inverse prefactor of $[h]^{\alpha}_m$ and any inverse post-factor of $[k]^n_{\beta}$.

Since the vertical rows of $[c]^n_m$ are connected with the vertical rows of $[a]^r_m$, they are connected with the vertical rows of $[h]^{\alpha}_m$, and we can write

$$[c]^n_m = [h]^{\alpha}_m [d]^n_{\alpha},$$

where

$$[d]^n_{\alpha} = \overline{\underline{H}}^{m}_{\alpha} [c]^n_m.$$

Again, since the horizontal rows of $[c]^n_m$ are connected with the horizontal rows of $[b]^n_s$, the horizontal rows of $[d]^n_{\alpha}$ are connected with the horizontal rows of $[b]^n_s$ and therefore with the horizontal rows of $[k]^n_{\beta}$, and we can write

$$[d]^n_{\alpha} = [c']^{\beta}_{\alpha} [k]^n_{\beta},$$

where

$$[c']^{\beta}_{\alpha} = [d]^n_{\alpha} \overline{\underline{K}}^{\beta}_{n}.$$

Accordingly we have

$$[c]^n_m = [h]^{\alpha}_m [c']^{\beta}_{\alpha} [k]^n_{\beta}, \qquad \text{.......................................}(5)$$

where

$$[c']^{\beta}_{\alpha} = \overline{\underline{H}}^{m}_{\alpha} [c]^n_m \overline{\underline{K}}^{\beta}_{n}.$$

Substituting from (4) and (5) we see by § 84 that the equation (C) is satisfied when and only when

$$[a']^r_{\alpha} [x]^s_r [b']^{\beta}_s = [c']^{\beta}_{\alpha}. \qquad \text{..................................}(c')$$

Thus the general solution of (C) is the same as the general solution of the irreducible equation (c′) which is given by a formula corresponding to (c_1).

The equation (5) shows that $[c']^{\beta}_{\alpha}$ has the same rank γ as $[c]^n_m$.

4. *Symmetric solutions of the symmetric matrix $A'XA = C$.*

This equation will be called symmetric when the given matrix C is symmetric, and the given matrices A and A' are mutually conjugate. We will first prove the following lemma:

Lemma D. *The general symmetric solution of the irreducible symmetric equation*

$$\overline{\underline{a}}\,{}^{r}_{a}\,[x]^{r}_{r}\,[a]^{a}_{r} = [c]^{a}_{a}, \qquad \text{.................................(d)}$$

in which $(a)^{a}_{a} \neq 0$, *and* $[c]^{a}_{a}$ *is symmetric, can be expressed in the form*

$$[x]^{r}_{r} = \overline{\underline{A}}\,{}^{r}_{r} \begin{bmatrix} c, & u \\ u', & w \end{bmatrix}^{a,\, r-a}_{a,\, r-a} [A]^{r}_{r}, \qquad \text{..........................(d}_1\text{)}$$

where $[u]^{r-a}_{a}$ *and* $[u']^{a}_{r-a}$ *are two arbitrary mutually conjugate matrices,* $[w]^{r-a}_{r-a}$ *is an arbitrary self-conjugate matrix, and* $[A]^{r}_{r}$ *is the inverse of any particular undegenerate square matrix* $[a]^{r}_{r}$ *formed by adding* $r - a$ *final vertical rows to* $[a]^{a}_{r}$.

By Lemma C the general solution of the equation (d) is

$$[x]^{r}_{r} = \overline{\underline{A}}\,{}^{r}_{r} \begin{bmatrix} c, & u \\ v, & w \end{bmatrix}^{a,\, r-a}_{a,\, r-a} [A]^{r}_{r},$$

where the elements of $[u]^{r-a}_{a}$, $[v]^{a}_{r-a}$ and $[w]^{r-a}_{r-a}$ are arbitrary.

This solution is symmetric or self-conjugate, i.e. we have $[x]^{r}_{r} = \overline{\underline{x}}\,{}^{r}_{r}$, when and only when the middle factor matrix on the right is self-conjugate, i.e. when and only when

$$[v]^{a}_{r-a} = \overline{\underline{u}}\,{}^{a}_{r-a}, \qquad [w]^{r-a}_{r-a} = \overline{\underline{w}}\,{}^{r-a}_{r-a}.$$

Using Lemma D we can prove the following theorem:

Theorem IV. *If* $[a]^{r}_{m}$ *is a given matrix of rank* α, *and* $[c]^{m}_{m}$ *is a given symmetric matrix of rank* γ, *and if the equation*

$$\overline{\underline{a}}\,{}^{r}_{m}\,[x]^{r}_{r}\,[a]^{m}_{r} = [c]^{m}_{m} \qquad \text{..........................(D)}$$

admits of finite solutions, then it has symmetric solutions of rank ρ *when and only when* ρ *is an integer satisfying the conditions*

$$\rho \nless \gamma, \quad \rho \ngtr r, \quad \rho + 2\alpha \ngtr \gamma + 2r. \qquad \text{.................(D}'\text{)}$$

The equation (D) admits of finite solutions when and only when $\begin{bmatrix} a \\ c \end{bmatrix}^{m}_{r,\, m}$ has the same rank α as $[a]^{m}_{r}$, i.e. when and only when the horizontal rows of $[c]^{m}_{m}$ are connected with the horizontal rows of $[a]^{m}_{r}$.

SPECIAL CASE. *When the equation* (D) *is irreducible and* $(a)_m^m \neq 0$.

In this case we have $m = \alpha$, and (D) is the equation (d) of the lemma. The general symmetric solution is given by (d_1), and the possible ranks of the symmetric solutions $[x]_r^r$ are the same as the possible ranks of the middle factor matrix on the right in (d_1). Since this is a symmetric matrix of the form $[c]_r^r$ which contains a given symmetric diagonal minor matrix $[c]_\alpha^\alpha$ of rank γ and whose remaining elements are arbitrary, it follows by Theorem II of § 106 that the possible ranks of this matrix, and therefore the possible ranks of the symmetric solutions $[x]_r^r$, are the values of ρ satisfying the conditions (D′).

GENERAL CASE. *When the equation* (D) *is finitely solvable but not necessarily irreducible.*

In this general case the equation (D) can be reduced to an irreducible equation of the form (d), the matrix $[x]_r^r$ in (d) being a symmetric derangement of the matrix $[x]_r^r$ in (D), and therefore as in the special case the possible values of the ranks of the symmetric solutions are the values of ρ given by (D′).

§ 133. Restrictions on the rank of any matrix product.

If $[a]_m^p$, $[b]_p^q$, $[c]_q^r$, ... $[k]_v^w$, $[l]_w^n$ are matrices of given orders having respectively ranks α, β, γ, ... κ, λ, and if the product matrix

$$[x]_m^n = [a]_m^p [b]_p^q [c]_q^r \dots [k]_v^w [l]_w^n \quad \text{.....................(1)}$$

has rank ρ, we will show that the integer ρ must satisfy the conditions

$$\rho \not> \alpha,\ \rho \not> \beta,\ \rho \not> \gamma,\ \dots \rho \not> \kappa,\ \rho \not> \lambda; \quad \text{..................(A)}$$

$$\rho + p + q + r + \dots + v + w \not< \alpha + \beta + \gamma + \dots + \kappa + \lambda. \quad \text{......(B)}$$

These results are equivalent to the following theorem:

Theorem. *In every matrix product the ranks of the factor matrices and the product matrix must satisfy the following conditions:*

(1) *The rank of the product matrix cannot exceed the rank of any factor matrix.*

(2) *The sum of the rank and the passivities of the product cannot be less than the sum of the ranks of the factor matrices.*

That the conditions (A) must be satisfied has been proved in Theorem IV of § 71. It remains to prove the condition (B), i.e. to show that the second part of the above theorem is true.

From Theorems I and II of § 132 we see that the second part of the theorem is true for all products of two matrices.

Let $$[y]_p^n = [b]_p^q [c]_q^r \dots [k]_v^w [l]_w^n, \quad \dots\dots\dots\dots(2)$$

so that $$[x]_m^n = [a]_m^p [y]_p^n, \quad \dots\dots\dots\dots(3)$$

and let $[y]_p^n$ have rank σ. If the second part of the theorem is true for the product (2), then since it is certainly true for the product (3) we have

$$\sigma + q + r + \dots + v + w \not< \beta + \gamma + \dots + \kappa + \lambda,$$

$$\rho + p \not< \alpha + \sigma.$$

By addition it follows that the inequality (B) is true.

Thus if the second part of the theorem is true for the product (2), it is true for the product (1); and if it is true for all products of t matrices, where t is any positive integer not less than 2, then it is true for all products of $t+1$ matrices. Since it is certainly true for all products of 2 matrices, it follows in succession that it is true for all products of 2, 3, ... t matrices, i.e. it is true for all matrix products.

Ex. i. *Direct proof of the theorem of the text.*

We can prove the theorem of the text without using the results of § 132.

It has been shown above that if the theorem is true for all products of two matrices, then it must be true generally. Accordingly it will be established if we prove the following result.

Theorem. *If* $[a]_m^r$ *and* $[b]_r^n$ *are two matrices having ranks* α *and* β, *and if the product matrix* $[x]_m^n = [a]_m^r [b]_r^n$ *has rank* ρ, *then the integers* α, β, ρ *must satisfy the condition*

$$\rho + r \not< \alpha + \beta.$$

We will now give a direct proof of this latter theorem.

First special case: When one of the factor matrices $[a]_m^r$, $[b]_r^n$ *has rank* r.

By § 131 the product matrix has the same rank as the other factor matrix. Therefore in this case $\rho + r = \alpha + \beta$, and the theorem is true.

Second special case: When the factor matrices $[a]_m^r$, $[b]_r^n$ *have ranks* m, n.

If ρ is the rank of $[x]_m^n$, then by § 90 there exists a matrix $[h]_n^{n-\rho}$ of rank $n-\rho$ such that $[x]_m^n [h]_n^{n-\rho} = 0$, and we have

$$[a]_m^r [k]_r^{n-\rho} = 0,$$

where $$[k]_r^{n-\rho} = [b]_r^n [h]_n^{n-\rho}.$$

Since the matrix $[a]_m^r$ has rank m, therefore by § 90 or Ex. vi of § 82 there cannot be more than $r-m$ unconnected connections between its vertical rows.

Consequently the rank of $[k]_r^{n-\rho}$ cannot exceed $r-m$.

But by § 131 or the first special case the matrix $[k]_r^{n-\rho}$ has rank $n-\rho$. Therefore we must have

$$n-\rho \not> r-m, \quad \text{or} \quad \rho+r \not< m+n.$$

Thus the theorem is true in this case.

General case: When the factor matrices $[a]_m^r$, $[b]_r^n$ *have any ranks* α, β.

In this case we have $\alpha \not> r$, $\alpha \not> m$, $\beta \not> r$, $\beta \not> n$. By Ex. i of § 115 there exist matrices $[h]_m^\alpha$, $[p]_\alpha^r$ of rank α, and matrices $[q]_r^\beta$, $[k]_\beta^n$ of rank β such that

$$[a]_m^r = [h]_m^\alpha [p]_\alpha^r, \quad [b]_r^n = [q]_r^\beta [k]_\beta^n,$$

and therefore $$[x]_m^n = [a]_m^r [b]_r^n = [h]_m^\alpha . [p]_\alpha^r [q]_r^\beta . [k]_\beta^n .$$

By § 131 or the first special case the matrix $[x]_m^n$ has the same rank as the product $[p]_\alpha^r [q]_r^\beta$, and if this rank is ρ, it follows from the second special case that $\rho+r \not< \alpha+\beta$.

Thus the theorem is true in all cases.

Ex. ii. *Alternative proof of the general case in Ex.* i.

Let $(a_{pu})_\alpha^\alpha$, $(b_{vq})_\beta^\beta$ be non-vanishing derived determinants of $[a]_m^r$, $[b]_r^n$ of ranks α, β. Then by Theorem VI of § 71 the matrix $[x]_m^n$ has the same rank as its derived matrix

$$[x_{pq}]_\alpha^\beta = [a_{p1}]_\alpha^r [b_{1q}]_r^\beta,$$

and if this rank is ρ, we see by the second special case that $\rho+r \not< \alpha+\beta$.

Ex. iii. The results given in Exs. v—vii of § 71, Ex. iv of § 73, and Theorem I of § 131 are all particular cases of the theorem of Ex. i.

Ex. iv. *If* $[a]_m^r$, $[b]_r^s$, $[c]_s^n$ *are matrices having ranks* α, β, γ, *and if the product matrix* $[x]_m^n = [a]_m^r [b]_r^s [c]_s^n$ *has rank* ρ, *then the integers* α, β, γ, ρ *must satisfy the condition*

$$\rho+r+s \not< \alpha+\beta+\gamma.$$

This result follows from Theorem III of § 132 as well as from the theorem of the text. It can also be deduced from the theorem of Ex. i.

§ 134. Possible ranks of the product matrix and the factor matrices in any product of two or three matrices.

1. *Products of two matrices.*

If $[a]_m^r$ and $[b]_r^n$ are matrices of given orders having ranks α and β, and if the product matrix

$$[x]_m^n = [a]_m^r [b]_r^n \quad \text{..............................(A)}$$

has rank ρ, we see from § 133 that the integers α, β, ρ must satisfy the conditions

$$\rho \not< 0, \quad \rho \not> \alpha, \quad \rho \not> \beta, \quad \rho+r \not< \alpha+\beta; \quad \text{..............(B)}$$

and $\alpha \ngtr m, \quad \beta \ngtr n;$ ....(C)

which include the conditions

$$\alpha \nless 0, \quad \alpha \ngtr r, \quad \alpha \ngtr m; \quad \beta \nless 0, \quad \beta \ngtr r, \quad \beta \ngtr n; \quad \text{.........(D)}$$

and $\rho \nless 0, \quad \rho \ngtr m, \quad \rho \ngtr n, \quad \rho \ngtr r.$ ....(E)

These conditions reduce to (E) when we eliminate α and β, and they reduce to (D) when we eliminate ρ.

We will now prove the theorems which follow.

Theorem I a. *If the factor matrices* $[a]_m^r$, $[b]_r^n$ *in the product* (A) *have any assigned ranks* α, β *consistent with the necessary conditions* (D), *and if their elements are otherwise arbitrary, then the possible values of the rank* ρ *of the product matrix* $[x]_m^n$ *are those consistent with the conditions* (B), *i.e.* ρ *can have any value consistent with the necessary conditions* (B) *and* (C).

We know that ρ must satisfy the conditions (B). It remains to show that ρ can have any value which is consistent with them. To see this let ρ be any integer satisfying the conditions (B). Then we obtain matrices $[a]_m^r$, $[b]_r^n$ of ranks α, β such that the product $[a]_m^r [b]_r^n$ has rank ρ by putting

$$[a]_m^r = \begin{bmatrix} 1, & 0, & 0, & 0 \\ 0, & 1, & 0, & 0 \\ 0, & 0, & 0, & 0 \end{bmatrix}_{\rho,\ \alpha-\rho,\ m-\alpha}^{\rho,\ \alpha-\rho,\ \beta-\rho,\ r+\rho-\alpha-\beta}, \quad [b]_r^n = \begin{bmatrix} 1, & 0, & 0 \\ 0, & 0, & 0 \\ 0, & 1, & 0 \\ 0, & 0, & 0 \end{bmatrix}_{\rho,\ \alpha-\rho,\ \beta-\rho,\ r+\rho-\alpha-\beta}^{\rho,\ \beta-\rho,\ n-\beta},$$

so that

$$[a]_m^r [b]_r^n = \begin{bmatrix} 1, & 0, & 0 \\ 0, & 0, & 0 \\ 0, & 0, & 0 \end{bmatrix}_{\rho,\ \alpha-\rho,\ m-\alpha}^{\rho,\ \beta-\rho,\ n-\beta}.$$

The theorem also follows immediately from Theorem II *a*.

The following examples will be used in proving the next theorems.

Ex. i. *If* $[x]_m^n$ *is a given matrix of rank* ρ, *we can construct a matrix* $[a]_m^r$ *of rank* α *which is such that the equation* $[a]_m^r [b]_r^n = [x]_m^n$ *admits of solution for* $[b]_r^n$ *when and only when* α *is an integer consistent with the conditions* (B) *and* (C), *i.e. when and only when*

$$\alpha \nless \rho, \quad \alpha \ngtr r, \quad \alpha \ngtr m. \quad \text{..................................(a)}$$

The matrix $[a]_m^r$ must be such that the vertical rows of $[x]_m^n$ are connected with the vertical rows of $[a]_m^r$, and the conditions (a) are clearly necessary. Suppose them to be satisfied so that ρ, α, r are in ascending order of magnitude; let $[\xi]_m^\rho$ be any undegenerate vertical minor of $[x]_m^n$ of rank ρ; and let $[\xi, \eta]_m^{\rho,\ \alpha-\rho}$ be undegenerate and have rank α. Then the matrix $[a]_m^r = [\xi, \eta, 0]_m^{\rho,\ \alpha-\rho,\ r-\alpha}$ has the required properties. A general formula for all such matrices is

$$[a]_m^r = [\xi, \eta]_m^{\rho,\ \alpha-\rho} [k]_\alpha^r,$$

where $[k]_\alpha^r$ has rank α.

Ex. ii. *If* $[x]_m^n$ *is a given matrix of rank* ρ, *we can construct a matrix* $[b]_r^n$ *of rank* β *which is such that the equation* $[a]_m^r [b]_r^n = [x]_m^n$ *admits of solution for* $[a]_m^r$ *when and only when* β *is an integer consistent with the conditions* (B) *and* (C), *i.e. when and only when*

$$\beta \not< \rho, \quad \beta \not> r, \quad \beta \not> n. \quad \text{.................................(b)}$$

If $[\xi]_\rho^n$ is any undegenerate horizontal minor of $[x]_m^n$, and if $\begin{bmatrix} \xi \\ \eta \end{bmatrix}_{\rho,\, \beta-\rho}^n$ has rank β, a general formula for all such matrices is

$$[b]_r^n = [h]_r^\beta \begin{bmatrix} \xi \\ \eta \end{bmatrix}_{\rho,\, \beta-\rho}^n,$$

where $[h]_r^\beta$ has rank β.

Ex. iii. *If* $[a]_m^r$ *is a given matrix of rank* α, *we can construct a matrix* $[x]_m^n$ *of rank* ρ *which is such that the equation* $[a]_m^r [b]_r^n = [x]_m^n$ *admits of solution for* $[b]_r^n$ *when and only when* ρ *is an integer consistent with the conditions* (B) *and* (C), *i.e. when and only when*

$$\rho \not< 0, \quad \rho \not> \alpha, \quad \rho \not> n. \quad \text{.................................(c)}$$

The matrix $[x]_m^n$ must be such that its vertical rows are connected with the vertical rows of $[a]_m^r$, and the conditions (c) are clearly necessary. Suppose them to be satisfied, and let $[c]_m^\alpha$ be any undegenerate vertical minor of $[a]_m^r$ of rank α. Then a general formula for all matrices having the required properties is

$$[x]_m^n = [c]_m^\alpha [k]_\alpha^n,$$

where $[k]_\alpha^n$ has rank ρ.

Ex. iv. *If* $[b]_r^n$ *is a given matrix of rank* β, *we can construct a matrix* $[x]_m^n$ *of rank* ρ *which is such that the equation* $[a]_m^r [b]_r^n = [x]_m^n$ *admits of solution for* $[a]_m^r$ *when and only when* ρ *is an integer consistent with the conditions* (B) *and* (C), *i.e. when and only when*

$$\rho \not< 0, \quad \rho \not> \beta, \quad \rho \not> m. \quad \text{.................................(d)}$$

If $[c]_\beta^n$ is any undegenerate horizontal minor of $[b]_r^n$ of rank β, a general formula for all such matrices is

$$[x]_m^n = [h]_m^\beta [c]_\beta^n,$$

where $[h]_m^\beta$ has rank ρ.

Theorem II a. *If the factor matrices* $[a]_m^r$, $[b]_r^n$ *in the product* (A) *have any assigned ranks* α, β *consistent with the necessary conditions* (D), *and if either one of them is a given matrix, we can determine the elements of the other so that the product matrix* $[x]_m^n$ *has rank* ρ *when and only when* ρ *is an integer satisfying the conditions* (B).

Suppose that $[a]_m^r$ is a given matrix having any rank α consistent with the necessary conditions $\alpha \not< 0$, $\alpha \not> r$, $\alpha \not> m$. Then the equation (A) admits of finite solutions for $[b]_r^n$ when and only when the vertical rows of $[x]_m^n$ are connected with the vertical rows of

$[a]_m^r$. The possible values of the rank ρ of $[x]_m^n$ when this is so are those consistent with the conditions

$$\rho \nless 0, \quad \rho \ngtr a, \quad \rho \ngtr n. \qquad \text{.........(c)}$$

Again when $[x]_m^n$ is any given matrix whose vertical rows are connected with the vertical rows of $[a]_m^r$ and which has any rank ρ consistent with the necessary conditions (c), it follows from Theorem I of § 132 that we can determine a matrix $[b]_r^n$ of rank β satisfying the equation (A) when and only when

$$\beta \nless \rho, \quad \beta \ngtr n, \quad \beta + a \ngtr \rho + r. \qquad \text{.........(1)}$$

Thus when $[a]_m^r$ is given and has rank a, we can determine matrices $[b]_r^n$, $[x]_m^n$ of ranks β, ρ satisfying the equation (A) when and only when β and ρ are integers satisfying the conditions (c) and (1). Omitting the superfluous condition $\rho \ngtr n$, and eliminating ρ from (c) and (1), we see that these conditions include $\beta \nless 0$, $\beta \ngtr r$, and are equivalent to

$$\beta \nless 0, \quad \beta \ngtr r, \quad \beta \ngtr n; \quad \rho \nless 0, \quad \rho \ngtr a, \quad \rho \ngtr \beta, \quad \rho + r \nless a + \beta. \qquad \text{.........(2)}$$

Accordingly β can have any value consistent with the first three inequalities in (2), and when the value of β is assigned, the possible values of ρ are those consistent with the conditions (B).

This proves the theorem for the case in which $[a]_m^r$ is given; and the case in which $[b]_r^n$ is given can be treated in a similar way.

Theorem III a. *If $[x]_m^n$ is a given matrix having any assigned rank ρ consistent with the necessary conditions* (B) *and* (C), *i.e. consistent with the necessary conditions* (E), *we can determine matrices $[a]_m^r$, $[b]_r^n$ of ranks a, β satisfying the equation* (A) *when and only when a and β are integers consistent with the necessary conditions* (B) *and* (C).

The equation (A) admits of finite solutions for $[b]_r^n$ when and only when $[a]_m^r$ is such that the vertical rows of $[x]_m^n$ are connected with the vertical rows of $[a]_m^r$. The possible ranks a of $[a]_m^r$ when this is so are those consistent with the conditions

$$a \nless \rho, \quad a \ngtr r, \quad a \ngtr m. \qquad \text{.........(a)}$$

Again if $[a]_m^r$ is any given matrix so chosen that the vertical rows of $[x]_m^n$ are connected with its vertical rows and having any rank a satisfying the conditions (a), it follows from Theorem I of § 132 that we can determine a matrix $[b]_r^n$ of rank β satisfying the equation (A) when and only when

$$\beta \nless \rho, \quad \beta \ngtr n, \quad \beta + a \ngtr \rho + r. \qquad \text{.........(1)}$$

Thus we can determine matrices $[a]_m^r$, $[b]_r^n$ of ranks a, β satisfying the equation (A) when and only when a and β are integers consistent with the conditions (a) and (1), which, since $\rho \nless 0$, are equivalent to the conditions (B) and (C).

Since the elimination of a and β from (B) and (C) leads to (E), we see that:

If $[x]_m^n$ is a given matrix of rank ρ, there exist matrices $[a]_m^r$, $[b]_r^n$ satisfying the equation (A) *when and only when ρ satisfies the conditions* (D).

Theorem IV a. *If both* $[x]_m^n$ *and* $[a]_m^r$ *are given matrices having any assigned ranks* ρ *and* α *consistent with the necessary conditions* (B) *and* (C), *and (as is possible) so chosen that the equation* (A) *admits of finite solutions for* $[b]_r^n$, *then we can determine a solution* $[b]_r^n$ *of rank* β *when and only when* β *is an integer consistent with the necessary conditions* (B) *and* (C).

This theorem remains true when we interchange $[a]_m^r$, $[b]_r^n$ *and* α, β.

Let $[x]_m^n$ and $[a]_m^r$ have any assigned ranks ρ and α consistent with (B) and (C). Eliminating β from (B) and (C) we see that the possible values of ρ and α are those consistent with the conditions

$$\alpha \nless 0, \quad \alpha \ngtr r, \quad \alpha \ngtr m; \quad \rho \nless 0, \quad \rho \ngtr \alpha, \quad \rho \ngtr n. \quad \text{..................(3)}$$

Now if $[a]_m^r$ is any given matrix of rank α we can determine a matrix $[x]_m^n$ of rank ρ whose vertical rows are connected with the vertical rows of $[a]_m^r$ when and only when $\rho \nless 0$, $\rho \ngtr \alpha$, $\rho \ngtr n$. Consequently we can always determine matrices $[x]_m^n$, $[a]_m^r$ which have any assigned ranks consistent with the conditions (3), i.e. with the conditions (B) and (C), and are such that the equation (A) admits of finite solutions for $[b]_r^n$.

When $[x]_m^n$ and $[a]_m^r$ are any given matrices determined in this way, we see from Theorem I of § 132 that the possible ranks β of the solutions $[b]_r^n$ of the equation (A) are given by

$$\beta \nless \rho, \quad \beta \ngtr n, \quad \beta + \alpha \ngtr \rho + r, \quad \text{................................(1)}$$

and these are the values of β consistent with the conditions (B) and (C) when ρ and α have assigned values.

Thus the theorem is true when $[x]_m^n$ and $[a]_m^r$ are given; and in a similar way we can show that it is true when $[x]_m^n$ and $[b]_r^n$ are given.

We will now give a summary of all the foregoing theorems.

Summary A. *In every product of two matrices of the form* (A) *the product matrix and the factor matrices can have any ranks consistent with their orders and with the necessary conditions of* § 133. *This is the case even under the following circumstances:*

(1) *When one of the factor matrices is given.*

(2) *When the product matrix is given.*

(3) *When both the product matrix and one of the factor matrices are given and are so chosen that the equation* (A) *admits of solution for the other factor matrix.*

Ex. v. *If the matrix* $[b]_r^n$ *has rank* n, *the rank of the product matrix* $[c]_m^n = [a]_m^r [b]_r^n$ *is equal to* $n - q$, *where* q *is the number of unconnected solutions of the equation* $[a]_m^r \overline{\underline{x}}_r = 0$ *connected with the active (or vertical) rows of* $[b]_r^n$.

If the matrix $[a]^r_m$ *has rank m, the rank of the product matrix* $[c]^n_m = [a]^r_m [b]^n_r$ *is equal to* $m-p$, *where p is the number of unconnected solutions of the equation* $\overline{\underline{b}}{}^r_n\, \overline{\underline{x}}{}_r = 0$ *connected with the active (or horizontal) rows of* $[a]^r_m$

It will be sufficient to prove the first result. We will denote the rank of $[c]^n_m$ by $n-t$, and show that $t=q$.

Since the equation $[c]^n_m\, \overline{\underline{x}}{}_n = 0$ has t unconnected solutions, there exists a matrix $\overline{\underline{k}}{}^t_n$ of rank t such that

$$[c]^n_m\, \overline{\underline{k}}{}^t_n = [a]^r_m [b]^n_r\, \overline{\underline{k}}{}^t_n = 0,$$

or

$$[a]^r_m\, \overline{\underline{\beta}}{}^t_r = 0,$$

where

$$\overline{\underline{\beta}}{}^t_r = [b]^n_r\, \overline{\underline{k}}{}^t_n.$$

By § 131 the matrix $\overline{\underline{\beta}}{}^t_r$ has the same rank t as $\overline{\underline{k}}{}^t_n$. Therefore the vertical rows of $\overline{\underline{\beta}}{}^t_r$ are t unconnected solutions of the equation $[a]^r_m\, \overline{\underline{x}}{}_r = 0$ connected with the active rows of $[b]^n_r$.

It follows that

$$q \not< t.$$

Again by supposition there exists a matrix $\overline{\underline{\beta}}{}^q_r$ of rank q and a matrix $\overline{\underline{k}}{}^q_n$ (which by § 131 must also have rank q) such that

$$[a]^r_m\, \overline{\underline{\beta}}{}^q_r = 0, \quad \overline{\underline{\beta}}{}^q_r = [b]^n_r\, \overline{\underline{k}}{}^q_n.$$

The last two equations show that

$$[a]^r_m [b]^n_r\, \overline{\underline{k}}{}^q_n = [c]^n_m\, \overline{\underline{k}}{}^q_n = 0.$$

Thus the equation $[c]^n_m\, \overline{\underline{x}}{}_n = 0$ has q unconnected solutions and therefore (or by Theorem I) the rank of $[c]^n_m$ cannot exceed $n-q$, i.e. we must have $n-t \not> n-q$, or

$$q \not> t.$$

It follows that $t=q$.

Ex. vi. **Theorem.** *If the matrices* $[a]^r_m$ *and* $[b]^n_r$ *have respectively ranks* α *and* β, *and if the product matrix* $[c]^n_m = [a]^r_m [b]^n_r$ *has rank* ρ, *then*

$$\rho = \alpha - p = \beta - q,$$

where p is the greatest number of unconnected solutions of the equation $\overline{\underline{b}}{}^r_n\, \overline{\underline{x}}{}_r = 0$ *connected with the active (or horizontal) rows of* $[a]^r_m$, *and q is the greatest number of unconnected solutions of the equation* $[a]^r_m\, \overline{\underline{x}}{}_r = 0$ *connected with the active (or vertical) rows of* $[b]^n_r$.

Let $[b_{1v}]^{\beta}_{r}$ be a vertical minor of $[b]^{n}_{r}$ of rank β. Then there exists a matrix $[k]^{n}_{\beta}$ of rank β such that

$$[b]^{n}_{r}=[b_{1v}]^{\beta}_{r}\,[k]^{n}_{\beta}, \quad [c]^{n}_{m}=[a]^{r}_{m}\,[b_{1v}]^{\beta}_{r}\,[k]^{n}_{\beta}.$$

By § 131 the matrix $[c]^{n}_{m}$ has the same rank as $[a]^{r}_{m}\,[b_{1v}]^{\beta}_{r}$, and by Ex. iii the last product matrix has rank $\beta-q$, where q is the number of unconnected solutions of the equation $[a]^{r}_{m}\,\overline{\underline{x}}_{r}=0$ connected with the vertical rows of $[b_{1v}]^{\beta}_{r}$, i.e. with the vertical rows of $[b]^{n}_{r}$.

Thus we have $\rho=\beta-q$; and similarly we can show that $\rho=\alpha-p$.

2. *Products of three matrices.*

If $[a]^{r}_{m}$, $[b]^{s}_{r}$, $[c]^{n}_{s}$ are matrices of given orders having ranks α, β, γ, and if the product matrix

$$[x]^{n}_{m}=[a]^{r}_{m}\,[b]^{s}_{r}\,[c]^{n}_{s} \quad \text{.........................(A}')$$

has rank ρ, we see from § 133 that the integers α, β, γ, ρ must satisfy the conditions

$$\rho\nless 0,\quad \rho\ngtr\alpha,\quad \rho\ngtr\beta,\quad \rho\ngtr\gamma,\quad \rho+r+s\nless\alpha+\beta+\gamma; \quad \text{......(B}')$$

$$\alpha\nless 0,\ \ \alpha\ngtr m,\ \ \alpha\ngtr r;\quad \beta\nless 0,\ \ \beta\ngtr r,\ \ \beta\ngtr s;\quad \gamma\nless 0,\ \ \gamma\ngtr s,\ \ \gamma\ngtr n; \quad \text{...(C}')$$

where the conditions $\alpha\nless 0$, $\beta\nless 0$, $\gamma\nless 0$ in (C′) are superfluous when we take (B′) and (C′) together.

The conditions (B′) and (C′) include

$$\rho\nless 0,\quad \rho\ngtr m,\quad \rho\ngtr r,\quad \rho\ngtr s,\quad \rho\ngtr n. \quad \text{..............(D}')$$

They reduce to (D′) when we eliminate α, β and γ; and they reduce to (C′) when we eliminate ρ.

When we eliminate α, β, γ respectively we obtain

$$\beta\nless 0,\ \beta\ngtr r,\ \beta\ngtr s;\quad \gamma\nless 0,\ \gamma\ngtr s,\ \gamma\ngtr n;\quad \rho\nless 0,\ \rho\ngtr\beta,\ \rho\ngtr\gamma,\ \rho\ngtr m; \quad \text{......(a}')$$

$$\alpha\nless 0,\ \alpha\ngtr m,\ \alpha\ngtr r;\quad \gamma\nless 0,\ \gamma\ngtr s,\ \gamma\ngtr n;\quad \rho\nless 0,\ \rho\ngtr\alpha,\ \rho\ngtr\gamma; \quad \text{...(b}')$$

$$\alpha\nless 0,\ \alpha\ngtr m,\ \alpha\ngtr r;\quad \beta\nless 0,\ \beta\ngtr r,\ \beta\ngtr s;\quad \rho\nless 0,\ \rho\ngtr\alpha,\ \rho\ngtr\beta,\ \rho\ngtr n. \quad \text{......(c}')$$

When we eliminate α and γ we obtain

$$\beta\nless 0,\quad \beta\ngtr r,\quad \beta\ngtr s;\quad \rho\nless 0,\quad \rho\ngtr\beta,\quad \rho\ngtr m,\quad \rho\ngtr n, \quad \text{......(d}')$$

where the condition $\beta\nless 0$ is superfluous.

Here m, r, s, n are any given positive integers which are not less than 1. We may however consider that the value 0 is permissible for each of these quantities when we regard a matrix one of whose orders is 0 as non-existent and replaceable as a factor by the scalar number 0.

We will now prove the theorems which follow.

Theorem I b. *If* $[x]_m^n$ *is a given matrix having any assigned rank* ρ *consistent with the necessary conditions* (B′) *and* (C′), *i.e. with the conditions* (D′), *we can determine matrices* $[a]_m^r$, $[b]_r^s$, $[c]_s^n$ *of assigned ranks* α, β, γ *satisfying the equation* (A′) *when and only when* α, β, γ *are integers consistent with the necessary conditions* (B′) *and* (C′), *i.e. with the conditions*

$$\alpha \nless \rho,\ \ \alpha \ngtr r,\ \ \alpha \ngtr m;\quad \beta \nless \rho,\ \ \beta \ngtr r,\ \ \beta \ngtr s;\quad \gamma \nless \rho,\ \ \gamma \ngtr s,\ \ \gamma \ngtr n;\quad \alpha+\beta+\gamma \ngtr \rho+r+s.$$

By Exs. i and ii we can determine matrices $[a]_m^r$ of rank α and $[c]_s^n$ of rank γ such that the vertical rows of $[x]_m^n$ are connected with the vertical rows of $[a]_m^r$ and the horizontal rows of $[x]_m^n$ are connected with the horizontal rows of $[c]_s^n$ when and only when

$$\alpha \nless \rho,\ \ \alpha \ngtr r,\ \ \alpha \ngtr m;\quad \gamma \nless \rho,\ \ \gamma \ngtr s,\ \ \gamma \ngtr n. \qquad \text{......................(5)}$$

Again when $[a]_m^r$ and $[c]_s^n$ are any given matrices thus determined having any ranks α and β consistent with (5), the equation (A′) admits of solution for $[b]_r^s$, and by § 132.3 it has solutions of rank β when and only when

$$\beta \nless \rho,\ \ \beta \ngtr r,\ \ \beta \ngtr s,\ \ \alpha+\beta+\gamma \ngtr \rho+r+s. \qquad \text{........................(6)}$$

Since the conditions (D′), (5) and (6) are together equivalent to the conditions (B′) and (C′), it follows that the equation (A′) can be satisfied by matrices $[a]_m^r$, $[b]_r^s$, $[c]_s^n$ having any ranks α, β, γ consistent with the necessary conditions (B′) and (C′).

COROLLARY 1. *We can determine matrices* $[a]_m^r$, $[b]_r^s$, $[c]_s^n$, $[x]_m^n$ *satisfying the equation* (A′) *and having any assigned ranks* α, β, γ, ρ *consistent with the necessary conditions* (B′) *and* (C′).

COROLLARY 2. *When* α, β, γ *have any assigned values consistent with the necessary conditions* (C′), *the possible ranks of the product matrix* $[x]_m^n$ *are those consistent with the conditions* (B′).

Theorem II b. *If* $[x]_m^n$, $[a]_m^r$, $[c]_s^n$ *are given matrices having any assigned ranks* ρ, α, γ *consistent with the necessary conditions* (B′) *and* (C′), *i.e. with the conditions* (b′), *and (as is possible) so chosen that the equation* (A′) *admits of solution for* $[b]_r^s$, *then the possible ranks of the solutions* $[b]_r^s$ *are the values of* β *consistent with the necessary conditions* (B′) *and* (C′), *i.e. with the conditions*

$$\beta \nless \rho,\ \ \beta \ngtr r,\ \ \beta \ngtr s,\ \ \alpha+\beta+\gamma \ngtr \rho+r+s. \qquad \text{........................(6)}$$

Since the conditions (D′) and (5) in Theorem I *b* are together equivalent to the conditions (b′), we see that the condition that (A′) shall admit of solution for $[b]_r^s$ imposes no restriction on the possible ranks of $[a]_m^r$, $[c]_s^n$, $[x]_m^n$ which is not already imposed by the necessary conditions (B′) and (C′). For (b′) are the conditions that it shall be possible to determine matrices $[x]_m^n$, $[a]_m^r$, $[c]_s^n$ of ranks ρ, α, γ and some matrix $[b]_r^s$ satisfying the equation (A′). Accordingly, as shown in the proof of Theorem I *b*, the above theorem is true.

Theorem III b. *If* $[x]_m^n$ *and* $[b]_r^s$ *are given matrices having any assigned ranks* ρ *and* β *consistent with the necessary conditions* (B′) *and* (C′), *i.e. with the conditions* (d′), *we can determine matrices* $[a]_m^r$, $[c]_s^n$ *of ranks* α, γ *satisfying the equation* (A′) *when and only when* α *and* γ *are integers consistent with the necessary conditions* (B′) *and* (C′), *i.e. with the conditions*

$$\alpha \nless \rho, \quad \alpha \ngtr r, \quad \alpha \ngtr m; \quad \gamma \nless \rho, \quad \gamma \ngtr s, \quad \gamma \ngtr n; \quad \alpha+\beta+\gamma \ngtr \rho+r+s. \quad \text{......(7)}$$

Using Ex. i of § 115 we can write

$$[b]_r^s = [p]_r^\beta [q]_\beta^s, \quad [u]_m^\beta = [a]_m^r [p]_r^\beta, \quad [v]_\beta^n = [q]_\beta^s [c]_s^n,$$

where $[p]_r^\beta$ and $[q]_\beta^s$ are given matrices of ranks β.

By Theorem III *a* we can determine matrices $[u]_m^\beta$, $[v]_\beta^n$ of ranks h, k satisfying the equation $[u]_m^\beta [v]_\beta^n = [x]_m^n$ when and only when

$$\rho \nless 0, \quad \rho \ngtr h, \quad \rho \ngtr k, \quad \rho+\beta \nless h+k; \quad h \ngtr m, \quad k \ngtr n. \quad \text{...............(8)}$$

Again when $[u]_m^\beta$, $[v]_\beta^n$ are any two given matrices determined in this manner having any ranks h and k consistent with (8), the equations

$$[a]_m^r [p]_r^\beta = [u]_m^\beta, \quad [q]_\beta^s [c]_s^n = [v]_\beta^n$$

admit of solutions for $[a]_m^r$ and $[c]_s^n$ respectively, and it follows from Theorems I and II of § 132 that we can determine solutions $[a]_m^r$, $[c]_s^n$ of ranks α, γ when and only when

$$\alpha \nless h, \quad \alpha \ngtr m, \quad \alpha+\beta \ngtr h+r; \quad \gamma \nless k, \quad \gamma \ngtr n, \quad \beta+\gamma \ngtr k+s. \quad \text{............(9)}$$

The inequalities obtained by eliminating h and k from (8) and (9) are equivalent to (B′) and (C′), i.e. to (d′) and (7); and this establishes the theorem.

All the foregoing results can be summarised in the following way.

Summary B. *In every product of three matrices of the form* (A′) *the product matrix and the factor matrices can have any ranks consistent with their orders and with the necessary conditions of* § 133 *for that product. This is the case even under the following circumstances:*

(1) *When the product matrix is given.*

(2) *When any one of the factor matrices is given.*

(3) *When both the product matrix and the middle factor matrix are given.*

(4) *When the product matrix and one or both of the extreme factor matrices are given, provided that when either extreme factor matrix is given it is so chosen that the equation* (A) *admits of solution for the product of the remaining two factor matrices.*

The following examples deal with the case in which $[x]_m^n$, $[b]_r^s$, $[c]_s^n$ are given. We have similar results when $[x]_m^n$, $[a]_m^r$, $[b]_r^s$ are given.

Ex. vii. Let $[a]_m^r$, $[b]_r^s$, $[c]_s^n$, $[x]_m^n$ be matrices of ranks a, β, γ, ρ satisfying the equation (A′), and let $[b]_r^s [c]_s^n$ have rank σ. Then if we write

$$[b]_r^s [c]_s^n = [y]_r^n, \quad [a]_m^r [y]_r^n = [x]_m^n, \qquad \text{.........................(10)}$$

we see by applying § 133 to these last two products that the integers a, β, γ, ρ, σ must satisfy the conditions

$$\rho \nless 0, \quad \rho \ngtr a, \quad \rho \ngtr \sigma, \quad \rho + r \nless a + \sigma, \quad a \ngtr m, \qquad \text{.....................(11)}$$

$$\sigma \ngtr \beta, \quad \sigma \ngtr \gamma, \quad \sigma + s \nless \beta + \gamma, \quad \beta \ngtr r, \quad \gamma \ngtr n. \qquad \text{.....................(12)}$$

When we eliminate σ from (11) and (12) we obtain (B′) and (C′).

If $[x]_m^n$ is a given matrix having any rank ρ consistent with these conditions, i.e. with the conditions (D′), we can by Theorem III *a* determine matrices $[a]_m^r$, $[y]_r^n$ of ranks a, σ satisfying the second of the equations (10) when and only when the conditions (11) are satisfied; and if $[a]_m^r$, $[y]_r^n$ are any two given matrices determined in this way, we can determine matrices $[b]_r^s$, $[c]_s^n$ of ranks β, γ satisfying the first of the equations (10) when and only when the conditions (12) are satisfied.

Thus the equation (A′) *can be satisfied when* a, β, γ, ρ, σ *have any assigned values satisfying the necessary conditions* (11) *and* (12), *and this is true even when* $[x]_m^n$ *is a given matrix.*

Leaving σ arbitrary and eliminating a we deduce that:

When the equation (A′) *admits of solution for* $[a]_m^r$, *the integers* β, γ, ρ, σ *can have any assigned values consistent with the necessary conditions* (11) *and* (12), *and this is true even when* $[x]_m^n$ *is a given matrix.*

Ex. viii. When the equation (A′) is satisfied or admits of solution for $[a]_m^r$, the possible values of ρ and σ are those consistent with the conditions

$$\rho \nless 0, \quad \rho \ngtr m; \quad \sigma \nless \rho, \quad \sigma \ngtr r, \quad \sigma \ngtr s, \quad \sigma \ngtr n. \qquad \text{..................(13)}$$

When ρ and σ have any given values consistent with these conditions the possible values of β and γ are those consistent with the conditions (11) and (12).

Ex. ix. When the equation (A′) is satisfied or admits of solution for $[a]_m^r$, the possible values of ρ, β, γ are those consistent with the conditions (11) and (12), i.e. those consistent with the conditions (B′) and (C′); and when ρ, β, γ are given the possible values of σ are those consistent with the conditions

$$\sigma \nless \rho, \quad \sigma \ngtr \beta, \quad \sigma \ngtr \gamma, \quad \rho + r \nless a + \sigma, \quad \sigma + s \nless \beta + \gamma.$$

Thus the condition that (A′) shall admit of solution for $[a]_m^r$ imposes no restrictions on the values of ρ, β, γ in addition to those given by (B′) and (C′).

Ex. x. *If* $[x]_m^n$, $[b]_r^s$, $[c]_s^n$ *are three given matrices of ranks* ρ, β, γ *so determined that the equation* (A′) *admits of solution for* $[a]_m^r$, *and if* $[b]_r^s [c]_s^n$ *has rank* σ, *the possible ranks* a *of the solutions* $[a]_m^r$ *are those consistent with the conditions*

$$a \nless \rho, \quad a \ngtr m, \quad a + \sigma \ngtr \rho + r,$$

i.e. those consistent with the conditions (B′) *and* (C′).

This follows from Theorem IV a. Thus the possible values of a are the same when $[x]_m^n$, $[b]_r^s$, $[c]_s^n$ are given as when only ρ and σ are given.

§ 135. Possible ranks of the product matrix and the factor matrices in any matrix product.

If $[a]_m^p$, $[b]_p^q$, $[c]_q^r$, ... $[k]_v^w$, $[l]_w^n$ are matrices of given orders having ranks $\alpha, \beta, \gamma, \ldots \kappa, \lambda$, and if the product matrix

$$[x]_m^n = [a]_m^p [b]_p^q [c]_q^r \ldots [k]_v^w [l]_w^n \ldots\ldots\ldots\ldots\ldots\ldots\ldots(A)$$

has rank ρ, we see from § 133 that the integers $\alpha, \beta, \gamma, \ldots \kappa, \lambda, \rho$ must satisfy the conditions

$$\rho \nless 0,\quad \rho \ngtr \alpha,\quad \rho \ngtr \beta,\quad \rho \ngtr \gamma, \ldots \rho \ngtr \kappa,\quad \rho \ngtr \lambda; \ldots\ldots\ldots\ldots(B)$$

$$\rho + p + q + r + \ldots + v + w \nless \alpha + \beta + \gamma + \ldots + \kappa + \lambda; \quad \ldots\ldots(C)$$

$$\alpha \nless 0,\quad \alpha \ngtr m,\quad \alpha \ngtr p;\quad \beta \nless 0,\quad \beta \ngtr p,\quad \beta \ngtr q, \ldots \lambda \nless 0,\quad \lambda \ngtr w,\quad \lambda \ngtr n. \quad \ldots\ldots(D)$$

The conditions (B), (C), (D) include

$$\rho \nless 0,\quad \rho \ngtr m,\quad \rho \ngtr p,\quad \rho \ngtr q, \ldots \rho \ngtr w,\quad \rho \ngtr n. \ldots\ldots\ldots\ldots(E)$$

They reduce to (E) when we eliminate $\alpha, \beta, \gamma, \ldots \kappa, \lambda$; and they reduce to (D) when we eliminate ρ.

We will now prove the theorems which follow.

Theorem I. *If* $[x]_m^n$ *is a given matrix having any assigned rank* ρ *consistent with the necessary conditions* (B), (C), (D), *i.e. with the conditions* (E), *we can determine matrices* $[a]_m^p$, $[b]_p^q$, $[c]_q^r$, ... $[k]_v^w$, $[l]_w^n$ *of assigned ranks* $\alpha, \beta, \gamma, \ldots \kappa, \lambda$ *satisfying the equation* (A) *when and only when* $\alpha, \beta, \gamma, \ldots \kappa, \lambda$ *are integers consistent with the necessary conditions* (B), (C), (D).

It has been shown in § 134 that the theorem is true for products of two and three matrices. We will now suppose that there are more than three factor matrices.

Let
$$[b]_p^q [c]_q^r \ldots [k]_v^w = [y]_p^w, \quad \ldots\ldots\ldots\ldots\ldots\ldots\ldots\ldots(1)$$

so that
$$[a]_m^p [y]_p^w [l]_w^n = [x]_m^n, \quad \ldots\ldots\ldots\ldots\ldots\ldots\ldots\ldots(2)$$

and let $[y]_p^w$ have rank σ. We will suppose that the theorem is true for the product (1), and show that it must then be true for the product (A).

By Exs. i and ii of § 134 we can determine matrices $[a]_m^p$, $[l]_w^n$ of ranks α, λ which are such that (2) admits of solution for $[y]_p^w$ when and only when

$$\alpha \nless \rho,\quad \alpha \ngtr p,\quad \alpha \ngtr m;\quad \lambda \nless \rho,\quad \lambda \ngtr w,\quad \lambda \ngtr n. \ldots\ldots\ldots(3)$$

Let $[a]_m^p$, $[l]_w^n$ be any given matrices determined in this way and having any ranks α, λ consistent with (3). Then by § 132.3 the equation (2) has solutions $[y]_p^w$ of rank σ when and only when

$$\sigma \nless \rho, \quad \sigma \ngtr p, \quad \sigma \ngtr w, \quad \alpha+\sigma+\lambda \ngtr \rho+p+w,$$

and therefore it has solutions of rank σ when σ is any integer satisfying the conditions

$$\sigma \nless \rho, \quad \sigma \ngtr p, \quad \sigma \ngtr q, \dots \sigma \ngtr v, \quad \alpha+\sigma+\lambda \ngtr \rho+p+w, \quad \dots\dots(4)$$

which are consistent with one another by virtue of (E) and (3).

Let $[y]_p^w$ be any given solution of (2) having any rank σ consistent with (4). Then by the hypothesis we can determine matrices $[b]_p^q$, $[c]_q^r$, ... $[k]_v^w$ satisfying the equation (1) and having any ranks β, γ, ... κ consistent with the conditions

$$\sigma \nless 0, \ \sigma \ngtr \beta, \ \sigma \ngtr \gamma, \dots \sigma \ngtr \kappa; \dots\dots\dots\dots\dots\dots\dots\dots(\text{B}')$$

$$\sigma + q + r + \dots + v \nless \beta + \gamma + \dots + \kappa; \quad \dots\dots\dots\dots\dots(\text{C}')$$

$$\beta \ngtr p, \ \beta \ngtr q; \ \gamma \ngtr q, \ \gamma \ngtr r; \dots \kappa \ngtr v, \ \kappa \ngtr w. \quad \dots\dots\dots(\text{D}')$$

Thus the equation (A) can be satisfied when α, β, γ, ... κ, λ, σ have any integral values consistent with (B′), (C′), (D′), (3) and (4). When we eliminate σ from these five sets of conditions, observing that $\rho \nless 0$ and in (4) all inequalities except the first and last are superfluous, we obtain the conditions (B), (C), (D). This shows that we can determine matrices $[a]_m^p$, $[b]_p^q$, $[c]_q^r$, ... $[k]_v^w$, $[l]_w^n$ of ranks α, β, γ, ... κ, λ satisfying the equation (A) when α, β, γ, ... κ, λ are any integers satisfying the necessary conditions (B), (C), (D).

We have now proved that if the theorem is true for the product (1), it is true for the product (A). If then it is true for all products of t matrices, where t is any positive integer not less than 2, it is true for all products of $t+2$ matrices. Since it is true for all products of two and three matrices, it must be true for every matrix product.

COROLLARY 1. *We can determine matrices* $[x]_m^n$, $[a]_m^p$, $[b]_p^q$, ... $[l]_w^n$ *which satisfy the equation* (A) *and have the assigned ranks* ρ, α, β, ... λ *when and only when* ρ, α, β, ... λ *are integers satisfying the necessary conditions* (B), (C), (D).

COROLLARY 2. *If* $[a]_m^p$, $[b]_p^q$, ... $[l]_w^n$ *are matrices having the assigned ranks* α, β, ... λ *whose elements are arbitrary, the possible values of the rank* ρ *of their product* (A) *are those consistent with the necessary conditions* (B), (C), (D).

COROLLARY 3. *If* $[x]_m^n$ *is a given matrix having any assigned rank* ρ *consistent with the necessary conditions* (B), (C), (D), *i.e. with the conditions* (E), *we can determine matrices* $[b]_p^q, [c]_q^r, \ldots [k]_v^w$ *which have the assigned ranks* $\beta, \gamma, \ldots \kappa$, *and are such that the equation* (A) *admits of solutions for* $[a]_m^p, [l]_w^n$ *when and only when* $\beta, \gamma, \ldots \kappa$ *are integers consistent with the necessary conditions* (B), (C), (D), *i.e. when and only when they satisfy the conditions* (B′), (C′), (D′).

For it is possible to determine integers α and λ satisfying the conditions (B), (C), (D) when and only when the integers $\beta, \gamma, \ldots \kappa$ are chosen in this way.

Theorem II. *If in* (A) *the product matrix* $[x]_m^n$ *and every alternate factor matrix beginning from the first or second are given having any assigned ranks consistent with the necessary conditions* (B), (C), (D), *then we can determine the remaining alternate factor matrices so that they satisfy the equation* (A) *and have any assigned ranks which are consistent with the necessary conditions* (B), (C), (D), *provided only that:*

(1) *When the first factor matrix is given, the vertical rows of* $[x]_m^n$ *are connected with the vertical rows of* $[a]_m^p$.

(2) *When the last factor matrix is given, the horizontal rows of* $[x]_m^n$ *are connected with the horizontal rows of* $[l]_w^n$.

It has been shown in § 134 that this theorem is true for products of two and three matrices. Knowing this we can prove the theorem by induction. We consider separately the following three cases which include all possible cases.

CASE I. *When the first factor matrix is not given.*

In this case $[b]_p^q$ is a given matrix of rank β and can be expressed in the form

$$[b]_p^q = [H]_p^\beta [b']_\beta^q,$$

where the two matrices on the right are given undegenerate matrices of rank β, and we can write

$$[x]_m^n = [S]_m^\beta [T]_\beta^n, \qquad \text{..........(5)}$$

where

$$[S]_m^\beta = [a]_m^p [H]_p^\beta, \qquad \text{..........(6)}$$

and

$$[T]_\beta^n = [b']_\beta^q [c]_q^r \ldots [k]_v^w [l]_w^n. \qquad \text{..........(7)}$$

We will make the hypothesis that the theorem is true for the product (7), and we will show that it must then be true for the product (A).

By Theorem I we can determine matrices $[S]_m^\beta$, $[T]_\beta^n$ of ranks σ, τ satisfying the equation (5) when and only when

$$\rho \not< 0,\quad \rho \not> \sigma,\quad \rho \not> \tau,\quad \rho+\beta \not< \sigma+\tau,\quad \sigma \not> m,\quad \tau \not> n. \quad \text{..............(8)}$$

Again if $[S]_m^\beta$, $[T]_\beta^n$ are any two given matrices determined in this way and having any ranks σ, τ consistent with the necessary conditions (8), it follows from Theorem IV *a* of § 134 that we can determine a matrix $[a]_m^p$ of rank a satisfying the equation (6) when and only when

$$\sigma \not< 0,\quad \sigma \not> a,\quad \sigma \not> \beta,\quad \sigma+p \not< a+\beta,\quad a \not> m; \quad \text{..................(9)}$$

and it follows from the hypothesis that we can determine the unknown factor matrices occurring on the right in (7), which are the alternate factor matrices beginning with $[c]_q^r$, so that they satisfy the equation (7) and have ranks γ, ϵ, ... when and only when

$$\left.\begin{array}{l} \tau \not< 0,\ \tau \not> \beta,\ \tau \not> \gamma,\ \dots\ \tau \not> \kappa,\ \tau \not> \lambda; \\ \tau+q+r+\dots+w \not< \beta+\gamma+\kappa+\dots+\lambda; \\ \beta \not> q;\ \gamma \not> q,\ \gamma \not> r;\ \dots\ \kappa \not> v,\ \kappa \not> w;\ \lambda \not> w,\ \lambda \not> n. \end{array}\right\} \text{..............(10)}$$

It is here assumed that when $[l]_w^n$ is known, the horizontal rows of $[x]_m^n$ are connected with the horizontal rows of $[l]_w^n$. When this is so, the horizontal rows of $[T]_\beta^n$ are connected with the horizontal rows of $[l]_w^n$, for by prefixing an inverse prefactor of $[S]_m^\beta$ on both sides of (5) we see that the horizontal rows of $[T]_\beta^n$ are connected with the horizontal rows of $[x]_m^n$.

Thus we can determine the unknown factor matrices $[a]_m^p$, $[c]_q^r$, ... in (A) so that they satisfy the equation (A) and have the assigned ranks a, γ, ... when and only when we can determine integers σ and τ so that the conditions (8), (9) and (10) are satisfied. Now when we eliminate σ and τ from (8), (9) and (10) we obtain the conditions (B), (C) and (D). Therefore this is possible when and only when a, γ, ... satisfy the conditions (B), (C), (D), where ρ, β, δ, ... are given integers consistent with those conditions.

It follows that if the theorem is true for the product (7), then it is true for the product (A).

CASE II. *When the last factor matrix is not given.*

In this case $[k]_v^w$ is a given matrix of rank κ, and we can proceed in a similar way. We express $[k]_v^w$ in the form

$$[k]_v^w = [k']_v^\kappa [K]_\kappa^w.$$

Then writing $$[x]_m^n = [T]_m^\kappa [S]_\kappa^n,$$

where $$[S]_\kappa^n = [K]_\kappa^w [l]_w^n,$$

and $$[T]_m^\kappa = [a]_m^p [b]_p^q [c]_q^r \dots [k']_v^\kappa, \quad \text{............................(11)}$$

we show that if the theorem is true for the product (11), then it is true for the product (A).

CASE III. *When both the first and the last factor matrices are given.*

In this case $[a]_m^p$ and $[l]_w^n$ are given matrices of ranks a and λ. Also the vertical rows of $[x]_m^n$ are connected with the vertical rows of $[a]_m^p$, and the horizontal rows of $[x]_m^n$ are connected with the horizontal rows of $[l]_w^n$.

We define $[y]_p^w$ as in (1), make the hypothesis that the theorem is true for the product (1), and show that it must then be true for the product (A).

By Theorem II *b* of § 134 we can determine matrices $[a]_m^p$, $[y]_p^w$, $[l]_w^n$ of ranks a, σ, λ satisfying the equation (2) when and only when

$$\left.\begin{array}{l}\rho \nless 0,\ \rho \ngtr a,\ \rho \ngtr \sigma,\ \rho \ngtr \lambda;\ \rho+p+w \nless a+\sigma+\lambda; \\ a \ngtr m,\ a \ngtr p;\ \sigma \ngtr p,\ \sigma \ngtr w,\ \lambda \ngtr w,\ \lambda \ngtr n.\end{array}\right\} \qquad (12)$$

Again when $[a]_m^p$, $[y]_p^w$, $[l]_w^n$ are any three given matrices determined in this way, it follows from the hypothesis that we can determine the unknown factor matrices in (1), i.e. the alternate factor matrices beginning with $[b]_p^q$ and ending with $[k]_v^w$, so that they satisfy the equation (1) and have ranks β, δ, ... κ when and only when

$$\left.\begin{array}{l}\sigma \nless 0,\ \sigma \ngtr \beta,\ \sigma \ngtr \gamma,\ \ldots\ \sigma \ngtr \kappa; \\ \sigma+q+r+\ldots+v \nless \beta+\gamma+\ldots+\kappa; \\ \beta \ngtr p,\ \beta \ngtr q;\ \gamma \ngtr q,\ \gamma \ngtr r;\ \ldots\ \kappa \ngtr v,\ \kappa \ngtr w.\end{array}\right\} \qquad (13)$$

Thus we can determine the unknown factor matrices $[b]_p^q$, ... $[k]_v^w$ in (A) so that they satisfy the equation (A) and have the assigned ranks β, δ, ... κ when and only when we can determine an integer σ so that the conditions (12) and (13) are satisfied. Now when we eliminate σ from these conditions, we obtain the conditions (B), (C), (D). Therefore this is possible when and only when β, δ, ... κ satisfy the conditions (B), (C), (D), where ρ, a, γ, ... λ are given integers consistent with those conditions.

It follows that if the theorem is true for the product (1), then it is true for the product (A).

From the results proved in Cases I, II and III we conclude that if Theorem II is true for all products of t and $t+1$ matrices, where t is any integer which is not less than 2, then it is true for all products of $t+2$ matrices. Since it is true for all products of two and three matrices, it follows in succession that it is true for all products of 2, 3, ... t matrices, i.e. it is true for all matrix products.

We now give a summary of the results obtained.

Summary. *In every matrix product of the form* (A) *the product matrix and the factor matrices can have any ranks consistent with their orders and the necessary conditions of* § 133 *for that product. This is the case even when the product matrix or any one of the factor matrices is given; and even when the product matrix and any number of the factor matrices, no two of which are consecutive, are given; provided that when an extreme factor matrix is given, the equation* (A) *admits of solution for the complementary factor.*

Ex. The equation (A) admits of solution whenever at least one factor matrix is unknown, provided that when the product matrix and an extreme factor matrix are both given the equation admits of solution for the complementary factor on the right. To obtain the possible ranks of the solutions we replace every set of consecutive given factor matrices by its product matrix. We then obtain an equation of the same form in which no two consecutive factor matrices are given, and the possible ranks of the unknown matrices are then given by Theorem II applied to this new equation.

§ 136. Possible ranks of the product matrix and the factor matrices in any product of two or three matrices which is self-conjugate in form.

In this article we shall make use of some of the results proved in Chapter XVII.

1. *Symmetric products of two matrices.*

If $[a]_r^m$ is a matrix of given orders having rank α, and if the self-conjugate product matrix

$$[x]_m^m = \overline{\underline{a}}{}_m^r\,[a]_r^m \quad \text{.............................(A)}$$

has rank ρ, we see from § 133 that the integers α and ρ must satisfy the conditions

$$\rho \nless 0, \quad \rho \ngtr \alpha, \quad \rho + r \nless 2\alpha; \quad \text{.......................(B)}$$

and

$$\alpha \ngtr m; \quad \text{..................................(C)}$$

which include the conditions

$$\alpha \nless 0, \quad \alpha \ngtr r, \quad \alpha \ngtr m; \quad \text{.......................(D)}$$

and

$$\rho \nless 0, \quad \rho \ngtr m, \quad \rho \ngtr r. \quad \text{.........................(E)}$$

These conditions reduce to (D) when we eliminate α; and they reduce to (C) when we eliminate ρ.

We will now prove the theorems which follow.

Theorem I a. *If the matrix* $[a]_r^m$ *has any assigned rank* a *consistent with the necessary conditions* (D), *and if its elements are otherwise arbitrary, then the possible values of the rank* ρ *of the product* (A) *are those consistent with the conditions* (B), *i.e.* ρ *can have any value consistent with the necessary conditions* (B) *and* (C).

We know that ρ must satisfy the conditions (B). It remains to show that ρ can have any value which is consistent with them. To see this let ρ be any integer satisfying the conditions (B). Then if $i=\sqrt{-1}$, we obtain a matrix $[a]_r^m$ which has rank a and is such that the product $\overline{\underline{a}}{}_m^r\,[a]_r^m$ has rank ρ when

$$[a]_r^m = \begin{bmatrix} 1, & 0, & 0 \\ 0, & 1, & 0 \\ 0, & i, & 0 \\ 0, & 0, & 0 \end{bmatrix}_{\rho,\ \alpha-\rho,\ \alpha-\rho,\ r+\rho-2\alpha}^{\rho,\ \alpha-\rho,\ m-\alpha}, \qquad \overline{\underline{a}}{}_m^r\,[a]_r^m = \begin{bmatrix} 1, & 0, & 0 \\ 0, & 0, & 0 \\ 0, & 0, & 0 \end{bmatrix}_{\rho,\ \alpha-\rho,\ m}^{\rho,\ \alpha-\rho,\ m}$$

NOTE. In Ex. xiv of § 160 it is shown that a general formula for all matrices $[a]_r^m$ of rank a which are such that the product $\overline{a}_{m}^{r}\ [a]_r^m$ has rank ρ is

$$[a]_r^m = [l]_r^r \begin{bmatrix} 1, & 0, & 0 \\ 0, & 1, & 0 \\ 0, & i, & 0 \\ 0, & 0, & 0 \end{bmatrix}_{\rho,\ a-\rho,\ a-\rho,\ r+\rho-2a}^{\rho,\ a-\rho,\ m-a} [k]_m^m,$$

where $[k]_m^m$ is any undegenerate square matrix of order m, and $[l]_r^r$ is any matrix such that

$$[l]_r^r\ \overline{l}_r^{r} = \overline{l}_r^{r}\ [l]_r^r = [1]_r^r.$$

Ex. i. If $[a]_r^m$ is undegenerate and has rank r, then:

(1) The product $[a]_r^m\ \overline{a}_m^{r}$ can have rank ρ when and only when

$$\rho \nless 0, \quad \rho \ngtr r, \quad \rho \nless 2r-m.$$

(2) The product $\overline{a}_m^{r}\ [a]_r^m$ must have rank r.

Theorem II a. *If $[x]_m^m$ is a given self-conjugate matrix having any assigned rank ρ, we can determine a matrix $[a]_r^m$ of rank a satisfying the equation* (A) *when and only when a is an integer consistent with the conditions*

$$a \nless \rho, \quad a \ngtr m, \quad 2a \ngtr r+\rho, \quad \text{..............................(F)}$$

which include the conditions $\rho \ngtr r$, $\rho \ngtr m$, $a \ngtr r$.

We know from (B) and (C) that the conditions (F) must be satisfied. Conversely if a is any integer satisfying these conditions, it is shown in § 160.4 that we can determine a matrix $[a]_r^m$ of rank a satisfying the equation (A); and this establishes the theorem.

In fact the formula of the Note gives solutions when

$$\overline{k}_m^{\rho}\ [k]_\rho^m = [x]_m^m,$$

and it is shown in Ex. xvi of § 147 that this last equation admits of solutions for $[k]_\rho^m$.

Since ρ must be such that $\rho \nless 0$, the conditions (F) are equivalent to the necessary conditions (B) and (C). Eliminating a from (F) or from (B) and (C) we see that:

There exists a matrix $[a]_r^m$ satisfying the equation (A) *when and only when the rank ρ of $[x]_m^m$ satisfies the necessary conditions*

$$\rho \nless 0, \quad \rho \ngtr m, \quad \rho \ngtr r. \quad \text{..................................(E)}$$

Ex. ii. We can deduce Theorem I *a* from Theorem II *a*. For the conditions (F) are equivalent to (B) and (C) or to (D) and (B), and Theorem II *a* shows that the equation (A) can be satisfied when a has any value consistent with the orders of $[a]_r^m$, and ρ has any value consistent with the conditions (B).

We can summarise the foregoing two theorems in the following way.

Summary A. *In every product of two mutually conjugate matrices the product matrix and the factor matrices can have any ranks consistent with their orders and with the necessary conditions of* § 133. *This is the case even when the product matrix is given.*

Ex. iii. *If the matrix* $[a]_r^m$ *has rank* α, *then the product matrix* $\overline{\underline{a}}{}_m^r\,[a]_r^m$ *has rank* $\alpha - s$, *where* s *is the greatest number of unconnected solutions of the equation* $\overline{\underline{a}}{}_m^r\;\overline{\underline{x}}{}_r = 0$ *connected with the vertical rows of* $[a]_r^m$.

This follows from Ex. v of § 134.

Ex. iv. *If* $\overline{\underline{a}}{}_n^r$ *is undegenerate and has rank* r, *then the product matrix* $[a]_r^n\,\overline{\underline{a}}{}_n^r$ *has rank* $r-s$, *where* s *is the rank of the greatest spacelet which is contained in and is orthogonal with the spacelet* $\omega_r \equiv \overline{\underline{a}}{}_n^r$.

2. *Symmetric products of three matrices.*

If $[a]_r^m$ is a matrix of given orders having rank α, if $[b]_r^r$ is a symmetric matrix of rank β, and if the self-conjugate product matrix

$$[x]_m^m = \overline{\underline{a}}{}_m^r\,[b]_r^r\,[a]_r^m \qquad \text{.........................(A}'\text{)}$$

has rank ρ, we see from § 133 that the integers α, β, ρ must satisfy the conditions

$$\rho \not< 0, \quad \rho \not> \alpha, \quad \rho \not> \beta, \quad \rho + 2r \not< 2\alpha + \beta; \qquad \text{...........(B}'\text{)}$$

and

$$\alpha \not< 0, \quad \alpha \not> m, \quad \alpha \not> r; \quad \beta \not< 0, \quad \beta \not> r; \qquad \text{..............(C}'\text{)}$$

where the conditions $\alpha \not< 0$, $\beta \not< 0$, $\alpha \not> r$ in (C′) are superfluous when we take (B′) and (C′) together.

The conditions (B′) and (C′) include

$$\rho \not< 0, \quad \rho \not> m, \quad \rho \not> r. \qquad \text{.......................(D}'\text{)}$$

They reduce to (D′) when we eliminate α and β; and they reduce to (C′) when we eliminate ρ.

When we eliminate α, β respectively, they reduce to

$$\beta \not< 0, \quad \beta \not> r; \quad \rho \not< 0, \quad \rho \not> \beta, \quad \rho \not> m; \qquad \text{..............(a)}$$

$$\alpha \not< 0, \quad \alpha \not> m, \quad \alpha \not> r; \quad \rho \not< 0, \quad \rho \not> \alpha. \qquad \text{..............(b)}$$

We will now prove the theorems which follow.

Theorem I b. *If* $[x]_m^m$ *is a given self-conjugate matrix having any assigned rank* ρ *consistent with the necessary conditions* (B′) *and* (C′), *i.e. with the conditions* (D′), *we can* *determine a matrix* $[a]_r^m$ *and a self-conjugate matrix* $[b]_r^r$ *which satisfy the equation* (A′) *and have the assigned ranks* a *and* β *when and only when* a *and* β *are integers consistent with the necessary conditions* (B′) *and* (C′).

By Ex. ii of § 134 we can determine a matrix $[a]_r^m$ of rank a such that the horizontal rows of $[x]_m^m$ are connected with the horizontal rows of $[a]_r^m$ when and only when

$$a \nless \rho, \quad a \ngtr r, \quad a \ngtr m. \qquad \text{.................................(1)}$$

Again when $[a]_r^m$ is any given matrix determined in this way and having any rank a consistent with the conditions (1), the equation (A′) admits of solution for $[b]_r^r$ and by Theorem IV of § 132 it has symmetric solutions $[b]_r^r$ of rank β when and only when

$$\beta \nless \rho, \quad \beta \ngtr r, \quad \beta + 2a \ngtr \rho + 2r. \qquad \text{.............................(2)}$$

Since the conditions (1) and (2) are together equivalent to the conditions (B′) and (C′), it follows that we can determine a matrix $[a]_r^m$ of rank a and a symmetric matrix $[b]_r^r$ of rank β which satisfy the equation (A′) when a and β are any assigned integers consistent with the necessary conditions (B′) and (C′).

COROLLARY 1. *We can determine a matrix* $[a]_r^m$ *of rank* a *and symmetric matrices* $[b]_r^r$, $[x]_m^m$ *of ranks* β, ρ *which satisfy the equation* (A′) *when* a, β, ρ *are any assigned integers consistent with the necessary conditions* (B′) *and* (C′).

COROLLARY 2. *When the matrix* $[a]_r^m$ *and the symmetric matrix* $[b]_r^r$ *have any assigned ranks* a *and* β *consistent with the necessary conditions* (C′), *we can determine their elements so that the symmetric product matrix* $[x]_m^m$ *has any assigned rank* ρ *consistent with the necessary conditions* (B′).

COROLLARY 3. *We can determine a symmetric matrix* $[x]_m^m$ *of rank* ρ *and a matrix* $[a]_r^m$ *of rank* a *which are such that the equation* (A′) *admits of solution for* $[b]_r^r$ *when* ρ *and* a *are any assigned integers consistent with the necessary conditions* (B′) *and* (C′), *i.e. with the necessary conditions* (b).

For the conditions (D′) and (1) in Theorem I *b* are equivalent to the conditions (b). Thus the conditions that $[x]_m^m$ shall be symmetric and that (A′) shall admit of solution for $[b]_r^r$ impose no restriction on the possible values of ρ and a which is not already imposed by the necessary conditions (B′) and (C′).

Theorem II b. *If* $[x]_m^m$ *and* $[a]_r^m$ *are a given symmetric matrix and a given matrix having any ranks* ρ *and* a *consistent with the necessary conditions* (B′) *and* (C′), *i.e. with the conditions* (b), *and (as is possible) so chosen that the equation* (A′) *admits of solution for* $[b]_r^r$, *then the possible ranks of the symmetric solutions* $[b]_r^r$ *are the values of* β *consistent with the necessary conditions* (B′) *and* (C′), *i.e. with the conditions* (2).

By Corollary 3 to Theorem I *b* the given integers ρ and a can have any values consistent with (b) when $[x]_m^m$ is symmetric and (A′) admits of solution for $[b]_r^r$. Accordingly it is shown in the proof of Theorem I *b* that the above theorem is true.

Theorem III b. *If* $[x]_m^m$ *and* $[b]_r^r$ *are given symmetric matrices having any assigned ranks* ρ *and* β *consistent with the necessary conditions* (B′) *and* (C′), *i.e. with the conditions* (a), *we can determine a matrix* $[a]_r^m$ *of rank* α *satisfying the equation* (A′) *when and only when* α *is an integer consistent with the necessary conditions* (B′) *and* (C′), *i.e. with the conditions*

$$\alpha \not< \rho, \quad \alpha \not> m, \quad 2\alpha+\beta \not> \rho+2r. \qquad \text{.........(3)}$$

The equation in this case is equivalent to that considered in § 162.

As shown in §§ 160 and 145 we can express the given symmetric matrix $[b]_r^r$ in the form $[b]_r^r = \overline{\underline{c}}{}_r^{\beta}\,[c]_\beta^r$, where $[c]_\beta^r$ is a given undegenerate matrix of rank β; and the equation (A′) is then satisfied when and only when

$$\overline{\underline{y}}{}_m^{\beta}\,[y]_\beta^m = [x]_m^m, \qquad \text{.........(4)}$$

where

$$[y]_\beta^m = [c]_\beta^r\,[a]_r^m. \qquad \text{.........(4′)}$$

By Theorem II a or § 161 we can determine a matrix $[y]_\beta^m$ of rank σ satisfying the equation (4) when and only when

$$\sigma \not< \rho, \quad \sigma \not> m, \quad 2\sigma \not> \beta+\rho. \qquad \text{.........(5)}$$

Again when $[y]_\beta^m$ is any matrix of rank β determined in this way and having any rank σ consistent with (5), it follows from Theorem I of § 132 or from § 134 that we can determine a matrix $[a]_r^m$ of rank α satisfying the equation (4′) when and only when

$$\alpha \not< \sigma, \quad \alpha \not> m, \quad \alpha+\beta \not> \sigma+r. \qquad \text{.........(6)}$$

When we eliminate σ from (5) and (6), we obtain the conditions (3). This shows that we can determine a matrix $[a]_r^m$ of rank α satisfying the equation (A′) when and only when α is an integer satisfying the conditions (3). Since the conditions (a) and (3) are together equivalent to (B′) and (C′), the possible values of α are those consistent with the conditions (B′) and (C′).

The foregoing results can be summarised as follows.

Summary B. *In every symmetric product of three matrices of the form* (A′) *the product matrix and the factor matrices can have any ranks consistent with their orders and with the necessary conditions of* § 133 *for that product. This is the case even under the following circumstances:*

(1) *When the product matrix is given.*

(2) *When any one of the factor matrices is given.*

(3) *When both the product matrix and the middle factor matrix are given.*

(4) *When the product matrix and the two extreme factor matrices are given, provided that the extreme factor matrices are so chosen that the equation* (A′) *admits of solution for the middle factor matrix.*

Ex. v. *If* $[b]_r^r$ *is a given undegenerate symmetric matrix, and if* $[a]_r^m$ *is a given matrix having the assigned rank* a *whose elements are otherwise arbitrary, then the possible ranks of the product* $[x]_m^m = \overline{\underline{a}}{}_m^{r}\,[b]_r^r\,[a]_r^m$ *are the same as the possible ranks of the product* $\overline{\underline{a}}{}_m^{r}\,[a]_r^m$.

We can see this directly by expressing $[b]_r^r$ in the form $[b]_r^r = \overline{\underline{c}}{}_r^{r}\,[c]_r^r$, where $[c]_r^r$ is a given undegenerate square matrix. We then have

$$[x]_m^m = \overline{\underline{d}}{}_m^{r}\,[d]_r^m,$$

where
$$[d]_r^m = [c]_r^r\,[a]_r^m.$$

When $[a]_r^m$ has rank a and its elements receive all possible values subject to this condition, then $[d]_r^m$ has rank a and its elements receive all possible values subject to this condition.

§ 137. Possible ranks of the product matrix and the factor matrices in any matrix product which is self-conjugate in form.

A self-conjugate or symmetric matrix product containing an odd number of factors has the form

$$[x]_m^m = \overline{\underline{a}}{}_m^{p}\,\overline{\underline{b}}{}_p^{q}\,\overline{\underline{c}}{}_q^{r}\,\ldots\,\overline{\underline{k}}{}_u^{v}\,[l]_v^v\,[k]_v^u\,\ldots\,[c]_r^q\,[b]_q^p\,[a]_p^m, \quad \ldots\ldots\text{(A)}$$

where $[l]_v^v$ and $[x]_m^m$ are symmetric matrices. If $\alpha, \beta, \gamma, \ldots \kappa, \lambda, \rho$ are the ranks of $[a]_p^m, [b]_q^p, [c]_r^q, \ldots [k]_v^u, [l]_v^v, [x]_m^m$, then these integers must satisfy the conditions

$$\left.\begin{array}{l} \rho \nless 0,\ \rho \ngtr \alpha,\ \rho \ngtr \beta,\ \rho \ngtr \gamma,\ \ldots\ \rho \ngtr \kappa,\ \rho \ngtr \lambda; \\ \rho + 2p + 2q + 2r + \ldots + 2v \nless 2\alpha + 2\beta + 2\gamma + \ldots + 2\kappa + \lambda; \\ \alpha \ngtr m,\ \alpha \ngtr p;\ \beta \ngtr p,\ \beta \ngtr q;\ \ldots\ \kappa \ngtr u,\ \kappa \ngtr v;\ \lambda \ngtr v. \end{array}\right\} \ldots\ldots\text{(A}')$$

These are the restrictions imposed on the ranks by the orders of the matrices and by the necessary conditions of § 133 for the product (A). They are the same as the corresponding conditions for the product obtained by prefixing and postfixing $[1]_m^m$ in the product on the right in (A).

A self-conjugate or symmetric matrix product containing an even number of factors has the form

$$[x]_m^m = \overline{\underline{a}}{}_m^{p}\,\overline{\underline{b}}{}_p^{q}\,\overline{\underline{c}}{}_q^{r}\,\ldots\,\overline{\underline{k}}{}_u^{v}\,\overline{\underline{l}}{}_v^{w}\,[l]_w^v\,[k]_v^u\,\ldots\,[c]_r^q\,[b]_q^p\,[a]_p^m, \quad \ldots\text{(B)}$$

where $[x]_m^m$ is a symmetric matrix. If $\alpha, \beta, \gamma, \ldots \kappa, \lambda, \rho$ are the ranks of $[a]_p^m, [b]_q^p, [c]_r^q, \ldots [k]_v^u, [l]_w^v, [x]_m^m$, then these integers must satisfy the conditions

$$\left.\begin{aligned}&\rho \not< 0,\ \rho \not> \alpha,\ \rho \not> \beta,\ \rho \not> \gamma,\ \ldots\ \rho \not> \kappa,\ \rho \not> \lambda;\\ &\rho + 2p + 2q + \ldots + 2v + w \not< 2\alpha + 2\beta + \ldots + 2\kappa + 2\lambda;\\ &\alpha \not> m,\ \alpha \not> p;\ \beta \not> p,\ \beta \not> q;\ \ldots\ \kappa \not> u,\ \kappa \not> v;\ \lambda \not> v,\ \lambda \not> w.\end{aligned}\right\} \quad \ldots(\mathrm{B}')$$

These are the conditions imposed on the ranks by the orders of the matrices and by the necessary conditions of § 133 for the product (B). They are the same as the corresponding conditions for the product obtained by inserting the middle factor $[1]_m^m$ in the product on the right in (B).

In both the products (A) and (B) the following condition must also be satisfied:

The two extreme factor matrices must be such that the horizontal rows of $[x]_m^m$ *are connected with the horizontal rows of* $[a]_p^m$. ....(C)

We will now prove the theorems which follow.

Theorem I a. *If in* (A) *the matrix* $[x]_m^m$ *is a given symmetric matrix having any rank* ρ *consistent with the necessary conditions* (A′), *i.e. with the necessary conditions*

$$\rho \not< 0,\ \rho \not> m,\ \rho \not> p,\ \rho \not> q,\ \ldots\ \rho \not> u,\ \rho \not> v, \quad \ldots\ldots\ldots..(1)$$

then we can determine the factor matrices on the right so that they satisfy the equation (A) *and have any assigned ranks which are consistent with the necessary conditions* (A′).

It is to be understood that the middle factor matrix $[l]_v^v$ is to be so determined as to be symmetric.

It has been shown in § 136 that the theorem is true for symmetric products of three matrices. Let there be at least five factor matrices in (A), and let

$$[y]_p^p = \underline{\overline{b}}{}_p^q\, \underline{\overline{c}}{}_q^r \ldots \underline{\overline{k}}{}_u^v\, [l]_v^v\, [k]_v^u \ldots [c]_r^q\, [b]_q^p, \quad \ldots\ldots\ldots..(2)$$

so that $$[x]_m^m = \underline{\overline{a}}{}_m^p\, [y]_p^p\, [a]_p^m. \quad \ldots\ldots\ldots\ldots\ldots\ldots\ldots..(3)$$

We will make the hypothesis that the theorem is true for the product (2), and show that it must then be true for the product (A).

By § 136 we can determine a symmetric matrix $[y]_p^p$ of rank σ and a matrix $[a]_p^m$ of rank α satisfying (3) when and only when

$$\left.\begin{array}{l} \rho \not< 0,\ \rho \not> \alpha,\ \rho \not> \sigma;\ \ \rho + 2p \not< 2\alpha + \sigma; \\ \alpha \not> m,\ \alpha \not> p;\ \ \sigma \not> p. \end{array}\right\} \qquad \text{..............(4)}$$

Again if $[y]_p^p$ and $[a]_p^m$ are any two given matrices determined in this way, it follows from the hypothesis that we can determine a symmetric matrix $[l]_v^v$ of rank λ and matrices $[b]_q^p$, $[c]_r^q$, ... $[k]_v^u$ of ranks $\beta, \gamma, \ldots \kappa$ satisfying (2) when and only when

$$\left.\begin{array}{l} \sigma \not< 0,\ \sigma \not> \beta,\ \sigma \not> \gamma,\ \ldots\ \sigma \not> \kappa,\ \sigma \not> \lambda; \\ \sigma + 2q + 2r + \ldots + 2v \not< 2\beta + 2\gamma + \ldots + 2\kappa + \lambda; \\ \beta \not> p,\ \beta \not> q;\ \ \gamma \not> q,\ \gamma \not> r;\ \ldots\ \kappa \not> u,\ \kappa \not> v;\ \ \lambda \not> v. \end{array}\right\} \qquad \text{......(5)}$$

When we eliminate σ from (4) and (5), we obtain the conditions (A′). We conclude that it is possible to determine the factor matrices in (A) so that they have assigned ranks and satisfy the equation (A) when and only when their ranks are consistent with (4) and (5), i.e. with (A′).

Thus if the theorem is true for the symmetric product (2), it is true for the symmetric product (A); and if it is true for all symmetric products of $2t+1$ matrices, then it must be true for all symmetric products of $2t+3$ matrices. Since it is true for all symmetric products of three matrices, it follows that it is true for all symmetric products of the form (A).

COROLLARY 1 *a*. *We can determine symmetric matrices* $[x]_m^m$, $[l]_v^v$ *of ranks* ρ, λ *and matrices* $[a]_p^m$, $[b]_q^p$, ... $[k]_v^u$ *of ranks* $\alpha, \beta, \ldots \kappa$ *which satisfy the equation* (A) *when and only when* $\rho, \alpha, \beta, \ldots \kappa, \lambda$ *are integers which satisfy the necessary conditions* (A′).

COROLLARY 2 *a*. *If* $[a]_p^m$, $[b]_q^p$, ... $[k]_v^u$, $[l]_v^v$ *are matrices having the assigned ranks* $\alpha, \beta, \ldots \kappa, \lambda$ *whose elements are arbitrary subject to the condition that* $[l]_v^v$ *is symmetric, the possible values of the rank* ρ *of the symmetric product* (A) *are those consistent with the necessary conditions* (A′).

COROLLARY 3 *a*. *If* $[x]_m^m$ *is a given symmetric matrix having any assigned rank* ρ *consistent with the necessary conditions* (1), *we can determine a symmetric matrix* $[l]_v^v$ *of rank* λ *and matrices* $[b]_q^p$, $[c]_r^q$, ... $[k]_v^u$ *of ranks* $\beta, \gamma, \ldots \kappa$ *which are such that the equation* (A) *admits of solution for* $[a]_p^m$ *when and only when* $\beta, \gamma, \ldots \kappa, \lambda$ *are integers consistent with the necessary conditions* (A′).

Theorem I b. *If in* (B) *the matrix* $[x]_m^m$ *is a given symmetric matrix having any rank* ρ *consistent with the necessary conditions* (B′), *i.e. with the necessary conditions*

$$\rho \nless 0,\quad \rho \ngtr p,\quad \rho \ngtr q,\quad \dots\ \rho \ngtr u,\quad \rho \ngtr v,\quad \rho \ngtr w, \qquad \text{(1′)}$$

then we can determine the factor matrices on the right so that they satisfy the equation (B) *and have any assigned ranks consistent with the necessary conditions* (B′).

It has been shown in § 136 that the theorem is true for symmetric products of two matrices. Let there be at least four factors in (B), and let

$$[y]_p^p = \overline{\underline{b}}{}_p^q\, \overline{\underline{c}}{}_q^r \dots \overline{\underline{k}}{}_u^v\, \overline{\underline{l}}{}_v^w\, [l]_w^v\, [k]_v^u \dots [c]_r^q\, [b]_q^p, \qquad \text{(2′)}$$

so that

$$[x]_m^m = \overline{\underline{a}}{}_m^p\, [y]_p^p\, [a]_p^m. \qquad \text{(3′)}$$

Then proceeding as in the proof of Theorem I *a* we can show that if the theorem is true for the product (2′), it is also true for the product (B); i.e. if it is true for all symmetric products of $2t$ matrices, then it must be true for all symmetric products of $2t+2$ matrices. Since it is true for all symmetric products of two matrices, it follows that it is true for all matrix products of the form (B).

COROLLARY 1 *b*. *We can determine a symmetric matrix* $[x]_m^m$ *of rank* ρ *and matrices* $[a]_p^m, [b]_q^p, \dots [l]_w^v$ *of ranks* $\alpha, \beta, \dots \lambda$ *which satisfy the equation* (B) *when and only when* $\rho, \alpha, \beta, \dots \lambda$ *are integers which satisfy the necessary conditions* (B′).

COROLLARY 2 *b*. *If* $[a]_p^m, [b]_q^p, \dots [l]_w^v$ *are matrices having the assigned ranks* $\alpha, \beta, \dots \lambda$ *whose elements are arbitrary, the possible values of the rank* ρ *of the symmetric product* (B) *are those consistent with the necessary conditions* (B′).

COROLLARY 3 *b*. *If* $[x]_m^m$ *is a given symmetric matrix having any assigned rank* ρ *consistent with the necessary conditions* (1′), *we can determine matrices* $[b]_q^p, [c]_r^q, \dots [l]_w^v$ *of ranks* $\beta, \gamma, \dots \lambda$ *which are such that the equation* (B) *admits of solution for* $[a]_p^m$ *when and only when* $\beta, \gamma, \dots \lambda$ *are integers consistent with the necessary conditions* (B′).

Theorem II a. *If in the product* (A) *the product matrix* $[x]_m^m$ *and every alternate factor matrix starting with the first or second at either end are given and have any assigned ranks consistent with the necessary conditions* (A′), *and if the necessary condition* (C) *is satisfied when the two extreme factor matrices* *are given, then we can determine the remaining alternate factor matrices so that they satisfy the equation* (A) *and have any assigned ranks which are consistent with the necessary conditions* (A′).

It has been shown in § 136 that the theorem is true for symmetric products of three matrices. Using this fact we can prove the theorem by induction. We will suppose that there are at least five factor matrices in (A), and consider separately two cases which include all possible cases.

CASE I. *When the two extreme factor matrices are given.*

Let
$$[y]_p^p = \overline{\underline{b}}^{\,q}_{\,p}\, \overline{\underline{c}}^{\,r}_{\,q} \ldots \overline{\underline{k}}^{\,w}_{\,v}\, [l]_w^w\, [k]_w^v \ldots [c]_r^q\, [b]_q^p, \qquad (6)$$
so that
$$[x]_m^m = \overline{\underline{a}}^{\,p}_{\,m}\, [y]_p^p\, [a]_p^m. \qquad (7)$$

We will make the hypothesis that the theorem is true for the product (6), and show that it must then be true for the product (A).

By Theorem II *b* of § 136 we can determine a symmetric matrix $[y]_p^p$ of rank σ satisfying the equation (7) when and only when
$$\left.\begin{array}{l} \rho \not< 0,\ \ \rho \not> a,\ \ \rho \not> \sigma;\ \ \rho + 2p \not< 2a + \sigma; \\ a \not> m,\ \ a \not> p;\ \ \sigma \not> p. \end{array}\right\} \qquad (8)$$

If $[y]_p^p$ is any given symmetric matrix determined in this way, it follows from the hypothesis that we can determine the alternate unknown factor matrices in (6), which include $[b]_q^p$, so that the equation (6) is satisfied when and only when their ranks satisfy the conditions
$$\left.\begin{array}{l} \sigma \not> \beta,\ \ \sigma \not> \gamma,\ \ldots\ \sigma \not> \kappa,\ \ \sigma \not> \lambda;\ \ \sigma + 2q + 2r + \ldots + 2w \not< 2\beta + 2\gamma + \ldots + 2\kappa + \lambda; \\ \beta \not> p,\ \ \beta \not> q;\ \ \gamma \not> q,\ \ \gamma \not> r;\ \ldots\ \kappa \not> v,\ \ \kappa \not> w;\ \ \lambda \not> w. \end{array}\right\} \ldots (9)$$
When we eliminate σ from (8) and (9) we obtain the conditions (A′). This shows that we can determine the unknown factor matrices in (A) so that they satisfy the equation (A) and have assigned ranks when and only when their ranks are consistent with (8) and (9), i.e. when and only when they are consistent with the necessary conditions (A′). This is possible since the ranks of the known factor matrices are consistent with (A′).

Thus if the theorem is true for the product (6), it must be true for the product (A).

CASE II. *When the two extreme factor matrices are not given.*

In this case we can express the known factor matrix $[b]_q^p$ in the form
$$[b]_q^p = [b']_q^\beta\, [H]_\beta^p,$$
where the matrices on the right are given undegenerate matrices of rank β. Then if
$$[y]_\beta^\beta = \overline{\underline{b'}}^{\,q}_{\,\beta}\, \overline{\underline{c}}^{\,r}_{\,q} \ldots \overline{\underline{k}}^{\,w}_{\,v}\, [l]_w^w\, [k]_w^v \ldots [c]_r^q\, [b']_q^\beta, \qquad (10)$$
and
$$[T]_\beta^m = [H]_\beta^p\, [a]_p^m, \qquad (11)$$
we have
$$[x]_m^m = \overline{\underline{T}}^{\,\beta}_{\,m}\, [y]_\beta^\beta\, [T]_\beta^m. \qquad (12)$$

We will make the hypothesis that the theorem is true for the product (10), and show that it must then be true for the product (A).

By Theorem I *b* of § 136 we can determine a symmetric matrix $[y]_\beta^\beta$ of rank σ and a matrix $[T]_\beta^m$ of rank τ satisfying the equation (12) when and only when

$$\left.\begin{array}{l} \rho \not< 0,\ \rho \not> \tau,\ \rho \not> \sigma;\ \rho+2\beta \not< 2\tau+\sigma; \\ \tau \not> m,\ \tau \not> \beta;\ \sigma \not> \beta. \end{array}\right\} \quad \ldots\ldots\ldots\ldots\ldots\ldots\ldots(13)$$

If $[y]_\beta^\beta$, $[T]_\beta^m$ are any two given matrices determined in this way, it follows from Theorem IV *a* of § 134 that we can determine a matrix $[a]_p^m$ of rank a satisfying the equation (11) when and only when

$$\left.\begin{array}{l} \tau \not> \beta,\ \tau \not> a;\ \tau+p \not< a+\beta; \\ \beta \not> p;\ a \not> p,\ a \not> m; \end{array}\right\} \quad \ldots\ldots\ldots\ldots\ldots\ldots\ldots(14)$$

and it follows from the hypothesis that we can determine the unknown factor matrices in (10), which do not include $[b']_q^\beta$, when and only when their ranks satisfy the conditions

$$\left.\begin{array}{l} \beta \not> \gamma,\ \ldots\ \beta \not> \kappa,\ \beta \not> \lambda;\ 2q+2r+\ldots+2w \not< \beta+2\gamma+\ldots+2\kappa+\lambda; \\ \beta \not> q;\ \gamma \not> q,\ \gamma \not> r;\ \ldots\ \kappa \not> v,\ \kappa \not> w;\ \lambda \not> w. \end{array}\right\} \quad \ldots\ldots(15)$$

When we eliminate σ and τ from (13), (14) and (15), we obtain the conditions (A′).

Consequently we can determine the unknown factor matrices in (A) so that they satisfy the equation (A) and have assigned ranks, when and only when these ranks are consistent with (13), (14) and (15), i.e. when and only when they are consistent with the necessary conditions (A′).

This is possible since the ranks of the known factor matrices are consistent with (A′).

Thus if the theorem is true for the product (10), it must be true for the product (A)

We conclude that if the theorem is true for all symmetric products of $2t+1$ matrices, then it is true for all products of $2t+3$ matrices. Since it is true for all symmetric products of three matrices, it is true for all matrix products of the form (A).

Theorem II b. *If in the product* (B) *the product matrix* $[x]_m^m$ *and every alternate factor matrix before and after the two middle factor matrices (but not including these two) are given and have any assigned ranks consistent with the necessary conditions* (B′), *and if the necessary condition* (C) *is satisfied when the two extreme factor matrices are given, then we can determine the remaining factor matrices so that they satisfy the equation* (B) *and have any assigned ranks which are consistent with the necessary conditions* (B′).

We can deduce this theorem from Theorem II *a* by inserting on the right in (B) the additional middle factor matrix $[1]_w^w$.

The corresponding result when the two middle factor matrices in (B) are given is obtained by replacing them by their product matrix and then applying Theorem II *a*.

The foregoing results can be all summarised as follows.

Summary. *In every symmetric matrix product the product matrix and the factor matrices can have any assigned ranks which are consistent with their orders and with the necessary conditions of* § 133 *for that product. This is the case even when the product matrix or any factor matrix is given, and even when the product matrix and any number of factor matrices, no two of which are consecutive, are given; provided that the necessary condition* (C) *is satisfied when the two extreme factor matrices are given.*

§ 138. Equivalences of two matrices and equivalent systems of linear algebraic equations.

1. *Equivalences of two matrices.*

Any two matrices A and B will be said to have *horizontal equivalence* or to be *horizontally equivalent* to one another when they have the same number of vertical rows, and when also every horizontal row of A is connected with the horizontal rows of B and every horizontal row of B is connected with the horizontal rows of A. Thus the matrices $A=[a]_r^n$, $B=[b]_s^n$ are horizontally equivalent when and only when there exist relations of both the forms

$$[a]_r^n=[p]_r^s\,[b]_s^n, \qquad [b]_s^n=[q]_s^r\,[a]_r^n.$$

Applying Theorem III of § 71 we see that these two matrices cannot be horizontally equivalent unless their ranks are the same; and applying Theorem VI of § 71 we see that they are horizontally equivalent when and only when $[a]_r^n$, $[b]_s^n$, $\begin{bmatrix} a \\ b \end{bmatrix}_{r,s}^n$ all have the same rank.

Again any two matrices A and B will be said to have *vertical equivalence* or to be *vertically equivalent* to one another when they have the same number of horizontal rows, and when also every vertical row of A is connected with the vertical rows of B and every vertical row of B is connected with the vertical rows of A. Thus the matrices $A=[a]_m^r$, $B=[b]_m^s$ are vertically equivalent when and only when there exist relations of both the forms

$$[a]_m^r=[b]_m^s\,[p]_s^r, \qquad [b]_m^s=[a]_m^r\,[q]_r^s.$$

These two matrices cannot be vertically equivalent unless their ranks are the same; and they are vertically equivalent when and only when $[a]_m^r$, $[b]_m^s$, $[a, b]_m^{r,s}$ all have the same rank.

We now enunciate two theorems:

Theorem I. *If* $s \not< r$, *the two matrices* $[a]_r^n$, $[b]_s^n$ *are horizontally equivalent when and only when there exists a relation of the form*

$$[b]_s^n=[h]_s^r\,[a]_r^n, \qquad \text{...(A)}$$

in which $[h]_s^r$ *is an undegenerate matrix of rank* r.

Theorem II. *If* $s \not< r$, *the two matrices* $[a]_m^r$, $[b]_m^s$ *are vertically equivalent when and only when there exists a relation of the form*

$$[b]_m^s=[a]_m^r\,[k]_r^s, \qquad \text{...(B)}$$

in which $[k]_r^s$ *is an undegenerate matrix of rank* r.

We prove Theorem I only, the proof of Theorem II being similar.

First suppose that there exists a relation (A) in which $[h]_s^r$ has rank r. Then by Ex. v of § 82 the matrix $[h]_s^r$ has an inverse prefactor $\overline{\underline{H}}_r^s$, and prefixing this on both sides of (A) we obtain

$$[a]_r^n = \overline{\underline{H}}_r^s \; [b]_s^n \; ; \quad \text{.......................................(A')}$$

and equations (A) and (A′) show that $[a]_r^n$ and $[b]_s^n$ are horizontally equivalent.

Next suppose that $[a]_r^n$ and $[b]_s^n$ are horizontally equivalent. Then they necessarily have the same rank, which we will denote by t, and the matrix $\begin{bmatrix} a \\ b \end{bmatrix}_{r,s}^n$ has the same rank t as $[a]_r^n$. Consequently the equation (A) admits of finite solutions for $[h]_s^r$, and by Theorem II of § 132 it has solutions of ranks $t, t+1, t+2, \dots r$. Accordingly we can determine a matrix $[h]_s^r$ of rank r satisfying the equation (A), i.e. there exists a relation (A) in which $[h]_s^r$ has rank r.

Thus Theorem I is proved; and in a similar way we can prove Theorem II.

The definitions of horizontal and vertical equivalence given above can be used concurrently with the definition of (unqualified) equivalence of similar undegenerate matrices given in § 113.

Ex. i. Two square matrices of the same order may have neither horizontal nor vertical equivalence, or horizontal equivalence only, or vertical equivalence only, or both horizontal and vertical equivalences.

The matrices

$$\begin{bmatrix} 1 & 2 & 3 \\ 3 & 1 & 4 \\ 4 & 3 & 7 \end{bmatrix} = \begin{bmatrix} 1 & 0 \\ 0 & 1 \\ 1 & 1 \end{bmatrix} \begin{bmatrix} 1 & 2 & 3 \\ 3 & 1 & 4 \end{bmatrix}, \quad \begin{bmatrix} -2, & 1, & -1 \\ 9, & 8, & 17 \\ 15, & 10, & 25 \end{bmatrix} = \begin{bmatrix} 1, & -1 \\ 3, & 2 \\ 3, & 4 \end{bmatrix} \begin{bmatrix} 1 & 2 & 3 \\ 3 & 1 & 4 \end{bmatrix}$$

have the same rank 2, and are horizontally but not vertically equivalent.

The matrices

$$\begin{bmatrix} 1 & 2 & 3 \\ 3 & 2 & 5 \\ 4 & 4 & 8 \end{bmatrix} = \begin{bmatrix} 1 & 0 \\ 0 & 1 \\ 1 & 1 \end{bmatrix} \begin{bmatrix} 1 & 2 & 3 \\ 3 & 2 & 5 \end{bmatrix} = \begin{bmatrix} 1 & 1 \\ 3 & 1 \\ 4 & 2 \end{bmatrix} \begin{bmatrix} 1 & 0 & 1 \\ 0 & 2 & 2 \end{bmatrix},$$

$$\begin{bmatrix} 1 & 2 & 3 \\ 2 & 0 & 2 \\ 3 & 2 & 5 \end{bmatrix} = \begin{bmatrix} 1, & 0 \\ -1, & 1 \\ 0, & 1 \end{bmatrix} \begin{bmatrix} 1 & 2 & 3 \\ 3 & 2 & 5 \end{bmatrix} = \frac{1}{2} \begin{bmatrix} 1 & 1 \\ 3 & 1 \\ 4 & 2 \end{bmatrix} \begin{bmatrix} 1, & -2, & -1 \\ 1, & 6, & 7 \end{bmatrix}$$

have the same rank 2, and are both horizontally and vertically equivalent.

The matrices $\begin{bmatrix} 1 & 1 \\ 2 & 2 \end{bmatrix}, \begin{bmatrix} 2 & 2 \\ 4 & 4 \end{bmatrix}$ have the same rank 1, and are both horizontally and vertically equivalent.

The matrices $\begin{bmatrix} 1 & 2 \\ 2 & 4 \end{bmatrix}, \begin{bmatrix} 2 & 3 \\ 6 & 9 \end{bmatrix}$ have the same rank 1, but have neither horizontal nor vertical equivalence.

Ex. ii. Two undegenerate square matrices of the same order have both horizontal and vertical equivalences.

Ex. iii. The matrices $\overline{a}^{\,r}_{n}$, $\overline{b}^{\,s}_{n}$ represent the same spacelet ω_t of rank t when and only when they are vertically equivalent and have the common rank t.

Ex. iv. *If A and X are vertically equivalent matrices, and also B and Y are vertically equivalent matrices, then A and B are vertically equivalent when and only when X and Y are vertically equivalent.*

This result remains true when we replace 'vertically' by 'horizontally.'

Ex. v. Let $A=[a]_n^p$, $B=[b]_n^q$ be two matrices having the same number n of horizontal rows.

Then if A and B have the same rank r, they are vertically equivalent when and only when the vertical rows of A are connected with the vertical rows of B.

This is true both when p is greater and when it is not greater than q.

We can write
$$[a]_n^p=[a]_n^r\,[h]_r^p, \quad [b]_n^q=[\beta]_n^r\,[k]_r^q,$$
where all the factor matrices on the right have rank r.

First suppose that A and B are vertically equivalent. Then the vertical rows of A are necessarily connected with the vertical rows of B.

Next suppose that the vertical rows of A are connected with the vertical rows of B. Then there exists a relation of the form
$$[a]_n^p=[b]_n^q\,[\lambda]_q^p,$$
or
$$[a]_n^r\,[h]_r^p=[\beta]_n^r\,[k]_r^q\,[\lambda]_q^p.$$

Postfixing on both sides an inverse postfactor of $[h]_r^p$, we obtain an equation of the form
$$[a]_n^r=[\beta]_n^r\,[u]_r^r,$$
and since $[a]_n^r$ has rank r, $[u]_r^r$ must be an undegenerate square matrix. Thus $[a]_n^r$ and $[\beta]_n^r$ are vertically equivalent, and therefore by Ex. iv the matrices A and B are vertically equivalent.

A similar theorem is true for the matrices $A=[a]_p^n$, $B=[b]_q^n$ when we interchange the terms 'horizontal' and 'vertical.'

2. *Equivalent systems of linear algebraic equations.*

The two homogeneous systems of equations
$$[a]_r^n\,\overline{x}_n=0, \qquad [b]_s^n\,\overline{x}_n=0 \quad \text{.................................(1)}$$
are mutually equivalent, i.e. they have identical solutions, when and only when the matrices $[a]_r^n$, $[b]_s^n$ have horizontal equivalence.

The two systems of finitely solvable equations
$$[a]_r^{n+1}\,\overline{\begin{matrix} x \\ 1 \end{matrix}}_{n,1}=0, \qquad [b]_s^{n+1}\,\overline{\begin{matrix} x \\ 1 \end{matrix}}_{n,1}=0 \quad \text{..........................(2)}$$

are mutually equivalent when and only when the matrices $[a]_r^{n+1}$, $[b]_s^{n+1}$ have horizontal equivalence.

If each of the two systems of equations (2) has infinite solutions but no finite solutions, then the two systems are mutually equivalent when and only when the matrices $[a]_r^n$, $[b]_s^n$ have horizontal equivalence.

These results follow from Theorem I of § 96.

NOTE 1. *Sign of equivalence.*

The horizontal equivalence of two matrices $[a]_r^n$ and $[b]_s^n$ will often be denoted by $[a]_r^n \equiv [b]_s^n$, this notation indicating that if $s \nless r$ there exist two mutually inverse matrices $[h]_s^r$, $\overline{\underline{H}}_r^s$ of rank r such that

$$[b]_s^n = [h]_s^r \, [a]_r^n, \quad [a]_r^n = \overline{\underline{H}}_r^s \, [b]_s^n, \quad \overline{\underline{H}}_r^s \, [h]_s^r = [1]_r^r .$$

Also the vertical equivalence of two matrices $[a]_m^r$, $[b]_m^s$ will often be denoted by $[a]_m^r \equiv [b]_m^s$, this notation indicating that if $s \nless r$ there exist two mutually inverse matrices $[k]_r^s$, $\overline{\underline{K}}_s^r$ of rank r such that

$$[b]_m^s = [a]_m^r \, [k]_r^s, \quad [a]_m^r = [b]_m^s \, \overline{\underline{K}}_s^r, \quad [k]_r^s \, \overline{\underline{K}}_s^r = [1]_r^r .$$

This use of the sign $\equiv$, which is an extension of that given in Note 2 of § 113, is ambiguous when both matrices are square and not necessarily undegenerate.

NOTE 2. *Spacelets represented by matrices which are not necessarily undegenerate.*

In Note 3 of § 113 a system of mutually equivalent similar undegenerate matrices is regarded as representing a spacelet which is completely defined by any one of them. More generally any system of matrices, undegenerate or degenerate, all of which have vertical (or horizontal) equivalence with one another represent a spacelet which is completely defined by any one of them. For if $\overline{\underline{a}}_n^r$ is a matrix of rank a and if $\overline{\underline{b}}_n^a$ is an undegenerate matrix of the same rank a which is vertically equivalent to it, then any point $\omega_1 \equiv \overline{\underline{x}}_n$ lies in the spacelet $\omega_a \equiv \overline{\underline{b}}_n^a$ when and only when $\overline{\underline{x}}_n$ is connected with the vertical rows of $\overline{\underline{a}}_n^r$. Consequently the matrix $\overline{\underline{a}}_n^r$ represents the same spacelet ω_a as $\overline{\underline{b}}_n^a$.

We will use the notation

$$\omega_a \equiv \overline{\underline{a}}_n^r \equiv \overline{\underline{b}}_n^s \equiv \overline{\underline{c}}_n^t \cdots$$

to indicate that $\overline{\underline{a}}_n^r$, $\overline{\underline{b}}_n^s$, $\overline{\underline{c}}_n^t$, ... are vertically equivalent matrices of rank a, and that ω_a is the spacelet of rank a defined by any one of them or by any other matrix vertically equivalent to them.

The notation $\boxed{a}^{r}_{n} \equiv \boxed{b}^{s}_{n}$ here indicates that $\boxed{a}^{r}_{n}$ and $\boxed{b}^{s}_{n}$ are vertically equivalent matrices. It therefore indicates that if $s \not< r$ there exist two mutually inverse undegenerate matrices $\boxed{k}^{s}_{r}$, $[K]^{r}_{s}$ of rank r such that

$$\boxed{b}^{s}_{n} = \boxed{a}^{r}_{n}\,\boxed{k}^{s}_{r}, \quad \boxed{a}^{r}_{n} = \boxed{b}^{s}_{n}\,[K]^{r}_{s}, \quad \boxed{k}^{s}_{r}\,[K]^{r}_{s} = [1]^{r}_{r}.$$

Whenever we use the notation $\omega_a \equiv \boxed{a}^{r}_{n}$, it will be understood that the matrix $\boxed{a}^{r}_{n}$ has rank a.

§ 139. Joins and intersections of spacelets in homogeneous space of $n-1$ dimensions or rank n.

1. *The join and the complete intersection of two spacelets.*

The join of any two spacelets ω_r and ω_s of ranks r and s in homogeneous space ω_n will be defined to be the (uniquely determinate) spacelet of smallest dimensions or smallest rank which contains all the points of ω_r and all the points of ω_s. If $\omega_r \equiv \boxed{a}^{r}_{n}$ and $\omega_s \equiv \boxed{b}^{s}_{n}$, so that the matrices on the right are undegenerate and have ranks r and s, and if the matrix $\phi = \boxed{a, b}^{r,s}_{n}$ has rank T, then the join of ω_r and ω_s is the spacelet ω_T of rank T given by

$$\omega_T \equiv \boxed{a, b}^{r,s}_{n} \equiv \boxed{x}^{T}_{n},$$

where $\boxed{x}^{T}_{n}$ is any undegenerate matrix vertically equivalent to ϕ, i.e. (see Ex. v of § 138) any matrix of rank T whose vertical rows are connected with the vertical rows of ϕ.

For in the first place the spacelet ω_T clearly contains all points of ω_r and ω_s. And in the second place if $\omega_u \equiv \boxed{y}^{u}_{n}$ is any spacelet of rank u which contains all the points of ω_r and ω_s, there must exist relations of the forms

$$\boxed{a}^{r}_{n} = \boxed{y}^{u}_{n}\,\boxed{h}^{r}_{u}, \quad \boxed{b}^{s}_{n} = \boxed{y}^{u}_{n}\,\boxed{k}^{s}_{u}, \quad \boxed{a, b}^{r,s}_{n} = \boxed{y}^{u}_{n}\,\boxed{h, k}^{r,s}_{u},$$

i.e. ω_u must contain ω_T. Thus every spacelet which contains all points of ω_r and ω_s must be either ω_T or some spacelet of greater dimensions than ω_T which contains ω_T. It follows that the join of ω_r and ω_s is uniquely determinate and is the spacelet ω_T.

Ex. i. If ω_r and ω_s are represented in the forms $\omega_r \equiv \boxed{a}^{p}_{n}$ and $\omega_s \equiv \boxed{b}^{q}_{n}$, where the matrices on the right have ranks r and s, and if the matrix $\lceil a, b \rfloor_n^{p,q}$ has rank T, then the join of ω_r and ω_s is the spacelet

$$\omega_T \equiv \lceil a, b \rfloor_n^{p,q}.$$

Ex. ii. *If two spacelets ω_u and ω_v both lie in a given spacelet ω_r, then every point of the join of ω_u and ω_v lies in ω_r.*

This has been proved in the text.

Ex. iii. *If $\lceil a \rfloor_n^r$ is a matrix of rank r, and if $\lceil x \rfloor_n^t$, where $t<r$, is a matrix of rank t whose vertical rows are connected with the vertical rows of $\lceil a \rfloor_n^r$, then we can determine a matrix $\lceil y \rfloor_n^{r-t}$ such that $\lceil x, y \rfloor_n^{t,r-t}$ is an undegenerate matrix of rank r vertically equivalent to $\lceil a \rfloor_n^r$.*

Let $\lceil x \rfloor_n^t = \lceil a \rfloor_n^r \lceil h \rfloor_r^t$, where $\lceil h \rfloor_r^t$ has rank t. Then as in § 104 we can determine a matrix $\lceil k \rfloor_r^{r-t}$ such that $\lceil h, k \rfloor_r^{t,r-t}$ is an undegenerate square matrix. This having been done, the required conditions are satisfied when

$$\lceil y \rfloor_n^{r-t} = \lceil a \rfloor_n^r \lceil k \rfloor_r^{r-t}, \quad \lceil x, y \rfloor_n^{t,r-t} = \lceil a \rfloor_n^r \lceil h, k \rfloor_r^{t,r-t}.$$

The *complete intersection* or simply the *intersection* of any two spacelets ω_r and ω_s of ranks r and s of homogeneous space ω_n will be defined to be the locus of all points common to ω_r and ω_s. It is the (uniquely determinate) spacelet of greatest dimensions or greatest rank which is contained both in ω_r and in ω_s.

To see this let t be the greatest possible rank of a spacelet which is contained both in ω_r and in ω_s, and let ω_t be a spacelet of rank t contained both in ω_r and ω_s. Then all points of ω_t are common to ω_r and ω_s. Further there cannot be any point which is common to both ω_r and ω_s and does not lie in ω_t. For if ω_1 were any such point, then by Ex. ii the join of ω_1 and ω_t would be a spacelet of rank $t+1$ contained in both ω_r and ω_s; and this by the hypothesis is impossible. Thus ω_t is the locus of all points common to ω_r and ω_s.

Two spacelets ω_r and ω_s of ranks r and s will be said to *intersect* when they have at least one point in common, and to be *non-intersecting* when they have no point in common, i.e. when their complete intersection has rank 0 or is non-existent. If $\omega_r \equiv \lceil a \rfloor_n^r$ and $\omega_s \equiv \lceil b \rfloor_n^s$, neither r nor s being zero, the two spacelets do or do not intersect according as the rank of the matrix

$\phi = \lceil a,\ b \rfloor_n^{r,s}$ is less than or equal to $r + s$. For ϕ has rank less than $r + s$ when and only when there exists a non-zero matrix $[\lambda,\ \mu]_{r,s}$ such that

$$\lceil a,\ b \rfloor_n^{r,s} \begin{bmatrix} \lambda \\ \mu \end{bmatrix}_{r,s} = 0,$$

i.e. when and only when there exists a relation of the form

$$\lceil a \rfloor_n^r \lceil \lambda \rfloor_r = - \lceil b \rfloor_n^s \lceil \mu \rfloor_s \cdot$$

in which both $\lceil \lambda \rfloor_r$ and $\lceil \mu \rfloor_s$ are non-zero matrices, i.e. when and only when there exists a point

$$\lceil x \rfloor_n = \lceil a \rfloor_n^r \lceil \lambda \rfloor_r = - \lceil b \rfloor_n^s \lceil \mu \rfloor_s$$

common to ω_r and ω_s.

Two spacelets, one of which has rank 0, are necessarily non-intersecting.

Ex. iv. The two spacelets $\omega_r \equiv \lceil a \rfloor_n^p$ and $\omega_s \equiv \lceil b \rfloor_n^q$ are non-intersecting when and only when the matrix $\lceil a,\ b \rfloor_n^{p,q}$ has rank $r + s$. In this case their join is the spacelet

$$\omega_{r+s} \equiv \lceil a,\ b \rfloor_n^{p,q}.$$

If either r or s is zero, the two spacelets are necessarily non-intersecting.

Ex. v. *If ω_u lies in ω_r, and ω_v does not intersect ω_r, then ω_u is the complete intersection of ω_r with the join of ω_u and ω_v, i.e. the only points of the join of ω_u and ω_v which lie in ω_r are those which lie in ω_u.*

Because ω_v does not intersect ω_r, and ω_u lies in ω_r, therefore ω_v does not intersect ω_u.

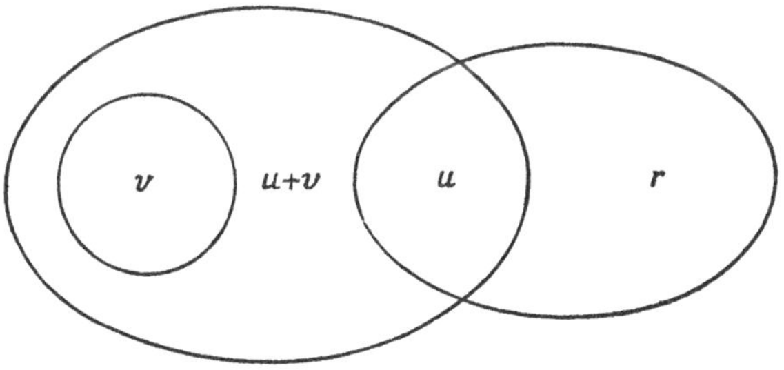

It follows that the join of ω_u and ω_v has rank $u + v$, and may be denoted by ω_{u+v}.

By Ex. iii we can write

$$\omega_u \equiv \lceil c \rfloor_n^u, \quad \omega_r \equiv \lceil c,\ b \rfloor_n^{u,\, r-u}, \quad \omega_v \equiv \lceil a \rfloor_n^v, \quad \omega_{u+v} \equiv \lceil c,\ a \rfloor_n^{u,\, v},$$

the matrices on the right being undegenerate and having ranks u, r, v, $u+v$. Then since ω_v does not intersect ω_r, the matrix $\phi = \left[c, a, b\right]_n^{u, v, r-u}$ is undegenerate and has rank $r+v$.

If ω_u is not the complete intersection of ω_{u+v} and ω_r, then some point of ω_{u+v} not lying in ω_u must be a point of ω_r not lying in ω_u, i.e. there must exist an equation of the form

$$\left[c, a\right]_n^{u, v} \left[\begin{matrix} h \\ \lambda \end{matrix}\right]_{u, v} = \left[c, b\right]_n^{u, r-u} \left[\begin{matrix} k \\ \mu \end{matrix}\right]_{u, r-u}$$

in which $\left[\lambda\right]_v \neq 0$ and $\left[\mu\right]_{r-u} \neq 0$. Writing this last equation in the form

$$\left[a\right]_n^v \left[\lambda\right]_v = \left[c, b\right]_n^{u, r-u} \left[\begin{matrix} \nu \\ \mu \end{matrix}\right]_{u, r-u},$$

where

$$\left[\nu\right]_u = \left[k\right]_u - \left[h\right]_u,$$

we see that there must be some connection between the vertical rows of ϕ, which is not the case.

Consequently ω_u is the complete intersection of ω_{u+v} and ω_r.

Ex. vi. *If* $\omega_t \equiv \left[c\right]_n^t$ *is the complete intersection of the two spacelets* $\omega_r \equiv \left[a\right]_n^r$ *and* $\omega_s \equiv \left[\beta\right]_n^s$, *we can represent* ω_r *and* ω_s *in the forms*

$$\omega_r \equiv \left[c, a\right]_n^{t, r-t} \quad \text{and} \quad \omega_s \equiv \left[c, b\right]_n^{t, s-t}. \qquad \text{.........................(1)}$$

The matrix $\phi = \left[c, a, b\right]_n^{t, r-t, s-t}$ *is then undegenerate and has rank* $r+s-t$, *and therefore* $\omega_{r+s-t} \equiv \left[c, a, b\right]_n^{t, r-t, s-t}$ *is the join of* ω_r *and* ω_s.

By Ex. iii we can represent ω_r and ω_s in the forms (1) where the matrices on the right are undegenerate and have ranks r and s. Then ω_r is the join of the two non-intersecting spacelets $\omega_t \equiv \left[c\right]_n^t$ and $\omega_{r-t} \equiv \left[a\right]_n^{r-t}$. Since ω_t is the complete intersection of ω_r and ω_s, all points common to ω_r and ω_s lie in ω_t, and no point of ω_{r-t} lies in ω_s. Consequently ω_{r-t} and ω_s do not intersect, and the matrix $\left[a, \beta\right]_n^{r-t, s}$ is undegenerate and has rank $r+s-t$. It follows that the equivalent matrix ϕ is undegenerate and has rank $r+s-t$.

Finally since ϕ has rank $r+s-t$, it follows from the definition of a join that ω_{r+s-t} is the join of ω_r and ω_s.

That the matrix ϕ has rank $r+s-t$ can also be seen as follows.

Suppose that there exists a relation of the form

$$\left[c, a, b\right]_n^{t, r-t, s-t} \left[\begin{matrix} \nu \\ \lambda \\ \mu \end{matrix}\right]_{t, r-t, s-t} = 0. \qquad \text{..............................(2)}$$

If $[\lambda]_{r-t} \neq 0$, the point $[x]_n = [a]_n^{r-t}[\lambda]_{r-t} = -[c,\, b]_n^{t,\, s-t}\begin{bmatrix}\nu\\ \mu\end{bmatrix}_{t,\, s-t}$ is common to ω_r and ω_s, and must lie in ω_t.

Therefore there must exist a relation of the form

$$[a]_n^{r-t}[\lambda]_{r-t} = -[c]_n^{t}[h]_t, \quad \text{or} \quad [c,\, a]_n^{t,\, r-t}\begin{bmatrix}h\\ \lambda\end{bmatrix}_{t,\, r-t} = 0,$$

and this is impossible because the matrix $[c,\, a]_n^{t,\, r-t}$ has rank r.

If $[\mu]_{s-t} \neq 0$, the point $[x]_n = [b]_n^{s-t}[\mu]_{s-t} = -[c,\, a]_n^{t,\, r-t}\begin{bmatrix}\nu\\ \lambda\end{bmatrix}_{t,\, r-t}$ is common to ω_r and ω_s, and must lie in ω_t.

Therefore there must exist a relation of the form

$$[b]_n^{s-t}[\mu]_{s-t} = -[c]_n^{t}[k]_t, \quad \text{or} \quad [c,\, b]_n^{t,\, s-t}\begin{bmatrix}k\\ \mu\end{bmatrix}_{t,\, s-t} = 0,$$

and this is impossible because the matrix $[c,\, b]_n^{t,\, s-t}$ has rank s.

Thus in (2) we must have $[\lambda]_{r-t} = 0$ and $[\mu]_{s-t} = 0$. The relation then reduces to $[c]_n^{t}[\nu]_t = 0$, and because $[c]_n^{t}$ has rank t, we must have $[\nu]_t = 0$.

Thus in every relation of the form (2) we must have $[\nu]_t = 0$, $[\lambda]_{r-t} = 0$, $[\mu]_{s-t} = 0$. Consequently there is no connection between the vertical rows of ϕ, and the matrix ϕ must be undegenerate and have rank $r+s-t$.

Ex. vii. *If two spacelets ω_r and ω_s are represented in the forms* (1) *where the matrix $\phi = [c,\, a,\, b]_n^{t,\, r-t,\, s-t}$ is undegenerate and has rank $r+s-t$, then*

$$\omega_t \equiv [c]_n^{t} \quad \text{and} \quad \omega_{r+s-t} \equiv [c,\, a,\, b]_n^{t,\, r-t,\, s-t}$$

are respectively the complete intersection and the join of ω_r and ω_s.

By the definition of a join, the join of ω_r and ω_s is the spacelet

$$\omega_{r+s-t} \equiv [c,\, a,\, c,\, b]_n^{t,\, r-t,\, t,\, s-t} \equiv [c,\, a,\, b]_n^{t,\, r-t,\, s-t}$$

If $\omega_t \equiv [c]_n^{t}$ is not the complete intersection of ω_r and ω_s, there must be some point of ω_r not lying in ω_t which is also a point of ω_s not lying in ω_t. Therefore there must exist an equation of the form

$$[c,\, a]_n^{t,\, r-t}\begin{bmatrix}h\\ \lambda\end{bmatrix}_{t,\, r-t} = [c,\, b]_n^{t,\, s-t}\begin{bmatrix}k\\ \mu\end{bmatrix}_{t,\, s-t}$$

in which $[\lambda]_{r-t} \neq 0$ and $[\mu]_{s-t} \neq 0$. Writing this equation in the form

$$[a]_n^{r-t}\,[\lambda]_{r-t} = [c,\ b]_n^{t,\ s-t}\left[\begin{matrix}\nu\\ \mu\end{matrix}\right]_{t,\ s-t}, \quad \text{where } [\nu]_t = [k]_t - [h]_t,$$

we see that there must exist some connection between the vertical rows of the matrix ϕ, which by hypothesis is not the case. Thus ω_t must be the complete intersection of ω_r and ω_s.

Ex. viii. *If* $\omega_r \equiv [c,\ a]_n^{t,\ r-t}$ *and* $\omega_s \equiv [c,\ b]_n^{t,\ s-t}$ *are two spacelets of ranks r and s whose complete intersection is the spacelet* $\omega_t \equiv [c]_n^t$, *then:*

(1) *Any point* $[x]_n \equiv [c,\ a]_n^{t,\ r-t}\left[\begin{matrix}h\\ \lambda\end{matrix}\right]_{t,\ r-t}$ *of* ω_r *lies in* ω_s *when and only when* $[\lambda]_{r-t} = 0$.

(2) *Any point* $[x]_n \equiv [c,\ b]_n^{t,\ s-t}\left[\begin{matrix}k\\ \mu\end{matrix}\right]_{t,\ s-t}$ *of* ω_s *lies in* ω_r *when and only when* $[\mu]_{s-t} = 0$.

This theorem is equivalent to and can be deduced from Ex. v.

To prove (1) directly we observe that if the point $[x]_n$ of ω_r lies in ω_s, then it lies in ω_t, i.e. there must exist an equation of the form

$$[c,\ a]_n^{t,\ r-t}\left[\begin{matrix}h\\ \lambda\end{matrix}\right]_{t,\ r-t} = [c]_n^t\,[k]_t .$$

Writing this equation in the form

$$[c,\ a]_n^{t,\ r-t}\left[\begin{matrix}\nu\\ \lambda\end{matrix}\right]_{t,\ r-t} = 0,$$

where

$$[\nu]_t = [h]_t - [k]_t,$$

it follows, since $[c,\ a]_n^{t,\ r-t}$ has rank r, that $[\nu]_t = 0$ and $[\lambda]_{r-t} = 0$; i.e. we must have $[\lambda]_{r-t} = 0$ and $[h]_t = [k]_t$.

We can prove (2) in a similar way.

The foregoing examples enable us to enunciate the two important theorems which follow.

Theorem I. *Any two spacelets* ω_r *and* ω_s *of ranks r and s in homogeneous space* ω_n *can be represented in the forms*

$$\omega_r \equiv [c,\ a]_n^{t,\ r-t} \quad \textit{and} \quad \omega_s \equiv [c,\ b]_n^{t,\ s-t}, \ \ldots\ldots\ldots\ldots\ldots\ldots\text{(A)}$$

where the matrix $\phi = \begin{bmatrix} c, a, b \end{bmatrix}^{t,\, r-t,\, s-t}_{n}$ *is undegenerate and has rank*

$$T = r + s - t.$$

Whenever they are represented in these forms, the complete intersection and the join of ω_r *and* ω_s *are respectively the spacelets* ω_t *and* ω_T *of ranks t and* $r+s-t$ *given by*

$$\omega_t \equiv \begin{bmatrix} c \end{bmatrix}^{t}_{n}, \quad \omega_T = \begin{bmatrix} c, a, b \end{bmatrix}^{t,\, r-t,\, s-t}_{n} \quad \text{......................(B)}$$

In the figure, which is of course merely schematic, a closed area marked with a number x represents a spacelet ω_x of rank x. The spacelet ω_r is the join of the two non-intersecting spacelets $\omega_t \equiv \begin{bmatrix} c \end{bmatrix}^{t}_{n}$ and $\omega_{r-t} \equiv \begin{bmatrix} a \end{bmatrix}^{r-t}_{n}$; the spacelet ω_s is the join of the two non-intersecting spacelets $\omega_t \equiv \begin{bmatrix} c \end{bmatrix}^{t}_{n}$ and $\omega_{s-t} \equiv \begin{bmatrix} b \end{bmatrix}^{s-t}_{n}$; and the spacelet ω_T is the join of the three unconnected spacelets $\omega_t, \omega_{r-t}, \omega_{s-t}$, which is also the join of ω_r and ω_s.

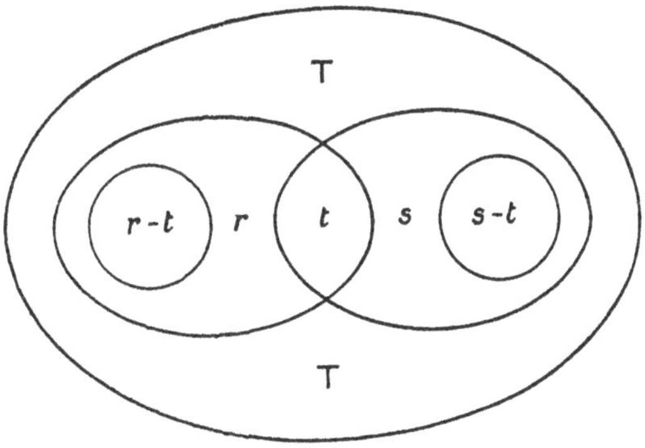

The first part of the theorem has been proved in Ex. vi, and the second part of the theorem has been proved in Ex. vii.

Theorem II. *If* ω_r *and* ω_s *are any two spacelets of ranks r and s in homogeneous space* ω_n, *and if t and T are respectively the ranks of the complete intersection and the join of* ω_r *and* ω_s, *then*

$$t + T = r + s. \quad \text{..............................(C)}$$

If the complete intersection of ω_r and ω_s has rank t, then by Theorem I the join of ω_r and ω_s has rank $r+s-t$. Writing $T = r+s-t$, we have $t + T = r + s$.

Ex. ix. *Possible ranks of the intersection and join of two spacelets of given ranks.*

If ω_r *and* ω_s *are two arbitrary spacelets of homogeneous space* ω_n *of given ranks r and s, the possible values of the rank t of their complete intersection are those consistent with the conditions*

$$t \nless 0, \quad t \ngtr r, \quad t \ngtr s, \quad t \nless r+s-n, \quad \text{..............................(3)}$$

which include $\quad r \nless 0, \quad r \ngtr n, \quad s \nless 0, \quad s \ngtr n.$

It is clear that t must satisfy all the conditions (3), the last of which is equivalent to $T \ngtr n$. Conversely when t is any integer satisfying these conditions, we see from Theorem I that there exist spacelets ω_r and ω_s of ranks r and s whose complete intersection has rank t.

The possible values of the rank T of the join of ω_r and ω_s are those consistent with the conditions

$$T \not> r+s, \quad T \not< r, \quad T \not< s, \quad T \not> n. \qquad \text{.........................(3')}$$

Ex. x. *If the spacelets ω_r and ω_s do not intersect, we must have $r+s \not> n$.*

For in this case $T=r+s$.

Ex. xi. *If $r+s>n$, the two spacelets ω_r and ω_s necessarily intersect, and they have at least $r+s-n$ unconnected points in common.*

For in this case we have $t \not< r+s-n$.

Ex. xii. If ω_u and ω_v are any two spacelets both of which lie in the spacelet $\omega_r \equiv \boxed{a}_{n}^{r}$, they can be represented in the forms

$$\omega_u \equiv \boxed{a}_{n}^{r}\,\boxed{\gamma,\ \alpha}_{r}^{t,\ u-t}, \quad \omega_v \equiv \boxed{a}_{n}^{r}\,\boxed{\gamma,\ \beta}_{n}^{t,\ v-t}, \qquad \text{....................}(A_1)$$

where the matrix $\boxed{\gamma,\ \alpha,\ \beta}_{r}^{t,\ u-t,\ v-t}$ is undegenerate and has rank $T=u+v-t$. Their complete intersection and join are then respectively the spacelets

$$\omega_t \equiv \boxed{a}_{n}^{r}\,\boxed{\gamma}_{r}^{t}, \quad \omega_T \equiv \boxed{a}_{n}^{r}\,\boxed{\gamma,\ \alpha,\ \beta}_{r}^{t,\ u-t,\ v-t} \qquad \text{....................}(B_1)$$

Ex. xiii. If ω_u and ω_v are any two spacelets both of which contain the spacelet $\omega_r \equiv \boxed{a}_{n}^{r}$, they can be represented in the forms

$$\omega_u \equiv \boxed{a,\ \gamma,\ \alpha}_{n}^{r,\ t-r,\ u-t}, \quad \omega_v \equiv \boxed{a,\ \gamma,\ \beta}_{n}^{r,\ t-r,\ v-t}, \qquad \text{....................}(A_2)$$

where the matrix $\boxed{a,\ \gamma,\ \alpha,\ \beta}_{n}^{r,\ t-r,\ u-t,\ v-t}$ is undegenerate and has rank $T=u+v-t$. Their complete intersection and join are then respectively the spacelets

$$\omega_t \equiv \boxed{a,\ \gamma}_{n}^{r,\ t-r}, \quad \omega_T \equiv \boxed{a,\ \gamma,\ \alpha,\ \beta}_{n}^{r,\ t-r,\ u-t,\ v-t}. \qquad \text{..................}(B_2)$$

Ex. xiv. Let $\omega_r \equiv \boxed{a}_{n}^{r}$ and $\omega_s \equiv \boxed{b}_{n}^{s} \equiv \boxed{a}_{n}^{r}\,\boxed{k}_{r}^{s}$ be two spacelets of ranks r and s of homogeneous space ω_n such that ω_r contains ω_s. Then if ω_u and ω_v are any two spacelets of ranks u and v which both lie in ω_r and both contain ω_s, they can be represented in the forms

$$\omega_u \equiv \boxed{a}_{n}^{r}\,\boxed{k,\ \gamma,\ \alpha}_{r}^{s,\ t-s,\ u-t}, \quad \omega_v \equiv \boxed{a}_{n}^{r}\,\boxed{k,\ \gamma,\ \beta}_{r}^{s,\ t-s,\ v-t}, \qquad \text{...............}(A_3)$$

where the matrix $\boxed{k,\ \gamma,\ \alpha,\ \beta}_{r}^{s,\ t-s,\ u-t,\ v-t}$ is undegenerate and has rank $T=u+v-t$. The complete intersection and the join of ω_u and ω_v are then respectively the spacelets ω_t and ω_T of ranks t and T given by

$$\omega_t \equiv \boxed{a}_{n}^{r}\,\boxed{k,\ \gamma}_{r}^{s,\ t-s}, \quad \omega_T \equiv \boxed{a}_{n}^{r}\,\boxed{k,\ \gamma,\ \alpha,\ \beta}_{r}^{s,\ t-s,\ u-t,\ v-t}. \qquad \text{...............}(B_3)$$

We deduce this result from Exs. xii and xiii when we observe that if

$$\omega_u \equiv [x]_n^u \equiv [a]_n^r [\xi]_r^u, \quad \omega_v \equiv [y]_n^v \equiv [a]_n^r [\eta]_r^v, \quad \omega_w \equiv [z]_n^w \equiv [a]_n^r [\zeta]_r^w, \ldots$$

are any number of spacelets lying in ω_r, then the matrices

$$[x, y, z, \ldots]_n^{u, v, w, \ldots}, \quad [\xi, \eta, \zeta, \ldots]_r^{u, v, w, \ldots}$$

have equal ranks, and the vertical rows of $[b]_n^s$ are connected with the vertical rows of $[x]_n^u$ when and only when the vertical rows of $[k]_r^s$ are connected with the vertical rows of $[\xi]_r^u$.

Ex. xv. *Possible ranks of the intersection and join of two spacelets of given ranks which both lie in a given spacelet ω_r.*

If ω_u and ω_v are two arbitrary spacelets of given ranks u and v which both lie in a given spacelet ω_r of rank r, the possible values of the rank t of their complete intersection are those consistent with the conditions

$$t \nless 0, \quad t \ngtr u, \quad t \ngtr v, \quad t \nless u+v-r, \qquad \text{.........................(4)}$$

which include $\qquad u \nless 0, \quad u \ngtr r, \quad v \nless 0, \quad v \ngtr r, \quad t \ngtr r.$

This follows from Ex. xii. Since $t+T=u+v$, we see also that:

The possible values of the rank T of the join of ω_u and ω_v are those consistent with the conditions

$$T \ngtr u+v, \quad T \nless u, \quad T \nless v, \quad T \ngtr r. \qquad \text{.........................(4')}$$

If ω_u and ω_v do not intersect, we must have $u+v \ngtr r$.

If $u+v>r$, then ω_u and ω_v necessarily intersect and have at least $u+v-r$ unconnected points in common.

Ex. xvi. *Possible ranks of the intersection and join of two spacelets of given ranks which both contain a given spacelet ω_r.*

If ω_u and ω_v are two arbitrary spacelets of given ranks u and v which both contain a given spacelet ω_r of rank r, the possible values of the rank t of their complete intersection are those consistent with the conditions

$$t \nless r, \quad t \ngtr u, \quad t \ngtr v, \quad t \nless u+v-n, \qquad \text{.........................(5)}$$

which include $\qquad u \nless r, \quad u \ngtr n, \quad v \nless r, \quad v \ngtr n.$

This follows from Ex. xiii. Since $t+T=u+v$, we see also that:

The possible values of the rank T of the join of ω_u and ω_v are those consistent with the conditions

$$T \ngtr u+v-r, \quad T \nless u, \quad T \nless v, \quad T \ngtr n. \qquad \text{.........................(5')}$$

If ω_r is the complete intersection of ω_u and ω_v, we must have $u+v \ngtr r+n$.

If $u+v>r+n$, then ω_u and ω_v have at least $u+v-n$ unconnected points in common, and ω_r is not their complete intersection.

Ex. xvii. *Possible ranks of the intersection and join of two spacelets of given ranks which both lie in a given spacelet ω_r and both contain a given spacelet ω_s.*

If the given spacelet ω_r contains the given spacelet ω_s, and if ω_u and ω_v are two arbitrary spacelets of given ranks u and v which both lie in ω_r and both contain ω_s, then the possible values of the rank t of their complete intersection are those consistent with the conditions

$$t \nless s, \quad t \ngtr u, \quad t \ngtr v, \quad t \nless u+v-r, \quad \ldots\ldots\ldots\ldots(6)$$

which include

$$u \nless s, \quad u \ngtr r, \quad v \nless s, \quad v \ngtr r, \quad t \ngtr r.$$

This follows from Ex. xiv. Since $t+T=u+v$, we see also that:

The possible values of the rank T of the join of ω_u and ω_v are those consistent with the conditions

$$T \ngtr u+v-s, \quad T \nless u, \quad T \nless v, \quad T \ngtr r. \quad \ldots\ldots\ldots\ldots(6')$$

If ω_s is the complete intersection of ω_u and ω_v, we must have $u+v \ngtr r+s$.

If $u+v > r+s$, then ω_u and ω_v have at least $u+v-r$ unconnected points in common, and ω_s is not the complete intersection of ω_u and ω_v.

If ω_r is the join of ω_u and ω_v, we must have $u+v \nless r+s$.

If $u+v < r+s$, then the rank of the join of ω_u and ω_v cannot exceed $u+v-s$, and is a spacelet of smaller rank than ω_r contained in ω_r.

Ex. xviii. *If ω_u is a given spacelet of rank u which lies in a given spacelet ω_p of rank p, and if ω_x is any spacelet of rank x which lies in ω_p and does not intersect ω_u, then the possible values of x are those consistent with the conditions*

$$x \nless 0, \quad x+u \ngtr p.$$

For if $\omega_p \equiv [a]_n^p$, $\omega_u \equiv [a]_n^p [h]_p^u$, we have $\omega_x \equiv [a]_n^p [k]_p^x$, where $[k]_p^x$ is any matrix so chosen that $[h, k]_p^{u, x}$ has rank $x+u$. It is to be here understood that the spacelet ω_x is non-existent when $x=0$.

In particular the possible ranks x of a spacelet which does not intersect a given spacelet ω_r of rank r are given by

$$x \nless 0, \quad x+r \ngtr n.$$

Ex. xix. *If ω_r lies in ω_p and ω_u lies in ω_r, then ω_x will lie in ω_p and have ω_u for its complete intersection with ω_r when and only when it is the join of ω_u with a spacelet ω_{x-u} which lies in ω_p and does not intersect ω_r.*

First let ω_x lie in ω_p and have ω_u for its complete intersection with ω_r. Then by Ex. iii we can express ω_x as the join of ω_u with a spacelet ω_{x-u} which does not intersect ω_u. Because ω_x lies in ω_p, therefore ω_{x-u} lies in ω_p; and because ω_u is the complete intersection of ω_x and ω_r, therefore no point of ω_{x-u} lies in ω_r, i.e. ω_{x-u} does not intersect ω_r.

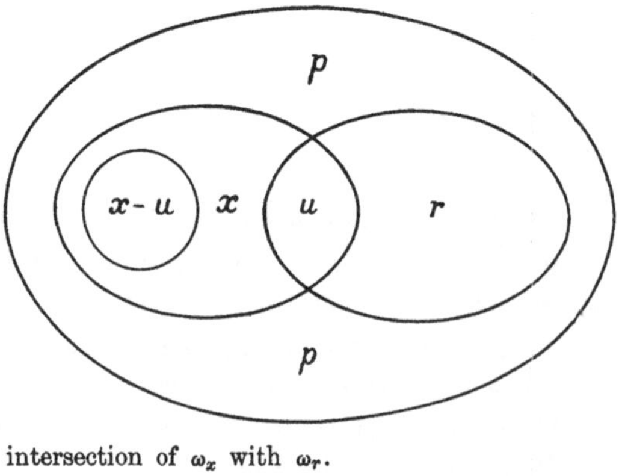

Again let ω_{x-u} be any spacelet which lies in ω_p and does not intersect ω_r. Then by Ex. ii the join of ω_u and ω_{x-u} is a spacelet ω_x which lies in ω_p; and it follows from Ex. v that ω_u is the complete intersection of ω_x with ω_r.

Hence if ω_x lies in ω_p and has ω_u for its complete intersection with ω_r, the possible values of x are those consistent with the conditions

$$x \nless u, \quad x+r-u \ngtr p.$$

For by Ex. xviii the possible values of $x-u$ are given by $x-u \nless 0$, $(x-u)+r \ngtr p$.

2. *Complementary spacelets.*

Two spacelets ω_r and ω_s of homogeneous space ω_n will be said to be mutually *complementary* when their join is the complete space ω_n. This cannot be the case unless $r+s \nless n$. If t and T are the ranks of the complete intersection and the join of two complementary spacelets ω_r and ω_s, we have

$$T=n, \quad t=r+s-n.$$

More generally if ω_r is any given spacelet of ω_n, two spacelets ω_u and ω_v will be said to be mutually *complementary in* ω_r when they both lie in ω_r and have ω_r for their join. This cannot be the case unless $u+v \nless r$. If t and T are the ranks of the complete intersection and join of two spacelets ω_u and ω_v which are complementary to one another in ω_r, we have

$$T=r, \quad t=u+v-r.$$

3. *Mutually incident spacelets.*

Two spacelets ω_r and ω_s of homogeneous space ω_n are sometimes said to be mutually *incident* or to *incide* when either one of them contains the other. This is the case when and only when the larger spacelet is the join of the two, and when and only when the smaller spacelet is the complete intersection of the two.

Ex. xx. Let t and T be the ranks of the complete intersection and the join of ω_r and ω_s, so that $t+T=r+s$. Then ω_r and ω_s are

(1) *incident* when $T=r$ or $T=s$, i.e. when $t=s$ or $t=r$;

(2) *non-incident* when $T\neq r$ and $T\neq s$, i.e. when $t\neq s$ and $t\neq r$;

(3) *co-incident* when $T=r=s=t$.

4. *The join and the complete intersection of any number of spacelets.*

If $\omega_\alpha \equiv \overline{\underline{\,a\,}}{}^{p}_{n}$, $\omega_\beta \equiv \overline{\underline{\,b\,}}{}^{q}_{n}$, ... $\omega_\gamma \equiv \overline{\underline{\,c\,}}{}^{r}_{n}$ are any number of spacelets of ranks $\alpha, \beta, \ldots \gamma$ of homogeneous space ω_n, their *join* is the spacelet of smallest dimensions or smallest rank which contains all of them. If the matrix $\phi = \overline{\underline{\,a, b, \ldots c\,}}{}^{p, q, \ldots r}_{n}$ has rank T, then their join is the uniquely determinate spacelet ω_T of rank T given by

$$\omega_T \equiv \overline{\underline{\,a, b, \ldots c\,}}{}^{p, q, \ldots r}_{n} \equiv \overline{\underline{\,x\,}}{}^{T}_{n},$$

where $\overline{\underline{\,x\,}}{}^{T}_{n}$ is any undegenerate matrix vertically equivalent to ϕ, i.e. (see Ex. v of § 138) any matrix of rank T whose vertical rows are connected with the vertical rows of ϕ.

The spacelets $\omega_\alpha, \omega_\beta, \ldots \omega_\gamma$ are *unconnected* (see § 140) when and only when their join has rank $\alpha+\beta+\ldots+\gamma$, i.e. when and only when no one of the spacelets has a point in common with the join of the others.

The *complete common intersection* or simply the *intersection* of $\omega_\alpha, \omega_\beta, \ldots \omega_\gamma$ is the locus of all points which are common to all these spacelets. It is the (uniquely determinate) spacelet of greatest dimensions or greatest rank which is contained in all the spacelets $\omega_\alpha, \omega_\beta, \ldots \omega_\gamma$; and it will often be a spacelet of zero rank, i.e. be non-existent.

Ex. xxi. If $\omega_u \equiv [a]_n^p\,[a]_p^u$, $\omega_v \equiv [a]_n^p\,[\beta]_p^v$, ... are any number of spacelets of ranks $u, v, \ldots$ lying in a given spacelet ω_p of homogeneous space ω_n, and if we write

$$\Omega_u \equiv [a]_p^u, \quad \Omega_v \equiv [\beta]_p^v, \ldots,$$

we can regard $\Omega_u, \Omega_v, \ldots$ as spacelets of ranks $u, v, \ldots$ lying in a homogeneous space Ω_p of rank p. There is a one-one correspondence between the spacelets $\omega_u, \omega_v, \ldots$ and the spacelets $\Omega_u, \Omega_v, \ldots$, any two corresponding spacelets having the same rank. Also the joins and intersections of $\omega_u, \omega_v, \ldots$ and $\Omega_u, \Omega_v, \ldots$ have the same ranks.

Hence from any general theorem relating to spacelets lying in homogeneous space ω_n which refers only to ranks, intersections, joins and incidences, we can deduce a corresponding theorem for spacelets lying in a given spacelet ω_p by simply substituting p for n.

5. *Spacelets which do not intersect either of two given spacelets.*

Theorem III. *If ω_u and ω_v are two given spacelets of ranks u and v which both lie in a given spacelet ω_r of rank r of homogeneous space ω_n, then there exist spacelets ω_x of rank x which lie in ω_r and intersect neither ω_u nor ω_v when and only when the integer x satisfies the conditions*

$$x \nless 0, \quad u+x \ngtr r, \quad v+x \ngtr r. \qquad \text{.....................(7)}$$

It is to be understood that the spacelet ω_x is non-existent when $x=0$.

Thus (see Ex. xviii) *if ω_r contains a spacelet of rank x which does not intersect ω_u and a spacelet of rank x which does not intersect ω_v, then it contains a spacelet of rank x which intersects neither ω_u nor ω_v.*

Let ω_t, ω_T be the complete intersection and the join of ω_u and ω_v, so that $T=u+v-t$. Then by Ex. iii we can write

$$\omega_t \equiv [c]_n^t,$$

$$\omega_u \equiv [a,\ c]_n^{u-t,\ t},$$

$$\omega_v \equiv [b,\ c]_n^{v-t,\ t},$$

$$\omega_T \equiv [a,\ b,\ c]_n^{u-t,\ v-t,\ t},$$

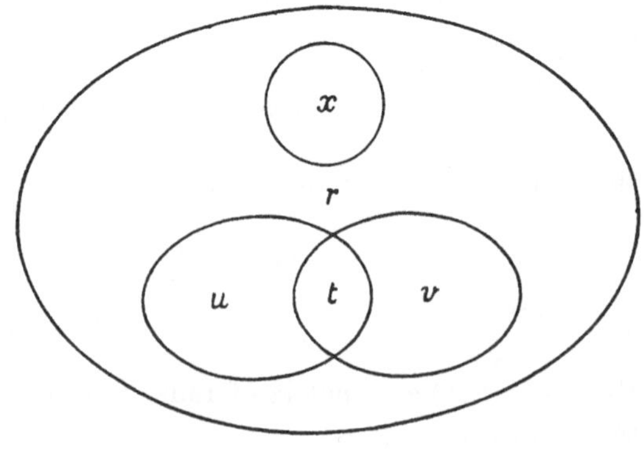

and $$\omega_r \equiv \lceil e \rfloor_n^{r}, \quad \text{where} \quad \lceil e \rfloor_n^{r} = \lceil a,\ b,\ c,\ d \rfloor_n^{u-t,\ v-t,\ t,\ r-T},$$

the vertical rows of all matrices on the right being unconnected.

A general formula for any spacelet ω_x of rank x lying in ω_r is

$$\omega_x \equiv \lceil e \rfloor_n^{r} \left\lceil \begin{matrix} \lambda \\ \mu \\ \nu \\ \pi \end{matrix} \right\rfloor_{u-t,\ v-t,\ t,\ r-T}^{x}, \quad \text{.......................(8)}$$

where the postfactor on the right is undegenerate and has rank x.

The spacelet ω_x given by (8) will be one which intersects neither ω_u nor ω_v when and only when the join of ω_x, ω_u has rank $x+u$, and the join of ω_x, ω_v has rank $x+v$.

Now the join of ω_x, ω_u and the join of ω_x, ω_v have respectively the same ranks as the matrices

$$\phi = \left\lceil \begin{matrix} \lambda, & 1, & 0 \\ \mu, & 0, & 0 \\ \nu, & 0, & 1 \\ \pi, & 0, & 0 \end{matrix} \right\rfloor_{u-t,\ v-t,\ t,\ r-T}^{x,\ u-t,\ t} \quad \text{and} \quad \psi = \left\lceil \begin{matrix} \lambda, & 0, & 0 \\ \mu, & 1, & 0 \\ \nu, & 0, & 1 \\ \pi, & 0, & 0 \end{matrix} \right\rfloor_{u-t,\ v-t,\ t,\ r-T}^{x,\ v-t,\ t};$$

for they are the spacelets represented by $\lceil e \rfloor_n^{r}\,\phi$ and $\lceil e \rfloor_n^{r}\,\psi$.

Again by Theorem IV of § 105 the matrices ϕ and ψ have ranks $x+u$ and $x+v$ when and only when the matrices

$$[\alpha]_{r-u}^{x} = \left\lceil \begin{matrix} \mu \\ \pi \end{matrix} \right\rfloor_{v-t,\ r-T}^{x} \quad \text{and} \quad [\beta]_{r-v}^{x} = \left\lceil \begin{matrix} \lambda \\ \pi \end{matrix} \right\rfloor_{u-t,\ r-T}^{x}$$

both have rank x; and clearly we can determine the postfactor on the right in (8) so that these last conditions are satisfied when and only when x satisfies the conditions (7). This establishes the theorem.

Ex. xxii. In determining a spacelet ω_x having the properties mentioned in the theorem, we can choose the postfactor on the right in (8) so as to be real; and at the same time we so choose it that $\lceil \nu \rfloor_t^{x} = 0$. Again if ω_u, ω_v, ω_r are all real, we can choose $\lceil e \rfloor_n^{r}$ to be real.

Hence if ω_u, ω_v, ω_r are all real, there exist real spacelets of rank x which lie in ω_r and intersect neither ω_u nor ω_v when and only when x satisfies the conditions (7).

Ex. xxiii. We can determine points ω_1 which lie in ω_r but lie neither in ω_u nor in ω_v when and only when

$$u < r \quad \text{and} \quad v < r. \quad \text{.......................................(9)}$$

Ex. xxiv. We can deduce Theorem III from Ex. xxiii.

First suppose that ω_r contains a spacelet ω_x which intersects neither ω_u nor ω_v. Then by Ex. xviii the conditions (7) must be satisfied.

Next suppose that x is an integer greater than 0 which satisfies the conditions (7).

If $u+1 \not> r$ and $v+1 \not> r$, then by Ex. xxiii we can determine a point P_1 which lies in ω_r and intersects neither ω_u nor ω_v. The joins of P_1, ω_u and of P_1, ω_v are then spacelets ω_{u+1} and ω_{v+1} which lie in ω_r.

If $u+2 \not> r$ and $v+2 \not> r$, then by Ex. xxiii we can determine a point P_2 which lies in ω_r and intersects neither ω_{u+1} nor ω_{v+1}. The joins of P_2, ω_{u+1} and of P_2, ω_{v+1}, i.e. the joins of P_1, P_2, ω_u and of P_1, P_2, ω_v, are then spacelets ω_{u+2} and ω_{v+2} which lie in ω_r; and the join of P_1, P_2 is a spacelet ω_2 which lies in ω_r and intersects neither ω_u nor ω_v.

If $u+3 \not> r$ and $v+3 \not> r$, then by Ex. xxiii we can determine a point P_3 which lies in ω_r and intersects neither ω_{u+2} nor ω_{v+2}. The joins of P_1, P_2, P_3, ω_u and of P_1, P_2, P_3, ω_v are then spacelets ω_{u+3} and ω_{v+3} which lie in ω_r; and the join of P_1, P_2, P_3 is a spacelet ω_3 which lies in ω_r and intersects neither ω_u nor ω_v.

Proceeding in this way we see that when x satisfies the conditions (7) we can determine a spacelet ω_x which lies in ω_r and intersects neither ω_u nor ω_v.

In the particular case when ω_r is the complete space ω_n Theorem III assumes the following form:

Theorem III a. *If ω_p and ω_q are two given spacelets of ranks p and q of homogeneous space ω_n, there exist spacelets ω_x of rank x which intersect neither ω_p nor ω_q when and only when the integer x satisfies the conditions*

$$x \not< 0, \quad p+x \not> n, \quad q+x \not> n. \quad \text{.....................(7')}$$

Thus if there exists a spacelet of rank x which does not intersect ω_p and a spacelet of rank x which does not intersect ω_q, then there exists a spacelet of rank x which intersects neither ω_p nor ω_q.

Ex. xxv. There exist points which lie neither in ω_p nor in ω_q when and only when

$$p<n \text{ and } q<n. \quad \text{.......................................(9')}$$

Ex. xxvi. We can deduce Theorem III *a* from Ex. xxv.

6. *Joins and intersections of matrices.*

First let $A = \underset{n}{\overset{p}{a}}$, $B = \underset{n}{\overset{q}{b}}$, ... $C = \underset{n}{\overset{r}{c}}$ be a number of matrices all having the same number n of horizontal rows.

Then any matrix $X = \underset{n}{\overset{u}{x}}$ will be called a *vertical join* of the matrices $A, B, \ldots C$ when the spacelet represented by X is the join of the spacelets represented by $A, B, \ldots C$. This is the case when and only when X is a matrix of the smallest possible rank having the property that all the vertical rows of $A, B, \ldots C$ are connected with the vertical rows of X. All the vertical joins of $A, B, \ldots C$ have the same rank and are vertically equivalent to one another; and they are all known when any one of them is known. One of the vertical joins of $A, B, \ldots C$ is the matrix

$$\phi = \overline{\underline{a, b, \ldots c}}^{\,p, q, \ldots r}_{\,n} ;$$

and the other vertical joins are the other matrices vertically equivalent to ϕ.

Also any matrix $Y = \overline{\underline{y}}^{\,v}_{\,n}$ will be called a *vertical intersection* of the matrices $A, B, \ldots C$ when the spacelet represented by Y is the intersection of the spacelets represented by $A, B, \ldots C$. This is the case when and only when Y is a matrix of the greatest possible rank having the property that all the vertical rows of Y are connected with the vertical rows of each one of the matrices $A, B, \ldots C$. All the vertical intersections of $A, B, \ldots C$ have the same rank and are vertically equivalent to one another; and they are all known when any one of them is known.

Next let $A' = [a]^n_p$, $B' = [b]^n_q$, $\ldots$ $C' = [c]^n_r$ be a number of matrices all having the same number n of horizontal rows.

Then any matrix $X' = [x]^n_u$ will be called a *horizontal join* of the matrices $A', B', \ldots C'$ when X' is a matrix of the smallest possible rank having the property that all the horizontal rows of $A', B', \ldots C'$ are connected with the horizontal rows of X'. This is the case when and only when the matrix $X = \overline{\underline{x}}^{\,u}_{\,n}$ is a vertical join of the matrices $A = \overline{\underline{a}}^{\,p}_{\,n}$, $B = \overline{\underline{b}}^{\,q}_{\,n}$, $\ldots$ $C = \overline{\underline{c}}^{\,r}_{\,n}$. All the horizontal joins of $A, B, \ldots C$ have the same rank and are horizontally equivalent to one another; and they are all known when any one of them is known.

Also any matrix $Y' = [y]^n_v$ will be called a *horizontal intersection* of the matrices $A', B', \ldots C'$ when Y' is a matrix of the greatest possible rank having the property that all the horizontal rows of Y' are connected with the horizontal rows of each one of the matrices $A', B', \ldots C'$. This is the case when and only when the matrix $Y = \overline{\underline{y}}^{\,v}_{\,n}$ is a vertical intersection of the matrices $A = \overline{\underline{a}}^{\,p}_{\,n}$, $B = \overline{\underline{b}}^{\,q}_{\,n}$, $\ldots$ $C = \overline{\underline{c}}^{\,r}_{\,n}$. All the horizontal intersections of $A', B', \ldots C'$ have the same rank and are horizontally equivalent to one another; and they are all known when any one of them is known.

§ 140. Connections between matrices and between spacelets.

1. *Horizontal and vertical connections between matrices.*

We will now give definitions of connections between matrices which can be regarded as parallel simple minors of some one matrix. These correspond to the definitions of connections between the horizontal and vertical rows of a matrix given in § 69.

We will define a *vertical connection* between the matrices

$$A=\overline{\underline{a}}{}^{p}_{n},\quad B=\overline{\underline{b}}{}^{q}_{n},\ \dots\ C=\overline{\underline{c}}{}^{r}_{n},\quad D=\overline{\underline{d}}{}^{s}_{n},\ \dots \qquad \dots\dots\dots(1)$$

to be a relation of the form

$$\overline{\underline{a}}{}^{p}_{n}\,\overline{\underline{\alpha}}{}_{p}+\overline{\underline{b}}{}^{q}_{n}\,\overline{\underline{\beta}}{}_{q}+\dots+\overline{\underline{c}}{}^{r}_{n}\,\overline{\underline{\gamma}}{}_{r}+\overline{\underline{d}}{}^{s}_{n}\,\overline{\underline{\delta}}{}_{s}+\dots=0 \qquad \dots\dots\dots(2)$$

in which at least one of the terms on the left does not vanish, and therefore at least two of the terms on the left do not vanish. Thus a vertical connection between these matrices is a connection between the vertical rows of all of them which is not deducible from the connections between the vertical rows of the individual matrices. The matrices (1) will be said to be vertically connected or vertically unconnected according as there is or is not a connection between them. They are therefore *vertically unconnected* when and only when in every relation of the form (2) every term on the left vanishes separately.

Ex. i. Every vertical connection between the matrices (1) is also a connection between the vertical rows of those matrices; but a connection between the vertical rows of the matrices is not necessarily a vertical connection between the matrices except when the matrices are all undegenerate.

Ex. ii. A connection between the vertical rows of a matrix is not necessarily a connection between the matrices of the vertical rows except when no vertical row is a row of 0's; but every connection between the matrices of the individual vertical rows is also a connection between the vertical rows.

Ex. iii. If every one of the matrices (1) has zero rank, then the matrices are unconnected; but the vertical rows of the matrices are connected.

Ex. iv. If some but not all of the matrices (1) have zero rank, then the vertical connections between all the matrices are the same as the vertical connections between those of the matrices whose ranks are not zero. In this case the matrices are unconnected when and only when those matrices whose ranks are not zero are unconnected.

The matrix A will be said to *have a vertical connection with* the remaining matrices $B, \dots C, D, \dots$ when there exists a relation of the form (2) in which the first term does not vanish. If A has zero rank, then it has no connection with the remaining matrices, although all the vertical rows of A are connected with the vertical rows of the remaining matrices.

More generally the matrices $A, B, \dots$ will be said to *have a vertical connection with* the matrices $C, D, \dots$, or there will be said to *be a vertical connection between* the matrices $A, B, \dots$ and the matrices $C, D, \dots$ when there exists a relation of the form

$$\overline{\underline{a}}{}^{p}_{n}\,\overline{\underline{\alpha}}{}_{p}+\overline{\underline{b}}{}^{q}_{n}\,\overline{\underline{\beta}}{}_{q}+\dots=\overline{\underline{c}}{}^{r}_{n}\,\overline{\underline{\gamma}}{}_{r}+\overline{\underline{d}}{}^{s}_{n}\,\overline{\underline{\delta}}{}_{s}+\dots \qquad \dots\dots\dots(2')$$

in which the left-hand side does not vanish or (which is the same thing) neither side vanishes.

The matrix A will be said to *be vertically connected with* (or to have complete vertical connection with) the remaining matrices $B, \dots C, D, \dots$ when there exists a relation of the form

$$\overline{\underline{a}}{}^{p}_{n}=\overline{\underline{b}}{}^{q}_{n}\,\overline{\underline{\beta}}{}_{q}+\dots+\overline{\underline{c}}{}^{r}_{n}\,\overline{\underline{\gamma}}{}_{r}+\overline{\underline{d}}{}^{s}_{n}\,\overline{\underline{\delta}}{}_{s}+\dots, \qquad \dots\dots\dots(3)$$

and when moreover $\overline{\underline{a}}{}^{p}_{n}\neq 0$. This is the case when the rank of A is not zero, and when moreover every vertical row of A is connected with the vertical rows of the remaining matrices $B, \ldots C, D, \ldots$. If A has rank 0, then it is not vertically connected with the remaining matrices, although every vertical row of A is connected with the vertical rows of the remaining matrices.

We can in a similar way define *horizontal connections* between a number of matrices of the forms

$$A' = [a]_p^n, \quad B' = [b]_q^n, \ldots C' = [c]_r^n, \quad D' = [d]_s^n, \ldots.$$

Ex. v. If

$$\phi = \left[a, b, \ldots c, d, \ldots \right]_n^{p, q, \ldots r, s, \ldots}, \quad \phi' = \left[b, \ldots c, d, \ldots \right]_n^{q, \ldots r, s, \ldots},$$

so that ϕ is a join of $A, B, \ldots C, D, \ldots$ and ϕ' is a join of $B, \ldots C, D, \ldots$, the equations (2) and (3) are

$$\phi \begin{bmatrix} \alpha \\ \beta \\ \vdots \\ \gamma \\ \delta \\ \vdots \end{bmatrix}_{p, q, \ldots r, s, \ldots} = 0, \quad \text{and} \quad [a]_n^p = \phi' \begin{bmatrix} \beta \\ \vdots \\ \gamma \\ \delta \\ \vdots \end{bmatrix}_{q, \ldots r, s, \ldots}.$$

Ex. vi. The matrices $A, B, \ldots C, D, \ldots$ are vertically unconnected or vertically connected according as the rank of their join ϕ is or is not equal to (i.e. according as it is equal to or less than) the sum of the ranks of $A, B, \ldots C, D, \ldots$.

In the particular case when $A, B, \ldots C, D, \ldots$ are all undegenerate, they are vertically unconnected or vertically connected according as their join ϕ is undegenerate or degenerate.

Ex. vii. The matrix A has a vertical connection with the remaining matrices $B, \ldots C, D, \ldots$ when and only when it has a connection with the join ϕ' of $B, \ldots C, D, \ldots$; also it is vertically connected with the remaining matrices $B, \ldots C, D, \ldots$ when and only when it is vertically connected with the join ϕ' of $B, \ldots C, D, \ldots$.

Ex. viii. There is a vertical connection between the matrices $A, B, \ldots$ and the matrices $C, D, \ldots$ when and only when there is a vertical connection between the joins of $A, B, \ldots$ and the joins of $C, D, \ldots$.

2. *Connections between spacelets.*

If
$$A = [a]_n^p, \quad B = [b]_n^q, \ldots C = [c]_n^r, \quad D = [d]_n^s, \ldots$$

are a number of matrices of ranks $\alpha, \beta, \ldots \gamma, \delta, \ldots$, the connections between the spacelets

$$\omega_\alpha \equiv [a]_n^p, \quad \omega_\beta \equiv [b]_n^q, \ldots \omega_\gamma \equiv [c]_n^r, \quad \omega_\delta \equiv [d]_n^s, \ldots \qquad \text{(4)}$$

will be defined to be the same as the vertical connections between the matrices $A, B, \ldots C, D, \ldots$. If some of the spacelets have zero ranks, then the connections between them are the same as the connections between those whose ranks are not zero. If all the spacelets have zero ranks, then they are unconnected. In all cases the spacelets $\omega_\alpha, \omega_\beta, \ldots \omega_\gamma, \omega_\delta, \ldots$ are *unconnected* when and only when their join has rank

$$\alpha + \beta + \ldots + \gamma + \delta + \ldots.$$

We can give geometrical interpretations to the definitions of connection.

There is a *connection between the spacelets* (4) when and only when one of the spacelets has a point in common with (or intersects) the join of the remaining spacelets ; and these spacelets are *unconnected* when and only when no one of them has a point in common with the join of the others.

The spacelet ω_a *has a connection with* the remaining spacelets when and only when it has a point in common with the join of the remaining spacelets, and this is only possible when ω_a is not a spacelet of zero rank; and the spacelet ω_a *is connected with* (or has complete connection with) the remaining spacelets when and only when its rank is not 0 and all its points lie in the join of the remaining spacelets. If ω_a has rank 0, i.e. if $a=0$, then ω_a is not connected with the join of the remaining spacelets, although it necessarily lies in that join.

There is a connection between the spacelets $\omega_\alpha, \omega_\beta, \ldots$ and the spacelets $\omega_\gamma, \omega_\delta, \ldots$ when and only when the join of $\omega_\alpha, \omega_\beta, \ldots$ and the join of $\omega_\gamma, \omega_\delta, \ldots$ have a point or points in common, i.e. when and only when these two joins intersect.

Ex. ix. Two spacelets ω_α and ω_β are unconnected when and only when they have no point in common, i.e. when and only when they do not intersect. If one of them has zero rank, then they are necessarily unconnected.

Ex. x. The three spacelets $\omega_p, \omega_q, \omega_r$ are unconnected when and only when ω_q and ω_r are two non-intersecting spacelets whose join does not intersect ω_p.

Ex. xi. *Let ω_r and ω_s be two spacelets which lie in ω_p, and let ω_u be a spacelet which lies in ω_r and does not intersect ω_s. Then ω_x will be a spacelet lying in ω_p which does not intersect ω_s and has ω_u for its complete intersection with ω_r when and only when it is the join of ω_u with a spacelet ω_{x-u} lying in ω_p which does not intersect ω_r and also does not intersect the join ω_{u+s} of ω_u and ω_s.*

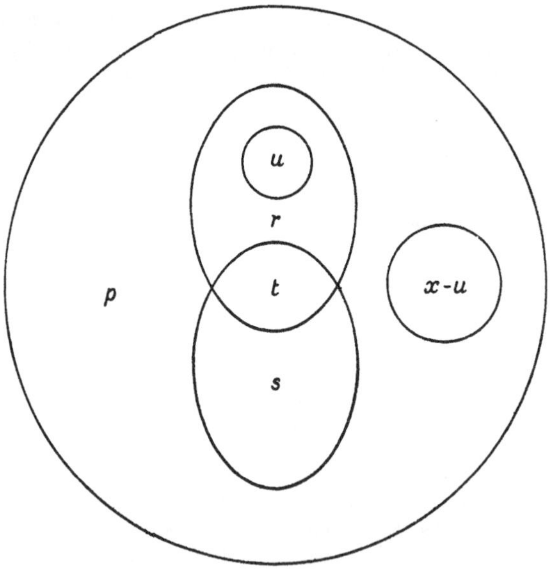

First let ω_x be a spacelet lying in ω_p which does not intersect ω_s and has ω_u for its complete intersection with ω_r. Then we can represent ω_x as the join of ω_u with a spacelet ω_{x-u} which does not intersect ω_u. Because ω_x lies in ω_p, therefore ω_{x-u} lies in ω_p;

because ω_u is the complete intersection of ω_x and ω_r, therefore ω_{x-u} does not intersect ω_r; and because ω_x does not intersect ω_s, therefore ω_u, ω_{x-u}, ω_s are unconnected, i.e. ω_{x-u} does not intersect the join of ω_u and ω_s.

Next let ω_{x-u} be any spacelet lying in ω_p which does not intersect ω_r and does not intersect the join of ω_u and ω_s. Then the join of the two non-intersecting spacelets ω_u and ω_{x-u} is a spacelet ω_x lying in ω_p. From Ex. v of § 139 we see that ω_u is the complete intersection of ω_x with ω_r; and because ω_{x-u} does not intersect the join of the two non-intersecting spacelets ω_u and ω_s, therefore ω_{x-u}, ω_u, ω_s are unconnected, i.e. ω_x does not intersect ω_s.

Ex. xii. *If ω_r and ω_s are two spacelets lying in ω_p whose complete intersection is ω_t, and if ω_x is a spacelet lying in ω_p which does not intersect ω_s and has a complete intersection ω_u of rank u with ω_r, then the possible values of u are those consistent with the conditions*

$$u \nless 0, \quad u+t \ngtr r;$$

and when u or ω_u is given, the possible values of x are those consistent with the conditions

$$x \nless u, \quad x+r-u \ngtr p, \quad x+s \ngtr p.$$

For in Ex. xi we see from Theorem III of § 139 that when ω_u is given, the possible values of $x-u$ or x are those consistent with the conditions

$$x-u \nless 0, \quad (x-u)+r \ngtr p, \quad (x-u)+(u+s) \ngtr p.$$

CHAPTER XVI

EQUIGRADENT TRANSFORMATIONS OF A MATRIX WHOSE ELEMENTS ARE CONSTANTS

[The transformations of this chapter correspond to the ordinary linear transformations of bilinear and quadratic forms. In § 141 equigradent transformations of any matrix are defined, and some of the general properties of such transformations are described. In §§ 142—5 certain special unitary and non-unitary equigradent transformations of matrices with constant elements are considered, the most general of these transformations being those given in §§ 143 and 144. The remaining articles deal with the reduction of matrices with constant elements to standard forms by equigradent transformations, symmetric matrices being specially considered in § 147, real symmetric matrices in §§ 148 and 149, and skew-symmetric matrices in § 150. The signants and signature of a real symmetric matrix are defined in § 148, and in § 149 such a matrix is called indefinite or definite according as it has or has not both positive and negative signants.]

§ 141. Equigradent transformations of a matrix.

1. *Equigradent transformations.*

Any equation of the form

$$[b]_m^n = [h]_m^r [a]_r^s [k]_s^n \quad \text{..........................(A)}$$

in which $[h]_m^r$ and $[k]_s^n$ are undegenerate matrices *with constant elements* having ranks r and s equal to their respective passivities will be called an *equigradent transformation* converting the matrix $[a]_r^s$ into the matrix $[b]_m^n$. This definition holds good both when $[a]_r^s$ and $[b]_m^n$ are matrices whose elements are constants and when they are matrices whose elements are rational integral functions of certain variables. It will be shown in a later chapter that an equigradent transformation of a matrix whose elements are rational integral functions of certain variables leaves the minimum degrees of connection of the matrix unaltered; and it is on account of this fact that the term 'equigradent' is applied to the transformations (A).

An equigradent transformation of the form (A) can only exist when $m \nless r$ and $n \nless s$.

When (A) is an equigradent transformation, we can determine an inverse prefactor $\overline{\underline{H}}{}_r^m$ of $[h]_m^r$ and an inverse postfactor $\overline{\underline{K}}{}_n^s$ of $[k]_s^n$, these being undegenerate matrices of ranks r and s with constant elements such that $\overline{\underline{H}}{}_r^m [h]_m^r = [1]_r^r$, $[k]_s^n \overline{\underline{K}}{}_n^s = [1]_s^s$; and we deduce from (A) the inverse transformation

$$[a]_r^s = \overline{\underline{H}}{}_r^m [b]_m^n \overline{\underline{K}}{}_n^s, \quad \text{.........................(A')}$$

which (unless $r = m$ and $s = n$) is not equigradent.

When the elements of $[h]_m^r$ and $[k]_s^n$ all lie in a domain of rationality Ω, we shall call (A) a *transformation in* Ω. We can in this case choose the inverse matrices $\overline{\underline{H}}{}_r^m$ and $\overline{\underline{K}}{}_n^s$ so that their elements also all lie in Ω, and then (A′) is also a transformation in Ω. In particular (A) is a *real transformation* when all the elements of $[h]_m^r$ and $[k]_s^n$ are real, and in this case we can choose the inverse matrices so that (A′) is also a real transformation.

2. *Equigradent matrices.*

Two matrices will be said to be *equigradent* with one another when both of them can be converted into the same matrix by equigradent transformations; in particular they are equigradent with one another when one of them can be converted into the other by an equigradent transformation. Thus the two matrices $[a]_r^s$ and $[b]_m^n$ are equigradent with one another when and only when there exists an equation of the form

$$[p]_u^m [b]_m^n [q]_n^v = [h]_u^r [a]_r^s [k]_s^v, \quad \text{.....................(B)}$$

where $[p]_u^m$, $[q]_n^v$, $[h]_u^r$, $[k]_s^v$ are undegenerate matrices with constant elements having ranks m, n, r, s equal to their respective passivities. This definition also holds good both when the elements of $[a]_r^s$ and $[b]_m^n$ are constants and when they are rational integral functions of certain variables.

Two equigradent matrices necessarily have the same rank.

If the elements of $[a]_r^s$ and $[b]_m^n$ are constants, these two matrices are equigradent (see Ex. ii of § 144) when and only when there exists an equation of the form (B) in which u is the larger of the two numbers m and r, and v is the larger of the two numbers n and s; also when and only when they have the same rank.

3. *Symmetric equigradent transformations.*

An equation of the form

$$[b]_m^m = \overline{\underline{h}}{}_m^r [a]_r^r [h]_r^m \quad \text{.........................(C)}$$

in which $[h]^m_r$ and $\overline{\underline{h}}{}^r_m$ are two mutually conjugate undegenerate matrices of rank r with constant elements will be called a *symmetric equigradent transformation* converting the square matrix $[a]^r_r$ into the square matrix $[b]^m_m$. Such an equation can only exist when $m \not< r$. If $\overline{\underline{H}}{}^r_m$ is an inverse postfactor of $[h]^m_r$, so that $[h]^m_r \overline{\underline{H}}{}^r_m = [1]^r_r$, we deduce from (C) the inverse transformation

$$[a]^r_r = [H]^m_r [b]^m_m \overline{\underline{H}}{}^r_m, \quad \text{..........................(C}')$$

which is equigradent when and only when $r = m$.

When (C) is a symmetric equigradent transformation in Ω, we can choose the inverse matrix $\overline{\underline{H}}{}^r_m$ so that (C′) also is a transformation in Ω.

If $m \not< r$, two square matrices $A = [a]^r_r$, $B = [b]^m_m$ will be said to be *symmetrically equigradent* when there exists a symmetric equigradent transformation of the form (C) converting A into B. This definition does not assert that A and B are symmetrically equigradent whenever they can both be converted into the same matrix by symmetric equigradent transformations.

NOTE 1. *Equigradent transformations of a matrix into a similar matrix.*

An equigradent transformation converting $[a]^n_m$ into a similar matrix $[b]^n_m$ has the form

$$[b]^n_m = [h]^m_m [a]^n_m [k]^n_n, \quad \text{..................................(D)}$$

where $[h]^m_m$ and $[k]^n_n$ are undegenerate square matrices with constant elements; and if $\overline{\underline{H}}{}^m_m$ and $\overline{\underline{K}}{}^n_n$ are the inverses of $[h]^m_m$ and $[k]^n_n$, we deduce from (D) the equivalent inverse transformation

$$[a]^n_m = \overline{\underline{H}}{}^m_m [b]^n_m \overline{\underline{K}}{}^n_n, \quad \text{..................................(D}')$$

which is an equigradent transformation converting $[b]^n_m$ into $[a]^n_m$.

When (D) is a transformation in Ω, Ω being any domain of rationality, then (D′) also is a transformation in Ω.

If the elements of the two similar matrices $[a]^n_m$ and $[b]^n_m$ are constants, these two matrices are equigradent when and only when there exists an equigradent transformation of the form (D).

Two similar matrices whose elements are constants are equigradent when and only when they are equipotent according to the definitions of Chapter XXII. We might therefore, as in Ex. xii of § 71, call them equipotent matrices. But we shall in future reserve the term 'equipotent' for functional matrices, and not apply it to matrices whose elements are constants.

A symmetric equigradent transformation converting the square matrix $[a]_m^m$ into the similar square matrix $[b]_m^m$ has the form

$$[b]_m^m = \underline{\overline{h}}{}_m^m [a]_m^m [h]_m^m, \quad \text{......(E)}$$

where $[h]_m^m$ is an undegenerate square matrix with constant elements. If $\underline{\overline{H}}{}_m^m$ is the inverse of $[h]_m^m$, we deduce from (E) the equivalent inverse transformation

$$[a]_m^m = [H]_m^m [b]_m^m \underline{\overline{H}}{}_m^m, \quad \text{......(E')}$$

which is a symmetric equigradent transformation converting $[b]_m^m$ into $[a]_m^m$. When (E) is a transformation in the domain of rationality Ω, then (E′) is also a transformation in Ω.

If the elements of $[a]_m^m$ and $[b]_m^m$ are constants, these two matrices are symmetrically equigradent when and only when there exists a symmetric equigradent transformation of the form (E).

NOTE 2. *Composition of equigradent transformations.*

Any number of equigradent transformations applied in succession to a matrix are together equivalent to a single resultant equigradent transformation applied to that matrix. For if

$$[a']_r^s = [h']_r^u [a]_u^v [k']_v^s, \quad [a'']_p^q = [h'']_p^r [a']_r^s [k'']_s^q, \quad [a''']_m^n = [h''']_m^p [a'']_p^q [k''']_q^n$$

are equigradent transformations converting

$$[a]_u^v \text{ into } [a']_r^s, \quad [a']_r^s \text{ into } [a'']_p^q, \quad [a'']_p^q \text{ into } [a''']_m^n$$

respectively, we have

$$[a''']_m^n = [h]_m^u [a]_u^v [k]_v^n, \quad \text{......(1)}$$

where $$[h]_m^u = [h''']_m^p [h'']_p^r [h']_r^u, \quad [k]_v^n = [k']_v^s [k'']_s^q [k''']_q^n.$$

By Theorem I of § 131 the matrix $[h]_m^u$ has the same rank as $[h'']_p^r [h']_r^u$, and this again has the same rank as $[h']_r^u$. Therefore $[h]_m^u$ has rank u. Similarly $[k]_v^n$ has rank v. Consequently (1) is an equigradent transformation converting $[a]_u^v$ into $[a''']_m^n$.

Clearly any number of symmetric equigradent transformations applied in succession to a matrix are together equivalent to a single resultant symmetric equigradent transformation applied to that matrix.

If all the component equigradent transformations are transformations in Ω, Ω being any domain of rationality, then the resultant equigradent transformation is also a transformation in Ω. In particular if the component equigradent transformations are all real, then the resultant equigradent transformation is real.

NOTE 3. *Elementary equigradent transformations.*

The following will be called elementary equigradent transformations of a matrix:

(a) The insertion of an additional horizontal or vertical row of 0's in any position.

(b) The interchange of any two parallel rows.

(c) The addition to any row of any parallel row multiplied by a scalar constant.

(d) The multiplication of all the elements of any horizontal or vertical row by a non-zero scalar constant.

That transformations of the type (a) are equigradent will be clear when we observe that

$$\begin{bmatrix} a_1 & 0 & b_1 & c_1 \\ a_2 & 0 & b_2 & c_2 \end{bmatrix} = \begin{bmatrix} a_1 & b_1 & c_1 \\ a_2 & b_2 & c_2 \end{bmatrix} \begin{bmatrix} 1 & 0 & 0 & 0 \\ 0 & 0 & 1 & 0 \\ 0 & 0 & 0 & 1 \end{bmatrix}, \quad \begin{bmatrix} a_1 & b_1 & c_1 \\ 0 & 0 & 0 \\ a_2 & b_2 & c_2 \end{bmatrix} = \begin{bmatrix} 1 & 0 \\ 0 & 0 \\ 0 & 1 \end{bmatrix} \begin{bmatrix} a_1 & b_1 & c_1 \\ a_2 & b_2 & c_2 \end{bmatrix}.$$

All transformations of the types (b), (c), (d) are of the form (D) in Note 1, and convert any matrix into a similar matrix.

A transformation of the type (b) which consists in the interchange of the ith and jth horizontal rows has the form $[b]_m^n = [h]_m^m [a]_m^n$, where $[h]_m^m$ is formed from $[1]_m^m$ by the interchange of its ith and jth horizontal (or vertical) rows, and the inverse transformation is $[a]_m^n = [h]_m^m [b]_m^n$.

A transformation of the type (b) which consists in the interchange of the ith and jth vertical rows has the form $[b]_m^n = [a]_m^n [k]_n^n$, where $[k]_n^n$ is formed from $[1]_n^n$ by the interchange of its ith and jth vertical (or horizontal) rows, and the inverse transformation is $[a]_m^n = [b]_m^n [k]_n^n$.

A transformation of the type (c) has one of the forms

$$[b]_m^n = [h]_m^m [a]_m^n, \quad [b]_m^n = [a]_m^n [k]_n^n,$$

where $[h]_m^m$ is formed from $[1]_m^m$ by replacing one of its zero elements by a scalar constant σ, and $[k]_n^n$ is formed from $[1]_n^n$ by replacing one of its zero elements by a scalar constant τ. The inverse transformations have the same forms, and to obtain them we replace σ by $-\sigma$ and τ by $-\tau$ respectively, and interchange $[a]_m^n$ and $[b]_m^n$.

A transformation of the type (d) has one of the forms

$$[b]_m^n = [h]_m^m [a]_m^n, \quad [b]_m^n = [a]_m^n [k]_n^n,$$

where $[h]_m^m$ and $[k]_n^n$ are formed from $[1]_m^m$ and $[1]_n^n$ respectively by replacing one of the diagonal elements 1 by a non-zero scalar constant κ. The inverse transformations have the same forms, and to obtain them we replace κ by $\frac{1}{\kappa}$, and interchange $[a]_m^n$ and $[b]_m^n$.

NOTE 4. *Derangements, unitary transformations and quasi-scalar transformations.*

The resultant of any number of transformations of the type (a) in Note 3 is an equigradent transformation of the form $[b]_m^n = [h]_m^r [a]_r^s [k]_s^n$, where $[h]_m^r$ is derived from $[1]_r^r$ by the insertion of $m-r$ horizontal rows of 0's, and $[k]_s^n$ is derived from $[1]_s^s$ by the insertion of $n-s$ vertical rows of 0's.

The resultant of any number of successive transformations of the type (b) in Note 3 will be called a *derangement*. Since all such transformations convert $[a]_m^n$ into one of its derangements, and all derangements of $[a]_m^n$ can be obtained by a succession of such transformations, we see that a derangement is any equigradent transformation of the form $[b]_m^n = [h]_m^m [a]_m^n [k]_n^n$ in which $[h]_m^m$ and $[k]_n^n$ are derangements of $[1]_m^m$ and $[1]_n^n$.

respectively. In this case the inverses of $[h]^m_m$ and $[k]^n_n$ are identical with their conjugates. Consequently the inverse transformation is $[a]^n_m = \underline{\overline{h}}^m_m \, [b]^n_m \, \underline{\overline{k}}^n_n$, and is also a derangement. If $[a_{uv}]^n_m$ is any derangement of $[a]^n_m$, so that $[u_1 u_2 \dots u_m]$ and $[v_1 v_2 \dots v_n]$ are derangements of $[1\,2 \dots m]$ and $[1\,2 \dots n]$ respectively, we have the two mutually inverse transformations

$$[a_{uv}]^n_m = [h]^m_m [a]^n_m [k]^n_n, \quad [a]^n_m = \underline{\overline{h}}^m_m [a_{uv}]^n_m \underline{\overline{k}}^n_n,$$

where the 1st, 2nd, ... mth horizontal rows of $[h]^m_m$ are the u_1th, u_2th, ... u_mth horizontal rows of $[1]^m_m$, and the 1st, 2nd, ... nth vertical rows of $[k]^n_n$ are the v_1th, v_2th, ... v_nth vertical rows of $[1]^n_n$. The resultant of any number of successive derangements is also a derangement.

By a *unitary equigradent transformation* will be meant one which is a resultant of a number of successive transformations of the type (c) in Note 3. The inverse of such a transformation and the resultant of any number of such transformations are again unitary equigradent transformations.

If $[b]^n_m = [h]^m_m [a]^n_m [k]^n_n$ is a unitary equigradent transformation, we must have $(h)^m_m = 1$ and $(k)^n_n = 1$.

Conversely if both $(h)^m_m$ and $(k)^n_n$ have one of the values 1 and -1, then (see Ex. xi of § 146) at least one of the transformations $[b]^n_m = \pm [h]^m_m [a]^n_m [k]^n_n$ is compounded of derangements and unitary equigradent transformations.

A transformation of $[a]^n_m$ is the resultant of a number of successive transformations of the type (d) in Note 3 when and only when it has the form

$$[b]^n_m = {}^1[h]_m [a]^n_m \, {}^1[k]_n,$$

where ${}^1[h]_m$ and ${}^1[k]_n$ are quasi-scalar matrices whose diagonal elements $h_1, h_2, \dots h_m$ and $k_1, k_2, \dots k_n$ are all non-zero constants. Such transformations will be called *undegenerate quasi-scalar transformations*. The inverse transformation is obtained by replacing $h_1, h_2, \dots h_m$ and $k_1, k_2, \dots k_n$ by their reciprocals and interchanging $[a]^n_m$ and $[b]^n_m$; and it is also an undegenerate quasi-scalar transformation. The resultant of any number of such transformations is also of the same form.

In the particular case when $h_1 = h_2 = \dots = h_m$ and $k_1 = k_2 = \dots = k_n$, the above transformation becomes

$$[b]^n_m = \rho [a]^n_m,$$

where ρ is a non-zero constant. This may be called a *scalar transformation*.

When $h_1, h_2, \dots h_m$ and $k_1, k_2, \dots k_n$ are not all non-zero quantities, we have a *degenerate quasi-scalar transformation*.

NOTE 5. *Equigradent transformations of a compartite matrix.*

$$\text{Let} \qquad \phi = \begin{bmatrix} a, & 0, & \dots & 0 \\ 0, & b, & \dots & 0 \\ \dots & \dots & \dots & \dots \\ 0, & 0, & \dots & c \end{bmatrix}^{p,\, q,\, \dots\, r}_{l,\, m,\, \dots\, n}, \quad \psi = \begin{bmatrix} \alpha, & 0, & \dots & 0 \\ 0, & \beta, & \dots & 0 \\ \dots & \dots & \dots & \dots \\ 0, & 0, & \dots & \gamma \end{bmatrix}^{P,\, Q,\, \dots\, R}_{L,\, M,\, \dots\, N}$$

be two compartite matrices having the same number of parts. Then if

$$[a]_L^P = [h]_L^l\,[a]_l^p\,[k]_p^P,\quad [\beta]_M^Q = [h']_M^m\,[b]_m^q\,[k']_q^Q,\quad \ldots\quad [\gamma]_N^R = [h'']_N^n\,[c]_n^r\,[k'']_r^R \quad \ldots(2)$$

are equigradent transformations converting the successive parts of ϕ into the corresponding parts of ψ,

$$\psi = \begin{bmatrix} h, & 0, & \ldots & 0 \\ 0, & h', & \ldots & 0 \\ \cdots & \cdots & \cdots & \cdots \\ 0, & 0, & \ldots & h'' \end{bmatrix}_{L,\,M,\,\ldots\,N}^{l,\,m,\,\ldots\,n} \phi \begin{bmatrix} k, & 0, & \ldots & 0 \\ 0, & k', & \ldots & 0 \\ \cdots & \cdots & \cdots & \cdots \\ 0, & 0, & \ldots & k'' \end{bmatrix}_{p,\,q,\,\ldots\,r}^{P,\,Q,\,\ldots\,R} \quad \ldots\ldots\ldots\ldots\ldots(3)$$

is an equigradent transformation converting ϕ into ψ.

Conversely if (3) is an equigradent transformation converting ϕ into ψ, then all the transformations (2) are equigradent.

These results follow from the theorem given in § 100.

Moreover (3) is a derangement or a unitary transformation or a quasi-scalar transformation when and only when (2) are all derangements, or all unitary transformations or all quasi-scalar transformations respectively.

Thus if ϕ and ψ are compartite matrices having the same number of parts, and if each part of ϕ can be converted into a corresponding part of ψ by an equigradent transformation in Ω, Ω being any domain of rationality, then ϕ can be converted into ψ by an equigradent transformation in Ω.

Similarly if ϕ and ψ are square compartite matrices whose parts are all square and the same in number, and if each part of ϕ can be converted into a corresponding part of ψ by a symmetric equigradent transformation in Ω, then ϕ can be converted into ψ by a symmetric equigradent transformation in Ω.

NOTE 6. *Correspondences between equigradent transformations of a matrix with constant elements and linear transformations of an algebraic form.*

Every equigradent transformation of a matrix with constant elements corresponds to a linear transformation of an algebraic bilinear form.

For if

$$[b]_m^n = \overline{\underline{h}}_m^p\,[a]_p^q\,[k]_q^n$$

is an equigradent transformation converting $[a]_p^q$ into $[b]_m^n$, then

$$\overline{\underline{x}}_p = [h]_p^m\,\overline{\underline{u}}_m,\quad \overline{\underline{y}}_q = [k]_q^n\,\overline{\underline{v}}_n$$

are linear transformations converting the bilinear form

$$S = \det [x]_p\,[a]_p^q\,\overline{\underline{y}}_q = \Sigma a_{ij} x_i y_j$$

into

$$T = \det [u]_m\,[b]_m^n\,\overline{\underline{v}}_n = \Sigma b_{ij} u_i v_j.$$

This correspondence is reversible when the two matrices are similar.

Again every symmetric equigradent transformation of a symmetric matrix with constant elements corresponds to a linear transformation of an algebraic quadratic form.

For if
$$[b]_m^m = \overline{\underline{h}}_m^p \, [a]_p^p \, [h]_p^m$$
is a symmetric equigràdent transformation converting the symmetric matrix $[a]_p^p$ into the symmetric matrix $[b]_m^m$, then
$$\overline{\underline{x}}_p = [h]_p^m \, \overline{\underline{y}}_m$$
is a linear transformation of variables converting the quadratic form
$$S = \det [x]_p \, [a]_p^p \, \overline{\underline{x}}_p = \Sigma a_{ii} x_i^2 + 2\Sigma a_{ij} x_i x_j$$
into
$$T = \det [y]_m \, [b]_m^m \, \overline{\underline{y}}_m = \Sigma b_{ii} y_i^2 + 2\Sigma b_{ij} y_i y_j,$$
where $j \neq i$.

This correspondence is reversible when the two symmetric matrices have the same order.

NOTE 7. *Completion of any two mutually inverse matrices.*

Let $[h]_r^m$ and $[k]_m^r$ be two mutually inverse undegenerate matrices of rank r whose elements are constants lying in a domain of rationality Ω, so that
$$[h]_r^m \, [k]_m^r = [1]_r^r .$$

Then if $r < m$, we can by the addition of final active rows to $[h]_r^m$ and $[k]_m^r$ form two mutually inverse undegenerate square matrices $[h]_m^m$ and $[k]_m^m$ whose elements are constants lying in Ω.

Since $[h]_r^m$ has rank r, we can determine a matrix $[v]_m^{m-r}$ of rank $m-r$ whose elements are constants in Ω, and which satisfies the equation
$$[h]_r^m \, [v]_m^{m-r} = 0,$$
as is evident when we solve this equation for $[v]_m^{m-r}$.

Then the square matrix $[k]_m^m = [k, v]_m^{r,\, m-r}$ is undegenerate; for if there exists a relation of the form
$$[k, v]_m^{r,\, m-r} \, \overline{\underline{\begin{matrix}\lambda \\ \mu\end{matrix}}}_{r,\, m-r} = 0,$$
we see by prefixing $[h]_r^m$ that
$$[h]_r^m \, [k]_m^r \, \overline{\underline{\lambda}}_r = 0, \quad \text{i.e. } [1]_r^r \, \overline{\underline{\lambda}}_r = 0, \text{ or } \overline{\underline{\lambda}}_r = 0,$$
and this shows that the relation has the form $[v]_m^{m-r} \, \overline{\underline{\mu}}_{m-r} = 0$, which can only be satisfied when $\overline{\underline{\mu}}_{m-r} = 0$.

Having formed $[k]_m^m$ in this manner, we can determine a matrix $[u]_{m-r}^m$ whose elements are constants lying in Ω, and which satisfies the equation
$$[u]_{m-r}^m \, [k]_m^m = [0, 1]_{m-r}^{r,\, m-r}.$$

Then writing
$$[h]^m_m = \begin{bmatrix} h \\ u \end{bmatrix}^m_{r,\, m-r},$$

we have
$$[h]^m_m [k]^m_m = \begin{bmatrix} h \\ u \end{bmatrix}^m_{r,\, m-r} [k,\ v]^{r,\, m-r}_m = \begin{bmatrix} 1, & 0 \\ 0, & 1 \end{bmatrix}^{r,\, m-r}_{r,\, m-r}$$
$$= [1]^m_m = [k]^m_m [h]^m_m.$$

NOTE 8. *Conversion of any equigradent transformation and its inverse into two mutually inverse equigradent transformations of similar matrices.*

Let
$$[h]^r_m [e]^s_r [k]^n_s = [a]^n_m, \quad \overline{\underline{H}}{}^m_r [a]^n_m \overline{\underline{K}}{}^s_n = [e]^s_r \quad \ldots\ldots\ldots\ldots\ldots\ldots\ldots(4)$$

be respectively an equigradent transformation and its inverse, so that $[h]^r_m$, $\overline{\underline{H}}{}^m_r$ and $[k]^n_s$, $\overline{\underline{K}}{}^s_n$ are two pairs of mutually inverse undegenerate matrices whose elements lie in the domain of rationality Ω, and which are such that
$$\overline{\underline{H}}{}^m_r [h]^r_m = [1]^r_r, \quad [k]^n_s \overline{\underline{K}}{}^s_n = [1]^s_s.$$

If $[h]^r_m$ and $[k]^n_s$ are given, $\overline{\underline{H}}{}^m_r$ may be any inverse prefactor of $[h]^r_m$ which lies in Ω, and $\overline{\underline{K}}{}^s_n$ may be any inverse postfactor of $[k]^n_s$ which lies in Ω.

Then from (4) *we can derive two mutually inverse equigradent transformations in* Ω *of the forms*
$$[h]^m_m \begin{bmatrix} e, & 0 \\ 0, & 0 \end{bmatrix}^{s,\, n-s}_{r,\, m-r} [k]^n_n = [a]^n_m, \quad \overline{\underline{H}}{}^m_m [a]^n_m \overline{\underline{K}}{}^n_n = \begin{bmatrix} e, & 0 \\ 0, & 0 \end{bmatrix}^{s,\, n-s}_{r,\, m-r}, \quad \ldots\ldots\ldots(5)$$

where $[h]^m_m$, $[k]^n_n$ *and* $\overline{\underline{H}}{}^m_m$, $\overline{\underline{K}}{}^n_n$ *are formed respectively by adding final passive rows to* $[h]^r_m$, $[k]^n_s$ *and final active rows to* $\overline{\underline{H}}{}^m_r$, $\overline{\underline{K}}{}^s_n$.

By Note 7 we can form two mutually inverse undegenerate square matrices $\overline{\underline{H}}{}^m_m$, $[h]^m_m$ lying in Ω by adding final active rows to $\overline{\underline{H}}{}^m_r$, $[h]^r_m$, and two mutually inverse undegenerate square matrices $[k]^n_n$, $\overline{\underline{K}}{}^n_n$ lying in Ω by adding final active rows to $[k]^n_s$, $\overline{\underline{K}}{}^s_n$. Then the first of the equations (5) is obviously true, and we deduce the second by prefixing $\overline{\underline{H}}{}^m_m$ and postfixing $\overline{\underline{K}}{}^n_n$.

If we write $\overline{\underline{H}}{}^m_m = \overline{\underline{\begin{matrix} H \\ U \end{matrix}}}{}^m_{r,\, m-r}$, $\overline{\underline{K}}{}^n_n = \overline{\underline{K,\ V}}{}^{s,\, n-s}_n$, it follows from the second of the equations (5) that the rows which have been added to $\overline{\underline{H}}{}^m_r$ and $\overline{\underline{K}}{}^s_n$ are such that
$\overline{\underline{U}}{}^m_{m-r} [a]^n_m = 0$, $[a]^n_m \overline{\underline{V}}{}^{n-s}_n = 0$. These results follow also from the equations $\overline{\underline{U}}{}^m_{m-r} [h]^r_m = 0$, $[k]^n_s \overline{\underline{V}}{}^{n-s}_n = 0$ derived from $\overline{\underline{H}}{}^m_m [h]^m_m = [1]^m_m$, $[k]^n_n \overline{\underline{K}}{}^n_n = [1]^n_n$.

When the first of the equations (4) is a given equigradent transformation in Ω, and the values of $\overline{\underline{H}}^{m}_{r}$ and $\overline{\underline{K}}^{s}_{n}$ are not prescribed, we can use a simpler method. For if $[h]_m^m$ and $[k]_n^n$ are any two undegenerate square matrices in Ω formed by adding final vertical and horizontal rows to $[h]_m^r$ and $[k]_s^n$, and if $\overline{\underline{H}}^{m}_{m}$ and $\overline{\underline{K}}^{n}_{n}$ are their inverses, then both the equations (4) are true.

In the following examples Ω is any domain of rationality. When it is the domain of all real numbers, quantities lying in Ω and transformations in Ω are real.

Ex. i. *If* $[a]_m^n$ *can be converted into the similar matrix* $[b]_m^n$ *by an equigradent transformation in* Ω, *then* $[b]_m^n$ *can be converted into* $[a]_m^n$ *by an equigradent transformation in* Ω.

This follows from Note 1.

Ex. ii. *If the matrix A can be converted into the matrix B and the matrix B into the matrix C by equigradent transformations in* Ω, *then A can be converted into C by an equigradent transformation in* Ω.

This follows from Note 2.

Ex. iii. *If* $[a]_m^n$ *can be converted into* $[b]_\mu^\nu$ *by an equigradent transformation in* Ω, *then it can be converted into* $\begin{bmatrix} b, & 0 \\ 0, & 0 \end{bmatrix}_{\mu,\rho}^{\nu,\sigma}$ *by an equigradent transformation in* Ω.

For we have $\mu \not< m$, $\nu \not< n$, and if

$$[h]_\mu^m [a]_m^n [k]_n^\nu = [b]_\mu^\nu$$

is an equigradent transformation in Ω, then

$$\begin{bmatrix} h \\ 0 \end{bmatrix}_{\mu,\rho}^{m} [a]_m^n [k,\ 0]_n^{\nu,\sigma} = \begin{bmatrix} b, & 0 \\ 0, & 0 \end{bmatrix}_{\mu,\rho}^{\nu,\sigma}$$

is an equigradent transformation in Ω.

Ex. iv. *If* $[a]_m^n$ *can be converted into* $[b]_\mu^\nu$ *by an equigradent transformation in* Ω, *so that* $m \not> \mu$ *and* $n \not> \nu$, *and if* $m+r \not> \mu$, $n+s \not> \nu$, *then* $\begin{bmatrix} a, & 0 \\ 0, & 0 \end{bmatrix}_{m,r}^{n,s}$ *can be converted into* $[b]_\mu^\nu$ *by an equigradent transformation in* Ω.

For if
$$[h]_\mu^m [a]_m^n [k]_n^\nu = [b]_\mu^\nu$$

is an equigradent transformation in Ω, we can as in § 104 by inserting additional final vertical rows in $[h]_\mu^m$ and additional final horizontal rows in $[k]_n^\nu$ form matrices $[h]_\mu^{m+r}$ and $[k]_{n+s}^\nu$ of ranks $m+r$ and $n+s$ whose elements all lie in Ω, and then

$$[h]_\mu^{m+r} \begin{bmatrix} a, & 0 \\ 0, & 0 \end{bmatrix}_{m,r}^{n,s} [k]_{n+s}^\nu = [b]_\mu^\nu$$

is an equigradent transformation in Ω.

Ex. v. *If* $[a]_r^r$ *and* $[b]_r^r$ *are undegenerate square matrices, and if we can convert* $\begin{bmatrix} a, & 0 \\ 0, & 0 \end{bmatrix}_{r,m}^{r,n}$ *into* $\begin{bmatrix} b, & 0 \\ 0, & 0 \end{bmatrix}_{r,\mu}^{r,\nu}$ *by an equigradent transformation in* Ω, *then we can convert each of the matrices* $[a]_r^r$ *and* $[b]_r^r$ *into the other by an equigradent transformation in* Ω.

We have $\mu \not< m$, $\nu \not< n$, and

$$[h]_{r+\mu}^{r+m} \begin{bmatrix} a, & 0 \\ 0, & 0 \end{bmatrix}_{r,m}^{r,n} [k]_{r+n}^{r+\nu} = \begin{bmatrix} b, & 0 \\ 0, & 0 \end{bmatrix}_{r,\mu}^{r,\nu},$$

where the prefactor and postfactor on the left are undegenerate matrices of ranks $r+m$ and $r+n$ whose elements are constants lying in Ω. It follows by the properties of active and passive rows that

$$[b]_r^r = [h]_r^r\,[a]_r^r\,[k]_r^r.$$

Since $[b]_r^r$ is undegenerate, the matrices $[h]_r^r$ and $[k]_r^r$ are undegenerate, and the last equation is an equigradent transformation in Ω converting $[a]_r^r$ into $[b]_r^r$.

Ex. vi. *If the square matrix* $[a]_m^m$ *can be converted into the similar square matrix* $[b]_m^m$ *by a symmetric equigradent transformation in* Ω, *then* $[b]_m^m$ *can be converted into* $[a]_m^m$ *by a symmetric equigradent transformation in* Ω.

This follows from Note 1.

Ex. vii. *If the square matrix A can be converted into the square matrix B, and the square matrix B into the square matrix C by symmetric equigradent transformations in* Ω, *then A can be converted into C by a symmetric equigradent transformation in* Ω.

This follows from Note 2.

Ex. viii. *If the square matrix* $[a]_m^m$ *can be converted into the square matrix* $[b]_\mu^\mu$ *by a symmetric equigradent transformation in* Ω, *then it can be converted into the square matrix* $\begin{bmatrix} b, & 0 \\ 0, & 0 \end{bmatrix}_{\mu,\rho}^{\mu,\rho}$ *by a symmetric equigradent transformation in* Ω.

For we have $\mu \not< m$, and if

$$\overline{\underline{h}}_{\mu}^{m}\,[a]_m^m\,[h]_m^\mu = [b]_\mu^\mu$$

is a symmetric equigradent transformation in Ω, then

$$\begin{array}{c} h \\ 0 \end{array}\Bigg|_{\mu,\rho}^{m} [a]_m^m\,[h,\ 0]_m^{\mu,\rho} = \begin{bmatrix} b, & 0 \\ 0, & 0 \end{bmatrix}_{\mu,\rho}^{\mu,\rho}$$

is a symmetric equigradent transformation in Ω.

This result also follows directly from Ex. vii.

Ex. ix. *If the square matrix* $[a]_m^m$ *can be converted into the square matrix* $[b]_\mu^\mu$ *by a symmetric equigradent transformation in* Ω, *so that* $m \not> \mu$, *and if* $m+r \not> \mu$, *then the square matrix* $\begin{bmatrix} a, & 0 \\ 0, & 0 \end{bmatrix}_{m,r}^{m,r}$ *can be converted into* $[b]_\mu^\mu$ *by a symmetric equigradent transformation in* Ω.

For if
$$\overline{\underline{h}}{}^{m}_{\mu}\,[a]^{m}_{m}\,[h]^{\mu}_{m}=[b]^{\mu}_{\mu}$$
is a symmetric equigradent transformation in Ω, we can, as in § 104, by inserting additional final horizontal rows in $[h]^{\mu}_{m}$ form a matrix $[h]^{\mu}_{m+r}$ of rank $m+r$ whose elements all lie in Ω, and then
$$\overline{\underline{h}}{}^{m+r}_{\mu}\begin{bmatrix}a, & 0\\ 0, & 0\end{bmatrix}^{m,\,r}_{m,\,r}[h]^{\mu}_{m+r}=[b]^{\mu}_{\mu}$$
is a symmetric equigradent transformation in Ω.

Ex. x. *If we can convert $\phi=[a]^{m}_{m}$ into $\psi=[b]^{\mu}_{\mu}$ by a symmetric equigradent transformation in Ω, so that $m \not> \mu$, and if $m+r \not> \mu+\rho$, then we can convert $\Phi=\begin{bmatrix}a, & 0\\ 0, & 0\end{bmatrix}^{m,\,r}_{m,\,r}$ into $\Psi=\begin{bmatrix}b, & 0\\ 0, & 0\end{bmatrix}^{\mu,\,\rho}_{\mu,\,\rho}$ by a symmetric equigradent transformation in Ω.*

For by Exs. viii and ix we can convert ϕ into Ψ, and then Φ into Ψ, by symmetric equigradent transformations in Ω.

If the transformation converting ϕ into ψ is $\overline{\underline{h}}{}^{m}_{\mu}\,[a]^{m}_{m}\,[h]^{\mu}_{m}=[b]^{\mu}_{\mu}$, then the transformation converting Φ into Ψ obtained in this way has the form
$$\overline{\underline{\begin{matrix}h, & u\\ 0, & v\end{matrix}}}{}^{m,\,r}_{\mu,\,\rho}\begin{bmatrix}a, & 0\\ 0, & 0\end{bmatrix}^{m,\,r}_{m,\,r}\begin{bmatrix}h, & 0\\ u, & v\end{bmatrix}^{\mu,\,\rho}_{m,\,r}=\begin{bmatrix}b, & 0\\ 0, & 0\end{bmatrix}^{\mu,\,\rho}_{\mu,\,\rho},$$
where $\begin{bmatrix}h, & 0\\ u, & v\end{bmatrix}^{\mu,\,\rho}_{m,\,r}$ is a matrix of rank $m+r$ which can be determined by Theorem V of § 105.

Ex. xi. *If $[a]^{r}_{r}$ and $[b]^{r}_{r}$ are undegenerate square matrices, and if we can convert the square matrix $\begin{bmatrix}a, & 0\\ 0, & 0\end{bmatrix}^{r,\,m}_{r,\,m}$ into the square matrix $\begin{bmatrix}b, & 0\\ 0, & 0\end{bmatrix}^{r,\,\mu}_{r,\,\mu}$ by a symmetric equigradent transformation in Ω, then we can convert each of the matrices $[a]^{r}_{r}$ and $[b]^{r}_{r}$ into the other by a symmetric equigradent transformation in Ω.*

We have $\mu \not< m$, and
$$\overline{\underline{h}}{}^{r+m}_{r+\mu}\begin{bmatrix}a, & 0\\ 0, & 0\end{bmatrix}^{r,\,m}_{r,\,m}[h]^{r+\mu}_{r+m}=\begin{bmatrix}b, & 0\\ 0, & 0\end{bmatrix}^{r,\,\mu}_{r,\,\mu},$$
where $[h]^{r+\mu}_{r+m}$ is an undegenerate matrix of rank $r+m$ whose elements are constants lying in Ω. It follows by the properties of active and passive rows that
$$[b]^{r}_{r}=\overline{\underline{h}}{}^{r}_{r}\,[a]^{r}_{r}\,[h]^{r}_{r}.$$

Since $[b]^{r}_{r}$ is undegenerate, the matrix $[h]^{r}_{r}$ is undegenerate, and the last equation is a symmetric equigradent transformation in Ω converting $[a]^{r}_{r}$ into $[b]^{r}_{r}$.

Ex. xii. *If $[b]^{r}_{r}$ and $[c]^{r}_{r}$ are undegenerate square matrices, and if the square matrix $[a]^{p}_{p}$ can be converted*
$$\textit{both into } \begin{bmatrix}b, & 0\\ 0, & 0\end{bmatrix}^{r,\,m-r}_{r,\,m-r} \qquad \textit{and into } \begin{bmatrix}c, & 0\\ 0, & 0\end{bmatrix}^{r,\,\mu-r}_{r,\,\mu-r}$$

by symmetric equigradent transformations in Ω, *then each of the square matrices* $[b]_r^r$ *and* $[c]_r^r$ *can be converted into the other by a symmetric equigradent transformation in* Ω.

We have $r \not< p$, $p \not< m$, $r \not< m$, and we may suppose that $\mu \not< m$.

Let $$A=\begin{bmatrix} a, & 0 \\ 0, & 0 \end{bmatrix}_{p,\,\mu-p}^{p,\,\mu-p}, \quad \phi=\begin{bmatrix} b, & 0 \\ 0, & 0 \end{bmatrix}_{r,\,\mu-r}^{r,\,\mu-r}, \quad \psi=\begin{bmatrix} c, & 0 \\ 0, & 0 \end{bmatrix}_{r,\,\mu-r}^{r,\,\mu-r}.$$

By Exs. x and vi we can convert each of the matrices A and ϕ into the other and also each of the matrices A and ψ into the other by symmetric equigradent transformations in Ω. Therefore by Ex. vii we can convert each of the matrices ϕ and ψ into the other by a symmetric equigradent transformation in Ω, and the theorem now follows from Ex. xi.

Ex. xiii. The transformation $[b]_m^n=[h]_m^m[a]_m^n[k]_n^n$ in which $[h]_m^m$ and $[k]_n^n$ are matrices with constant elements having the forms

$$[h]_m^m = \begin{bmatrix} 1 & 0 & 0 & 0 & \dots & 0 \\ a_{21} & 1 & 0 & 0 & \dots & 0 \\ a_{31} & a_{32} & 1 & 0 & \dots & 0 \\ a_{41} & a_{42} & a_{43} & 1 & \dots & 0 \\ \dots & \dots & \dots & \dots & \dots & \dots \\ a_{m1} & a_{m2} & a_{m3} & a_{m4} & \dots & 1 \end{bmatrix}, \quad [k]_n^n = \begin{bmatrix} 1 & b_{12} & b_{13} & b_{14} & \dots & b_{1n} \\ 0 & 1 & b_{23} & b_{24} & \dots & b_{2n} \\ 0 & 0 & 1 & b_{34} & \dots & b_{3n} \\ 0 & 0 & 0 & 1 & \dots & b_{4n} \\ \dots & \dots & \dots & \dots & \dots & \dots \\ 0 & 0 & 0 & 0 & \dots & 1 \end{bmatrix}$$

occurs frequently in the following articles. Referring to Exs. xiii and xv of § 101, it will be seen that this transformation can be expressed as the resultant of a number of transformations of the type (c) in Note 3, and is therefore a *unitary equigradent transformation*. The inverse of such a transformation, and the resultant of any number of such transformations, are again transformations of the same form.

Ex. xiv. The results stated in Ex. xiii remain true when $[h]_m^m$ and $[k]_n^n$ have the forms

$$[h]_m^m = \begin{bmatrix} 1, & 0, & 0, & \dots & 0, & 0 \\ \sigma, & 1, & 0, & \dots & 0, & 0 \\ \sigma, & \sigma, & 1, & \dots & 0, & 0 \\ \dots & \dots & \dots & \dots & \dots & \dots \\ \sigma, & \sigma, & \sigma, & \dots & 1, & 0 \\ \sigma, & \sigma, & \sigma, & \dots & \sigma, & 1 \end{bmatrix}_{\alpha,\,\beta,\,\gamma,\,\dots\,\kappa,\,\lambda}^{\alpha,\,\beta,\,\gamma,\,\dots\,\kappa,\,\lambda}, \quad [k]_n^n = \begin{bmatrix} 1, & \tau, & \tau, & \dots & \tau, & \tau \\ 0, & 1, & \tau, & \dots & \tau, & \tau \\ 0, & 0, & 1, & \dots & \tau, & \tau \\ \dots & \dots & \dots & \dots & \dots & \dots \\ 0, & 0, & 0, & \dots & 1, & \tau \\ 0, & 0, & 0, & \dots & 0, & 1 \end{bmatrix}_{\alpha,\,\beta,\,\gamma,\,\dots\,\kappa,\,\lambda}^{\alpha,\,\beta,\,\gamma,\,\dots\,\kappa,\,\lambda},$$

where in the first matrix $\alpha, \beta, \gamma, \dots \kappa, \lambda$ are any non-zero positive integers whose sum is m; in the second matrix $\alpha, \beta, \gamma, \dots \kappa, \lambda$ are any non-zero positive integers whose sum is n; and a constituent matrix denoted by $[\sigma]_p^q$ or $[\tau]_p^q$ is any matrix with constant elements whose orders are p and q.

Ex. xv. *The inverses of the derangements of* $\begin{bmatrix} 1, & 0 \\ \sigma, & 1 \end{bmatrix}_{r,\,m-r}^{r,\,m-r}$ *and* $\begin{bmatrix} 1, & \tau \\ 0, & 1 \end{bmatrix}_{r,\,m-r}^{r,\,m-r}$.

Let $[p]_r$, $[\mu]_{m-r}$ and $[q]_r$, $[\nu]_{m-r}$ be two pairs of complementary minors of the sequence $[1\,2\,\dots\,m]$.

Then if $[a]_m^m$ *is a square matrix in which*

$$[a_{pq}]_r^r=[1]_r^r, \quad [a_{p\nu}]_r^{m-r}=0, \quad [a_{\mu q}]_{m-r}^r=[\sigma]_{m-r}^r, \quad [a_{\mu\nu}]_{m-r}^{m-r}=[1]_{m-r}^{m-r},$$

its inverse is the matrix $[b]_m^m$ *and therefore the matrix* $\overline{\underline{A}}_m^m$ *in which*

$$[b_{qp}]_r^r = [1]_r^r, \quad [b_{q\mu}]_r^{m-r} = 0, \quad [b_{\nu p}]_{m-r}^r = -[\sigma]_{m-r}^r, \quad [b_{\nu\mu}]_{m-r}^{m-r} = [1]_{m-r}^{m-r},$$

$$\overline{\underline{A_{pq}}}_r^r = [1]_r^r, \quad \overline{\underline{A_{\mu q}}}_r^{m-r} = 0, \quad \overline{\underline{A_{p\nu}}}_{m-r}^r = -[\sigma]_{m-r}^r, \quad \overline{\underline{A_{\mu\nu}}}_{m-r}^{m-r} = [1]_{m-r}^{m-r}.$$

We have
$$\begin{bmatrix} 1, & 0 \\ \sigma, & 1 \end{bmatrix}_{r,\, m-r}^{r,\, m-r} = \begin{bmatrix} a_{pq}, & a_{p\nu} \\ a_{\mu q}, & a_{\mu\nu} \end{bmatrix}_{r,\, m-r}^{r,\, m-r} = [h]_m^m [a]_m^m [k]_m^m,$$

where $[h]_m^m$ is the matrix whose 1st, 2nd, ... rth, $(r+1)$th, ... mth horizontal rows are the p_1th, p_2th, ... p_rth, μ_1th, ... μ_{m-r}th horizontal rows of $[1]_m^m$,

and $[k]_m^m$ is the matrix whose 1st, 2nd, ... rth, $(r+1)$th, ... mth vertical rows are the q_1th, q_2th, ... q_rth, ν_1th, ... ν_{m-r}th vertical rows of $[1]_m^m$.

Since $\begin{bmatrix} 1, & 0 \\ -\sigma, & 1 \end{bmatrix}_{r,\, m-r}^{r,\, m-r}$ is the inverse of $\begin{bmatrix} 1, & 0 \\ \sigma, & 1 \end{bmatrix}_{r,\, m-r}^{r,\, m-r}$, it follows that when $[b]_m^m$ is the inverse of $[a]_m^m$ we have

$$\begin{bmatrix} 1, & 0 \\ -\sigma, & 1 \end{bmatrix}_{r,\, m-r}^{r,\, m-r} = \overline{\underline{k}}_m^m [b]_m^m \overline{\underline{h}}_m^m = \begin{bmatrix} b_{qp}, & b_{q\mu} \\ b_{\nu p}, & b_{\nu\mu} \end{bmatrix}_{r,\, m-r}^{r,\, m-r};$$

for	the 1st, ... rth, $(r+1)$th, ...	mth horizontal rows of $\overline{\underline{k}}_m^m$	
are	the q_1th, ... q_rth,	ν_1th, ... ν_{m-r}th horizontal rows of $[1]_m^m$,	
and	the 1st, ... rth, $(r+1)$th, ...	mth vertical rows of $\overline{\underline{h}}_m^m$	
are	the p_1th, ... p_rth,	μ_1th, ... μ_{m-r}th vertical rows of $[1]_m^m$.	

Similarly if $[a]_m^m$ *is a square matrix in which*

$$[a_{pq}]_r^r = [1]_r^r, \quad [a_{p\nu}]_r^{m-r} = [\tau]_r^{m-r}, \quad [a_{\mu q}]_{m-r}^r = 0, \quad [a_{\mu\nu}]_{m-r}^{m-r} = [1]_{m-r}^{m-r},$$

its inverse is the matrix $[b]_m^m$ *and therefore the matrix* $\overline{\underline{A}}_m^m$ *in which*

$$[b_{qp}]_r^r = [1]_r^r, \quad [b_{q\mu}]_r^{m-r} = -[\tau]_r^{m-r}, \quad [b_{\nu p}]_{m-r}^r = 0, \quad [b_{\nu\mu}]_{m-r}^{m-r} = [1]_{m-r}^{m-r},$$

$$\overline{\underline{A_{pq}}}_r^r = [1]_r^r, \quad \overline{\underline{A_{\mu q}}}_r^{m-r} = -[\tau]_r^{m-r}, \quad \overline{\underline{A_{p\nu}}}_{m-r}^r = 0, \quad \overline{\underline{A_{\mu\nu}}}_{m-r}^{m-r} = [1]_{m-r}^{m-r}.$$

§ 142. Some special unitary equigradent transformations of a matrix with constant elements.

Throughout this article Ω denotes any domain of rationality. When it is the domain of all real numbers, scalar quantities lying in Ω are real, and transformations in Ω are real transformations.

Theorem I. *If* $\phi = [a]_m^n$ *is a matrix of rank* ρ *whose elements are constants lying in a domain of rationality* Ω, *and if* $a_{11} \neq 0$, *we can convert* ϕ *by unitary equigradent transformations in* Ω *into the compartite matrix*

$$\psi = \begin{bmatrix} a_{11}, & 0 \\ 0, & b \end{bmatrix}_{1,\, m-1}^{1,\, n-1},$$

where $[b]_{m-1}^{n-1}$ *is the matrix of rank* $\rho - 1$ *in which*

$$a_{11} b_{uv} = \begin{pmatrix} 1, 1+v \\ a \\ 1, 1+u \end{pmatrix}, \qquad a_{11} (b_{uv})_s^s = \begin{pmatrix} 1, 1+v_1, 1+v_2, \dots 1+v_s \\ a \\ 1, 1+u_1, 1+u_2, \dots 1+u_s \end{pmatrix} \quad \text{.......(1)}$$

The first equation in (1) completely defines the matrix $[b]_{m-1}^{n-1}$, and the second equation follows from the first by the identity of § 110.

First proof of Theorem I.

Putting

$$\sigma_{i1} = \frac{a_{1+i,1}}{a_{11}}, \qquad \tau_{1j} = \frac{a_{1,1+j}}{a_{11}}, \qquad (i = 1, 2, \dots m-1;\ j = 1, 2, \dots n-1),$$

we will perform on ϕ in succession and in any order the following two sets of operations:

(1) *We will add to the* $(1+i)$*th horizontal row the first horizontal row multiplied by* $-\sigma_{i1}$, *doing this for the values* 1, 2, ... $m-1$ *of* i.

(2) *We will add to the* $(1+j)$*th vertical row the first vertical row multiplied by* $-\tau_{1j}$, *doing this for the values* 1, 2, ... $n-1$ *of* j.

These are unitary equigradent transformations in Ω which are together equivalent to the single resultant transformation

$$\begin{bmatrix} 1 & , 0, 0, \dots 0 \\ -\sigma_{11} & , 1, 0, \dots 0 \\ -\sigma_{21} & , 0, 1, \dots 0 \\ \dots & \dots \\ -\sigma_{m-1,1}, & 0, 0, \dots 1 \end{bmatrix} [a]_m^n \begin{bmatrix} 1, & -\tau_{11}, & -\tau_{12}, & \dots & -\tau_{1,n-1} \\ 0, & 1, & 0, & \dots & 0 \\ 0, & 0, & 1, & \dots & 0 \\ \dots & \dots & \dots & \dots & \dots \\ 0, & 0, & 0, & \dots & 1 \end{bmatrix} = \begin{bmatrix} a_{11}, & 0 \\ 0, & b \end{bmatrix}_{1,\, m-1}^{1,\, n-1},$$

or

$$\begin{bmatrix} 1, & 0 \\ -\sigma, & 1 \end{bmatrix}_{1,\, m-1}^{1,\, m-1} [a]_m^n \begin{bmatrix} 1, & -\tau \\ 0, & 1 \end{bmatrix}_{1,\, n-1}^{1,\, n-1} = \begin{bmatrix} a_{11}, & 0 \\ 0, & b \end{bmatrix}_{1,\, m-1}^{1,\, n-1}, \quad \text{..................(A)}$$

where $$b_{i-1,j-1} = a_{ij} - \sigma_{i-1,1} a_{1j} - \tau_{1,j-1} a_{i1} + \sigma_{i-1,1} \tau_{1,j-1} a_{11} = a_{ij} - \frac{a_{i1} a_{1j}}{a_{11}},$$

i.e. $$a_{11} b_{i-1,j-1} = \begin{vmatrix} a_{11} & a_{1j} \\ a_{i1} & a_{ij} \end{vmatrix} = \begin{pmatrix} 1j \\ a \\ 1i \end{pmatrix}.$$

Thus the matrix $[b]_{m-1}^{n-1}$ in (A) is that defined by the first equation in (1). We can see this more easily by equating correspondingly formed minor determinants of orders 2 and $s+1$ on both sides of (A), the determinants of the right-hand side being those which contain a_{11}; for on doing this we obtain both the equations (1).

That $[b]_{m-1}^{n-1}$ has rank $\rho - 1$ follows from the theorem of § 100 when we observe that the matrix on the right in (A) has the same rank ρ as $[a]_m^n$. It also follows from the second equation in (1); for this equation shows that all minor determinants of $[b]_{m-1}^{n-1}$ of order ρ vanish, and that if $\begin{pmatrix} 1, 1+v_1, \dots 1+v_{\rho-1} \\ a \\ 1, 1+u_1, \dots 1+u_{\rho-1} \end{pmatrix}$ is one of those non-vanishing minor determinants of ϕ of order ρ which contain a_{11}, then $(b_{uv})_{\rho-1}^{\rho-1}$ is a non-vanishing minor determinant of $[b]_{m-1}^{n-1}$ of order $\rho-1$.

By prefixing $\begin{bmatrix} 1, & 0 \\ \sigma, & 1 \end{bmatrix}_{1, m-1}^{1, m-1}$ and postfixing $\begin{bmatrix} 1, & \tau \\ 0, & 1 \end{bmatrix}_{1, n-1}^{1, n-1}$ on both sides of (A) we obtain the inverse equation

$$[a]_m^n = \begin{bmatrix} 1, & 0 \\ \sigma, & 1 \end{bmatrix}_{1, m-1}^{1, m-1} \begin{bmatrix} a_{11}, & 0 \\ 0, & b \end{bmatrix}_{1, m-1}^{1, n-1} \begin{bmatrix} 1, & \tau \\ 0, & 1 \end{bmatrix}_{1, n-1}^{1, n-1}. \quad \text{..................(A')}$$

Equation (A) is a unitary equigradent transformation in Ω which converts ϕ into ψ; and equation (A') is a unitary equigradent transformation in Ω which converts ψ into ϕ.

SECOND PROOF OF THEOREM I.

If $Q_{uv} = \begin{pmatrix} 1, 1+v \\ a \\ 1, 1+u \end{pmatrix}$, $b_{uv} = \frac{1}{a_{11}} Q_{uv}$, so that $[b]_{m-1}^{n-1}$ is the matrix defined in the enunciation, we have by § 116 the equations

$$a_{11} [a]_m^n = [a]_m^1 [a]_1^n + \begin{bmatrix} 0, & 0 \\ 0, & Q \end{bmatrix}_{1, m-1}^{1, n-1}, \quad \text{..........................(a)}$$

$$[a]_m^n = \frac{1}{a_{11}} [a]_m^1 [a]_1^n + \begin{bmatrix} 0, & 0 \\ 0, & b \end{bmatrix}_{1, m-1}^{1, n-1}; \quad \text{.....................(a')}$$

and by Ex. viii of § 116 the matrices $[Q]_{m-1}^{n-1}$ and $[b]_{m-1}^{n-1}$ have rank $\rho - 1$.

Writing $$\frac{1}{a_{11}} [a]_m^1 = \begin{bmatrix} 1 \\ \sigma \end{bmatrix}_{1, m-1}^1, \qquad \frac{1}{a_{11}} [a]_1^n = [1, \tau]_1^{1, n-1},$$

so that σ_{i1} and τ_{1j} have the same values as before, we can replace (a') by the equation

$$[a]_m^n = a_{11} \begin{bmatrix} 1 \\ \sigma \end{bmatrix}_{1, m-1}^1 [1, \tau]_1^{1, n-1} + \begin{bmatrix} 0 \\ 1 \end{bmatrix}_{1, m-1}^{m-1} [b]_{m-1}^{n-1} [0, 1]_{n-1}^{1, n-1},$$

and by the property of passive rows given in § 43.9 this is the same as

$$[a]_m^n = \begin{bmatrix} 1, & 0 \\ \sigma, & 1 \end{bmatrix}_{1, m-1}^{1, m-1} \begin{bmatrix} a_{11}, & 0 \\ 0, & b \end{bmatrix}_{1, m-1}^{1, n-1} \begin{bmatrix} 1, & \tau \\ 0, & 1 \end{bmatrix}_{1, n-1}^{1, n-1}. \quad \text{..................(A')}$$

Prefixing $\begin{bmatrix} 1, & 0 \\ -\sigma, & 1 \end{bmatrix}_{1, m-1}^{1, m-1}$ and postfixing $\begin{bmatrix} 1, & -\tau \\ 0, & 1 \end{bmatrix}_{1, n-1}^{1, n-1}$ on both sides of (A') we obtain the inverse transformation

$$\begin{bmatrix} 1, & 0 \\ -\sigma, & 1 \end{bmatrix}_{1, m-1}^{1, m-1} [a]_m^n \begin{bmatrix} 1, & -\tau \\ 0, & 1 \end{bmatrix}_{1, n-1}^{1, n-1} = \begin{bmatrix} a_{11}, & 0 \\ 0, & b \end{bmatrix}_{1, m-1}^{1, n-1} \quad \text{..................(A)}$$

converting ϕ into ψ, which is the resultant of the elementary unitary transformations in Ω described in the first proof.

Ex. i. As particular cases of the second equation in (1) we have the equations

$$a_{11}b_{11}=(a)^2_2,\ a_{11}(b)^2_2=(a)^3_3,\ \dots\ a_{11}(b)^{\rho-1}_{\rho-1}=(a)^{\rho}_{\rho},\ \dots\dots\dots\dots(1')$$

which we can also obtain by equating correspondingly formed minor determinants of both sides of (A).

Ex. ii. *If a_{uv} is any non-vanishing element of ϕ, we can convert ϕ by unitary equigradent transformations in Ω into the matrix $[b]^n_m$ in which $b_{uv}=a_{uv}$, all other elements of the uth horizontal row and the vth vertical row vanish, and*

$$a_{uv}b_{ij}=\begin{pmatrix} vj \\ a \\ ui \end{pmatrix}=\begin{vmatrix} a_{uv} & a_{uj} \\ a_{iv} & a_{ij} \end{vmatrix}\quad \textit{when } i \neq u \textit{ and } j \neq v.$$

As in Ex. ix we write $\sigma_{iu}=\frac{a_{iv}}{a_{uv}}$, $\tau_{vj}=\frac{a_{uj}}{a_{uv}}$, and add to the ith horizontal row the uth horizontal row multiplied by $-\sigma_{iu}$, and to the jth vertical row the vth vertical row multiplied by $-\tau_{vj}$, doing this in succession for all the values 1, 2, ... m of i except u and for all the values 1, 2, ... n of j except v.

The notation here used is different from that of the text, for when $a_{uv}=a_{11}$ we now have $\sigma_{i1}=\frac{a_{i1}}{a_{11}}$, $\tau_{1j}=\frac{a_{1j}}{a_{11}}$.

Theorem II. *If $\phi=[a]^n_m$ is a matrix of rank ρ whose elements are constants lying in a domain of rationality Ω, and if $\Delta=(a)^r_r \neq 0$ is a non-vanishing diagonal minor determinant of ϕ, we can convert ϕ by unitary equigradent transformations in Ω into the compartite matrix*

$$\psi=\begin{bmatrix} a, & 0 \\ 0, & b \end{bmatrix}^{r,\,n-r}_{r,\,m-r}$$

where $[b]^{n-r}_{m-r}$ is the matrix of rank $\rho-r$ in which

$$\Delta b_{uv}=\begin{pmatrix} 1, 2, \dots r, r+v \\ a \\ 1, 2, \dots r, r+u \end{pmatrix},\quad \Delta(b_{uv})^s_s=\begin{pmatrix} 1, 2, \dots r, r+v_1, \dots r+v_s \\ a \\ 1, 2, \dots r, r+u_1, \dots r+u_s \end{pmatrix}.\ \dots(2)$$

The first of the equations (2) completely defines the matrix $[b]^{n-r}_{m-r}$, and the second equation follows from the first by the identity of § 110. The second equation shows that $[b]^{n-r}_{m-r}$ has rank $\rho-r$; for it shows that all minor determinants of $[b]^{n-r}_{m-r}$ of order greater than $\rho-r$ vanish, and that to every one of the non-vanishing minor determinants of ϕ of order ρ which contains Δ there corresponds a non-vanishing minor determinant of $[b]^{n-r}_{m-r}$ of order $\rho-r$.

FIRST PROOF OF THEOREM II.

Let $[A]^r_r$ be the reciprocal of $[a]^r_r$; let $Q_{uv}=\begin{pmatrix} 1, 2, \dots r, r+v \\ a \\ 1, 2, \dots r, r+u \end{pmatrix}$; and let

$b_{uv} = \frac{1}{\Delta} Q_{uv}$, $[b]_{m-r}^{n-r} = \frac{1}{\Delta}[Q]_{m-r}^{n-r}$, so that $[b]_{m-r}^{n-r}$ is the matrix defined in the enunciation. Then by § 116 we have

$$\Delta [a]_m^n = [a]_m^r \, \overline{\underline{A}}_r^r \, [a]_r^n + \begin{bmatrix}0, & 0\\ 0, & Q\end{bmatrix}_{r,\, m-r}^{r,\, n-r}, \quad \dots\dots\dots\dots\dots(\text{b})$$

$$[a]_m^n = \frac{1}{\Delta} [a]_m^r \, \overline{\underline{A}}_r^r \, [a]_r^n + \begin{bmatrix}0, & 0\\ 0, & b\end{bmatrix}_{r,\, m-r}^{r,\, n-r}; \quad \dots\dots\dots\dots(\text{b}')$$

and by Ex. viii of § 116 the matrices $[Q]_{m-r}^{n-r}$ and $[b]_{m-r}^{n-r}$ have rank $\rho - r$.

We can replace (b′) by

$$[a]_m^n = \frac{1}{\Delta^2} [a]_m^r \, \overline{\underline{A}}_r^r \, . [a]_r^r \, . \, \overline{\underline{A}}_r^r \, [a]_r^n + \begin{bmatrix}0, & 0\\ 0, & b\end{bmatrix}_{r,\, m-r}^{r,\, n-r} . \quad \dots\dots\dots(3)$$

Using a notation analogous to that given in §§ 102 and 115 let

$$[a]_m^r \, \overline{\underline{A}}_r^r = \Delta \begin{bmatrix}1\\ \sigma\end{bmatrix}_{r,\, m-r}^{r}, \quad \overline{\underline{A}}_r^r \, [a]_r^n = \Delta \, [1, \ \tau]_r^{r,\, n-r}, \quad \dots\dots\dots(4)$$

so that $[\sigma]_{m-r}^r$ and $[\tau]_r^{n-r}$ are the matrices in which:

$\sigma_{iu} = \frac{1}{\Delta}$ times the primary of Δ formed when we replace the uth horizontal row of Δ by the $(r+i)$th horizontal row of $[a]_m^r$, or when in Δ we replace the horizontal suffix u by $r+i$;

$\tau_{vj} = \frac{1}{\Delta}$ times the primary of Δ formed when we replace the vth vertical row of Δ by the $(r+j)$th vertical row of $[a]_r^n$, or when in Δ we replace the vertical suffix v by $r+j$.

Then the equation (3) is equivalent to

$$[a]_m^n = \begin{bmatrix}1\\ \sigma\end{bmatrix}_{r,\, m-r}^{r} [a]_r^r \, [1, \ \tau]_r^{r,\, n-r} + \begin{bmatrix}0\\ 1\end{bmatrix}_{r,\, m-r}^{m-r} [b]_{m-r}^{n-r} \, [0, \ 1]_{n-r}^{r,\, n-r},$$

and by the property of passive rows given in § 43.9 this is the same as

$$[a]_m^n = \begin{bmatrix}1, & 0\\ \sigma, & 1\end{bmatrix}_{r,\, m-r}^{r,\, m-r} \begin{bmatrix}a, & 0\\ 0, & b\end{bmatrix}_{r,\, m-r}^{r,\, n-r} \begin{bmatrix}1, & \tau\\ 0, & 1\end{bmatrix}_{r,\, n-r}^{r,\, n-r} \quad \dots\dots\dots\dots(\text{B}')$$

Prefixing $\begin{bmatrix}1, & 0\\ -\sigma, & 1\end{bmatrix}_{r,\, m-r}^{r,\, m-r}$ and postfixing $\begin{bmatrix}1, & -\tau\\ 0, & 1\end{bmatrix}_{r,\, n-r}^{r,\, n-r}$ on both sides of (B′), we obtain the equivalent inverse equation

$$\begin{bmatrix}1, & 0\\ -\sigma, & 1\end{bmatrix}_{r,\, m-r}^{r,\, m-r} [a]_m^n \begin{bmatrix}1, & -\tau\\ 0, & 1\end{bmatrix}_{r,\, n-r}^{r,\, n-r} = \begin{bmatrix}a, & 0\\ 0, & b\end{bmatrix}_{r,\, m-r}^{r,\, n-r} . \quad \dots\dots\dots(\text{B})$$

Equation (B′) is a unitary equigradent transformation in Ω converting the matrix ψ of the theorem into ϕ, and equation (B) is a unitary equigradent transformation in Ω converting ϕ into ψ.

To effect the transformation (B) we perform on ϕ in succession and in any order the following two sets of transformations:

(1) *We add to the* $(r+i)$*th horizontal row the* 1*st*, 2*nd*, ... r*th horizontal rows multiplied by* $-\sigma_{i1}, -\sigma_{i2}, \dots -\sigma_{ir}$ *respectively, doing this for all the values* 1, 2, ... $m-r$ *of* i.

(2) *We add to the* $(r+j)$*th vertical row the* 1*st*, 2*nd*, ... r*th vertical rows multiplied by* $-\tau_{1j}, -\tau_{2j}, \dots -\tau_{rj}$ *respectively, doing this for all the values* 1, 2, ... $n-r$ *of* j.

Thus (B) is the resultant of $2r$ elementary unitary equigradent transformations in Ω; and this is also true of (B′).

The fact that $[b]_{m-r}^{n-r}$ has rank $\rho - r$ also follows from the theorem of § 100 when we observe that in (B) the compartite matrix on the right has the same rank ρ as $[a]_m^n$. Further the equations (2) of the enunciation can be deduced from (B) by equating correspondingly formed minor determinants of orders $r+1$ and $r+s$ on both sides, the determinants of the right-hand side being those which contain $[a]_r^r$.

SECOND PROOF OF THEOREM II.

Since all horizontal rows of $[a]_m^r$ are connected with the horizontal rows of $[a]_r^r$, and all vertical rows of $[a]_r^n$ are connected with the vertical rows of $[a]_r^r$, there exist equations of the forms

$$[-\sigma,\ 1]_{m-r}^{r,\,m-r}[a]_m^r = 0,\quad [a]_r^n\begin{bmatrix}-\tau\\ 1\end{bmatrix}_{r,\,n-r}^{n-r} = 0. \quad \dots\dots\dots(4')$$

Solving for $[\sigma]_{m-r}^r$ and $[\tau]_r^{n-r}$, we obtain

$$\Delta[\sigma]_{m-r}^r = \begin{bmatrix} a_{r+1,1} & \dots & a_{r+1,r}\\ \dots & \dots & \dots\\ a_{m1} & \dots & a_{mr}\end{bmatrix}\overline{\underline{A}}_r^r,\quad \Delta[\tau]_r^{n-r} = \overline{\underline{A}}_r^r\begin{bmatrix} a_{1,r+1} & \dots & a_{1n}\\ \dots & \dots & \dots\\ a_{r,r+1} & \dots & a_{rn}\end{bmatrix},$$

from which the equations (4) follow, this showing that σ_{ui} and τ_{vj} are the same quantities as before.

From (4′) we obtain at once the transformation

$$\begin{bmatrix}1, & 0\\ -\sigma, & 1\end{bmatrix}_{r,\,m-r}^{r,\,m-r}[a]_m^n\begin{bmatrix}1, & -\tau\\ 0, & 1\end{bmatrix}_{r,\,n-r}^{r,\,n-r} = \begin{bmatrix}a, & 0\\ 0, & b\end{bmatrix}_{r,\,m-r}^{r,\,n-r}, \quad \dots\dots\dots(B)$$

where $$[b]_{m-r}^{n-r} = [-\sigma, 1]_{m-r}^{r, m-r} [a]_m^n \begin{bmatrix} -\tau \\ 1 \end{bmatrix}_{r, n-r}^{n-r}$$

$$= [-\sigma, 1]_{m-r}^{r, m-r} \begin{bmatrix} a_{1, r+1} \dots a_{1n} \\ \dots\dots\dots\dots \\ a_{m, r+1} \dots a_{mn} \end{bmatrix} = \begin{bmatrix} a_{r+1, 1} \dots a_{r+1, n} \\ \dots\dots\dots\dots \\ a_{m1} \quad \dots a_{mn} \end{bmatrix} \begin{bmatrix} -\tau \\ 1 \end{bmatrix}_{r, n-r}^{n-r}.$$

It follows that

$$b_{uv} = -(\sigma_{u1} a_{1, r+v} + \sigma_{u2} a_{2, r+v} + \dots + \sigma_{ur} a_{r, r+v}) + a_{r+u, r+v},$$

$$\Delta b_{uv} = -\det[a_{r+u, 1} \dots a_{r+u, r}] \overline{\underline{A}}_r^r \begin{bmatrix} a_{1, r+v} \\ \vdots \\ a_{r, r+v} \end{bmatrix} + \Delta a_{r+u, r+v} = \begin{pmatrix} 1, 2, \dots r, r+v \\ a \\ 1, 2, \dots r, r+u \end{pmatrix}.$$

Thus $[b]_{m-r}^{n-r}$ is the same matrix as before; and since the matrix ψ on the right in (B) must have the same rank ρ as $[a]_m^n$, it follows by the theorem of § 100 that $[b]_{m-r}^{n-r}$ has rank $\rho - r$.

Prefixing $\begin{bmatrix} 1, & 0 \\ \sigma, & 1 \end{bmatrix}_{r, m-r}^{r, m-r}$ and postfixing $\begin{bmatrix} 1, & \tau \\ 0, & 1 \end{bmatrix}_{r, n-r}^{r, n-r}$ on both sides of (B), we obtain (B′).

Ex. iii. Particular cases of the second equation in (2) are

$$\Delta b_{11} = (a)_{r+1}^{r+1}, \quad \Delta (b)_2^2 = (a)_{r+2}^{r+2}, \dots \Delta (b)_{\rho-r}^{\rho-r} = (a)_\rho^\rho. \quad\dots\dots\dots\dots(2').$$

We can obtain these equations by equating correspondingly formed minor determinants of both sides of (B′).

Ex. iv. When $\Delta = a_{11} \neq 0$, we have as in Theorem I

$$\sigma_{i-1,1} = \frac{a_{i1}}{a_{11}}, \quad \tau_{1,j-1} = \frac{a_{1j}}{a_{11}}, \quad a_{11} b_{i-1,j-1} = \begin{pmatrix} 1j \\ a \\ 1i \end{pmatrix}.$$

Ex. v. When $\Delta = (a)_2^2 \neq 0$, we have

$$\Delta\sigma_{i-2,1} = \begin{pmatrix} 12 \\ a \\ i2 \end{pmatrix}, \quad \Delta\sigma_{i-2,2} = \begin{pmatrix} 12 \\ a \\ 1i \end{pmatrix}; \quad \Delta\tau_{1,j-2} = \begin{pmatrix} j2 \\ a \\ 12 \end{pmatrix}, \quad \Delta\tau_{2,j-2} = \begin{pmatrix} 1j \\ a \\ 12 \end{pmatrix};$$

and $$\Delta b_{i-2,j-2} = \begin{pmatrix} 12j \\ a \\ 12i \end{pmatrix}.$$

Ex. vi. When $a_{11} = a_{22} = 0$, and $\Delta = (a)_2^2 = -a_{12}a_{21} \neq 0$, we have

$$\sigma_{i-2,1} = \frac{a_{i2}}{a_{12}}, \quad \sigma_{i-2,2} = \frac{a_{i1}}{a_{21}}; \quad \tau_{1,j-2} = \frac{a_{2j}}{a_{21}}, \quad \tau_{2,j-2} = \frac{a_{1j}}{a_{12}},$$

and $$\Delta b_{i-2,j-2} = a_{12} a_{i1} a_{2j} + a_{21} a_{i2} a_{1j} - a_{12} a_{21} a_{ij}.$$

Ex. vii. *If* $\phi = [a]_m^m$ *is a symmetric or skew-symmetric matrix in Theorem II, then the transformations* (B) *and* (B′) *are symmetric, and the reduced matrix* ψ *is symmetric or skew-symmetric according as* ϕ *is symmetric or skew-symmetric.*

In these cases we have $n=m$.

If $[a]_m^m$ is symmetric, then $[A]_r^r$ is symmetric, and $\overline{\underline{a}}_r^m=[a]_r^m$. When therefore we equate the conjugates of both sides in the equation

$$\Delta\begin{bmatrix}1\\ \sigma\end{bmatrix}_{r,\,m-r}^{r}=[a]_m^r\,\overline{\underline{A}}_r^r,$$

we obtain

$$\Delta\,\overline{\underline{1,\ \sigma}}_r^{r,\,m-r}=[A]_r^r\,\overline{\underline{a}}_r^m=\overline{\underline{A}}_r^r\,[a]_r^m=\Delta[1,\ \tau]_r^{r,\,m-r}.$$

Thus $$\overline{\underline{\sigma}}_r^{m-r}=[\tau]_r^{m-r},\quad [\sigma]_{m-r}^r=\overline{\underline{\tau}}_{m-r}^r,$$

and the equations (B) and (B′) assume the symmetric forms

$$\overline{\underline{\begin{matrix}1, & 0\\ -\tau, & 1\end{matrix}}}_{r,\,m-r}^{r,\,m-r}\,[a]_m^m\begin{bmatrix}1, & -\tau\\ 0, & 1\end{bmatrix}_{r,\,m-r}^{r,\,m-r}=\begin{bmatrix}a, & 0\\ 0, & b\end{bmatrix}_{r,\,m-r}^{r,\,m-r}, \quad\ldots\ldots\ldots\ldots(\text{B}_1)$$

$$[a]_m^m=\overline{\underline{\begin{matrix}1, & 0\\ \tau, & 1\end{matrix}}}_{r,\,m-r}^{r,\,m-r}\begin{bmatrix}a, & 0\\ 0, & b\end{bmatrix}_{r,\,m-r}^{r,\,m-r}\begin{bmatrix}1, & \tau\\ 0, & 1\end{bmatrix}_{r,\,m-r}^{r,\,m-r}. \quad\ldots\ldots\ldots\ldots(\text{B}_1')$$

If $[a]_m^m$ is skew-symmetric, then r and ρ are necessarily even integers; moreover $[A]_r^r$ is skew-symmetric, and $\overline{\underline{a}}_r^m=-[a]_r^m$. Proceeding as before, we again obtain the same results, and the equations (B) and (B′) again assume the symmetric forms (B_1) and (B_1').

Equating the conjugates of both sides in (B_1), we see that the matrix ψ on the right is symmetric when $[a]_m^m$ is symmetric, and skew-symmetric when $[a]_m^m$ is skew-symmetric.

To effect the transformation (B_1) we perform on ϕ in succession and in any order the following operations for all the values $1, 2, \ldots\, m-r$ of i:

We add to the $(r+i)$*th horizontal and vertical rows the* 1*st*, 2*nd*, ... r*th parallel rows multiplied by* $-\tau_{1i}, -\tau_{2i}, \ldots -\tau_{ri}$ *respectively.*

NOTE 1. *Identities corresponding to the equations* (B) *and* (B′).

We can replace (b) by the identity

$$\Delta^2[a]_m^m=[a]_m^r\,\overline{\underline{A}}_r^r\,.\,[a]_r^r\,.\,\overline{\underline{A}}_r^r\,[a]_r^n+\Delta\begin{bmatrix}0, & 0\\ 0, & Q\end{bmatrix}_{r,\,m-r}^{r,\,n-r}.$$

Let $$[a]_m^r\,\overline{\underline{A}}_r^r=\begin{bmatrix}\Delta\\ a\end{bmatrix}_{r,\,m-r}^{r},\quad \overline{\underline{A}}_r^r\,[a]_r^n=[\Delta,\ \beta]_r^{r,\,n-r},$$

so that a_{iu} is the primary of Δ formed when in Δ we replace the horizontal suffix u by $r+i$, and β_{vj} is the primary of Δ formed when in Δ we replace the vertical suffix v by $r+j$.

Then we have

$$\Delta^2[a]_m^n=\begin{bmatrix}\Delta\\ a\end{bmatrix}_{r,\,m-r}^{r}[a]_r^r\,[\Delta,\ \beta]_r^{r,\,n-r}+\Delta\begin{bmatrix}0\\ 1\end{bmatrix}_{r,\,m-r}^{m-r}[Q]_{m-r}^{n-r}\,[0,\ 1]_{n-r}^{r,\,n-r},$$

or $$\Delta^2[a]_m^n=\begin{bmatrix}\Delta, & 0\\ a, & 1\end{bmatrix}_{r,\,m-r}^{r,\,m-r}\begin{bmatrix}a, & 0\\ 0, & \Delta Q\end{bmatrix}_{r,\,m-r}^{r,\,n-r}\begin{bmatrix}\Delta, & \beta\\ 0, & 1\end{bmatrix}_{r,\,n-r}^{r,\,n-r}. \quad\ldots\ldots\ldots\ldots(\text{B}_2')$$

The inverse transformation is

$$\begin{bmatrix} 1, & 0 \\ -a, & \Delta \end{bmatrix}_{r,\, m-r}^{r,\, m-r} [a]_m^n \begin{bmatrix} 1, & -\beta \\ 0, & \Delta \end{bmatrix}_{r,\, n-r}^{r,\, n-r} = \begin{bmatrix} a, & 0 \\ 0, & \Delta Q \end{bmatrix}_{r,\, m-r}^{r,\, n-r}. \quad \text{............}(B_2)$$

The equations (B_2) and (B_2') are identities in the elements of $[a]_m^n$.

Theorem III. *Let $\phi = [a]_m^n$ be a matrix of rank ρ whose elements are constants lying in a domain of rationality Ω, and let $\Delta = (a_{pq})_r^r \neq 0$ be any non-vanishing minor determinant of ϕ. Then if $[\mu]_{m-r}$ and $[\nu]_{n-r}$ are the corranged complements of $[p]_r$ in $[1\,2 \ldots m]$ and of $[q]_r$ in $[1\,2 \ldots n]$ respectively, and if $P_{\mu_i \nu_j} = \begin{pmatrix} q_1 q_2 \ldots q_r \nu_j \\ a \\ p_1 p_2 \ldots p_r \mu_i \end{pmatrix}$, we can convert ϕ by unitary equigradent transformations in Ω into the compartite matrix $\psi = [b]_m^n$ in which*

(1) $[b_{pq}]_r^r = [a_{pq}]_r^r$; $[b_{\mu\nu}]_{m-r}^{n-r} = \frac{1}{\Delta}[P_{\mu\nu}]_{m-r}^{n-r}$, *and has rank $\rho - r$.*

(2) *All other elements are 0's.*

It follows from Ex. viii of § 116 that the matrices $[P_{\mu\nu}]_{m-r}^{n-r}$ and $[b_{\mu\nu}]_{m-r}^{n-r}$ defined in the enunciation have rank $\rho - r$.

The theorem can be deduced from Theorem II by derangements of horizontal and vertical rows. We give below a direct and independent proof.

PROOF OF THEOREM III.

If $[A_{pq}]_r^r$ is the reciprocal of $[a_{pq}]_r^r$, we have by § 116

$$\Delta [a]_m^n = [a_{1q}]_m^r \boxed{A_{pq}}_r^r [a_{p1}]_r^n + [P]_m^n$$

$$= \frac{1}{\Delta}[a_{1q}]_m^r \boxed{A_{pq}}_r^r \cdot [a_{pq}]_r^r \cdot \boxed{A_{pq}}_r^r [a_{p1}]_r^n + [P]_m^n. \quad \text{............}(5)$$

Using a notation analogous to that of §§ 102 and 115 we will write

$$[h_{1p}]_m^r = \frac{1}{\Delta}[a_{1q}]_m^r \boxed{A_{pq}}_r^r, \qquad [k_{q1}]_r^n = \frac{1}{\Delta}\boxed{A_{pq}}_r^r [a_{p1}]_r^n,$$

$$[\sigma_{\mu p}]_{m-r}^r = \frac{1}{\Delta}[a_{\mu q}]_{m-r}^r \boxed{A_{pq}}_r^r, \quad [\tau_{q\nu}]_r^{n-r} = \frac{1}{\Delta}\boxed{A_{pq}}_r^r [a_{p\nu}]_r^{n-r},$$

so that

$$[h_{pp}]_r^r = [1]_r^r, \quad [h_{\mu p}]_{m-r}^r = [\sigma_{\mu p}]_{m-r}^r;$$

$$[k_{qq}]_r^r = [1]_r^r, \quad [k_{q\nu}]_r^{n-r} = [\tau_{q\nu}]_r^{n-r};$$

$\sigma_{iu} = \frac{1}{\Delta}$ times the horizontal primary of Δ formed when we replace the uth horizontal row of $[a_{1q}]_m^r$ (occurring in Δ) by the ith horizontal row of $[a_{1q}]_m^r$ (not occurring in Δ), or when in Δ we replace the horizontal suffix u by i;

$\tau_{vj} = \frac{1}{\Delta}$ times the vertical primary of Δ formed when we replace the vth vertical row of $[a_{p1}]_r^n$ (occurring in Δ) by the jth vertical row of $[a_{p1}]_r^n$ (not occurring in Δ), or when in Δ we replace the vertical suffix v by j.

We can then replace (5) by

$$[a]_m^n = [h_{1p}]_m^r [a_{pq}]_r^r [k_{q1}]_r^n + \frac{1}{\Delta}[P]_m^n . \qquad (6)$$

By adding $m-r$ suitably placed vertical rows to $[h_{1p}]_m^r$ and $n-r$ suitably placed horizontal rows to $[k_{q1}]_r^n$ we will form the square matrices $[h]_m^m$ and $[k]_n^n$ in which

$$[h_{pp}]_r^r = [1]_r^r, \quad [h_{p\mu}]_r^{m-r} = 0, \quad [h_{\mu p}]_{m-r}^r = [\sigma_{\mu p}]_{m-r}^r, \quad [h_{\mu\mu}]_{m-r}^{m-r} = [1]_{m-r}^{m-r};$$

$$[k_{qq}]_r^r = [1]_r^r, \quad [k_{q\nu}]_r^{n-r} = [\tau_{q\nu}]_r^{n-r}, \quad [k_{\nu q}]_{n-r}^r = 0, \quad [k_{\nu\nu}]_{n-r}^{n-r} = [1]_{n-r}^{n-r}.$$

Then if we write $[h_{1\mu}]_m^{m-r} [P_{\mu\nu}]_{m-r}^{n-r} [\beta_{\nu 1}]_{n-r}^n = [x]_m^n$, we have

$$[x_{p1}]_r^n = 0, \quad [x_{1q}]_m^r = 0, \quad [x_{\mu\nu}]_{m-r}^{n-r} = [P_{\mu\nu}]_{m-r}^{n-r}, \quad \text{i.e. } [x]_m^n = [P]_m^n.$$

Accordingly equation (6) is

$$[a]_m^n = [h_{1p}]_m^r [a_{pq}]_r^r [k_{q1}]_r^n + \frac{1}{\Delta}[h_{1\mu}]_m^{m-r} [P_{\mu\nu}]_{m-r}^{n-r} [k_{\nu 1}]_{n-r}^n ,$$

which by the properties of passive rows is the same as

$$[a]_m^n = [h]_m^m [b]_m^n [k]_n^n, \qquad (C')$$

where $[b]_m^n$ is the matrix defined in the enunciation; and this is a unitary equigradent transformation in Ω converting ψ into ϕ.

By Ex. xv of § 141 the inverse transformation is

$$[h']_m^m [a]_m^n [k']_n^n = [b]_m^n, \qquad (C)$$

where $[h']_m^m$ and $[k']_n^n$ are the matrices in which

$$[h'_{pp}]_r^r = [1]_r^r, \quad [h'_{p\mu}]_r^{m-r} = 0, \quad [h'_{\mu p}]_{m-r}^r = -[\sigma_{\mu p}]_{m-r}^r, \quad [h'_{\mu\mu}]_{m-r}^{m-r} = [1]_{m-r}^{m-r};$$

$$[k'_{qq}]_r^r = [1]_r^r, \quad [k'_{q\nu}]_r^{n-r} = -[\tau_{q\nu}]_r^{n-r}, \quad [k'_{\nu q}]_{n-r}^r = 0, \quad [k'_{\nu\nu}]_{n-r}^{n-r} = [1]_{n-r}^{n-r}.$$

Equation (C) is a unitary equigradent transformation in Ω converting ϕ into ψ. We effect this transformation by performing on ϕ in succession and in any order the following two sets of operations:

(1) *We add to the μth horizontal row the p_1th, p_2th, ... p_rth horizontal rows multiplied by $-\sigma_{\mu p_1}, -\sigma_{\mu p_2}, \dots -\sigma_{\mu p_r}$ respectively, doing this for all the values $\mu_1, \mu_2, \dots \mu_{m-r}$ of μ.*

(2) *We add to the νth vertical row the q_1th, q_2th, ... q_rth vertical rows multiplied by $-\tau_{q_1\nu}, -\tau_{q_2\nu}, \dots -\tau_{q_r\nu}$ respectively, doing this for all the values $\nu_1, \nu_2, \dots \nu_{n-r}$ of ν.*

At each step we add to a horizontal or vertical row which does not occur in Δ the parallel rows which do occur in Δ multiplied by certain constants each of which is $-\frac{1}{\Delta}$ times one of the primaries of Δ.

Ex. viii. From the definition of $[b]_m^n$ in the enunciation, we have

$$\Delta\, b_{uv} = \begin{pmatrix} q_1 q_2 \dots q_r v \\ a \\ p_1 p_2 \dots p_r u \end{pmatrix}, \quad \Delta\,(b_{uv})_s^s = \begin{pmatrix} q_1 q_2 \dots q_r v_1 v_2 \dots v_s \\ a \\ p_1 p_2 \dots p_r u_1 u_2 \dots u_s \end{pmatrix}, \quad \dots\dots\dots\dots(7)$$

the first equation being true whenever u is not an element of the sequence $[p_1 p_2 \dots p_r]$ and v is not an element of the sequence $[q_1 q_2 \dots q_r]$, and the second equation being true whenever the sequences $[u]_s$ and $[p]_s$ have no element in common and the sequences $[v]_s$ and $[q]_s$ have no element in common.

Ex. ix. *Special case when* $\Delta = a_{uv} \neq 0$.

In this case we have $r=1$, $p_1=u$, $q_1=v$, and

$$\sigma_{iu} = \frac{a_{iv}}{a_{uv}}, \quad \tau_{vj} = \frac{a_{uj}}{a_{uv}}, \quad a_{uv}\, b_{ij} = \begin{pmatrix} v\,j \\ a \\ u\,i \end{pmatrix}$$

This is the case considered in Ex. ii.

Ex. x. *Special case when* $\Delta = \begin{pmatrix} q\,v \\ a \\ p\,u \end{pmatrix} \neq 0$.

In this case $r=2$, $p_1=p$, $p_2=u$, $q_1=q$, $q_2=v$;

$$\Delta\,\sigma_{ip} = \begin{pmatrix} q\,v \\ a \\ i\,u \end{pmatrix}, \quad \Delta\,\sigma_{iu} = \begin{pmatrix} q\,v \\ a \\ p\,i \end{pmatrix}, \quad \Delta\,\tau_{qj} = \begin{pmatrix} j\,v \\ a \\ p\,u \end{pmatrix}, \quad \Delta\,\tau_{vj} = \begin{pmatrix} q\,j \\ a \\ p\,u \end{pmatrix};$$

and

$$\Delta\, b_{ij} = \begin{pmatrix} q\,v\,j \\ a \\ p\,u\,i \end{pmatrix};$$

it being understood that $i \neq p$, $i \neq u$, $j \neq q$, $j \neq v$.

We add to the ith horizontal row the pth and uth horizontal rows multiplied by $-\sigma_{ip}$ and $-\sigma_{iu}$, and we add to the jth vertical row the qth and vth vertical rows multiplied by $-\tau_{qj}$ and $-\tau_{vj}$, doing this in succession for all the values $1, 2, \dots m$ of i except p and u, and for all the values $1, 2, \dots n$ of j except q and v.

Ex. xi. *Special case when* $a_{uu} = a_{vv} = 0$, $\Delta = \begin{pmatrix} u\,v \\ a \\ u\,v \end{pmatrix} = -a_{uv}\,a_{vu} \neq 0$.

We have
$$\sigma_{iu} = \frac{a_{iv}}{a_{uv}}, \quad \sigma_{iv} = \frac{a_{iu}}{a_{vu}}, \quad \tau_{uj} = \frac{a_{vj}}{a_{vu}}, \quad \tau_{vj} = \frac{a_{uj}}{a_{uv}},$$

and
$$b_{ij} = a_{ij} - \frac{a_{uj}\,a_{iv} + a_{vj}\,a_{iu}}{a_{uv}\,a_{vu}},$$

when neither i nor j is equal to u or v.

We add to the ith horizontal row the uth and vth horizontal rows multiplied by $-\sigma_{iu}$ and $-\sigma_{iv}$, and to the jth vertical row the uth and vth vertical rows multiplied by $-\tau_{uj}$ and $-\tau_{vj}$.

Ex. xii. *If* $\phi = [a]_m^m$ *is a symmetric or skew-symmetric matrix in Theorem III, and if* $\Delta = (a_{pp})_r^r \neq 0$ *is a non-vanishing diagonal minor determinant of* ϕ, *then the transformations* (C) *and* (C′) *are symmetric, and the reduced matrix* $\psi = [b]_m^m$ *is symmetric or skew-symmetric according as* ϕ *is symmetric or skew-symmetric.*

In these cases we have $n=m$, $[q]_r=[p]_r$, and $[\nu]_r=[\mu]_r$ = corranged complement of $[p]_r$ in $[1\ 2 \ldots m]$.

If $[a]_m^m$ is symmetric, then $[A_{pp}]_r^r$ is symmetric, and $\ulcorner a_{\mu p} \lrcorner_r^{m-r} = [a_{p\mu}]_r^{m-r}$. When therefore we equate the conjugates of both sides in the equation

$$[\sigma_{\mu p}]_{m-r}^r = \frac{1}{\Delta}[a_{\mu\nu}]_{m-r}^r \ulcorner A_{pp} \lrcorner_r^r,$$

we obtain

$$\ulcorner \sigma_{\mu p} \lrcorner_r^{m-r} = \frac{1}{\Delta}[A_{pp}]_r^r \ulcorner a_{\mu p} \lrcorner_r^{m-r} = \frac{1}{\Delta}\ulcorner A_{pp} \lrcorner_r^r [a_{p\mu}]_r^{m-r} = [\tau_{p\mu}]_r^{m-r}.$$

Therefore $\quad [\sigma_{\mu p}]_{m-r}^r = \ulcorner \tau_{p\mu} \lrcorner_{m-r}^r, \quad [h]_m^m = \ulcorner k \lrcorner_m^m, \quad [h']_m^m = \ulcorner k' \lrcorner_m^m;$

and the equations (C) and (C′) assume the symmetric forms

$$\ulcorner k' \lrcorner_m^m [a]_m^m [k']_m^m = [b]_m^m, \quad \text{.................................(C}_1\text{)}$$

$$[a]_m^m = \ulcorner k \lrcorner_m^m [b]_m^m [k]_m^m, \quad \text{.................................(C}_1'\text{)}$$

where $[k]_m^m$ and $[k']_m^m$ are defined as before.

If $[a]_m^m$ is skew-symmetric, then r and ρ are necessarily even integers; moreover $[A_{pp}]_r^r$ is skew-symmetric, and $\ulcorner a_{\mu p} \lrcorner_r^{m-r} = -[a_{p\mu}]_r^{m-r}$. Proceeding as before we again obtain the same results, and the equations (C) and (C′) again assume the symmetric forms (C_1) and (C_1').

Equating the conjugates of both sides in (C_1) we see that the matrix ψ on the right is symmetric when ϕ is symmetric, and skew-symmetric when ϕ is skew-symmetric.

To effect the transformation (C_1) we perform on ϕ in succession and in any order the following operations for all the values $\mu_1, \mu_2, \ldots \mu_{m-r}$ of μ.

We add to the μth horizontal and vertical rows the p_1th, p_2th, ... p_rth parallel rows multiplied by $-\tau_{p_1\mu}, -\tau_{p_2\mu}, \ldots -\tau_{p_r\mu}$ respectively.

§ 143. A more general unitary equigradent transformation.

Theorem. *Let $\phi=[a]_m^n$ be a matrix of rank ρ whose elements are constants lying in a domain of rationality Ω, and let $\alpha, \beta, \gamma, \ldots \kappa, \lambda, \mu, \ldots \pi, \rho$ be a series of constantly increasing positive integers, the first of which is greater than 0, and the last of which is equal to the rank of ϕ. Then if none of the leading diagonal minor determinants*

$$\Delta_0=1, \quad \Delta_\alpha=(a)_\alpha^\alpha, \quad \Delta_\beta=(a)_\beta^\beta, \quad \ldots \quad \Delta_\lambda=(a)_\lambda^\lambda, \quad \ldots \quad \Delta_\rho=(a)_\rho^\rho$$

of ϕ vanish, we can convert ϕ by a unitary equigradent transformation in Ω into a certain compartite matrix

$$\psi = \begin{bmatrix} a, & 0, & 0, & \dots & 0, & \dots & 0, & 0 \\ 0, & b, & 0, & \dots & 0, & \dots & 0, & 0 \\ 0, & 0, & c, & \dots & 0, & \dots & 0, & 0 \\ \cdots & & & & & & & \\ 0, & 0, & 0, & \dots & l, & \dots & 0, & 0 \\ \cdots & & & & & & & \\ 0, & 0, & 0, & \dots & 0, & \dots & r, & 0 \\ 0, & 0, & 0, & \dots & 0, & \dots & 0, & 0 \end{bmatrix}_{\alpha,\ \beta-\alpha,\ \gamma-\beta,\ \dots\ \lambda-\kappa,\ \dots\ \rho-\pi,\ m-\rho}^{\alpha,\ \beta-\alpha,\ \gamma-\beta,\ \dots\ \lambda-\kappa,\ \dots\ \rho-\pi,\ n-\rho}$$

in which $A=(a)_\alpha^\alpha \neq 0$, $B=(b)_{\beta-\alpha}^{\beta-\alpha}\neq 0, \dots L=(l)_{\lambda-\kappa}^{\lambda-\kappa}\neq 0, \dots R=(r)_{\rho-\pi}^{\rho-\pi}\neq 0$, the part $[a]_\alpha^\alpha$ being the leading diagonal minor of order α of ϕ, and the parts $[b]_{\beta-\alpha}^{\beta-\alpha}, [c]_{\gamma-\beta}^{\gamma-\beta}, \dots [l]_{\lambda-\kappa}^{\lambda-\kappa}, [m]_{\mu-\lambda}^{\mu-\lambda}, \dots [r]_{\rho-\pi}^{\rho-\pi}$ being leading diagonal minors of the matrices

$$[b]_{m-\alpha}^{n-\alpha},\ [c]_{m-\beta}^{n-\beta}, \dots [l]_{m-\kappa}^{n-\kappa},\ [m]_{m-\lambda}^{n-\lambda}, \dots [r]_{m-\pi}^{n-\pi}$$

of ranks $\rho-\alpha, \rho-\beta, \dots \rho-\kappa, \rho-\lambda, \dots \rho-\pi$ whose elements are defined in succession by the equations

$$Ab_{uv} = \begin{pmatrix} 1, 2, \dots \alpha, \alpha+v \\ a \\ 1, 2, \dots \alpha, \alpha+u \end{pmatrix}, \quad Bc_{uv} = \begin{pmatrix} 1, 2, \dots \beta-\alpha, \beta-\alpha+v \\ b \\ 1, 2, \dots \beta-\alpha, \beta-\alpha+u \end{pmatrix},$$

$$Cd_{uv} = \begin{pmatrix} 1, 2, \dots \gamma-\beta, \gamma-\beta+v \\ c \\ 1, 2, \dots \gamma-\beta, \gamma-\beta+u \end{pmatrix}$$

and so on, the determinants on the right being primary superdeterminants of $A, B, C, \dots L, \dots P$.

If we assume provisionally that $A, B, C, \dots$ do not vanish, we deduce in succession from the equations defining the elements $b_{uv}, c_{uv}, d_{uv}, \dots$ by means of the identity of § 110 the equations

$$A\,(b_{uv})_s^s = \begin{pmatrix} 1, 2, \dots \alpha, \alpha+v_1, \dots \alpha+v_s \\ a \\ 1, 2, \dots \alpha, \alpha+u_1, \dots \alpha+u_s \end{pmatrix}, \qquad A\,(b)_s^s = (a)_{\alpha+s}^{\alpha+s},$$

$$B\,(c_{uv})_s^s = \begin{pmatrix} 1, 2, \dots \beta-\alpha, \beta-\alpha+v_1, \dots \beta-\alpha+v_s \\ b \\ 1, 2, \dots \beta-\alpha, \beta-\alpha+u_1, \dots \beta-\alpha+u_s \end{pmatrix}, \quad B\,(c)_s^s = (b)_{\beta-\alpha+s}^{\beta-\alpha+s},$$

$$C\,(d_{uv})_s^s = \begin{pmatrix} 1, 2, \dots \gamma-\beta, \gamma-\beta+v_1, \dots \gamma-\beta+v_s \\ c \\ 1, 2, \dots \gamma-\beta, \gamma-\beta+u_1, \dots \gamma-\beta+u_s \end{pmatrix}, \quad C\,(d)_s^s = (c)_{\gamma-\beta+s}^{\gamma-\beta+s},$$

$$\dots\dots\dots\dots\dots\dots \qquad (1);$$

and from these we deduce in succession the equations

$$A\,(b_{uv})_s^s = \begin{pmatrix} 1, 2, \ldots \alpha, \alpha+v_1, \ldots \alpha+v_s \\ a \\ 1, 2, \ldots \alpha, \alpha+u_1, \ldots \alpha+u_s \end{pmatrix}, \qquad A\,(b)_s^s = (a)_{\alpha+s}^{\alpha+s},$$

$$AB\,(c_{uv})_s^s = \begin{pmatrix} 1, 2, \ldots \beta, \beta+v_1, \ldots \beta+v_s \\ a \\ 1, 2, \ldots \beta, \beta+u_1, \ldots \beta+u_s \end{pmatrix}, \qquad AB\,(c)_s^s = (a)_{\beta+s}^{\beta+s},$$

$$ABC\,(d_{uv})_s^s = \begin{pmatrix} 1, 2, \ldots \gamma, \gamma+v_1, \ldots \gamma+v_s \\ a \\ 1, 2, \ldots \gamma, \gamma+u_1, \ldots \gamma+u_s \end{pmatrix}, \qquad ABC\,(d)_s^s = (a)_{\gamma+s}^{\gamma+s},$$

.... (2)

Since $A = \Delta_\alpha \neq 0$, the first pairs of equations in (1) and (2) are true, and they show that $AB = \Delta_\beta \neq 0$. Therefore B does not vanish, and the second pairs of equations in (1) and (2) are true.

These show that $ABC = \Delta_\gamma \neq 0$. Therefore C does not vanish and the third pairs of equations in (1) and (2) are true.

These show that $ABCD = \Delta_\delta \neq 0$. Therefore D does not vanish and the fourth pairs of equations in (1) and (2) are true.

Proceeding in this way we see that

$$A = \Delta_\alpha,\ AB = \Delta_\beta,\ ABC = \Delta_\gamma, \ldots ABC\ldots L = \Delta_\lambda, \ldots ABC\ldots R = \Delta_\rho. \ldots (3)$$

Thus none of the determinants $A, B, C, \ldots$ vanish; all the equations (1) and (2) are true; and we can determine matrices

$$\phi_\beta = [b]_{m-\alpha}^{n-\alpha}, \quad \phi_\gamma = [c]_{m-\beta}^{n-\beta}, \ldots \phi_\lambda = [l]_{m-\kappa}^{n-\kappa}, \ldots \phi_\rho = [r]_{m-\pi}^{n-\pi}$$

whose elements are defined as in the theorem.

The equations (2) show in succession that $\phi_\beta, \phi_\gamma, \ldots \phi_\lambda, \ldots \phi_\rho$ have non-vanishing minor determinants $(b)_{\rho-\alpha}^{\rho-\alpha}, (c)_{\rho-\beta}^{\rho-\beta}, \ldots (l)_{\rho-\kappa}^{\rho-\kappa}, \ldots (r)_{\rho-\pi}^{\rho-\pi}$ of orders $\rho-\alpha, \rho-\beta, \ldots \rho-\kappa, \ldots \rho-\pi$, and that all their minor determinants of higher orders vanish. Consequently these matrices have ranks $\rho-\alpha$, $\rho-\beta, \ldots \rho-\kappa, \ldots \rho-\pi$ respectively.

From the equations (2) we see that

$$\Delta_\alpha\,(b_{uv})_s^s = \begin{pmatrix} 1, 2, \ldots \alpha, \alpha+v_1, \ldots \alpha+v_s \\ a \\ 1, 2, \ldots \alpha, \alpha+u_1, \ldots \alpha+u_s \end{pmatrix}, \qquad \Delta_\alpha\, b_{uv} = \begin{pmatrix} 1, 2, \ldots \alpha, \alpha+v \\ a \\ 1, 2, \ldots \alpha, \alpha+u \end{pmatrix},$$

$$\Delta_\beta\,(c_{uv})_s^s = \begin{pmatrix} 1, 2, \ldots \beta, \beta+v_1, \ldots \beta+v_s \\ a \\ 1, 2, \ldots \beta, \beta+u_1, \ldots \beta+u_s \end{pmatrix}, \qquad \Delta_\beta\, c_{uv} = \begin{pmatrix} 1, 2, \ldots \beta, \beta+v \\ a \\ 1, 2, \ldots \beta, \beta+u \end{pmatrix},$$

....

$$\Delta_\kappa\,(l_{uv})_s^s = \begin{pmatrix} 1, 2, \ldots \kappa, \kappa+v_1, \ldots \kappa+v_s \\ a \\ 1, 2, \ldots \kappa, \kappa+u_1, \ldots \kappa+u_s \end{pmatrix}, \qquad \Delta_\kappa\, l_{uv} = \begin{pmatrix} 1, 2, \ldots \kappa, \kappa+v \\ a \\ 1, 2, \ldots \kappa, \kappa+u \end{pmatrix},$$

.... (4)

We shall regard λ as any integer of the series $\alpha, \beta, \gamma, \ldots \kappa, \lambda, \mu, \ldots \pi, \rho$; κ as that integer of the series which immediately precedes λ; and μ as that integer of the series which immediately follows λ. When λ is the first integer of the series, we must consider that $\kappa = 0$, $\Delta_\kappa = \Delta_0 = 1$, $K = 1$. When λ is the last integer of the series, we must consider that $\Delta_\mu = 0$, $\phi_\mu = 0$.

Proof of the theorem.

Let the reciprocals of

$$[a]^{\alpha}_{\alpha}, \quad [b]^{\beta-\alpha}_{\beta-\alpha}, \quad [c]^{\gamma-\beta}_{\gamma-\beta}, \ldots [l]^{\lambda-\kappa}_{\lambda-\kappa}, \ldots [r]^{\rho-\pi}_{\rho-\pi}$$

be respectively

$$[A]^{\alpha}_{\alpha}, \quad [B]^{\beta-\alpha}_{\beta-\alpha}, \quad [C]^{\gamma-\beta}_{\gamma-\beta}, \ldots [L]^{\lambda-\kappa}_{\lambda-\kappa}, \ldots [R]^{\rho-\pi}_{\rho-\pi};$$

let $\phi_\alpha = [a]^{n}_{m}, \quad \phi_\beta = [b]^{n-\alpha}_{m-\alpha}, \quad \phi_\gamma = [c]^{n-\beta}_{m-\beta}, \ldots \phi_\lambda = [l]^{n-\kappa}_{m-\kappa}, \ldots \phi_\rho = [r]^{n-\pi}_{m-\pi};$

and let

$$\Phi_\lambda = \begin{bmatrix} 0, & 0 \\ 0, & l \end{bmatrix}^{\kappa,\, n-\kappa}_{\kappa,\, m-\kappa} = \begin{bmatrix} 0 \\ 1 \end{bmatrix}^{m-\kappa}_{\kappa,\, m-\kappa} [l]^{n-\kappa}_{m-\kappa} [0,\ 1]^{\kappa,\, n-\kappa}_{n-\kappa} \quad \ldots\ldots\ldots\ldots(5)$$

$$T_\lambda = \frac{1}{L} \begin{bmatrix} 0 \\ l \end{bmatrix}^{\lambda-\kappa}_{\kappa,\, m-\kappa} \overline{\underline{L}}^{\lambda-\kappa}_{\lambda-\kappa} [0,\ l]^{\kappa,\, n-\kappa}_{\lambda-\kappa}$$

$$= \frac{1}{L^2} \begin{bmatrix} 0 \\ l \end{bmatrix}^{\lambda-\kappa}_{\kappa,\, m-\kappa} \overline{\underline{L}}^{\lambda-\kappa}_{\lambda-\kappa} \cdot [l]^{\lambda-\kappa}_{\lambda-\kappa} \cdot \overline{\underline{L}}^{\lambda-\kappa}_{\lambda-\kappa} [0,\ l]^{\kappa,\, n-\kappa}_{\lambda-\kappa}; \quad \ldots\ldots\ldots\ldots(6)$$

so that Φ_λ is the matrix similar to $[a]^{n}_{m}$ formed by adding initial horizontal and vertical rows of 0's to ϕ_λ.

The corresponding expressions for $\Phi_\alpha, \Phi_\beta, \Phi_\gamma, \ldots \Phi_\rho$ and $T_\alpha, T_\beta, T_\gamma, \ldots T_\rho$ are obtained by replacing (λ, κ, l, L) by $(\alpha, 0, a, A)$, (β, α, b, B), (γ, β, c, C), $\ldots (\rho, \pi, r, R)$; and in particular we have

$$\Phi_\alpha = \phi_\alpha = \phi,$$

$$T_\alpha = \frac{1}{A} [a]^{\alpha}_{m} \overline{\underline{A}}^{\alpha}_{\alpha} [a]^{n}_{\alpha} = \frac{1}{A^2} [a]^{\alpha}_{m} \overline{\underline{A}}^{\alpha}_{\alpha} \cdot [a]^{\alpha}_{\alpha} \cdot \overline{\underline{A}}^{\alpha}_{\alpha} [a]^{n}_{\alpha}.$$

The identity (A′) of § 116 gives the equation

$$(l)^{\lambda-\kappa}_{\lambda-\kappa} [l]^{n-\kappa}_{m-\kappa} = [l]^{\lambda-\kappa}_{m-\kappa} \overline{\underline{L}}^{\lambda-\kappa}_{\lambda-\kappa} [l]^{n-\kappa}_{\lambda-\kappa} + (l)^{\lambda-\kappa}_{\lambda-\kappa} \begin{bmatrix} 0, & 0 \\ 0, & m \end{bmatrix}^{\lambda-\kappa,\, n-\lambda}_{\lambda-\kappa,\, m-\lambda},$$

or

$$[l]^{n-\kappa}_{m-\kappa} = \frac{1}{L} [l]^{\lambda-\kappa}_{m-\kappa} \overline{\underline{L}}^{\lambda-\kappa}_{\lambda-\kappa} [l]^{n-\kappa}_{\lambda-\kappa} + \begin{bmatrix} 0, & 0 \\ 0, & m \end{bmatrix}^{\lambda-\kappa,\, n-\lambda}_{\lambda-\kappa,\, m-\lambda}, \quad \ldots\ldots\ldots\ldots(7)$$

or

$$\Phi_\lambda = T_\lambda + \Phi_\mu. \quad \ldots\ldots\ldots\ldots(8)$$

From the equations

$$\phi = T_\alpha + \Phi_\beta, \quad \Phi_\beta = T_\beta + \Phi_\gamma, \quad \Phi_\gamma = T_\gamma + \Phi_\delta, \ldots \Phi_\rho = T_\rho$$

obtained in this way we deduce such equations as

$$\phi = T_\alpha + T_\beta + T_\gamma + \ldots + T_\lambda + \Phi_\mu, \quad \ldots\ldots\ldots\ldots(9)$$

the last of which is

$$[a]_m^n = T_\alpha + T_\beta + T_\gamma + \ldots + T_\kappa + T_\lambda + T_\mu + \ldots + T_\pi + T_\rho. \quad \text{..............(A)}$$

Now let

$$h_\lambda = \frac{1}{L}\begin{bmatrix}0\\ l\end{bmatrix}_{\kappa,\, m-\kappa}^{\lambda-\kappa} \overline{L}_{\lambda-\kappa}^{\lambda-\kappa} = \begin{bmatrix}0\\ 1\\ \lambda'\end{bmatrix}_{\kappa,\, \lambda-\kappa,\, m-\lambda}^{\lambda-\kappa}, \quad \text{..................(10)}$$

and

$$k_\lambda = \frac{1}{L}\overline{L}_{\lambda-\kappa}^{\lambda-\kappa}[0,\ l]_{\lambda-\kappa}^{\kappa,\, n-\kappa} = [0,\ 1,\ \lambda]_{\lambda-\kappa}^{\kappa,\, \lambda-\kappa,\, n-\lambda}, \quad \text{...................(11)}$$

where $\lambda'_{iu} = \frac{1}{L}$ times the determinant formed from $(l)_{\lambda-\kappa}^{\lambda-\kappa}$ when we replace its uth horizontal row by the $(\lambda-\kappa+i)$th horizontal row of $[l]_{m-\kappa}^{\lambda-\kappa}$, and $\lambda_{vj} = \frac{1}{L}$ times the determinant formed from $(l)_{\lambda-\kappa}^{\lambda-\kappa}$ when we replace its vth vertical row by the $(\lambda-\kappa+j)$th vertical row of $[l]_{\lambda-\kappa}^{n-\kappa}$.

We obtain the corresponding expressions for $h_\alpha, h_\beta, h_\gamma, \ldots h_\rho$ and $k_\alpha, k_\beta, k_\gamma, \ldots k_\rho$ by replacing (λ, κ, l, L) by $(\alpha, 0, a, A)$, (β, α, b, B), (γ, β, c, C), ... (ρ, π, r, R); and in particular we have

$$h_\alpha = \frac{1}{A}[a]_m^\alpha \overline{A}_\alpha^\alpha = \begin{bmatrix}1\\ a'\end{bmatrix}_{\alpha,\, m-\alpha}^{\alpha}, \quad k_\alpha = \frac{1}{A}\overline{A}_\alpha^\alpha [a]_\alpha^n = [1,\ a]_\alpha^{\alpha,\, n-\alpha}.$$

Further let

$$u_{m-\rho} = \begin{bmatrix}0\\ 1\end{bmatrix}_{\rho,\, m-\rho}^{m-\rho}, \quad v_{n-\rho} = [0,\ 1]_{n-\rho}^{\rho,\, n-\rho}. \quad \text{.......................(12)}$$

Then the equation (A) can be written in the form

$$[a]_m^n = h_\alpha [a]_\alpha^\alpha k_\alpha + h_\beta [b]_{\beta-\alpha}^{\beta-\alpha} k_\beta + \ldots + h_\lambda [l]_{\lambda-\kappa}^{\lambda-\kappa} k_\lambda + \ldots$$

$$+ h_\rho [r]_{\rho-\pi}^{\rho-\pi} k_\rho + 0 \,.\, u_{m-\rho} v_{n-\rho}. \quad \text{....................(A}_1\text{)}$$

By the property of passive rows given in § 43.9 the equation (A_1) is equivalent to

$$\phi = [h]_m^m \psi [k]_n^n, \quad \text{.......................................(B)}$$

where ψ is the matrix defined in the enunciation;

$[h]_m^m$ is the matrix whose successive vertical rows are the successive vertical rows of the matrices $h_\alpha, h_\beta, h_\gamma, \ldots h_\rho, u_{m-\rho}$ taken in this order;

$[k]_n^n$ is the matrix whose successive horizontal rows are the successive horizontal rows of the matrices $k_\alpha, k_\beta, k_\gamma, \ldots k_\rho, v_{n-\rho}$ taken in this order.

We may define $[h]_m^m$ by saying that its last $m-\rho$ vertical rows are $u_{m-\rho}$, and its $(\kappa+1)$th, $(\kappa+2)$th, ... λth vertical rows are the 1st, 2nd, ... $(\lambda-\kappa)$th vertical rows of h_λ;

and we may define $[k]_n^n$ by saying that its last $n-\rho$ horizontal rows are $v_{n-\rho}$, and its $(\kappa+1)$th, $(\kappa+2)$th, ... λth horizontal rows are the 1st, 2nd, ... $(\lambda-\kappa)$th horizontal rows of k_λ.

The matrices $[h]_m^m$ and $[k]_n^n$ have the general forms

$$[h]_m^m = \begin{bmatrix} 1, & 0, & 0, & \dots & 0, & 0 \\ \sigma, & 1, & 0, & \dots & 0, & 0 \\ \sigma, & \sigma, & 1, & \dots & 0, & 0 \\ \dots & \dots & \dots & \dots & \dots & \dots \\ \sigma, & \sigma, & \sigma, & \dots & 1, & 0 \\ \sigma, & \sigma, & \sigma, & \dots & \sigma, & 1 \end{bmatrix}_{\alpha,\, \beta-\alpha,\, \gamma-\beta,\, \dots\, \rho-\pi,\, m-\rho}^{\alpha,\, \beta-\alpha,\, \gamma-\beta,\, \dots\, \rho-\pi,\, m-\rho},$$

$$[k]_n^n = \begin{bmatrix} 1, & \tau, & \tau, & \dots & \tau, & \tau \\ 0, & 1, & \tau, & \dots & \tau, & \tau \\ 0, & 0, & 1, & \dots & \tau, & \tau \\ \dots & \dots & \dots & \dots & \dots & \dots \\ 0, & 0, & 0, & \dots & 1, & \tau \\ 0, & 0, & 0, & \dots & 0, & 1 \end{bmatrix}_{\alpha,\, \beta-\alpha,\, \gamma-\beta,\, \dots\, \rho-\pi,\, n-\rho}^{\alpha,\, \beta-\alpha,\, \gamma-\beta,\, \dots\, \rho-\pi,\, n-\rho},$$

the constituent matrices where a σ or τ occurs being determined in the manner shown above.

It follows that the equation (B) is a unitary equigradent transformation in Ω converting ψ into ϕ; and if $\overline{\underline{H}}_m^m$ and $\overline{\underline{K}}_n^n$ are the inverse matrices of $[h]_m^m$ and $[k]_n^n$, having the same general forms as $[h]_m^m$ and $[k]_n^n$ respectively, the inverse equation

$$\overline{\underline{H}}_m^m \psi \overline{\underline{K}}_n^n = \phi \quad \dots\dots\dots\dots\dots\dots\dots\dots (B')$$

is a unitary equigradent transformation in Ω converting ϕ into ψ.

We have thus proved the theorem.

NOTE 1. *Resolution of* (B) *and* (B′) *into simpler unitary transformations.*

Let p_λ be the matrix formed from $[1]_m^m$ when we replace the $\lambda-\kappa$ vertical rows immediately following the κth vertical row by h_λ; let q_λ be the matrix formed from $[1]_n^n$ when we replace the $\lambda-\kappa$ horizontal rows immediately following the κth horizontal row by k_λ; let P_λ be the matrix formed from p_λ when we replace $[\lambda']_{m-\lambda}^{\lambda-\kappa}$ by $-[\lambda']_{m-\lambda}^{\lambda-\kappa}$; let Q_λ be the matrix formed from q_λ when we replace $[\lambda]_{\lambda-\kappa}^{n-\lambda}$ by $-[\lambda]_{\lambda-\kappa}^{n-\lambda}$; so that P_λ is the inverse of p_λ, and Q_λ is the inverse of q_λ; and let $p_\alpha, p_\beta, \dots p_\rho, q_\alpha, q_\beta, \dots q_\rho, P_\alpha, P_\beta, \dots P_\rho, Q_\alpha, Q_\beta, \dots Q_\rho$ be similarly defined.

In particular we form p_α by replacing the first α vertical rows of $[1]_m^m$ by h_α, and we form q_α by replacing the first α horizontal rows of $[1]_n^n$ by k_α.

Then by Ex. xiii of § 101 we have

$$[h]_m^m = p_\alpha p_\beta \dots p_\rho, \quad [k]_n^n = q_\rho \dots q_\beta q_\alpha,$$

and therefore

$$\overline{\underline{H}}{}^m_m = P_\rho \dots P_\beta P_\alpha, \quad \overline{\underline{K}}{}^n_n = Q_\alpha Q_\beta \dots Q_\rho.$$

Thus we can replace the equations (B) and (B′) by

$$\phi = p_\alpha p_\beta \dots p_\rho \,.\, \psi \,.\, q_\rho \dots q_\beta q_\alpha, \quad \dots\dots\dots\dots(\text{B}_1)$$

$$\psi = P_\rho \dots P_\beta P_\alpha \,.\, \phi \,.\, Q_\alpha Q_\beta \dots Q_\rho, \quad \dots\dots\dots\dots(\text{B}_1{}')$$

and so express each of them as the resultant of a number of simpler unitary transformations in Ω. Each of these simpler transformations can be expressed as the resultant of a number of elementary unitary equigradent transformations in Ω.

NOTE 2. *First alternative proof of the theorem.*

Let

$$\psi_\alpha = \begin{bmatrix} a, & 0 \\ 0, & b \end{bmatrix}^{\alpha,\, n-\alpha}_{\alpha,\, m-\alpha}, \quad \psi_\beta = \begin{bmatrix} a, & 0, & 0 \\ 0, & b, & 0 \\ 0, & 0, & c \end{bmatrix}^{\alpha,\, \beta-\alpha,\, n-\beta}_{\alpha,\, \beta-\alpha,\, m-\beta},$$

$$\psi_\gamma = \begin{bmatrix} a, & 0, & 0, & 0 \\ 0, & b, & 0, & 0 \\ 0, & 0, & c, & 0 \\ 0, & 0, & 0, & d \end{bmatrix}^{\alpha,\, \beta-\alpha,\, \gamma-\beta,\, n-\gamma}_{\alpha,\, \beta-\alpha,\, \gamma-\beta,\, m-\gamma}, \quad \dots \quad \psi_\rho = \psi. \quad \dots\dots\dots\dots(13)$$

By the properties of passive rows we see as in the first proof of Theorem II of § 142 that the successive equations (7) are equivalent to

$$[a]^n_m = \begin{bmatrix} 1, & 0 \\ a', & 1 \end{bmatrix}^{\alpha,\, m-\alpha}_{\alpha,\, m-\alpha} \begin{bmatrix} a, & 0 \\ 0, & b \end{bmatrix}^{\alpha,\, n-\alpha}_{\alpha,\, m-\alpha} \begin{bmatrix} 1, & a \\ 0, & 1 \end{bmatrix}^{\alpha,\, n-\alpha}_{\alpha,\, n-\alpha},$$

$$[b]^{n-\alpha}_{m-\alpha} = \begin{bmatrix} 1, & 0 \\ \beta', & 1 \end{bmatrix}^{\beta-\alpha,\, m-\beta}_{\beta-\alpha,\, m-\beta} \begin{bmatrix} b, & 0 \\ 0, & c \end{bmatrix}^{\beta-\alpha,\, n-\beta}_{\beta-\alpha,\, m-\beta} \begin{bmatrix} 1, & \beta \\ 0, & 1 \end{bmatrix}^{\beta-\alpha,\, n-\beta}_{\beta-\alpha,\, n-\beta},$$

$$\dots\dots\dots\dots\dots\dots\dots\dots\dots\dots\dots\dots$$

$$[l]^{n-\kappa}_{m-\kappa} = \begin{bmatrix} 1, & 0 \\ \lambda', & 1 \end{bmatrix}^{\lambda-\kappa,\, m-\lambda}_{\lambda-\kappa,\, m-\lambda} \begin{bmatrix} l, & 0 \\ 0, & m \end{bmatrix}^{\lambda-\kappa,\, n-\lambda}_{\lambda-\kappa,\, m-\lambda} \begin{bmatrix} 1, & \lambda \\ 0, & 1 \end{bmatrix}^{\lambda-\kappa,\, n-\lambda}_{\lambda-\kappa,\, n-\lambda},$$

$$\dots\dots\dots\dots\dots\dots\dots\dots\dots\dots\dots\dots$$

$$[r]^{n-\pi}_{m-\pi} = \begin{bmatrix} 1, & 0 \\ \rho', & 1 \end{bmatrix}^{\rho-\pi,\, m-\rho}_{\rho-\pi,\, m-\rho} \begin{bmatrix} r, & 0 \\ 0, & 0 \end{bmatrix}^{\rho-\pi,\, n-\rho}_{\rho-\pi,\, m-\rho} \begin{bmatrix} 1, & \rho \\ 0, & 1 \end{bmatrix}^{\rho-\pi,\, n-\rho}_{\rho-\pi,\, n-\rho}. \quad \dots\dots\dots\dots(14)$$

By Note 5 of § 141 we deduce from (14) the successive equations

$$\phi = p_\alpha \psi_\alpha q_\alpha, \quad \psi_\alpha = p_\beta \psi_\beta q_\beta, \quad \dots \quad \psi_\kappa = p_\lambda \psi_\lambda q_\lambda, \quad \dots \quad \psi_\pi = p_\rho \psi_\rho q_\rho. \quad \dots\dots\dots(15)$$

From the equations (15) we deduce the resultant transformation (B_1) which is equivalent to (B), and from (B) we deduce (B′). Thus we can prove the theorem by means of the successive transformations (14) which can be obtained as in the first proof of Theorem II in § 142; i.e. we can prove the theorem by successive applications of formula (B′) of § 142.

NOTE 3. *Second alternative proof of the theorem.*

The successive equations (14) are equivalent to the successive equations

$$\begin{bmatrix} 1, & 0 \\ -a', & 1 \end{bmatrix}_{a,\, m-a}^{a,\, m-a} [a]_m^n \begin{bmatrix} 1, & -a \\ 0, & 1 \end{bmatrix}_{a,\, n-a}^{a,\, n-a} = \begin{bmatrix} a, & 0 \\ 0, & b \end{bmatrix}_{a,\, m-a}^{a,\, n-a},$$

$$\begin{bmatrix} 1, & 0 \\ -\beta', & 1 \end{bmatrix}_{\beta-a,\, m-\beta}^{\beta-a,\, m-\beta} [b]_{m-a}^{n-a} \begin{bmatrix} 1, & -\beta \\ 0, & 1 \end{bmatrix}_{\beta-a,\, n-\beta}^{\beta-a,\, n-\beta} = \begin{bmatrix} b, & 0 \\ 0, & c \end{bmatrix}_{\beta-a,\, m-\beta}^{\beta-a,\, n-\beta},$$

....

$$\begin{bmatrix} 1, & 0 \\ -\lambda', & 1 \end{bmatrix}_{\lambda-\kappa,\, m-\lambda}^{\lambda-\kappa,\, m-\lambda} [l]_{m-\kappa}^{n-\kappa} \begin{bmatrix} 1, & -\lambda \\ 0, & 1 \end{bmatrix}_{\lambda-\kappa,\, n-\lambda}^{\lambda-\kappa,\, n-\lambda} = \begin{bmatrix} l, & 0 \\ 0, & m \end{bmatrix}_{\lambda-\kappa,\, m-\lambda}^{\lambda-\kappa,\, n-\lambda},$$

....

$$\begin{bmatrix} 1, & 0 \\ -\rho', & 1 \end{bmatrix}_{\rho-\pi,\, m-\rho}^{\rho-\pi,\, m-\rho} [r]_{m-\pi}^{n-\pi} \begin{bmatrix} 1, & -\rho \\ 0, & 1 \end{bmatrix}_{\rho-\pi,\, n-\rho}^{\rho-\pi,\, n-\rho} = \begin{bmatrix} r, & 0 \\ 0, & 0 \end{bmatrix}_{\rho-\pi,\, m-\rho}^{\rho-\pi,\, n-\rho}. \quad \text{.......(14')}$$

By Note 5 of § 141 we deduce from (14′) the successive equations

$$P_a\phi Q_a = \psi_a,\quad P_\beta\psi_a Q_\beta = \psi_\beta,\ \dots\ P_\lambda\psi_\kappa Q_\lambda = \psi_\lambda,\ \dots\ P_\rho\psi_\pi Q_\rho = \psi_\rho. \quad \text{.......(15')}$$

From the equations (15′) we deduce the resultant transformation (B_1') which is equivalent to (B′), and from (B′) we deduce (B).

Thus we can prove the theorem by means of the successive transformations (14′) which can be obtained as in the second proof of Theorem II of § 142; i.e. we can prove the theorem by successive applications of formula (B) of § 142.

NOTE 4. *Transformation represented by the equation* (9).

By Ex. xiii of § 101 we see that $p_a p_\beta \dots p_\lambda$ is the square matrix of order m whose successive vertical rows are the successive vertical rows of the matrices $h_a, h_\beta, \dots h_\lambda$, $\begin{bmatrix} 0 \\ 1 \end{bmatrix}_{\lambda,\, m-\lambda}^{m-\lambda}$ taken in this order; and that $q_\kappa \dots q_\beta q_a$ is the square matrix of order n whose successive horizontal rows are the successive horizontal rows of the matrices $k_a, k_\beta, \dots k_\lambda, [0,\ 1]_{n-\lambda}^{\lambda,\, n-\lambda}$.

Hence when we write (9) in the form

$$[a]_m^n = h_a [a]_a^a k_a + h_\beta [b]_{\beta-a}^{\beta-a} k_\beta + \dots + h_\lambda [l]_{\lambda-\kappa}^{\lambda-\kappa} k_\lambda + \begin{bmatrix} 0 \\ 1 \end{bmatrix}_{\lambda,\, m-\lambda}^{m-\lambda} [m]_{m-\lambda}^{n-\lambda} [0,\ 1]_{n-\lambda}^{\lambda,\, n-\lambda}, \quad \text{...(9')}$$

we see that it is equivalent to the transformations

$$\phi = p_a p_\beta \dots p_\lambda \psi_\lambda q_\lambda \dots q_\beta q_a,\quad \psi_\lambda = P_\lambda \dots P_\beta P_a \phi Q_a Q_\beta \dots Q_\lambda. \quad \text{.........(9'')}$$

Ex. i. *Special case when* ϕ *is symmetric.*

In this case we have $n=m$, the matrices $\phi_\beta, \phi_\gamma, \dots \phi_\rho$ are all symmetric, the matrices $[l]_{\lambda-\kappa}^{\lambda-\kappa}$ and $[L]_{\lambda-\kappa}^{\lambda-\kappa}$ are symmetric, and h_λ and k_λ are mutually conjugate. Consequently the compartite matrix ψ is symmetric, and the transformations (B) and (B′) are symmetric.

Ex. ii. *Special case when ϕ is skew-symmetric.*

In this case we have $n=m$, the matrices $\phi_\beta, \phi_\gamma, \dots \phi_\rho$ are all skew-symmetric, the matrices $[l]_{\lambda-\kappa}^{\lambda-\kappa}$ and $[L]_{\lambda-\kappa}^{\lambda-\kappa}$ are skew-symmetric, and h_λ and k_λ are mutually conjugate. Consequently the compartite matrix ψ is skew-symmetric and the transformations (B) and (B′) are symmetric.

§ 144. A corresponding non-unitary equigradent transformation.

Theorem. *Let $\alpha, \beta, \gamma, \dots \kappa, \lambda, \mu, \dots \pi, \rho$ be a series of constantly increasing non-zero positive integers, and let $\phi=[a]_m^n$ be a matrix of rank ρ in which none of the leading diagonal minor determinants*

$$\Delta_0=1,\ \Delta_\alpha=(a)_\alpha^\alpha,\ \Delta_\beta=(a)_\beta^\beta, \dots \Delta_\kappa=(a)_\kappa^\kappa,\ \Delta_\lambda=(a)_\lambda^\lambda,\ \Delta_\mu=(a)_\mu^\mu, \dots \Delta_\rho=(a)_\rho^\rho$$

vanish. Then we can express ϕ in the form

$$[a]_m^n = T_\alpha + T_\beta + T_\gamma + \dots + T_\kappa + T_\lambda + T_\mu + \dots + T_\pi + T_\rho, \quad \dots(\mathrm{A})$$

where

$$T_\lambda = \frac{1}{\Delta_\kappa \Delta_\lambda^2} \begin{bmatrix} 0 \\ \mathbf{L}' \end{bmatrix}_{\kappa,\, m-\kappa}^{\lambda-\kappa} [l]_{\lambda-\kappa}^{\lambda-\kappa} [0,\ \mathbf{L}]_{\lambda-\kappa}^{\kappa,\, n-\kappa}; \quad l_{uv} = \begin{pmatrix} 1, 2, \dots \kappa, \kappa+v \\ a \\ 1, 2, \dots \kappa, \kappa+u \end{pmatrix};$$

and $\mathbf{L}'_{iu}$, $\mathbf{L}_{vj}$ are the determinants formed from Δ_λ when we respectively replace the horizontal suffix $\kappa+u$ by $\kappa+i$, the vertical suffix $\kappa+v$ by $\kappa+j$.

According to these definitions we have

$$\mathbf{L}'_{i1} = \begin{pmatrix} 1, 2, \dots \kappa, \kappa+1, \kappa+2, \dots \lambda \\ a \\ 1, 2, \dots \kappa, \kappa+i, \kappa+2, \dots \lambda \end{pmatrix}, \quad \mathbf{L}_{1j} = \begin{pmatrix} 1, 2, \dots \kappa, \kappa+j, \kappa+2, \dots \lambda \\ a \\ 1, 2, \dots \kappa, \kappa+1, \kappa+2, \dots \lambda \end{pmatrix}.$$

We obtain $T_\alpha, T_\beta, T_\gamma, \dots T_\rho$ from T_λ by replacing $(\lambda, \kappa, l, \mathbf{L}', \mathbf{L})$ by $(\alpha, 0, a, \mathbf{A}', \mathbf{A})$, $(\beta, \alpha, b, \mathbf{B}', \mathbf{B})$, $(\gamma, \beta, c, \mathbf{C}', \mathbf{C})$, ... $(\rho, \pi, r, \mathbf{R}', \mathbf{R})$ respectively; and we obtain $(a_{uv}, \mathbf{A}_{uv}, \mathbf{A}'_{uv})$, $(b_{uv}, \mathbf{B}_{uv}, \mathbf{B}'_{uv})$, $(c_{uv}, \mathbf{C}_{uv}, \mathbf{C}'_{uv})$, ... $(r_{uv}, \mathbf{R}_{uv}, \mathbf{R}'_{uv})$ from $(l_{uv}, \mathbf{L}_{uv}, \mathbf{L}'_{uv})$ by replacing κ by $0, \alpha, \beta, \dots \rho$ respectively.

In particular a_{uv} is an element of ϕ, $T_\alpha = \dfrac{1}{\Delta_0 \Delta_\alpha^2} [\mathbf{A}]_m^\alpha [a]_\alpha^\alpha [\mathbf{A}']_\alpha^n$, and $\mathbf{A}'_{iu}$, $\mathbf{A}_{vj}$ are the determinants formed from Δ_α when we respectively replace the horizontal suffix u by i, the vertical suffix v by j.

Direct proof of the theorem. Let

$$\phi_\beta=[b]_{m-\alpha}^{n-\alpha},\quad \phi_\gamma=[c]_{m-\beta}^{n-\beta}, \dots \phi_\lambda=[l]_{m-\kappa}^{n-\kappa},\quad \phi_\mu=[m]_{m-\lambda}^{n-\lambda}, \dots \phi_\rho=[r]_{m-\pi}^{n-\pi}$$

be the matrices of ranks $\rho-\alpha, \rho-\beta, \dots \rho-\kappa, \rho-\lambda, \dots \rho-\pi$ whose elements are defined as in the enunciation by the equations

$$b_{uv}=\begin{pmatrix} 1, 2, \dots \alpha, \alpha+v \\ a \\ 1, 2, \dots \alpha, \alpha+u \end{pmatrix}, \quad c_{uv}=\begin{pmatrix} 1, 2, \dots \beta, \beta+v \\ a \\ 1, 2, \dots \beta, \beta+u \end{pmatrix}, \dots l_{uv}=\begin{pmatrix} 1, 2, \dots \kappa, \kappa+v \\ a \\ 1, 2, \dots \kappa, \kappa+u \end{pmatrix}, \dots,$$

and let
$$L=(l)_{\lambda-\kappa}^{\lambda-\kappa}, \quad \Phi_\lambda=\begin{bmatrix} 0, & 0 \\ 0, & l \end{bmatrix}_{\kappa,\, m-\kappa}^{\kappa,\, n-\kappa},$$

the corresponding expressions for (A, Φ_a), (B, Φ_β), (C, Φ_γ), ... (R, Φ_ρ) being obtained respectively from (L, Φ_λ) by replacing (λ, κ, l) by $(a, 0, a)$, (β, a, b), (γ, β, c), ... (ρ, π, r).

In particular we have $\qquad A = (a)_a^a = \Delta_a, \quad \Phi_a = \phi_a = \phi.$

From the identity of § 110 we see that

$$A = \Delta_0^{a-1}\,\Delta_a, \quad B = \Delta_a^{\beta-a-1}\,\Delta_\beta, \; \ldots \; L = \Delta_\kappa^{\lambda-\kappa-1}\,\Delta_\lambda, \; \ldots \; R = \Delta_\pi^{\rho-\pi-1}\,\Delta_\rho, \qquad \ldots\ldots(1)$$

and all these are non-vanishing determinants.

From the same identity we obtain such equations as

$$(l_{uv})_s^s = \Delta_\kappa^{s-1}\begin{pmatrix} 1, 2, \ldots \kappa, \kappa+v_1, \ldots \kappa+v_s \\ a \\ 1, 2, \ldots \kappa, \kappa+u_1, \ldots \kappa+u_s \end{pmatrix}, \quad \begin{pmatrix} 1, 2, \ldots \lambda-\kappa, \lambda-\kappa+v \\ l \\ 1, 2, \ldots \lambda-\kappa, \lambda-\kappa+u \end{pmatrix} = \Delta_\kappa^{\lambda-\kappa} m_{uv}. \qquad \ldots\ldots(2)$$

The equations (1) and (2) show that the matrices $\phi_\beta, \phi_\gamma, \ldots \phi_\lambda, \phi_\mu, \ldots \phi_\rho$ have ranks $\rho - a$, $\rho - \beta$, ... $\rho - \kappa$, $\rho - \lambda$, ... $\rho - \pi$.

Denoting the reciprocal of $[l]_{\lambda-\kappa}^{\lambda-\kappa}$ by $[L]_{\lambda-\kappa}^{\lambda-\kappa}$ and applying the identity (A′) of § 116, we obtain the equation

$$(l)_{\lambda-\kappa}^{\lambda-\kappa}[l]_{m-\kappa}^{n-\kappa} = [l]_{m-\kappa}^{\lambda-\kappa}\,\overline{\underline{L}}_{\lambda-\kappa}^{\lambda-\kappa}\,[l]_{\lambda-\kappa}^{n-\kappa} + \Delta_\kappa^{\lambda-\kappa}\begin{bmatrix} 0, & 0 \\ 0, & m \end{bmatrix}_{\lambda-\kappa,\, m-\lambda}^{\lambda-\kappa,\, n-\lambda}. \qquad \ldots\ldots\ldots\ldots(3)$$

When we write $\qquad \overline{\underline{L}}_{\lambda-\kappa}^{\lambda-\kappa} = \dfrac{1}{L}\,\overline{\underline{L}}_{\lambda-\kappa}^{\lambda-\kappa}\,[l]_{\lambda-\kappa}^{\lambda-\kappa}\,\overline{\underline{L}}_{\lambda-\kappa}^{\lambda-\kappa},$

$$[l]_{m-\kappa}^{\lambda-\kappa}\,\overline{\underline{L}}_{\lambda-\kappa}^{\lambda-\kappa} = [\lambda']_{m-\kappa}^{\lambda-\kappa}, \qquad \overline{\underline{L}}_{\lambda-\kappa}^{\lambda-\kappa}\,[l]_{\lambda-\kappa}^{n-\kappa} = [\lambda]_{\lambda-\kappa}^{n-\kappa},$$

so that λ'_{iu}, λ_{vj} are the determinants formed from L when we replace the horizontal suffix u by i, the vertical suffix v by j, we can replace (3) by

$$[l]_{m-\kappa}^{n-\kappa} = \frac{1}{L^2}[\lambda']_{m-\kappa}^{\lambda-\kappa}[l]_{\lambda-\kappa}^{\lambda-\kappa}[\lambda]_{\lambda-\kappa}^{n-\kappa} + \frac{\Delta_\kappa}{\Delta_\lambda}\begin{bmatrix} 0, & 0 \\ 0, & m \end{bmatrix}_{\lambda-\kappa,\, m-\lambda}^{\lambda-\kappa,\, n-\lambda}. \qquad \ldots\ldots\ldots\ldots(4)$$

Now if $\mathbf{A}'_{iu}, \mathbf{B}'_{iu}, \ldots \mathbf{L}'_{iu}, \ldots \mathbf{R}'_{iu}$ and $\mathbf{A}_{vj}, \mathbf{B}_{vj}, \ldots \mathbf{L}_{vj}, \ldots \mathbf{R}_{vj}$ are defined as in the enunciation of the theorem, it follows from (2) that

$$\lambda'_{iu} = \Delta_\kappa^{\lambda-\kappa-1}\,\mathbf{L}'_{iu} = \frac{L}{\Delta_\kappa}\mathbf{L}'_{iu}, \qquad \lambda_{vj} = \Delta_\kappa^{\lambda-\kappa-1}\,\mathbf{L}_{vj} = \frac{L}{\Delta_\lambda}\mathbf{L}_{vj},$$

and the equation (4) is equivalent to

$$\frac{1}{\Delta_\kappa}[l]_{m-\kappa}^{n-\kappa} = \frac{1}{\Delta_\kappa\Delta_\lambda^2}[\mathbf{L}']_{m-\kappa}^{\lambda-\kappa}[l]_{\lambda-\kappa}^{\lambda-\kappa}[\mathbf{L}]_{\lambda-\kappa}^{n-\kappa} + \frac{1}{\Delta_\lambda}\begin{bmatrix} 0, & 0 \\ 0, & m \end{bmatrix}_{\lambda-\kappa,\, m-\lambda}^{\lambda-\kappa,\, n-\lambda},$$

or

$$\frac{1}{\Delta_\kappa}\Phi_\lambda = T_\lambda + \frac{1}{\Delta_\lambda}\Phi_\mu. \qquad \ldots\ldots\ldots\ldots\ldots\ldots\ldots\ldots\ldots\ldots\ldots\ldots(5)$$

From the equations

$$\frac{1}{\Delta_0}\phi = T_a + \frac{1}{\Delta_a}\Phi_\beta, \quad \frac{1}{\Delta_a}\Phi_\beta = T_\beta + \frac{1}{\Delta_\beta}\Phi_\gamma, \; \ldots \; \frac{1}{\Delta_\pi}\Phi_\rho = T_\rho$$

obtained in this way we deduce such equations as

$$\phi = T_a + T_\beta + T_\gamma + \ldots + T_\lambda + \frac{1}{\Delta_\lambda}\Phi_\mu, \qquad \ldots\ldots\ldots\ldots\ldots\ldots\ldots\ldots\ldots(6)$$

the last of which is the equation (A).

Alternative proof of the theorem. We can deduce the theorem from the equation (A) of § 143. Defining a_{uv}, b_{uv}, c_{uv}, ... l_{uv}, ... r_{uv} and A, B, C, ... L, ... R as in § 143, we have

$$[a]_m^n = T_\alpha + T_\beta + T_\gamma + \ldots + T_\kappa + T_\lambda + T_\mu + \ldots + T_\pi + T_\rho, \quad \ldots\ldots(7)$$

where
$$T_\lambda = \frac{1}{L^2} \begin{bmatrix} 0 \\ l \end{bmatrix}_{\kappa,\, m-\kappa}^{\lambda-\kappa} \underline{\overline{L}}_{\lambda-\kappa}^{\lambda-\kappa} \cdot [l]_{\lambda-\kappa}^{\lambda-\kappa} \cdot \underline{\overline{L}}_{\lambda-\kappa}^{\lambda-\kappa} [0,\, l]_{\lambda-\kappa}^{\kappa,\, n-\kappa}.$$

Writing
$$[l]_{m-\kappa}^{\lambda-\kappa} \underline{\overline{L}}_{\lambda-\kappa}^{\lambda-\kappa} = [\lambda']_{m-\kappa}^{\lambda-\kappa}, \quad \underline{\overline{L}}_{\lambda-\kappa}^{\lambda-\kappa} [l]_{\lambda-\kappa}^{n-\kappa} = [\lambda]_{\lambda-\kappa}^{n-\kappa},$$

we have
$$T_\lambda = \frac{1}{L^2} \begin{bmatrix} 0 \\ \lambda' \end{bmatrix}_{\kappa,\, m-\kappa}^{\lambda-\kappa} [l]_{\lambda-\kappa}^{\lambda-\kappa} [0,\, \lambda]_{\lambda-\kappa}^{\kappa,\, n-\kappa}, \quad \ldots\ldots(8)$$

where λ'_{iu}, λ_{vj} are the determinants formed from the determinant $L = (l)_{\lambda-\kappa}^{\lambda-\kappa}$ when we replace the horizontal suffix u by i, the vertical suffix v by j.

If $\mathbf{A}'_{iu}$, $\mathbf{B}'_{iu}$, ... $\mathbf{L}'_{iu}$, ... $\mathbf{R}'_{iu}$ and $\mathbf{A}_{vj}$, $\mathbf{B}_{vj}$, ... $\mathbf{L}_{vj}$, ... $\mathbf{R}_{vj}$ are defined as in the enunciation of the theorem, it follows from the equations (4) and (3) of § 143 that

$$\Delta_\kappa \lambda'_{iu} = \mathbf{L}'_{iu}, \quad \Delta_\kappa \lambda_{vj} = \mathbf{L}_{vj}, \quad L = \frac{\Delta_\lambda}{\Delta_\kappa}.$$

Substituting these values for λ'_{iu}, λ_{vj}, L in (8) we obtain

$$T_\lambda = \frac{1}{\Delta_\lambda^2} \begin{bmatrix} 0 \\ \mathbf{L}' \end{bmatrix}_{\kappa,\, m-\kappa}^{\lambda-\kappa} [l]_{\lambda-\kappa}^{\lambda-\kappa} [0,\, \mathbf{L}]_{\lambda-\kappa}^{\kappa,\, n-\kappa}. \quad \ldots\ldots(9)$$

Now let
$$\mathbf{l}_{uv} = \begin{pmatrix} 1, 2, \ldots \kappa, \kappa+v \\ a \\ 1, 2, \ldots \kappa, \kappa+u \end{pmatrix} = \Delta_\kappa l_{uv}.$$

Substituting this value of l_{uv} in (9) we see that

$$T_\lambda = \frac{1}{\Delta_\kappa \Delta_\lambda^2} \begin{bmatrix} 0 \\ \mathbf{L}' \end{bmatrix}_{\kappa,\, m-\kappa}^{\lambda-\kappa} [\mathbf{l}]_{\lambda-\kappa}^{\lambda-\kappa} [0,\, \mathbf{L}]_{\lambda-\kappa}^{\kappa,\, n-\kappa}. \quad \ldots\ldots(10)$$

Equation (7) when we express T_λ in the form (10) is equivalent to the equation (A) in the theorem of this article.

If Φ_λ is defined as in § 143, and if

$$\Psi_\lambda = \begin{bmatrix} 0, & 0 \\ 0, & \mathbf{1} \end{bmatrix}_{\kappa,\, m-\kappa}^{\kappa,\, n-\kappa}, \quad \text{so that} \quad \Delta_\kappa \Phi_\lambda = \Psi_\lambda,$$

we see from the equations (8) and (9) of § 143 that

$$\frac{1}{\Delta_\kappa} \Psi_\lambda = T_\lambda + \frac{1}{\Delta_\lambda} \Psi_\mu, \quad \ldots\ldots(11)$$

$$\phi = T_\alpha + T_\beta + T_\gamma + \ldots + T_\lambda + \frac{1}{\Delta_\lambda} \Psi_\mu. \quad \ldots\ldots(12)$$

Equations (11) and (12) are equivalent to the equations (5) and (6) in the first proof.

Reverting to the notation of the theorem and the first proof, we see from the properties of passive rows that the equation (A) is equivalent to the equation

$$\phi = [h]_m^m \psi [k]_n^n, \quad \ldots\ldots(B)$$

where ψ is a compartite matrix of standard form whose successive parts are

$$\frac{1}{\Delta_0\Delta_\alpha^2}[a]_\alpha^\alpha,\quad \frac{1}{\Delta_\alpha\Delta_\beta^2}[b]_{\beta-\alpha}^{\beta-\alpha},\ldots \frac{1}{\Delta_\kappa\Delta_\lambda^2}[l]_{\lambda-\kappa}^{\lambda-\kappa},\ldots \frac{1}{\Delta_\pi\Delta_\rho^2}[r]_{\rho-\pi}^{\rho-\pi},\quad [0]_{m-\rho}^{n-\rho};$$

$[h]_m^m$ is the matrix whose successive vertical rows are the successive vertical rows of the matrices

$$[\mathbf{A}']_m^\alpha,\quad \begin{bmatrix}0\\ \mathbf{B}'\end{bmatrix}_{\alpha,\,m-\alpha}^{\beta-\alpha},\ldots \begin{bmatrix}0\\ \mathbf{L}'\end{bmatrix}_{\kappa,\,m-\kappa}^{\lambda-\kappa},\ldots \begin{bmatrix}0\\ \mathbf{R}'\end{bmatrix}_{\pi,\,m-\pi}^{\rho-\pi},\quad \begin{bmatrix}0\\ 1\end{bmatrix}_{\rho,\,m-\rho}^{m-\rho};$$

$[k]_n^n$ is the matrix whose successive horizontal rows are the successive horizontal rows of the matrices

$$[\mathbf{A}]_\alpha^n,\quad [0,\mathbf{B}]_{\beta-\alpha}^{\alpha,\,n-\alpha},\ldots [0,\mathbf{L}]_{\lambda-\kappa}^{\kappa,\,n-\kappa},\ldots [0,\mathbf{R}]_{\rho-\pi}^{\pi,\,n-\pi},\quad [0,1]_{n-\rho}^{\rho,\,n-\rho}.$$

Since

$$[\mathbf{A}']_\alpha^\alpha=[\mathbf{A}]_\alpha^\alpha=\Delta_\alpha[1]_\alpha^\alpha,\ldots [\mathbf{L}']_{\lambda-\kappa}^{\lambda-\kappa}=[\mathbf{L}]_{\lambda-\kappa}^{\lambda-\kappa}=\Delta_\kappa[1]_{\lambda-\kappa}^{\lambda-\kappa},\ldots,$$

none of the first ρ leading diagonal elements of $[h]_m^m$ and $[k]_n^n$ vanish. The compartite matrix ψ has the same general form as the matrix ψ in § 143.

If $\overline{H}_m^m$ and $\overline{K}_n^n$ are the inverses of $[h]_m^m$ and $[k]_n^n$, we deduce from (B) the inverse equation

$$\overline{H}_m^m\,\phi\,\overline{K}_n^n=\psi. \qquad\ldots\ldots\ldots\ldots\ldots\ldots(\mathrm{B}')$$

Equations (B) and (B′) are equigradent transformations in Ω which respectively convert ψ into ϕ and ϕ into ψ, and which correspond to the transformations (B) and (B′) of § 143. They are not unitary, but they have the property that the elements of $[h]_m^m$ and $[k]_n^n$ which are not 0's are all minor determinants of ϕ.

Ex. i. *Special case when* $[\alpha\beta\gamma\ldots\rho]=[1\,2\,3\ldots\rho]$.

In this case $\kappa=\lambda-1$, $\mathbf{L}_{i1}=l_{i1}$, $\mathbf{L}_{1j}=l_{1j}$, $l_{11}=\Delta_\lambda$, and the equation (A) becomes

$$[a]_m^n=\frac{1}{\Delta_0\Delta_1}[a]_m^1[a]_1^n+\frac{1}{\Delta_1\Delta_2}\begin{bmatrix}0\\ b\end{bmatrix}_{1,\,m-1}^1[0,b]_1^{1,\,n-1}+\frac{1}{\Delta_2\Delta_3}\begin{bmatrix}0\\ c\end{bmatrix}_{2,\,m-2}^1[0,c]_1^{2,\,n-2}$$

$$+\ldots+\frac{1}{\Delta_{\rho-1}\Delta_\rho}\begin{bmatrix}0\\ r\end{bmatrix}_{\rho-1,\,m-\rho+1}^1[0,r]_1^{\rho-1,\,n-\rho+1}. \qquad\ldots\ldots\ldots\ldots(\mathrm{C})$$

Ex. ii. *Special case when* ϕ *is symmetric.*

In this case we have $n=m$, the matrices $\phi_\beta,\ \phi_\gamma,\ \ldots\ \phi_\rho$ are all symmetric, and the matrix $[L]_{\lambda-\kappa}^{\lambda-\kappa}$ is symmetric. It follows in succession that the two matrices of the pairs $[\lambda']_{m-\kappa}^{\lambda-\kappa}$, $[\lambda]_{\lambda-\kappa}^{m-\kappa}$; $[\mathbf{L}']_{m-\kappa}^{\lambda-\kappa}$, $[\mathbf{L}]_{\lambda-\kappa}^{m-\kappa}$; $[h]_m^m$, $[k]_m^m$ are mutually conjugate. Consequently the compartite matrix ψ is symmetric, and the transformations (B) and (B′) are symmetric.

Ex. iii. *Special case when* ϕ *is skew-symmetric.*

In this case we have $n=m$; the matrices $\phi_\beta, \phi_\gamma, \dots \phi_\rho$ are all skew-symmetric; and the matrix $[L]^{\lambda-\kappa}_{\lambda-\kappa}$ is skew-symmetric. It follows as in Ex. ii that $[h]^m_m$ and $[k]^m_m$ are mutually conjugate. Consequently the compartite matrix ψ is skew-symmetric, and the transformations (B) and (B′) are symmetric.

§ 145. The corresponding equigradent transformations of symmetric and skew-symmetric matrices whose elements are constants.

The proofs of the following theorems are included as particular cases in the proofs of the c rresponding more general theorems of §§ 142—4.

Theorem I. *If* $\phi=[a]^m_m$ *is a symmetric or skew-symmetric matrix of rank* ρ *whose elements are constants lying in a domain of rationality* Ω, *and if* $\Delta=(a)^r_r \neq 0$ *is a non-vanishing leading diagonal minor determinant of* ϕ, *then there exists a symmetric unitary equigradent transformation in* Ω *of the form*

$$[H]^m_m [a]^n_m \overline{H}^m_m = \begin{bmatrix} a, & 0 \\ 0, & b \end{bmatrix}^{r,\, m-r}_{r,\, m-r}, \quad \dots\dots\dots\dots\dots\dots\text{(A)}$$

where $[b]^{m-r}_{m-r}$ *is the matrix of rank* $\rho-r$ *in which*

$$\Delta b_{uv} = \begin{pmatrix} 1, 2, \dots r, r+v \\ a \\ 1, 2, \dots r, r+u \end{pmatrix}, \quad \Delta\,(b_{uv})^s_s = \begin{pmatrix} 1, 2, \dots r, r+v_1, \dots r+v_s \\ a \\ 1, 2, \dots r, r+u_1, \dots r+u_s \end{pmatrix}.$$

The matrix $[b]^{m-r}_{m-r}$ *and the reduced matrix on the right in* (A) *are symmetric when* ϕ *is symmetric, and skew-symmetric when* ϕ *is skew-symmetric.*

If $[h]^m_m$ is the inverse matrix of $\overline{H}^m_m$, the transformation inverse to (A) is

$$[a]^m_m = \overline{h}^m_m \begin{bmatrix} a, & 0 \\ 0, & b \end{bmatrix}^{r,\, m-r}_{r,\, m-r} [h]^m_m. \quad \dots\dots\dots\dots\dots\dots\text{(A′)}$$

By Ex. vii of § 142 we have transformations of these forms when

$$[h]^m_m = \begin{bmatrix} 1, & \tau \\ 0, & 1 \end{bmatrix}^{r,\, m-r}_{r,\, m-r}, \quad \overline{H}^m_m = \begin{bmatrix} 1, & -\tau \\ 0, & 1 \end{bmatrix}^{r,\, m-r}_{r,\, m-r},$$

where $\tau_{ui} = \dfrac{1}{\Delta}$ times the determinant formed when we replace the uth vertical row of Δ by the $(r+i)$th vertical row of $[a]^m_r$. To effect the transformation (A), we perform on ϕ in succession and in any order the following operations for all the values $1, 2, \dots m-r$ of i:

We add to the $(r+i)$*th horizontal and vertical rows the* 1*st*, 2*nd*, ... r*th parallel rows multiplied by* $-\tau_{1i}, -\tau_{2i}, \dots -\tau_{ri}$ *respectively.*

We can prove the theorem by obtaining the equation (A′) from which (A) can be deduced. If $[A]^r_r$ is the reciprocal of $[a]^r_r$, the identity of § 116 gives the equation

$$[a]^m_m = \frac{1}{\Delta}[a]^r_m \,\overline{\underline{A}}^{\,r}_{\,r}\, [a]^m_r + \begin{bmatrix} 0, & 0 \\ 0, & b \end{bmatrix}^{r,\,m-r}_{r,\,m-r}.$$

When ϕ is symmetric or skew-symmetric this is equivalent to

$$[a]^m_m = \frac{1}{\Delta^2}\,\overline{\underline{a}}^{\,r}_{\,m}\,[A]^r_r \,.\, [a]^r_r \,.\, \overline{\underline{A}}^{\,r}_{\,r}\,[a]^m_r + \begin{bmatrix} 0, & 0 \\ 0, & b \end{bmatrix}^{r,\,m-r}_{r,\,m-r}$$

$$= \overline{\underline{\begin{matrix} 1 \\ \tau \end{matrix}}}^{\,r}_{\,r,\,m-r}\,[a]^r_r\,[1,\,\tau]^{r,\,m-r}_r + \overline{\underline{\begin{matrix} 0 \\ 1 \end{matrix}}}^{\,m-r}_{\,r,\,m-r}\,[b]^{m-r}_{m-r}\,[0,\,1]^{r,\,m-r}_{m-r},$$

i.e. to (A′).

Ex. i. *Special case when ϕ is symmetric and $\Delta = a_{11} \neq 0$.*

In this case
$$\tau_{1,\,i-1} = \frac{a_{1i}}{a_{11}} = \frac{a_{i1}}{a_{11}}, \quad a_{11}b_{i-1,\,j-1} = \begin{vmatrix} a_{11} & a_{1j} \\ a_{i1} & a_{ij} \end{vmatrix}.$$

We add to the $(1+i)$th horizontal and vertical rows the 1st parallel row multiplied by $-\tau_{1i}$, doing this in succession for all the values 1, 2, ... $m-1$ of i.

Ex. ii. *Special case when ϕ is skew-symmetric and $\Delta = (a)^2_2 \neq 0$.*

In this case $a_{11} = a_{22} = 0$, and we can write $a_{12} = -a_{21} = a \neq 0$. Then

$$\Delta = a^2, \quad \tau_{1,\,i-2} = -\frac{1}{a}a_{2i} = \frac{1}{a}a_{i2}, \quad \tau_{2,\,i-2} = \frac{1}{a}a_{1i} = -\frac{1}{a}a_{i1},$$

and
$$b_{i-2,\,j-2} = \frac{1}{\Delta}\begin{vmatrix} 0, & a, & a_{1j} \\ -a, & 0, & a_{2j} \\ a_{i1}, & a_{i2}, & a_{ij} \end{vmatrix} = a_{ij} - \frac{1}{a}(a_{1i}a_{2j} - a_{1j}a_{2i}).$$

We add to the $(2+i)$th horizontal and vertical rows the 1st and 2nd parallel rows multiplied by $-\tau_{1i}$ and $-\tau_{2i}$ respectively, doing this in succession for all the values 1, 2, ... $m-2$ of i.

Ex. iii. *Special case when ϕ is symmetric, $a_{11} = a_{22} = 0$, and $\Delta = (a)^2_2 \neq 0$.*

Writing $a_{21} = a_{12} = a \neq 0$, so that $\Delta = -a^2$, we have

$$\tau_{1,\,i-2} = \frac{1}{a}a_{2i} = \frac{1}{a}a_{i2}, \quad \tau_{2,\,i-2} = \frac{1}{a}a_{1i} = \frac{1}{a}a_{i1},$$

and
$$b_{i-2,\,j-2} = \frac{1}{\Delta}\begin{vmatrix} 0 & a & a_{1j} \\ a & 0 & a_{2j} \\ a_{i1} & a_{i2} & a_{ij} \end{vmatrix} = a_{ij} - \frac{1}{a}(a_{1i}a_{2j} + a_{1j}a_{2i}).$$

We add to the $(2+i)$th horizontal and vertical rows the 1st and 2nd parallel rows multiplied by $-\tau_{1i}$ and $-\tau_{2i}$ respectively, doing this in succession for all the values 1, 2, ... $m-2$ of i.

Ex. iv. *In the special case when* ϕ *is symmetric,*

$$a_{11}=0,\quad a_{21}=a_{12}=a\neq 0,\quad and\quad \Delta=(a)_2^2=-a^2\neq 0,$$

there exists a symmetric unitary equigradent transformation in Ω *of the form*

$$\underline{\overline{h}}_m^m [a]_m^m [h]_m^m = \begin{bmatrix} 0 & a & 0 \\ a & 0 & 0 \\ 0 & 0 & b \end{bmatrix}_{1,1,m-2}^{1,1,m-2}, \qquad \text{.....(1)}$$

where $[b]_{m-2}^{m-2}$ *is the symmetric matrix of rank* $\rho-2$ *in which*

$$\Delta b_{i-2,j-2} = \begin{pmatrix} 12j \\ a \\ 12i \end{pmatrix} = \begin{vmatrix} 0 & a & a_{1j} \\ a & a_{22} & a_{2j} \\ a_{i1} & a_{i2} & a_{ij} \end{vmatrix}.$$

Writing
$$\tau_{1,i-2}=\frac{1}{\Delta}\begin{vmatrix} a_{1i} & a \\ a_{2i} & a_{22}\end{vmatrix},\quad \tau_{2,i-2}=\frac{1}{\Delta}\begin{vmatrix} 0 & a_{1i} \\ a & a_{2i}\end{vmatrix},$$

or
$$\tau_{1,i-2}=\frac{1}{a^2}(aa_{2i}-a_{22}a_{1i}),\quad \tau_{2,i-2}=\frac{1}{a}a_{1i},$$

we have by Theorem I the unitary equigradent transformation

$$\underline{\overline{\begin{matrix} 1, & 0 \\ -\tau, & 1 \end{matrix}}}_{2,m-2}^{2,m-2} [a]_m^m \begin{bmatrix} 1, & -\tau \\ 0, & 1 \end{bmatrix}_{2,m-2}^{2,m-2} = \begin{bmatrix} 0, & a, & 0 \\ a, & a_{22}, & 0 \\ 0, & 0, & b \end{bmatrix}_{1,1,m-2}^{1,1,m-2} \qquad \text{.....(2)}$$

If $[p]_2^2$ is a matrix so chosen that $\begin{bmatrix} x \\ y \end{bmatrix}=[p]_2^2\begin{bmatrix} \xi \\ \eta \end{bmatrix}$ is one of the transformations which convert $2axy+a_{22}y^2$ into $2a\xi\eta$, or

$$[x\ y]\begin{bmatrix} 0 & a \\ a & a_{22}\end{bmatrix}\begin{bmatrix} x \\ y \end{bmatrix} \text{ into } [\xi\ \eta]\begin{bmatrix} 0 & a \\ a & 0\end{bmatrix}\begin{bmatrix} \xi \\ \eta \end{bmatrix},$$

and if we postfix $\begin{bmatrix} p, & 0 \\ 0, & 1 \end{bmatrix}_{2,m-2}^{2,m-2}$ and prefix its conjugate on both sides of (2), we obtain an equation of the form (1).

Consequently we obtain a unitary equigradent transformation of the form (1) when we write

$$\beta=-\frac{a_{22}}{2a},\quad [p]_2^2=\begin{bmatrix} 1 & \beta \\ 0 & 1\end{bmatrix},\quad [h]_m^m=\begin{bmatrix} p, & -\tau \\ 0, & 1 \end{bmatrix}_{2,m-2}^{2,m-2}. \qquad \text{.....(3)}$$

Then to effect the transformation (1) we perform the following operations in succession on $[a]_m^m$:

(1) We add to the $(2+i)$th horizontal and vertical rows the 1st and 2nd parallel rows multiplied by $-\tau_{1i}$ and $-\tau_{2i}$ respectively, doing this in succession for all the values $1, 2, \dots m-2$ of i.

(2) We then add to the 2nd horizontal row the 1st horizontal row multiplied by β, and to the 2nd vertical row the 1st vertical row multiplied by β.

Ex. v. *In the special case of Ex.* iv *there also exists an equigradent transformation in* Ω *of the form*

$$\underline{\overline{k}}_m^m [a]_m^m [k]_m^m = \begin{bmatrix} 0, & 1, & 0 \\ 1, & 0, & 0 \\ 0, & 0, & b \end{bmatrix}_{1,1,m-2}^{1,1,m-2}, \qquad \text{.....(4)}$$

where $[b]_{m-2}^{m-2}$ *has the same value as before.*

We obtain a resultant transformation of this form when we first convert $[a]_m^m$ into the matrix on the right in (1), and then divide the first horizontal and vertical rows by a, or the second horizontal and vertical rows by a.

Consequently we obtain (4) when we write

$$[q]_2^2 = \frac{1}{a}\begin{bmatrix} 1, & -\frac{1}{2}a_{22} \\ 0, & a \end{bmatrix}, \quad [k]_m^m = \begin{bmatrix} q, & -\tau \\ 0, & 1 \end{bmatrix}_{2,\, m-2}^{2,\, m-2}.$$

Here $[q]_2^2$ is a matrix so chosen that $\begin{bmatrix} x \\ y \end{bmatrix} = [q]_2^2 \begin{bmatrix} \xi \\ \eta \end{bmatrix}$ is one of the transformations which converts $2axy + a_{22}y^2$ into $2\xi\eta$.

Theorem II. *Let $\phi = [a]_m^m$ be a symmetric or skew-symmetric matrix of rank ρ whose elements are constants lying in a domain of rationality Ω, and let $\Delta = (a_{pp})_r^r \neq 0$ be any non-vanishing diagonal minor determinant of ϕ. Then if $[\mu]_{m-r}$ is the corranged complement of $[p]_r$ in $[1\ 2 \ldots m]$, and if $P_{\mu_i \nu_j} = \begin{pmatrix} q_1\, q_2 \ldots q_r\, \nu_j \\ a \\ p_1\, p_2 \ldots p_r\, \mu_i \end{pmatrix}$, there exists a symmetric unitary equigradent transformation Ω of the form*

$$[H]_m^m [a]_m^m \overline{\underline{H}}_m^m = [b]_m^m, \quad \text{.........................(B)}$$

where $[b]_m^m$ is the compartite matrix in which:

(1) $[b_{pp}]_r^r = [a_{pp}]_r^r$; $[b_{\mu\mu}]_{m-r}^{m-r} = \frac{1}{\Delta}[P_{\mu\mu}]_{m-r}^{m-r}$, *and has rank $\rho - r$.*

(2) *All other elements are 0's.*

The reduced matrix $[b]_m^m$ is symmetric when ϕ is symmetric and skew-symmetric when ϕ is skew-symmetric.

If $[h]_m^m$ is the inverse matrix of $\overline{\underline{H}}_m^m$, the transformation inverse to (B) is

$$[a]_m^m = \overline{\underline{h}}_m^m [b]_m^m [h]_m^m. \quad \text{.........................(B')}$$

By Ex. xii of § 142 we have transformations of these forms when

$$[h_{pp}]_r^r = [1]_r^r, \quad [h_{p\mu}]_r^{m-r} = [\tau_{p\mu}]_r^{m-r}, \quad [h_{\mu p}]_{m-r}^r = 0, \quad [h_{\mu\mu}]_{m-r}^{m-r} = [1]_{m-r}^{m-r},$$

$$[H_{pp}]_r^r = [1]_r^r, \quad [H_{p\mu}]_r^{m-r} = 0, \quad [H_{\mu p}]_{m-r}^r = -\overline{\underline{\tau_{p\mu}}}_{m-r}^r, \quad [H_{\mu\mu}]_{m-r}^{m-r} = [1]_{m-r}^{m-r},$$

where $\tau_{ui} = \frac{1}{\Delta}$ times the determinant formed from Δ when we replace the uth vertical row of $[a_{p1}]_r^m$ (occurring in Δ) by the ith vertical row of $[a_{p1}]_r^n$, or when in Δ we replace the vertical suffix u by i.

To effect the transformation (B), we perform on ϕ in succession and in any order the following operations for all the values $\mu_1, \mu_2, \ldots \mu_{m-r}$ of μ:

We add to the μth horizontal and vertical rows the p_1th, p_2th, ... p_rth parallel rows multiplied by $-\tau_{p_1\mu}, -\tau_{p_2\mu}, \ldots -\tau_{p_r\mu}$ respectively.

Ex. vi. *Special case when ϕ is symmetric and $\Delta = a_{uu} \neq 0$.*

In this case $$\tau_{ui} = \frac{a_{ui}}{a_{uu}} = \frac{a_{iu}}{a_{uu}}.$$

We add to the ith horizontal and vertical rows the uth parallel row multiplied by $-\tau_{ui}$, doing this in succession for all the values 1, 2, ... m of i except u.

Ex. vii. *Special case when ϕ is skew-symmetric and $\Delta = \begin{pmatrix} u\,v \\ a \\ u\,v \end{pmatrix} \neq 0$.*

In this case $$a_{uu} = a_{vv} = 0, \quad a_{uv} = -a_{vu} = a \neq 0, \quad \Delta = a^2 \neq 0,$$

and $$\tau_{ui} = \frac{1}{\Delta}\begin{pmatrix} i\,v \\ a \\ u\,v \end{pmatrix} = -\frac{1}{a}a_{vi} = \frac{1}{a}a_{iv}, \quad \tau_{vi} = \frac{1}{\Delta}\begin{pmatrix} u\,i \\ a \\ u\,v \end{pmatrix} = \frac{1}{a}a_{ui} = -\frac{1}{a}a_{iu}.$$

We add to the ith horizontal and vertical rows the uth and vth parallel rows multiplied by $-\tau_{ui}$ and $-\tau_{vi}$ respectively, doing this in succession for all the values 1, 2, ... m of i except u and v.

Ex. viii. *Special case when ϕ is symmetric, $a_{uu} = a_{vv} = 0$, and $\Delta = \begin{pmatrix} u\,v \\ a \\ u\,v \end{pmatrix} \neq 0$.*

Writing $a_{vu} = a_{uv} = a$, so that $\Delta = -a^2$, we have

$$\tau_{ui} = \frac{1}{a}a_{vi} = \frac{1}{a}a_{iv}, \quad \tau_{vi} = \frac{1}{a}a_{ui} = \frac{1}{a}a_{iu}.$$

We add to the ith horizontal and vertical rows the uth and vth parallel rows multiplied by $-\tau_{ui}$ and $-\tau_{vi}$ respectively, doing this in succession for all the values 1, 2, ... m of i except u and v.

Ex. ix. *In the special case when ϕ is symmetric, $a_{uu} = 0$, $a_{vu} = a_{uv} = a \neq 0$, and*

$$\Delta = \begin{pmatrix} u\,v \\ a \\ u\,v \end{pmatrix} = -a^2 \neq 0,$$

there exists a symmetric unitary equigradent transformation in Ω of the form (B), *where $[b]_m^m$ is the symmetric matrix in which $\begin{bmatrix} b_{uu} & b_{uv} \\ b_{vu} & b_{vv} \end{bmatrix} = \begin{bmatrix} 0 & a \\ a & 0 \end{bmatrix}$, all other elements of the uth and vth horizontal and vertical rows are 0's, and $b_{ij} = \frac{1}{\Delta}\begin{pmatrix} u\,v\,j \\ a \\ u\,v\,i \end{pmatrix}$ when neither i nor j is equal to u or v.*

In this case we have $$\tau_{ui} = \frac{1}{a^2}(aa_{vi} - a_{vv}a_{ui}), \quad \tau_{vi} = \frac{1}{a}a_{ui}.$$

We obtain a resultant transformation of the character stated when we perform on ϕ in succession the following two sets of operations:

(1) We add to the ith horizontal and vertical rows in succession the uth and vth parallel rows multiplied by $-\tau_{ui}$ and $-\tau_{vi}$ respectively, doing this in succession for all the values 1, 2, ... m of i except u and v.

(2) We then add to the vth horizontal row the uth horizontal row multiplied by $-\frac{a_{vv}}{2a}$, and to the vth vertical row the uth vertical row multiplied by the same quantity.

The transformations (1) convert $[a]_m^m$ into a symmetric matrix $[b]_m^m$ in which $\begin{bmatrix} b_{uu} & b_{uv} \\ b_{vu} & b_{vv} \end{bmatrix} = \begin{bmatrix} 0 & a \\ a & a_{vv} \end{bmatrix}$, all other elements of the uth and vth horizontal and vertical rows are 0's, and the remaining elements have the values stated in the enunciation; and the transformations (2) convert $[b]_m^m$ into the matrix derived from it by putting $b_{vv}=0$.

Ex. x. *In the special case of Ex.* ix *there also exists a symmetric equigradent transformation in* Ω *of the form* (B), *where* $[b]_m^m$ *is the symmetric matrix in which* $\begin{bmatrix} b_{uu} & b_{uv} \\ b_{vu} & b_{vv} \end{bmatrix} = \begin{bmatrix} 0 & 1 \\ 1 & 0 \end{bmatrix}$, *all other elements of the uth and vth horizontal and vertical rows are 0's, and the remaining elements have the same values as in Ex.* ix.

We first reduce ϕ to the matrix $[b]_m^m$ of Ex. ix, and then divide the uth horizontal and vertical rows by a, or the vth horizontal and vertical rows by a.

Theorem III. *Let $\phi = [a]_m^m$ be a symmetric or skew-symmetric matrix of rank ρ whose elements are constants lying in a domain of rationality Ω, and let $\alpha, \beta, \gamma, \ldots \kappa, \lambda, \mu, \ldots \pi, \rho$ be a series of constantly increasing non-zero positive integers, the last of which is equal to the rank of ϕ. Then if none of the leading diagonal minor determinants*

$$\Delta_0 = 1, \quad \Delta_\alpha = (a)_\alpha^\alpha, \quad \Delta_\beta = (a)_\beta^\beta, \ \ldots \ \Delta_\lambda = (a)_\lambda^\lambda, \ \ldots \ \Delta_\rho = (a)_\rho^\rho$$

of ϕ vanish, we can convert ϕ by a symmetric unitary equigradent transformation in Ω into a certain compartite matrix

$$\psi = \begin{bmatrix} a, & 0, & 0, & \ldots & 0, & \ldots & 0, & 0 \\ 0, & b, & 0, & \ldots & 0, & \ldots & 0, & 0 \\ 0, & 0, & c, & \ldots & 0, & \ldots & 0, & 0 \\ \cdots & & & & & & & \cdots \\ 0, & 0, & 0, & \ldots & l, & \ldots & 0, & 0 \\ \cdots & & & & & & & \cdots \\ 0, & 0, & 0, & \ldots & 0, & \ldots & r, & 0 \\ 0, & 0, & 0, & \ldots & 0, & \ldots & 0, & 0 \end{bmatrix}_{\alpha,\ \beta-\alpha,\ \gamma-\beta,\ \ldots\ \lambda-\kappa,\ \ldots\ \rho-\pi,\ m-\rho}^{\alpha,\ \beta-\alpha,\ \gamma-\beta,\ \ldots\ \lambda-\kappa,\ \ldots\ \rho-\pi,\ m-\rho}$$

in which

$$A = (a)_\alpha^\alpha \neq 0, \quad B = (b)_{\beta-\alpha}^{\beta-\alpha} \neq 0, \ \ldots \ L = (l)_{\lambda-\kappa}^{\lambda-\kappa} \neq 0, \ \ldots \ R = (r)_{\rho-\pi}^{\rho-\pi} \neq 0,$$

the parts $\quad [a]_\alpha^\alpha, \quad [b]_{\beta-\alpha}^{\beta-\alpha}, \quad [c]_{\gamma-\beta}^{\gamma-\beta}, \ \ldots \ [l]_{\lambda-\kappa}^{\lambda-\kappa}, \quad [m]_{\mu-\lambda}^{\mu-\lambda}, \ \ldots \ [r]_{\rho-\pi}^{\rho-\pi}$

of ψ being leading diagonal minors of the matrices

$$[a]_m^m, \quad [b]_{m-\alpha}^{m-\alpha}, \quad [c]_{m-\beta}^{m-\beta}, \ \ldots \ [l]_{m-\kappa}^{m-\kappa}, \quad [m]_{m-\lambda}^{m-\lambda}, \ \ldots \ [r]_{m-\pi}^{m-\pi}$$

of ranks $\rho, \rho-\alpha, \rho-\beta, \ldots \rho-\kappa, \rho-\lambda, \ldots \rho-\pi$ whose elements are defined in succession by such equations as

$$Lm_{uv} = \begin{pmatrix} 1, 2, \ldots \lambda-\kappa, \lambda-\kappa+v \\ l \\ 1, 2, \ldots \lambda-\kappa, \lambda-\kappa+u \end{pmatrix}, \quad \textit{or} \quad \Delta_\lambda m_{uv} = \begin{pmatrix} 1, 2, \ldots \lambda, \lambda+v \\ a \\ 1, 2, \ldots \lambda, \lambda+u \end{pmatrix},$$

the part $[a]_\alpha^\alpha$ in particular being the leading diagonal minor of ϕ of order α.

All the parts of ψ *and the matrix* ψ *itself are symmetric when* ϕ *is symmetric, and skew-symmetric when* ϕ *is skew-symmetric.*

We deduce this theorem from the general theorem of § 143.

Let

$$h_\lambda = \frac{1}{L} \overline{\underline{L}}_{\lambda-\kappa}^{\lambda-\kappa} [0,\ l]_{\lambda-\kappa}^{\kappa,\ m-\kappa} = [0,\ 1,\ \lambda]_{\lambda-\kappa}^{\kappa,\ \lambda-\kappa,\ m-\lambda}, \quad u_{m-\rho} = [0,\ 1]_{m-\rho}^{\rho,\ m-\rho},$$

so that $\lambda_{ui} = \frac{1}{L}$ times the determinant formed from the determinant $L = (l)_{\lambda-\kappa}^{\lambda-\kappa}$ when we replace the vertical suffix u by $\lambda - \kappa + i$.

Then the transformations converting ψ into ϕ and ϕ into ψ are

$$[a]_m^m = \overline{\underline{h}}_m^m\, \psi\, [h]_m^m, \quad [H]_m^m [a]_m^m \overline{\underline{H}}_m^m = \psi, \qquad \text{(C)}$$

where $\overline{\underline{H}}_m^m$ is the inverse matrix of $[h]_m^m$, and $[h]_m^m$ is the matrix whose successive horizontal rows are the successive horizontal rows of the matrices $h_\alpha, h_\beta, h_\gamma, \ldots h_\lambda, \ldots h_\rho, u_{m-\rho}$ taken in this order.

Ex. xi. *Direct proof of the theorem.*

Let $h_\alpha', h_\beta', h_\gamma', \ldots h_\lambda', \ldots h_\rho'$ be the conjugates of $h_\alpha, h_\beta, h_\gamma, \ldots h_\lambda, \ldots h_\rho$, and let

$$\Phi_\lambda = \begin{bmatrix} 0, & 0 \\ 0, & l \end{bmatrix}_{\kappa,\ m-\kappa}^{\kappa,\ m-\kappa}, \qquad \text{(5)}$$

$$T_\lambda = \frac{1}{L} \overline{\underline{\begin{matrix} 0 \\ l \end{matrix}}}_{\kappa,\ m-\kappa}^{\lambda-\kappa} \overline{\underline{L}}_{\lambda-\kappa}^{\lambda-\kappa} [0,\ l]_{\lambda-\kappa}^{\kappa,\ m-\kappa} = h_\lambda' [l]_{\lambda-\kappa}^{\lambda-\kappa} h_\lambda. \qquad \text{(6)}$$

Then if we assume that $[l]_{m-\kappa}^{m-\kappa}$ is symmetric or skew-symmetric the identity (A′) of § 116 gives the equation

$$(l)_{\lambda-\kappa}^{\lambda-\kappa} [l]_{m-\kappa}^{m-\kappa} = [l]_{m-\kappa}^{\lambda-\kappa} \overline{\underline{L}}_{\lambda-\kappa}^{\lambda-\kappa} [l]_{\lambda-\kappa}^{m-\kappa} + (l)_{\lambda-\kappa}^{\lambda-\kappa} \begin{bmatrix} 0, & 0 \\ 0, & m \end{bmatrix}_{\lambda-\kappa,\ m-\lambda}^{\lambda-\kappa,\ m-\lambda}$$

or

$$[l]_{m-\kappa}^{m-\kappa} = \frac{1}{L^2} \overline{\underline{l}}_{m-\kappa}^{\lambda-\kappa} [L]_{\lambda-\kappa}^{\lambda-\kappa} [l]_{\lambda-\kappa}^{\lambda-\kappa} \overline{\underline{L}}_{\lambda-\kappa}^{\lambda-\kappa} [l]_{\lambda-\kappa}^{m-\kappa} + \begin{bmatrix} 0, & 0 \\ 0, & m \end{bmatrix}_{\lambda-\kappa,\ m-\lambda}^{\lambda-\kappa,\ m-\lambda}.$$

This equation shows that $[m]_{m-\lambda}^{m-\lambda}$ is symmetric or skew-symmetric according as $[l]_{m-\kappa}^{m-\kappa}$ is symmetric or skew-symmetric, and is equivalent to

$$\Phi_\lambda = T_\lambda + \Phi_\mu. \qquad \text{(7)}$$

From the successive equations $\phi = T_\alpha + \Phi_\beta$, $\Phi_\beta = T_\beta + \Phi_\gamma$, ... $\Phi_\rho = T_\rho$ obtained in this way we deduce the equation

$$[a]_m^m = T_\alpha + T_\beta + T_\gamma + \ldots + T_\kappa + T_\lambda + T_\mu + \ldots + T_\pi + T_\rho, \qquad \text{(D)}$$

which is equivalent to the first of the equations (C). We see also that the matrices $[b]_{m-\alpha}^{m-\alpha}$, $[c]_{m-\beta}^{m-\beta}$, ... $[r]_{m-\pi}^{m-\pi}$ are all symmetric or all skew-symmetric according as ϕ is symmetric or skew-symmetric.

Theorem IV. *The matrix ϕ of Theorem III can be expressed in the form*

$$[a]_m^m = T_\alpha + T_\beta + T_\gamma + \ldots + T_\kappa + T_\lambda + T_\mu + \ldots + T_\pi + T_\rho, \quad \ldots\ldots\text{(E)}$$

where

$$T_\lambda = \frac{1}{\Delta_\kappa \Delta_\lambda^2} \overline{\begin{matrix} 0 \\ \mathbf{L} \end{matrix}}_{\kappa,\, m-\kappa}^{\lambda-\kappa} [l]_{\lambda-\kappa}^{\lambda-\kappa} [0, \mathbf{L}]_{\lambda-\kappa}^{\kappa,\, m-\kappa}, \quad l_{uv} = \begin{pmatrix} 1, 2, \ldots \kappa, \kappa+v \\ a \\ 1, 2, \ldots \kappa, \kappa+u \end{pmatrix},$$

and $\mathbf{L}_{ui}$ is the determinant formed from Δ_λ when we replace the vertical suffix $\kappa + u$ by $\kappa + i$.

The matrix $[l]_{\lambda-\kappa}^{\lambda-\kappa}$ is symmetric or skew-symmetric according as ϕ is symmetric or skew-symmetric.

Here we obtain T_α, T_β, ... T_ρ from T_λ by replacing $(\kappa, \lambda, l, \mathbf{L})$ by $(0, \alpha, a, \mathbf{A}), (\alpha, \beta, b, \mathbf{B}), \ldots (\pi, \rho, r, \mathbf{R})$; and we obtain $(a_{uv}, \mathbf{A}_{ui}), (b_{uv}, \mathbf{B}_{ui}), \ldots (r_{uv}, \mathbf{R}_{ui})$ from $(l_{uv}, \mathbf{L}_{ui})$ by replacing (κ, λ) by $(0, \alpha), (\alpha, \beta), \ldots (\pi, \rho)$.

We can deduce Theorem IV from the general theorem of § 144.

From the equation (E) we obtain symmetric equigradent transformations of the forms (C) in which ψ is a compartite matrix whose successive parts are

$$\frac{1}{\Delta_0 \Delta_\alpha^2} [a]_\alpha^\alpha, \quad \frac{1}{\Delta_\alpha \Delta_\beta^2} [b]_{\beta-\alpha}^{\beta-\alpha}, \ldots \frac{1}{\Delta_\kappa \Delta_\lambda^2} [l]_{\lambda-\kappa}^{\lambda-\kappa}, \ldots \frac{1}{\Delta_\pi \Delta_\rho^2} [r]_{\rho-\pi}^{\rho-\pi}, \quad [0]_{m-\rho}^{m-\rho},$$

and $[h]_m^m$ is the matrix whose successive horizontal rows are the successive horizontal rows of the matrices

$$[\mathbf{A}]_\alpha^m, \quad [0, \mathbf{B}]_{\beta-\alpha}^{\alpha,\, m-\alpha}, \ldots [0, \mathbf{L}]_{\lambda-\kappa}^{\kappa,\, m-\kappa}, \ldots [0, \mathbf{R}]_{\rho-\pi}^{\pi,\, m-\pi}, \quad [0, 1]_{m-\rho}^{\rho,\, m-\rho}$$

taken in this order.

The compartite matrix ψ is symmetric or skew-symmetric according as ϕ is symmetric or skew-symmetric.

We obtain (E) directly as in § 144 by writing

$$\Phi_\lambda = \begin{bmatrix} 0, & 0 \\ 0, & l \end{bmatrix}_{\kappa,\, m-\kappa}^{\kappa,\, m-\kappa},$$

where $[l]_{m-\kappa}^{m-\kappa}$ is symmetric or skew-symmetric according as ϕ is symmetric or skew-symmetric, and proving the equation

$$\frac{1}{\Delta_\kappa} \Phi_\lambda = T_\lambda + \frac{1}{\Delta_\lambda} \Phi_\mu$$

by means of the identity of § 116. We can also deduce (E) from (D) as in the alternative proof of the theorem in § 144.

§ 146. Reduction of any matrix with constant elements to standard forms by equigradent transformations.

1. *Reduction by unrestricted equigradent transformations.*

Theorem I. *Any matrix* $\phi = [a]^n_m$ *of rank r whose elements are constants lying in a domain of rationality* Ω *can be reduced by equigradent transformations in* Ω *to the standard form*

$$\psi = \begin{bmatrix} 1, & 0 \\ 0, & 0 \end{bmatrix}^{r,\, n-r}_{r,\, m-r}$$

As in Ex. i of § 115 we can determine matrices $[h]^r_m$ and $[k]^n_r$ of rank r with constant elements lying in Ω such that

$$[a]^n_m = [h]^r_m\, [k]^n_r = [h]^r_m\, [1]^r_r\, [k]^n_r\,. \quad \text{.................(1)}$$

Let $[h]^r_m$, $[k]^n_r$ be converted into undegenerate square matrices $[h]^m_m$ and $[k]^n_n$ whose elements are constants in Ω by the insertion of additional final vertical and horizontal rows, and let $\overline{\underline{H}}{}^m_m$ and $\overline{\underline{K}}{}^n_n$ be the inverses of $[h]^m_m$ and $[k]^n_n$. Then it follows from (1) that

$$[a]^n_m = [h]^m_m \begin{bmatrix} 1, & 0 \\ 0, & 0 \end{bmatrix}^{r,\, n-r}_{r,\, m-r} [k]^n_n, \quad \overline{\underline{H}}{}^m_m\, [a]^n_m\, \overline{\underline{K}}{}^n_n = \begin{bmatrix} 1, & 0 \\ 0, & 0 \end{bmatrix}^{r,\, n-r}_{r,\, m-r}; \quad \text{...(A)}$$

and the last equation shows that the theorem is true.

Ex. i. We can determine two pairs of mutually inverse undegenerate matrices $[h]^r_m$, $\overline{\underline{H}}{}^m_r$ and $[k]^n_r$, $\overline{\underline{K}}{}^r_n$ of rank r whose elements are constants in Ω such that

$$[a]^n_m = [h]^r_m\, [1]^r_r\, [k]^n_r, \quad [1]^r_r = \overline{\underline{H}}{}^m_r\, [a]^n_m\, \overline{\underline{K}}{}^r_n\,. \quad \text{..................(A')}$$

The first of the equations (A') is an equigradent transformation in Ω converting $[1]^r_r$ into $[a]^n_m$ and the second can be derived from it by prefixing $\overline{\underline{H}}{}^m_r$ and postfixing $\overline{\underline{K}}{}^r_n$.

As in Note 8 of § 141 we can derive either of the two pairs of equations (A) and (A') from the other.

Ex. ii. *Necessary and sufficient condition that two matrices with constant elements may be equigradent.*

Theorem. *Any two matrices* $[a]^n_m$ *and* $[b]^\nu_\mu$ *with constant elements have the same rank when and only when there exists a relation of the form*

$$[h]^m_u\, [a]^n_m\, [k]^v_n = [p]^\mu_u\, [b]^\nu_\mu\, [q]^v_\nu \quad \text{..............................(2)}$$

where the prefactors and postfactors are undegenerate matrices with constant elements whose ranks are equal to their passivities.

This theorem remains true when we restrict u to be the larger of the two integers m and μ, and v to be the larger of the two integers n and ν.

If there exists a relation of the form (2), then by Theorem II of § 131 the two matrices $[a]_m^n$ and $[b]_\mu^\nu$ have the same rank. Conversely if these two matrices have the same rank r, there exist undegenerate square matrices $[h]_m^m$, $[k]_n^n$, $[p]_\mu^\mu$, $[q]_\nu^\nu$ such that

$$[h]_m^m [a]_m^n [k]_n^n = \begin{bmatrix} 1, & 0 \\ 0, & 0 \end{bmatrix}_{r,\, m-r}^{r,\, n-r}, \quad [p]_\mu^\mu [b]_\mu^\nu [q]_\nu^\nu = \begin{bmatrix} 1, & 0 \\ 0, & 0 \end{bmatrix}_{r,\, \mu-r}^{r,\, \nu-r};$$

and if u is any integer not less than either m or μ, and v any integer not less than either n or ν, it follows that

$$\begin{bmatrix} h \\ 0 \end{bmatrix}_{m,\, u-m}^{m} [a]_m^n [k, 0]_n^{n,\, v-n} = \begin{bmatrix} p \\ 0 \end{bmatrix}_{\mu,\, u-\mu}^{\mu} [b]_\mu^\nu [q, 0]_\nu^{\nu,\, v-\nu}$$
$$= \begin{bmatrix} 1, & 0 \\ 0, & 0 \end{bmatrix}_{r,\, u-r}^{r,\, v-r},$$

where the first equality has the form (2).

The theorem in its more general form simply states that:

Two matrices with constant elements are equigradent when and only when they have the same rank.

2. *Reduction by derangements and unitary equigradent transformations.*

Theorem II a. *If $\phi = [a]_m^n$ is a matrix of rank r whose elements are constants lying in a domain of rationality Ω, and if none of the leading diagonal minor determinants*

$$\Delta_0 = 1, \quad \Delta_1 = a_{11}, \quad \Delta_2 = (a)_2^2, \quad \Delta_3 = (a)_3^3, \ldots \Delta_r = (a)_r^r$$

vanish, then we can convert ϕ by unitary equigradent transformations in Ω into the quasi-scalar matrix

$$\psi = \begin{bmatrix} e, & 0 \\ 0, & 0 \end{bmatrix}_{r,\, m-r}^{r,\, n-r}, \quad \textit{where } [e]_r^r = \begin{bmatrix} e_1 & 0 & \ldots & 0 \\ 0 & e_2 & \ldots & 0 \\ \cdots & \cdots & \cdots & \cdots \\ 0 & 0 & \ldots & e_r \end{bmatrix},$$

and $e_1, e_2, \ldots e_r$ are the non-zero scalar quantities lying in Ω which are defined by the equations

$$e_1 = \Delta_1, \quad e_1 e_2 = \Delta_2, \quad e_1 e_2 e_3 = \Delta_3, \ldots e_1 e_2 e_3 \ldots e_r = \Delta_r.$$

This theorem is a particular case of the more general theorem of § 143, and the proofs of that more general theorem include corresponding proofs of the present theorem. When we follow the proofs given in § 143, we in this particular case use matrices $[b]_{m-1}^{n-1}$, $[c]_{m-2}^{n-2}$, ... $[l]_{m-r+1}^{n-r+1}$ of ranks $r-1, r-2, \ldots 1$ whose elements are defined by the successive equations

$$a_{11} b_{uv} = \begin{pmatrix} 1, 1+v \\ a \\ 1, 1+u \end{pmatrix}, \quad b_{11} c_{uv} = \begin{pmatrix} 1, 1+v \\ b \\ 1, 1+u \end{pmatrix}, \quad c_{11} d_{uv} = \begin{pmatrix} 1, 1+v \\ c \\ 1, 1+u \end{pmatrix}, \ldots, \quad \ldots(3)$$

so that $\quad a_{11} = \Delta_1, \quad a_{11} b_{11} = \Delta_2, \quad a_{11} b_{11} c_{11} = \Delta_3, \ldots a_{11} b_{11} \ldots l_{11} = \Delta_r,$

and

$$a_{11} (b_{uv})_s^s = \begin{pmatrix} 1, 1+v_1, \ldots 1+v_s \\ a \\ 1, 1+u_1, \ldots 1+u_s \end{pmatrix}, \quad \Delta_1 (b_{uv})_s^s = \begin{pmatrix} 1, 1+v_1, \ldots 1+v_s \\ a \\ 1, 1+u_1, \ldots 1+u_s \end{pmatrix},$$

$$b_{11} (c_{uv})_s^s = \begin{pmatrix} 1, 1+v_1, \ldots 1+v_s \\ b \\ 1, 1+u_1, \ldots 1+u_s \end{pmatrix}, \quad \Delta_2 (c_{uv})_s^s = \begin{pmatrix} 1, 2, 2+v_1, \ldots 2+v_s \\ a \\ 1, 2, 2+u_1, \ldots 2+u_s \end{pmatrix},$$

$$c_{11} (d_{uv})_s^s = \begin{pmatrix} 1, 1+v_1, \ldots 1+v_s \\ c \\ 1, 1+u_1, \ldots 1+u_s \end{pmatrix}, \quad \Delta_3 (d_{uv})_s^s = \begin{pmatrix} 1, 2, 3, 3+v_1, \ldots 3+v_s \\ a \\ 1, 2, 3, 3+u_1, \ldots 3+u_s \end{pmatrix},$$

....

Then $a_{11}, b_{11}, \ldots l_{11}$ are the quantities $e_1, e_2, \ldots e_r$ of the theorem.

The resultant unitary transformations in Ω which convert ψ into ϕ and ϕ into ψ have the forms

$$[a]_m^n = [h]_m^m \begin{bmatrix} e, & 0 \\ 0, & 0 \end{bmatrix}_{r,\, m-r}^{r,\, n-r} [k]_n^n, \quad \overline{H}_m^m [a]_m^n \overline{K}_n^n = \begin{bmatrix} e, & 0 \\ 0, & 0 \end{bmatrix}_{r,\, m-r}^{r,\, n-r}. \quad \ldots(B)$$

If

$$h_1 = \frac{1}{a_{11}} [a]_m^1, \quad h_2 = \frac{1}{b_{11}} \begin{bmatrix} 0 \\ b \end{bmatrix}_{1,\, m-1}^1, \ldots h_r = \frac{1}{l_{11}} \begin{bmatrix} 0 \\ l \end{bmatrix}_{r-1,\, m-r+1}^1, \quad u_{m-r} = \begin{bmatrix} 0 \\ 1 \end{bmatrix}_{r,\, m-r}^{m-r},$$

$$k_1 = \frac{1}{a_{11}} [a]_1^n, \quad k_2 = \frac{1}{b_{11}} [0,\ b]_1^{1,\, n-1}, \ldots k_r = \frac{1}{l_{11}} [0,\ l]_1^{r-1,\, n-r+1},$$

$$v_{n-r} = [0,\ 1]_{n-r}^{r,\, n-r},$$

then $[h]_m^m$ is the matrix whose 1st, 2nd, ... rth, last $m-r$ vertical rows are $h_1, h_2, \ldots h_r, u_{m-r}$; $[k]_n^n$ is the matrix whose 1st, 2nd, ... rth, last $n-r$ horizontal rows are $k_1, k_2, \ldots k_r, v_{n-r}$; $\overline{H}_m^m$ is the inverse of $[h]_m^m$ and has the same general form as $[h]_m^m$; $\overline{K}_n^n$ is the inverse of $[k]_n^n$ and has the same general form as $[k]_n^n$.

Proceeding as in Note 2 or Note 3 of § 143, we can deduce the present theorem from Theorem I of § 142. In Note 1 below we give in brief the direct proof of § 143 for this particular case.

NOTE 1. *Direct proof of Theorem II a.*

From the identity (A′) of § 116 we obtain such equations as

$$[c]_{m-2}^{n-2}=\frac{1}{c_{11}}[c]_{m-2}^{1}[c]_{1}^{n-2}+\begin{bmatrix}0, & 0\\ 0, & d\end{bmatrix}_{1,\,m-3}^{1,\,n-3},$$

i.e. we obtain the successive equations

$$[a]_{m}^{n}=\frac{1}{a_{11}}[a]_{m}^{1}[a]_{1}^{n}+\begin{bmatrix}0, & 0\\ 0, & b\end{bmatrix}_{1,\,m-1}^{1,\,n-1},$$

$$\begin{bmatrix}0, & 0\\ 0, & b\end{bmatrix}_{1,\,m-1}^{1,\,n-1}=\frac{1}{b_{11}}\begin{bmatrix}0\\ b\end{bmatrix}_{1,\,m-1}^{1}[0,\ b]_{1}^{1,\,n-1}+\begin{bmatrix}0, & 0\\ 0, & c\end{bmatrix}_{2,\,m-2}^{2,\,n-2},$$

$$\begin{bmatrix}0, & 0\\ 0, & c\end{bmatrix}_{2,\,m-2}^{2,\,n-2}=\frac{1}{c_{11}}\begin{bmatrix}0\\ c\end{bmatrix}_{2,\,m-2}^{1}[0,\ c]_{1}^{2,\,n-2}+\begin{bmatrix}0, & 0\\ 0, & d\end{bmatrix}_{3,\,m-3}^{3,\,n-3},$$

....

$$\begin{bmatrix}0, & 0\\ 0, & l\end{bmatrix}_{r-1,\,m-r+1}^{r-1,\,n-r+1}=\frac{1}{l_{11}}\begin{bmatrix}0\\ l\end{bmatrix}_{r-1,\,m-r+1}^{1}[0,\ l]_{1}^{r-1,\,n-r+1}. \quad \text{......................(4)}$$

From equations (4) we deduce the equation

$$[a]_{m}^{n}=\frac{1}{a_{11}}[a]_{m}^{1}[a]_{1}^{n}+\frac{1}{b_{11}}\begin{bmatrix}0\\ b\end{bmatrix}_{1,\,m-1}^{1}[0,\ b]_{1}^{1,\,n-1}+\frac{1}{c_{11}}\begin{bmatrix}0\\ c\end{bmatrix}_{2,\,m-2}^{1}[0,\ c]_{1}^{2,\,n-2}$$

$$+\ldots+\frac{1}{l_{11}}\begin{bmatrix}0\\ l\end{bmatrix}_{r-1,\,m-r+1}^{1}[0,\ l]_{1}^{r-1,\,n-r+1}. \quad \text{........................(C)}$$

The last equation can be replaced by

$$[a]_{m}^{n}=a_{11}h_1k_1+b_{11}h_2k_2+c_{11}h_3k_3+\ldots+l_{11}h_rk_r+0\,u_{m-r}\,v_{n-r}, \quad \text{............(D)}$$

where h_1, h_2, ... k_1, k_2, ... are the matrices defined in the text, and this equation is by the properties of passive rows equivalent to

$$[a]_{m}^{n}=[h]_{m}^{r}[e]_{r}^{r}[k]_{r}^{n}, \text{ and } [a]_{m}^{n}=[h]_{m}^{m}\begin{bmatrix}e, & 0\\ 0, & 0\end{bmatrix}_{r,\,m-r}^{r,\,n-r}[k]_{n}^{n}.$$

We have thus obtained the first of the equations (B) from which the second equation can be deduced.

NOTE 2. The equation (C) can be written in the form

$$[a]_{m}^{n}=\frac{\Delta_0}{\Delta_1}[a]_{m}^{1}[a]_{1}^{n}+\frac{\Delta_1}{\Delta_2}\begin{bmatrix}0\\ b\end{bmatrix}_{1,\,m-1}^{1}[0,\ b]_{1}^{1,\,n-1}+\frac{\Delta_2}{\Delta_3}\begin{bmatrix}0\\ c\end{bmatrix}_{2,\,m-2}^{1}[0,\ c]_{1}^{2,\,n-2}$$

$$+\ldots+\frac{\Delta_{r-1}}{\Delta_r}\begin{bmatrix}0\\ l\end{bmatrix}_{r-1,\,m-r+1}^{1}[0,\ l]_{1}^{r-1,\,n-r+1}, \quad \text{........................(C′)}$$

from which we derive in the same way non-unitary equigradent transformations in Ω of the forms (B) in which

$$e_1=\frac{\Delta_0}{\Delta_1},\quad e_2=\frac{\Delta_1}{\Delta_2},\ \ldots\ e_r=\frac{\Delta_{r-1}}{\Delta_r}.$$

If $\quad \mathbf{a}_{uv}=a_{uv},\quad \mathbf{b}_{uv}=\begin{pmatrix}1,\ 1+v\\ a\\ 1,\ 1+u\end{pmatrix},\quad \mathbf{c}_{uv}=\begin{pmatrix}1,\ 2,\ 2+v\\ a\\ 1,\ 2,\ 2+u\end{pmatrix},\ \ldots\ \mathbf{l}_{uv}=\begin{pmatrix}1,\ 2,\ \ldots\ r-1,\ r-1+v\\ a\\ 1,\ 2,\ \ldots\ r-1,\ r-1+u\end{pmatrix},$

and if we substitute in (C′) for a_{uv}, b_{uv}, c_{uv}, ... l_{uv} the values given by

$$\Delta_0 a_{uv}=\mathbf{a}_{uv},\quad \Delta_1 b_{uv}=\mathbf{b}_{uv},\quad \Delta_2 c_{uv}=\mathbf{c}_{uv},\ \ldots\ \Delta_{r-1} l_{uv}=\mathbf{l}_{uv},$$

we obtain the equation (E) of Note 3.

NOTE 3. *The matrix ϕ of Theorem IIa can be expressed in the form*

$$[a]_m^n=\frac{1}{\Delta_0\Delta_1}[a]_m^1[a]_1^n+\frac{1}{\Delta_1\Delta_2}\begin{bmatrix}0\\ b\end{bmatrix}_{1,\ m-1}^1[0,\ b]_1^{1,\ n-1}+\frac{1}{\Delta_2\Delta_3}\begin{bmatrix}0\\ c\end{bmatrix}_{2,\ m-2}^1[0,\ c]_1^{2,\ n-2}$$
$$+\ldots+\frac{1}{\Delta_{r-1}\Delta_r}\begin{bmatrix}0\\ l\end{bmatrix}_{r-1,\ m-r+1}^1[0,\ l]_1^{r-1,\ n-r+1},\ \ldots\ldots\ldots\ldots\ldots\ldots\text{(E)}$$

where $\quad b_{uv}=\begin{pmatrix}1,\ 1+v\\ a\\ 1,\ 1+u\end{pmatrix},\quad c_{uv}=\begin{pmatrix}1,\ 2,\ 2+v\\ a\\ 1,\ 2,\ 2+u\end{pmatrix},\ \ldots\ l_{uv}=\begin{pmatrix}1,\ 2,\ \ldots\ r-1,\ r-1+v\\ a\\ 1,\ 2,\ \ldots\ r-1,\ r-1+u\end{pmatrix}.$

This result is a particular case of the general theorem of § 144, and has been proved indirectly in Note 2.

To prove it directly we observe that now by the identity of § 110

$$\begin{pmatrix}1,\ 1+v\\ b\\ 1,\ 1+u\end{pmatrix}=\Delta_1 c_{uv},\quad \begin{pmatrix}1,\ 1+v\\ c\\ 1,\ 1+u\end{pmatrix}=\Delta_2 d_{uv},\quad \begin{pmatrix}1,\ 1+v\\ d\\ 1,\ 1+u\end{pmatrix}=\Delta_3 e_{uv},\ \ldots.$$

Let $\quad \Phi_2=\begin{bmatrix}0,&0\\ 0,&b\end{bmatrix}_{1,\ m-1}^{1,\ n-1},\quad \Phi_3=\begin{bmatrix}0,&0\\ 0,&c\end{bmatrix}_{2,\ m-2}^{2,\ n-2},\ \ldots\ \Phi_r=\begin{bmatrix}0,&0\\ 0,&l\end{bmatrix}_{r-1,\ m-r+1}^{r-1,\ n-r+1}.$

Then from the identity (A′) of § 116 we have such equations as

$$c_{11}[c]_{m-2}^{n-2}=[c]_{m-2}^1[c]_1^{n-2}+\Delta_2\begin{bmatrix}0,&0\\ 0,&d\end{bmatrix}_{1,\ m-3}^{1,\ n-3},$$

i.e. we have the successive equations

$$\frac{1}{\Delta_0}\phi\ =\frac{1}{\Delta_0\Delta_1}[a]_m^1[a]_1^n+\frac{1}{\Delta_1}\Phi_2,$$

$$\frac{1}{\Delta_1}\Phi_2\ =\frac{1}{\Delta_1\Delta_2}\begin{bmatrix}0\\ b\end{bmatrix}_{1,\ m-1}^1[0,\ b]_1^{1,\ n-1}+\frac{1}{\Delta_2}\Phi_3,$$

....

$$\frac{1}{\Delta_{r-1}}\Phi_r=\frac{1}{\Delta_{r-1}\Delta_r}\begin{bmatrix}0\\ l\end{bmatrix}_{r-1,\ m-r+1}^1[0,\ l]_1^{r-1,\ n-r+1},$$

which lead to the equation (E).

The equation (E) is equivalent to equigradent transformations in Ω of the forms (B) in which :

(1) $e_1=\dfrac{1}{\Delta_0\Delta_1},\quad e_2=\dfrac{1}{\Delta_1\Delta_2},\quad e_3=\dfrac{1}{\Delta_2\Delta_3},\ \ldots\ e_{r-1}=\dfrac{1}{\Delta_{r-1}\Delta_r}.$

(2) $[h]_m^m$ is the matrix in which the 1st, 2nd, ... rth, last $m-r$ vertical rows are

$$[a]_m^1,\quad \begin{bmatrix}0\\ b\end{bmatrix}_{1,\ m-1}^1,\ \ldots\ \begin{bmatrix}0\\ l\end{bmatrix}_{r-1,\ m-r+1}^1,\quad \begin{bmatrix}0\\ 1\end{bmatrix}_{r,\ m-r}^{m-r}.$$

(3) $[k]_n^n$ is the matrix in which the 1st, 2nd, ... rth, last $n-r$ horizontal rows are

$$[a]_1^n,\quad [0,\ b]_1^{1,\ n-1},\ \ldots\ [0,\ l]_1^{r-1,\ n-r+1},\quad [0,\ 1]_{n-r}^{r,\ n-r}.$$

These transformations are not unitary, but all the elements of $[h]^m_m$ and $[k]^n_n$ which are not 0's are minor determinants of $[a]^n_m$.

Ex. iii. From the equations (B) we deduce the equations

$$[a]^n_m = [h]^r_m [e]^r_r [k]^n_r, \quad \overline{\underline{H}}^m_r [a]^n_m \overline{\underline{K}}^r_n = [e]^r_r;$$

$$[a]^r_r = [h]^r_r [e]^r_r [k]^r_r, \quad \overline{\underline{H}}^r_r [a]^r_r \overline{\underline{K}}^r_r = [e]^r_r;$$

the last equation following from the forms of $\overline{\underline{H}}^m_r$ and $\overline{\underline{K}}^r_n$. In these equations we have

$$\overline{\underline{H}}^m_r [h]^r_m = \overline{\underline{H}}^r_r [h]^r_r = [1]^r_r, \quad [k]^n_r \overline{\underline{K}}^r_n = [k]^r_r \overline{\underline{K}}^r_r = [1]^r_r.$$

Theorem II b. *If $\phi = [a]^n_m$ is any matrix of rank r whose elements are constants lying in a domain of rationality Ω, we can convert ϕ by derangements and unitary equigradent transformations in Ω into a quasi-scalar matrix of the form*

$$\psi = \begin{bmatrix} e, & 0 \\ 0, & 0 \end{bmatrix}^{r,\, n-r}_{r,\, m-r}, \quad \textit{where } [e]^r_r = \begin{bmatrix} e_1 & 0 & \dots & 0 \\ 0 & e_2 & \dots & 0 \\ \dots & \dots & \dots & \dots \\ 0 & 0 & \dots & e_r \end{bmatrix},$$

and $e_1, e_2, \dots e_r$ are non-zero constants lying in Ω.

By Ex. i of § 116 we can convert ϕ by derangements of its horizontal and vertical rows into a similar matrix $[b]^n_m$ in which

$$b_{11} \neq 0, \quad (b)^2_2 \neq 0, \quad (b)^3_3 \neq 0, \dots (b)^r_r \neq 0.$$

We can then as in Theorem II *a* convert $[b]^n_m$ by unitary equigradent transformations in Ω into a matrix ψ of the form stated in the theorem. Consequently we can convert ϕ into a matrix ψ of that form by derangements and unitary equigradent transformations in Ω. In the particular case when ϕ is real, all these transformations are real.

We see then that there exist two mutually inverse equigradent transformations

$$[a]^n_m = [h]^m_m \begin{bmatrix} e, & 0 \\ 0, & 0 \end{bmatrix}^{r,\, n-r}_{r,\, m-r} [k]^n_n, \quad \overline{\underline{H}}^m_m [a]^n_m \overline{\underline{K}}^n_n = \begin{bmatrix} e, & 0 \\ 0, & 0 \end{bmatrix}^{r,\, n-r}_{r,\, m-r}, \quad \dots\text{(F)}$$

each of which is the resultant of derangements and unitary equigradent transformations in Ω.

Ex. iv. From (F) we deduce the two transformations

$$[a]^n_m = [h]^r_m \, [e]^r_r \, [k]^n_r, \quad \overline{\underline{H}}{}^m_r \, [a]^n_m \, \overline{\underline{K}}{}^r_n = [e]^r_r. \quad \text{.....................(F}'\text{)}$$

The first of the equations (F′) is an equigradent transformation in Ω converting $[e]^r_r$ into $[a]^n_m$, and since

$$\overline{\underline{H}}{}^m_r \, [h]^r_m = [k]^n_r \, \overline{\underline{K}}{}^r_n = [1]^r_r,$$

we can deduce the second equation from the first by prefixing $\overline{\underline{H}}{}^m_r$ and postfixing $\overline{\underline{K}}{}^r_n$.

As in Note 8 of § 141 we can deduce each of the pairs of transformations (F) and (F′) from the other.

Ex. v. In (F) the matrices $[h]^m_m$ and $[k]^n_n$ have the forms

$$[h]^m_m = [u]^m_m \begin{bmatrix} 1 & 0 & \ldots & 0 & 0 & \ldots & 0 \\ \sigma & 1 & \ldots & 0 & 0 & \ldots & 0 \\ \ldots & & & & & & \ldots \\ \sigma & \sigma & \ldots & 1 & 0 & \ldots & 0 \\ \sigma & \sigma & \ldots & \sigma & 1 & \ldots & 0 \\ \ldots & & & & & & \ldots \\ \sigma & \sigma & \ldots & \sigma & 0 & \ldots & 1 \end{bmatrix}, \quad [k]^n_n = \begin{bmatrix} 1 & \tau & \ldots & \tau & \tau & \ldots & \tau \\ 0 & 1 & \ldots & \tau & \tau & \ldots & \tau \\ \ldots & & & & & & \ldots \\ 0 & 0 & \ldots & 1 & \tau & \ldots & \tau \\ 0 & 0 & \ldots & 0 & 1 & \ldots & 0 \\ \ldots & & & & & & \ldots \\ 0 & 0 & \ldots & 0 & 0 & \ldots & 1 \end{bmatrix} [v]^n_n,$$

where $[u]^m_m$ and $[v]^n_n$ are derangements of $[1]^m_m$ and $[1]^n_n$ respectively, and each σ or τ merely denotes some element which is not necessarily 0 or 1. The σ's occur in the first r vertical rows and the τ's in the first r horizontal rows of their respective matrices.

Ex. vi. Any matrix $[a]^n_m$ whose elements are constants lying in Ω can be converted into the standard form of Theorem I by three successive equigradent transformations in Ω which are respectively a derangement, a unitary transformation, and a quasi-scalar transformation. For the matrix ψ of Theorem II *b* can be converted into the matrix ψ of Theorem I by an undegenerate quasi-scalar transformation in Ω.

Ex. vii. If $[a]^n_m$ and $[b]^n_m$ are two similar matrices of the same rank whose elements are constants in Ω, there exists an equigradent transformation in Ω of the form

$$[b]^n_m = [h]^m_m \, [a]^n_m \, [k]^n_n$$

which is the resultant of a derangement, a unitary transformation, a quasi-scalar transformation, a unitary transformation and a derangement applied in succession, the unitary transformations having the special form shown in Ex. xiii of § 141.

Ex. viii. *If* $\phi = {}^1[a]_m$ *and* $\psi = {}^1[b]_m$ *are two undegenerate quasi-scalar matrices whose diagonal elements are non-zero constants lying in the domain of rationality* Ω, *and whose determinants are equal in value, then we can convert* ϕ *into* ψ *by unitary equigradent transformations in* Ω.

There are many ways in which we can convert any one diagonal element of ϕ which is not the last into the corresponding element of ψ without altering the values of any of the preceding elements of ϕ, and by a succession of such transformations we can convert ϕ into ψ.

For example if a_i is any diagonal element of ϕ which is not the last, and if we

add the $(i+1)$th horizontal row to the ith,

add $\frac{b_i - a_i}{a_{i+1}}$ times the $(i+1)$th vertical row to the ith,

add $\frac{a_i - b_i}{b_i}$ times the ith horizontal row to the $(i+1)$th,

add $-\frac{a_{i+1}}{b_i}$ times the ith vertical row to the $(i+1)$th,

performing these operations on ϕ in succession, we convert ϕ into the matrix derived from it by replacing a_i by b_i and a_{i+1} by $\frac{a_i a_{i+1}}{b_i}$. If we perform these operations in succession for the values 1, 2, ... $m-1$ of i, we convert ϕ into the matrix derived from it by replacing $a_1, a_2, \ldots a_{m-1}$ by $b_1, b_2, \ldots b_{m-1}$ and a_m by $\frac{a_1 a_2 \ldots a_m}{b_1 b_2 \ldots b_{m-1}}$ or b_m, i.e. we convert ϕ into ψ.

Ex. ix. *Under the same circumstances we can convert ϕ into ψ by symmetric unitary equigradent transformations which are not necessarily transformations in Ω.*

If we first add in succession to the ith horizontal and vertical rows of ϕ the $(i+1)$th parallel row multiplied by κ, choosing κ so that

$$a_i + \kappa^2 a_{i+1} = b_i,$$

and then add in succession to the $(i+1)$th horizontal and vertical rows the ith parallel row multiplied by λ, choosing λ so that

$$\kappa a_{i+1} + \lambda b_i = 0,$$

we convert ϕ into the matrix derived from it by replacing a_i by b_i and a_{i+1} by $\frac{a_i a_{i+1}}{b_i}$.

Performing these operations in succession for the values 1, 2, ... $m-1$ of i we convert ϕ into ψ.

Ex. x. *If $\phi = [a]_m^m$ and $\psi = [b]_m^m$ are two similar undegenerate square matrices whose elements are constants lying in Ω, and whose determinants are equal in value, then we can convert ϕ either into ψ or into $-\psi$ by derangements and unitary equigradent transformations in Ω.*

Let $(a)_m^m = (b)_m^m = \Delta$, and let ϕ and ψ be converted into quasi-scalar matrices $[\alpha]_m^m$ and $[\beta]_m^m$ by derangements and unitary equigradent transformations in Ω.

Then $(\alpha)_m^m = \epsilon\Delta$, $(\beta)_m^m = \epsilon'\Delta$, where ϵ is either 1 or -1 and ϵ' is either 1 or -1.

If $\epsilon = \epsilon'$, then by Ex. viii we can convert $[\alpha]_m^m$ into $[\beta]_m^m$ by unitary equigradent transformations in Ω. Thus we can convert ϕ into $[\alpha]_m^m$ by derangements and unitary equigradent transformations in Ω, $[\alpha]_m^m$ into $[\beta]_m^m$ by unitary equigradent transformations in Ω, and $[\beta]_m^m$ into ψ by derangements and unitary equigradent transformations in Ω. It follows that we can convert ϕ into ψ by derangements and unitary equigradent transformations in Ω.

If $\epsilon = -\epsilon'$, we can convert $[\alpha]_m^m$ into $-[\beta]_m^m$ by unitary equigradent transformations in Ω, and it follows in the same way that we can convert ϕ into $-\psi$ by derangements and unitary equigradent transformations in Ω.

When the matrices ϕ and ψ are symmetric, then (using Theorem IIb) we can convert ϕ either into ψ or into $-\psi$ by symmetric derangements and symmetric unitary equigradent transformations which are not necessarily transformations in Ω.

Ex. xi. *If $[h]_m^m$ is a matrix whose elements are constants lying in the domain of rationality Ω, and if $(h)_m^m=1$ or -1, then we can convert $[h]_m^m$ into one or both of the matrices $[1]_m^m$, $-[1]_m^m$ by derangements and unitary equigradent transformations in Ω.*

This is a particular case of Ex. x.

It follows that we have $[h]_m^m=\pm[p]_m^m[u]_m^m[q]_m^m$ where $[p]_m^m$ and $[q]_m^m$ are derangements of $[1]_m^m$ and $[u]_m^m$ is the matrix of a unitary transformation.

§ 147. Reduction of a symmetric matrix with constant elements to standard forms by symmetric equigradent transformations.

1. *Reduction by derangements and unitary equigradent transformations.*

Theorem Ia. *If $\phi=[a]_m^m$ is a symmetric matrix of rank ρ whose elements are constants lying in a domain of rationality Ω, and whose rows are so arranged that none of the leading diagonal minor determinants*

$$\Delta_0=1,\quad \Delta_1=a_{11},\quad \Delta_2=(a)_2^2,\ldots \Delta_\lambda=(a)_\lambda^\lambda,\ldots \Delta_\rho=(a)_\rho^\rho$$

vanish, then we can convert ϕ by a symmetric unitary equigradent transformation in Ω into a quasi-scalar matrix of the standard form

$$\psi=\begin{bmatrix} e, & 0 \\ 0, & 0\end{bmatrix}_{\rho,\, m-\rho}^{\rho,\, m-\rho},\quad \text{where } [e]_\rho^\rho=\begin{bmatrix} e_1 & 0 & \ldots & 0 \\ 0 & e_2 & \ldots & 0 \\ \ldots & \ldots & \ldots & \ldots \\ 0 & 0 & \ldots & e_\rho\end{bmatrix},$$

and $e_1, e_2, \ldots e_\rho$ are the non-zero scalar quantities lying in Ω which are given by the equations

$$e_1=\Delta_1,\quad e_1e_2=\Delta_2,\quad e_1e_2e_3=\Delta_3,\ \ldots\ e_1e_2e_3\ldots e_\rho=\Delta_\rho.$$

This theorem is a particular case of Theorem III of § 145 and Theorem II*a* of § 146.

We define symmetric matrices

$$[b]_{m-1}^{m-1},\quad [c]_{m-2}^{m-2},\ldots [l]_{m-\lambda+1}^{m-\lambda+1},\ldots [r]_{m-\rho+1}^{m-\rho+1}$$

of ranks $\rho-1, \rho-2, \ldots \rho-\lambda+1, \ldots 1$ by the equations

$$a_{11}b_{uv}=\begin{pmatrix} 1, 1+v \\ a \\ 1, 1+u\end{pmatrix},\quad b_{11}c_{uv}=\begin{pmatrix} 1, 1+v \\ b \\ 1, 1+u\end{pmatrix},\quad c_{11}d_{uv}=\begin{pmatrix} 1, 1+v \\ c \\ 1, 1+u\end{pmatrix},\ \ldots$$

so that

$$a_{11}=\Delta_1,\quad a_{11}b_{11}=\Delta_2,\quad a_{11}b_{11}c_{11}=\Delta_3,\ \ldots\ a_{11}b_{11}c_{11}\ldots r_{11}=\Delta_\rho;$$

$$\Delta_1 b_{uv}=\begin{pmatrix}1,\,1+v\\ a\\ 1,\,1+u\end{pmatrix},\quad \Delta_2 c_{uv}=\begin{pmatrix}1,\,2,\,2+v\\ a\\ 1,\,2,\,2+u\end{pmatrix},\ \ldots\ \Delta_{\lambda-1} l_{uv}=\begin{pmatrix}1,\,2,\ldots\lambda-1,\,\lambda-1+v\\ a\\ 1,\,2,\ldots\lambda-1,\,\lambda-1+u\end{pmatrix},\ \ldots.$$

Then $a_{11}, b_{11}, \ldots l_{11}, \ldots r_{11}$ are the quantities $e_1, e_2, \ldots e_\lambda, \ldots e_\rho$ of the theorem; and if we write

$$h_\lambda=\frac{1}{l_{11}}[0,\ l]_1^{\lambda-1,\,m-\lambda+1}=[0,\ 1,\ \lambda]_1^{\lambda-1,\,1,\,m-\lambda}$$

and proceed as in the proof of Theorem II *a* of § 146, we obtain resultant symmetric unitary equigradent transformations in Ω of the forms

$$[a]_m^m=\overline{\underline{h}}_m^m\begin{bmatrix}e, & 0\\ 0, & 0\end{bmatrix}_{\rho,\,m-\rho}^{\rho,\,m-\rho}[h]_m^m,\quad [H]_m^m[a]_m^m\overline{\underline{H}}_m^m=\begin{bmatrix}e, & 0\\ 0, & 0\end{bmatrix}_{\rho,\,m-\rho}^{\rho,\,m-\rho},\qquad (A)$$

where $\overline{\underline{H}}_m^m$ is the inverse of $[h]_m^m$, and $[h]_m^m$ is the undegenerate square matrix whose 1st, 2nd, ... λth, ... ρth, last $m-\rho$ horizontal rows are

$$h_1,\ h_2,\ldots h_\lambda,\ldots h_\rho,\quad [0,\ 1]_{m-\rho}^{\rho,\,m-\rho}$$

respectively.

Here we obtain $h_1, h_2, h_3, \ldots h_\rho$ from h_λ by replacing (λ, l, l_{11}) by $(1, a, a_{11}), (2, b, b_{11}), (3, c, c_{11}), \ldots (\rho, r, r_{11})$.

When the matrix ϕ is real, the transformations (A) are real.

Ex. i. When $m=6$ and $\rho=3$, we have

$$[h]_6^6=\begin{bmatrix}1 & a_{11} & a_{12} & a_{13} & a_{14} & a_{15}\\ 0 & 1 & \beta_{11} & \beta_{12} & \beta_{13} & \beta_{14}\\ 0 & 0 & 1 & \gamma_{11} & \gamma_{12} & \gamma_{13}\\ 0 & 0 & 0 & 1 & 0 & 0\\ 0 & 0 & 0 & 0 & 1 & 0\\ 0 & 0 & 0 & 0 & 0 & 1\end{bmatrix},\quad \overline{\underline{H}}_6^6=\begin{bmatrix}1 & \mathrm{A}_{11} & \mathrm{A}_{12} & \mathrm{A}_{13} & \mathrm{A}_{14} & \mathrm{A}_{15}\\ 0 & 1 & \mathrm{B}_{11} & \mathrm{B}_{12} & \mathrm{B}_{13} & \mathrm{B}_{14}\\ 0 & 0 & 1 & \Gamma_{11} & \Gamma_{12} & \Gamma_{13}\\ 0 & 0 & 0 & 1 & 0 & 0\\ 0 & 0 & 0 & 0 & 1 & 0\\ 0 & 0 & 0 & 0 & 0 & 1\end{bmatrix},$$

where
$$a_{1,i-1}=\frac{a_{1i}}{a_{11}},\quad \beta_{1,i-1}=\frac{b_{1i}}{b_{11}},\quad \gamma_{1,i-1}=\frac{c_{1i}}{c_{11}},$$

and $\overline{\underline{H}}_6^6$ is the inverse of $[h]_6^6$. The elements of $\overline{\underline{H}}_6^6$ are completely and uniquely determinate; and in particular we have

$$\mathrm{A}_{11}=-a_{11},\quad \mathrm{B}_{11}=-\beta_{11},\quad \Gamma_{11}=-\gamma_{11},\quad \Gamma_{12}=-\gamma_{12},\quad \Gamma_{13}=-\gamma_{13}.$$

Here
$$a_{11}b_{uv}=\begin{pmatrix}1,\,1+v\\ a\\ 1,\,1+u\end{pmatrix},\quad b_{11}c_{uv}=\begin{pmatrix}1,\,1+v\\ b\\ 1,\,1+u\end{pmatrix}$$

or
$$\Delta_1 b_{uv}=\begin{pmatrix}1,\,1+v\\ a\\ 1,\,1+u\end{pmatrix},\quad \Delta_2 c_{uv}=\begin{pmatrix}1,\,2,\,2+v\\ a\\ 1,\,2,\,2+u\end{pmatrix}.$$

Ex. ii. *Direct proof of Theorem Ia.*

Let h_λ' be the conjugate of h_λ for the values 1, 2, ... ρ of λ, and let

$$\Phi_\lambda = \begin{bmatrix} 0, & 0 \\ 0, & l \end{bmatrix}_{\lambda-1,\, m-\lambda+1}^{\lambda-1,\, m-\lambda+1}, \quad T_\lambda = \frac{1}{l_{11}} \begin{bmatrix} 0 \\ l \end{bmatrix}_{\lambda-1,\, m-\lambda+1}^{1} [0, \ l]_1^{\lambda-1,\, m-\lambda+1},$$

so that
$$T_\lambda = l_{11} h_\lambda' h_\lambda = h_\lambda' [l_{11}] h_\lambda.$$

From the identity (A′) of § 116 we obtain such equations as

$$c_{11} [c]_{m-2}^{m-2} = \begin{bmatrix} c \end{bmatrix}_{m-2}^{1} [c]_1^{m-2} + c_{11} \begin{bmatrix} 0, & 0 \\ 0, & d \end{bmatrix}_{3,\, m-3}^{3,\, m-3}, \quad \text{or} \quad \Phi_3 = T_3 + \Phi_4.$$

From the equations

$$\phi = T_1 + \Phi_2, \ \ldots \ \Phi_\lambda = T_\lambda + \Phi_{\lambda+1}, \ \ldots \ \Phi_\rho = T_\rho \quad \text{.........................(1)}$$

obtained in this way we deduce the equation

$$[a]_m^m = T_1 + T_2 + T_3 + \ldots + T_\lambda + \ldots + T_\rho \quad \text{..................................}(A_1)$$

or
$$[a]_m^m = a_{11} h_1' h_1 + b_{11} h_2' h_2 + \ldots + l_{11} h_\lambda' h_\lambda + \ldots + r_{11} h_\rho' h_\rho. \quad \text{...............}(A_2)$$

By the properties of passive rows (A_2) is equivalent to the first of the equations (A), and the second of the equations (A), which follows from the first, shows that the theorem is true.

By expressing T_λ in other forms we can obtain in the same way other symmetric equigradent transformations of ϕ which are not unitary.

In (A_2) and ψ we have

$$a_{11} = \frac{\Delta_1}{\Delta_0}, \quad b_{11} = \frac{\Delta_2}{\Delta_1}, \quad c_{11} = \frac{\Delta_3}{\Delta_2}, \ \ldots \ l_{11} = \frac{\Delta_\lambda}{\Delta_{\lambda-1}}, \ \ldots \ r_{11} = \frac{\Delta_\rho}{\Delta_{\rho-1}}.$$

Ex. iii. Let $\mathbf{l}_{uv} = \begin{pmatrix} 1, 2, \ldots \lambda-1, \lambda-1+v \\ a \\ 1, 2, \ldots \lambda-1, \lambda-1+u \end{pmatrix} = \Delta_{\lambda-1} l_{uv}$, so that $\mathbf{l}_{11} = \Delta_\lambda$. Then in ($A_1$) we have

$$T_\lambda = \frac{1}{\Delta_{\lambda-1}\Delta_\lambda} \begin{bmatrix} 0 \\ \mathbf{1} \end{bmatrix}_{\lambda-1,\, m-\lambda+1}^{1} [0, \ \mathbf{1}]_1^{\lambda-1,\, m-\lambda+1}$$

This proves the results of Ex. iv.

Ex. iv. *A corresponding non-unitary symmetric equigradent transformation.*

Let $[b]_{m-1}^{m-1}$, $[c]_{m-2}^{m-2}$, ... $[l]_{m-\lambda+1}^{m-\lambda+1}$, ... $[r]_{m-\rho+1}^{m-\rho+1}$ be the symmetric matrices of ranks $\rho-1, \rho-2, \ldots \rho-\lambda+1, \ldots 1$ whose elements are defined by such equations as

$$l_{uv} = \begin{pmatrix} 1, 2, \ldots \lambda-1, \lambda-1+v \\ a \\ 1, 2, \ldots \lambda-1, \lambda-1+u \end{pmatrix},$$

where, when $\lambda = 1, 2, 3, \ldots \rho$, we have $l = a, b, c, \ldots r$.

Then the symmetric matrix ϕ of Theorem Ia can be expressed in the form

$$[a]_m^m = T_1 + T_2 + T_3 + \ldots + T_\lambda + \ldots + T_\rho, \quad \text{.........................}(A_1)$$

where
$$T_\lambda = \frac{1}{\Delta_{\lambda-1}\Delta_\lambda} \begin{bmatrix} 0 \\ l \end{bmatrix}_{\lambda-1,\, m-\lambda+1}^{1} [0, \ l]_1^{\lambda-1,\, m-\lambda+1}$$

Here $T_1, T_2, T_3, \ldots T_\rho$ are obtained from T_λ by replacing (λ, l) by $(1, a)$, $(2, b)$, $(3, c)$, ... (ρ, r).

This result is a particular case of Theorem IV of § 145 and Note 3 of § 146; and it has already been proved in Ex. iii. To prove it directly let

$$\Phi_\lambda = \begin{bmatrix} 0, & 0 \\ 0, & l \end{bmatrix}_{\lambda-1,\, m-\lambda+1}^{\lambda-1,\, m-\lambda+1}, \quad \text{so that} \quad \Phi_1 = \phi.$$

Then the identity (A′) of § 116 furnishes such equations as

$$c_{11}\,[c]_{m-2}^{m-2} = \Delta_3\,[c]_{m-2}^{m-2} = \overline{c}\,{}^{1}_{m-2}\,[c]_1^{m-2} + \Delta_2 \begin{bmatrix} 0, & 0 \\ 0, & d \end{bmatrix}_{1,\, m-3}^{1,\, m-3},$$

or

$$\frac{1}{\Delta_2}\Phi_3 = T_3 + \frac{1}{\Delta_3}\Phi_4.$$

From the equations

$$\frac{1}{\Delta_0}\phi = T_1 + \frac{1}{\Delta_1}\Phi_2, \ \ldots \ \frac{1}{\Delta_{\lambda-1}}\Phi_\lambda = T_\lambda + \frac{1}{\Delta_\lambda}\Phi_{\lambda+1}, \ \ldots \ \frac{1}{\Delta_{\rho-1}}\Phi_\rho = T_\rho \quad \text{............(2)}$$

obtained in this way we deduce such equations as

$$[a]_m^m = T_1 + T_2 + T_3 + \ldots + T_\lambda + \frac{1}{\Delta_\lambda}\Phi_{\lambda+1},$$

and finally the equation (A_1).

The equation (A_1) with this new notation is equivalent to symmetric non-unitary transformations in Ω of the forms (A) in which

$$e_i = \frac{1}{\Delta_{i-1}\Delta_i}, \quad (i = 1, 2, \ldots \rho),$$

and $[h]_m^m$ is the matrix whose 1st, 2nd, ... λth, ... ρth, last $m-\rho$ horizontal rows are

$$[a]_1^m, \quad [0, b]_1^{1,\, m-1}, \ldots [0, l]_1^{\lambda-1,\, m-\lambda+1}, \ldots [0, r]_1^{\rho-1,\, m-\rho+1}, \quad [0, 1]_{m-\rho}^{\rho,\, m-\rho}.$$

Those elements of $[h]_m^m$ which are not 0's are minor determinants of $[a]_m^m$, and therefore these transformations lie in Ω.

Theorem I b. *Let $\phi = [a]_m^m$ be a symmetric matrix of rank ρ whose elements are constants lying in a domain of rationality Ω, and whose rows are so arranged that the series of leading diagonal minor determinants*

$$\Delta_0 = 1, \quad \Delta_1 = a_{11}, \quad \Delta_2 = (a)_2^2, \quad \Delta_3 = (a)_3^3, \ \ldots \ \Delta_\rho = (a)_\rho^\rho \quad \text{......(3)}$$

has the following properties:

(1) *The first and last determinants Δ_0 and Δ_ρ do not vanish.*

(2) *No two consecutive determinants both vanish.*

(3) *If $\Delta_{\lambda-1} = 0$ (where $\lambda - 1 < r$, and $\lambda - 1 > 0$), then all those diagonal minor determinants of ϕ of order $\lambda - 1$ which contain $\Delta_{\lambda-2}$ vanish, but*

$$\delta_{\lambda-1} = \begin{pmatrix} 1, 2, \ldots \lambda-2, \lambda \\ a \\ 1, 2, \ldots \lambda-2, \lambda-1 \end{pmatrix} = \begin{pmatrix} 1, 2, \ldots \lambda-2, \lambda-1 \\ a \\ 1, 2, \ldots \lambda-2, \lambda \end{pmatrix} \neq 0. \quad \text{...............(4)}$$

Then we can convert ϕ by a symmetric unitary equigradent transformation in Ω into a similar symmetric matrix

$$\psi = \begin{bmatrix} e, & 0 \\ 0, & 0 \end{bmatrix}_{\rho,\, m-\rho}^{\rho,\, m-\rho},$$

where $[e]_\rho^\rho$ is a certain symmetric compartite matrix of standard form whose parts correspond one by one to the successive non-vanishing determinants of the series $\Delta_1, \Delta_2, \ldots \Delta_\rho$. If Δ_λ is any one of these determinants which does not vanish, the part of $[e]_\rho^\rho$ corresponding to Δ_λ is formed as follows:

(1) *When $\Delta_{\lambda-1} \neq 0$, it consists of the single element l_{11} given by the equation*

$$\Delta_{\lambda-1} l_{11} = \Delta_\lambda \neq 0. \quad \text{.............................(5)}$$

(2) *When $\Delta_{\lambda-1} = 0$, it is the symmetric matrix $[l]_2^2 = \begin{bmatrix} 0 & l_{12} \\ l_{12} & 0 \end{bmatrix}$, where*

$$\Delta_{\lambda-2} l_{12} = \delta_{\lambda-1} \neq 0. \quad \text{.............................(6)}$$

By Note 2 of § 126 every symmetric matrix of order m and rank ρ whose elements are constants can be reduced by a symmetric derangement to a symmetric matrix of the character described above.

The transformation which we shall determine remains valid in the more general case when the third property of the series (3) is that of the four minor determinants of ϕ of order $\lambda - 1$ which contain $\Delta_{\lambda-2}$ and are contained in $\Delta_{\lambda-1}$ those two which are diagonal minor determinants vanish, but those two which are non-diagonal minor determinants (and are mutually conjugate and equal in value) do not vanish.

We can deduce this theorem from Theorem III of § 145.

Let
$$\Delta_\alpha, \Delta_\beta, \Delta_\gamma, \ldots \Delta_\kappa, \Delta_\lambda, \Delta_\mu, \ldots \Delta_\pi, \Delta_\rho$$

be the successive non-vanishing determinants of the series $\Delta_1, \Delta_2, \ldots \Delta_\rho$; let

$$[a]_m^m, \quad [b]_{m-\alpha}^{m-\alpha}, \quad [c]_{m-\beta}^{m-\beta}, \ldots [l]_{m-\kappa}^{m-\kappa}, \quad [m]_{m-\lambda}^{m-\lambda}, \ldots [r]_{m-\pi}^{m-\pi}$$

be the symmetric matrices of ranks $\rho, \rho-\alpha, \rho-\beta, \ldots \rho-\kappa, \rho-\lambda, \ldots \rho-\pi$ whose elements are defined as in the theorem of § 143, the first of these being identical with ϕ; and let ϕ be reduced to the compartite matrix ψ of Theorem III of § 145, the same notation being used as in § 143. In this particular case each one of the differences $\alpha - 0, \beta - \alpha, \gamma - \beta, \ldots \lambda - \kappa, \mu - \lambda, \ldots \rho - \pi$ is either 1 or 0.

The part of ψ corresponding to Δ_λ is $[l]_{\lambda-\kappa}^{\lambda-\kappa}$, where

$$\Delta_\kappa l_{uv} = \begin{pmatrix} 1, 2, \ldots \kappa, \kappa+v \\ a \\ 1, 2, \ldots \kappa, \kappa+u \end{pmatrix}.$$

When $\Delta_{\lambda-1} \neq 0$, so that $\kappa = \lambda - 1$, we have $[l]_{\lambda-\kappa}^{\lambda-\kappa} = [l]_1^1 = [l_{11}]$, where

$$\Delta_{\lambda-1} l_{11} = \Delta_\lambda \neq 0.$$

When $\Delta_{\lambda-1}=0$, so that $\Delta_{\lambda-2}\neq 0$ and $\kappa=\lambda-2$, we have $[l]_{\lambda-\kappa}^{\lambda-\kappa}=[l]_2^2$, where

$$\Delta_{\lambda-2}l_{11}=\Delta_{\lambda-2}l_{22}=0, \text{ and } \Delta_{\lambda-2}l_{12}=\Delta_{\lambda-2}l_{21}=\delta_{\lambda-1}\neq 0.$$

Thus we have $\psi=\begin{bmatrix} e, & 0 \\ 0, & 0\end{bmatrix}_{\rho,\, m-\rho}^{\rho,\, m-\rho}$, where $[e]_\rho^\rho$ is the matrix defined in the enunciation.

We have thus established the theorem and proved the existence of equations of the forms

$$[a]_m^m=\underline{\overline{h}}_m^m\begin{bmatrix} e, & 0 \\ 0, & 0\end{bmatrix}_{\rho,\, m-\rho}^{\rho,\, m-\rho}[h]_m^m, \quad [H]_m^m[a]_m^m\underline{\overline{H}}_m^m=\begin{bmatrix} e, & 0 \\ 0, & 0\end{bmatrix}_{\rho,\, m-\rho}^{\rho,\, m-\rho}, \quad \text{...(B)}$$

where $[h]_m^m$ and $\underline{\overline{H}}_m^m$ are two mutually inverse undegenerate square matrices whose elements are constants lying in Ω.

If as in Theorem III of § 145 we write

$$L=(l)_{\lambda-\kappa}^{\lambda-\kappa}, \quad h_\lambda=\frac{1}{L}\underline{\overline{L}}_{\lambda-\kappa}^{\lambda-\kappa}[0,\, l]_{\lambda-\kappa}^{\kappa,\, m-\kappa}=[0,\, 1,\, \lambda]_{\lambda-\kappa}^{\kappa,\, \lambda-\kappa,\, m-\lambda},$$

the corresponding expressions for $A, B, C, \ldots, h_\alpha, h_\beta, h_\gamma, \ldots$ being obtained by replacing (λ, κ, l) by $(\alpha, 0, a)$, (β, α, b), (γ, β, c), ..., the successive horizontal rows of $[h]_m^m$ are the successive horizontal rows of the matrices $h_\alpha, h_\beta, h_\gamma, \ldots h_\rho, [0,\, 1]_{m-\rho}^{\rho,\, m-\rho}$ taken in this order.

When $\kappa=\lambda-1$, i.e. when $\Delta_{\lambda-1}\neq 0$, we have $L=l_{11}$, $\underline{\overline{L}}_{\lambda-\kappa}^{\lambda-\kappa}=[1]$, and

$$h_\lambda=\frac{1}{l_{11}}[0,\, l]_1^{\kappa,\, m-\kappa}=[0,\, 1,\, \lambda]_1^{\kappa,\, 1,\, m-\lambda},$$

where
$$\lambda_{1,\, i-1}=\frac{l_{1i}}{l_{11}};$$

and in this case the λth horizontal row of $[h]_m^m$ is h_λ.

When $\kappa=\lambda-2$, i.e. when $\Delta_{\lambda-1}=0$, we have

$$L=-l_{12}^2, \quad \underline{\overline{L}}_{\lambda-\kappa}^{\lambda-\kappa}=\begin{bmatrix} 0\,, & -l_{12} \\ -l_{12}, & 0\end{bmatrix},$$

and
$$h_\lambda=\frac{1}{l_{12}}\begin{bmatrix} 0, & 1 \\ 1, & 0\end{bmatrix}[0,\, l]_2^{\kappa,\, m-\kappa}=[0,\, 1,\, \lambda]_2^{\kappa,\, 2,\, m-\lambda},$$

where
$$\lambda_{1,\, i-2}=\frac{l_{2i}}{l_{12}}, \quad \lambda_{2,\, i-2}=\frac{l_{1i}}{l_{12}};$$

and in this case the $(\lambda-1)$th and $(\lambda-2)$th horizontal rows of $[h]_m^m$ are h_λ.

The equations (B) are deduced from the equivalent equation

$$[a]_m^m = T_\alpha + T_\beta + T_\gamma + \dots + T_\kappa + T_\lambda + T_\mu + \dots + T_\pi + T_\rho, \qquad \dots(B_1)$$

where
$$T_\lambda = \frac{1}{L}\,\underset{\kappa,\, m-\kappa}{\overset{\lambda-\kappa}{\begin{matrix}0\\ l\end{matrix}}}\quad \underset{\lambda-\kappa}{\overset{\lambda-\kappa}{L}}\,[0,\ l]_{\lambda-\kappa}^{\kappa,\, m-\kappa} = h_\lambda'\,[l]_{\lambda-\kappa}^{\lambda-\kappa}\,h_\lambda.$$

When $\kappa = \lambda - 1$, i.e. when $\Delta_{\lambda-1} \neq 0$, we have

$$T_\lambda = \frac{1}{l_{11}}\,\underset{\kappa,\, m-\kappa}{\overset{1}{\begin{matrix}0\\ l\end{matrix}}}\quad [0,\ l]_1^{\kappa,\, m-\kappa}, \qquad \dots\dots\dots\dots(7)$$

and when $\kappa = \lambda - 2$, i.e. when $\Delta_{\lambda-1} = 0$, we have

$$T_\lambda = \frac{1}{l_{12}}\,\underset{\kappa,\, m-\kappa}{\overset{2}{\begin{matrix}0\\ l\end{matrix}}}\quad \begin{bmatrix}0 & 1\\ 1 & 0\end{bmatrix}[0,\ l]_2^{\kappa,\, m-\kappa}$$

$$= \frac{1}{2l_{12}}\,\underset{\kappa,\, m-\kappa}{\overset{2}{\begin{matrix}0\\ l\end{matrix}}}\quad \begin{bmatrix}1, & 1\\ 1, & -1\end{bmatrix}\cdot\begin{bmatrix}1, & 0\\ 0, & -1\end{bmatrix}\cdot\begin{bmatrix}1, & 1\\ 1, & -1\end{bmatrix}[0,\ l]_2^{\kappa,\, m-\kappa}. \qquad \dots(8)$$

Ex. v. If $m=7$, $\rho=5$; and if $\Delta_2, \Delta_4, \Delta_5$ are the non-vanishing determinants of the series $\Delta_1, \Delta_2, \Delta_3, \Delta_4, \Delta_5$, we have

$$[h]_7^7 = \begin{bmatrix}1 & 0 & \alpha_{11} & \alpha_{12} & \alpha_{13} & \alpha_{14} & \alpha_{15}\\ 0 & 1 & \alpha_{21} & \alpha_{22} & \alpha_{23} & \alpha_{24} & \alpha_{25}\\ 0 & 0 & 1 & 0 & \beta_{11} & \beta_{12} & \beta_{13}\\ 0 & 0 & 0 & 1 & \beta_{21} & \beta_{22} & \beta_{23}\\ 0 & 0 & 0 & 0 & 1 & \gamma_{11} & \gamma_{12}\\ 0 & 0 & 0 & 0 & 0 & 1 & 0\\ 0 & 0 & 0 & 0 & 0 & 0 & 1\end{bmatrix}, \qquad \underset{7}{\overset{7}{H}} = \begin{bmatrix}1 & 0 & A_{11} & A_{12} & A_{13} & A_{14} & A_{15}\\ 0 & 1 & A_{21} & A_{22} & A_{23} & A_{24} & A_{25}\\ 0 & 0 & 1 & 0 & B_{11} & B_{12} & B_{13}\\ 0 & 0 & 0 & 1 & B_{21} & B_{22} & B_{23}\\ 0 & 0 & 0 & 0 & 1 & \Gamma_{11} & \Gamma_{12}\\ 0 & 0 & 0 & 0 & 0 & 1 & 0\\ 0 & 0 & 0 & 0 & 0 & 0 & 1\end{bmatrix},$$

where
$$\alpha_{1,i-2} = \frac{a_{2i}}{a_{12}},\quad \alpha_{2,i-2} = \frac{a_{1i}}{a_{12}},\quad \beta_{1,i-2} = \frac{b_{2i}}{b_{12}},\quad \beta_{2,i-2} = \frac{b_{1i}}{b_{12}},\quad \gamma_{1,i-1} = \frac{c_{1i}}{c_{11}}.$$

In this case equation (B_1) is

$$[a]_7^7 = T_2 + T_4 + T_5,$$

where

$$T_2 = \frac{1}{a_{12}}\,\underset{7}{\overset{2}{a}}\begin{bmatrix}0 & 1\\ 1 & 0\end{bmatrix}[a]_2^7 = \frac{1}{2a_{12}}\,\underset{7}{\overset{2}{a}}\begin{bmatrix}1, & 1\\ 1, & -1\end{bmatrix}\cdot\begin{bmatrix}1, & 0\\ 0, & -1\end{bmatrix}\cdot\begin{bmatrix}1, & 1\\ 1, & -1\end{bmatrix}[a]_2^7,$$

$$T_4 = \frac{1}{b_{12}}\,\underset{2,\,5}{\overset{2}{\begin{matrix}0\\ b\end{matrix}}}\quad \begin{bmatrix}0 & 1\\ 1 & 0\end{bmatrix}[0,\ b]_2^{2,\,5} = \frac{1}{2b_{12}}\,\underset{2,\,5}{\overset{2}{\begin{matrix}0\\ b\end{matrix}}}\quad \begin{bmatrix}1, & 1\\ 1, & -1\end{bmatrix}\cdot\begin{bmatrix}1, & 0\\ 0, & -1\end{bmatrix}\cdot\begin{bmatrix}1, & 1\\ 1, & -1\end{bmatrix}[0,\ b]_2^{2,\,5},$$

$$T_5 = \frac{1}{c_{11}}\,\underset{4,\,3}{\overset{1}{\begin{matrix}0\\ c\end{matrix}}}\quad [0,\ c]_1^{4,\,3}.$$

Here
$$(a)^2_2\, b_{uv}=\begin{pmatrix}1,\,2,\,2+u\\ a\\ 1,\,2,\,2+u\end{pmatrix},\qquad (b)^2_2\, c_{uv}=\begin{pmatrix}1,\,2,\,2+u\\ b\\ 1,\,2,\,2+u\end{pmatrix};$$
or
$$\Delta_2 b_{uv}=\begin{pmatrix}1,\,2,\,2+u\\ a\\ 1,\,2,\,2+u\end{pmatrix},\qquad \Delta_4 c_{uv}=\begin{pmatrix}1,\,2,\,3,\,4,\,4+u\\ a\\ 1,\,2,\,3,\,4,\,4+u\end{pmatrix}.$$

Ex. vi. *Direct proof of Theorem I b.*

Let $\Delta_\alpha, \Delta_\beta, \Delta_\gamma, \ldots \Delta_\kappa, \Delta_\lambda, \Delta_\mu, \ldots \Delta_\pi, \Delta_\rho$ be the successive non-vanishing determinants of the series $\Delta_1, \Delta_2, \ldots \Delta_\rho$, and let

$$a_{11} b_{uv}=\begin{pmatrix}1,\,1+v\\ a\\ 1,\,1+u\end{pmatrix} \text{ or } (a)^2_2\, b_{uv}=\begin{pmatrix}1,\,2,\,2+v\\ a\\ 1,\,2,\,2+u\end{pmatrix} \text{ according as } a_{11}\neq 0 \text{ or } a_{11}=0,$$

i.e. according as $\alpha=1$ or $\alpha=2$;

$$b_{11} c_{uv}=\begin{pmatrix}1,\,1+v\\ b\\ 1,\,1+u\end{pmatrix} \text{ or } (b)^2_2\, c_{uv}=\begin{pmatrix}1,\,2,\,2+v\\ b\\ 1,\,2,\,2+u\end{pmatrix} \text{ according as } b_{11}\neq 0 \text{ or } b_{11}=0,$$

i.e. according as $\beta=\alpha+1$ or $\beta=\alpha+2$;

....

$$l_{11} m_{uv}=\begin{pmatrix}1,\,1+v\\ l\\ 1,\,1+u\end{pmatrix} \text{ or } (l)^2_2\, l_{uv}=\begin{pmatrix}1,\,2,\,2+v\\ l\\ 1,\,2,\,2+u\end{pmatrix} \text{ according as } l_{11}\neq 0 \text{ or } l_{11}=0,$$

i.e. according as $\lambda=\kappa+1$ or $\lambda=\kappa+2$.

....

Then the determinants $A=(a)^\alpha_\alpha$, $B=(b)^{\beta-\alpha}_{\beta-\alpha}$, ... $L=(l)^{\lambda-\kappa}_{\lambda-\kappa}$, ... do not vanish, and we have

$$A=\Delta_\alpha,\quad AB=\Delta_\beta,\quad ABC=\Delta_\gamma,\quad \ldots\quad ABC\ldots L=\Delta_\lambda,\ \ldots$$

and
$$\Delta_\kappa l_{uv}=\begin{pmatrix}1,\,2,\,\ldots\,\kappa,\,\kappa+v\\ a\\ 1,\,2,\,\ldots\,\kappa,\,\kappa+u\end{pmatrix}.$$

If $\Delta_{\lambda-1}\neq 0$, i.e. if $\kappa=\lambda-1$, we have $\Delta_{\lambda-1} l_{11}=\Delta_\lambda\neq 0$; and if $\Delta_{\lambda-1}=0$, i.e. if $\kappa=\lambda-2$, we have $l_{11}=l_{22}=0$, $\Delta_{\lambda-2} l_{12}=\Delta_{\lambda-2} l_{21}=\delta_{\lambda-1}\neq 0$, where $\delta_{\lambda-1}$ is defined as in (4).

Let
$$h_\lambda=\frac{1}{l_{11}}[0,\ l]_1^{\kappa,\,m-\kappa},\qquad h_\lambda=\frac{1}{l_{12}}\begin{bmatrix}0, & 1\\ 1, & 0\end{bmatrix}[0,\ l]_2^{\kappa,\,m-\kappa}$$

according as $\kappa=\lambda-1$ or $\kappa=\lambda-2$; let h_λ' be the conjugate of h_λ; and let

$$\Phi_\lambda=\begin{bmatrix}0, & 0\\ 0, & l\end{bmatrix}^{\kappa,\,m-\kappa}_{\kappa,\,m-\kappa},\qquad T_\lambda=h_\lambda'\,[l]^{\lambda-\kappa}_{\lambda-\kappa}\,h_\lambda,$$

so that T_λ has the value (7) or (8) according as $\kappa=\lambda-1$ or $\kappa=\lambda-2$.

If $\kappa=\lambda-1$, i.e. if $\Delta_{\lambda-1}\neq 0$, the identity of § 116 gives the equation

$$l_{11}\,[l]^{m-\kappa}_{m-\kappa}=\overline{\underline{l}}^{\,1}_{\;m-\kappa}\,[l]^{m-\kappa}_1+l_{11}\begin{bmatrix}0, & 0\\ 0, & m\end{bmatrix}^{1,\,m-\lambda}_{1,\,m-\lambda},$$

or
$$\Phi_\lambda=T_\lambda+\Phi_\mu\,. \qquad\qquad \text{(9)}$$

If $\kappa=\lambda-2$, i.e. if $\Delta_{\lambda-1}=0$, the same identity gives the equation

$$(l)^2_2\,[l]^{m-\kappa}_{m-\kappa}=\overline{\underline{l}}^{\,2}_{\;m-\kappa}\begin{bmatrix}0 & , -l_{12}\\ -l_{12}, & 0\end{bmatrix}[l]^{m-\kappa}_2+(l)^2_2\begin{bmatrix}0, & 0\\ 0, & m\end{bmatrix}^{2,\,m-\lambda}_{2,\,m-\lambda},$$

or
$$-l^2_{12}\,[l]^{m-\kappa}_{m-\kappa}=-l_{12}\,\overline{\underline{l}}^{\,2}_{\;m-\kappa}\begin{bmatrix}0 & 1\\ 1 & 0\end{bmatrix}[l]^{m-\kappa}_2-l^2_{12}\begin{bmatrix}0, & 0\\ 0, & m\end{bmatrix}^{2,\,m-\lambda}_{2,\,m-\lambda},$$

or
$$\Phi_\lambda = T_\lambda + \Phi_\mu. \qquad \text{...(9)}$$

We obtain the same equation (9) in both cases.

From the equations

$$\phi = T_\alpha + \Phi_\beta, \quad \Phi_\beta = T_\beta + \Phi_\gamma, \ \ldots\ \Phi_\kappa = T_\kappa + \Phi_\lambda, \ \ldots\ \Phi_\rho = T_\rho$$

obtained in this way we deduce the equation (B_1) in which

$$T_\lambda = h_\lambda' [l_{11}] h_\lambda \quad \text{or} \quad T_\lambda = h_\lambda' \begin{bmatrix} 0 & l_{12} \\ l_{12} & 0 \end{bmatrix} h_\lambda$$

according as $\Delta_{\lambda-1} \neq 0$ or $\Delta_{\lambda-1} = 0$; and this last equation is equivalent to the first of the transformations (B) from which the second transformation follows.

Ex. vii. From (7) and (8) we see that we can convert ϕ by a non-unitary equigradent transformation in Ω into a matrix of the same form as ψ in the theorem where the part of $[e]_\rho^\rho$ corresponding to Δ_λ is $[l_{11}]$ when $\Delta_{\lambda-1} \neq 0$ and $\begin{bmatrix} 1, & 0 \\ 0, & -1 \end{bmatrix}$ when $\Delta_{\lambda-1} = 0$.

Ex. viii. When $\kappa = \lambda - 2$, we have $\Delta_{\lambda-2} l_{12} = \delta_{\lambda-1}$ and $\Delta_{\lambda-2} (l)_2^2 = \Delta_\lambda$,

i.e.
$$\Delta_{\lambda-2} l_{12} = \delta_{\lambda-1}, \quad \Delta_{\lambda-2} l_{12}^2 = -\Delta_\lambda.$$

Therefore the quantity $\delta_{\lambda-1}$ or $\delta_{\kappa+1}$ defined by (4) satisfies the equation

$$\Delta_{\lambda-2}\Delta_\lambda = -\delta_{\lambda-1}^2, \quad \text{or} \quad \Delta_\kappa \Delta_{\kappa+2} = -\delta_{\kappa+1}^2. \qquad \text{........................(10)}$$

This also follows by § 110 from the equation

$$\begin{vmatrix} \begin{pmatrix} 1, 2, \ldots \kappa, \kappa+1 \\ a \\ 1, 2, \ldots \kappa, \kappa+1 \end{pmatrix} & \begin{pmatrix} 1, 2, \ldots \kappa, \kappa+2 \\ a \\ 1, 2, \ldots \kappa, \kappa+1 \end{pmatrix} \\ \begin{pmatrix} 1, 2, \ldots \kappa, \kappa+1 \\ a \\ 1, 2, \ldots \kappa, \kappa+2 \end{pmatrix} & \begin{pmatrix} 1, 2, \ldots \kappa, \kappa+2 \\ a \\ 1, 2, \ldots \kappa, \kappa+2 \end{pmatrix} \end{vmatrix} = \begin{vmatrix} 0, & \delta_{\kappa+1} \\ \delta_{\kappa+1}, & 0 \end{vmatrix} = \Delta_\kappa \Delta_{\kappa+2}.$$

Ex. ix. If the symmetric matrix ϕ of Theorem I*b* is real, and if $\Delta_{\lambda-1} = 0$, then $\Delta_{\lambda-2}$ and Δ_λ have opposite signs.

Ex. x. Let $\mathbf{l}_{uv} = \begin{pmatrix} 1, 2, \ldots \kappa, \kappa+v \\ a \\ 1, 2, \ldots \kappa, \kappa+u \end{pmatrix} = \Delta_\kappa l_{uv}$. Then when $\kappa = \lambda - 1$, we have $\mathbf{l}_{11} = \Delta_\lambda$; and when $\kappa = \lambda - 2$, we have $\mathbf{l}_{11} = \mathbf{l}_{22} = 0$, $\mathbf{l}_{12} = \mathbf{l}_{21} = \delta_{\lambda-1} = \delta_{\kappa+1}$. Hence when we make the substitution $l_{uv} = \frac{1}{\Delta_\kappa} \mathbf{l}_{uv}$ in (7) and (8) we see that in (B_1) we have

$$T_\lambda = \frac{1}{\Delta_\kappa \Delta_{\kappa+1}} \begin{bmatrix} 0 \\ \mathbf{1} \end{bmatrix}^{1}_{\kappa,\, m-\kappa} [0, \mathbf{1}]_1^{\kappa,\, m-\kappa}$$

or
$$T_\lambda = \frac{1}{\Delta_\kappa \delta_{\kappa+1}} \begin{bmatrix} 0 \\ \mathbf{1} \end{bmatrix}^{2}_{\kappa,\, m-\kappa} \begin{bmatrix} 0 & 1 \\ 1 & 0 \end{bmatrix} [0, \mathbf{1}]_2^{\kappa,\, m-\kappa}$$

$$= \frac{1}{2\Delta_\kappa \delta_{\kappa+1}} \begin{bmatrix} 0 \\ \mathbf{1} \end{bmatrix}^{2}_{\kappa,\, m-\kappa} \begin{bmatrix} 1, & 1 \\ 1, & -1 \end{bmatrix} \cdot \begin{bmatrix} 1, & 0 \\ 0, & -1 \end{bmatrix} \cdot \begin{bmatrix} 1, & 1 \\ 1, & -1 \end{bmatrix} [0, \mathbf{1}]_2^{\kappa,\, m-\kappa}$$

according as $\kappa = \lambda - 1$ or $\kappa = \lambda - 2$.

In this way we deduce from (B_1) the equation (B_2) of Ex. xi.

Ex. xi. *A corresponding non-unitary symmetric equigradent transformation.*

Let

$$\Delta_\alpha,\ \Delta_\beta,\ \Delta_\gamma,\ \ldots\ \Delta_\kappa,\ \Delta_\lambda,\ \Delta_\mu,\ \ldots\ \Delta_\pi,\ \Delta_\rho$$

be the successive non-vanishing determinants of the series $\Delta_1, \Delta_2, \ldots \Delta_\rho$, and let

$$[a]_m^m,\quad [b]_{m-\alpha}^{m-\alpha},\quad [c]_{m-\beta}^{m-\beta},\ \ldots\ [l]_{m-\kappa}^{m-\kappa},\quad [m]_{m-\lambda}^{m-\lambda},\ \ldots\ [r]_{m-\pi}^{m-\pi}$$

be the symmetric matrices of ranks $\rho, \rho-\alpha, \rho-\beta, \ldots \rho-\kappa, \rho-\lambda, \ldots \rho-\pi$ whose elements are defined by such equations as

$$l_{uv} = \begin{pmatrix} 1, 2, \ldots \kappa, \kappa+v \\ a \\ 1, 2, \ldots \kappa, \kappa+u \end{pmatrix},$$

the matrix $[a]_m^m$ being identical with ϕ. Also let

$$\delta_{\kappa+1} = \begin{pmatrix} 1, 2, \ldots \kappa, \kappa+2 \\ a \\ 1, 2, \ldots \kappa, \kappa+1 \end{pmatrix} = \begin{pmatrix} 1, 2, \ldots \kappa, \kappa+1 \\ a \\ 1, 2, \ldots \kappa, \kappa+2 \end{pmatrix} \quad \text{when } \lambda-\kappa=2.$$

Then the matrix ϕ of Theorem I b can be expressed in the form

$$[a]_m^m = T_\alpha + T_\beta + T_\gamma + \ldots + T_\kappa + T_\lambda + T_\mu + \ldots + T_\pi + T_\rho, \quad \ldots\ldots\ldots\ldots\ldots(\mathrm{B}_2)$$

where

$$T_\lambda = \frac{1}{\Delta_\kappa \Delta_{\kappa+1}} \left\lceil\begin{matrix}0\\ l\end{matrix}\right\rfloor_{\kappa,\, m-\kappa}^{1} [0,\ l]_1^{\kappa,\, m-\kappa} \quad \ldots\ldots\ldots\ldots\ldots\ldots\ldots\ldots\ldots\ldots\ldots\ldots(11)$$

or

$$T_\lambda = \frac{1}{\Delta_\kappa \delta_{\kappa+1}} \left\lceil\begin{matrix}0\\ l\end{matrix}\right\rfloor_{\kappa,\, m-\kappa}^{2} \begin{bmatrix} 0 & 1 \\ 1 & 0 \end{bmatrix} [0,\ l]_2^{\kappa,\, m-\kappa}$$

$$= \frac{1}{2\Delta_\kappa \delta_{\kappa+1}} \left\lceil\begin{matrix}0\\ l\end{matrix}\right\rfloor_{\kappa,\, m-\kappa}^{2} \begin{bmatrix} 1, & 1 \\ 1, & -1 \end{bmatrix} \cdot \begin{bmatrix} 1, & 0 \\ 0, & -1 \end{bmatrix} \cdot \begin{bmatrix} 1, & 1 \\ 1, & -1 \end{bmatrix} [0,\ l]_2^{\kappa,\, m-\kappa} \quad \ldots\ldots\ldots(12)$$

according as $\kappa=\lambda-1$ or $\kappa=\lambda-2$, i.e. according as $\Delta_{\lambda-1} \neq 0$ or $\Delta_{\lambda-1}=0$.

We have deduced this result from (B_1) in Ex. x. To prove it directly let

$$\Phi_\lambda = \begin{bmatrix} 0, & 0 \\ 0, & l \end{bmatrix}_{\kappa,\, m-\kappa}^{\kappa,\, m-\kappa},$$

so that in particular $\Phi_\alpha = \phi$.

If $\kappa=\lambda-1$, we have $l_{11} = \Delta_\lambda \neq 0$, $\begin{pmatrix} 1,\ 1+v \\ l \\ 1,\ 1+u \end{pmatrix} = \Delta_\kappa m_{uv}$, and the identity of § 116 gives the equation

$$l_{11} [l]_{m-\kappa}^{m-\kappa} = \Delta_\lambda [l]_{m-\kappa}^{m-\kappa} = \left\lceil l \right\rfloor_{m-\kappa}^{1} [l]_1^{m-\kappa} + \Delta_\kappa \begin{bmatrix} 0, & 0 \\ 0, & m \end{bmatrix}_{1,\, m-\lambda}^{1,\, m-\lambda}.$$

Dividing both sides by $\Delta_\kappa \Delta_\lambda$ we obtain the equation

$$\frac{1}{\Delta_\kappa} \Phi_\lambda = T_\lambda + \frac{1}{\Delta_\lambda} \Phi_\mu. \quad \ldots\ldots\ldots\ldots\ldots\ldots\ldots\ldots\ldots\ldots\ldots(13)$$

If $\kappa=\lambda-2$, we have $l_{11}=l_{22}=0$, $l_{12}=l_{21}=\delta_{\kappa+1} \neq 0$, $\begin{pmatrix} 1, 2, 2+v \\ l \\ 1, 2, 2+u \end{pmatrix} = \Delta_\kappa^2 m_{uv}$, and the identity of § 116 gives the equation

$$(l)_2^2 [l]_{m-\kappa}^{m-\kappa} = -l_{12}^2 [l]_{m-\kappa}^{m-\kappa} = \left\lceil l \right\rfloor_{m-\kappa}^{2} \begin{bmatrix} 0\ , & -l_{12} \\ -l_{12}, & 0 \end{bmatrix} [l]_2^{m-\kappa} + \Delta_\kappa^2 \begin{bmatrix} 0, & 0 \\ 0, & m \end{bmatrix}_{2,\, m-\lambda}^{2,\, m-\lambda},$$

or

$$-\delta_{\kappa+1}^2 [l]_{m-\kappa}^{m-\kappa} = -\delta_{\kappa+1} \left\lceil l \right\rfloor_{m-\kappa}^{2} \begin{bmatrix} 0 & 1 \\ 1 & 0 \end{bmatrix} [l]_2^{m-\kappa} + \Delta_\kappa^2 \begin{bmatrix} 0, & 0 \\ 0, & m \end{bmatrix}_{2,\, m-\lambda}^{2,\, m-\lambda}.$$

It has been shown in Ex. viii that $\Delta_\kappa\Delta_{\kappa+2}=\Delta_\kappa\Delta_\lambda=-\delta^2_{\kappa+1}$.

Hence when we divide both sides of the last equation by $-\Delta_\kappa\delta^2_{\kappa+1}$, we obtain

$$\frac{1}{\Delta_\kappa}[l]^{m-\kappa}_{m-\kappa}=\frac{1}{\Delta_\kappa\delta_{\kappa+1}}\overline{l}^{\,2}_{m-\kappa}\begin{bmatrix}0 & 1\\ 1 & 0\end{bmatrix}[l]^{m-\kappa}_2+\frac{1}{\Delta_\lambda}\begin{bmatrix}0, & 0\\ 0, & m\end{bmatrix}^{2,\,m-\lambda}_{2,\,m-\lambda},$$

or again

$$\frac{1}{\Delta_\kappa}\Phi_\lambda=T_\lambda+\frac{1}{\Delta_\lambda}\Phi_\mu. \qquad \text{(13)}$$

In both cases we obtain the same equation (13), and from the equations

$$\frac{1}{\Delta_0}\phi=T_a+\frac{1}{\Delta_a}\Phi_\beta,\quad \frac{1}{\Delta_a}\Phi_\beta=T_\beta+\frac{1}{\Delta_\beta}\Phi_\gamma,\ \ldots\ \frac{1}{\Delta_\pi}\Phi_\rho=T_\rho$$

obtained in this way we deduce such equations as

$$\phi=T_a+T_\beta+T_\gamma+\ldots+T_\kappa+\frac{1}{\Delta_\kappa}\Phi_\lambda,$$

the last of which is the equation (B_2).

The equation (B_2) is equivalent to non-unitary symmetric equigradent transformations in Ω of the forms (B) in which $[e]^\rho_\rho$ is a symmetric compartite matrix of standard form having one part corresponding to each one of the non-vanishing determinants $\Delta_a, \Delta_\beta, \ldots \Delta_\kappa, \Delta_\lambda, \ldots \Delta_\rho$, the part corresponding to T_λ being

$$\frac{1}{\Delta_\kappa\Delta_{\kappa+1}}[1] \quad\text{or}\quad \frac{1}{2\Delta_\kappa\delta_{\kappa+1}}\begin{bmatrix}1, & 0\\ 0, & -1\end{bmatrix}$$

according as $\kappa=\lambda-1$ or $\kappa=\lambda-2$.

If

$$h_\lambda=[0,\ l]^{\kappa,\,m-\kappa}_1 \quad\text{or}\quad h_\lambda=\begin{bmatrix}1, & 1\\ 1, & -1\end{bmatrix}[0,\ l]^{\kappa,\,m-\kappa}_2$$

according as $\kappa=\lambda-1$ or $\kappa=\lambda-2$, i.e. according as $\Delta_{\lambda-1}\neq 0$ or $\Delta_{\lambda-1}=0$, then $[h]^m_m$ is now the matrix whose successive horizontal rows are the successive horizontal rows of the matrices $h_a, h_\beta, \ldots h_\rho, [0,\ 1]^{\rho,\,m-\rho}_{m-\rho}$ taken in this order. Those of its elements which do not vanish are minor determinants of $[a]^m_m$.

Ex. xii. We can deduce the equation (B_2) from the equation (E) in Theorem IV of § 145. Let l_{uv} be defined as in Ex. xi and $\mathbf{L}_{ui}$ as in Theorem IV of § 145.

Then when $\kappa=\lambda-1$, we have $\mathbf{L}_{1i}=l_{1i}$; and when $\kappa=\lambda-2$, we have by § 110 the equations

$$\begin{vmatrix}l_{1i} & \delta_{\kappa+1}\\ l_{2i} & 0\end{vmatrix}=\Delta_\kappa\mathbf{L}_{1i},\quad \begin{vmatrix}0 & l_{1i}\\ \delta_{\kappa+1} & l_{2i}\end{vmatrix}=\Delta_\kappa\mathbf{L}_{2i};$$

and when we make use of (10), it follows that

$$\frac{\mathbf{L}_{1i}}{l_{2i}}=\frac{\mathbf{L}_{2i}}{l_{1i}}=-\frac{\delta_{\kappa+1}}{\Delta_\kappa}=\frac{\Delta_{\kappa+2}}{\delta_{\kappa+1}}.$$

Ex. xiii. When we prefix $[x]_m$ and postfix $\overline{x}_m$ on both sides of (B_2), we obtain a reduction of the quadratic form $S=\det[x]_m[a]^m_m\overline{x}_m$ to an algebraic sum of ρ squares of linear expressions.

Theorem I c. *The symmetric matrix* $\phi = [a]_m^m$ *of Theorem I b whose rank is* ρ *can be converted by a symmetric unitary equigradent transformation in* Ω *into a symmetric matrix* $[c]_m^m$ *of the same order and rank in which none of the leading diagonal minor determinants* $c_{11}, (c)_2^2, (c)_3^3, \ldots (c)_\rho^\rho$ *vanish.*

Suppose that $\Delta_s = 0$ (where $s < \rho$, and $s > 0$), and let

$$A_{11} = \Delta_s = \begin{pmatrix} 1, 2, \ldots s-1, s \\ a \\ 1, 2, \ldots s-1, s \end{pmatrix}, \qquad A_{22} = \begin{pmatrix} 1, 2, \ldots s-1, s+1 \\ a \\ 1, 2, \ldots s-1, s+1 \end{pmatrix},$$

$$A_{12} = \begin{pmatrix} 1, 2, \ldots s-1, s+1 \\ a \\ 1, 2, \ldots s-1, s \end{pmatrix}, \quad A_{21} = \begin{pmatrix} 1, 2, \ldots s-1, s \\ a \\ 1, 2, \ldots s-1, s+1 \end{pmatrix},$$

so that $$A_{11} = 0, \quad A_{22} = 0, \quad A_{12} = A_{21} \neq 0.$$

Then if we convert $[a]_m^m$ into a similar symmetric matrix $[b]_m^m$ by adding in succession and in either order the $(s+1)$th horizontal row to the sth horizontal row and the $(s+1)$th vertical row to the sth vertical row, we have

$$[b]_m^m = \overline{h}_m^m \, [a]_m^m \, [h]_m^m, \quad \ldots\ldots\ldots\ldots\ldots\ldots\ldots\ldots \text{(C)}$$

where
$$[h]_m^m = \begin{bmatrix} 1, & 0, & 0, & 0 \\ 0, & 1, & 0, & 0 \\ 0, & 1, & 1, & 0 \\ 0, & 0, & 0, & 1 \end{bmatrix}_{s-1,\,1,\,1,\,m-s-1}^{s-1,\,1,\,1,\,m-s-1}$$

It follows from this equation that, when $i \neq s$ and $j \neq s$,

$$b_{ij} = a_{ij}, \qquad b_{ss} = a_{ss} + 2a_{s,\,s+1} + a_{s+1,\,s+1},$$

$$b_{is} = a_{is} + a_{i,\,s+1}, \quad b_{sj} = a_{sj} + a_{s+1,\,j}.$$

Let $\quad \Delta_0' = 1, \quad \Delta_1' = b_{11}, \quad \Delta_2' = (b)_2^2, \quad \Delta_3' = (b)_3^3, \ldots \Delta_m' = (b)_m^m.$

Then it also follows from the above equation that

$$\Delta_i' = \Delta_i \text{ when } i \neq s,$$

$$\Delta_s' = A_{11} + 2A_{12} + A_{22} = 2A_{12} \neq 0.$$

These two elementary unitary equigradent transformations have converted $[a]_m^m$ into another symmetric matrix of the same form in which $\Delta_s \neq 0$, and the rest of the determinants $\Delta_1, \Delta_2, \ldots \Delta_m$ retain the same values as before. By a succession of such double operations, i.e. by a series of symmetric unitary equigradent transformations in Ω, we can convert $[a]_m^m$ into a symmetric matrix $[c]_m^m$ having the properties specified in the theorem.

Since any symmetric matrix $[a]_m^m$ can be converted into one having the form considered in Theorem I b by a symmetric derangement, we see that any symmetric matrix $[a]_m^m$ can be converted into a symmetric matrix $[c]_m^m$ having the form considered in Theorem I a by a symmetric unitary equigradent transformation $\overline{\underline{h}}_m^m [a]_m^m [h]_m^m = [c]_m^m$ in which $(h)_m^m = 1$, and every element of $[h]_m^m$ is either 0 or 1. This transformation lies in every domain of rationality however restricted.

Theorem II. *Every symmetric matrix $\phi = [a]_m^m$ of rank r whose elements are constants lying in a domain of rationality Ω can be converted by a symmetric derangement followed by a symmetric unitary equigradent transformation in Ω into a quasi-scalar matrix of the standard form*

$$\psi = \begin{bmatrix} e, & 0 \\ 0, & 0 \end{bmatrix}_{r,\, m-r}^{r,\, m-r}, \quad \textit{where } [e]_r^r = \begin{bmatrix} e_1 & 0 & \dots & 0 \\ 0 & e_2 & \dots & 0 \\ \dots & \dots & \dots & \dots \\ 0 & 0 & \dots & e_r \end{bmatrix},$$

and $e_1, e_2, \dots e_r$ are non-zero constants lying in Ω.

By Note 2 of § 126 we can convert ϕ into a similar symmetric matrix ϕ' whose rows are arranged as in Theorem I b by a symmetric derangement; then by Theorem I c we can convert ϕ' into a similar matrix ϕ'' whose rows are arranged as in Theorem I a by a symmetric unitary equigradent transformation in Ω; and finally by Theorem I a we can convert ϕ'' into a matrix having the form assigned to ψ in the enunciation by a further symmetric unitary equigradent transformation in Ω.

The theorem also follows from Ex. vii.

The resultant transformations converting ψ into ϕ and ϕ into ψ have the forms

$$[a]_m^m = \overline{\underline{h}}_m^m \begin{bmatrix} e, & 0 \\ 0, & 0 \end{bmatrix}_{r,\, m-r}^{r,\, m-r} [h]_m^m, \quad [H]_m^m [a]_m^m \overline{\underline{H}}_m^m = \begin{bmatrix} e, & 0 \\ 0, & 0 \end{bmatrix}_{r,\, m-r}^{r,\, m-r}, \quad \dots\text{(D)}$$

where $[h]_m^m$ and $\overline{\underline{H}}_m^m$ are two mutually inverse undegenerate matrices whose elements are constants lying in Ω.

Ex. xiv. From (D) we deduce that

$$[a]_m^m = \overline{\underline{h}}_m^r [e]_r^r [h]_r^m, \quad [e]_r^r = [H]_r^m [a]_m^m \overline{\underline{H}}_m^r, \quad \dots\dots\dots\dots\text{(D')}$$

where

$$[h]_r^m \overline{\underline{H}}_m^r = [1]_r^r.$$

The first of the equations (D′) is a symmetric equigradent transformation in Ω converting $[e]_r^r$ into ϕ.

When equations of the form (D′) are given, we can (see Note 8 of § 141) re-construct equations of the form (D).

2. *Reduction by unrestricted symmetric equigradent transformations.*

Theorem III. *If $\phi = [a]_m^m$ is a symmetric matrix of rank r with constant elements, we can convert it by a symmetric equigradent transformation into the standard form*

$$\psi = \begin{bmatrix} 1, & 0 \\ 0, & 0 \end{bmatrix}_{r,\, m-r}^{r,\, m-r}.$$

By Theorem II we can determine a symmetric equigradent transformation

$$[K]_m^m [a]_m^m \overline{\underline{K}}_m^m = \begin{bmatrix} e, & 0 \\ 0, & 0 \end{bmatrix}_{r,\, m-r}^{r,\, m-r}, \quad \text{where } [e]_r^r = \begin{bmatrix} e_1 & 0 & \dots & 0 \\ 0 & e_2 & \dots & 0 \\ \dots & \dots & \dots & \dots \\ 0 & 0 & \dots & e_r \end{bmatrix},$$

and $e_1, e_2, \dots e_r$ are non-zero constants; and we can complete the reduction by dividing the successive horizontal and vertical rows of the matrix on the right by $\sqrt{e_1}, \sqrt{e_2}, \dots \sqrt{e_r}, 1, 1, \dots 1$.

Let

$$\overline{\underline{H}}_m^m = \overline{\underline{K}}_m^m \begin{bmatrix} e^{-\frac{1}{2}}, & 0 \\ 0\ , & 1 \end{bmatrix}_{r,\, m-r}^{r,\, m-r}, \quad \text{where } [e^{-\frac{1}{2}}]_r^r = \begin{bmatrix} e_1^{-\frac{1}{2}} & 0 & \dots & 0 \\ 0 & e_2^{-\frac{1}{2}} & \dots & 0 \\ \dots & \dots & \dots & \dots \\ 0 & 0 & \dots & e_r^{-\frac{1}{2}} \end{bmatrix}.$$

Then if $[h]_m^m$ is the inverse matrix of $\overline{\underline{H}}_m^m$, we have

$$[a]_m^m = \overline{\underline{h}}_m^m \begin{bmatrix} 1, & 0 \\ 0, & 0 \end{bmatrix}_{r,\, m-r}^{r,\, m-r} [h]_m^m, \quad [H]_m^m [a]_m^m \overline{\underline{H}}_m^m = \begin{bmatrix} 1, & 0 \\ 0, & 0 \end{bmatrix}_{r,\, m-r}^{r,\, m-r}, \quad \text{(E)}$$

and the second of these equations is a symmetric equigradent transformation converting ϕ into ψ which is compounded of a symmetric derangement, a symmetric unitary equigradent transformation, and an undegenerate quasi-scalar transformation.

If the elements of ϕ all lie in a restricted domain of rationality Ω, the transformations (E) are not necessarily transformations in Ω, for the component quasi-scalar transformation does not in general lie in Ω. In particular when ϕ is real, it may not be possible to find real transformations of the forms (E).

Ex. xv. From (E) we deduce that

$$[a]_m^m = \overline{\underline{h}}{}_m^r\,[1]_r^r\,[h]_r^m = \overline{\underline{h}}{}_m^r\,[h]_r^m, \quad [H]_r^m\,[a]_m^m\,\overline{\underline{H}}{}_m^r = [1]_r^r, \qquad \text{............(E}')$$

where

$$[h]_r^m\,\overline{\underline{H}}{}_m^r = [1]_r^r.$$

By Note 8 of § 141 we can also deduce (E) from (E′).

Ex. xvi. The matrix $[h]_r^m$ in Ex. xv is a solution of rank r of the equation

$$\overline{\underline{x}}{}_m^r\,[x]_r^m = [a]_m^m.$$

Ex. xvii. If $e_1, e_2, \ldots e_r$ are all non-zero scalar quantities, and

$$[e]_r^r = \begin{bmatrix} e_1 & 0 & \ldots & 0 \\ 0 & e_2 & \ldots & 0 \\ \ldots & \ldots & \ldots & \ldots \\ 0 & 0 & \ldots & e_r \end{bmatrix}, \qquad [\sqrt{e}]_r^r = \begin{bmatrix} \sqrt{e_1} & 0 & \ldots & 0 \\ 0 & \sqrt{e_2} & \ldots & 0 \\ \ldots & \ldots & \ldots & \ldots \\ 0 & 0 & \ldots & \sqrt{e_r} \end{bmatrix},$$

then

$$\begin{bmatrix} e, & 0 \\ 0, & 0 \end{bmatrix}_{r,\,m-r}^{r,\,m-r} = \begin{bmatrix} \sqrt{e}, & 0 \\ 0, & 1 \end{bmatrix}_{r,\,m-r}^{r,\,m-r} \begin{bmatrix} 1, & 0 \\ 0, & 0 \end{bmatrix}_{r,\,m-r}^{r,\,m-r} \begin{bmatrix} \sqrt{e}, & 0 \\ 0, & 1 \end{bmatrix}_{r,\,m-r}^{r,\,m-r}$$

and its inverse are symmetric equigradent transformations.

Ex. xviii. If $e_1, e_2, \ldots e_\pi$ are real non-zero positive quantities, and $f_1, f_2, \ldots f_\nu$ are real non-zero negative quantities, and if

$$[e]_\pi^\pi = \begin{bmatrix} e_1 & 0 & \ldots & 0 \\ 0 & e_2 & \ldots & 0 \\ \ldots & \ldots & \ldots & \ldots \\ 0 & 0 & \ldots & e_\pi \end{bmatrix}, \qquad [\sqrt{e}]_\pi^\pi = \begin{bmatrix} \sqrt{e_1} & 0 & \ldots & 0 \\ 0 & \sqrt{e_2} & \ldots & 0 \\ \ldots & \ldots & \ldots & \ldots \\ 0 & 0 & \ldots & \sqrt{e_\pi} \end{bmatrix},$$

$$[f]_\nu^\nu = \begin{bmatrix} f_1 & 0 & \ldots & 0 \\ 0 & f_2 & \ldots & 0 \\ \ldots & \ldots & \ldots & \ldots \\ 0 & 0 & \ldots & f_\nu \end{bmatrix}, \qquad [\sqrt{-f}]_\nu^\nu = \begin{bmatrix} \sqrt{-f_1} & 0 & \ldots & 0 \\ 0 & \sqrt{-f_2} & \ldots & 0 \\ \ldots & \ldots & \ldots & \ldots \\ 0 & 0 & \ldots & \sqrt{-f_\nu} \end{bmatrix},$$

then

$$\begin{bmatrix} e, & 0, & 0 \\ 0, & f, & 0 \\ 0, & 0, & 0 \end{bmatrix}_{\pi,\nu,\kappa}^{\pi,\nu,\kappa} = \begin{bmatrix} \sqrt{e}, & 0, & 0 \\ 0, & \sqrt{-f}, & 0 \\ 0, & 0, & 1 \end{bmatrix}_{\pi,\nu,\kappa}^{\pi,\nu,\kappa} \begin{bmatrix} 1, & 0, & 0 \\ 0, & -1, & 0 \\ 0, & 0, & 0 \end{bmatrix}_{\pi,\nu,\kappa}^{\pi,\nu,\kappa} \begin{bmatrix} \sqrt{e}, & 0, & 0 \\ 0, & \sqrt{-f}, & 0 \\ 0, & 0, & 1 \end{bmatrix}_{\pi,\nu,\kappa}^{\pi,\nu,\kappa},$$

and its inverse are real symmetric equigradent transformations.

Theorem IV. *If* $A = [a]_u^u$ *and* $B = [b]_m^m$ *are two symmetric matrices with constant elements, and if* $m \nless u$, *we can convert* A *into* B *by a symmetric equigradent transformation when and only when* A *and* B *have the same rank.*

First suppose that A and B have the same rank r, and let

$$\phi = \begin{bmatrix} 1, & 0 \\ 0, & 0 \end{bmatrix}_{r,\,u-r}^{r,\,u-r}, \qquad \Phi = \begin{bmatrix} 1, & 0 \\ 0, & 0 \end{bmatrix}_{r,\,m-r}^{r,\,m-r}$$

Then making use of Theorem III we see that we can convert A into ϕ, ϕ into Φ, and Φ into B by symmetric equigradent transformations. Therefore there exists a symmetric equigradent transformation

$$\overline{\underline{h}}{}^{u}_{m}\,[a]^{u}_{u}\,[h]^{m}_{u} = [b]^{m}_{m} \quad \text{.........................(F)}$$

converting A into B.

Next suppose that there exists an equigradent transformation (F). Then by Theorem II of § 131 the matrices A and B have the same rank.

Thus two symmetric matrices with constant elements are symmetrically equigradent when and only when they have the same rank.

Ex. xix. If we can convert the symmetric matrices A and B into the same symmetric matrix by symmetric equigradent transformations, then there exists a symmetric equigradent transformation of the form (F).

§ 148. The signants and signature of a real symmetric matrix.

Before proving the theorem which leads to the definitions of these quantities we will establish the lemma given below.

Lemma. *If* $p+n=\pi+\nu=r$, *there cannot exist an equation of the form*

$$\overline{\underline{h}}{}^{r}_{r}\begin{bmatrix}1, & 0\\ 0, & -1\end{bmatrix}^{p,\,n}_{p,\,n}[h]^{r}_{r} = \overline{\underline{k}}{}^{r}_{r}\begin{bmatrix}1, & 0\\ 0, & -1\end{bmatrix}^{\pi,\,\nu}_{\pi,\,\nu}[k]^{r}_{r} \quad \text{.....................(A)}$$

in which $[h]^{r}_{r}$ *and* $[k]^{r}_{r}$ *are real undegenerate matrices except when* $p=\pi$ *and* $n=\nu$.

The lemma asserts that we can convert $\begin{bmatrix}1, & 0\\ 0, & -1\end{bmatrix}^{p,\,n}_{p,\,n}$ into $\begin{bmatrix}1, & 0\\ 0, & -1\end{bmatrix}^{\pi,\,\nu}_{\pi,\,\nu}$ by a real symmetric equigradent transformation only when $p=\pi$ and $n=\nu$.

Suppose that $p<\pi$, and therefore $n>\nu$.

Writing $\overline{\underline{h}}{}^{r}_{r} = \overline{\underline{a,\ b}}{}^{p,\,n}_{r}$, $\overline{\underline{k}}{}^{r}_{r} = \overline{\underline{a,\ \beta}}{}^{\pi,\,\nu}_{r}$, equation (A) is equivalent to

$$\overline{\underline{a}}{}^{p}_{r}\,[a]^{r}_{p} - \overline{\underline{b}}{}^{n}_{r}\,[b]^{r}_{n} = \overline{\underline{a}}{}^{\pi}_{r}\,[a]^{r}_{\pi} - \overline{\underline{\beta}}{}^{\nu}_{r}\,[\beta]^{r}_{\nu}$$

or

$$\overline{\underline{a,\ \beta}}{}^{p,\,\nu}_{r}\begin{bmatrix}a\\ \beta\end{bmatrix}^{r}_{p,\,\nu} = \overline{\underline{b,\ a}}{}^{n,\,\pi}_{r}\begin{bmatrix}b\\ a\end{bmatrix}^{r}_{n,\,\pi}. \quad \text{...........................(1)}$$

Since $p+\nu<r$, there exist *real* quantities $\lambda_1, \lambda_2, \ldots \lambda_r$, not all zero, such that

$$[\lambda_1\,\lambda_2\ldots\lambda_r]\,\overline{\underline{a,\ \beta}}{}^{p,\,\nu}_{r} = 0, \quad \text{..................................(2)}$$

as can be seen by solving the equation for $\lambda_1, \lambda_2, \ldots \lambda_r$.

From (1) and (2) we deduce that

$$[\lambda]_r \,\overline{\underline{\lvert\, b,\ a \,\rvert}}_r^{\,n,\,\pi} \cdot \begin{bmatrix} b \\ a \end{bmatrix}_{n,\,\pi}^{r} \overline{\underline{\lvert\,\lambda\,\rvert}}_r = 0; \quad \text{.............................(3)}$$

and from (3) it follows by § 72 that

$$[\lambda_1\, \lambda_2 \ldots \lambda_r]\, \overline{\underline{\lvert\, b,\ a \,\rvert}}_r^{\,n,\,\pi} = 0. \quad \text{..................................(4)}$$

From (2) and (4) we see that

$$[\lambda_1\, \lambda_2 \ldots \lambda_r]\, \overline{\underline{\lvert\, a,\ b \,\rvert}}_r^{\,p,\,n} = [\lambda_1\, \lambda_2 \ldots \lambda_r]\, \overline{\underline{\lvert\, h \,\rvert}}_r^{\,r} = 0.$$

But this is impossible, since $\overline{\underline{\lvert\, h \,\rvert}}_r^{\,r}$ is undegenerate.

We see then that p cannot be less than π; and we can show in a similar way that π cannot be less than p.

Thus when there exists an equation of the form (A), we must have

$$p = \pi, \quad \text{and therefore also} \quad n = \nu.$$

We can deduce (4) from (3) by observing that (3) and (4) are respectively

$$B_1^2 + B_2^2 + \ldots + B_n^2 + \mathrm{A}_1^2 + \mathrm{A}_2^2 + \ldots + \mathrm{A}_\pi^2 = 0, \quad \text{......................(3')}$$

and

$$[B_1\, B_2 \ldots B_n\, \mathrm{A}_1\, \mathrm{A}_2 \ldots \mathrm{A}_\pi] = 0, \quad \text{...............................(4')}$$

where

$$B_i = \lambda_1 b_{i1} + \lambda_2 b_{i2} + \ldots + \lambda_r b_{ir}, \quad \mathrm{A}_i = \lambda_1 a_{i1} + \lambda_2 a_{i2} + \ldots + \lambda_r a_{ir}.$$

Since every term in (3') is essentially positive, it follows from (3') that $B_1 = 0$, $B_2 = 0$, ... $\mathrm{A}_1 = 0$, $\mathrm{A}_2 = 0$, ...; and these equations are equivalent to (4').

Theorem I. *Let $\phi = [a]_m^m$ be a real symmetric matrix of rank r whose elements are constants, and let ϕ be converted by a real symmetric equigradent transformation into the quasi-scalar matrix*

$$\psi = \begin{bmatrix} e, & 0 \\ 0, & 0 \end{bmatrix}_{r,\,m-r}^{r,\,m-r}, \quad \textit{where} \quad [e]_r^r = \begin{bmatrix} e_1 & 0 & \ldots & 0 \\ 0 & e_2 & \ldots & 0 \\ \ldots & \ldots & \ldots & \ldots \\ 0 & 0 & \ldots & e_r \end{bmatrix},$$

and $e_1, e_2, \ldots e_r$ are real non-zero quantities.

Then if π and ν are the numbers of the quantities $e_1, e_2, \ldots e_r$ which are positive and negative respectively, so that $\pi + \nu = r$, the integers π and ν are always the same for all such transformations.

We know by Theorem II of § 147 that such transformations do exist; and there will clearly be no loss of generality in assuming that the first π of the quantities $e_1, e_2, \ldots e_r$ are positive, and that the remaining ν of them are negative.

Suppose then that

$$\overline{h}^{m}_{m}\,[a]^{m}_{m}\,[h]^{m}_{m} = \begin{bmatrix} e, & 0 \\ 0, & 0 \end{bmatrix}^{r,\,m-r}_{r,\,m-r}, \qquad \overline{h'}^{m}_{m}\,[a]^{m}_{m}\,[h']^{m}_{m} = \begin{bmatrix} \epsilon, & 0 \\ 0, & 0 \end{bmatrix}^{r,\,m-r}_{r,\,m-r},$$

where
$$[e]^{r}_{r} = \begin{bmatrix} e_1 & 0 & \ldots & 0 \\ 0 & e_2 & \ldots & 0 \\ \cdots & \cdots & \cdots & \cdots \\ 0 & 0 & \ldots & e_r \end{bmatrix}, \qquad [\epsilon]^{r}_{r} = \begin{bmatrix} \epsilon_1 & 0 & \ldots & 0 \\ 0 & \epsilon_2 & \ldots & 0 \\ \cdots & \cdots & \cdots & \cdots \\ 0 & 0 & \ldots & \epsilon_r \end{bmatrix};$$

$e_1, e_2, \ldots e_r$ are real non-zero quantities of which the first π are positive and the remaining ν are negative; and $\epsilon_1, \epsilon_2, \ldots \epsilon_r$ are real non-zero quantities of which the first p are positive and the remaining n are negative; so that $\pi + \nu = p + n = r$; the matrices $[h]^{m}_{m}$ and $[h']^{m}_{m}$ being real and undegenerate.

The theorem will be established by showing that $p = \pi$, $n = \nu$.

By Ex. xii of § 141 there exists a real undegenerate matrix $[u]^{r}_{r}$ such that

$$[\epsilon]^{r}_{r} = \overline{u}^{r}_{r}\,[e]^{r}_{r}\,[u]^{r}_{r}, \quad \ldots\ldots\ldots\ldots\ldots\ldots(5)$$

and by such transformations as that given in Ex. xviii of § 147 we deduce from (5) an equation of the form

$$\begin{bmatrix} 1, & 0 \\ 0, & -1 \end{bmatrix}^{p,\,n}_{p,\,n} = \overline{v}^{r}_{r} \begin{bmatrix} 1, & 0 \\ 0, & -1 \end{bmatrix}^{\pi,\,\nu}_{\pi,\,\nu} [v]^{r}_{r}, \quad \ldots\ldots\ldots\ldots(5')$$

where $[v]^{r}_{r}$ is a real undegenerate matrix. Applying the lemma proved above, we see from (5′) that $p = \pi$ and $n = \nu$.

The numbers π and ν defined in Theorem I will be called respectively the *positive* and *negative signants* of the real symmetric matrix $[a]^{m}_{m}$; and the number s defined by the equation $s = \pi - \nu$ will be called the *signature* of that matrix. Thus if r is the rank, s the signature, π the positive signant, and ν the negative signant of any real self-conjugate matrix with constant elements, we have

$$\pi + \nu = r, \quad \pi - \nu = s, \quad \pi = \tfrac{1}{2}(r + s), \quad \nu = \tfrac{1}{2}(r - s).$$

When r is given, π can have any integral value from r to 0, ν can have any integral value from 0 to r, and s can have any one of the values $r, r-2, r-4, \ldots 4-r, 2-r, -r$.

Two real self-conjugate matrices with constant elements will be said to be *equisignant* when they have the same positive and negative signants. This is the case when and only when they have the same rank and the same signature.

Theorem II. *If $\phi = [a]^m_m$ is a real symmetric matrix whose elements are constants and whose positive and negative signants are π and ν, then there exists a real symmetric equigradent transformation of the form*

$$\overline{h}^{\,m}_{\,m}\,[a]^m_m\,[h]^m_m = \begin{bmatrix} 1, & 0, & 0 \\ 0, & -1, & 0 \\ 0, & 0, & 0 \end{bmatrix}^{\pi,\,\nu,\,m-\pi-\nu}_{\pi,\,\nu,\,m-\pi-\nu} \qquad \ldots\ldots\ldots\ldots\ldots(\mathrm{B})$$

By Theorem II of § 147 we can convert ϕ by derangements and real unitary equigradent transformations, the same for horizontal and vertical rows, into the form ψ of Theorem I, where $r = \pi + \nu$. Then π of the real non-zero quantities $e_1, e_2, \ldots e_r$ are positive, and ν of them are negative; and by further derangements, the same for horizontal and vertical rows, we can convert ψ into a matrix ψ' of the same form in which the first π of the quantities $e_1, e_2, \ldots e_r$ are positive, and the remaining ν are negative. Finally dividing the ith horizontal and vertical rows of ψ' by $\sqrt{e_i}$ or $\sqrt{-e_i}$ according as e_i is positive or negative, doing this for all the values $1, 2, \ldots r$ of i, we convert ψ' by real quasi-scalar transformations into the matrix on the right in (B). In this way we obtain a transformation of the form (B) compounded of equigradent transformations, all of which are real and symmetric.

Ex. i. If $[a]^m_m$ is a real symmetric matrix with constant elements whose positive and negative signants are π and ν, and if $r = \pi + \nu$, there exist equations of the form

$$\overline{h}^{\,r}_{\,m}\begin{bmatrix} 1, & 0 \\ 0, & -1 \end{bmatrix}^{\pi,\,\nu}_{\pi,\,\nu}[h]^m_r = [a]^m_m, \quad [H]^m_r\,[a]^m_m\,\overline{H}^{\,r}_{\,m} = \begin{bmatrix} 1, & 0 \\ 0, & -1 \end{bmatrix}^{\pi,\,\nu}_{\pi,\,\nu}, \qquad \ldots\ldots\ldots(\mathrm{B}')$$

where $[h]^m_r$ and $\overline{H}^{\,r}_{\,m}$ are two mutually inverse real undegenerate matrices with constant elements.

Ex. ii. *For the real symmetric matrix* $\begin{bmatrix} 0 & 1 \\ 1 & 0 \end{bmatrix}$ *we have* $\pi = 1$, $\nu = 1$.

For we have the real symmetric equigradent transformation

$$\frac{1}{2}\begin{bmatrix} 1, & 1 \\ 1, & -1 \end{bmatrix}\begin{bmatrix} 0, & 1 \\ 1, & 0 \end{bmatrix}\begin{bmatrix} 1, & 1 \\ 1, & -1 \end{bmatrix} = \begin{bmatrix} 1, & 0 \\ 0, & -1 \end{bmatrix}.$$

Theorem III. *If $A = [a]^u_u$ and $B = [b]^m_m$ are two real symmetric matrices with constant elements, and if $m \nless u$, we can convert A into B by a real symmetric equigradent transformation when and only when A and B are equisignant.*

First suppose that A and B have the same positive signant π and the same negative signant ν, so that they have the same rank $r = \pi + \nu$.

Let $$\phi = \begin{bmatrix} 1, & 0, & 0 \\ 0, & -1, & 0 \\ 0, & 0, & 0 \end{bmatrix}^{\pi,\,\nu,\,u-r}_{\pi,\,\nu,\,u-r}, \quad \Phi = \begin{bmatrix} 1, & 0, & 0 \\ 0, & -1, & 0 \\ 0, & 0, & 0 \end{bmatrix}^{\pi,\,\nu,\,m-r}_{\pi,\,\nu,\,m-r}.$$

Then, making use of Theorem II, we can convert A into ϕ, ϕ into Φ, and Φ into B by real symmetric equigradent transformations. Therefore there exists a real symmetric equigradent transformation

$$\overline{\underline{h}}_m^u \; [a]_u^u \; [h]_u^m = [b]_m^m \quad \ldots\ldots\ldots\ldots\ldots\ldots\ldots\ldots\text{(C)}$$

converting A into B.

Next suppose that there exists a real symmetric equigradent transformation of the form (C), so that A and B have the same rank, which will be denoted by r.

Let the positive and negative signants of A be π and ν, and the positive and negative signants of B be p and n, so that $p + n = \pi + \nu = r$; and let

$$\phi = \begin{bmatrix} 1, & 0, & 0 \\ 0, & -1, & 0 \\ 0, & 0, & 0 \end{bmatrix}_{\pi,\, \nu,\, u-r}^{\pi,\, \nu,\, u-r}, \quad \Phi = \begin{bmatrix} 1, & 0, & 0 \\ 0, & -1, & 0 \\ 0, & 0, & 0 \end{bmatrix}_{p,\, n,\, m-r}^{p,\, n,\, m-r}.$$

Then we can convert ϕ into A, A into B, and B into Φ by real symmetric equigradent transformations. Therefore we can convert ϕ into Φ by a real symmetric equigradent transformation. It follows from Ex. xi of § 141 that there exists a real symmetric equigradent transformation of the form

$$\begin{bmatrix} 1, & 0 \\ 0, & -1 \end{bmatrix}_{p,\, n}^{p,\, n} = \overline{\underline{k}}_r^r \begin{bmatrix} 1, & 0 \\ 0, & -1 \end{bmatrix}_{\pi,\, \nu}^{\pi,\, \nu} [k]_r^r.$$

Applying the lemma given at the commencement of this article, we see from the last equation that $p = \pi$ and $n = \nu$, i.e. A and B are equisignant.

Theorem III is equivalent to the following statement:

Two real symmetric matrices with constant elements are symmetrically equigradent when and only when they are equisignant.

Ex. iii. *The positive and negative signants of a real symmetric compartite matrix* Φ *of the standard form*

$$\Phi = \begin{bmatrix} a, & 0, & \ldots & 0 \\ 0, & b, & \ldots & 0 \\ \ldots & \ldots & \ldots & \ldots \\ 0, & 0, & \ldots & c \end{bmatrix}_{p,\, q,\, \ldots\, r}^{p,\, q,\, \ldots\, r}$$

whose elements are constants, and whose successive parts are symmetric matrices, are respectively the sums of the positive and negative signants of its successive parts.

Let the positive and negative signants of the successive parts $\Phi' = [a]_p^p$, $\Phi'' = [b]_q^q$, ... be (π', ν'), (π'', ν''), ... Then we can convert Φ', Φ'', ... into the forms

$$\phi' = \begin{bmatrix} 1, & 0, & 0 \\ 0, & -1, & 0 \\ 0, & 0, & 0 \end{bmatrix}_{\pi',\, \nu',\, \kappa'}^{\pi',\, \nu',\, \kappa'}, \quad \phi'' = \begin{bmatrix} 1, & 0, & 0 \\ 0, & -1, & 0 \\ 0, & 0, & 0 \end{bmatrix}_{\pi'',\, \nu'',\, \kappa''}^{\pi'',\, \nu'',\, \kappa''}, \ldots$$

by real symmetric equigradent transformations. Hence if ϕ is the compartite matrix of standard form whose successive parts are ϕ', ϕ'', ..., we can convert Φ into ϕ by a real symmetric equigradent transformation. The matrix Φ has therefore the same signants as ϕ; and ϕ is a quasi-scalar matrix whose positive and negative signants π and ν are

$$\pi = \pi' + \pi'' + \dots, \quad \nu = \nu' + \nu'' + \dots.$$

We can also obtain this result by converting the compartite matrix ψ of standard form whose successive parts are

$$\psi' = \begin{bmatrix} 1, & 0 \\ 0, & -1 \end{bmatrix}^{\pi', \nu'}_{\pi', \nu'}, \quad \psi'' = \begin{bmatrix} 1, & 0 \\ 0, & -1 \end{bmatrix}^{\pi'', \nu''}_{\pi'', \nu''}, \dots$$

into Φ by a real symmetric equigradent transformation.

§ 149. Definite and indefinite matrices.

A *real symmetric matrix* $A = [a]^m_m$ will be said to be *definite* when one of its two signants vanishes, and to be *indefinite* when neither of them vanishes. If we reduce A to the standard form given in Theorem II of § 147, then it is definite or indefinite according as the real quantities $e_1, e_2, \dots e_r$ have or have not all the same sign. In the particular case when A has rank 0, it is definite.

When the real symmetric matrix $[a]^m_m$ is definite, the two equations

$$[x]_m [a]^m_m \overline{\underline{x}}_m = 0, \quad [\grave{a}]^m_m \overline{\underline{x}}_m = 0 \quad \dots\dots\dots\dots\dots\dots(1)$$

have the same real solutions.

For if $[a]^m_m$ has rank r, we can write

$$[a]^m_m = \overline{\underline{h}}^r_m \begin{bmatrix} e_1 & 0 & \dots & 0 \\ 0 & e_2 & \dots & 0 \\ \dots & \dots & \dots & \dots \\ 0 & 0 & \dots & e_r \end{bmatrix} [h]^m_r$$

where $[h]^m_r$ is a real undegenerate matrix of rank r, and $e_1, e_2, \dots e_r$ are real non-zero quantities all having the same sign; and the first of the equations (1) is equivalent to

$$e_1 X_1^2 + e_2 X_2^2 + \dots + e_r X_r^2 = 0, \quad \dots\dots\dots\dots\dots\dots(2)$$

where $\quad X_i = h_{i1} x_1 + h_{i2} x_2 + \dots + h_{im} x_m.$

If $x_1, x_2, \dots x_m$ are real, $X_1, X_2, \dots X_r$ are real, and all terms on the left of (2) which do not vanish have the same sign. Consequently the equation (2) is satisfied when and only when every term on the left vanishes, i.e. when and only when $X_1 = X_2 = \dots = X_r = 0$, or $[X_1 X_2 \dots X_r] = [h]^m_r \overline{\underline{x}}_m = 0$, i.e. when and only when $[a]^m_m \overline{\underline{x}}_m = 0$.

As a particular case of the above theorem we have the following result:

When the real symmetric matrix $[a]_m^m$ *is undegenerate and definite, the only real solution of the equation* $[x]_m\,[a]_m^m\,\overline{x}_m = 0$ *is* $x_1 = x_2 = \dots = x_m = 0$.

Ex. i. *If* $A = [a]_m^m$ *is a real symmetric matrix of rank r with constant elements, and if* $S = \det\,[x]_m\,[a]_m^m\,\overline{x}_m = \Sigma a_{ii}x_i^2 + 2\Sigma a_{ij}x_ix_j$, *where* $j \neq i$, *then:*

(1) *If A is definite and has positive signature or only positive signants, the quadratic expression S cannot be negative for real values of the variables* $x_1, x_2, \dots x_m$.

(2) *If A is definite and has negative signature or only negative signants, the quadratic expression S cannot be positive for real values of the variables.*

(3) *If A is indefinite, and has therefore both positive and negative signants, the quadratic expression S can be both positive and negative for real values of the variables.*

The first two properties follow at once from equation (2). It only remains to prove the third property.

Let A be indefinite, and let its positive and negative signants be π and ν respectively, so that $\pi + \nu = r$, and let $m - r = \kappa$.

Then by Theorem II of § 148 there exists a real undegenerate matrix

$$\overline{h}_m^m = \overline{\alpha,\ \beta,\ \gamma}_m^{\pi,\ \nu,\ \kappa}$$

such that

$$[a]_m^m = \overline{h}_m^m \begin{bmatrix} 1, & 0, & 0 \\ 0, & -1, & 0 \\ 0, & 0, & 0 \end{bmatrix}_{\pi,\ \nu,\ \kappa}^{\pi,\ \nu,\ \kappa} [h]_m^m,$$

and if we write

$$X_i = \alpha_{i1}x_1 + \alpha_{i2}x_2 + \dots + \alpha_{im}x_m, \quad (i = 1, 2, \dots \pi),$$

$$Y_i = \beta_{i1}x_1 + \beta_{i2}x_2 + \dots + \beta_{im}x_m, \quad (i = 1, 2, \dots \nu),$$

$$Z_i = \gamma_{i1}x_1 + \gamma_{i2}x_2 + \dots + \gamma_{im}x_m, \quad (i = 1, 2, \dots \kappa),$$

we have

$$S = (X_1^2 + X_2^2 + \dots + X_\pi^2) - (Y_1^2 + Y_2^2 + \dots + Y_\nu^2),$$

$$[\alpha]_\pi^m\,\overline{x}_m = \overline{X}_\pi, \quad [\beta]_\nu^m\,\overline{x}_m = \overline{Y}_\nu, \quad [\gamma]_\kappa^m\,\overline{x}_m = \overline{Z}_\kappa.$$

First let $\overline{x}_m$ be a real non-zero solution of the equation $\begin{bmatrix}\beta \\ \gamma\end{bmatrix}_{\nu,\ \kappa}^m \overline{x}_m = 0$. Then we cannot have $[\alpha]_\pi^m\,\overline{x}_m = 0$, for we cannot have $[h]_m^m\,\overline{x}_m = 0$. Thus for the real values of $x_1, x_2, \dots x_m$ thus determined we have $Y_1 = Y_2 = \dots = Y_\nu = 0$, $Z_1 = Z_2 = \dots = Z_\kappa = 0$, but $X_1, X_2, \dots X_\pi$ do not all vanish and are all real. It follows that S has a positive non-zero value.

Next let $\overline{x}_m$ be a real non-zero solution of the equation $\begin{bmatrix}\alpha \\ \gamma\end{bmatrix}_{\pi,\ \kappa}^m \overline{x}_m = 0$. Then we cannot have $[\beta]_\nu^m\,\overline{x}_m = 0$, for we cannot have $[h]_m^m\,\overline{x}_m = 0$. Thus for the real values of $x_1, x_2, \dots x_m$ thus determined we have $X_1 = X_2 = \dots = X_\pi = 0$, $Z_1 = Z_2 = \dots = Z_\kappa = 0$, but

$Y_1, Y_2, \ldots Y_\nu$ do not all vanish and are all real. It follows that S has a negative non-zero value.

Ex. ii. If a real symmetric compartite matrix of standard form with constant elements whose successive parts are all square has $\begin{bmatrix} 0 & 1 \\ 1 & 0 \end{bmatrix}$ as one of its parts, it must be indefinite.

This follows from Exs. ii and iii of § 148.

Ex. iii. *If the real symmetric matrix $\phi = [a]^m_m$ is definite, and if the diagonal element a_{ii} vanishes, then every element of the ith horizontal and vertical rows must vanish.*

For if $a_{ii} = 0$, and $a_{ij} = a_{ji} \neq 0$, then by Ex. x of § 145 we can convert ϕ by a real symmetric equigradent transformation into a symmetric compartite matrix having two parts, both of which are square, and one of which is $\begin{bmatrix} 0 & 1 \\ 1 & 0 \end{bmatrix}$. By Ex. ii this is only possible when ϕ is indefinite.

Ex. iv. Let $[h]^m_r$ be some real undegenerate matrix of rank r, and let ϵ be either 1 or -1.

Then every real symmetric matrix $\phi = [a]^m_m$ of rank r can be expressed in the form

$$[a]^m_m = \epsilon\, \underline{\overline{h}}{}^r_m\, [h]^m_r = \underline{\overline{h}}{}^r_m\, [\epsilon]^r_r\, [h]^m_r \qquad \text{.........................(3)}$$

where ϵ is 1 or -1 according as the signature of ϕ is positive or negative, i.e. according as the signature is r or $-r$. Conversely every matrix given by (3) is a real and definite symmetric matrix of rank r whose signature is positive or negative according as ϵ is 1 or -1.

For the real symmetric matrix ϕ has rank r and is definite when and only when $[a]^m_m$ is equisignant with $[\epsilon]^r_r$, where ϵ is 1 or -1 according as the signature of ϕ is positive or negative.

We obtain Ex. iii by observing that $a_{ii} = \epsilon\,(h^2_{1i} + h^2_{2i} + \ldots + h^2_{ri})$, which shows that $a_{ii} = 0$ when and only when $h_{1i}, h_{2i}, \ldots h_{ri}$ all vanish. We see further that:

When ϕ is definite, those of its diagonal elements which do not vanish all have the same sign, viz. the sign of the signature of ϕ.

Ex. v. If $[a]^n_m$ *is any real matrix of rank r, and if ϵ is either 1 or -1, then the matrix*

$$\phi = \epsilon\, [a]^n_m\, \underline{\overline{a}}{}^m_n$$

is a real and definite symmetric matrix of rank r whose signature is positive or negative according as ϵ is 1 or -1.

By § 72.1 the matrix ϕ is a real symmetric matrix of rank r. Therefore by Theorem II of § 147 we can determine a real undegenerate matrix $[h]^m_m$ such that

$$[h]^m_m\, \phi\, \underline{\overline{h}}{}^m_m = \epsilon\, [h]^m_m\, [a]^n_m\, \underline{\overline{a}}{}^m_n\, \underline{\overline{h}}{}^m_m = \begin{bmatrix} e, & 0 \\ 0, & 0 \end{bmatrix}^{r,\, m-r}_{r,\, m-r}$$

where $[e]^r_r = {}^1[e]_r$ is an undegenerate quasi-scalar matrix whose diagonal elements $e_1, e_2, \ldots e_r$ are all non-zero scalar quantities. Writing $[h]^m_m\, [a]^n_m = [b]^n_m$, so that $[b]^n_m$ is real, we have

$$e_i = \epsilon\,(b^2_{i1} + b^2_{i2} + \ldots + b^2_{in}), \text{ where } i = 1, 2, \ldots r.$$

Since e_i does not vanish, it must have the same sign as ϵ. Thus all the diagonal elements $e_1, e_2, \ldots e_r$ of the quasi-scalar matrix $[e]_r^r$ have the same sign, and are positive or negative according as ϵ is 1 or -1. This shows that ϕ is definite, and that its signature is positive or negative according as ϵ is 1 or -1.

Ex. vi. *If the real symmetric matrix* $\phi=[a]_m^m$ *is undegenerate and definite, then none of its diagonal elements vanish, and they all have the same sign, viz. the sign of the signature of* ϕ.

By Ex. iii none of the diagonal elements can vanish. The rest of the theorem then follows from the last result in Ex. iv.

Ex. vii. *If the real symmetric matrix* $\phi=[a]_m^m$ *is definite, then every symmetrically formed complete matrix* $\Phi=[A]_\mu^\mu$ *of its minor determinants of order s is definite.*

Let ϕ have rank r. Then if $s > r$, we have $\Phi=0$. If $s \ngtr r$, we see by equating correspondingly formed complete matrices of both sides in the equation of Ex. iv that

$$[A]_\mu^\mu = \epsilon^s \underline{\overline{H}}_\mu^\rho [H]_\rho^\mu = \epsilon^s \underline{\overline{H}}_\mu^\rho [1]_\rho^\rho [H]_\rho^\mu,$$

where $[H]_\rho^\mu$ is a real undegenerate matrix of rank ρ. This shows that Φ is equisignant with $\epsilon^s[1]_\rho^\rho$.

Applying Exs. iii, iv and vi to the matrix $[A]_\mu^\mu$, we obtain the following further properties of the real symmetric matrix $\phi=[a]_m^m$.

(1) *If* ϕ *is definite and the diagonal minor determinant* $(a_{uu})_s^s$ *vanishes, then all the simple minor determinants of* $[a_{u1}]_s^m$ *and* $[a_{1u}]_m^s$ *vanish.*

(2) *When* ϕ *is definite, those of its diagonal minor determinants of order s which do not vanish all have the same sign, viz. the sign of* ϵ^s.

(3) *When* ϕ *is undegenerate and definite, then none of its diagonal minor determinants of order s vanish, and they all have the same sign, viz. the sign of* ϵ^s.

NOTE. *Semi-definite matrices.*

The term 'definite' is sometimes reserved for a real symmetric matrix which is undegenerate and definite according to the definitions of the text. Then a real symmetric matrix which is degenerate and definite according to the definitions of the text is said to be *semi-definite.*

§ 150. Reduction of a skew-symmetric matrix with constant elements to standard forms by symmetric equigradent transformations.

1. *Reduction by derangements and unitary equigradent transformations.*

Theorem I. *If* $\phi=[a]_m^m$ *is a skew-symmetric matrix of (even) rank* ρ *whose elements are constants lying in a domain of rationality* Ω, *and if none of the successive leading diagonal minor determinants*

$$\Delta_0 = 1, \quad \Delta_2 = (a)_2^2, \quad \Delta_4 = (a)_4^4, \ldots \Delta_\lambda = (a)_\lambda^\lambda, \ldots \Delta_\rho = (a)_\rho^\rho$$

of even order vanish, then we can convert ϕ *by a symmetric unitary equigradent transformation in* Ω *into a skew-symmetric compartite matrix of the standard form*

$$\psi = \begin{bmatrix} e, & 0 \\ 0, & 0 \end{bmatrix}_{\rho,\, m-\rho}^{\rho,\, m-\rho}, \quad \textit{where } [e]_\rho^\rho = \begin{bmatrix} a, & 0, \dots & 0, \dots & 0 \\ 0, & b, \dots & 0, \dots & 0 \\ \dots & \dots & \dots & \dots \\ 0, & 0, \dots & l, \dots & 0 \\ \dots & \dots & \dots & \dots \\ 0, & 0, \dots & 0, \dots & r \end{bmatrix}_{2, 2, \dots 2, \dots 2}^{2, 2, \dots 2, \dots 2},$$

the successive parts of $[e]_\rho^\rho$ *being*

$$[a]_2^2 = \begin{bmatrix} 0 & , a_{12} \\ -a_{12}, & 0 \end{bmatrix}, \quad [b]_2^2 = \begin{bmatrix} 0 & , b_{12} \\ -b_{12}, & 0 \end{bmatrix}, \dots$$

$$[l]_2^2 = \begin{bmatrix} 0 & , l_{12} \\ -l_{12}, & 0 \end{bmatrix}, \dots [r]_2^2 = \begin{bmatrix} 0 & , r_{12} \\ -r_{12}, & 0 \end{bmatrix},$$

where

$$\Delta_2 b_{12} = \begin{pmatrix} 124 \\ a \\ 123 \end{pmatrix}, \quad \Delta_4 c_{12} = \begin{pmatrix} 12346 \\ a \\ 12345 \end{pmatrix}, \dots \Delta_{\lambda-2} l_{12} = \begin{pmatrix} 1, 2, \dots \lambda-2, \lambda \\ a \\ 1, 2, \dots \lambda-2, \lambda-1 \end{pmatrix}, \dots .$$

This theorem is a particular case of Theorem III of § 145.

Writing $\quad A = (a)_2^2, \quad B = (b)_2^2, \dots L = (l)_2^2, \dots R = (r)_2^2$

we define skew-symmetric matrices

$$[b]_{m-2}^{m-2}, \quad [c]_{m-4}^{m-4}, \dots [l]_{m-\lambda+2}^{m-\lambda+2}, \dots [r]_{m-\rho+2}^{m-\rho+2}$$

of ranks $\rho - 2, \rho - 4, \dots \rho - \lambda + 2, \dots 2$ by the equations

$$A\, b_{uv} = \begin{pmatrix} 1, 2, 2+v \\ a \\ 1, 2, 2+u \end{pmatrix}, \quad B\, c_{uv} = \begin{pmatrix} 1, 2, 2+v \\ b \\ 1, 2, 2+u \end{pmatrix}, \quad C\, d_{uv} = \begin{pmatrix} 1, 2, 2+v \\ c \\ 1, 2, 2+u \end{pmatrix}, \dots$$

so that $A = \Delta_2$, $AB = \Delta_4$, $ABC = \Delta_6$, $ABCD = \Delta_8, \dots$, and

$$\Delta_2 b_{uv} = \begin{pmatrix} 1, 2, 2+v \\ a \\ 1, 2, 2+u \end{pmatrix}, \quad \Delta_4 c_{uv} = \begin{pmatrix} 1, 2, \dots 4, 4+v \\ a \\ 1, 2, \dots 4, 4+u \end{pmatrix}, \dots \Delta_{\lambda-2} l_{uv} = \begin{pmatrix} 1, 2, \dots \lambda-2, \lambda-2+v \\ a \\ 1, 2, \dots \lambda-2, \lambda-2+u \end{pmatrix}, \dots .$$

Then $a_{12}, b_{12}, c_{12}, \dots l_{12}, \dots r_{12}$ are the quantities defined in the enunciation, and if we write

$$h_\lambda = \frac{1}{L} \begin{bmatrix} 0, & -l_{12} \\ l_{12}, & 0 \end{bmatrix} [0, \; l]_2^{\lambda-2,\, m-\lambda+2} = \frac{1}{l_{12}} \begin{bmatrix} 0, & -1 \\ 1, & 0 \end{bmatrix} [0, \; l]_2^{\lambda-2,\, m-\lambda+2}$$

$$= [0, \; 1, \; \lambda]_2^{\lambda-2,\, 2,\, m-\lambda},$$

and proceed as in § 143 we obtain resultant symmetric unitary equigradent transformations in Ω of the forms

$$[a]_m^m = \overleftarrow{h}_m^m \begin{bmatrix} e, & 0 \\ 0, & 0 \end{bmatrix}_{\rho,\, m-\rho}^{\rho,\, m-\rho} [h]_m^m, \quad [H]_m^m [a]_m^m \overleftarrow{H}_m^m = \begin{bmatrix} e, & 0 \\ 0, & 0 \end{bmatrix}_{\rho,\, m-\rho}^{\rho,\, m-\rho}, \quad \text{(A)}$$

where $\overline{\underline{H}}^{m}_{m}$ is the inverse matrix of $[h]^{m}_{m}$, and $[h]^{m}_{m}$ is the matrix whose 1st and 2nd, 3rd and 4th, ...$(\lambda-1)$th and λth, ...$(\rho-1)$th and ρth, last $m-\rho$ horizontal rows are $h_2, h_4, \ldots h_\lambda, \ldots h_\rho, [0, 1]^{\rho, m-\rho}_{m-\rho}$ respectively.

Ex. i. When $m=7$ and $\rho=4$ we have

$$[h]^{7}_{7}=\begin{bmatrix} 1 & 0 & a_{11} & a_{12} & a_{13} & a_{14} & a_{15} \\ 0 & 1 & a_{21} & a_{22} & a_{23} & a_{24} & a_{25} \\ 0 & 0 & 1 & 0 & \beta_{11} & \beta_{12} & \beta_{13} \\ 0 & 0 & 0 & 1 & \beta_{21} & \beta_{22} & \beta_{23} \\ 0 & 0 & 0 & 0 & 1 & 0 & 0 \\ 0 & 0 & 0 & 0 & 0 & 1 & 0 \\ 0 & 0 & 0 & 0 & 0 & 0 & 1 \end{bmatrix}, \quad \overline{\underline{H}}^{7}_{7}=\begin{bmatrix} 1 & 0 & A_{11} & A_{12} & A_{13} & A_{14} & A_{15} \\ 0 & 1 & A_{21} & A_{22} & A_{23} & A_{24} & A_{25} \\ 0 & 0 & 1 & 0 & B_{11} & B_{12} & B_{13} \\ 0 & 0 & 0 & 1 & B_{21} & B_{22} & B_{23} \\ 0 & 0 & 0 & 0 & 1 & 0 & 0 \\ 0 & 0 & 0 & 0 & 0 & 1 & 0 \\ 0 & 0 & 0 & 0 & 0 & 0 & 1 \end{bmatrix}$$

where $$a_{1, i-2}=-\frac{a_{2i}}{a_{12}}, \quad a_{2, i-2}=\frac{a_{1i}}{a_{12}}; \quad \beta_{1, i-2}=-\frac{b_{2i}}{b_{12}}, \quad \beta_{2, i-2}=\frac{b_{1i}}{b_{12}},$$

and $$\Delta_2\beta_{1, i-2}=\begin{pmatrix} 12i \\ a \\ 123 \end{pmatrix}, \quad \Delta_2\beta_{2, i-2}=\begin{pmatrix} 12i \\ a \\ 124 \end{pmatrix}.$$

Ex. ii. *Direct proof of Theorem I.*

Let $$\Phi_\lambda=\begin{bmatrix} 0, & 0 \\ 0, & l \end{bmatrix}^{\lambda-2, m-\lambda+2}_{\lambda-2, m-\lambda+2}, \quad T_\lambda=\frac{1}{l^{2}_{12}}\,\overline{\underline{\begin{matrix} 0 \\ l \end{matrix}}}^{2}_{\lambda-2, m-\lambda+2}\begin{bmatrix} 0 & , l_{12} \\ -l_{12}, & 0 \end{bmatrix}[0, l]^{\lambda-2, m-\lambda+2}_{2}.$$

For $\lambda=2, 4, 6, \ldots \rho$ we put $l=a, b, c, \ldots r$.

The identity of § 116 gives the equation

$$(c)^{2}_{2}[c]^{m-4}_{m-4}=c^{2}_{12}[c]^{m-4}_{m-4}=[c]^{2}_{m-4}\begin{bmatrix} 0 & , -c_{12} \\ c_{12}, & 0 \end{bmatrix}[c]^{m-4}_{2}+(c)^{2}_{2}\begin{bmatrix} 0, & 0 \\ 0, & d \end{bmatrix}^{2, m-6}_{2, m-6}$$

$$=\overline{\underline{c}}^{2}_{m-4}\begin{bmatrix} 0 & , c_{12} \\ -c_{12}, & 0 \end{bmatrix}[c]^{m-4}_{2}+c^{2}_{12}\begin{bmatrix} 0, & 0 \\ 0, & d \end{bmatrix}^{2, m-6}_{2, m-6}.$$

Dividing both sides by c^{2}_{12}, we obtain $\Phi_6=T_6+\Phi_8$.

From the equations $\phi=T_2+\Phi_4, \ldots \Phi_{\lambda-2}=T_{\lambda-2}+\Phi_\lambda, \ldots \Phi_\rho=T_\rho$ obtained in this way we deduce the equation

$$[a]^{m}_{m}=T_2+T_4+\ldots+T_\lambda+\ldots+T_\rho. \qquad \text{(B)}$$

The equation (B) is equivalent to the first of the equations (A), and the second of those equations follows from the first.

Ex. iii. Let $$\mathbf{l}_{uv}=\begin{pmatrix} 1, 2, \ldots \lambda-2, \lambda-2+v \\ a \\ 1, 2, \ldots \lambda-2, \lambda-2+u \end{pmatrix}, \text{ so that } \Delta_{\lambda-2}l_{uv}=\mathbf{l}_{uv},$$

and let $$\delta_{\lambda-1}=\begin{pmatrix} 1, 2, \ldots \lambda-2, \lambda \\ a \\ 1, 2, \ldots \lambda-2, \lambda-1 \end{pmatrix}=-\begin{pmatrix} 1, 2, \ldots \lambda-2, \lambda-1 \\ a \\ 1, 2, \ldots \lambda-2, \lambda \end{pmatrix}=\mathbf{l}_{12}. \qquad (1)$$

Then in equation (B) we have

$$T_\lambda=\frac{1}{\Delta_{\lambda-2}\,\delta_{\lambda-1}}\,\overline{\underline{\begin{matrix} 0 \\ \mathbf{1} \end{matrix}}}^{2}_{\lambda-2, m-\lambda+2}\begin{bmatrix} 0, & 1 \\ -1, & 0 \end{bmatrix}[0, \mathbf{1}]^{\lambda-2, m-\lambda+2}_{2}.$$

This proves the results of Ex. vi.

Ex. iv. Since $\Delta_{\lambda-2} l_{12}=\delta_{\lambda-1}$, and $\Delta_{\lambda-2}(l)_2^2=\Delta_\lambda$,

we have $\Delta_{\lambda-2} l_{12}=\delta_{\lambda-1}$, $\Delta_{\lambda-2} l_{12}^2=\Delta_\lambda$.

Therefore the quantity $\delta_{\lambda-1}$ defined in Ex. iii satisfies the equation

$$\Delta_{\lambda-2}\Delta_\lambda=\delta_{\lambda-1}^2. \quad\text{......(2)}$$

This result also follows by § 110 from the equation

$$\begin{vmatrix} \begin{pmatrix} 1,2,\dots\lambda-2,\lambda-1 \\ a \\ 1,2,\dots\lambda-2,\lambda-1 \end{pmatrix}\begin{pmatrix} 1,2,\dots\lambda-2,\lambda \\ a \\ 1,2,\dots\lambda-2,\lambda-1 \end{pmatrix} \\ \begin{pmatrix} 1,2,\dots\lambda-2,\lambda-1 \\ a \\ 1,2,\dots\lambda-2,\lambda \end{pmatrix}\begin{pmatrix} 1,2,\dots\lambda-2,\lambda \\ a \\ 1,2,\dots\lambda-2,\lambda \end{pmatrix} \end{vmatrix} = \begin{vmatrix} 0, & \delta_{\lambda-1} \\ -\delta_{\lambda-1}, & 0 \end{vmatrix} = \Delta_{\lambda-2}\Delta_\lambda.$$

Ex. v. Denoting the conjugate of h_λ by h_λ', and using for T_λ in (B) the form

$$T_\lambda=h_\lambda'\begin{bmatrix} 0, & 1 \\ -1, & 0\end{bmatrix}h_\lambda, \quad\text{where } h_\lambda=\frac{1}{l_{12}}\begin{bmatrix} 0, & -l_{12} \\ 1, & 0\end{bmatrix}[0,\ l]_2^{\lambda-2,\ m-\lambda+2}$$

we can deduce symmetric equigradent transformations in Ω of the forms (A) in which all the successive parts of $[e]_\rho^\rho$ have the common value $\begin{bmatrix} 0, & 1 \\ -1, & 0\end{bmatrix}$.

Ex. vi. *A corresponding non-unitary symmetric equigradent transformation.*

Let $[b]_{m-2}^{m-2}$, $[c]_{m-4}^{m-4}$, ... $[l]_{m-\lambda+2}^{m-\lambda+2}$, ... $[r]_{m-\rho+2}^{m-\rho+2}$ be the skew-symmetric matrices of ranks $\rho-2$, $\rho-4$, ... $\rho-\lambda+2$, ... 2, whose elements are defined by such equations as

$$l_{uv}=\begin{pmatrix} 1,2,\dots\lambda-2,\lambda-2+v \\ a \\ 1,2,\dots\lambda-2,\lambda-2+u \end{pmatrix}, \text{ where } \lambda \text{ is even};$$

and let
$$\delta_{\lambda-1}=\begin{pmatrix} 1,2,\dots\lambda-2,\lambda \\ a \\ 1,2,\dots\lambda-2,\lambda-1 \end{pmatrix}=-\begin{pmatrix} 1,2,\dots\lambda-2,\lambda-1 \\ a \\ 1,2,\dots\lambda-2,\lambda \end{pmatrix}.$$

Then the matrix ϕ of Theorem I can be expressed in the form

$$[a]_m^m=T_2+T_4+\dots+T_\lambda+\dots+T_\rho, \quad\text{......(B)}$$

where
$$T_\lambda=\frac{1}{\Delta_{\lambda-2}\delta_{\lambda-1}}\overline{\underline{\begin{matrix}0\\ l\end{matrix}}}^{\,2}_{\,\lambda-2,\ m-\lambda+2}\begin{bmatrix} 0, & 1 \\ -1, & 0\end{bmatrix}[0,\ l]_2^{\lambda-2,\ m-\lambda+2}.$$

We have already proved this result indirectly in Ex. iii. To prove it directly let

$$\Phi_\lambda=\begin{bmatrix} 0, & 0 \\ 0, & l\end{bmatrix}_{\lambda-2,\ m-\lambda+2}^{\lambda-2,\ m-\lambda+2}.$$

When $\lambda=2, 4, 6, \dots \rho$ we have $l=a, b, c, \dots r$; and with this new notation we have

$$l_{11}=l_{22}=0,\ l_{12}=-l_{21}=\delta_{\lambda-1};\ a_{12}=\delta_1,\ b_{12}=\delta_3,\ c_{12}=\delta_5, \dots l_{12}=\delta_{\lambda-1}, \dots;$$

and
$$\begin{pmatrix} 1,2,2+v \\ b \\ 1,2,2+u\end{pmatrix}=\Delta_2^2 c_{uv}, \quad \begin{pmatrix} 1,2,2+v \\ c \\ 1,2,2+u\end{pmatrix}=\Delta_4^2 d_{uv}, \dots.$$

The identity of § 116 gives the equation

$$(c)_2^2[c]_{m-4}^{m-4}=[c]_{m-4}^2\begin{bmatrix} 0, & -c_{12} \\ c_{12}, & 0\end{bmatrix}[c]_2^{m-4}+\Delta_4^2\begin{bmatrix} 0, & 0 \\ 0, & d\end{bmatrix}_{2,\ m-6}^{2,\ m-6}$$

or
$$\delta_5^2[c]_{m-4}^{m-4}=\delta_5\overline{\underline{c}}^{\,2}_{\,m-4}\begin{bmatrix} 0, & 1 \\ -1, & 0\end{bmatrix}[c]_2^{m-4}+\Delta_4^2\begin{bmatrix} 0, & 0 \\ 0, & d\end{bmatrix}_{2,\ m-6}^{2,\ m-6}.$$

Dividing both sides by $\Delta_4 \delta_5^2$ and using Ex. iv we deduce that

$$\frac{1}{\Delta_4}\Phi_6 = T_6 + \frac{1}{\Delta_6}\Phi_8.$$

In this way we obtain the successive equations

$$\frac{1}{\Delta_0}\phi = T_2 + \frac{1}{\Delta_2}\Phi_4,\quad \frac{1}{\Delta_2}\Phi_4 = T_4 + \frac{1}{\Delta_4}\Phi_6, \ldots \frac{1}{\Delta_{\rho-2}}\Phi_\rho = T_\rho,$$

which lead to the equation (B).

The equation (B) is with this new notation equivalent to transformations of the form (A) in which $[e]_\rho^\rho$ is a compartite matrix in standard form whose successive parts are

$$\frac{1}{\Delta_0\delta_1}\begin{bmatrix} 0, & 1 \\ -1, & 0 \end{bmatrix},\quad \frac{1}{\Delta_2\delta_3}\begin{bmatrix} 0, & 1 \\ -1, & 1 \end{bmatrix},\quad \frac{1}{\Delta_4\delta_5}\begin{bmatrix} 0, & 1 \\ -1, & 0 \end{bmatrix}, \ldots \frac{1}{\Delta_{\rho-2}\delta_{\rho-1}}\begin{bmatrix} 0, & 1 \\ -1, & 0 \end{bmatrix},$$

and $[h]_m^m$ is a matrix in which the $(\lambda-1)$th, and λth horizontal rows are $[0, l]_2^{\lambda-2,\, m-\lambda+2}$, and the last $m-\rho$ horizontal rows are $[0, 1]_{m-\rho}^{\rho,\, m-\rho}$. All the non-vanishing elements of $[h]_m^m$ are minor determinants of $[a]_m^m$. These are non-unitary symmetric equigradent transformations in Ω.

Theorem II. *Every skew-symmetric matrix $\phi = [a]_m^m$ of rank r whose elements are constants lying in a domain of rationality Ω can be converted by a symmetric derangement followed by a symmetric unitary equigradent transformation in Ω into a skew-symmetric matrix of the standard form*

$$\psi = \begin{bmatrix} e, & 0 \\ 0, & 0 \end{bmatrix}_{r,\, m-r}^{r,\, m-r}, \quad \textit{where } [e]_r^r = \begin{bmatrix} a', & 0, \ldots & 0 \\ 0, & b', \ldots & 0 \\ \ldots & \ldots & \ldots \\ 0, & 0, \ldots & l' \end{bmatrix}_{2,2,\ldots 2}^{2,2,\ldots 2},$$

the successive parts of $[e]_r^r$ being

$$[a']_2^2 = \begin{bmatrix} 0\ , & \alpha \\ -\alpha, & 0 \end{bmatrix},\quad [b']_2^2 = \begin{bmatrix} 0\ , & \beta \\ -\beta, & 0 \end{bmatrix},\ \ldots [l']_2^2 = \begin{bmatrix} 0\ , & \lambda \\ -\lambda, & 0 \end{bmatrix},$$

and $\alpha, \beta, \ldots \lambda$ being non-zero constants lying in Ω.

By Note 2 of § 127 we can convert ϕ by a symmetric derangement into a skew-symmetric matrix $[b]_m^m$ of the same order and rank whose rows are so arranged that none of the leading diagonal minor determinants $(b)_2^2, (b)_4^4, \ldots (b)_r^r$ of even order vanish, and by Theorem I we can then convert $[b]_m^m$ into the above form by a symmetric unitary equigradent transformation in Ω.

2. *Reduction by unrestricted symmetric equigradent transformations.*

Theorem III. *Every skew-symmetric matrix* $\phi = [a]_m^m$ *of rank* r *whose elements are constants lying in a domain of rationality* Ω *can be converted by a symmetric equigradent transformation in* Ω *into a similar skew-symmetric compartite matrix* ψ *of the standard form*

$$\psi = \begin{bmatrix} e, & 0 \\ 0, & 0 \end{bmatrix}_{r,\, m-r}^{r,\, m-r}, \quad \text{where } [e]_r^r = \begin{bmatrix} \omega, & 0, & \dots & 0 \\ 0, & \omega, & \dots & 0 \\ \dots & \dots & \dots & \dots \\ 0, & 0, & \dots & \omega \end{bmatrix}_{2,2,\dots 2}^{2,2,\dots 2},$$

the successive parts of $[e]_r^r$ *having the common value*

$$[\omega]_2^2 = \begin{bmatrix} 0, & 1 \\ -1, & 0 \end{bmatrix}.$$

We first convert ϕ into a matrix ϕ' having the standard form of Theorem II. Then if we divide the 1st horizontal and vertical rows of ϕ' by α, the 3rd horizontal and vertical rows by $\beta, \dots$ the $(r-1)$th horizontal and vertical rows by λ, we convert ϕ' into the matrix ψ of the above theorem. In effecting this conversion we have applied in succession a symmetric derangement, a symmetric unitary equigradent transformation in Ω, and an undegenerate quasi-scalar transformation in Ω.

We can also deduce Theorem III from Ex. v.

Theorem IV. *If* $A = [a]_u^u$ *and* $B = [b]_m^m$ *are two skew-symmetric matrices whose elements are constants in* Ω, *and if* $m \not> u$, *we can convert* A *into* B *by a symmetric equigradent transformation in* Ω *of the form*

$$\overline{h}_m^{\,u} [a]_u^u [h]_u^m = [b]_m^m \quad \text{..........................(C)}$$

when and only when A *and* B *have the same rank.*

We deduce this theorem from Theorem III, the proof being similar to that of Theorem IV in § 147.

Thus two skew-symmetric matrices with constant elements are symmetrically equigradent when and only when they have the same rank.

Ex. vii. If we can convert the skew-symmetric matrices A and B into the same skew-symmetric matrix by symmetric equigradent transformations, then there exists a symmetric equigradent transformation of the form (C).

CHAPTER XVII

SOME MATRIX EQUATIONS OF THE SECOND DEGREE

[The equations $XY = AB$, $XY = C$ are first considered in §§ 152—4. The rest of the chapter deals with equations which are symmetric in form. In § 155 we show how all solutions of symmetric equations of the form $X'X = I$, where I is a unit matrix, can be found, and we call these solutions semi-unit matrices. In §§ 156—160 we determine the general solutions of all symmetric equations of the forms $X'X = A'A$, $X'X = C$; and in § 161 we show how all solutions of the symmetric equation $X'AX = C$ can be found. The remaining articles deal with certain special equations of the foregoing types.]

§ 151. Matrix equations of the second degree.

In Chapter XV we have considered equations, and in particular symmetric equations, of the form

$$X_1 X_2 \ldots X_n = C, \quad \text{.............................(1)}$$

where C is a known matrix, and $X_1 X_2 \ldots X_n$ is a standard product of matrices of given orders some but not all of which are known, whilst in the rest all the elements are unknown; and we have shown how all possible values of the unknown matrices which satisfy the equation can be found. Such an equation may be called a matrix equation of the rth degree when exactly r of the factor matrices on the left are unknown.

The most general matrix equation of the second degree having this form is

$$AXMYB = C, \quad \text{................................(2)}$$

where A, M, B, C are known matrices and X and Y are two unknown matrices all of whose elements are to be so determined that the equation is satisfied.

The most general symmetric matrix equation of the second degree having this form is

$$A'X'MXA = C, \quad \text{.............................(3)}$$

where M and C are given symmetric matrices, A and A' are two given mutually conjugate matrices, and X and X' are two unknown mutually conjugate matrices all of whose elements are to be so determined that the equation is satisfied.

In the present chapter we shall consider in greater detail certain equations of the forms (2) and (3); and we shall consider more particularly equations of the form

$$XY = C, \qquad \text{.....................................(4)}$$

and symmetric equations of the form

$$X'X = C \qquad \text{.....................................(5)}$$

to which (2) and (3) are always reducible.

§ 152. Solutions of some special equations of the form $XY=AB$.

1. *The equation* $$\underline{\overline{x}}_m\,[y]_n = \underline{\overline{a}}_m\,[b]_n, \qquad \text{............................(A)}$$

i.e.

$$\begin{bmatrix} x_1 \\ x_2 \\ \vdots \\ x_m \end{bmatrix} [y_1\, y_2 \dots y_n] = \begin{bmatrix} a_1 \\ a_2 \\ \vdots \\ a_m \end{bmatrix} [b_1\, b_2 \dots b_n].$$

We shall distinguish two cases.

CASE I. *When both the given matrices on the right have rank* 1.

We may suppose that $a_i \neq 0$, and $b_j = 0$. Then by equating corresponding elements of the product matrices on the two sides of the given equation, we see that when it is satisfied we must have $x_i y_j = a_i b_j \neq 0$. It follows that $x_i \neq 0$, $y_j \neq 0$; and when we equate the ith horizontal rows and also the jth vertical rows of the two product matrices, we obtain

$$x_i\,[y_1\, y_2 \dots y_n] = a_i\,[b_1\, b_2 \dots b_n], \quad y_j\,[x_1\, x_2 \dots x_m] = b_j\,[a_1\, a_2 \dots a_m]. \;\text{...(1)}$$

If we write $\dfrac{b_j}{y_j} = \lambda$, $\dfrac{a_i}{x_i} = \mu$, we have $\lambda\mu = 1$; and we conclude that:

The general solution of the equation (A) *in Case I is*

$$\left.\begin{array}{l} [x_1\, x_2 \dots x_m] = \lambda\,[a_1\, a_2 \dots a_m] \\ [y_1\, y_2 \dots y_n] \;= \mu\,[b_1\, b_2 \dots b_n] \end{array}\right\}, \; \textit{where } \lambda\mu = 1. \qquad \text{............(A}_1\text{)}$$

For (1) shows that every solution must have this form, and conversely every pair of values of $[x]_m$ and $[y]_n$ given by (A_1) clearly constitutes a solution of the equation.

CASE II. *When one at least of the given matrices on the right has rank* 0.

In this case the given equation assumes the form $\underline{\overline{x}}_m\,[y]_n = 0$.

It is clearly satisfied when one at least of the matrices $[x]_m$ and $[y]_n$ has rank 0 or vanishes, and unless this is so, it cannot be satisfied; for if $[x]_m$

and $[y]_n$ both have rank 1, we may suppose that $x_i \neq 0$, $y_j \neq 0$; and then the product matrix on the left contains the non-vanishing element $x_i y_j$, and cannot be a zero matrix.

Thus in Case II the equation (A) *is satisfied when and only when*

$$\textit{either}\quad [x_1\, x_2 \ldots x_m] = 0, \quad \textit{or} \quad [y_1\, y_2 \ldots y_n] = 0. \quad \ldots\ldots\ldots\ldots\ldots (A_2)$$

Ex. i. *The equation* $p\,\overline{\underline{x}}_m\,[y]_n = q\,\overline{\underline{a}}_m\,[b]_n$ *in which* $p \neq 0$ *and* $q \neq 0$.

This equation can be reduced to the equation (A) by putting

$$p\,\overline{\underline{x}}_m = \overline{\underline{\xi}}_m, \quad q\,\overline{\underline{a}}_m = \overline{\underline{a}}_m.$$

When both the given matrices on the right have rank 1, *the general solution is*

$$\left.\begin{array}{l}[x_1\, x_2 \ldots x_m] = \lambda\,[a_1\, a_2 \ldots a_m] \\ [y_1\, y_2 \ldots y_n] = \mu\,[b_1\, b_2 \ldots b_n]\end{array}\right\}, \quad \textit{where}\ \ p\lambda\mu = q.$$

When one of the given matrices on the right has rank 0, *the equation is satisfied when and only when*

$$\textit{either}\quad [x_1\, x_2 \ldots x_m] = 0, \qquad \textit{or}\quad [y_1\, y_2 \ldots y_n] = 0.$$

Ex. ii. *The equation* $\begin{bmatrix} x_1 & y_1 \\ x_2 & y_2 \\ \cdots & \cdots \\ x_m & y_m \end{bmatrix} \begin{bmatrix} y_1\, y_2 \ldots y_m \\ x_1\, x_2 \ldots x_m \end{bmatrix} = 0.$

By the properties of passive rows this is the same as $\overline{\underline{x}}_m\,[y]_m + \overline{\underline{y}}_m\,[x]_m = 0$. Supposing that neither of the matrices $[x]_m$ and $[y]_m$ vanishes, it follows from Ex. i that the equation can only be satisfied when

$$[x]_m = \lambda\,[y]_m, \quad [y]_m = \mu\,[x]_m, \quad \text{where } \lambda\mu = -1.$$

This requires that $[x]_m = -[x]_m$, which is impossible.

Thus the equation is satisfied when and only when

$$\textit{either}\quad [x_1\, x_2 \ldots x_m] = 0, \qquad \textit{or}\quad [y_1\, y_2 \ldots y_m] = 0.$$

The equation is therefore the condition that one at least of the long rows of the matrix $\begin{bmatrix} x_1\, x_2 \ldots x_m \\ y_1\, y_2 \ldots y_m \end{bmatrix}$ shall be a row of 0's.

2. *The equation* $\qquad [x]_m^r\,[y]_r^n = [a]_m^r\,[b]_r^n \quad \ldots\ldots\ldots\ldots\ldots\ldots\ldots\ldots (B)$

in which both the given matrices on the right have rank r.

When both the given matrices $[a]_m^r$ and $[b]_r^n$ have rank r, the product matrix on the right of the equation (B) has rank r, and therefore in every solution both the factor matrices $[x]_m^r$ and $[y]_r^n$ on the left must have rank r.

Whenever the equation is satisfied, we see by postfixing an inverse postfactor of $[y]_r^n$ on both sides, and by prefixing an inverse prefactor of $[x]_m^r$ on both sides, that there must exist relations of the form

$$[x]_m^r = [a]_m^r\,[h]_r^r, \quad [y]_r^n = [k]_r^r\,[b]_r^n, \qquad \text{.................(2)}$$

where $[h]_r^r$ and $[k]_r^r$ are necessarily undegenerate.

NOTE. We can deduce this result from the first sub-article.

When we equate correspondingly formed complete matrices of the minor determinants of order r on both sides of (A), we obtain

$$\underline{\overline{\xi}}_\mu\,[\eta]_\nu = \underline{\overline{a}}_\mu\,[\beta]_\nu, \qquad \text{..................................(3)}$$

where $\mu = \binom{m}{r}$, $\nu = \binom{n}{r}$; $\xi_1, \xi_2, \ldots \xi_\mu$ and $\eta_1, \eta_2, \ldots \eta_\nu$ are the distinct simple minor determinants of $[x]_m^r$ and $[y]_r^n$ respectively; and $a_1, a_2, \ldots a_\mu$ and $\beta_1, \beta_2, \ldots \beta_\nu$ are the similarly formed distinct simple minor determinants of $[a]_m^r$ and $[b]_r^n$ respectively. Since both the factor matrices on the right of (3) have rank 1, therefore by sub-article 1 the general solution of the equation (3) is

$$\left.\begin{array}{l}[\xi_1\,\xi_2 \ldots \xi_\mu] = p\,[a_1\,a_2 \ldots a_\mu]\\ [\eta_1\,\eta_2 \ldots \eta_\nu] = q\,[\beta_1\,\beta_2 \ldots \beta_\nu]\end{array}\right\}, \quad \text{where } pq = 1.$$

It follows from § 113 that when the equation (A) is satisfied, there must exist relations of the form (2) in which $[h]_r^r$ and $[k]_r^r$ are undegenerate square-matrices.

Substituting the values given by (2) in (B), we see that the equation (B) is satisfied if and only if

$$[a]_m^r\,[h]_r^r\,[k]_r^r\,[b]_r^n = [a]_m^r\,[1]_r^r\,[b]_r^n.$$

Since the extreme factor matrices have rank r, we can cancel them by § 84, and the above condition becomes

$$[h]_r^r\,[k]_r^r = [1]_r^r.$$

Thus the general solution of the equation (B) *is given by*

$$[x]_m^r = [a]_m^r\,[h]_r^r, \quad [y]_r^n = [k]_r^r\,[b]_r^n, \quad \text{where } [h]_r^r\,[k]_r^r = [1]_r^r. \qquad (\mathrm{B}_1)$$

Here $[h]_r^r$ *and* $[k]_r^r$ *are any two mutually inverse undegenerate square matrices of order r, one of which can be chosen arbitrarily.*

Ex. iii. If $\lambda \neq 0$ and $\mu \neq 0$ the general solution of the equation

$$\lambda\,[x]_m^r\,[y]_r^n = \mu\,[a]_m^r\,[b]_r^n,$$

in which both the factor matrices on the right have rank r, is given by

$$\sqrt{\lambda}\,[x]_m^r=\sqrt{\mu}\,[a]_m^r\,[h]_r^r,\quad \sqrt{\lambda}\,[y]_r^n=\sqrt{\mu}\,[k]_r^r\,[b]_r^n,$$

or

$$\lambda\,[x]_m^r=\mu\,[a]_m^r\,[h]_r^r,\quad [y]_r^n=[k]_r^r\,[b]_r^n,$$

where

$$[h]_r^r\,[k]_r^r=[1]_r^r.$$

§ 153. Solutions of all equations of the form $XY=AB$.

1. *The equation* $$[x]_m^s\,[y]_s^n=[a]_m^r\,[b]_r^n, \qquad \text{(A)}$$

in which both the given matrices on the right have rank r.

Since the product matrix on the right is a given matrix of rank r, it follows from Theorem III a of § 134 that the equation admits of solution when and only when

$$r \nless 0,\quad r \ngtr m,\quad r \ngtr n;\quad r \ngtr s. \qquad \text{(1)}$$

If r is any integer satisfying the necessary conditions (1), there are solutions in which $[x]_m^s$ and $[y]_s^n$ have respectively ranks α and β when and only when α and β satisfy the conditions

$$\alpha \nless r,\quad \alpha \ngtr m;\quad \beta \nless r,\quad \beta \ngtr n;\quad \alpha+\beta \ngtr r+s. \qquad \text{(2)}$$

When the equation (A) is satisfied, we see by postfixing an inverse postfactor of $[b]_r^n$ on both sides that the vertical rows of $[a]_m^r$ must be connected with the vertical rows of $[x]_m^s$; and when this is so, the vertical rows of $[a]_m^r\,[b]_r^n$ are connected with the vertical rows of $[x]_m^s$, and the equation therefore admits of solution for $[y]_s^n$. Thus there are solutions for $[y]_s^n$ when and only when $[x]_m^s$ satisfies a relation of the form

$$[x]_m^s\,[l]_s^r=[a]_m^r,\quad \text{or}\quad [x]_m^s\,[l]_s^s=[a,\ a']_m^{r,\,s-r},$$

where $[l]_s^s$ is undegenerate. It follows that there are solutions in which $[x]_m^s$ has rank α when and only when $[x]_m^s$ has the form

$$[x]_m^s=[a,\ a']_m^{r,\,\alpha-r}\,[\xi]_\alpha^s, \qquad \text{(3)}$$

where both matrices on the right have rank α.

Similarly there are solutions in which $[y]_s^n$ has rank β when and only when $[y]_s^n$ has the form

$$[y]_s^n=[\eta]_s^\beta\begin{bmatrix} b \\ b' \end{bmatrix}_{r,\,\beta-r}^n, \qquad \text{(4)}$$

where both matrices on the right have rank β.

The values of $[x]_m^s$ and $[y]_s^n$ given by (3) and (4) satisfy the equation (A) when and only when

$$[a,\ a']_m^{r,\ \alpha-r}\,[\xi]_\alpha^s\,[\eta]_s^\beta\begin{bmatrix} b \\ b' \end{bmatrix}_{r,\ \beta-r}^n = [a,\ a']_m^{r,\ \alpha-r}\begin{bmatrix} 1, & 0 \\ 0, & 0 \end{bmatrix}_{r,\ \alpha-r}^{r,\ \beta-r}\begin{bmatrix} b \\ b' \end{bmatrix}_{r,\ \beta-r}^n .$$

or

$$[\xi]_\alpha^s\,[\eta]_s^\beta = \begin{bmatrix} 1, & 0 \\ 0, & 0 \end{bmatrix}_{r,\ \alpha-r}^{r,\ \beta-r} = \begin{bmatrix} 1 \\ 0 \end{bmatrix}_{r,\ \alpha-r}^r\,[1,\ 0]_r^{r,\ \beta-r}. \qquad \text{.........(5)}$$

Thus all solutions of the equation (A) *in which* $[x]_m^s$ *and* $[y]_s^n$ *have ranks* α *and* β *are given by* (3) *and* (4) *where* $[\xi]_\alpha^s$ *and* $[\eta]_s^\beta$ *are undegenerate matrices of ranks* α *and* β *satisfying the equation* (5).

Here $[\xi]_\alpha^s$ can be chosen arbitrarily subject to the condition that it has rank α, and the matrices $[\eta]_s^\beta$ of rank β which satisfy (5) can then be found as in § 132.1; or we can choose $[\eta]_s^\beta$ arbitrarily subject to the condition that it has rank β, and then solve (5) for $[\xi]_\alpha^s$ as in § 132.2.

Ex. i. *Undegenerate solutions of ranks* α *and* β *of the equation*

$$[x]_\alpha^s\,[y]_s^\beta = \begin{bmatrix} 1, & 0 \\ 0, & 0 \end{bmatrix}_{r,\ \alpha-r}^{r,\ \beta-r} = \begin{bmatrix} 1 \\ 0 \end{bmatrix}_{r,\ \alpha-r}^r\,[1,\ 0]_r^{r,\ \beta-r}. \qquad \text{..................(B)}$$

We have reduced the solution of the equation (A) to the solution of (B).

The equation (B) admits of solution when and only when

$$r \nless 0,\quad r \ngtr \alpha,\quad r \ngtr \beta;\quad r \ngtr s.$$

These conditions being satisfied, it has undegenerate solutions $[x]_\alpha^s$, $[y]_s^\beta$ of ranks α, β when and only when

$$\alpha+\beta \ngtr r+s.$$

Any given matrix $[x]_\alpha^s$ in which $(x)_\alpha^\alpha \neq 0$ can be expressed in the form

$$[x]_\alpha^s = [h]_\alpha^\alpha\,[1,\ p]_\alpha^{\alpha,\ s-\alpha},$$

where $[h]_\alpha^\alpha$ is undegenerate; and when $[x]_\alpha^s$ has this value, it follows from Ex. iv of § 132 that the general formula for a matrix $[y]_s^\beta$ of rank β satisfying (B) is

$$[y]_s^\beta = \begin{bmatrix} k, & -p \\ 0, & 1 \end{bmatrix}_{\alpha,\ s-\alpha}^{r,\ s-\alpha}\begin{bmatrix} 1, & 0 \\ u, & v \end{bmatrix}_{r,\ s-\alpha}^{r,\ \beta-r},$$

where $[v]_{s-\alpha}^{\beta-r}$ has rank $\beta-r$, $[u]_{s-\alpha}^r$ is entirely arbitrary, and $[k]_\alpha^\alpha$ is the inverse of $[h]_\alpha^\alpha$.

Thus all solutions of the equation (B) *in which* $[x]_\alpha^s$ *and* $[y]_s^\beta$ *have ranks* α *and* β *are given by*

$$[x]_\alpha^s = [h]_\alpha^\alpha\,[1,\ p]_\alpha^{\alpha,\ s-\alpha}\,[\omega]_s^s, \qquad \text{......................................(6)}$$

$$[y]_s^\beta = \underline{\overline{\omega}}_s^s\begin{bmatrix} k, & -p \\ 0, & 1 \end{bmatrix}_{\alpha,\ s-\alpha}^{r,\ s-\alpha}\begin{bmatrix} 1, & 0 \\ u, & v \end{bmatrix}_{r,\ s-\alpha}^{r,\ \beta-r}, \qquad \text{........................(7)}$$

where $[\omega]_s^s$ *is a derangement of* $[1]_s^s$, $[h]_a^a$ *is undegenerate,* $[k]_a^a$ *is the inverse of* $[h]_a^a$, *and* $[v]_{s-a}^{\beta-r}$ *has rank* $\beta - r$.

We can replace (7) by

$$[y]_s^\beta = \overline{\underline{\omega}}_s^s \begin{bmatrix} k, & -p \\ 0, & 1 \end{bmatrix}_{a,\, s-a}^{a,\, s-a} \begin{bmatrix} 1, & 0 \\ 0, & 0 \\ u, & v \end{bmatrix}_{r,\, a-r,\, s-a}^{r,\, \beta-r}.$$

Ex. ii. If

$$[\xi]_m^s = \begin{bmatrix} 1, & 0, & 0, & 0 \\ 0, & 1, & 0, & 0 \\ 0, & 0, & 0, & 0 \end{bmatrix}_{r,\, a-r,\, m-a}^{r,\, a-r,\, \beta-r,\, r+s-a-\beta}, \qquad [\eta]_s^n = \begin{bmatrix} 1, & 0, & 0 \\ 0, & 0, & 0 \\ 0, & 1, & 0 \\ 0, & 0, & 0 \end{bmatrix}_{r,\, a-r,\, \beta-r,\, r+s-a-\beta}^{r,\, \beta-r,\, n-\beta},$$

then solutions of the equation (A) of ranks a and β are given by

$$[x]_m^s = [a,\ a']_m^{r,\, m-r} [\xi]_m^s [h]_s^s, \qquad [y]_s^n = [k]_s^s [\eta]_s^n \begin{bmatrix} b \\ b' \end{bmatrix}_{r,\, n-r}^n,$$

where $[h]_s^s\ [k]_s^s = [1]_s^s$, and the matrices $[a,\ a']_m^{r,\, m-r}$ and $\begin{bmatrix} b \\ b' \end{bmatrix}_{r,\, n-r}^n$ are undegenerate.

Ex. iii. If

$$[\xi]_a^s = \begin{bmatrix} 1, & 0, & 0, & 0 \\ 0, & 1, & 0, & 0 \end{bmatrix}_{r,\, a-r}^{r,\, a-r,\, \beta-r,\, r+s-a-\beta}, \qquad [\eta]_s^\beta = \begin{bmatrix} 1, & 0 \\ 0, & 0 \\ 0, & 1 \\ 0, & 0 \end{bmatrix}_{r,\, a-r,\, \beta-r,\, r+s-a-\beta}^{r,\, \beta-r},$$

then solutions of the equation (B) of ranks a and β are given by

$$[x]_a^s = \begin{bmatrix} 1, & p \\ 0, & u \end{bmatrix}_{r,\, a-r}^{r,\, a-r} [\xi]_a^s [h]_s^s, \qquad [y]_s^\beta = [k]_s^s\ [\eta]_s^\beta \begin{bmatrix} 1, & 0 \\ q, & v \end{bmatrix}_{r,\, \beta-r}^{r,\, \beta-r},$$

where the prefactors and postfactors are undegenerate square matrices, and

$$[h]_s^s\ [k]_s^s = [1]_s^s.$$

2. *The equation* $[x]_m^s\ [y]_s^n = [a]_m^r\ [b]_r^n$ *in which the given matrices on the right have any ranks.*

We treat this equation as in § 154, writing it in the form

$$[x]_m^s\ [y]_s^n = [c]_m^n, \text{ where } [c]_m^n = [a]_m^r\ [b]_r^n.$$

The rank ρ of the given product matrix $[c]_m^n$ must satisfy the conditions $\rho \nless 0$, $\rho \ngtr m$, $\rho \ngtr n$, $\rho \ngtr r$; and the equation admits of solution only when we also have $\rho \ngtr s$.

§ 154. Solutions of equations of the form $XY = C$.

1. *The special equation*

$$\lceil x \rfloor_m [y]_n = [c]_m^n \quad \text{..............................(A)}$$

in which $[c]_m^n$ *is a given matrix whose rank does not exceed* 1.

Written in full the equation is

$$\begin{bmatrix} x_1 \\ x_2 \\ \vdots \\ x_m \end{bmatrix} [y_1\, y_2 \cdots y_n] = \begin{bmatrix} c_{11} & c_{12} & \cdots & c_{1n} \\ c_{21} & c_{22} & \cdots & c_{2n} \\ \cdots & \cdots & \cdots & \cdots \\ c_{m1} & c_{m2} & \cdots & c_{mn} \end{bmatrix} = [c]_m^n.$$

Clearly the equation does not admit of solution when the rank of $[c]_m^n$ exceeds 1, for the rank of the product matrix on the left cannot exceed 1; and we have therefore only two cases to consider.

Case I. *When* $[c]_m^n$ *has rank* 1.

Since $[c]_m^n$ has at least one non-vanishing element, we may suppose that $c_{ij} \neq 0$. Then by Ex. v of § 115 or Ex. vi of § 116 the equation (A) is equivalent to

$$c_{ij} \begin{bmatrix} x_1 \\ x_2 \\ \vdots \\ x_m \end{bmatrix} [y_1\, y_2 \cdots y_n] = \begin{bmatrix} c_{1j} \\ c_{2j} \\ \vdots \\ c_{mj} \end{bmatrix} [c_{i1}\, c_{i2} \ldots c_{in}],$$

where each factor matrix on the right contains the non-vanishing element c_{ij} and has rank 1. Hence by Ex. i of § 152 we see that:

If $c_{ij} \neq 0$, *the general solution of the equation* (A) *in Case I is*

$$\left.\begin{aligned} [x_1\, x_2 \ldots x_m] &= \lambda\, [c_{1j}\, c_{2j} \ldots c_{mj}] \\ [y_1\, y_2 \ldots y_n] &= \mu\, [c_{i1}\, c_{i2} \ldots c_{in}] \end{aligned}\right\}, \ \textit{where } \lambda\mu c_{ij} = 1. \quad \text{.........}(A_1)$$

Again if $\lceil x \rfloor_m = \lceil a \rfloor_m$, $[y]_n = [b]_n$ is any particular solution, the equation can be replaced by $\lceil x \rfloor_m [y]_n = \lceil a \rfloor_m [b]_n$, where the factor matrices on the right have rank 1. Solving this equation by § 152.1 we see that:

If $\lceil x \rfloor_m = \lceil a \rfloor_m$, $[y]_n = [b]_n$ *is any particular solution of* (A), *the general solution is*

$$\lceil x \rfloor_m = \lambda \lceil a \rfloor_m, \quad [y]_n = \mu\, [b]_n, \ \textit{where } \lambda\mu = 1. \quad \text{...........}(A_2)$$

CASE II. *When* $[c]_m^n$ *has rank* 0.

The equation is now that considered in Case II of § 152.1.

Thus in Case II the equation (A) *is satisfied when and only when*

$$\text{either}\quad [x_1 x_2 \dots x_m] = 0, \quad \text{or}\quad [y_1 y_2 \dots y_n] = 0. \quad\text{............}(A_3)$$

Ex. i. If $p \neq 0$ and $q \neq 0$, the general solution of the equation

$$p \,\overline{\underline{x}}_m\, [y]_n = q\,[c]_m^n$$

when $[c]_m^n$ has rank 1 and $c_{ij} \neq 0$ is given by

$$\left.\begin{aligned}[x_1 x_2 \dots x_m] &= \lambda\,[c_{1j}\, c_{2j} \dots c_{mj}]\\ [y_1 y_2 \dots y_n] &= \mu\,[c_{i1}\, c_{i2} \dots c_{in}]\end{aligned}\right\},\quad \text{where } p\lambda\mu c_{ij} = q.$$

Ex. ii. The general solution of the equation

$$\begin{bmatrix} x \\ y \\ z \end{bmatrix} [p\, q\, r\, s] = \begin{bmatrix} 1, & 2, & -1, & 3 \\ 0, & 0, & 0, & 0 \\ 2, & 4, & -2, & 6 \end{bmatrix}$$

is given by

$$\left.\begin{aligned}[x\, y\, z] &= \lambda\,[1\;0\;2]\\ [p\, q\, r\, s] &= \mu\,[1,\; 2,\; -1,\; 3]\end{aligned}\right\},\quad \text{where } \lambda\mu = 1.$$

2. *The special equation*

$$[x]_m^r\, [y]_r^n = [c]_m^n, \quad\text{.............................}(B)$$

in which $[c]_m^n$ *is a given matrix of rank r.*

Clearly in every solution the matrices $[x]_m^r$ and $[y]_r^n$ must be undegenerate and have rank r; and it has been shown in Ex. i of § 115 and in § 146 how particular solutions can be obtained. If $[x]_m^r = [a]_m^r$, $[y]_r^n = [b]_r^n$ is any particular solution, the equation can be written in the form

$$[x]_m^r\, [y]_r^n = [a]_m^r\, [b]_r^n, \quad\text{..........................}(1)$$

and the general solution can then be obtained as in § 152.2.

Thus if $[x]_m^r = [a]_m^r$, $[y]_r^n = [b]_r^n$ *is any particular solution of the equation* (B) *the general solution of the equation is*

$$[x]_m^r = [a]_m^r\, [h]_r^r, \quad [y]_r^n = [k]_r^r\, [b]_r^n, \quad \text{where } [h]_r^r\, [k]_r^r = [1]_r^r. \quad\text{...}(B_1)$$

To obtain a more precise formula for the general solution let $\Delta = (c_{pq})_r^r$ be any non-vanishing derived determinant of $[c]_m^n$ of order r, and let $[C_{pq}]_r^r$ be the reciprocal of $[c_{pq}]_r^r$. Then by § 115 or § 116 the equation (B) is equivalent to

$$(c_{pq})_r^r\, [x]_m^r\, [y]_r^n = [c_{1q}]_m^r\, \overline{\underline{C_{pq}}}_r^r\, [c_{p1}]_r^n. \quad\text{...................}(2)$$

By postfixing on both sides of (2) an inverse postfactor of $[y]_r^n$, and by prefixing on both sides an inverse prefactor of $[x]_m^r$, we see that when the equation (2) is satisfied, there must exist relations of the form

$$[x]_m^r = [c_{1q}]_m^r \, [h]_r^r, \quad [y]_r^n = [k]_r^r \, [c_{p1}]_r^n, \quad \text{..................(3)}$$

where $[h]_r^r$ and $[k]_r^r$ are undegenerate square matrices.

NOTE. When we equate correspondingly formed complete matrices of the minor determinants of order r on both sides of (2) we obtain the equation

$$\Delta \ulcorner\xi\lrcorner_\mu \, [\eta]_\nu = \ulcorner a \lrcorner_\mu \, [\beta]_\nu, \quad \text{..................................(4)}$$

where the elements of $\ulcorner a \lrcorner_\mu$, $[\beta]_\nu$ are the distinct simple minor determinants of the undegenerate matrices $[c_{1q}]_m^r$, $[c_{p1}]_r^n$; and the elements of $\ulcorner\xi\lrcorner_\mu$, $[\eta]_\nu$ are the correspondingly formed simple minor determinants of the similar undegenerate matrices $[x]_m^r$, $[y]_r^n$. Applying Ex. i of § 152 to (4) we see that

$$\left.\begin{matrix}[\xi_1 \xi_2 \ldots \xi_\mu] = \lambda \, [a_1 a_2 \ldots a_\mu] \\ [\eta_1 \eta_2 \ldots \eta_\nu] = \mu \, [\beta_1 \beta_2 \ldots \beta_\nu]\end{matrix}\right\}, \text{ where } \Delta\lambda\mu = 1. \quad \text{....................(5)}$$

From the equations (5) we deduce by § 113 that there must exist relations of the form (3).

Substituting the values (3) of $[x]_m^r$ and $[y]_r^n$ in (2), we see that the equation (2) or (B) is satisfied when and only when $(c_{pq})_r^r \, [h]_r^r \, [k]_r^r = \ulcorner C_{pq} \lrcorner_r^r$, i.e. when and only when $[h]_r^r \, [k]_r^r$ is equal to the inverse matrix of $[c_{pq}]_r^r$.

Thus when $[c]_m^n$ *has rank* r *and* $(c_{pq})_r^r \neq 0$, *the general solution of the equation* (B) *is given by*

$$[x]_m^r = [c_{1q}]_m^r \, [h]_r^r, \quad [y]_r^n = [k]_r^r \, [c_{p1}]_r^n, \quad \text{..................(B}_2\text{)}$$

where

$$(c_{pq})_r^r \cdot [h]_r^r [k]_r^r = \ulcorner C_{pq} \lrcorner_r^r.$$

We can regard either one of the matrices $[h]_r^r$, $[k]_r^r$ as an arbitrary undegenerate square matrix of order r, and the other matrix is then completely determinate.

Ex. iii. We will consider the equation

$$[x]_3^2 \, [y]_2^4 = \begin{bmatrix} 5 & 3 & 1 & 2 \\ 7 & 5 & 2 & 3 \\ 0 & 4 & 3 & 1 \end{bmatrix} = [c]_3^4,$$

in which $[c]_3^4$ has rank 2.

Here $\begin{bmatrix}5 & 3\\ 7 & 5\end{bmatrix}$ is an undegenerate derived matrix whose inverse is $\frac{1}{4}\begin{bmatrix}5, & -3\\ -7, & 5\end{bmatrix}$. Accordingly the general solution is

$$[x]_3^2 = \begin{bmatrix}5 & 3\\ 7 & 5\\ 0 & 4\end{bmatrix} [h]_2^2, \quad [y]_2^4 = [k]_2^2 \begin{bmatrix}5 & 3 & 1 & 2\\ 7 & 5 & 2 & 3\end{bmatrix},$$

where

$$[h]_2^2 [k]_2^2 = \frac{1}{4}\begin{bmatrix}5, & -3\\ -7, & 5\end{bmatrix}.$$

Putting $[h]_2^2 = \frac{1}{2}\begin{bmatrix}1, & 0\\ 0, & 1\end{bmatrix}$, we have $[k]_2^2 = \frac{1}{2}\begin{bmatrix}5, & -3\\ -7, & 5\end{bmatrix}$. Therefore a particular solution is

$$[x]_3^2 = \frac{1}{2}\begin{bmatrix}5 & 3\\ 7 & 5\\ 0 & 4\end{bmatrix}, \quad [y]_2^4 = \frac{1}{2}\begin{bmatrix}4, & 0, & -1, & 1\\ 0, & 4, & 3, & 1\end{bmatrix},$$

and the general solution is

$$[x]_3^2 = \frac{1}{2}\begin{bmatrix}5 & 3\\ 7 & 5\\ 0 & 4\end{bmatrix} [h]_2^2, \quad [y]_2^4 = \frac{1}{2}[k]_2^2\begin{bmatrix}4, & 0, & -1, & 1\\ 0, & 4, & 3, & 1\end{bmatrix}, \quad \text{where } [h]_2^2 [k]_2^2 = [1]_2^2.$$

To obtain a particular solution by the method described in § 146 we obtain in succession the equations

$$5\begin{bmatrix}5 & 3 & 1 & 2\\ 7 & 5 & 2 & 3\\ 0 & 4 & 3 & 1\end{bmatrix} = \begin{bmatrix}5\\ 7\\ 0\end{bmatrix}[5312] + \begin{bmatrix}0 & 0 & 0 & 0\\ 0 & 4 & 3 & 1\\ 0 & 20 & 15 & 5\end{bmatrix}, \quad 4\begin{bmatrix}4 & 3 & 1\\ 20 & 15 & 5\end{bmatrix} = \begin{bmatrix}4\\ 20\end{bmatrix}[431],$$

$$\begin{bmatrix}5 & 3 & 1 & 2\\ 7 & 5 & 2 & 3\\ 0 & 4 & 3 & 1\end{bmatrix} = \begin{bmatrix}5 & 0\\ 7 & 4\\ 0 & 20\end{bmatrix}\begin{bmatrix}\frac{1}{5} & 0\\ 0 & \frac{1}{20}\end{bmatrix}\begin{bmatrix}5 & 3 & 1 & 2\\ 0 & 4 & 3 & 1\end{bmatrix},$$

and thence the solution in which the successive vertical rows of $[x]_3^2$ and the successive horizontal rows of $[y]_2^4$ are

$$\frac{1}{\sqrt{5}}\begin{bmatrix}5\\ 7\\ 0\end{bmatrix}, \quad \frac{1}{\sqrt{20}}\begin{bmatrix}0\\ 4\\ 20\end{bmatrix} \quad \text{and} \quad \frac{1}{\sqrt{5}}[5312], \quad \frac{1}{\sqrt{20}}[0431].$$

3. *The general equation*

$$[x]_m^s [y]_s^n = [c]_m^n, \quad \text{..............................(C)}$$

in which $[c]_m^n$ *is a given matrix of rank r.*

By Theorem III a of § 134 the equation (C) admits of solution when and only when the rank r of $[c]_m^n$ satisfies the conditions

$$r \nless 0, \quad r \ngtr m, \quad r \ngtr n; \quad r \ngtr s; \quad \text{......................(6)}$$

and when this is so the equation has solutions in which $[x]^s_m$, $[y]^n_s$ have ranks α, β when and only when α and β satisfy the conditions

$$\alpha \nless r,\quad \alpha \ngtr m;\quad \beta \nless r,\quad \beta \ngtr n;\quad \alpha+\beta \ngtr r+s; \qquad\qquad (7)$$

which include $\alpha \ngtr s$ and $\beta \ngtr s$.

If we take $[x]^s_m$ to be any given matrix so chosen that the vertical rows of $[c]^n_m$ are connected with the vertical rows of $[x]^s_m$, the equation (C) admits of solution for $[y]^n_s$; and in this way all possible solutions of the equation can be found.

If we express $[c]^n_m$ in the form $[c]^n_m = [a]^r_m\,[b]^n_r$, the equation becomes the equation of § 153.2.

Ex. iv. Let $(c_{pq})^r_r$ be a non-vanishing minor determinant of $[c]^n_m$ of order r, and let $\underline{\overline{C_{pq}}}^{\,r}_{\,r}$ be the inverse of $[c_{pq}]^r_r$.

Then the solutions of (C) *in which* $[x]^s_m$, $[y]^n_s$ *have ranks* α, β *are given by*

$$[x]^s_m = [c_{1q},\ u]^{r,\ \alpha-r}_m\,[\xi]^s_\alpha,\quad [y]^n_s = [\eta]^\beta_s \begin{bmatrix} c_{p1} \\ v \end{bmatrix}^n_{r,\ \beta-r}, \qquad\qquad (8)$$

where the factor matrices on the right are undegenerate, and where

$$[\xi]^s_\alpha\,[\eta]^\beta_s = \underline{\overline{\begin{matrix} C_{pq}, & 0 \\ 0\ , & 0 \end{matrix}}}^{\,r,\ \beta-r}_{\,r,\ \alpha-r}. \qquad\qquad (9)$$

This appears when we substitute the matrices (8) in the equation

$$[x]^s_m\,[y]^n_s = [c_{1q}]^r_m\,\underline{\overline{C_{pq}}}^{\,r}_{\,r}\,[c_{p1}]^n_r,$$

which is equivalent to (C). Thus the solution of (C) is reduced to the determination of the undegenerate solutions of (9). These can be found as in § 132.

§ 155. Solutions of the equation $[x]^m_r\,\underline{\overline{x}}^{\,r}_{\,m} = [1]^r_r$. Semi-unit matrices.

1. *Semi-unit matrices.*

The equation

$$[x]^m_r\,\underline{\overline{x}}^{\,r}_{\,m} = [1]^r_r \qquad\qquad (1)$$

does not admit of solution when $r > m$, for then the rank of the product matrix on the left cannot exceed m and therefore cannot be equal to r. When $r \ngtr m$ the equation always has solutions some of which are real, and every solution has rank r. Assuming that $r \ngtr m$, any matrix $[x]^m_r$ or $\underline{\overline{x}}^{\,r}_{\,m}$ satisfying the equation (1) will be called a *semi-unit matrix* or a *unitary orthogonal matrix* of rank r. In particular any matrix $[x]^{m}_{m}$ or $\overline{\underline{x}}^{m}_{m}$ satisfying the equations $[x]^{m}_{m}\,\overline{\underline{x}}^{m}_{m} = \overline{\underline{x}}^{m}_{m}\,[x]^{m}_{m} = [1]^{m}_{m}$ will be called a *square semi-unit matrix.*

According to these definitions a matrix M is a semi-unit matrix when we can form with it and its conjugate matrix M' a product which is a unit matrix. If M is not square, the product which is a unit matrix must be that one of the two products MM' and $M'M$ in which long rows are active rows, for the other product is necessarily degenerate and cannot be a unit matrix. If M is a square matrix, it is a semi-unit matrix when either one of the products MM' or $M'M$ is a unit matrix; for if one of these products is a unit matrix, it has been shown in Ex. i of § 67 that the other also is a unit matrix.

Ex. i. *Every semi-unit matrix is undegenerate.*

Ex. ii. *Every derangement of a unit matrix is a square semi-unit matrix.*

Ex. iii. *If in every long row of a matrix one of the elements is* 1 *or* − 1 *and all other elements are* 0*'s, the matrix is a semi-unit matrix.*

Ex. iv. *Every derangement of a semi-unit matrix of rank r is also a semi-unit matrix of rank r.*

For if $[x]^{m}_{r}$ is a semi-unit matrix of rank r and if $[\xi]^{m}_{r} = [u]^{r}_{r}\,[x]^{m}_{r}\,[v]^{m}_{m}$, where $[u]^{r}_{r}$ and $[v]^{m}_{m}$ are derangements of $[1]^{r}_{r}$ and $[1]^{m}_{m}$, we have

$$[\xi]^{m}_{r}\,\overline{\underline{\xi}}^{r}_{m} = [u]^{r}_{r}\,[x]^{m}_{r}\,[v]^{m}_{m}\,\overline{\underline{v}}^{m}_{m}\,\overline{\underline{x}}^{r}_{m}\,\overline{\underline{u}}^{r}_{r}$$

$$= [u]^{r}_{r}\,[x]^{m}_{r}\,[1]^{m}_{m}\,\overline{\underline{x}}^{r}_{m}\,\overline{\underline{u}}^{r}_{r} = [u]^{r}_{r}\,[x]^{m}_{r}\,\overline{\underline{x}}^{r}_{m}\,\overline{\underline{u}}^{r}_{r}$$

$$= [u]^{r}_{r}\,[1]^{r}_{r}\,\overline{\underline{u}}^{r}_{r} = [u]^{r}_{r}\,\overline{\underline{u}}^{r}_{r} = [1]^{r}_{r}\,.$$

Ex. v. *If we change the sign of every element in any long or short row of a semi-unit matrix, we again obtain a semi-unit matrix.*

Ex. vi. *A semi-unit matrix and its conjugate (which is also a semi-unit matrix) are mutually inverse.*

Ex. vii. *Every long-cut simple minor of a semi-unit matrix is a semi-unit matrix.*

For if $[p_1\, p_2 \ldots p_s]$ is any minor of $[1\, 2 \ldots r]$ we deduce from

$$[x]^{m}_{r}\,\overline{\underline{x}}^{r}_{m} = [1]^{r}_{r} \quad \text{that} \quad [x_{p1}]^{m}_{s}\,\overline{\underline{x_{p1}}}^{s}_{m} = [1]^{s}_{s}\,.$$

Ex. viii. *If a product of two mutually conjugate matrices is an undegenerate quasi-scalar matrix, we can deduce from them two mutually conjugate semi-unit matrices.*

Let
$$[X]^{m}_{r}\,\overline{\underline{X}}^{r}_{m} = {}^{1}[k]_{r} = \begin{bmatrix} k_1 & 0 & \ldots & 0 \\ 0 & k_2 & \ldots & 0 \\ \ldots & \ldots & \ldots & \ldots \\ 0 & 0 & \ldots & k_r \end{bmatrix},$$

where $r \not> m$ and $k_1 \neq 0$, $k_2 \neq 0$, ... $k_r \neq 0$, and let

$$[x_{i1}\, x_{i2} \dots x_{im}] = \frac{1}{\sqrt{k_i}} [X_{i1}\, X_{i2} \dots X_{im}]$$

for the values 1, 2, ... r of i.

Then $$[x]_r^m \, \overline{\underline{x}}_{\,m}^{\,r} = [1]_r^r.$$

Thus $[x]_r^m$ and $\overline{\underline{x}}_{\,m}^{\,r}$ are two mutually conjugate semi-unit matrices. In particular when $[X]_r^m$ is a real matrix, then $k_1, k_2, \dots k_r$ are necessarily positive, and $[x]_r^m$ and its conjugate are real semi-unit matrices.

Ex. ix. If a square semi-unit matrix has either of the forms

$$\begin{bmatrix} a_{11} & 0 & 0 & \dots & 0 \\ a_{21} & a_{22} & 0 & \dots & 0 \\ a_{31} & a_{32} & a_{33} & \dots & 0 \\ \dots & \dots & \dots & \dots & \dots \\ a_{m1} & a_{m2} & a_{m3} & \dots & a_{mm} \end{bmatrix}, \quad \begin{bmatrix} a_{11} & a_{12} & a_{13} & \dots & a_{1m} \\ 0 & a_{21} & a_{22} & \dots & a_{2m} \\ 0 & 0 & a_{33} & \dots & a_{3m} \\ \dots & \dots & \dots & \dots & \dots \\ 0 & 0 & 0 & \dots & a_{mm} \end{bmatrix},$$

all elements on one side of the leading diagonal being 0's, then all its elements must be 0's except the elements $a_{11}, a_{22}, a_{33}, \dots a_{mm}$ of the leading diagonal, each one of which must be either 1 or -1.

2. *Determination of all real semi-unit matrices* $[x]_r^m$ *of rank* r.

We first observe that we can always determine a real matrix $[x_{11}\, x_{12} \dots x_{1m}]$ whose elements satisfy the equation

$$x_{11}^2 + x_{12}^2 + x_{13}^2 + \dots + x_{1m}^2 = 1.$$

Next assuming that $[x]_{r-1}^m$ is a real semi-unit matrix of rank $r-1$, we will show that, if $r-1 < m$, we can by adding to $[x]_{r-1}^m$ another final horizontal row $[x_{r1}\, x_{r2} \dots x_{rm}]$ form a real semi-unit matrix $[x]_r^m$ of rank r.

We have $$[x]_{r-1}^m \, \overline{\underline{x}}_{\,m}^{\,r-1} = [1]_{r-1}^{r-1}.$$

Since the real matrix $[x]_{r-1}^m$ has rank less than m, we can by § 94 determine a real non-zero solution $[x_{r1}\, x_{r2} \dots x_{rm}]$ of the equation

$$[x]_{r-1}^m \begin{bmatrix} x_{r1} \\ x_{r2} \\ \vdots \\ x_{rm} \end{bmatrix} = 0 \quad \dots\dots\dots\dots (2)$$

in which $m - r + 1$ of the elements $x_{r1}, x_{r2}, \dots x_{rm}$ have arbitrarily assigned real values. We then have

$$x_{r1}^2 + x_{r2}^2 + x_{r3}^2 + \dots + x_{rm}^2 = k_r^2,$$

where k_r is a real positive non-zero quantity. Dividing this solution by k_r we obtain a real unit solution $[x_{r1}\, x_{r2} \dots x_{rm}]$, i.e. a real solution in which $x_{r1}^2 + x_{r2}^2 + x_{r3}^2 + \dots + x_{rm}^2 = 1$. We then have

$$[x]_r^m \, \underline{\overline{x}}_m^r = [1]_r^r,$$

i.e. $[x]_r^m$ is a real semi-unit matrix of rank r.

It follows that if $r \not> m$, we can in succession determine real semi-unit matrices $[x]_1^m, [x]_2^m, [x]_3^m, \dots [x]_r^m$ of ranks 1, 2, 3, ... r, each matrix after the first being formed by the addition of a final horizontal row to the preceding matrix. All real semi-unit matrices of rank r can be formed in this way.

In practice it is usually more convenient to start with any real non-zero matrix $[X_{11}\, X_{12} \dots X_{1m}]$ and to then determine in succession any real non-zero solutions $[X_{21}\, X_{22} \dots X_{2m}]$, $[X_{31}\, X_{32} \dots X_{3m}]$, ... $[X_{r1}\, X_{r2} \dots X_{rm}]$ of the equations

$$[X]_m^1 \begin{bmatrix} X_{21} \\ X_{22} \\ \vdots \\ X_{2m} \end{bmatrix} = 0, \quad [X]_m^2 \begin{bmatrix} X_{31} \\ X_{32} \\ \vdots \\ X_{3m} \end{bmatrix} = 0, \ \dots \ [X]_m^{r-1} \begin{bmatrix} X_{r1} \\ X_{r2} \\ \vdots \\ X_{rm} \end{bmatrix} = 0. \quad \dots(3)$$

We then have

$$[X]_r^m \, \underline{\overline{X}}_m^r = \begin{bmatrix} k_1^2 & 0 & \dots & 0 \\ 0 & k_2^2 & \dots & 0 \\ \dots & \dots & \dots & \dots \\ 0 & 0 & \dots & k_r^2 \end{bmatrix}, \quad \dots\dots(4)$$

where $k_1, k_2, \dots k_r$ are real non-zero positive quantities.

If now we write

$$[x_{i1}\, x_{i2} \dots x_{im}] = \frac{1}{k_i} [X_{i1}\, X_{i2} \dots X_{im}],$$

for the values 1, 2, ... r of i, we have $[x]_r^m \, \underline{\overline{x}}_m^r = [1]_r^r$, i.e. $[x]_r^m$ is a real semi-unit matrix of rank r.

In the equation (4) all the m elements of $[X_{11}\, X_{12} \dots X_{1m}]$ are arbitrary real quantities, $m-1$ of the elements of $[X_{21}\, X_{22} \dots X_{2m}]$ are arbitrary real quantities, ... $m-r+1$ of the elements of $[X_{r1}\, X_{r2} \dots X_{rm}]$ are arbitrary real quantities. Therefore $rm - \frac{1}{2}r(r-1)$ of the elements of $[X]_r^m$ are arbitrary real quantities. Accordingly a real semi-unit matrix $[x]_r^m$ determined in this way may be regarded as containing $rm - \frac{1}{2}r(r+1)$ arbitrary real parameters; and a real square semi-unit matrix $[x]_m^m$ determined in this way may be regarded as containing $\frac{1}{2}m(m-1)$ arbitrary real parameters.

Ex. x. If $[x]_r^m$ is a real semi-unit matrix of rank r, then by § 94 we can determine sets of $m-r$ unconnected mutually orthogonal real unit solutions of the equation

$$[x]_r^m \overline{\underline{y}}_m = 0.$$

If $[y_1 y_2 \ldots y_m] = [x_{r+1,1} x_{r+1,2} \ldots x_{r+1,m}], \ldots [x_{m1} x_{m2} \ldots x_{mm}]$ is any one such set of $m-r$ solutions, then $[x]_m^m$ is a real square semi-unit matrix; and if $s > r$ and $\not> m$, then $[x]_s^m$ is a real semi-unit matrix of rank s formed by adding $s-r$ final horizontal rows to $[x]_r^m$.

Putting $r=1$, we can in this way determine all real semi-unit matrices $[x]_s^m$ of rank s.

Ex. xi. We will determine a particular real solution of the equation

$$\begin{bmatrix} l_1 & m_1 & n_1 \\ l_2 & m_2 & n_2 \\ l_3 & m_3 & n_3 \end{bmatrix} \begin{bmatrix} l_1 & l_2 & l_3 \\ m_1 & m_2 & m_3 \\ n_1 & n_2 & n_3 \end{bmatrix} = \begin{bmatrix} 1 & 0 & 0 \\ 0 & 1 & 0 \\ 0 & 0 & 1 \end{bmatrix}.$$

We first choose any real non-zero value of $[l_1 m_1 n_1]$. Let this be

$$[l_1 m_1 n_1] = [2\ 1\ 2].$$

We then determine any real non-zero solution of the equation

$$[2\ 1\ 2] \begin{bmatrix} l_2 \\ m_2 \\ n_2 \end{bmatrix} = 0.$$

The general solution is given by $2l_2 = -m_2 - 2n_2$. Putting $m_2 = 4$, $n_2 = 1$, we obtain the particular real non-zero solution

$$[l_2 m_2 n_2] = [-3,\ 4,\ 1].$$

Lastly we determine any real non-zero solution of the equation

$$\begin{bmatrix} 2, & 1, & 2 \\ -3, & 4, & 1 \end{bmatrix} \begin{bmatrix} l_3 \\ m_3 \\ n_3 \end{bmatrix} = 0.$$

The general solution is given by $11 \begin{bmatrix} l_3 \\ m_3 \end{bmatrix} = -\begin{bmatrix} 7 \\ 8 \end{bmatrix} n_3$. Putting $n_3 = 11$, we obtain the particular real non-zero solution

$$[l_3 m_3 n_3] = [-7,\ -8,\ 11].$$

We now see that

$$\begin{bmatrix} 2, & 1, & 2 \\ -3, & 4, & 1 \\ -7, & -8, & 11 \end{bmatrix} \begin{bmatrix} 2, & -3, & -7 \\ 1, & 4, & -8 \\ 2, & 1, & 11 \end{bmatrix} = \begin{bmatrix} 9 & 0 & 0 \\ 0 & 26 & 0 \\ 0 & 0 & 234 \end{bmatrix}.$$

Accordingly the matrix $[l\,m\,n]_{123}$ whose elements are given by

$$[l_1 m_1 n_1] = \frac{1}{3}[2\ 1\ 2], \quad [l_2 m_2 n_2] = \frac{1}{\sqrt{26}}[-3,\ 4,\ 1], \quad [l_3 m_3 n_3] = \frac{1}{\sqrt{234}}[-7,\ -8,\ 11]$$

is a particular real solution of the given equation.

By this process we can find the direction-cosines of all possible sets of three mutually perpendicular real directions in common 3-way space with reference to any set of three rectangular co-ordinate axes.

Other real solutions of the above equation are given by

$$\begin{bmatrix} l_1 & m_1 & n_1 \\ l_2 & m_2 & n_2 \\ l_3 & m_3 & n_3 \end{bmatrix} = \frac{1}{3}\begin{bmatrix} 1, & 2, & 3 \\ 2, & 1, & -2 \\ -2, & 2, & -1 \end{bmatrix}, \quad \begin{bmatrix} l_1 & m_1 & n_1 \\ l_2 & m_2 & n_2 \\ l_3 & m_3 & n_3 \end{bmatrix} = \frac{1}{\sqrt{6}}\begin{bmatrix} 1, & 1, & -2 \\ \sqrt{2}, & \sqrt{2}, & \sqrt{2} \\ \sqrt{3}, & -\sqrt{3}, & 0 \end{bmatrix}.$$

A particular solution which is not real is given in Ex. xvi.

Ex. xii. A particular real solution of the equation

$$\begin{bmatrix} l_1 & m_1 & n_1 & p_1 \\ l_2 & m_2 & n_2 & p_2 \\ l_3 & m_3 & n_3 & p_3 \\ l_4 & m_4 & n_4 & p_4 \end{bmatrix}\begin{bmatrix} l_1 & l_2 & l_3 & l_4 \\ m_1 & m_2 & m_3 & m_4 \\ n_1 & n_2 & n_3 & n_4 \\ p_1 & p_2 & p_3 & p_4 \end{bmatrix} = \begin{bmatrix} 1 & 0 & 0 & 0 \\ 0 & 1 & 0 & 0 \\ 0 & 0 & 1 & 0 \\ 0 & 0 & 0 & 1 \end{bmatrix}$$

can be derived from the equation

$$\begin{bmatrix} 1, & 0, & -1, & 0 \\ 1, & 1, & 1, & 0 \\ 1, & -2, & 1, & 1 \\ -1, & 2, & -1, & 6 \end{bmatrix}\begin{bmatrix} 1, & 1, & 1, & -1 \\ 0, & 1, & -2, & 2 \\ -1, & 1, & 1, & -1 \\ 0, & 0, & 1, & 6 \end{bmatrix} = \begin{bmatrix} 2, & 0, & 0, & 0 \\ 0, & 3, & 0, & 0 \\ 0, & 0, & 7, & 0 \\ 0, & 0, & 0, & 42 \end{bmatrix}$$

by putting

$$[l_1\, m_1\, n_1\, p_1] = \frac{1}{\sqrt{2}}[1,\ 0,\ -1,\ 0], \quad [l_2\, m_2\, n_2\, p_2] = \frac{1}{\sqrt{3}}[1,\ 1,\ 1,\ 0],$$

$$[l_3\, m_3\, n_3\, p_3] = \frac{1}{\sqrt{7}}[1,\ -2,\ 1,\ 1], \quad [l_4\, m_4\, n_4\, p_4] = \frac{1}{\sqrt{42}}[-1,\ 2,\ -1,\ 6].$$

Some particular solutions which are not real are given in Ex. xvii.

Ex. xiii. The matrix $\begin{bmatrix} l_1 & m_1 & n_1 & p_1 \\ l_2 & m_2 & n_2 & p_2 \end{bmatrix}$ of Ex. xii is a real semi-unit matrix of rank 2.

3. *Determination of all semi-unit matrices* $[x]_r^m$ *of rank r both real and not real.*

We proceed as in sub-article 2, now determining simply unit solutions or non-extravagant solutions instead of real unit solutions or real solutions.

We first observe that we can always determine a matrix $[x_{11}\, x_{12} \dots x_{1m}]$ whose elements (not necessarily real) satisfy the equation

$$x_{11}^2 + x_{12}^2 + x_{13}^2 + \dots + x_{1m}^2 = 1.$$

Next assuming that $[x]_{r-1}^m$ is a semi-unit matrix (not necessarily real) of rank $r-1$, we can by adding to $[x]_{r-1}^m$ another final horizontal row $[x_{r1}\, x_{r2} \dots x_{rm}]$ form a semi-unit matrix $[x]_r^m$ of rank r.

We have $$[x]_{r-1}^m \overline{\underline{x}}_m^{r-1} = [1]_{r-1}^{r-1}.$$

Since the matrix $[x]_{r-1}^{m}$ is non-extravagant and has rank less than m, we can by § 172 determine a non-extravagant solution $[x_{r1}\, x_{r2} \dots x_{rm}]$ of the equation (2); and we then have

$$x_{r1}^2 + x_{r2}^2 + x_{r3}^2 + \dots + x_{rm}^2 = k_r,$$

where $k_r \neq 0$. Dividing this solution by $\sqrt{k_r}$ we obtain a unit solution $[x_{r1}\, x_{r2} \dots x_{rm}]$, i.e. a solution in which

$$x_{r1}^2 + x_{r2}^2 + x_{r3}^2 + \dots + x_{rm}^2 = 1.$$

We then have $[x]_r^m \, \overline{\underline{x}}_m^{\,r} = [1]_r^r$, i.e. $[x]_r^m$ is a semi-unit matrix of rank r.

It follows that if $r \not> m$, we can in succession determine semi-unit matrices (not necessarily real) $[x]_1^m$, $[x]_2^m$, $[x]_3^m$, ... $[x]_r^m$ of ranks 1, 2, 3, ... r, each matrix after the first being formed by the addition of a final horizontal row to the preceding matrix. All semi-unit matrices of rank r can be formed in this way.

In practice it is usually more convenient to start with any non-extravagant matrix $[X_{11}\, X_{12} \dots X_{1m}]$ and to then determine in succession any non-extravagant solutions $[X_{21}\, X_{22} \dots X_{2m}]$, $[X_{31}\, X_{32} \dots X_{3m}]$, ... $[X_{r1}\, X_{r2} \dots X_{rm}]$ of the equations (3).

We then have

$$[X]_r^m \, \overline{\underline{X}}_m^{\,r} = {}^1[k]_r = \begin{bmatrix} k_1 & 0 & \dots & 0 \\ 0 & k_2 & \dots & 0 \\ \dots & \dots & \dots & \dots \\ 0 & 0 & \dots & k_r \end{bmatrix}, \quad \dots\dots\dots\dots(5)$$

where $k_1, k_2, \dots k_r$ are all non-zero quantities.

If now we write

$$[x_{i1}\, x_{i2} \dots x_{im}] = \frac{1}{\sqrt{k_i}} [X_{i1}\, X_{i2} \dots X_{im}]$$

for the values 1, 2, ... r of i, we have $[x]_r^m \, \overline{\underline{x}}_m^{\,r} = [1]_r^r$, i.e. $[x]_r^m$ is a semi-unit matrix of rank r. The semi-unit matrix $[x]_r^m$ determined in this way may be regarded as containing $rm - \frac{1}{2} r(r+1)$ arbitrary parameters.

Ex. xiv. If $[x]_r^m$ is any semi-unit matrix of rank r, it is necessarily non-extravagant, and by § 172 we can determine sets of $m-r$ unconnected mutually orthogonal unit solutions of the equation

$$[x]_r^m \, \overline{\underline{y}}_m = 0.$$

If $[y_1\, y_2 \ldots y_m]=[x_{r+1,1}\, x_{r+1,2} \ldots x_{r+1,m}],\ \ldots\ [x_{m1}\, x_{m2} \ldots x_{mm}]$ is any one such set of $m-r$ solutions, then $[x]_m^m$ is a square semi-unit matrix; and if $s>r$ and $\not> m$, then $[x]_s^m$ is a semi-unit matrix of rank s formed by adding $s-r$ final horizontal rows to $[x]_r^m$.

Putting $r=1$, we can in this way determine all semi-unit matrices $[x]_s^m$ of rank s.

Ex. xv. Every semi-unit matrix which is not square is a simple minor of a square semi-unit matrix.

Ex. xvi. From the equation

$$\begin{bmatrix} 1, & 1, & i \\ -1, & 2, & i \\ -i, & -2i, & 3 \end{bmatrix} \begin{bmatrix} 1, & -1, & -i \\ 1, & 2, & -2i \\ i, & i, & 3 \end{bmatrix} = \begin{bmatrix} 1, & 0, & 0 \\ 0, & 4, & 0 \\ 0, & 0, & 4 \end{bmatrix}$$

in which $i=\sqrt{-1}$, we derive the semi-unit matrix

$$\begin{bmatrix} l_1 & m_1 & n_1 \\ l_2 & m_2 & n_2 \\ l_3 & m_3 & n_3 \end{bmatrix} = \frac{1}{2} \begin{bmatrix} 2, & 2, & 2i \\ -1, & 2, & i \\ -i, & -2i, & 3 \end{bmatrix}.$$

Ex. xvii. Particular solutions of the equation of Ex. xii can be derived from the equations

$$\begin{bmatrix} 3, & 0, & 0, & 2i \\ 4, & 3, & 0, & 6i \\ 6, & 10, & i, & 9i \\ 6, & 10, & 55i, & 9i \end{bmatrix} \begin{bmatrix} 3, & 4, & 6, & 6 \\ 0, & 3, & 10, & 10 \\ 0, & 0, & i, & 55i \\ 2i, & 6i, & 9i, & 9i \end{bmatrix} = \begin{bmatrix} 5, & 0, & 0, & 0 \\ 0, & -11, & 0, & 0 \\ 0, & 0, & 54, & 0 \\ 0, & 0, & 0, & -2889 \end{bmatrix},$$

$$\begin{bmatrix} 1, & 5, & 3i, & 4i \\ 2, & 1, & i, & i \\ 5, & 17, & 18i, & 9i \\ 17, & 76, & 43i, & 67i \end{bmatrix} \begin{bmatrix} 1, & 2, & 5, & 17 \\ 5, & 1, & 17, & 76 \\ 3i, & i, & 18i, & 43i \\ 4i, & i, & 9i, & 67i \end{bmatrix} = \begin{bmatrix} 1, & 0, & 0, & 0 \\ 0, & 3, & 0, & 0 \\ 0, & 0, & -91, & 0 \\ 0, & 0, & 0, & -273 \end{bmatrix},$$

in which $i=\sqrt{-1}$.

Ex. xviii. *If a square semi-unit matrix* $[l]_r^r$ *has the form* $\begin{bmatrix} a, & ia \\ i\beta, & b \end{bmatrix}_{u,\,r-u}^{v,\,r-v}$, *where* $i=\sqrt{-1}$ *and* $[a]_u^v$, $[b]_{r-u}^{r-v}$, $[a]_u^{r-v}$, $[\beta]_{r-u}^{v}$ *are all real, then* $v=u$.

From the equation $[l]_r^r \, \underline{\overline{l}}_r^{\,r} = [1]_r^r$, it follows that

$$[a,\ ia]_u^{v,\,r-v}\ \underline{\overline{\begin{matrix} a \\ ia \end{matrix}}}_{v,\,r-v}^{\,u} = [1]_u^u, \quad [i\beta,\ b]_{r-u}^{v,\,r-v}\ \underline{\overline{\begin{matrix} i\beta \\ b \end{matrix}}}_{v,\,r-v}^{\,r-u} = [1]_{r-u}^{r-u},$$

or

$$[a]_u^v\, \underline{\overline{a}}_v^{\,u} = [1,\ a]_u^{u,\,r-v}\ \underline{\overline{\begin{matrix} 1 \\ a \end{matrix}}}_{u,\,r-v}^{\,u}, \quad [b]_{r-u}^{r-v}\, \underline{\overline{b}}_{r-v}^{\,r-u} = [1,\ \beta]_{r-u}^{r-u,\,v}\ \underline{\overline{\begin{matrix} 1 \\ \beta \end{matrix}}}_{r-u,\,v}^{\,r-u}. \quad \ldots\ldots(6)$$

In the first of the equations (6) the real matrix $[1,\ a]_u^{u,\,r-v}$ has rank u; therefore, by § 72.1 the product matrix on the right has rank u, and the real matrix $[a]_u^v$ has rank u. Thus $u \not> v$.

Similarly the second of the equations (6) shows that $[b]^{r-v}_{r-u}$ has rank $r-u$. Therefore $r-u \not> r-v$, i.e. $u \not< v$. Consequently we have $v=u$.

This theorem is illustrated in Ex. xvi.

Ex. xix. *Reciprocal of a square semi-unit matrix.*

Let $[L]^m_m$ be the reciprocal of the square semi-unit matrix $[l]^m_m$, and let $(l)^m_m=\Delta$.

Then $$[l]^m_m \overline{\underline{l}}{}^m_m=[1]^m_m, \quad [l]^m_m \overline{\underline{L}}{}^m_m=\Delta[1]^m_m.$$

Equating the determinants of both sides in the first equation, we obtain

$$\Delta^2=1, \quad \Delta=\pm 1.$$

If $\Delta=+1$, then $[l]^m_m \overline{\underline{L}}{}^m_m=[l]^m_m \overline{\underline{l}}{}^m_m$, and therefore by § 84

$$[L]^m_m=[l]^m_m.$$

If $\Delta=-1$, then $[l]^m_m \overline{\underline{L}}{}^m_m=-[l]^m_m \overline{\underline{l}}{}^m_m$, and therefore by § 84

$$[L]^m_m=-[l]^m_m.$$

Thus the reciprocal of $[l]^m_m$ *is* $\pm[l]^m_m$ *according as* $(l)^m_m=\pm 1$. In the former case the semi-unit matrix $[l]^m_m$ is self-reciprocal.

It may be observed that when $(l)^m_m=-1$, we can by changing the signs of all elements in any one of the rows of $[l]^m_m$ obtain a semi-unit matrix whose determinant has the value $+1$.

Ex. xx. *General solution of the equation*

$$\begin{bmatrix} l_1 & m_1 \\ l_2 & m_2 \end{bmatrix}\begin{bmatrix} l_1 & l_2 \\ m_1 & m_2 \end{bmatrix}=\begin{bmatrix} 1 & 0 \\ 0 & 1 \end{bmatrix}.$$

This equation is equivalent to the three scalar equations

$$l_1^2+m_1^2=1, \quad l_2^2+m_2^2=1, \quad l_1l_2+m_1m_2=0.$$

From the first of these equations we see that l_1 and m_1 are not both zero. Writing $l_1=\lambda$, $m_1=\mu$, we find that

$$\frac{l_2}{\mu}=-\frac{m_2}{\lambda}=k=\pm 1.$$

It follows that the general solution of the above equation can be expressed in either of the equivalent forms

$$\begin{bmatrix} l_1 & m_1 \\ l_2 & m_2 \end{bmatrix}=\begin{bmatrix} \lambda, & \mu \\ \mp\mu, & \pm\lambda \end{bmatrix}, \quad \begin{bmatrix} l_1 & m_1 \\ l_2 & m_2 \end{bmatrix}=\begin{bmatrix} \lambda, & \mp\mu \\ \mu, & \pm\lambda \end{bmatrix}, \quad \text{.....................(A)}$$

where $$\lambda^2+\mu^2=1.$$

It can also be expressed in either of the equivalent forms

$$\begin{bmatrix} l_1 & m_1 \\ l_2 & m_2 \end{bmatrix}=\begin{bmatrix} \cos\phi, & \sin\phi \\ \mp\sin\phi, & \pm\cos\phi \end{bmatrix}, \quad \begin{bmatrix} l_1 & m_1 \\ l_2 & m_2 \end{bmatrix}=\begin{bmatrix} \cos\phi, & \mp\sin\phi \\ \sin\phi, & \pm\cos\phi \end{bmatrix}, \quad \text{.........(B)}$$

where ϕ is not necessarily real.

The upper signs give the solutions in which $(lm)_{12}=+1$; and the lower signs the solutions in which $(lm)_{12}=-1$.

Ex. xxi. *General solution of the equation*

$$\begin{bmatrix} l_1 & m_1 & n_1 \\ l_2 & m_2 & n_2 \\ l_3 & m_3 & n_3 \end{bmatrix} \begin{bmatrix} l_1 & l_2 & l_3 \\ m_1 & m_2 & m_3 \\ n_1 & n_2 & n_3 \end{bmatrix} = \begin{bmatrix} 1 & 0 & 0 \\ 0 & 1 & 0 \\ 0 & 0 & 1 \end{bmatrix}.$$

The general solution in which $(lmn)_{123} = +1$ is given by

$$\begin{bmatrix} l_1 & m_1 & n_1 \\ l_2 & m_2 & n_2 \\ l_3 & m_3 & n_3 \end{bmatrix}$$

$$= \begin{bmatrix} \cos\phi\cos\psi, & \cos\theta\sin\psi + \sin\theta\sin\phi\cos\psi, & \sin\theta\sin\psi - \cos\theta\sin\phi\cos\psi \\ -\cos\phi\sin\psi, & \cos\theta\cos\psi - \sin\theta\sin\phi\sin\psi, & \sin\theta\cos\psi + \cos\theta\sin\phi\sin\psi \\ \sin\phi \quad , & -\sin\theta\cos\phi \quad , & \cos\theta\cos\phi \end{bmatrix}$$

$$= \begin{bmatrix} \cos\psi, & \sin\psi, & 0 \\ -\sin\psi, & \cos\psi, & 0 \\ 0 \; , & 0 \; , & 1 \end{bmatrix} \begin{bmatrix} \cos\phi, & 0, & -\sin\phi \\ 0 \; , & 1, & 0 \\ \sin\phi, & 0, & \cos\phi \end{bmatrix} \begin{bmatrix} 1, & 0 \; , & 0 \\ 0, & \cos\theta, & \sin\theta \\ 0, & -\sin\theta, & \cos\theta \end{bmatrix} \quad . \; \ldots.(\mathrm{C})$$

The general solution in which $(lmn)_{123} = -1$ is obtained by replacing the element 1 by -1 in any one of the factor matrices on the right, or by changing the sign of every element in any one of the horizontal or vertical rows in the product matrix on the right.

In these formulae θ, ϕ and ψ are not necessarily real. We obtain the real solutions of the equation when they are all real.

We find this general solution by the method described in the text.

First choosing $[l_3 m_3 n_3]$ to be an arbitrary unit matrix, we can write

$$[l_3 m_3 n_3] = [\sin\phi, \; -\sin\theta\cos\phi, \; \cos\theta\cos\phi].$$

We then determine $[l_2 m_2 n_2]$ to be an arbitrary matrix satisfying the conditions

$$l_2 l_3 + m_2 m_3 + n_2 n_3 = 0, \quad l_2^2 + m_2^2 + n_2^2 = 1.$$

We can ascribe to l_2 the arbitrary value $-\cos\phi\sin\psi$, and determine m_2 and n_2 by direct solution of the above two equations.

The matrix $[l_1 m_1 n_1]$ is then completely determinate, for we must have

$$[l_1 m_1 n_1] = \pm[(mn)_{23}, \; (nl)_{23}, \; (lm)_{23}].$$

The general solution in which $(lmn)_{123} = +1$ is also given by

$$\begin{bmatrix} l_1 & m_1 & n_1 \\ l_2 & m_2 & n_2 \\ l_3 & m_3 & n_3 \end{bmatrix}$$

$$= \begin{bmatrix} -\sin\phi\sin\psi + \cos\theta\cos\phi\cos\psi, & \sin\phi\cos\psi + \cos\theta\cos\phi\sin\psi, & -\sin\theta\cos\phi \\ -\cos\phi\sin\psi - \cos\theta\sin\phi\cos\psi, & \cos\phi\cos\psi - \cos\theta\sin\phi\sin\psi, & \sin\theta\sin\phi \\ \sin\theta\cos\psi \quad , & \sin\theta\sin\psi \quad , & \cos\theta \end{bmatrix}$$

$$= \begin{bmatrix} \cos\phi, & \sin\phi, & 0 \\ -\sin\phi, & \cos\phi, & 0 \\ 0 \; , & 0 \; , & 1 \end{bmatrix} \begin{bmatrix} \cos\theta, & 0, & -\sin\theta \\ 0 \; , & 1, & 0 \\ \sin\theta, & 0, & \cos\theta \end{bmatrix} \begin{bmatrix} \cos\psi, & \sin\psi, & 0 \\ -\sin\psi, & \cos\psi, & 0 \\ 0 \; , & 0 \; , & 1 \end{bmatrix} \quad . \; \ldots\ldots\ldots\ldots.(\mathrm{D})$$

The general solution in which $(lmn)_{123} = -1$ is then again obtained by replacing the element 1 by -1 in any one of the factor matrices on the right, or by changing the sign of every element in any one of the horizontal or vertical rows in the product matrix on the right.

First choosing $[l_3 m_3 n_3]$ to be an arbitrary unit matrix, we can write

$$[l_3 m_3 n_3] = [\sin\theta\cos\psi,\ \sin\theta\sin\psi,\ \cos\theta].$$

We then determine $[l_2\ m_2\ n_2]$, any unit matrix orthogonal with $[l_3\ m_3\ n_3]$, ascribing to n_2 the arbitrary value $\sin\theta\sin\phi$. The matrix $[l_1\ m_1\ n_1]$ is then completely determinate.

Ex. xxii. *The corresponding rotations of a rigid body in common 3-way space.*

If $[l\,m\,n]_{123}$ is any square semi-unit matrix of order 3 whose determinant has the value $+1$, then, as will be shown more fully in a later chapter, the equation

$$\begin{bmatrix} x' \\ y' \\ z' \end{bmatrix} = \begin{bmatrix} l_1 & l_2 & l_3 \\ m_1 & m_2 & m_3 \\ n_1 & n_2 & x_3 \end{bmatrix} \begin{bmatrix} x \\ y \\ z \end{bmatrix} \quad \text{.............................(E)}$$

represents the most general rotation of a rigid body about the origin in common 3-way space. Here (x, y, z) are the co-ordinates of any point P of the body referred to a set of three mutually perpendicular co-ordinate axes (OX, OY, OZ) fixed in space, and (x', y', z') are the co-ordinates with reference to the same axes of the point P' to which P is brought by the rotation. If the rotation would bring (OX, OY, OZ), when we regard these lines as fixed in the body, into the positions (OX_1, OY_1, OZ_1), then (l_1, m_1, n_1), (l_2, m_2, n_2), (l_3, m_3, n_3) are the direction-cosines of OX_1, OY_1, OZ_1 with reference to the axes (OX, OY, OZ). Hence if the co-ordinates of P and P' with reference to the axes (OX_1, OY_1, OZ_1) are (x_1, y_1, z_1) and (x_1', y_1', z_1'), we have

$$\begin{bmatrix} x \\ y \\ z \end{bmatrix} = \begin{bmatrix} l_1 & l_2 & l_3 \\ m_1 & m_2 & m_3 \\ n_1 & n_2 & n_3 \end{bmatrix} \begin{bmatrix} x_1 \\ y_1 \\ z_1 \end{bmatrix}, \quad \begin{bmatrix} x' \\ y' \\ z' \end{bmatrix} = \begin{bmatrix} l_1 & l_2 & l_3 \\ m_1 & m_2 & m_3 \\ n_1 & n_2 & n_3 \end{bmatrix} \begin{bmatrix} x_1' \\ y_1' \\ z_1' \end{bmatrix}. \quad \text{....(F)}$$

We can deduce (E) from the formulae of transformation (F); for by the nature of a rotation the co-ordinates of P' with reference to the axes (OX_1, OY_1, OZ_1) are the same as the co-ordinates of P with reference to the axes (OX, OY, OZ), i.e. we have $x_1' = x$, $y_1' = y$, $z_1' = z$.

First let

$$\Theta = \begin{bmatrix} 1, & 0\ , & 0 \\ 0, & \cos\theta, & -\sin\theta \\ 0, & \sin\theta, & \cos\theta \end{bmatrix}, \quad \Phi = \begin{bmatrix} \cos\phi, & 0, & \sin\phi \\ 0\ , & 1, & 0 \\ -\sin\phi, & 0, & \cos\phi \end{bmatrix}, \quad \Psi = \begin{bmatrix} \cos\psi, & -\sin\psi, & 0 \\ \sin\psi, & \cos\psi, & 0 \\ 0\ , & 0\ , & 1 \end{bmatrix};$$

and let

$$P = \begin{bmatrix} x \\ y \\ z \end{bmatrix}, \quad P' = \begin{bmatrix} x' \\ y' \\ z' \end{bmatrix}.$$

Then if we express $[l\,m\,n]_{123}$ in the form (C), the equation (E) is

$$P' = \Theta\Phi\Psi P. \quad \text{..(7)}$$

Thus the rotation (E) is the resultant of the three successive rotations

$$P' = \Psi P, \quad P' = \Phi P, \quad P' = \Theta P,$$

which are rotations through angles ψ, ϕ, θ about OZ, OY, OX when these are regarded as axes fixed in space. By Ex. xxiii it follows that the rotation (E) is also the resultant of three successive rotations through angles θ, ϕ, ψ about OX, OY, OZ when these are regarded as axes fixed in the body.

Accordingly formula (C) shows that :

Every rotation of a rigid body about a point O in common 3-way space is the resultant of three successive rotations about any three mutually perpendicular axes which pass through O and are either fixed in space or fixed in the body.

Next let the matrix $[l\,m\,n]_{123}$ be expressed in the form (D), and let

$$\Theta = \begin{bmatrix} \cos\theta, & 0, & \sin\theta \\ 0, & 1, & 0 \\ -\sin\theta, & 0, & \cos\theta \end{bmatrix}, \quad \Phi = \begin{bmatrix} \cos\phi, & -\sin\phi, & 0 \\ \sin\phi, & \cos\phi, & 0 \\ 0, & 0, & 1 \end{bmatrix}, \quad \Psi = \begin{bmatrix} \cos\psi, & -\sin\psi, & 0 \\ \sin\psi, & \cos\psi, & 0 \\ 0, & 0, & 1 \end{bmatrix}.$$

Then the equation (E) is

$$P' = \Psi\Theta\Phi P. \qquad \text{.............(8)}$$

Accordingly the rotation (E) is the resultant of three successive rotations through angles ϕ, θ, ψ about the axes OZ, OY, OZ regarded as fixed in space, and also the resultant of three successive rotations through angles ψ, θ, ϕ about the axes OZ, OY, OZ regarded as fixed in the body.

In this second case θ, ϕ, ψ are the three *Eulerian* angles defining the rotation.

Ex. xxiii. **Theorem.** *If OA, OB, OC are any three straight lines passing through O, successive rotations of a rigid body in common 3-way space through angles θ, ϕ, ψ about the axes OA, OB, OC regarded as fixed in space are equivalent to successive rotations of the body through angles ψ, ϕ, θ about the axes OC, OB, OA regarded as fixed in the body.*

Let (OX, OY, OZ) be a set of three mutually perpendicular co-ordinate axes which pass through O and are fixed in space ; let

$$P = \begin{bmatrix} x \\ y \\ z \end{bmatrix}, \quad P' = \begin{bmatrix} x' \\ y' \\ z' \end{bmatrix}, \quad P_1 = \begin{bmatrix} x_1 \\ y_1 \\ z_1 \end{bmatrix}, \quad P_2 = \begin{bmatrix} x_2 \\ y_2 \\ z_2 \end{bmatrix},$$

and so on; let the equations of the rotations θ about OA, ϕ about OB, ψ about OC referred to (OX, OY, OZ) as co-ordinate axes when OA, OB, OC are regarded as fixed in space be

$$P' = \Theta P, \quad P' = \Phi P, \quad P' = \Psi P, \qquad \text{.............(9)}$$

where Θ, Φ, Ψ are square semi-unit matrices of order 3 whose determinants all have the value $+1$; and let Θ', Φ', Ψ' be the inverses (or conjugates) of Θ, Φ, Ψ.

Then the successive rotations θ, ϕ, ψ about the axes OA, OB, OC regarded as fixed in space are equivalent to the resultant rotation

$$P' = \Psi\Phi\Theta P. \qquad \text{.............(10)}$$

Let the rotation ψ about OC bring (OX, OY, OZ, OB, OA) when we regard these lines as fixed in the body into the positions $(OX_1, OY_1, OZ_1, OB_1, OA_1)$; and let the rotation ϕ about OB_1 bring (OX_1, OY_1, OZ_1, OA_1) when we regard these lines as fixed in the body into the positions (OX_2, OY_2, OZ_2, OA_2). Then the equations of the rotations ψ about

OC, ϕ about OB_1, θ about OA_2 referred respectively to (OX, OY, OZ), (OX_1, OY_1, OZ_1), (OX_2, OY_2, OZ_2) as axes of co-ordinates are

$$P' = \Psi P, \quad P_1' = \Phi P_1, \quad P_2' = \Theta P_2. \qquad \text{.............................(11)}$$

If (x, y, z), (x_1, y_1, z_1), (x_2, y_2, z_2) are the co-ordinates of any assigned point of space with reference to the axes (OX, OY, OZ), (OX_1, OY_1, OZ_1), (OX_2, OY_2, OZ_2) respectively, we have as in the first of the equations (F)

$$P = \Psi P_1, \quad P_1 = \Psi' P; \quad P_1 = \Phi P_2, \quad P_2 = \Phi' P_1; \quad P = \Psi\Phi P_2, \quad P_2 = \Phi'\Psi' P;$$

and therefore the equations of the rotations ψ about OC, ϕ about OB_1, θ about OA_2 referred to (OX, OY, OZ) as axes of co-ordinates are

$$P' = \Psi P, \quad P' = \Psi\Phi\Psi' P, \quad P' = \Psi\Phi\Theta\Phi'\Psi' P. \qquad \text{........................(11')}$$

Now let the rotation ψ about OC change the point (x, y, z) into the point (x', y', z'), let the rotation ϕ about OB_1 change the point (x', y', z') into the point (x'', y'', z''), and let the rotation θ about OA_2 change the point (x'', y'', z'') into the point (x''', y''', z'''), the co-ordinates in each case being those with reference to the fixed axes of co-ordinates (OX, OY, OZ). Then from (11') we see that

$$P' = \Psi P, \quad P'' = \Psi\Phi\Psi' P' = \Psi\Phi P, \quad P''' = \Psi\Phi\Theta\Phi'\Psi' P'' = \Psi\Phi\Theta P.$$

Thus the successive rotations ψ, ϕ, θ about the axes OC, OB_1, OA_2 regarded as fixed in space, i.e. the successive rotations ψ, ϕ, θ about the axes OC, OB, OA regarded as fixed in the body, are equivalent to the resultant rotation

$$P' = \Psi\Phi\Theta P. \qquad \text{...(10')}$$

This being the same resultant rotation as (10), we see that the theorem is true.

In the same way we can show that:

If OA_1, OA_2, ... OA_{r-1}, OA_r are any r straight lines passing through O, successive rotations of a rigid body in common 3-way space through angles θ_1, θ_2, ... θ_{r-1}, θ_r about the axes OA_1, OA_2, ... OA_{r-1}, OA_r regarded as fixed in space are equivalent to successive rotations through angles θ_r, θ_{r-1}, ... θ_2, θ_1 about the axes OA_r, OA_{r-1}, ... OA_2, OA_1 regarded as fixed in the body.

The special case when there are only two axes of rotation is equivalent to *Rodrigues' Theorem* concerning finite rotations; and the general case follows by induction from this special case.

Ex. xxiv. *If the rotation ϕ about OB brings OA into the position OA_1, then successive rotations of a rigid body through angles $-\phi$, θ, ϕ about the axes OB, OA, OB regarded as fixed in space are equivalent to a rotation through an angle θ about OA_1.*

An analytical proof of this theorem is contained in Ex. xxiii, as we see by comparing the equation (10) with the second of the equations (11'). The theorem can also be proved as follows.

Let a sphere with centre O cut OA, OB, OA_1 in A, B, A_1. Then the rotations $-\phi$ about OB, θ about OA, ϕ about OB carry A_1 to A, leave A unaltered in position, carry A to A_1. Therefore the resultant of these rotations leaves A_1 unaltered in position, and OA_1 is the axis of the resultant rotation.

Again let the rotations θ about OA, ϕ about OB carry B to B_2, B_2 to B_3, so that the resultant of the above three rotations carries B to B_3. Then the spherical triangles A_1BB_3, ABB_2 are equal in all respects, since $A_1B = AB$, $BB_3 = BB_2$, $\angle A_1BB_3 = \angle ABB_2$. It follows that $\angle BA_1B_3 = \angle BAB_2 = \theta$, i.e. B is carried to B_3 by a rotation θ about OA_1.

§ 156. Some special symmetric equations of the form $X'X = A'A$, where A, A' and X, X' are pairs of mutually conjugate matrices.

1. *The equation* $\lceil x \rfloor_m [x]_m = \lceil a \rfloor_m [a]_m$, ...(A)

i.e.
$$\begin{bmatrix} x_1 \\ x_2 \\ \vdots \\ x_m \end{bmatrix} [x_1\, x_2 \ldots x_m] = \begin{bmatrix} a_1 \\ a_2 \\ \vdots \\ a_m \end{bmatrix} [a_1\, a_2 \ldots a_m].$$

We shall distinguish two cases.

CASE I. *When the given matrices on the right have rank* 1.

Let a_i be a non-vanishing element of $[a]_m$. Then by equating corresponding diagonal elements of the product matrices in (A) we see that $x_i^2 = a_i^2$, $x_i = \pm a_i \neq 0$. Further by equating the ith horizontal rows of the two product matrices we obtain

$$x_i [x_1\, x_2 \ldots x_m] = a_i [a_1\, a_2 \ldots a_m], \quad [x_1\, x_2 \ldots x_m] = \pm [a_1\, a_2 \ldots a_m],$$

and the last equation gives the only possible solutions of (A).

CASE II. *When the given matrices on the right have rank* 0.

In this case the right-hand side of (A) vanishes, and by equating the diagonal elements of both sides we see that the only solution is

$$[x_1\, x_2 \ldots x_m] = 0.$$

Thus in all cases the general solution of the equation (A) *is*

$$[x_1\, x_2 \ldots x_m] = \pm [a_1\, a_2 \ldots a_m]. \quad \text{.....................(A')}$$

Ex. i. *The equation* $p^2 \lceil x \rfloor_m [x]_m = q^2 \lceil a \rfloor_m [a]_m$, *where* $p \neq 0$.

If we write $p[x]_m = [\xi]_m$, $q[a]_m = [\alpha]_m$, the equation is equivalent to

$$\lceil \xi \rfloor_m [\xi]_m = \lceil \alpha \rfloor_m [\alpha]_m,$$

and is satisfied when and only when $[\xi]_m = \pm [\alpha]_m$.

Thus the general solution is given by

$$p[x_1\, x_2 \ldots x_m] = \pm q [a_1\, a_2 \ldots a_m].$$

Ex. ii. The general solution of the equation $\lceil x \rfloor_m [x]_m = -\lceil a \rfloor_m [a]_m$ is

$$[x_1\, x_2 \ldots x_m] = \pm \sqrt{-1}\, [a_1\, a_2 \ldots a_m].$$

Ex. iii. *The equation* $\overline{\underline{x}}_{m}\,[x]_{m} = \overline{\underline{a}}_{m}\,[b]_{m}$.

Case I. *When both the given matrices on the right have rank* 1.

By § 152.1 the equation can only be satisfied when

$$[x]_{m} = \lambda\,[a]_{m} = \mu\,[b]_{m}, \quad \text{where } \lambda\mu = 1.$$

Thus a necessary condition for the possibility of solution is the existence of a relation of the form

$$[b_1\, b_2 \dots b_m] = k\,[a_1\, a_2 \dots a_m], \quad \text{where } k \neq 0.$$

When this condition is satisfied, the general solution is

$$[x_1\, x_2 \dots x_m] = \pm\sqrt{k}\,[a_1\, a_2 \dots a_m].$$

Case II. *When one of the given matrices on the right has rank* 0.

In this case the only solution is $[x_1\, x_2 \dots x_m] = 0$.

Ex. iv. *The equation* $\begin{bmatrix} x_1 & x_1 \\ x_2 & x_2 \\ \dots & \dots \\ x_m & x_m \end{bmatrix} \begin{bmatrix} x_1 & x_2 \dots x_m \\ x_1 & x_2 \dots x_m \end{bmatrix} = k \begin{bmatrix} a_1 & a_1 \\ a_2 & a_2 \\ \dots & \dots \\ a_m & a_m \end{bmatrix} \begin{bmatrix} a_1 & a_2 \dots a_m \\ a_1 & a_2 \dots a_m \end{bmatrix}$.

By the properties of passive rows this equation is equivalent to

$$\overline{\underline{x}}_{m}\,[x]_{m} + \overline{\underline{x}}_{m}\,[x]_{m} = k\,\overline{\underline{a}}_{m}\,[a]_{m} + k\,\overline{\underline{a}}_{m}\,[a]_{m}, \quad \text{or} \quad \overline{\underline{x}}_{m}\,[x]_{m} = k\,\overline{\underline{a}}_{m}\,[a]_{m},$$

and the general solution is

$$[x_1\, x_2 \dots x_m] = \pm\sqrt{k}\,[a_1\, a_2 \dots a_m].$$

Ex. v. *The equation* $\begin{bmatrix} x_1 & y_1 \\ x_2 & y_2 \\ \dots & \dots \\ x_m & y_m \end{bmatrix} \begin{bmatrix} x_1 & x_2 \dots x_m \\ y_1 & y_2 \dots y_m \end{bmatrix} = 0$.

This equation is equivalent to $\overline{\underline{x}}_{m}\,[x]_{m} = -\overline{\underline{y}}_{m}\,[y]_{m}$, and by Ex. ii it is satisfied when and only when

$$[x_1\, x_2 \dots x_m] = \pm\sqrt{-1}\,[y_1\, y_2 \dots y_m],$$

or

$$\begin{bmatrix} x_1 & x_2 \dots x_m \\ y_1 & y_2 \dots y_m \end{bmatrix} = \begin{bmatrix} l \\ m \end{bmatrix} [\xi_1\, \xi_2 \dots \xi_m],$$

where $l^2 + m^2 = 0$, and $[\xi]_{m}$ is arbitrary.

The only real solution is $\begin{bmatrix} x_1 & x_2 \dots x_m \\ y_1 & y_2 \dots y_m \end{bmatrix} = 0$.

2. *The equation* $\overline{\underline{x}}^{\,r}_{m}\,[x]^{m}_{r} = \overline{\underline{a}}^{\,r}_{m}\,[a]^{m}_{r}$, ...(B)

in which $[a]^{m}_{r}$ *is a given matrix of rank* r.

Since the product matrix on the right has rank r, every solution $[x]^{m}_{r}$ of the equation (B) must have rank r. For when the equation is satisfied, the product matrix on the left must have rank r, and this is only possible when $[x]^{m}_{r}$ has rank r.

Again when the equation is satisfied, we see by prefixing an inverse prefactor of $\overline{\underline{x}}^{\,r}_{\,m}$ on both sides that there must exist a relation of the form

$$[x]^m_r = [l]^r_r\,[a]^m_r, \quad\text{.............................(1)}$$

where the square matrix $[l]^r_r$ is necessarily undegenerate. This can also be seen by equating correspondingly formed complete matrices of the minor determinants of order r of both sides, solving the resulting equation by sub-article 1, and making use of § 113 as in the note to § 152.2.

Substituting the value of $[x]^m_r$ given by (1) in (B), we see by § 84 that the equation is satisfied if and only if $\overline{\underline{l}}^{\,r}_{\,r}\,[l]^r_r = [1]^r_r$, i.e. if and only if $[l]^r_r$ is a square semi-unit matrix of rank r. All the possible values of $[l]^r_r$ can be found as in § 155.

Thus the general solution of the equation (B) *is*

$$[x]^m_r = [l]^r_r\,[a]^m_r, \quad\text{..........................(B')}$$

where $[l]^r_r$ *is any square semi-unit matrix of rank* r.

This general solution can also be deduced from § 152.2.

When $[a]^m_r$ has rank less than r, the equation (B) can be treated as in § 160.4.

Ex. vi. *The solutions of the equation* (B) *are all non-singular or singular according as* $[a]^m_r$ *is non-singular or singular.*

This is seen by equating the determinoids of both sides in (B').

Ex. vii. *The equation* $p\,\overline{\underline{x}}^{\,r}_{\,m}\,[x]^m_r = q\,\overline{\underline{a}}^{\,r}_{\,m}\,[a]^m_r$ *in which* $p \neq 0$, $q \neq 0$, *and* $[a]^m_r$ *has rank* r.

The general solution is given by

$$\sqrt{p}\,[x]^m_r = \sqrt{q}\,[l]^r_r\,[a]^m_r,$$

where $[l]^r_r$ is a square semi-unit matrix.

Ex. viii. The general solution of the equation

$$\begin{bmatrix} x_1 & x_2 \\ y_1 & y_2 \\ z_1 & z_2 \\ w_1 & w_2 \end{bmatrix}\begin{bmatrix} x_1 & y_1 & z_1 & w_1 \\ x_2 & y_2 & z_2 & w_2 \end{bmatrix} = \begin{bmatrix} a_1 & a_2 \\ b_1 & b_2 \\ c_1 & c_2 \\ d_1 & d_2 \end{bmatrix}\begin{bmatrix} a_1 & b_1 & c_1 & d_1 \\ a_2 & b_2 & c_2 & d_2 \end{bmatrix},$$

in which $[a\,b\,c\,d]_{12}$ is a given matrix of rank 2, is

$$\begin{bmatrix} x_1 & y_1 & z_1 & w_1 \\ x_2 & y_2 & z_2 & w_2 \end{bmatrix} = \begin{bmatrix} l_1 & l_2 \\ m_1 & m_2 \end{bmatrix}\begin{bmatrix} a_1 & b_1 & c_1 & d_1 \\ a_2 & b_2 & c_2 & d_2 \end{bmatrix}, \text{ where } \begin{bmatrix} l_1 & m_1 \\ l_2 & m_2 \end{bmatrix}\begin{bmatrix} l_1 & l_2 \\ m_1 & m_2 \end{bmatrix} = \begin{bmatrix} 1 & 0 \\ 0 & 1 \end{bmatrix}.$$

By Ex. xx of § 155 it can be expressed in either of the forms

$$\begin{bmatrix} x_1 & y_1 & z_1 & w_1 \\ x_2 & y_2 & z_2 & w_2 \end{bmatrix} = \begin{bmatrix} \lambda, & \mp\mu \\ \mu, & \pm\lambda \end{bmatrix} \begin{bmatrix} a_1 & b_1 & c_1 & d_1 \\ a_2 & b_2 & c_2 & d_2 \end{bmatrix}, \text{ where } \lambda^2+\mu^2=1;$$

$$\begin{bmatrix} x_1 & y_1 & z_1 & w_1 \\ x_2 & y_2 & z_2 & w_2 \end{bmatrix} = \begin{bmatrix} \cos\phi, & \mp\sin\phi \\ \sin\phi, & \pm\cos\phi \end{bmatrix} \begin{bmatrix} a_1 & b_1 & c_1 & d_1 \\ a_2 & b_2 & c_2 & d_2 \end{bmatrix}.$$

§ 157. Solutions of the equation $\overline{\underline{x}}{}^{s}_{m}\,[x]^{m}_{s} = \overline{\underline{a}}{}^{r}_{m}\,[a]^{m}_{r}$ in which $[a]^{m}_{r}$ is a given matrix of rank *r*.

1. *Distinct solutions.*

If $[x]^{m}_{s}$ is any matrix satisfying the equation

$$\overline{\underline{x}}{}^{s}_{m}\,[x]^{m}_{s} = \overline{\underline{a}}{}^{r}_{m}\,[a]^{m}_{r}, \qquad \text{(A)}$$

and if $[x']^{m}_{s}$ is any matrix derived from $[x]^{m}_{s}$ by deranging its horizontal rows only, i.e. if

$$[x']^{m}_{s} = [\omega]^{s}_{s}\,[x]^{m}_{s}, \qquad \text{(1)}$$

where $[\omega]^{s}_{s}$ is a derangement of the unit matrix $[1]^{s}_{s}$, then $[x']^{m}_{s}$ is also a matrix satisfying the equation (A). Two solutions $[x]^{m}_{s}$, $[x']^{m}_{s}$ will be regarded as distinct from one another when and only when each of these matrices is not merely a horizontal derangement of the other, i.e. when and only when there does not exist a relation of the form (1).

2. *Limits to the ranks of the solutions.*

If $[a]^{m}_{r}$ has rank r, the product matrix on the right in (A) is a given matrix of rank r. Hence if the equation has a solution $[x]^{m}_{s}$ of rank ρ, we know by § 133 that ρ must be an integer consistent with the necessary conditions

$$r \nless 0, \quad r \ngtr \rho, \quad r+s \nless 2\rho; \quad \rho \ngtr m \qquad \text{(2)}$$

which include $r \ngtr m$, $r \ngtr s$ and $\rho \ngtr s$.

Hence $s \nless r$ is a necessary condition for the existence of solutions; as is otherwise clear from the fact that when $s < r$, the rank of the product matrix on the left cannot exceed s and therefore cannot be equal to r, whereas the product matrix on the right has rank r.

Again if there is a solution of rank ρ, the integer ρ must satisfy the conditions

$$\rho \nless r, \quad \rho \ngtr m, \quad 2\rho \ngtr r+s, \qquad \text{(3)}$$

which include $s \nless r$, $\rho \ngtr s$.

NOTE. *Direct determination of the necessary condition* $2\rho \not> r+s$.

If the equation (A) has a solution $[x]^m_s$ of rank ρ, where necessarily $\rho \not> s$, it must have a solution $[x]^m_s$ of rank ρ in which $[x]^m_\rho$ has rank ρ. Let this be $[x]^m_s = \begin{bmatrix} x \\ i\xi \end{bmatrix}^m_{\rho,\, s-\rho}$, where $i = \sqrt{-1}$. Then by the properties of passive rows we see from (A) that

$$\overline{\underline{x}}^{\,\rho}_{m} [x]^m_\rho = \overline{\underline{a,\ \xi}}^{\,r,\, s-\rho}_{m} \begin{bmatrix} a \\ \xi \end{bmatrix}^m_{r,\, s-\rho}.$$

Since the product matrix on the left has rank ρ, the passivity of the product on the right cannot be less than ρ, i.e. we must have

$$r+s-\rho \not< \rho, \quad \text{or} \quad 2\rho \not> r+s.$$

3. *Limits to the ranks of real solutions.*

If $[x]^m_s$ is a matrix of rank ρ satisfying the equation (A), and if $\rho > r$, then from (A) we deduce by § 66 that

$$\overline{\underline{X}}^{\,\sigma}_{\mu} [X]^\mu_\sigma = 0,$$

where $[X]^\mu_\sigma$ is a complete matrix of the minor determinants of $[x]^m_s$ of order ρ. From this equation it follows that

$$X^2_{1i} + X^2_{2i} + \dots + X^2_{\sigma i} = 0$$

for the values $1, 2, \dots \mu$ of i. Consequently every undegenerate simple minor matrix of $[x]^m_s$ formed with ρ of its vertical rows is extravagant, and in particular $[x]^m_s$ cannot be real.

Thus (A) *cannot have a real solution whose rank is greater than* r.

4. *Some properties of the solutions.*

If $[a]^m_r$ has rank r and if the equation (A) is satisfied, we obtain by prefixing an inverse prefactor of $\overline{\underline{a}}^{\,r}_{m}$ on both sides an equation of the form $[a]^m_r = [h]^s_r [x]^m_s$.

Thus if $[a]^m_r$ *has rank* r, *and if* $[x]^m_s$ *is any solution of the equation* (A), *the horizontal rows of* $[a]^m_r$ *are connected with the horizontal rows of* $[x]^m_s$, *and the matrices* $[x]^m_s$ *and* $\begin{bmatrix} a \\ x \end{bmatrix}^m_{r,\, s}$ *have equal ranks.*

It follows that the solutions in which $[x]^m_s$ has rank ρ are identical with the solutions in which $\begin{bmatrix} a \\ x \end{bmatrix}^m_{r,\, s}$ has rank ρ.

Hence if $[x]_s^m$ is a solution of rank ρ, we can (see Theorem VI of § 71) select $\rho - r$ horizontal rows $[v]_{\rho-r}^m$ of $[x]_s^m$ in such a manner that the matrix $\begin{bmatrix} a \\ v \end{bmatrix}_{r,\,\rho-r}^m$ has rank ρ whilst the remaining $r+s-\rho$ horizontal rows of $[x]_s^m$, which we will denote by $[u]_{r+s-\rho}^m$, are connected with the horizontal rows of this matrix, i.e. we can write

$$[x]_s^m = [\omega]_s^s \begin{bmatrix} u \\ v \end{bmatrix}_{r+s-\rho,\,\rho-r}^m, \qquad \text{.......................(4)}$$

where $[\omega]_s^s$ is a derangement of $[1]_s^s$, $\begin{bmatrix} a \\ v \end{bmatrix}_{r,\,\rho-r}^m$ has rank ρ, and the horizontal rows of $[u]_{r+s-\rho}^m$ are connected with the horizontal rows of $\begin{bmatrix} a \\ v \end{bmatrix}_{r,\,\rho-r}^m$. If $i = \sqrt{-1}$, the equation (A) is then satisfied if and only if

$$\overline{\underline{u,\ v}}_m^{\,r+s-\rho,\,\rho-r} \begin{bmatrix} u \\ v \end{bmatrix}_{r+s-\rho,\,\rho-r}^m = \overline{\underline{a}}_m^{\,r} [a]_r^m,$$

or

$$\overline{\underline{a,\ iv}}_m^{\,r,\,\rho-r} \begin{bmatrix} a \\ iv \end{bmatrix}_{r,\,\rho-r}^m = \overline{\underline{u}}_m^{\,r+s-\rho} [u]_{r+s-\rho}^m \qquad \text{.........(5)}$$

Here the matrix $\overline{\underline{a,\ iv}}_m^{\,r,\,\rho-r}$ has rank ρ, and when we prefix one of its inverse prefactors on both sides of (5), we see that the horizontal rows of $[a]_r^m$ and $[v]_{\rho-r}^m$ are connected with the horizontal rows of $[u]_{r+s-\rho}^m$.

Thus if $[a]_r^m$ has rank r, and if any solution $[x]_s^m$ of rank ρ of the equation (A) is expressed in the form (4), where $\begin{bmatrix} a \\ v \end{bmatrix}_{r,\,\rho-r}^m$ has rank ρ, then all three of the matrices

$$\begin{bmatrix} a \\ u \\ v \end{bmatrix}_{r,\,r+s-\rho,\,\rho-r}^m, \quad \begin{bmatrix} a \\ v \end{bmatrix}_{r,\,\rho-r}^m, \quad [u]_{r+s-\rho}^m$$

have the same rank ρ.

5. *Solutions of lowest rank r.*

Theorem I. *When $[a]_r^m$ has rank r and the necessary condition $s \nless r$ is satisfied, the solutions of the equation (A) of rank r are the matrices $[x]_s^m$ given by the formula*

$$[x]_s^m = [l]_s^r\, [a]_r^m, \qquad \text{.........................(B)}$$

where $[l]_s^r$ is a semi-unit matrix of rank r.

By sub-article 4 the solutions of rank r are those in which $\begin{bmatrix} a \\ x \end{bmatrix}^m_{r,s}$ has rank r. When $[x]^m_s$ is such a solution, there must exist a relation of the form (B) in which $[l]^r_s$ is some matrix of rank r; and by § 84 this value of $[x]^m_s$ satisfies the equation (A) when and only when

$$\overline{\underline{l}}{}^{s}_{r}\,[l]^r_s = [1]^r_r .$$

Such solutions can be real when $[a]^m_r$ is real.

6. *Solutions of all possible ranks.*

Theorem II. *When $[a]^m_r$ has rank r, the equation* (A) *has solutions $[x]^m_s$ of rank ρ when and only when ρ is an integer consistent with the necessary conditions*

$$\rho \not< r, \quad \rho \not> m, \quad 2\rho \not> r+s, \qquad \text{.......................(3)}$$

which include the conditions $s \not< r$ and $\rho \not> s$.

When the conditions (3) *are satisfied the solutions of the equation* (A) *of rank ρ are the matrices $[x]^m_s$ given by*

$$\left.\begin{aligned} [x]^m_s &= [\omega]^s_s \begin{bmatrix} u \\ v \end{bmatrix}^m_{r+s-\rho,\, \rho-r}, \\ [u]^m_{r+s-\rho} = [l]^\rho_{r+s-\rho} &\begin{bmatrix} a \\ \xi \end{bmatrix}^m_{r,\, \rho-r}, \quad [v]^m_{\rho-r} = \sqrt{-1}\,[\xi]^m_{\rho-r}, \end{aligned}\right\} \text{......(C)}$$

where $[\omega]^s_s$ is any derangement of the unit matrix $[1]^s_s$, $[l]^\rho_{r+s-\rho}$ is a semi-unit matrix of rank ρ, and the elements of $[\xi]^m_{\rho-r}$ are arbitrary subject to the condition that the matrix $\begin{bmatrix} a \\ \xi \end{bmatrix}^m_{r,\, \rho-r}$ is undegenerate and has rank ρ.

To prove this theorem let $[x]^m_s$ be any solution of rank ρ, and let it be expressed in the form (4), where $\begin{bmatrix} a \\ v \end{bmatrix}^m_{r,\, \rho-r}$ has rank ρ. Then by sub-article 4 we see that the horizontal rows of the matrix $[u]^m_{r+s-\rho}$, whose rank is ρ, are connected with the horizontal rows of $\begin{bmatrix} a \\ v \end{bmatrix}^m_{r,\, \rho-r}$.

Writing $[v]^m_{\rho-r} = \sqrt{-1}\,[\xi]^m_{\rho-r}$, so that $\begin{bmatrix} a \\ \xi \end{bmatrix}^m_{r,\, \rho-r}$ has rank ρ, it follows that $[x]^m_s$ must be one of the matrices given by (C) when $[l]^\rho_{r+s-\rho}$ is some matrix, necessarily of rank ρ, whose orders only are specified.

Moreover every matrix $[x]_s^m$ given by (C) as in the theorem satisfies the equation (A) or the equivalent equation (5) when and only when

$$\overline{\underline{\,a,\ \xi\,}}_{m}^{r,\,\rho-r}\,[1]_\rho^\rho \begin{bmatrix} a \\ \xi \end{bmatrix}_{r,\,\rho-r}^{m} = \overline{\underline{\,a,\ \xi\,}}_{m}^{r,\,\rho-r}\,\overline{\underline{\,l\,}}_{\rho}^{r+s-\rho}\,[l]_{r+s-\rho}^{\rho}\begin{bmatrix} a \\ \xi \end{bmatrix}_{r,\,\rho-r}^{m},$$

or by § 84 when and only when

$$\overline{\underline{\,l\,}}_{\rho}^{r+s-\rho}\,[l]_{r+s-\rho}^{\rho} = [1]_\rho^\rho, \quad \text{......................(6)}$$

and it is then a solution of rank ρ.

Now by § 155 the equation (6) always admits of solution when $r+s-\rho \nless \rho$, or $2\rho \ngtr r+s$. Hence when the necessary conditions (3) are satisfied, we can determine solutions of the equation (A) of rank ρ, and these solutions are the matrices $[x]_s^m$ specified in Theorem II.

Ex. i. When $[a]_r^m$ has rank r, the equation (A) has solutions of ranks

$$r,\ r+1,\ r+2,\ \ldots\ r+p,\ \ldots\ k,$$

where k is the greatest integer which does not exceed either $\frac{1}{2}(r+s)$ or m.

It has solutions of rank $r+p$ when and only when

$$p \nless 0, \quad p \ngtr \tfrac{1}{2}(s-r), \quad p \ngtr m-r.$$

Ex. ii. Every matrix $[x]_s^m$ given by (C) when $[\omega]_s^s$ is a derangement of $[1]_s^s$, $[l]_{r+s-\rho}^{\rho}$ is a semi-unit matrix of rank ρ, and $[\xi]_{\rho-r}^m$ is arbitrary, is a solution of the equation (A). The matrix $[u]_{r+s-\rho}^m$ and the solution $[x]_s^m$ have the same rank as $\begin{bmatrix} a \\ \xi \end{bmatrix}_{r,\,\rho-r}^{m}$.

Ex. iii. If $[\tau]_{\rho-r} = [r+s-\rho+1,\ r+s-\rho+2,\ \ldots\ s]$, all distinct solutions of rank ρ can be found by determining those solutions in which $\begin{bmatrix} a \\ x_\tau \end{bmatrix}_{r,\,\rho-r}^{m}$ has rank ρ. When $[x]_s^m$ is such a solution we can write

$$[x]_{r+s-\rho}^m = [l]_{r+s-\rho}^{\rho}\begin{bmatrix} a \\ ix_\tau \end{bmatrix}_{r,\,\rho-r}^{m}, \quad \text{..........................(C')}$$

and writing equation (A) in the form

$$\overline{\underline{\,x\,}}_{m}^{r+s-\rho}\,[x]_{r+s-\rho}^{m} = \overline{\underline{\,a,\ ix_\tau\,}}_{m}^{r,\,\rho-r}\begin{bmatrix} a \\ ix_\tau \end{bmatrix}_{r,\,\rho-r}^{m},$$

where $i=\sqrt{-1}$, we see that it is satisfied when and only when $[l]_{r+s-\rho}^{\rho}$ is a semi-unit matrix of rank ρ.

Thus all distinct solutions $[x]_s^m$ of rank ρ are given by formula (C'), where $[l]_{r+s-\rho}^{\rho}$ is any semi-unit matrix of rank ρ and the elements of $[x_\tau]_{\rho-r}^m$ are arbitrary subject to the condition that $\begin{bmatrix} a \\ x_\tau \end{bmatrix}_{r,\,\rho-r}^{m}$ has rank ρ.

We will now express the solutions given by Theorem II in other forms.

Theorem III. *When* $[a]_r^m$ *has rank* r *and* ρ *is any integer consistent with the necessary conditions* (3), *the solutions* $[x]_s^m$ *of the equation* (A) *which have rank* ρ *are given by any one of the following formulae, in which* $i=\sqrt{-1}$, $[\omega]_s^s$ *is any derangement of* $[1]_s^s$, *and the elements of* $[\xi]_{\rho-r}^m$ *and* $[\xi]_{m-r}^m$ *are arbitrary subject respectively to the conditions that* $\begin{bmatrix} a \\ \xi \end{bmatrix}_{r,\ \rho-r}^{m}$ *has rank* ρ, *and* $\begin{bmatrix} a \\ \xi \end{bmatrix}_{r,\ m-r}^{m}$ *has rank* m:

$$(1)\quad [x]_s^m = [\omega]_s^s \begin{bmatrix} l, & \lambda \\ 0, & i \end{bmatrix}_{r+s-\rho,\ \rho-r}^{r,\ \rho-r} \begin{bmatrix} a \\ \xi \end{bmatrix}_{r,\ \rho-r}^{m}, \qquad \text{.........................}(D_1)$$

where $[l,\ \lambda]_{r+s-\rho}^{r,\ \rho-r}$ *is a semi-unit matrix of rank* ρ.

$$(2)\quad [x]_s^m = [l]_s^\rho \begin{bmatrix} a \\ \xi \end{bmatrix}_{r,\ \rho-r}^{m}, \qquad \text{...}(D_2)$$

where $[l]_s^\rho$ *is any solution of rank* ρ *of* $\underline{\overline{l}}_\rho^s\ [l]_s^\rho = \begin{bmatrix} 1, & 0 \\ 0, & 0 \end{bmatrix}_{r,\ \rho-r}^{r,\ \rho-r}$.

$$(3)\quad [x]_s^m = [\omega]_s^s \begin{bmatrix} l, & 0 \\ 0, & 1 \end{bmatrix}_{r+s-\rho,\ \rho-r}^{\rho,\ \rho-r} \begin{bmatrix} 1, & 0 \\ 0, & 1 \\ 0, & i \end{bmatrix}_{r,\ \rho-r,\ \rho-r}^{r,\ \rho-r} \begin{bmatrix} a \\ \xi \end{bmatrix}_{r,\ \rho-r}^{m}, \qquad \text{.........}(D_3)$$

where $[l]_{r+s-\rho}^{\rho}$ *is a semi-unit matrix of rank* ρ.

$$(4)\quad [x]_s^m = [l]_s^{2\rho-r} \begin{bmatrix} 1, & 0 \\ 0, & 1 \\ 0, & i \end{bmatrix}_{r,\ \rho-r,\ \rho-r}^{r,\ \rho-r} \begin{bmatrix} a \\ \xi \end{bmatrix}_{r,\ \rho-r}^{m}, \qquad \text{.........................}(D_4)$$

where $[l]_s^{2\rho-r}$ *is a semi-unit matrix of rank* $2\rho-r$.

$$(5)\quad [x]_s^m = [l]_s^s \begin{bmatrix} 1, & 0 \\ 0, & 1 \\ 0, & i \\ 0, & 0 \end{bmatrix}_{r,\ \rho-r,\ \rho-r,\ r+s-2\rho}^{r,\ \rho-r} \begin{bmatrix} a \\ \xi \end{bmatrix}_{r,\ \rho-r}^{m}, \qquad \text{.................}(D_5)$$

where $[l]_s^s$ *is a square semi-unit matrix of rank* s.

$$(6)\quad [x]^m_s = [l]^s_s \begin{bmatrix} 1, & 0, & 0 \\ 0, & 1, & 0 \\ 0, & i, & 0 \\ 0, & 0, & 0 \end{bmatrix}^{r,\ \rho-r,\ m-\rho}_{r,\ \rho-r,\ \rho-r,\ r+s-2\rho} \begin{bmatrix} a \\ \xi \end{bmatrix}^m_{r,\ m-r}, \quad \ldots\ldots\ldots\ldots\ldots(D_6)$$

where $[l]^s_s$ *is a square semi-unit matrix of rank s.*

Here (D_1) is clearly equivalent to the formulae of Theorem II.

Formula (D_2) includes all matrices given by (D_1), and every matrix $[x]^m_s$ given by (D_2) is a solution of (A) of rank ρ.

We obtain (D_3) from (D_1) by observing that

$$\begin{bmatrix} l, & \lambda \\ 0, & i \end{bmatrix}^{r,\ \rho-r}_{r+s-\rho,\ \rho-r} = \begin{bmatrix} l, & \lambda, & 0 \\ 0, & 0, & 1 \end{bmatrix}^{r,\ \rho-r,\ \rho-r}_{r+s-\rho,\ \rho-r} \begin{bmatrix} 1, & 0 \\ 0, & 1 \\ 0, & i \end{bmatrix}^{r,\ \rho-r}_{r,\ \rho-r,\ \rho-r}.$$

Formula (D_4) includes all matrices given by (D_3), and every matrix $[x]^m_s$ given by (D_4) is a solution of (A) of rank ρ.

Lastly formulae (D_5) and (D_6) clearly give the same matrices as formula (D_4).

Ex. iv. *The equation* $\overline{x}^{\,s}_{m}\,[x]^m_s = \overline{a}^{\,r}_{m}\,[a]^m_r$, *in which* $[a]^m_r$ *has rank r, has undegenerate solutions of rank m when and only when the condition*

$$s \not< 2m - r,$$

which includes $s \not< r$, $s \not< m$, *is satisfied.*

These solutions are given by the formulae of Theorem IV when we replace ρ by m.

Ex. v. *The same equation has undegenerate solutions of rank s when and only when* $s = r$.

The equation then has the form $\overline{x}^{\,r}_{m}\,[x]^m_r = \overline{a}^{\,r}_{m}\,[a]^m_r$, and has only solutions of rank r, these being given by

$$[x]^m_r = [l]^r_r\,[a]^m_r, \quad \text{where} \quad \overline{l}^{\,r}_{r}\,[l]^r_r = [1]^r_r.$$

Ex. vi. When $[a]^m_r$ has rank r, we can always determine an undegenerate square matrix $[a]^m_m$ such that $[a]^m_r = [1,\ 0]^{r,\ m-r}_r\,[a]^m_m$. Then the equation (A) is satisfied when and only when

$$[x]^m_s = [y]^m_s\,[a]^m_m, \quad \ldots\ldots\ldots\ldots\ldots\ldots\ldots\ldots\ldots\ldots\ldots\ldots(7)$$

where

$$\overline{y}^{\,s}_{m}\,[y]^m_s = \overline{\begin{matrix} 1 \\ 0 \end{matrix}}^{\,r}_{r,\ m-r}\,[1,\ 0]^{r,\ m-r}_r. \quad \ldots\ldots\ldots\ldots\ldots\ldots\ldots\ldots\ldots(8)$$

Thus there is a one-one correspondence between the solutions of the equations (A) and (8), corresponding solutions having equal ranks, and the general solution of the equation (A) can be expressed in the form (7), when $[y]_s^m$ is the general solution of the equation (8).

Ex. vii. *The equation*

$$\underline{\overline{x}}{}^{s}_{m}\,[x]_s^m = \begin{bmatrix}1, & 0\\ 0, & 0\end{bmatrix}_{r,\,m-r}^{r,\,m-r} = \underline{\overline{\begin{matrix}1\\0\end{matrix}}}{}^{r}_{r,\,m-r}\quad [1,\ 0]_r^{r,\,m-r}. \qquad \text{..................(E)}$$

If $r \nless 0$, $r \ngtr m$, this equation has solutions of rank ρ when and only when ρ is an integer consistent with the conditions

$$\rho \nless r, \quad \rho \ngtr m, \quad 2\rho \ngtr r+s,$$

which include the conditions $s \nless r$, $\rho \ngtr s$. The solutions of rank ρ are then given by the formulae of Theorem III when we replace

$$\begin{bmatrix}a\\ \xi\end{bmatrix}_{r,\,\rho-r}^{m}, \quad \begin{bmatrix}a\\ \xi\end{bmatrix}_{r,\,m-r}^{m} \quad \text{by} \quad \begin{bmatrix}1, & 0\\ \xi, & \eta\end{bmatrix}_{r,\,\rho-r}^{r,\,m-r}, \quad \begin{bmatrix}1, & 0\\ \xi, & \eta\end{bmatrix}_{r,\,m-r}^{r,\,m-r}$$

respectively, where $[\eta]_{\rho-r}^{m-r}$, $[\eta]_{m-r}^{m-r}$ have ranks $\rho-r$, $m-r$ and where $[\xi]_{\rho-r}^{r}$, $[\xi]_{m-r}^{r}$ are entirely arbitrary.

Ex. viii. The equation

$$\begin{bmatrix} x_1 & x_2 & x_3 & x_4 & x_5 & x_6 & x_7\\ y_1 & y_2 & y_3 & y_4 & y_5 & y_6 & y_7\\ z_1 & z_2 & z_3 & z_4 & z_5 & z_6 & z_7\\ w_1 & w_2 & w_3 & w_4 & w_5 & w_6 & w_7\end{bmatrix} \begin{bmatrix} x_1 & y_1 & z_1 & w_1\\ x_2 & y_2 & z_2 & w_2\\ x_3 & y_3 & z_3 & w_3\\ x_4 & y_4 & z_4 & w_4\\ x_5 & y_5 & z_5 & w_5\\ x_6 & y_6 & z_6 & w_6\\ x_7 & y_7 & z_7 & w_7\end{bmatrix} = \begin{bmatrix} a_1 & a_2\\ b_1 & b_2\\ c_1 & c_2\\ d_1 & d_2\end{bmatrix} \begin{bmatrix} a_1 & b_1 & c_1 & d_1\\ a_2 & b_2 & c_2 & d_2\end{bmatrix},$$

in which $[a\,b\,c\,d]_{12}$ has rank 2, has solutions of ranks 2, 3 and 4.

The distinct solutions of these several ranks are given respectively by the formulae

$$[x\,y\,z\,w]_{1234567} = [l]_7^2 \begin{bmatrix} a_1 & b_1 & c_1 & d_1\\ a_2 & b_2 & c_2 & d_2\end{bmatrix},$$

$$[x\,y\,z\,w]_{123456} = [l]_6^3 \begin{bmatrix} a_1 & b_1 & c_1 & d_1\\ a_2 & b_2 & c_2 & d_2\\ ix_7 & iy_7 & iz_7 & iw_7\end{bmatrix},$$

$$[x\,y\,z\,w]_{12345} = [l]_5^4 \begin{bmatrix} a_1 & b_1 & c_1 & d_1\\ a_2 & b_2 & c_2 & d_2\\ ix_6 & iy_6 & iz_6 & iw_6\\ ix_7 & iy_7 & iz_7 & iw_7\end{bmatrix},$$

where $[l]_7^2$, $[l]_6^3$, $[l]_5^4$ are semi-unit matrices of ranks 2, 3, 4; $i=\sqrt{-1}$; and the matrices on the right are undegenerate.

The first formula gives all distinct solutions of the lowest rank 2.

The second formula gives all distinct solutions of rank 3 when the postfactor on the right is undegenerate; and it gives solutions of rank 2 when the postfactor on the right has rank 2.

The third formula gives all distinct solutions of rank 4 when the postfactor on the right has rank 4; and it gives solutions of ranks 3 or 2 when the postfactor on the right has rank 3 or 2.

Ex. ix. The equation $\overline{\underline{x}}{}^{7}_{4}\,[x]_7^4 = \overline{\underline{a}}{}^{2}_{4}\,[a]_2^4$, in which $[a]_2^4$ has rank 2, has solutions of ranks 2, 3 and 4. All solutions of these several ranks are given respectively by the formulae

$$[x]_7^4 = [l]_7^2\,[a]_2^4, \qquad \text{where} \qquad \overline{\underline{l}}{}^{7}_{2}\,[l]_7^2 = [1]_2^2;$$

$$[x]_7^4 = [\omega]_7^7 \begin{bmatrix} l, & \lambda \\ 0, & i \end{bmatrix}_{6,1}^{2,1} \begin{bmatrix} a \\ \xi \end{bmatrix}_{2,1}^{4}, \qquad \text{where} \qquad \overline{\underline{\begin{matrix} l \\ \lambda \end{matrix}}}{}^{6}_{2,1}\,[l, \lambda]_6^{2,1} = [1]_3^3;$$

$$[x]_7^4 = [\omega]_7^7 \begin{bmatrix} l, & \lambda \\ 0, & i \end{bmatrix}_{5,2}^{2,2} \begin{bmatrix} a \\ \xi \end{bmatrix}_{2,2}^{4}, \qquad \text{where} \qquad \overline{\underline{\begin{matrix} l \\ \lambda \end{matrix}}}{}^{5}_{2,2}\,[l, \lambda]_5^{2,2} = [1]_4^4.$$

Here $[\omega]_7^7$ is a derangement of $[1]_7^7$, and the matrices $\begin{bmatrix} a \\ \xi \end{bmatrix}_{2,1}^{4}$, $\begin{bmatrix} a \\ \xi \end{bmatrix}_{2,2}^{4}$ are undegenerate.

All solutions of ranks 2, 3, and 4 are also given respectively by the formulae

$$[x]_7^4 = [l]_7^2\,[a]_2^4, \qquad \text{where} \qquad \overline{\underline{l}}{}^{7}_{2}\,[l]_7^2 = [1]_2^2;$$

$$[x]_7^4 = [l]_7^4 \begin{bmatrix} 1, & 0 \\ 0, & 1 \\ 0, & i \end{bmatrix}_{2,1,1}^{2,1} \begin{bmatrix} a \\ \xi \end{bmatrix}_{2,1}^{4}, \qquad \text{where} \qquad \overline{\underline{l}}{}^{7}_{4}\,[l]_7^4 = [1]_4^4;$$

$$[x]_7^4 = [l]_7^6 \begin{bmatrix} 1, & 0 \\ 0, & 1 \\ 0, & i \end{bmatrix}_{2,2,2}^{2,2} \begin{bmatrix} a \\ \xi \end{bmatrix}_{2,2}^{4}, \qquad \text{where} \qquad \overline{\underline{l}}{}^{7}_{6}\,[l]_7^6 = [1]_6^6;$$

the matrices $\begin{bmatrix} a \\ \xi \end{bmatrix}_{2,1}^{4}$ and $\begin{bmatrix} a \\ \xi \end{bmatrix}_{2,2}^{4}$ being again undegenerate.

§ 158. Solutions of the symmetric equation $X'X = 0$.

This equation, which is a particular case of that considered in § 157, may be taken to be

$$\overline{\underline{x}}{}^{s}_{m}\,[x]_s^m = 0. \qquad \text{..............................(A)}$$

If there is a solution $[x]_s^m$ of rank ρ, it follows from § 133 that the integer ρ must satisfy the conditions

$$\rho \nless 0, \quad \rho \ngtr m, \quad 2\rho \ngtr s; \qquad \text{.........................(1)}$$

and from § 72.1 it follows that:

The only real solution of the equation (A) *is* $[x]_s^m = 0$.

Any matrix $[x]_s^m$ of rank ρ can be expressed in the form

$$[x]_s^m = [\omega]_s^s \begin{bmatrix} u \\ v \end{bmatrix}_{s-\rho,\, \rho}^{m}, \qquad \text{.........................(2)}$$

where $[\omega]_s^s$ is a derangement of $[1]_s^s$, and $[v]_\rho^m$ has rank ρ.

If this matrix is a solution of (A) we must have

$$\underline{\overline{v}}_m^{\rho}\, [v]_\rho^m = -\, \underline{\overline{u}}_m^{s-\rho}\, [u]_{s-\rho}^m, \qquad \text{.........................(3)}$$

and when we prefix an inverse prefactor of $\underline{\overline{v}}_m^{\rho}$ on both sides we see that the horizontal rows of $[v]_\rho^m$ must be connected with the horizontal rows of $[u]_{s-\rho}^m$.

Thus if (2) *is a solution of rank* ρ *in which* $[v]_\rho^m$ *has rank* ρ, *then the matrices* $[x]_s^m$, $\begin{bmatrix} u \\ v \end{bmatrix}_{s-\rho,\, \rho}^{m}$, $[v]_\rho^m$, $[u]_{s-\rho}^m$ *all have the same rank* ρ.

Corresponding to Theorem II of § 157, we have the following theorem:

Theorem I. *The equation* (A) *has solutions of rank* ρ *when and only when*

$$\rho \nless 0, \quad \rho \ngtr m, \quad 2\rho \ngtr s. \qquad \text{.........................(1)}$$

When ρ *is an integer satisfying these conditions, the solutions of rank* ρ *are the matrices* $[x]_s^m$ *given by*

$$[x]_s^m = [\omega]_s^s \begin{bmatrix} u \\ v \end{bmatrix}_{s-\rho,\, \rho}^{m}, \quad [u]_{s-\rho}^m = [l]_{s-\rho}^\rho [\xi]_\rho^m, \quad [v]_\rho^m = \sqrt{-1}\,[\xi]_\rho^m, \quad \text{(B)}$$

where $[\omega]_s^s$ *is any derangement of* $[1]_s^s$, $[l]_{s-\rho}^\rho$ *is a semi-unit matrix of rank* ρ, *and* $[\xi]_\rho^m$ *has rank* ρ.

For if any solution $[x]_s^m$ of rank ρ is expressed in the form (2), and if we write $[v]_\rho^m = \sqrt{-1}\,[\xi]_\rho^m$, so that $[\xi]_\rho^m$ has rank ρ, there must exist a relation of the form $[u]^m_{s-\rho} = [l]^\rho_{s-\rho}\,[\xi]^m_\rho$, where $[l]^\rho_{s-\rho}$ is some matrix of rank ρ. The equation (A) or the equivalent equation (3) is then satisfied if and only if

$$\overline{\underline{l}}{}^{s-\rho}_{\rho}\,[l]^\rho_{s-\rho} = [1]^\rho_\rho;$$

and there exists a semi-unit matrix satisfying this equation whenever $s-\rho \nless \rho$, or $2\rho \ngtr s$.

Ex. i. Every matrix $[x]^m_s$ given by (B) when $[\omega]^s_s$ is a derangement of $[1]^s_s$, $[l]^\rho_{s-\rho}$ is a semi-unit matrix of rank ρ, and $[\xi]^m_\rho$ is arbitrary is a solution of the equation (A) having the same rank as $[\xi]^m_\rho$.

Ex. ii. If $[\tau]_\rho = [s-\rho+1, s-\rho+2, \ldots s]$, all *distinct* solutions of rank ρ can be found by determining those solutions in which $[x_\tau]^m_\rho$ has rank ρ. When $[x]^m_s$ is such a solution, we can write

$$[x]^m_{s-\rho} = \sqrt{-1}\,[l]^\rho_{s-\rho}\,[x_\tau]^m_\rho, \quad \ldots\ldots\ldots\ldots\ldots\ldots\ldots\ldots\text{(C)}$$

and the equation (A) or the equivalent equation $\overline{\underline{x}}{}^{s-\rho}_{m}\,[x]^m_{s-\rho} = -\overline{\underline{x_\tau}}{}^{\rho}_{m}\,[x_\tau]^m_\rho$ is then satisfied when and only when $[l]^\rho_{s-\rho}$ is a semi-unit matrix of rank ρ.

Thus all distinct solutions $[x]^m_s$ *of rank* ρ *are given by formula* (C), *where* $[l]^\rho_{s-\rho}$ *is a semi-unit matrix of rank* ρ, *and where the elements of* $[x]^m_s$ *occurring in* $[x_\tau]^m_\rho$ *are arbitrary subject to the condition that* $[x_\tau]^m_\rho$ *has rank* ρ.

From Theorem I we deduce the following equivalent theorem which corresponds to Theorem III of § 157.

Theorem II. *If* ρ *is any integer satisfying the necessary conditions* (1), *the solutions of the equation* (A) *of rank* ρ *are the matrices* $[x]^m_s$ *given by any one of the following formulae, in which* $i=\sqrt{-1}$, $[\omega]^s_s$ *is a derangement of* $[1]^s_s$, *and* $[\xi]^m_\rho$ *and* $[\xi]^m_m$ *are undegenerate having ranks* ρ *and* m:

(1) $$[x]^m_s = [\omega]^s_s \begin{bmatrix} l \\ i \end{bmatrix}^{\rho}_{s-\rho,\,\rho} [\xi]^m_\rho, \quad \ldots\ldots\ldots\ldots\ldots\ldots\ldots\ldots\text{(D}_1\text{)}$$

where $[l]^\rho_{s-\rho}$ *is a semi-unit matrix of rank* ρ.

(2) $$[x]^m_s = [l]^\rho_s\,[\xi]^m_\rho, \quad \ldots\ldots\ldots\ldots\ldots\ldots\ldots\ldots\text{(D}_2\text{)}$$

where $[l]^\rho_s$ *is any solution of rank* ρ *of* $\overline{\underline{l}}{}^{s}_{\rho}\,[l]^\rho_s = 0$.

(3) $$[x]^m_s = [\omega]^s_s \begin{bmatrix} l, & 0 \\ 0, & 1 \end{bmatrix}^{\rho,\,\rho}_{s-\rho,\,\rho} \begin{bmatrix} 1 \\ i \end{bmatrix}^{\rho}_{\rho,\,\rho} [\xi]^m_\rho, \quad \ldots\ldots\ldots\ldots\ldots\ldots\text{(D}_3\text{)}$$

where $[l]^\rho_{s-\rho}$ *is a semi-unit matrix of rank* ρ.

$$(4)\quad [x]_s^m = [l]_s^{2\rho} \begin{bmatrix} 1 \\ i \end{bmatrix}_{\rho,\,\rho}^{\rho} [\xi]_\rho^m, \quad \ldots\ldots\ldots\ldots\ldots\ldots(D_4)$$

where $[l]_s^{2\rho}$ *is a semi-unit matrix of rank* 2ρ.

$$(5)\quad [x]_s^m = [l]_s^s \begin{bmatrix} 1 \\ i \\ 0 \end{bmatrix}_{\rho,\,\rho,\,s-2\rho}^{\rho} [\xi]_\rho^m, \quad \ldots\ldots\ldots\ldots\ldots\ldots(D_5)$$

where $[l]_s^s$ *is a square semi-unit matrix of rank* s.

$$(6)\quad [x]_s^m = [l]_s^s \begin{bmatrix} 1, & 0 \\ i, & 0 \\ 0, & 0 \end{bmatrix}_{\rho,\,\rho,\,s-2\rho}^{\rho,\,m-\rho} [\xi]_m^m, \quad \ldots\ldots\ldots\ldots\ldots\ldots(D_6)$$

where $[l]_s^s$ *is a square semi-unit matrix of rank* s.

It will be shown in § 159 that when ρ satisfies the necessary conditions (1), the solutions of rank ρ of the equation $\underset{\rho}{\overline{x}}^{\,s} [x]_s^\rho = 0$ are given by the formula

$$[x]_s^\rho = [l]_s^{2\rho} \begin{bmatrix} 1 \\ i \end{bmatrix}_{\rho,\,\rho}^{\rho} [k]_\rho^\rho,$$

where $[k]_\rho^\rho$ is an undegenerate square matrix, and $[l]_s^{2\rho}$ is a *real* semi-unit matrix of rank 2ρ. Hence from (D_2) we see that:

In formulae (D_4), (D_5) *and* (D_6) *there is no loss of generality when we introduce the restrictions that the semi-unit matrices* $[l]_s^{2\rho}$, $[l]_s^s$ *are real.*

Ex. iii. *The equation* $\underset{m}{\overline{x}}^{\,s} [x]_s^m = 0$ *has undegenerate solutions of rank* m *when and only when* $s \nless 2m$.

These solutions are given by the formulae of Theorem III of § 157 when we replace ρ by m.

The equation has no undegenerate solutions of rank s.

Ex. iv. The equation

$$\begin{bmatrix} x_1 & x_2 & x_3 & x_4 & x_5 & x_6 \\ y_1 & y_2 & y_3 & y_4 & y_5 & y_6 \\ z_1 & z_2 & z_3 & z_4 & z_5 & z_6 \\ w_1 & w_2 & w_3 & w_4 & w_5 & w_6 \end{bmatrix} \begin{bmatrix} x_1 & y_1 & z_1 & w_1 \\ x_2 & y_2 & z_2 & w_2 \\ x_3 & y_3 & z_3 & w_3 \\ x_4 & y_4 & z_4 & w_4 \\ x_5 & y_5 & z_5 & w_5 \\ x_6 & y_6 & z_6 & w_6 \end{bmatrix} = 0$$

has solutions of ranks 0, 1, 2 and 3. The distinct solutions of these several ranks are given respectively by the formulae

$$[x\,y\,z\,w]_{123456} = 0\,;$$

$$[x\,y\,z\,w]_{12345} = \sqrt{-1}\,[l]_5^1\,[x\,y\,z\,w]_6\,, \quad \text{where } l_{11}^2 + l_{21}^2 + \ldots + l_{51}^2 = 1\,;$$

$$[x\,y\,z\,w]_{1234} = \sqrt{-1}\,[l]_4^2\,[x\,y\,z\,w]_{56}\,, \quad \text{where } \overline{\underline{l}}_2^{\,4}\,[l]_4^2 = [1]_2^2\,;$$

$$[x\,y\,z\,w]_{123} = \sqrt{-1}\,[l]_3^3\,[x\,y\,z\,w]_{456}, \quad \text{where } \overline{\underline{l}}_3^{\,3}\,[l]_3^3 = [1]_3^3\,.$$

The second formula gives all distinct solutions of rank 1 when $[x\,y\,z\,w]_6$ has rank 1, and gives solutions of rank 0 when this matrix has rank 0.

The third formula gives all distinct solutions of rank 2 when $[x\,y\,z\,w]_{56}$ has rank 2, and in other cases gives solutions having the same rank as $[x\,y\,z\,w]_{56}$.

The fourth formula gives all distinct solutions of rank 3 when $[x\,y\,z\,w]_{456}$ has rank 3, and in other cases gives solutions having the same rank as $[x\,y\,z\,w]_{456}$.

Ex. v. The equation $\overline{\underline{x}}_4^{\,6}\,[x]_6^4 = 0$ has solutions of ranks 0, 1, 2 and 3. All solutions of these several ranks are given respectively by the formulae

$$[x]_6^4 = 0\,;$$

$$[x]_6^4 = [\omega]_6^6 \begin{bmatrix} l \\ i \end{bmatrix}_{5,\,1}^{1} [\xi]_1^4, \quad \text{where } l_{11}^2 + l_{21}^2 + \ldots + l_{51}^2 = 1\,;$$

$$[x]_6^4 = [\omega]_6^6 \begin{bmatrix} l \\ i \end{bmatrix}_{4,\,2}^{2} [\xi]_2^4, \quad \text{where } \overline{\underline{l}}_2^{\,4}\,[l]_4^2 = [1]_2^2\,;$$

$$[x]_6^4 = [\omega]_6^6 \begin{bmatrix} l \\ i \end{bmatrix}_{3,\,3}^{3} [\xi]_3^4, \quad \text{where } \overline{\underline{l}}_3^{\,3}\,[l]_3^3 = [1]_3^3\,.$$

Here $[\omega]_6^6$ is any derangement of the unit matrix $[1]_6^6$, the matrices $[\xi]_1^4, [\xi]_2^4, [\xi]_3^4$ are undegenerate, and $i = \sqrt{-1}$.

All solutions of ranks 0, 1, 2 and 3 are also given respectively by the formulae

$$[x]_6^4 = 0\,;$$

$$[x]_6^4 = [l]_6^2 \begin{bmatrix} 1 \\ i \end{bmatrix}_{1,\,1}^{1} [\xi]_1^4 = [l]_6^2 \begin{bmatrix} \xi \\ i\xi \end{bmatrix}_{1,\,1}^{4}\,;$$

$$[x]_6^4 = [l]_6^4 \begin{bmatrix} 1 \\ i \end{bmatrix}_{2,\,2}^{2} [\xi]_2^4 = [l]_6^4 \begin{bmatrix} \xi \\ i\xi \end{bmatrix}_{2,\,2}^{4}\,;$$

$$[x]_6^4 = [l]_6^6 \begin{bmatrix} 1 \\ i \end{bmatrix}_{3,\,3}^{3} [\xi]_3^4 = [l]_6^6 \begin{bmatrix} \xi \\ i\xi \end{bmatrix}_{3,\,3}^{4}\,;$$

where $[l]_6^2, [l]_6^4, [l]_6^6$ are *real* semi-unit matrices of ranks 2, 4, 6, and $[\xi]_1^4, [\xi]_2^4, [\xi]_3^4$ are undegenerate.

Ex. vi. Another illustration is furnished by Ex. v of § 156.

§ 159. Solutions of rank r of the equation $[x]^n_r \lceil x \rfloor^r_n = 0$.

By the theorem of § 133 this equation cannot have undegenerate solutions of rank r unless $2r \not> n$. We will obtain independently of § 158 a general formula for all undegenerate solutions of rank r when this necessary condition is satisfied, and thereby show that such solutions exist. We shall then be able to deduce the general solutions of the equations $[x]^n_r \lceil x \rfloor^r_n = 0$, $\lceil x \rfloor^s_m [x]^m_s = 0$ in all cases without using the results of § 158.

The fundamental theorem to be proved is as follows:

Theorem. *If $[a]^n_r$ is an undegenerate matrix of rank r, and if $[a]^n_r \lceil a \rfloor^r_n = 0$, which is only possible when $2r \not> n$, then we can determine equivalent similar undegenerate matrices $[w]^n_r$ of the form*

$$[w]^n_r = [u]^n_r + i[v]^n_r = [1,\ i]^{r,\,r} \begin{bmatrix} u \\ v \end{bmatrix}^n_{r,\,r},$$

where $[u]^n_r$ and $[v]^n_r$ are real undegenerate matrices such that

$$\begin{bmatrix} u \\ v \end{bmatrix}^n_{r,\,r} \lceil u,\ v \rfloor^{r,\,r}_n = \begin{bmatrix} 1, & 0 \\ 0, & 1 \end{bmatrix}^{r,\,r}_{r,\,r} = [1]^{2r}_{2r},$$

or
$$[u]^n_r \lceil u \rfloor^r_n = [v]^n_r \lceil v \rfloor^r_n = [1]^r_r, \qquad [u]^n_r \lceil v \rfloor^r_n = [v]^n_r \lceil u \rfloor^r_n = 0.$$

Here 'equivalence' has the same meaning as in § 113, i.e. the similar undegenerate matrices $[a]^n_r$ and $[w]^n_r$ are equivalent when there exists an undegenerate square matrix $[k]^r_r$ such that

$$[a]^n_r = [k]^r_r [w]^n_r.$$

SPECIAL CASE. *When $r=1$ and $n \not< 2$.*

Let $[a_1\, a_2 \ldots a_n]$ be a matrix of rank 1 in which $a_1^2+a_2^2+\ldots+a_n^2=0$, i.e. let it be a non-vanishing one-rowed matrix which is extravagant. Then $a_1, a_2, \ldots a_n$ cannot be all real or all purely imaginary, and we can write

$$[a_1\, a_2 \ldots a_n] = [\alpha_1 + i\beta_1,\ \alpha_2 + i\beta_2,\ \ldots\ \alpha_n + i\beta_n]$$
$$= [\alpha_1\, \alpha_2 \ldots \alpha_n] + i[\beta_1\, \beta_2 \ldots \beta_n],$$

where $[\alpha]_n$ and $[\beta]_n$ are real non-vanishing matrices.

Since $(\alpha_1+i\beta_1)^2+(\alpha_2+i\beta_2)^2+\ldots+(\alpha_n+i\beta_n)^2=0$, we have

$$\alpha_1^2+\alpha_2^2+\ldots+\alpha_n^2=\beta_1^2+\beta_2^2+\ldots+\beta_n^2=\kappa^2, \quad \alpha_1\beta_1+\alpha_2\beta_2+\ldots+\alpha_n\beta_n=0,$$

where κ^2 is a real non-vanishing quantity, and is positive.

Putting $[\alpha_1\, \alpha_2 \ldots \alpha_n] = \kappa[u_1\, u_2 \ldots u_n]$, $[\beta_1\, \beta_2 \ldots \beta_n] = \kappa[v_1\, v_2 \ldots v_n]$,

we have $[a_1\, a_2 \ldots a_n] = \kappa[w_1\, w_2 \ldots w_n]$, where

$$[w_1\, w_2 \ldots w_n] = [u_1\, u_2 \ldots u_n] + i[v_1\, v_2 \ldots v_n] = [1,\ i]\begin{bmatrix} u_1 & u_2 \ldots u_n \\ v_1 & v_2 \ldots v_n \end{bmatrix},$$

the u's and v's being real quantities such that

$$u_1^2+u_2^2+\ldots+u_n^2=v_1^2+v_2^2+\ldots+v_n^2=1, \quad u_1v_1+u_2v_2+\ldots+u_nv_n=0,$$

or
$$\begin{bmatrix} u_1 & u_2 \ldots u_n \\ v_1 & v_2 \ldots v_n \end{bmatrix} \begin{bmatrix} u_1 & v_1 \\ u_2 & v_2 \\ \cdots & \cdots \\ u_n & v_n \end{bmatrix} = \begin{bmatrix} 1 & 0 \\ 0 & 1 \end{bmatrix}.$$

Thus the theorem is true in this special case.

GENERAL CASE. *When r and n are any positive integers such that $2r \not> n$.*

We now consider the matrix $\phi = [a]^n_r$ of the theorem.

Every horizontal row of ϕ and every one-rowed matrix of the form $[\lambda_1 \lambda_2 \dots \lambda_r][a]^n_r$ where $[\lambda]_r \neq 0$ is a non-vanishing one-rowed matrix which is extravagant, and is therefore a matrix to which the special case is applicable.

By the special case we can express the last horizontal row of ϕ in the form $\kappa\,[w_{11}\, w_{12} \dots w_{1n}]$, where

$$[w]^n_1 = [u]^n_1 + i[v]^n_1 = [1,\ i]\begin{bmatrix} u \\ v \end{bmatrix}^n_{1,\,1}, \qquad \begin{bmatrix} u \\ v \end{bmatrix}^n_{1,\,1} \overline{\underline{u,\ v}}^{1,\,1}_{n} = \begin{bmatrix} 1, & 0 \\ 0, & 1 \end{bmatrix},$$

and where the u's and v's are all real, and $\kappa \neq 0$.

Then replacing the last horizontal row of ϕ by $[w_{11}\, w_{12} \dots w_{1n}]$, we obtain the undegenerate matrix $\phi_1 = \begin{bmatrix} a \\ w \end{bmatrix}^n_{r-1,\,1}$ which is equivalent to ϕ. Using this fact, we can prove the theorem by induction.

If s is any integer which is less than r and not less than 1, we will assume that an undegenerate matrix $\phi_s = \begin{bmatrix} b \\ w \end{bmatrix}^n_{r-s,\,s}$ has been found which is equivalent to ϕ, and in which

$$[w]^n_s = [u]^n_s + i[v]^n_s = [1,\ i]^{s,\,s}_s \begin{bmatrix} u \\ v \end{bmatrix}^n_{s,\,s}, \quad \dots\dots\dots\dots(1)$$

the matrices $[u]^n_s$ and $[v]^n_s$ being real and such that

$$\begin{bmatrix} u \\ v \end{bmatrix}^n_{s,\,s} \overline{\underline{u,\ v}}^{s,\,s}_{n} = \begin{bmatrix} 1, & 0 \\ 0, & 1 \end{bmatrix}^{s,\,s}_{s,\,s} = [1]^{2s}_{2s}. \quad \dots\dots\dots\dots(1')$$

We will then show that we can determine an undegenerate matrix $\phi_{s+1} = \begin{bmatrix} c \\ w \end{bmatrix}^n_{r-s-1,\,s+1}$ which is equivalent to ϕ, and in which

$$[w]^n_{s+1} = [u]^n_{s+1} + i[v]^n_{s+1} = [1,\ i]^{s+1,\,s+1}_{s+1} \begin{bmatrix} u \\ v \end{bmatrix}^n_{s+1,\,s+1}, \quad \dots\dots\dots\dots(2)$$

the matrices $[u]^n_{s+1}$ and $[v]^n_{s+1}$ being real and such that

$$\begin{bmatrix} u \\ v \end{bmatrix}^n_{s+1,\,s+1} \overline{\underline{u,\ v}}^{s+1,\,s+1}_{n} = \begin{bmatrix} 1, & 0 \\ 0, & 1 \end{bmatrix}^{s+1,\,s+1}_{s+1,\,s+1} = [1]^{2s+2}_{2s+2}. \quad \dots\dots\dots\dots(2')$$

Since there exists a relation of the form $\phi_s = \begin{bmatrix} b \\ w \end{bmatrix}^n_{r-s,\,s} = [h]^r_r\,[a]^n_r$, where $[h]^r_r$ is an undegenerate square matrix, we have

$$\begin{bmatrix} b \\ w \end{bmatrix}^n_{r-s,\,s} \overline{\underline{b,\ w}}^{r-s,\,s}_{n} = 0,$$

and therefore

$$\phi_s\, \overline{\underline{w}}^{s}_{n} = 0, \quad \text{or} \quad \phi_s\, \overline{\underline{u,\ v}}^{s,\,s}_{n} \begin{bmatrix} 1 \\ i \end{bmatrix}^s_{s,\,s} = 0.$$

Thus there are s unconnected connections between the vertical rows of the matrix $\phi_s \; \overline{\underline{u, v}}_n^{s,s}$, and the rank of this matrix cannot exceed s. Since $s < r$, this matrix must be degenerate, and it is possible to determine a non-vanishing matrix $[\lambda, \mu]_{r-s, s}$ such that

$$[\lambda, \mu]_{r-s, s} \begin{bmatrix} b \\ w \end{bmatrix}_{r-s, s}^{n} \overline{\underline{u, v}}_n^{s,s} = 0. \quad \text{..............................(3)}$$

If $[\lambda]_{r-s} = 0$, so that $[\mu]_s \neq 0$, we should have

$$[\mu]_s \, [w]_s^n \, \overline{\underline{u, v}}_n^{s,s} = [\mu]_s \, [1, i]_s^{s,s} \begin{bmatrix} u \\ v \end{bmatrix}_{s,s}^{n} \overline{\underline{u, v}}_n^{s,s} = [\mu]_s \, [1, i]_s^{s,s} = 0,$$

which is impossible, since the last postfactor on the right is undegenerate. Hence in (3) we must have $[\lambda]_{r-s} \neq 0$, and we may assume that $\lambda_k \neq 0$.

The matrix $[\lambda, \mu]_{r-s, s} \begin{bmatrix} b \\ w \end{bmatrix}_{r-s, s}^{n}$ is a non-vanishing one-rowed matrix connected with the horizontal rows of ϕ_s or ϕ, and is therefore extravagant. By the special case it can be expressed in the form

$$[\lambda, \mu]_{r-s, s} \begin{bmatrix} b \\ w \end{bmatrix}_{r-s, s}^{n} = \kappa \, [w_{s+1,1} \, w_{s+1,2} \dots w_{s+1,n}],$$

where $\kappa \neq 0$, and

$$[w_{s+1,1} \, w_{s+1,2} \dots w_{s+1,n}] = [u_{s+1,1} \, u_{s+1,2} \dots u_{s+1,n}] + i \, [v_{s+1,1} \, v_{s+1,2} \dots v_{s+1,n}],$$

the u's and v's being real and such that

$$\begin{bmatrix} u_{s+1,1} & u_{s+1,2} \dots u_{s+1,n} \\ v_{s+1,1} & v_{s+1,2} \dots v_{s+1,n} \end{bmatrix} \begin{bmatrix} u_{s+1,1} & v_{s+1,1} \\ u_{s+1,2} & v_{s+1,2} \\ \dots\dots & \dots\dots \\ u_{s+1,n} & v_{s+1,n} \end{bmatrix} = \begin{bmatrix} 1, & 0 \\ 0, & 1 \end{bmatrix}. \quad \text{................(4)}$$

Then if in ϕ_s we replace the kth horizontal row of $[b]_{r-s}^n$ by $[w_{s+1,1} \, w_{s+1,2} \dots w_{s+1,n}]$ and then re-arrange the horizontal rows, we obtain (see Ex. iii of § 113) an equivalent matrix $\phi_{s+1} = \begin{bmatrix} c \\ w \end{bmatrix}_{r-s-1, s+1}^{n}$ in which $[c]_{r-s-1}^n$ is composed of the remaining $r-s-1$ horizontal rows of $[b]_{r-s}^n$. From (3) we see that both $[u_{s+1,1} \, u_{s+1,2} \dots u_{s+1,n}]$ and $[v_{s+1,1} \, v_{s+1,2} \dots v_{s+1,n}]$ are orthogonal with all the horizontal rows of $\begin{bmatrix} u \\ v \end{bmatrix}_{s,s}^{n}$. Hence from (4) we see that the equations (2) and (2′) are satisfied, i.e. we have determined a matrix ϕ_{s+1} which is equivalent to ϕ_s and therefore to ϕ, and has the required properties.

Since, as has been shown above, we can determine ϕ_1, it follows that we can in succession determine $\phi_1, \phi_2, \dots \phi_r$, and the last of these matrices has the form $[w]_s^n$ of the theorem.

Writing $\begin{bmatrix} u \\ v \end{bmatrix}_{r,r}^{n} = [l]_{2r}^n$ we deduce from the theorem the following two corollaries:

COROLLARY 1. *When the necessary condition $2r \not> n$ is satisfied, the equation $[x]_r^n \, \overline{\underline{x}}_n^r = 0$ has undegenerate solutions of rank r, and these are all given by the formula*

$$[x]_r^n = [k]_r^r \, [1, i]_r^{r,r} \, [l]_{2r}^n,$$

where $[k]_r^r$ is an undegenerate square matrix, and $[l]_{2r}^n$ is a real semi-unit matrix of rank $2r$.

COROLLARY 2. *When the necessary condition* $2m \not> s$ *is satisfied, the equation* $\overline{\underline{x}}{}^{s}_{m} \, [x]^{m}_{s} = 0$ *has undegenerate solutions of rank* m, *and these are all given by the formula*

$$[x]^{m}_{s} = [l]^{2m}_{s} \begin{bmatrix} 1 \\ i \end{bmatrix}^{m}_{m,\, m} [\xi]^{m}_{m},$$

where $[\xi]^{m}_{m}$ *is an undegenerate square matrix, and* $[l]^{2m}_{s}$ *is a real semi-unit matrix of rank* $2m$.

If $[x]^{n}_{r}$ is any solution of rank ρ of the equation $[x]^{n}_{r} \, \overline{\underline{x}}{}^{r}_{n} = 0$, we can write

$$[x]^{n}_{r} = [k]^{\rho}_{r} \, [y]^{n}_{\rho},$$

where both matrices on the right have rank ρ. By § 84 the equation is then satisfied if and only if $[y]^{n}_{\rho} \, \overline{\underline{y}}{}^{\rho}_{n} = 0$. Again, by the theorem of § 133, this last equation cannot have solutions of rank ρ unless $2\rho \not> n$; and when this condition is satisfied it has solutions of rank ρ, these being given by Corollary 1. Hence we have the following further corollaries:

COROLLARY 3. *The equation* $[x]^{n}_{r} \, \overline{\underline{x}}{}^{r}_{n} = 0$ *has solutions of rank* ρ *when and only when the conditions* $\rho \not< 0$, $\rho \not> r$, $2\rho \not> n$ *are satisfied, and they are then given by the formula*

$$[x]^{n}_{r} = [k]^{\rho}_{r} \, [1, \, i]^{\rho,\, \rho}_{\rho} \, [l]^{n}_{2\rho},$$

where $[k]^{\rho}_{r}$ *is an undegenerate matrix of rank* ρ, *and* $[l]^{n}_{2\rho}$ *is a real semi-unit matrix of rank* 2ρ.

COROLLARY 4. *The equation* $\overline{\underline{x}}{}^{s}_{m} \, [x]^{m}_{s} = 0$ *has solutions of rank* ρ *when and only when the conditions* $\rho \not< 0$, $\rho \not> m$, $2\rho \not> s$ *are satisfied, and they are then given by the formula*

$$[x]^{m}_{s} = [l]^{2\rho}_{s} \begin{bmatrix} 1 \\ i \end{bmatrix}^{\rho}_{\rho,\, \rho} [\xi]^{m}_{\rho},$$

where $[\xi]^{m}_{\rho}$ *is an undegenerate matrix of rank* ρ, *and* $[l]^{2\rho}_{s}$ *is a real semi-unit matrix of rank* 2ρ.

NOTE. Another proof of the theorem of the text is given in § 171.

§ 160. Solutions of symmetric equations of the form $X'X = C$.

1. *The equation* $\overline{\underline{x}}{}_{m} \, [x]_{m} = [c]^{m}_{m}$, ...(A)

in which $[c]^{m}_{m}$ *is a given symmetric matrix whose rank does not exceed* 1.

The equation clearly does not admit of solution when the rank of $[c]^{m}_{m}$ exceeds 1. We shall distinguish two cases.

Case I. *When the symmetric matrix* $[c]_m^m$ *has rank* 1.

In this case every solution must have rank 1. If $[x]_m = [a]_m$ is any particular solution, the equation can be written in the form

$$\underline{\overline{x}}_m [x]_m = \underline{\overline{a}}_m [a]_m,$$

and the general solution is then given by § 156.1.

Thus if $[x]_m = [a]_m$ *is any particular solution, the general solution is*

$$[x]_m = \pm [a]_m. \qquad \text{.............................}(A_1)$$

By § 125 the elements of the leading diagonal of $[c]_m^m$ do not all vanish, and if the diagonal element c_{ii} does not vanish, the equation can be written in the form

$$\begin{bmatrix} x_1 \\ x_2 \\ \vdots \\ x_m \end{bmatrix} [x_1 \, x_2 \dots x_m] = \begin{bmatrix} \sqrt{c_{11}} \\ \sqrt{c_{22}} \\ \vdots \\ \sqrt{c_{mm}} \end{bmatrix} [\sqrt{c_{11}} \, \sqrt{c_{22}} \dots \sqrt{c_{mm}}], \qquad \text{........}(1)$$

where the radicals are so chosen that

$$\sqrt{c_{ii}} [\sqrt{c_{11}} \, \sqrt{c_{22}} \dots \sqrt{c_{mm}}] = [c_{i1} \, c_{i2} \dots c_{im}]. \qquad \text{.............}(2)$$

Thus when $c_{ii} \neq 0$, *the general solution is*

$$[x_1 \, x_2 \dots x_m] = \pm [\sqrt{c_{11}} \, \sqrt{c_{22}} \dots \sqrt{c_{mm}}], \qquad \text{..................}(A_2)$$

where the radicals are chosen in accordance with (2).

The equation can also be solved in the following manner. When $c_{ii} \neq 0$, the equation can by Ex. v of § 115 or Ex. vi of § 116 be written in the form

$$c_{ii} \begin{bmatrix} x_1 \\ x_2 \\ \vdots \\ x_m \end{bmatrix} [x_1 \, x_2 \dots x_m] = \begin{bmatrix} c_{i1} \\ c_{i2} \\ \vdots \\ c_{im} \end{bmatrix} [c_{i1} \, c_{i2} \dots c_{im}]. \qquad \text{...........}(3)$$

Solving equation (3) by Ex. i of § 156, we see that

When $c_{ii} \neq 0$, *the general solution is given by*

$$\sqrt{c_{ii}} [x_1 \, x_2 \dots x_m] = \pm [c_{i1} \, c_{i2} \dots c_{im}]. \qquad \text{..................}(A_3)$$

The formulae (A_2) and (A_3) are equivalent in virtue of (2).

Case II. *When* $[c]_m^m$ *has rank* 0 *or vanishes.*

As in Case II of § 156.1 the only solution is

$$[x_1 \, x_2 \dots x_m] = 0. \qquad \text{..........................}(A_4)$$

The general solution is still given by (A_1) and (A_2) in this second case.

It has been shown in § 125 that when the symmetric matrix $[c]_m^m$ is real and has rank not exceeding 1, the elements of its leading diagonal which do not vanish are either all positive or all negative. Accordingly from (A_2) or by equating the elements of the leading diagonals on both sides of (A) we draw the following conclusion:

When the symmetric matrix $[c]_m^m$ is real and has rank not exceeding 1, the elements $x_1, x_2, \ldots x_m$ occurring in either of the solutions of the equation (A) are either all real or all purely imaginary.

Ex. i. *The equation $p \lceil x \rfloor_m [x]_m = q\,[c]_m^m$, in which $p \neq 0$, and $[c]_m^m$ is a symmetric matrix whose rank does not exceed* 1.

If $[c]_m^m$ has rank 0, the only solution is $[x_1\, x_2 \ldots x_m] = 0$.

If $[c]_m^m$ has rank 1 and $c_{ii} \neq 0$, the general solution is given by either of the formulae

$$\sqrt{p}\,[x_1\, x_2 \ldots x_m] = \pm\sqrt{q}\,[\sqrt{c_{11}}\,\sqrt{c_{22}} \ldots \sqrt{c_{mm}}], \qquad \ldots\ldots(4)$$

$$\sqrt{pc_{ii}}\,[x_1\, x_2 \ldots x_m] = \pm\sqrt{q}\,[c_{i1}\, c_{i2} \ldots c_{im}], \qquad \ldots\ldots(5)$$

the radicals in (5) being so chosen that the equation (2) is satisfied.

Ex. ii. *The equation* $\begin{bmatrix} x \\ y \\ z \\ w \end{bmatrix} [x\, y\, z\, w] = \begin{bmatrix} a & h & g & u \\ h & b & f & v \\ g & f & c & w \\ u & v & w & d \end{bmatrix} = \phi$, *when ϕ has rank* 1.

By § 125 the radicals $\sqrt{a}$, $\sqrt{b}$, $\sqrt{c}$, $\sqrt{d}$ can in all cases be so chosen that

$$\begin{bmatrix} a & h & g & u \\ h & b & f & v \\ g & f & c & w \\ u & v & w & d \end{bmatrix} = \begin{bmatrix} \sqrt{a} \\ \sqrt{b} \\ \sqrt{c} \\ \sqrt{d} \end{bmatrix} [\sqrt{a}\,\sqrt{b}\,\sqrt{c}\,\sqrt{d}].$$

When this is done the solutions of the equation are

$$[x\, y\, z\, w] = \pm[\sqrt{a}\,\sqrt{b}\,\sqrt{c}\,\sqrt{d}].$$

In the cases $a \neq 0$, $d \neq 0$ the equation can be written

$$a \begin{bmatrix} x \\ y \\ z \\ w \end{bmatrix} [x\, y\, z\, w] = \begin{bmatrix} a \\ h \\ g \\ u \end{bmatrix} [a\, h\, g\, u], \qquad d \begin{bmatrix} x \\ y \\ z \\ w \end{bmatrix} [x\, y\, z\, w] = \begin{bmatrix} u \\ v \\ w \\ d \end{bmatrix} [u\, v\, w\, d],$$

and its solutions are given by

$$\sqrt{a}\,[x\, y\, z\, w] = \pm[a\, h\, g\, u], \quad \sqrt{d}\,[x\, y\, z\, w] = \pm[u\, v\, w\, d].$$

The cases $b \neq 0$, $c \neq 0$ can be treated similarly.

Ex. iii. The solutions of the equations

$$\begin{bmatrix} x \\ y \\ z \\ w \end{bmatrix} [x\, y\, z\, w] = \begin{bmatrix} 4, & -6, & 2, & -4 \\ -6, & 9, & -3, & 6 \\ 2, & -3, & 1, & -2 \\ -4, & 6, & -2, & 4 \end{bmatrix}, \quad \begin{bmatrix} x \\ y \\ z \\ w \end{bmatrix} [x\, y\, z\, w] = \begin{bmatrix} -4, & 6, & -2, & 4 \\ 6, & -9, & 3, & -6 \\ -2, & 3, & -1, & 2 \\ 4, & -6, & 2, & -4 \end{bmatrix}$$

are respectively $[x\, y\, z\, w] = \pm[2, -3, 1, 2]$, $[x\, y\, z\, w] = \pm\sqrt{-1}\,[2, -3, 1, 2]$.

2. *Solutions of the equation*

$$\overline{x}_m^{\,r}\,[x]_r^m = [a]_m^m, \quad \text{.............................(B)}$$

in which $[a]_m^m$ *is a given symmetric matrix of rank* r.

That the equation (B) admits of solution when $[a]_m^m$ is a symmetric matrix of rank r has been shown in Ex. xv of § 147. In fact if the elements of $[a]_m^m$ are constants lying in any domain of rationality Ω, it has been shown in Theorem II of § 147 that we can determine an undegenerate square matrix $[h]_m^m$ and an undegenerate quasi-scalar matrix

$$[e]_r^r = \begin{bmatrix} e_1 & 0 & \dots & 0 \\ 0 & e_2 & \dots & 0 \\ \dots & \dots & \dots & \dots \\ 0 & 0 & \dots & e_r \end{bmatrix} \quad \text{...........................(6)}$$

which both lie in Ω and are such that

$$[a]_m^m = \overline{h}_m^{\,m} \begin{bmatrix} e, & 0 \\ 0, & 0 \end{bmatrix}_{r,\, m-r}^{r,\, m-r} [h]_m^m = \overline{h}_m^{\,r}\,[e]_r^r\,[h]_r^m, \quad \text{...........(7)}$$

and a particular solution of rank r of the equation (B) is given by

$$[x]_r^m = [\sqrt{e}]_r^r\,[h]_r^m, \text{ where } [\sqrt{e}]_r^r = \begin{bmatrix} \sqrt{e_1} & 0 & \dots & 0 \\ 0 & \sqrt{e_2} & \dots & 0 \\ \dots & \dots & \dots & \dots \\ 0 & 0 & \dots & \sqrt{e_r} \end{bmatrix}. \quad \text{...(B}_1\text{)}$$

Again if $[x]_r^m = [\alpha]_r^m$ is any one particular solution, where $[\alpha]_r^m$ necessarily has rank r, the equation (B) is equivalent to

$$\overline{x}_m^{\,r}\,[x]_r^m = \overline{\alpha}_m^{\,r}\,[\alpha]_r^m, \quad \text{..........................(8)}$$

and the general solution of this equation can be obtained as in § 156.2. We thus obtain the following theorem:

Theorem I. *When* $[a]_m^m$ *is a given symmetric matrix of rank* r, *the equation* (B) *always admits of solution, and every solution* $[x]_r^m$ *has rank* r.

If $[x]_r^m = [\alpha]_r^m$ *is any particular solution, the general solution is*

$$[x]_r^m = [l]_r^r\,[\alpha]_r^m, \qquad \text{.............................(B}_2\text{)}$$

where $[l]_r^r$ *is any square semi-unit matrix of rank* r.

The general solution is known when any one particular solution is known; and various ways of determining particular solutions have been described in § 147.

Ex. iv. With the notation used in (B_1) the general solution of the equation (B) is

$$[x]_r^m = [l]_r^r\,[\sqrt{e}]_r^r\,[h]_r^m, \qquad \text{.................................(B}_3\text{)}$$

where $[l]_r^r$ is any square semi-unit matrix of rank r.

Ex. v. If $\Delta = (a_{pp})_r^r$ is any one of the non-vanishing diagonal minor determinants of $[a]_m^m$ of order r, if $[A_{pp}]_r^r$ is the reciprocal of $[a_{pp}]_r^r$, and if $[u]_r^r$ and $[v]_r^r$ are any particular solutions of the equations

$$\overline{\underline{u}}{}_r^r\,[u]_r^r = [A_{pp}]_r^r, \quad \overline{\underline{v}}{}_r^r\,[v]_r^r = [a_{pp}]_r^r, \qquad \text{........................(9)}$$

then the general solution of the equation (B) is given by either of the formulae

$$\sqrt{\Delta}\,[x]_r^m = [l]_r^r\,[u]_r^r\,[a_{p1}]_r^m, \quad \Delta\,[x]_r^m = [l]_r^r\,[v]_r^r\,[A_{pp}]_r^r\,[a_{p1}]_r^m, \qquad \text{.......(B}_4\text{)}$$

where $[l]_r^r$ is any square semi-unit matrix of rank r.

For by § 116 we have the identities

$$\Delta\,[a]_m^m = \overline{\underline{a_{p1}}}{}_m^r\,[A_{pp}]_r^r\,[a_{p1}]_r^m, \quad \Delta^2\,[a]_m^m = \overline{\underline{a_{p1}}}{}_m^r\,\overline{\underline{A_{pp}}}{}_r^r\,[a_{pp}]_r^r\,[A_{pp}]_r^r\,[a_{p1}]_r^m,$$

and particular solutions of the equation (B) are given by

$$\sqrt{\Delta}\,[x]_r^m = [u]_r^r\,[a_{p1}]_r^m, \quad \Delta\,[x]_r^m = [v]_r^r\,[A_{pp}]_r^r\,[a_{p1}]_r^m.$$

Ex. vi. *Determination of a particular solution of the equation* (B).

A particular solution is best obtained by the transformation of Ex. xi in § 147.

If the symmetric matrix $[a]_m^m$ has rank 1, it has a non-vanishing diagonal element, which after a symmetric derangement may be supposed to be a_{11}. If the matrix has rank greater than 1, and if all its diagonal elements vanish, then it has a non-vanishing diagonal minor determinant of order 2, and after a symmetric derangement of $[a]_m^m$ we may suppose that $a_{12} \neq 0$. Consequently after a symmetric derangement of $[a]_m^m$ we can always apply to it one of the two following transformations.

First transformation. If $a_{11} \neq 0$, we have by § 116

$$a_{11}\,[a]_m^m = \overline{\underline{a}}{}_m^1\,[a]_1^m + \begin{bmatrix} 0, & 0 \\ 0, & b \end{bmatrix}_{1,\,m-1}^{1,\,m-1}, \quad \text{where} \quad b_{uv} = \begin{pmatrix} 1,\,1+v \\ a \\ 1,\,1+u \end{pmatrix};$$

and therefore $$[a]_m^m = \underset{m}{\overline{\underline{h}}}^1 [h]_1^m + \frac{1}{a_{11}} \begin{bmatrix} 0, & 0 \\ 0, & b \end{bmatrix}_{1,\, m-1}^{1,\, m-1}, \quad \text{.......................(10)}$$

where $$[h]_1^m = \frac{1}{\sqrt{a_{11}}} [a_{11}\, a_{12} \dots a_{1m}].$$

Second transformation. If all diagonal elements of $[a]_m^m$ vanish, and if $a_{12} \neq 0$, so that $(a)_2^2 = -a_{12}^2 \neq 0$, we have by § 116

$$(a)_2^2 [a]_m^m = \underset{m}{\overline{\underline{a}}}^2 \begin{bmatrix} 0 \;, & -a_{12} \\ -a_{12}, & 0 \end{bmatrix} [a]_2^m + \begin{bmatrix} 0, & 0 \\ 0, & b \end{bmatrix}_{2,\, m-2}^{2,\, m-2},$$

or $$[a]_m^m = \frac{1}{a_{12}} \underset{m}{\overline{\underline{a}}}^2 \begin{bmatrix} 0 & 1 \\ 1 & 0 \end{bmatrix} [a]_2^m + \frac{1}{(a)_2^2} \begin{bmatrix} 0, & 0 \\ 0, & b \end{bmatrix}_{2,\, m-2}^{2,\, m-2},$$

where $$b_{uv} = \begin{pmatrix} 1,\, 2,\, 2+v \\ a \\ 1,\, 2,\, 2+u \end{pmatrix} = a_{12} (a_{1,\, 2+u}\, a_{2,\, 2+v} + a_{1,\, 2+v}\, a_{2,\, 2+u}) - a_{12}^2\, a_{2+u,\, 2+v};$$

or writing $\begin{bmatrix} 0 & 1 \\ 1 & 0 \end{bmatrix} = \frac{1}{2} \begin{bmatrix} 1, & i \\ 1, & -i \end{bmatrix} \begin{bmatrix} 1, & 1 \\ i, & -i \end{bmatrix}$, where $i = \sqrt{-1}$, we have

$$[a]_m^m = \underset{m}{\overline{\underline{h}}}^2 [h]_2^m + \frac{1}{(a)_2^2} \begin{bmatrix} 0, & 0 \\ 0, & b \end{bmatrix}_{2,\, m-2}^{2,\, m-2}, \quad \text{.......................(11)}$$

where $$[h]_2^m = \begin{bmatrix} 1, & 1 \\ i, & -i \end{bmatrix} [a]_2^m$$

$$= \begin{bmatrix} (a_{11}+a_{21}), & (a_{12}+a_{22}), & \dots & (a_{1u}+a_{2u}), & \dots & (a_{1m}+a_{2m}) \\ i\,(a_{11}-a_{21}), & i\,(a_{12}-a_{22}), & \dots & i\,(a_{1u}-a_{2u}), & \dots & i\,(a_{1m}-a_{2m}) \end{bmatrix},$$

and $a_{11} = a_{22} = 0$, $a_{12} = a_{21} \neq 0$.

In the particular case when $[a]_m^m$ is real and definite only the first transformation can be used.

Whichever transformation is applicable, we can then treat $[b]_{m-1}^{m-1}$ or $[b]_{m-2}^{m-2}$ in the same manner; and proceeding in this way we finally express $[a]_m^m$ in the form

$$[a]_m^m = \underset{m}{\overline{\underline{\omega}}}^m \underset{m}{\overline{\underline{h}}}^r [h]_r^m [\omega]_m^m,$$

where $[\omega]_m^m$ is a derangement of the unit matrix $[1]_m^m$, and $[h]_r^m$ is a matrix in which none of the elements of the leading diagonal vanish. A particular solution of the equation (B) is then given by

$$[x]_r^m = [h]_r^m [\omega]_m^m. \quad \text{.......................................}(B_5)$$

If the symbol τ merely indicates the presence of a constituent matrix, the matrix $[h]_r^m$ has the form

$$[h]_r^m = \begin{bmatrix} \tau, & \tau, & \tau, & \dots & \tau, & \tau, & \tau \\ 0, & \tau, & \tau, & \dots & \tau, & \tau, & \tau \\ 0, & 0, & \tau, & \dots & \tau, & \tau, & \tau \\ \dots & \dots & \dots & \dots & \dots & \dots & \dots \\ 0, & 0, & 0, & \dots & \tau, & \tau, & \tau \\ 0, & 0, & 0, & \dots & 0, & \tau, & \tau \end{bmatrix}_{\alpha,\, \beta,\, \gamma,\, \dots\, \kappa,\, \lambda}^{\alpha,\, \beta,\, \gamma,\, \dots\, \kappa,\, \lambda,\, m-r},$$

where each of the integers $\alpha, \beta, \gamma, \dots \kappa, \lambda$ is either 1 or 2, and

$$\alpha + \beta + \gamma + \dots + \kappa + \lambda = r.$$

Ex. vii. *Particular solution of* (B) *when the rows of* $[a]_m^m$ *are arranged as in Theorem I b of* § 147.

We can always reduce the equation (B) to this form by applying a symmetric derangement to both sides.

Let Δ_κ, Δ_λ be any two consecutive non-vanishing determinants of the series

$$\Delta_0=1,\ \Delta_1=a_{11},\ \Delta_2=(a)_2^2,\ \Delta_3=(a)_3^3,\ \ldots\ \Delta_r=(a)_r^r,$$

so that $\lambda-\kappa$ is either 1 or 2, and let

$$l_{uv}=\begin{pmatrix}1,\,2,\,\ldots\,\kappa,\,\kappa+v\\ a\\ 1,\,2,\,\ldots\,\kappa,\,\kappa+u\end{pmatrix},\quad \delta_{\kappa+1}=\begin{pmatrix}1,\,2,\,\ldots\,\kappa,\,\kappa+2\\ a\\ 1,\,2,\,\ldots\,\kappa,\,\kappa+1\end{pmatrix}=l_{12}.$$

Then a particular solution of (B) is $[x]_r^m=[h]_r^m$, where:

(1) when $\kappa=\lambda-1$ (so that $l_{11}=\Delta_{\kappa+1}=\Delta_\lambda\neq0$), the λth horizontal row of $[h]_r^m$ is

$$\frac{1}{\sqrt{\Delta_\kappa\Delta_{\kappa+1}}}\,[0,\ l]_1^{\kappa,\,m-\kappa};$$

(2) when $\kappa=\lambda-2$ (so that $l_{11}=l_{22}=0$, and $l_{12}=l_{21}=\delta_{\kappa+1}\neq0$), the $(\lambda-1)$th and λth horizontal rows of $[h]_r^m$ are

$$\frac{1}{\sqrt{2\Delta_\kappa\delta_{\kappa+1}}}\begin{bmatrix}1, & 1\\ i, & -i\end{bmatrix}[0,\ l]_2^{\kappa,\,m-\kappa};$$

this being true for all the values of λ.

These results are equivalent to the equation (B_2) in Ex. xi of § 147.

Ex. viii. *Particular solution of the equation* $\overline{\underline{x}}_4^{\,r}\,[x]_r^4=[a]_4^4$, *when* $[a]_4^4$ *is a symmetric matrix of rank r whose rows are arranged as in Theorem I b of* § 147.

First suppose that $r=4$. Then by Ex. vii there is a particular solution whose successive horizontal rows in the five possible cases are as follows:

Case I. $\Delta_1\neq0,\ \Delta_2\neq0,\ \Delta_3\neq0,\ \Delta_4\neq0.$

$$\frac{1}{\sqrt{\Delta_0\Delta_1}}[a_{11}\,a_{12}\,a_{13}\,a_{14}],\quad \frac{1}{\sqrt{\Delta_1\Delta_2}}[0\,b_{11}\,b_{12}\,b_{13}],\quad \frac{1}{\sqrt{\Delta_2\Delta_3}}[0\,0\,c_{11}\,c_{12}],\quad \frac{1}{\sqrt{\Delta_3\Delta_4}}[0\,0\,0\,d_{11}],$$

where $$b_{uv}=\begin{pmatrix}1,\,1+v\\ a\\ 1,\,1+u\end{pmatrix},\quad c_{uv}=\begin{pmatrix}1,\,2,\,2+v\\ a\\ 1,\,2,\,2+u\end{pmatrix},\quad d_{uv}=\begin{pmatrix}1,\,2,\,3,\,3+v\\ a\\ 1,\,2,\,3,\,3+u\end{pmatrix}.$$

Case II. $\Delta_1=0,\ \Delta_2\neq0,\ \Delta_3\neq0,\ \Delta_4\neq0.$

$$\frac{1}{\sqrt{2\Delta_0\delta_1}}\begin{bmatrix}1, & 1\\ i, & -i\end{bmatrix}\begin{bmatrix}0 & a_{12} & a_{13} & a_{14}\\ a_{21} & 0 & a_{23} & a_{24}\end{bmatrix},\quad \frac{1}{\sqrt{\Delta_2\Delta_3}}[0\,0\,b_{11}\,b_{12}],\quad \frac{1}{\sqrt{\Delta_3\Delta_4}}[0\,0\,0\,c_{11}],$$

where $$b_{uv}=\begin{pmatrix}1,\,2,\,2+v\\ a\\ 1,\,2,\,2+u\end{pmatrix},\quad c_{uv}=\begin{pmatrix}1,\,2,\,3,\,3+v\\ a\\ 1,\,2,\,3,\,3+u\end{pmatrix},\quad \delta_1=a_{12}=a_{21}\neq0.$$

Case III. $\Delta_1\neq0,\ \Delta_2=0,\ \Delta_3\neq0,\ \Delta_4\neq0.$

$$\frac{1}{\sqrt{\Delta_0\Delta_1}}[a_{11}\,a_{12}\,a_{13}\,a_{14}],\quad \frac{1}{\sqrt{2\Delta_1\delta_2}}\begin{bmatrix}1, & 1\\ i, & -i\end{bmatrix}\begin{bmatrix}0 & 0 & b_{12} & b_{13}\\ 0 & b_{21} & 0 & b_{23}\end{bmatrix},\quad \frac{1}{\sqrt{\Delta_3\Delta_4}}[0\,0\,0\,c_{11}],$$

where $$b_{uv}=\begin{pmatrix}1,\,1+v\\ a\\ 1,\,1+u\end{pmatrix},\quad c_{uv}=\begin{pmatrix}1,\,2,\,3,\,3+v\\ a\\ 1,\,2,\,3,\,3+u\end{pmatrix},\quad \delta_2=b_{12}=b_{21}\neq0.$$

CASE IV. $\Delta_1 \neq 0$, $\Delta_2 \neq 0$, $\Delta_3 = 0$, $\Delta_4 \neq 0$.

$$\frac{1}{\sqrt{\Delta_0 \Delta_1}}[a_{11}\, a_{12}\, a_{13}\, a_{14}], \quad \frac{1}{\sqrt{\Delta_1 \Delta_2}}[0\, b_{11}\, b_{12}\, b_{13}], \quad \frac{1}{\sqrt{2\Delta_2 \delta_3}}\begin{bmatrix} 1, & 1 \\ i, & -i \end{bmatrix}\begin{bmatrix} 0 & 0 & 0 & c_{12} \\ 0 & 0 & c_{21} & 0 \end{bmatrix},$$

where
$$b_{uv} = \begin{pmatrix} 1,\ 1+v \\ a \\ 1,\ 1+u \end{pmatrix}, \quad c_{uv} = \begin{pmatrix} 1,\ 2,\ 2+v \\ a \\ 1,\ 2,\ 2+u \end{pmatrix}, \quad \delta_3 = c_{12} = c_{21} \neq 0.$$

CASE V. $\Delta_1 = 0$, $\Delta_2 \neq 0$, $\Delta_3 = 0$, $\Delta_4 \neq 0$.

$$\frac{1}{\sqrt{2\Delta_0 \delta_1}}\begin{bmatrix} 1, & 1 \\ i, & -i \end{bmatrix}\begin{bmatrix} 0 & a_{12} & a_{13} & a_{14} \\ a_{21} & 0 & a_{23} & a_{24} \end{bmatrix}, \quad \frac{1}{\sqrt{2\Delta_2 \delta_3}}\begin{bmatrix} 1, & 1 \\ i, & -i \end{bmatrix}\begin{bmatrix} 0 & 0 & 0 & b_{12} \\ 0 & 0 & b_{21} & 0 \end{bmatrix},$$

where
$$b_{uv} = \begin{pmatrix} 1,\ 2,\ 2+v \\ a \\ 1,\ 2,\ 2+u \end{pmatrix}, \quad \delta_1 = a_{12} = a_{21} \neq 0, \quad \delta_3 = b_{12} = b_{21} \neq 0.$$

When $r=3$, there are three possible cases derived from Cases I, II, III above by putting $\Delta_4 = 0$ and omitting the last horizontal row of $[h]_4^4$.

When $r=2$, there are two possible cases derived from Cases I, II above by putting $\Delta_3 = 0$, $\Delta_4 = 0$, and omitting the last two horizontal rows of $[h]_4^4$.

When $r=1$, there is only one possible case derived from Case I above by putting $\Delta_2 = 0$, $\Delta_3 = 0$, $\Delta_4 = 0$, and omitting the last three horizontal rows of $[h]_4^4$.

Ex. ix. *The equation* $\overline{\underline{x}}{}^{r}_{m}\, [x]_r^m = \overline{\underline{a}}{}^{r}_{m}\, [b]_r^r\, [a]_r^m$, *where* $[a]_r^m$ *and* $[b]_r^r$ *are given matrices of rank* r.

By equating the conjugates of both sides we see that the equation does not admit of solution unless $[b]_r^r$ is self-conjugate.

When $[b]_r^r$ is self-conjugate, the general solution is

$$[x]_r^m = [l]_r^r\, [\beta]_r^r\, [a]_r^m,$$

where $[\beta]_r^r$ is a particular solution of $\overline{\underline{\beta}}{}^{r}_{r}\, [\beta]_r^r = [b]_r^r$, and $[l]_r^r$ is any square semi-unit matrix of rank r.

Ex. x. *The equation* $\overline{\underline{x}}{}^{r}_{m}\, [b]_r^r\, [x]_r^m = \overline{\underline{a}}{}^{r}_{m}\, [a]_r^m$, *where* $[a]_r^m$ *and* $[b]_r^r$ *are given matrices of rank* r.

The equation does not admit of solution unless $[b]_r^r$ is self-conjugate.

When $[b]_r^r$ is self-conjugate, the general solution is given by

$$\sqrt{\Delta}\,[x]_r^m = \overline{\underline{\mathrm{B}}}{}^{r}_{r}\, [l]_r^r\, [a]_r^m,$$

where $\Delta = (b)_r^r$, $[\mathrm{B}]_r^r$ is the reciprocal of any particular solution $[\beta]_r^r$ of the equation $\overline{\underline{\beta}}{}^{r}_{r}\, [\beta]_r^r = [b]_r^r$, and $[l]_r^r$ is any square semi-unit matrix of rank r.

Ex. xi. *The equation* $\overline{\underline{x}}{}^{r}_{m}\, [b]_r^r\, [x]_r^m = \overline{\underline{a}}{}^{r}_{m}\, [c]_r^r\, [a]_r^m$, *where* $[a]_r^m$, $[b]_r^r$, $[c]_r^r$ *are given matrices of rank* r, *and where* $[b]_r^r$ *and* $[c]_r^r$ *are self-conjugate.*

The general solution is given by

$$\sqrt{\Delta}\,[x]_r^m = \overline{\underline{\mathrm{B}}}{}^{r}_{r}\, [l]_r^r\, [\gamma]_r^r\, [a]_r^m,$$

where $\Delta=(b)_{r}^{r}$, $[B]_{r}^{r}$ is the reciprocal of any particular solution $[\beta]_{r}^{r}$ of the equation $\overline{\underline{\beta}}{}_{r}^{r}\,[\beta]_{r}^{r}=[b]_{r}^{r}$, $[\gamma]_{r}^{r}$ is any particular solution of the equation $\overline{\underline{\gamma}}{}_{r}^{r}\,[\gamma]_{r}^{r}=[c]_{r}^{r}$, and $[l]_{r}^{r}$ is any square semi-unit matrix of rank r.

In §§ 162—4 we consider some special equations of this form.

3. *Semi-real solutions of the symmetric equation* $\overline{\underline{x}}{}_{m}^{r}\,[x]_{r}^{m}=[a]_{m}^{m}$ *when* $[a]_{m}^{m}$ *is a given real symmetric matrix of rank* r.

When $[a]_{m}^{m}$ is a *real* symmetric matrix of rank r, the matrices $[h]_{m}^{m}$ and $[e]_{r}^{r}$ in the particular solution (B_1) of the equation

$$\overline{\underline{x}}{}_{m}^{r}\,[x]_{r}^{m}=[a]_{m}^{m} \quad\text{..............................(B)}$$

are real. Consequently in the ith horizontal row of the particular solution $[x]_{r}^{m}=[\sqrt{e}]_{r}^{r}\,[h]_{r}^{m}$, i.e. in the matrix $\sqrt{e_i}\,[h_{i1}\,h_{i2}\ldots h_{im}]$, every element is real or every element is purely imaginary according as the real quantity e_i is positive or negative. Thus in this case the equation (B) has solutions such that the elements of each pair of corresponding passive rows in the product on the left are either all real or all purely imaginary. Such solutions will be called *semi-real solutions*. The particular solutions obtained in Exs. vi, vii and viii are semi-real when $[a]_{m}^{m}$ or $[a]_{4}^{4}$ is real.

If $[x]_{r}^{m}=[\alpha]_{r}^{m}$ is any semi-real solution in which π of the horizontal or passive rows are real, and the remaining ν horizontal or passive rows are purely imaginary, so that $\pi+\nu=r$, we can write

$$[\alpha]_{r}^{m}=[\omega]_{r}^{r}\begin{bmatrix}1, & 0\\ 0, & i\end{bmatrix}[k]_{r}^{m},$$

where $i=\sqrt{-1}$, $[\omega]_{r}^{r}$ is a derangement of $[1]_{r}^{r}$, and $[k]_{r}^{m}$ is a real undegenerate matrix of rank r. We then have

$$[a]_{m}^{m}=\overline{\underline{\alpha}}{}_{m}^{r}\,[\alpha]_{r}^{m}=\overline{\underline{k}}{}_{m}^{r}\begin{bmatrix}1, & 0\\ 0, & i\end{bmatrix}_{\pi,\,\nu}^{\pi,\,\nu}\begin{bmatrix}1, & 0\\ 0, & i\end{bmatrix}_{\pi,\,\nu}^{\pi,\,\nu}[k]_{r}^{m}$$

$$=\overline{\underline{k}}{}_{m}^{r}\begin{bmatrix}1, & 0\\ 0, & -1\end{bmatrix}_{\pi,\,\nu}^{\pi,\,\nu}[k]_{r}^{m}=\overline{\underline{k}}{}_{m}^{m}\begin{bmatrix}1, & 0, & 0\\ 0, & -1, & 0\\ 0, & 0, & 0\end{bmatrix}_{\pi,\,\nu,\,m-r}^{\pi,\,\nu,\,m-r}[k]_{m}^{m},$$

where $[k]_{m}^{m}$ is a real undegenerate square matrix.

By § 148 we conclude that π and ν are the positive and negative signants of $[a]_m^m$, and are therefore always the same.

We have thus proved the following theorem:

Theorem II. *When $[a]_m^m$ is a real symmetric matrix of rank r, the equation* (B) *has solutions $[x]_r^m$ such that the elements of each horizontal or passive row are either all real or all purely imaginary.*

In every such solution the numbers of real and purely imaginary horizontal or passive rows are always the same, being equal respectively to the positive and negative signants of $[a]_m^m$.

Ex. xii. *If $[a]_r^m$ and $[\beta]_r^m$ are two matrices of rank r in both of which the elements of each horizontal row are either all real or all purely imaginary, and if*

$$[\beta]_r^m=[l]_r^r\,[a]_r^m, \quad\text{.......................................(12)}$$

then any element l_{hk} of $[l]_r^r$ is real when the hth horizontal row of $[\beta]_r^m$ and the kth horizontal row of $[a]_r^m$ are both real or both purely imaginary, and is purely imaginary when one of those horizontal rows is real and the other is purely imaginary.

If the hth horizontal row of $[\beta]_r^m$ is real, then by equating the purely imaginary portions on both sides of the equation (5) we obtain

$$[\lambda_{h1}\,\lambda_{h2}\ldots\lambda_{hr}]\,[a]_r^m=0,$$

where λ_{hk} is the real portion or the purely imaginary portion of l_{hk} according as the kth horizontal row of $[a]_r^m$ is purely imaginary or real. Since there is no connection between the horizontal rows of $[a]_r^m$, it follows that

$$[\lambda_{h1}\,\lambda_{h2}\ldots\lambda_{hr}]=0, \quad \lambda_{hk}=0. \quad\text{..................................(13)}$$

The equation $\lambda_{hk}=0$ shows that in this case l_{hk} is real or purely imaginary according as the kth horizontal row of $[a]_r^m$ is real or purely imaginary.

Again if the hth horizontal row of $[\beta]_r^m$ is purely imaginary, then by equating the real portions on both sides of the equation (12) we obtain in the same way the equation (13), where now λ_{hk} is the purely imaginary portion or the real portion of l_{hk} according as the kth horizontal row of $[a]_r^m$ is purely imaginary or real. The equation $\lambda_{hk}=0$ shows in this case that l_{hk} is real or purely imaginary according as the kth horizontal row of $[a]_r^m$ is purely imaginary or real.

Ex. xiii. *Alternative proof of the last part of Theorem II.*

Let $[x]_r^m=[a]_r^m$ and $[x]_r^m=[\beta]_r^m$ be any two semi-real solutions of the equation (B), and let the respective matrices $[a]_r^m$ and $[\beta]_r^m$ contain exactly u and exactly v real horizontal rows.

The theorem states that $v=u$.

We may suppose without loss of generality that the first u horizontal rows of $[a]_r^m$ and the first v horizontal rows of $[\beta]_r^m$ are real, and we know that there exists a relation of the form

$$[\beta]_r^m=[l]_r^r\,[a]_r^m,$$

where $[l]_r^r$ is a square semi-unit matrix.

By Ex. xii the matrix $[l]_r^r$ has the form

$$[l]_r^r=\begin{bmatrix} P, & ip \\ iq, & Q \end{bmatrix}_{v,\,r-v}^{u,\,r-u},$$

where $i=\sqrt{-1}$, and $[P]_v^u$, $[Q]_{r-v}^{r-u}$, $[p]_v^{r-u}$, $[q]_{r-v}^u$ are all real.

It follows by Ex. xviii of § 155 that $v=u$.

4. *Solutions of the equation*

$$\overline{\underline{x}}_m^s\,[x]_s^m=[c]_m^m \quad \text{.............................(C)}$$

in which $[c]_m^m$ *is a given symmetric matrix of rank* r.

If we express $[c]_m^m$ in the form $\overline{\underline{a}}_m^r\,[a]_r^m$ by any of the methods described in sub-article 3 or in § 147, the equation (C) is equivalent to

$$\overline{\underline{x}}_m^s\,[x]_s^m=\overline{\underline{a}}_m^r\,[a]_r^m, \quad \text{..........................(C')}$$

where $[a]_r^m$ has rank r, and can be treated as in § 157. As there shown the equation admits of solution only when $s \nless r$, and it has solutions of rank ρ when and only when ρ is an integer consistent with the conditions

$$\rho \nless r, \quad \rho \ngtr m, \quad 2\rho \ngtr r+s,$$

which include the necessary condition $s \nless r$.

Ex. xiv. *General formulae for a matrix* $[x]_s^m$ *which has rank* ρ *and is such that the product* $\overline{\underline{x}}_m^s\,[x]_s^m$ *has rank* r.

The matrix $[x]_s^m$ has these properties when and only when it is a solution of rank ρ of some equation of the form (C′) in which $[a]_r^m$ is a matrix of rank r; and by § 157 this is possible when and only when ρ and r are integers satisfying the conditions

$$r \nless 0; \quad \rho \nless r, \quad \rho \ngtr m, \quad 2\rho \ngtr r+s,$$

which include $r \nless 0$, $r \ngtr m$, $r \ngtr s$ and $\rho \nless 0$, $\rho \ngtr s$, $\rho \ngtr m$.

When these conditions are satisfied, we obtain general formulae for all such matrices from the formulae given in Theorem III of § 157 by replacing $\begin{bmatrix} a \\ \xi \end{bmatrix}_{r,\,\rho-r}^m$ and $\begin{bmatrix} a \\ \xi \end{bmatrix}_{r,\,m-r}^m$

by arbitrary undegenerate matrices $[\xi]^m_\rho$ and $[\xi]^m_m$ having ranks ρ and m. In particular all such matrices are given by either of the following formulae in which $i=\sqrt{-1}$:

$$(1)\qquad [x]^m_s=[l]^{2\rho-r}_s \begin{bmatrix}1, & 0\\ 0, & 1\\ 0, & i\end{bmatrix}^{r,\ \rho-r}_{r,\ \rho-r,\ \rho-r} [\xi]^m_\rho, \qquad \text{.......................}(D_1)$$

where $[\xi]^m_\rho$ has rank ρ, and $[l]^{2\rho-r}_s$ is a semi-unit matrix of rank $2\rho-r$.

$$(2)\qquad [x]^m_s=[l]^s_s \begin{bmatrix}1, & 0, & 0\\ 0, & 1, & 0\\ 0, & i, & 0\\ 0, & 0, & 0\end{bmatrix}^{r,\ \rho-r,\ m-\rho}_{r,\ \rho-r,\ \rho-r,\ r+s-2\rho} [\xi]^m_m, \qquad \text{.................}(D_2)$$

where $[\xi]^m_m$ is an undegenerate square matrix, and $[l]^s_s$ is a semi-unit matrix of rank s.

Ex. xv. When $[c]^m_m$ is a symmetric matrix of rank r, it follows from § 147.2 that we can determine an undegenerate square matrix $[h]^m_m$ such that

$$[c]^m_m=\overline{h}^m_m \begin{bmatrix}1, & 0\\ 0, & 0\end{bmatrix}^{r,\ m-r}_{r,\ m-r} [h]^m_m .$$

The equation (C) is then satisfied when and only when

$$[x]^m_s=[y]^m_s\,[h]^m_m, \qquad \text{......................................}(14)$$

where

$$\overline{y}^s_m\,[y]^m_s=\begin{bmatrix}1, & 0\\ 0, & 0\end{bmatrix}^{r,\ m-r}_{r,\ m-r}=\overline{\begin{matrix}1\\ 0\end{matrix}}^{\,r}_{r,\ m-r}\,[1,\ 0]^{r,\ m-r}_r . \qquad \text{....................}(c)$$

Thus there is a one-one correspondence between the solutions of the equation (C) and the solutions of the equation (c), and the general solution of the equation (C) can be expressed in the form (14), where $[y]^m_s$ is the general solution of the equation (c), which is given in Ex. vii of § 157.

§ 161. Solutions of symmetric equations of the form $X'AX=C$.

Let m and r be any given positive integers, and let $[a]^r_r$ and $[c]^m_m$ be given symmetric matrices of ranks α and γ, so that α and γ are given integers satisfying the conditions

$$\alpha \nless 0,\quad \alpha \ngtr r;\quad \gamma \nless 0,\quad \gamma \ngtr m. \qquad \text{.....................}(1)$$

We shall always have $m \nless 1$, $r \nless 1$; but if we regard a matrix either of whose orders is 0 as non-existent and replaceable as a factor by the scalar number 0, there will be no necessity for excluding the value 0 of m or r.

If $[x]^m_r$ is a solution of rank ρ of the symmetric equation

$$\overline{x}^r_m\,[a]^r_r\,[x]^m_r=[c]^m_m, \qquad \text{..........................}(A)$$

we know by § 133 that the integers α, γ and ρ must satisfy the conditions

$$\gamma \nless 0,\quad \gamma \ngtr \alpha,\quad \gamma \ngtr \rho;\quad \gamma + 2r \nless 2\rho + \alpha;\quad \rho \ngtr m,\quad \rho \ngtr r;\quad \alpha \ngtr r; \quad \text{...(2)}$$

which are equivalent to (1) together with the conditions

$$\gamma \ngtr \alpha;\quad \rho \nless \gamma,\quad \rho \ngtr m,\quad 2\rho + \alpha \ngtr \gamma + 2r. \quad \text{..............(B)}$$

Thus the equation (A) does not admit of solution unless the condition $\gamma \ngtr \alpha$ is satisfied.

To solve the equation (A) we determine as in § 160.2 particular matrices $[p]^m_\gamma$, $[q]^r_\alpha$ of ranks γ and α such that

$$[c]^m_m = \lceil p \rfloor^{\gamma}_{m} [p]^m_\gamma, \quad [a]^r_r = \lceil q \rfloor^{\alpha}_{r} [q]^r_\alpha .$$

Then the equation (A) is satisfied when and only when

$$[q]^r_\alpha [x]^m_r = [y]^m_\alpha, \text{ where } \lceil y \rfloor^{\alpha}_{m} [y]^m_\alpha = \lceil p \rfloor^{\gamma}_{m} [p]^m_\gamma . \quad \text{........(3)}$$

By § 157 the second of the equations (3) admits of solution for $[y]^m_\alpha$ when and only when $\gamma \ngtr \alpha$, and it has solutions of rank t when and only when t satisfies the conditions

$$t \nless \gamma,\quad t \ngtr m,\quad 2t \ngtr \alpha + \gamma \quad \text{........................(4)}$$

which include $t \ngtr \alpha$, $\gamma \ngtr \alpha$. All solutions of rank t are then given by the formulae of Theorem III in § 157.

Again when $[y]^m_\alpha$ is any particular matrix of rank t thus determined, the first of the equations (3) admits of finite solutions for $[x]^m_r$, and by Theorem I of § 132 it has solutions $[x]^m_r$ of rank ρ when and only when ρ satisfies the conditions

$$\rho \nless t,\quad \rho \ngtr m,\quad \rho + \alpha \ngtr t + r. \quad \text{....................(5)}$$

Since the elimination of t from (4) and (5) leads to the conditions (B) we have the following theorem:

Theorem. *The equation* (A) *in which* $[c]^m_m$ *and* $[a]^r_r$ *are given symmetric matrices of ranks* γ *and* α *has solutions of rank* ρ *when and only when the conditions* (B) *are satisfied, i.e. when and only when the necessary conditions* (2) *are satisfied. It admits of solution when and only when* α *and* γ *satisfy the conditions*

$$\alpha \nless 0,\quad \alpha \ngtr m,\quad \beta \nless 0,\quad \beta \ngtr r;\quad \gamma \ngtr \alpha.$$

In particular the equation (A) has undegenerate solutions of rank m when and only when

$$\gamma \ngtr \alpha,\quad 2m + \alpha \ngtr \gamma + 2r;$$

and it has undegenerate solutions of rank r when and only when

$$\gamma = \alpha,\quad r \ngtr m.$$

NOTE 1. The theorem given above is equivalent to Theorem III *b* of § 136.

NOTE 2. *Alternative method of solving* (A) *and obtaining the conditions* (B).

By § 147 we can determine particular undegenerate square matrices $[h]_m^m$, $[k]_r^r$ such that

$$[c]_m^m = \overline{\underline{h}}{}_m^m \begin{bmatrix}1, & 0\\ 0, & 0\end{bmatrix}_{\gamma,\,m-\gamma}^{\gamma,\,m-\gamma} [h]_m^m, \quad [a]_r^r = \overline{\underline{k}}{}_r^r \begin{bmatrix}1, & 0\\ 0, & 0\end{bmatrix}_{\alpha,\,r-\alpha}^{\alpha,\,r-\alpha} [k]_r^r .$$

Then the equation (A) is satisfied when and only when

$$[k]_r^r\,[x]_r^m = [y]_r^m\,[h]_m^m,$$

where $$\overline{\underline{y}}{}_m^r \begin{bmatrix}1, & 0\\ 0, & 0\end{bmatrix}_{\alpha,\,r-\alpha}^{\alpha,\,r-\alpha} [y]_r^m = \overline{\underline{y}}{}_m^\alpha\,[y]_\alpha^m = \begin{bmatrix}1, & 0\\ 0, & 0\end{bmatrix}_{\gamma,\,m-\gamma}^{\gamma,\,m-\gamma} \qquad \text{.....(A}')$$

Thus there is a one-one correspondence between the solutions $[x]_r^m$ of the equation (A) and the solutions $[y]_r^m$ of the equation (A′), corresponding solutions have equal ranks; and the solution of (A) is reduced to the solution of (A′).

By Ex. vi of § 157 we can determine a matrix $[y]_\alpha^m$ of rank t satisfying the equation (A′) when and only when

$$t \nless \gamma, \quad t \ngtr m, \quad 2t \ngtr \alpha+\gamma; \qquad \text{.....(4)}$$

and, as shown in Theorem I *b* of § 104, we can by adding final horizontal rows to $[y]_\alpha^m$ form a matrix $[y]_r^m$ of rank ρ satisfying the equation (A′) when and only when

$$\rho \nless t, \quad \rho \ngtr m, \quad \rho+\alpha \ngtr t+r. \qquad \text{.....(5)}$$

Eliminating t from (4) and (5) we see that we can determine solutions $[y]_r^m$ of (A′) of rank ρ, and therefore also solutions $[x]_r^m$ of (A) of rank ρ, when and only when the conditions (B) are satisfied.

§ 162. Some special equations of the form $[x]_2^2\,[y]_2^2 = [c]_2^2$, where $[c]_2^2$ is a given symmetric matrix.

1. *The equation* $$\begin{bmatrix}x_1 & x_2\\ y_1 & y_2\end{bmatrix}\begin{bmatrix}x_1 & y_1\\ x_2 & y_2\end{bmatrix} = \begin{bmatrix}a & h\\ h & b\end{bmatrix} = \phi. \qquad \text{.....(A)}$$

(*a*) *Particular solutions.* Writing $\Delta = ab - h^2$, particular solutions in the special cases $a \neq 0$; $b \neq 0$; $a = b = 0$, $h \neq 0$ are

$$\begin{bmatrix}x_1 & y_1\\ x_2 & y_2\end{bmatrix} = \frac{1}{\sqrt{a}}\begin{bmatrix}a, & h\\ 0, & \sqrt{\Delta}\end{bmatrix}, \quad \frac{1}{\sqrt{b}}\begin{bmatrix}h, & b\\ \sqrt{\Delta}, & 0\end{bmatrix}, \quad \sqrt{\tfrac{1}{2}h}\begin{bmatrix}1, & 1\\ i, & -i\end{bmatrix}$$

where $i = \sqrt{-1}$. In these the horizontal rows can be interchanged.

(*b*) *General solution when* ϕ *has rank* 2.

If $[x\,y]_{12} = [\alpha\,\beta]_{12}$ is any particular solution, the general solution is

$$\begin{bmatrix}x_1 & y_1\\ x_2 & y_2\end{bmatrix} = \begin{bmatrix}l_1 & m_1\\ l_2 & m_2\end{bmatrix}\begin{bmatrix}\alpha_1 & \beta_1\\ \alpha_2 & \beta_2\end{bmatrix}, \qquad \text{.....(A}_1)$$

where $[l\,m]_{12}$ is an arbitrary square semi-unit matrix of rank 2. Inserting the particular solutions given in (*a*), and making use of Ex. xx of § 155 we can express the general solution in the following forms:

CASE I. $a \neq 0$. $$a\begin{bmatrix} x_1 & y_1 \\ x_2 & y_2 \end{bmatrix} = \begin{bmatrix} \lambda a, & \lambda h \pm \mu s \\ \mu a, & \mu h \mp \lambda s \end{bmatrix},$$

where $$s = \sqrt{ab - h^2} = \sqrt{\Delta}, \text{ and } \lambda^2 + \mu^2 = a.$$

CASE II. $b \neq 0$. $$b\begin{bmatrix} x_1 & y_1 \\ x_2 & y_2 \end{bmatrix} = \begin{bmatrix} \lambda h \pm \mu s, & \lambda b \\ \mu h \mp \lambda s, & \mu b \end{bmatrix},$$

where $$s = \sqrt{ab - h^2} = \sqrt{\Delta}, \text{ and } \lambda^2 + \mu^2 = b.$$

CASE III. $a = b = 0$, $h \neq 0$. $$\begin{bmatrix} x_1 & y_1 \\ x_2 & y_2 \end{bmatrix} = \begin{bmatrix} \lambda, & \mu \\ \pm i\lambda, & \mp i\mu \end{bmatrix},$$

where $$2\lambda\mu = h.$$

(*c*) *General solution when* ϕ *has rank* 1.

In this case we can write

$$\begin{bmatrix} a & h \\ h & b \end{bmatrix} = \begin{bmatrix} \alpha \\ \beta \end{bmatrix}[\alpha\,\beta], \quad \begin{bmatrix} x_1 & y_1 \\ x_2 & y_2 \end{bmatrix} = \begin{bmatrix} \lambda \\ \mu \end{bmatrix}[x\,y],$$

and replace the equation (A) by

$$(\lambda^2 + \mu^2)\begin{bmatrix} x \\ y \end{bmatrix}[x\,y] = \begin{bmatrix} \alpha \\ \beta \end{bmatrix}[\alpha\,\beta],$$

where λ, μ, x, y are to be determined.

Using Ex. i of § 160 the general solution will be seen to be

$$\begin{bmatrix} x_1 & y_1 \\ x_2 & y_2 \end{bmatrix} = \begin{bmatrix} l \\ m \end{bmatrix}[\sqrt{a}, \sqrt{b}], \text{ where } l^2 + m^2 = 1 \text{ and } \sqrt{a}\sqrt{b} = h. \quad \text{............}(A_2)$$

The formula (A_2) is included in formula (A_1).

(*d*) *General solution when* ϕ *has rank* 0.

This is $$[x_2\,y_2] = \pm\sqrt{-1}\,[x_1\,y_1], \text{ or } \begin{bmatrix} x_1 & y_1 \\ x_2 & y_2 \end{bmatrix} = \begin{bmatrix} l \\ m \end{bmatrix}[\xi\,\eta], \quad \text{................}(A_3)$$

where $l^2 + m^2 = 0$, and ξ and η are arbitrary.

(*e*) *Self-conjugate solutions.* If we write

$$\begin{bmatrix} x_1 & y_1 \\ x_2 & y_2 \end{bmatrix} = \begin{bmatrix} \lambda & \rho \\ \rho & \mu \end{bmatrix},$$

the equation (A) is satisfied when and only when

$$\lambda^2 + \rho^2 = a, \quad \mu^2 + \rho^2 = b, \quad (\lambda + \mu)\rho = h.$$

Except when $(a-b)^2 + 4h^2 = 0$ there are just two pairs of self-conjugate solutions given by

$$\frac{\lambda}{a \pm s} = \frac{\mu}{b \pm s} = \frac{\rho}{h} = \pm\frac{1}{\sqrt{(a+b) \pm 2s}}, \quad \text{..............................}(A_4)$$

where $s = \sqrt{ab - h^2}$, and the same sign is to be given to the radical s in all places.

Equations (A_4) lead to the identities

$$(a + b \pm 2s)\begin{bmatrix} a & h \\ h & b \end{bmatrix} = \begin{bmatrix} a \pm s, & h \\ h, & b \pm s \end{bmatrix}\begin{bmatrix} a \pm s, & h \\ h, & b \pm s \end{bmatrix}$$

where $s=\sqrt{ab-h^2}$, and either all the upper signs or all the lower signs are to be taken throughout.

One of these identities has the trivial form $0=0$ when

$$(a-b)^2+4h^2=(a+b+2s)(a+b-2s)=0,$$

and both have this trivial form when $a+b=0$ and $ab-h^2=0$. In other cases, i.e. when $(a-b)^2+4h^2\neq 0$, they are both non-trivial.

Ex. i. The general solution of the equation $\begin{bmatrix} x_1 & x_2 \\ y_1 & y_2 \end{bmatrix}\begin{bmatrix} x_1 & y_1 \\ x_2 & y_2 \end{bmatrix}=\begin{bmatrix} 1 & 0 \\ 0 & 1 \end{bmatrix}$ can be expressed in either of the two equivalent forms

$$\begin{bmatrix} x_1 & y_1 \\ x_2 & y_2 \end{bmatrix}=\begin{bmatrix} \lambda, & \mu \\ \pm\mu, & \mp\lambda \end{bmatrix},\quad \begin{bmatrix} x_1 & y_1 \\ x_2 & y_2 \end{bmatrix}=\begin{bmatrix} \lambda, & \pm\mu \\ \mu, & \mp\lambda \end{bmatrix},\ \text{where } \lambda^2+\mu^2=1.$$

The self-conjugate solutions are given by

$$\begin{bmatrix} x_1 & y_1 \\ x_2 & y_2 \end{bmatrix}=\begin{bmatrix} \pm 1, & 0 \\ 0, & \pm 1 \end{bmatrix},\ \text{and}\ \begin{bmatrix} x_1 & y_1 \\ x_2 & y_2 \end{bmatrix}=\begin{bmatrix} \pm\sqrt{1-\rho^2}, & \rho \\ \rho\ , & \mp\sqrt{1-\rho^2} \end{bmatrix},$$

where in the first case either sign may be taken in each place, and in the second case ρ is arbitrary.

Ex. ii. The general solution of the equation $\begin{bmatrix} x_1 & x_2 \\ y_1 & y_2 \end{bmatrix}\begin{bmatrix} x_1 & y_1 \\ x_2 & y_2 \end{bmatrix}=\begin{bmatrix} 0 & 1 \\ 1 & 0 \end{bmatrix}$ is

$$\begin{bmatrix} x_1 & y_1 \\ x_2 & y_2 \end{bmatrix}=\begin{bmatrix} \lambda, & \mu \\ \pm i\lambda, & \mp i\mu \end{bmatrix},\ \text{where } 2\lambda\mu=1.$$

The self-conjugate solutions are given by

$$\begin{bmatrix} x_1 & y_1 \\ x_2 & y_2 \end{bmatrix}=\frac{1}{2}\begin{bmatrix} 1\pm i, & 1\mp i \\ 1\mp i, & 1\pm i \end{bmatrix},\ \text{and}\ \begin{bmatrix} x_1 & y_1 \\ x_2 & y_2 \end{bmatrix}=-\frac{1}{2}\begin{bmatrix} 1\pm i, & 1\mp i \\ 1\mp i, & 1\pm i \end{bmatrix}.$$

Ex. iii. The following results are often of service:

$$2\begin{bmatrix} 1 & 0 \\ 0 & 1 \end{bmatrix}=\begin{bmatrix} 1, & 1 \\ i, & -i \end{bmatrix}\begin{bmatrix} 1, & -i \\ 1, & i \end{bmatrix}=\begin{bmatrix} i, & -i \\ 1, & 1 \end{bmatrix}\begin{bmatrix} -i, & 1 \\ i, & 1 \end{bmatrix}=\begin{bmatrix} 1, & -i \\ 1, & i \end{bmatrix}\begin{bmatrix} 1, & 1 \\ i, & -i \end{bmatrix}=\begin{bmatrix} i, & 1 \\ -i, & 1 \end{bmatrix}\begin{bmatrix} -i, & i \\ 1, & 1 \end{bmatrix}.$$

$$2\begin{bmatrix} 0 & 1 \\ 1 & 0 \end{bmatrix}=\begin{bmatrix} 1, & 1 \\ i, & -i \end{bmatrix}\begin{bmatrix} -i, & 1 \\ i, & 1 \end{bmatrix}=\begin{bmatrix} i, & -i \\ 1, & 1 \end{bmatrix}\begin{bmatrix} 1, & -i \\ 1, & i \end{bmatrix}=\begin{bmatrix} 1, & -i \\ 1, & i \end{bmatrix}\begin{bmatrix} 1, & 1 \\ -i, & i \end{bmatrix}=\begin{bmatrix} i, & 1 \\ -i, & 1 \end{bmatrix}\begin{bmatrix} i, & -i \\ 1, & 1 \end{bmatrix}.$$

We can change the sign of i in all of them, and they admit of many other transformations.

2. *The equation* $$\begin{bmatrix} x_1 & x_2 \\ y_1 & y_2 \end{bmatrix}\begin{bmatrix} x_2 & y_2 \\ x_1 & y_1 \end{bmatrix}=\begin{bmatrix} a & h \\ h & b \end{bmatrix}=\phi. \qquad\text{(B)}$$

(*a*) *Reduction to the equation of sub-article* 1. The solutions of the equations

$$\begin{bmatrix} x_1 & x_2 \\ y_1 & y_2 \end{bmatrix}\begin{bmatrix} x_2 & y_2 \\ x_1 & y_1 \end{bmatrix}=\begin{bmatrix} a & h \\ h & b \end{bmatrix},\quad \begin{bmatrix} \xi_1 & \xi_2 \\ \eta_1 & \eta_2 \end{bmatrix}\begin{bmatrix} \xi_1 & \eta_1 \\ \xi_2 & \eta_2 \end{bmatrix}=\begin{bmatrix} a & h \\ h & b \end{bmatrix}, \qquad\text{(1)}$$

are connected in pairs by the relations

$$\begin{bmatrix} x_1 & y_1 \\ x_2 & y_2 \end{bmatrix}=\sqrt{\frac{1}{2}}\begin{bmatrix} 1, & i \\ 1, & -i \end{bmatrix}\begin{bmatrix} \xi_1 & \eta_1 \\ \xi_2 & \eta_2 \end{bmatrix},\quad \begin{bmatrix} \xi_1 & \eta_1 \\ \xi_2 & \eta_2 \end{bmatrix}=\sqrt{\frac{1}{2}}\begin{bmatrix} 1, & 1 \\ -i, & i \end{bmatrix}\begin{bmatrix} x_1 & y_1 \\ x_2 & y_2 \end{bmatrix}, \qquad\text{(2)}$$

which lead to

$$\begin{bmatrix} x_1 & x_2 \\ y_1 & y_2 \end{bmatrix}\begin{bmatrix} x_2 & y_2 \\ x_1 & y_1 \end{bmatrix}=\begin{bmatrix} x_1 & x_2 \\ y_1 & y_2 \end{bmatrix}\begin{bmatrix} 0 & 1 \\ 1 & 0 \end{bmatrix}\begin{bmatrix} x_1 & y_1 \\ x_2 & y_2 \end{bmatrix}=\begin{bmatrix} \xi_1 & \xi_2 \\ \eta_1 & \eta_2 \end{bmatrix}\begin{bmatrix} \xi_1 & \eta_1 \\ \xi_2 & \eta_2 \end{bmatrix}.$$

Hence the solutions of the equation (B) can be deduced from the solutions of the second of the equations (1).

(*b*) *Particular solutions.* Writing $\Delta = ab - h^2$, particular solutions of (B) in the special cases $a \neq 0$; $b \neq 0$; $a = b = 0$, $h \neq 0$ are

$$\begin{bmatrix} x_1 & y_1 \\ x_2 & y_2 \end{bmatrix} = \frac{1}{\sqrt{2a}} \begin{bmatrix} a, & h \pm \sqrt{-\Delta} \\ a, & h \mp \sqrt{-\Delta} \end{bmatrix}, \quad \frac{1}{\sqrt{2b}} \begin{bmatrix} h \pm \sqrt{-\Delta}, & b \\ h \mp \sqrt{-\Delta}, & b \end{bmatrix}, \quad \sqrt{h} \begin{bmatrix} 1 & 0 \\ 0 & 1 \end{bmatrix}.$$

In these the horizontal rows can be interchanged.

(*c*) *General solution when ϕ has rank* 2.

If $[x\, y]_{12} = [\alpha\, \beta]_{12}$ is any particular solution, the general solution is

$$\begin{bmatrix} x_1 & y_1 \\ x_2 & y_2 \end{bmatrix} = \begin{bmatrix} l_1 & m_1 \\ l_2 & m_2 \end{bmatrix} \begin{bmatrix} \alpha_1 & \beta_1 \\ \alpha_2 & \beta_2 \end{bmatrix}, \quad \text{where } \begin{bmatrix} l_1 & l_2 \\ m_1 & m_2 \end{bmatrix} \begin{bmatrix} l_2 & m_2 \\ l_1 & m_1 \end{bmatrix} = \begin{bmatrix} 0 & 1 \\ 1 & 0 \end{bmatrix}.$$

Hence all solutions are given by the two formulae

$$\left.\begin{matrix} [x_1\, y_1] = \lambda\, [\alpha_1\, \beta_1] \\ [x_2\, y_2] = \mu\, [\alpha_2\, \beta_2] \end{matrix}\right\}, \quad \left.\begin{matrix} [x_1\, y_1] = \lambda\, [\alpha_2\, \beta_2] \\ [x_2\, y_2] = \mu\, [\alpha_1\, \beta_1] \end{matrix}\right\} \quad \text{..........................} (B_1)$$

where $$\lambda\mu = 1.$$

Using the particular solutions given in (*b*), we can express these general solutions as follows :

CASE I. $a \neq 0$. $\quad [x_1\, y_1] = \lambda\, [a,\ h \pm \sqrt{-\Delta}], \quad [x_2\, y_2] = \mu\, [a,\ h \mp \sqrt{-\Delta}],$

where $$2a\lambda\mu = 1.$$

CASE II. $b \neq 0$. $\quad [x_1\, y_1] = \lambda\, [h \pm \sqrt{-\Delta},\ b], \quad [x_2\, y_2] = \mu\, [h \mp \sqrt{-\Delta},\ b],$

where $$2b\lambda\mu = 1.$$

CASE III. $a = b = 0$, $h \neq 0$.

$$\left.\begin{matrix} [x_1\, y_1] = \lambda\, [0,\ 1] \\ [x_2\, y_2] = \mu\, [1,\ 0] \end{matrix}\right\}, \quad \left.\begin{matrix} [x_1\, y_1] = \lambda\, [1,\ 0] \\ [x_2\, y_2] = \mu\, [0,\ 1] \end{matrix}\right\},$$

where $$\lambda\mu = h.$$

(*d*) *General solution when ϕ has rank* 1. In this case we can write

$$\begin{bmatrix} a & h \\ h & b \end{bmatrix} = \begin{bmatrix} \alpha \\ \beta \end{bmatrix} [\alpha\, \beta], \quad \begin{bmatrix} x_1 & y_1 \\ x_2 & y_2 \end{bmatrix} = \begin{bmatrix} \lambda \\ \mu \end{bmatrix} [x\, y],$$

and replace the equation (B) by

$$2\lambda\mu \begin{bmatrix} x \\ y \end{bmatrix} [x\, y] = \begin{bmatrix} \alpha \\ \beta \end{bmatrix} [\alpha\, \beta],$$

where λ, μ, x, y are to be determined. Using Ex. i of § 160 the general solution will be seen to be

$$\begin{bmatrix} x_1 & y_1 \\ x_2 & y_2 \end{bmatrix} = \begin{bmatrix} l \\ m \end{bmatrix} [\sqrt{a},\ \sqrt{b}], \quad \text{..................................} (B_2)$$

where $$2lm = 1, \quad \text{and } \sqrt{a}\sqrt{b} = h.$$

(*e*) *General solution when ϕ has rank* 0.

In this case the equation (B) is satisfied when and only when one at least of the matrices $[x_1\, y_1]$, $[x_2\, y_2]$ vanishes.

(*f*) *Self-conjugate solutions.* If we write

$$\begin{bmatrix} x_1 & y_1 \\ x_2 & y_2 \end{bmatrix} = \begin{bmatrix} \lambda, & \rho \\ \rho, & \mu \end{bmatrix},$$

the equation (B) is satisfied when and only when

$$2\lambda\rho = a, \quad 2\mu\rho = b, \quad \lambda\mu + \rho^2 = h^2.$$

When neither a nor b is zero, so that $ab \neq 0$, we must have

$$4\rho^4 - 4h\rho^2 + ab = 0$$

and therefore

$$2\rho^2 = h \pm \sqrt{h^2 - ab}, \quad 2\rho = \pm(h + \sqrt{ab})^{\frac{1}{2}} \pm (h - \sqrt{ab})^{\frac{1}{2}}.$$

Thus except when $ab = 0$ there are just two pairs of self-conjugate solutions of (B) given by

$$\left.\begin{aligned} \frac{\lambda}{a} = \frac{\mu}{b} = \frac{\rho}{h + \sqrt{h^2 - ab}} = \pm \frac{1}{(h + \sqrt{ab})^{\frac{1}{2}} + (h - \sqrt{ab})^{\frac{1}{2}}} = \frac{1}{2\rho} \\ \frac{\lambda}{a} = \frac{\mu}{b} = \frac{\rho}{h - \sqrt{h^2 - ab}} = \pm \frac{1}{(h + \sqrt{ab})^{\frac{1}{2}} - (h - \sqrt{ab})^{\frac{1}{2}}} = \frac{1}{2\rho} \end{aligned}\right\} \quad \ldots\ldots\ldots\ldots(B_3)$$

From (B_3) we obtain the identities

$$2(h \pm s)\begin{bmatrix} a & h \\ h & b \end{bmatrix} = \begin{bmatrix} a, & h \pm s \\ h \pm s, & b \end{bmatrix}\begin{bmatrix} h \pm s, & b \\ a, & h \pm s \end{bmatrix}$$

where $s = \sqrt{h^2 - ab} = \sqrt{-\Delta}$, and either all the upper signs or all the lower signs are to be taken throughout.

If $ab = 0$, one of these identities has the trivial form $0 = 0$.

If $ab = 0$ and $h = 0$, both identities have this trivial form.

In other cases, i.e. when $ab \neq 0$, both the identities are non-trivial.

Ex. iv. The general solution of the equation $\begin{bmatrix} x_1 & x_2 \\ y_1 & y_2 \end{bmatrix}\begin{bmatrix} x_2 & y_2 \\ x_1 & y_1 \end{bmatrix} = \begin{bmatrix} 1 & 0 \\ 0 & 1 \end{bmatrix}$ can be expressed in either of the two equivalent forms

$$\begin{bmatrix} x_1 & y_1 \\ x_2 & y_2 \end{bmatrix} = \begin{bmatrix} \lambda, & \pm i\lambda \\ \mu, & \mp i\mu \end{bmatrix}, \quad \begin{bmatrix} x_1 & y_1 \\ x_2 & y_2 \end{bmatrix} = \begin{bmatrix} \pm i\lambda, & \lambda \\ \mp i\mu, & \mu \end{bmatrix}, \text{ where } 2\lambda\mu = 1.$$

The self-conjugate solutions are given by

$$\begin{bmatrix} x_1 & y_1 \\ x_2 & y_2 \end{bmatrix} = \frac{1}{2}\begin{bmatrix} 1 \pm i, & 1 \mp i \\ 1 \mp i, & 1 \pm i \end{bmatrix}, \text{ and } \begin{bmatrix} x_1 & y_1 \\ x_2 & y_2 \end{bmatrix} = -\frac{1}{2}\begin{bmatrix} 1 \pm i, & 1 \mp i \\ 1 \mp i, & 1 \pm i \end{bmatrix}.$$

Ex. v. All the solutions of the equation $\begin{bmatrix} x_1 & x_2 \\ y_1 & y_2 \end{bmatrix}\begin{bmatrix} x_2 & y_2 \\ x_1 & y_1 \end{bmatrix} = \begin{bmatrix} 0 & 1 \\ 1 & 0 \end{bmatrix}$ are given by the two formulae

$$\begin{bmatrix} x_1 & y_1 \\ x_2 & y_2 \end{bmatrix} = \begin{bmatrix} 0 & \lambda \\ \mu & 0 \end{bmatrix}, \quad \begin{bmatrix} x_1 & y_1 \\ x_2 & y_2 \end{bmatrix} = \begin{bmatrix} \lambda & 0 \\ 0 & \mu \end{bmatrix}, \text{ where } \lambda\mu = 1.$$

The self-conjugate solutions are given by

$$\begin{bmatrix} x_1 & y_1 \\ x_2 & y_2 \end{bmatrix} = \begin{bmatrix} 0, & \pm 1 \\ \pm 1, & 0 \end{bmatrix}, \text{ and } \begin{bmatrix} x_1 & y_1 \\ x_2 & y_2 \end{bmatrix} = \begin{bmatrix} \lambda & 0 \\ 0 & \mu \end{bmatrix} \text{ where } \lambda\mu = 1.$$

§ 163. Some special equations of the form $[x]_m^2 [y]_2^m = [a]_m^2 [b]_2^m$.

1. *The equation*

$$\begin{bmatrix} x_1 & y_1 \\ x_2 & y_2 \\ \cdots & \cdots \\ x_m & y_m \end{bmatrix}\begin{bmatrix} x_1 & x_2 \ldots x_m \\ y_1 & y_2 \ldots y_m \end{bmatrix} = \begin{bmatrix} a_1 & b_1 \\ a_2 & b_2 \\ \cdots & \cdots \\ a_m & b_m \end{bmatrix}\begin{bmatrix} a_1 & a_2 \ldots a_m \\ b_1 & b_2 \ldots b_m \end{bmatrix} \ldots\ldots\ldots\ldots\ldots(A)$$

CASE I. *When the product matrix on the right has rank* 2.

The factor matrices on both sides have rank 2, and the general solution is given by

$$\begin{bmatrix} x_1 & x_2 \dots x_m \\ y_1 & y_2 \dots y_m \end{bmatrix} = \begin{bmatrix} l_1 & m_1 \\ l_2 & m_2 \end{bmatrix} \begin{bmatrix} a_1 & a_2 \dots a_m \\ b_1 & b_2 \dots b_m \end{bmatrix}, \text{ where } \begin{bmatrix} l_1 & m_1 \\ l_2 & m_2 \end{bmatrix} \begin{bmatrix} l_1 & l_2 \\ m_1 & m_2 \end{bmatrix} = \begin{bmatrix} 1 & 0 \\ 0 & 1 \end{bmatrix}. \quad \dots(A_1)$$

It can also be expressed in the form

$$\begin{bmatrix} x_1 & x_2 \dots x_m \\ y_1 & y_2 \dots y_m \end{bmatrix} = \begin{bmatrix} \lambda, & \mu \\ \pm\mu, & \mp\lambda \end{bmatrix} \begin{bmatrix} a_1 & a_2 \dots a_m \\ b_1 & b_2 \dots b_m \end{bmatrix}, \text{ where } \lambda^2+\mu^2=1. \quad \dots\dots\dots(A_2)$$

CASE II. *When the product matrix on the right has rank* 1.

This case occurs when the factor matrices on the right have rank 1 and

$$[b_1\ b_2 \dots b_m] \neq \pm\sqrt{-1}\,[a_1\ a_2 \dots a_m].$$

Every solution has then rank 1, and we can put

$$\begin{bmatrix} x_1 & x_2 \dots x_m \\ y_1 & y_2 \dots y_m \end{bmatrix} = \begin{bmatrix} \xi \\ \eta \end{bmatrix} [z_1\ z_2 \dots z_m], \quad \begin{bmatrix} a_1 & a_2 \dots a_m \\ b_1 & b_2 \dots b_m \end{bmatrix} = \begin{bmatrix} \alpha \\ \beta \end{bmatrix} [c_1\ c_2 \dots c_m],$$

where all matrices on the right have rank 1, and $\alpha^2+\beta^2 \neq 0$.

We can then replace the equation (A) by

$$(\xi^2+\eta^2) \begin{bmatrix} z_1 \\ z_2 \\ \vdots \\ z_m \end{bmatrix} [z_1\ z_2 \dots z_m] = (\alpha^2+\beta^2) \begin{bmatrix} c_1 \\ c_2 \\ \vdots \\ c_m \end{bmatrix} [c_1\ c_2 \dots c_m],$$

where $\qquad \alpha^2+\beta^2 \neq 0, \quad \xi^2+\eta^2 \neq 0.$

Using Ex. i of § 160 we see that the general solution is given by

$$\begin{bmatrix} x_1 & x_2 \dots x_m \\ y_1 & y_2 \dots y_m \end{bmatrix} = \sqrt{\alpha^2+\beta^2} \begin{bmatrix} \lambda \\ \mu \end{bmatrix} [c_1\ c_2 \dots c_m], \text{ where } \lambda^2+\mu^2=1. \quad \dots\dots\dots\dots(A_3)$$

CASE III. *When the product matrix on the right has rank* 0.

This case occurs when $[b_1\, b_2 \dots b_m] = \pm\sqrt{-1}\,[a_1\, a_2 \dots a_m]$; and in this case the equation (A) is satisfied when and only when

$$[y_1\, y_2 \dots y_m] = \pm\sqrt{-1}\,[x_1 x_2 \dots x_m], \text{ or } \begin{bmatrix} x_1 & x_2 \dots x_m \\ y_1 & y_2 \dots y_m \end{bmatrix} = \begin{bmatrix} l \\ m \end{bmatrix} [\xi_1\ \xi_2 \dots \xi_m], \quad \dots(A_4)$$

where $l^2+m^2=0$, and $[\xi]_m$ is arbitrary.

2. *The equation*

$$\begin{bmatrix} x_1 & y_1 \\ x_2 & y_2 \\ \dots\dots & \\ x_m & y_m \end{bmatrix} \begin{bmatrix} y_1 & y_2 \dots y_m \\ x_1 & x_2 \dots x_m \end{bmatrix} = \begin{bmatrix} a_1 & b_1 \\ a_2 & b_2 \\ \dots\dots & \\ a_m & b_m \end{bmatrix} \begin{bmatrix} a_1 & a_2 \dots a_m \\ b_1 & b_2 \dots b_m \end{bmatrix}. \quad \dots\dots\dots(B)$$

CASE I. *When the product matrix on the right has rank* 2.

The factor matrices on both sides have rank 2, and the general solution is given by

$$\begin{bmatrix} x_1 & x_2 \dots x_m \\ y_1 & y_2 \dots y_m \end{bmatrix} = \begin{bmatrix} h_1 & k_1 \\ h_2 & k_2 \end{bmatrix} \begin{bmatrix} a_1 & a_2 \dots a_m \\ b_1 & b_2 \dots b_m \end{bmatrix}, \text{ where } \begin{bmatrix} h_1 & h_2 \\ k_1 & k_2 \end{bmatrix} \begin{bmatrix} h_2 & k_2 \\ h_1 & k_1 \end{bmatrix} = \begin{bmatrix} 1 & 0 \\ 0 & 1 \end{bmatrix}.$$

Using Ex. iv of § 162 we can express the general solution in the form

$$\begin{bmatrix} x_1 & x_2 \dots x_m \\ y_1 & y_2 \dots y_m \end{bmatrix} = \begin{bmatrix} \lambda, & \pm i\lambda \\ \mu, & \mp i\mu \end{bmatrix} \begin{bmatrix} a_1 & a_2 \dots & a_m \\ b_1 & b_2 \dots & b_m \end{bmatrix}, \quad \text{.........................}(B_1)$$

where $i = \sqrt{-1}$, and $2\lambda\mu = 1$.

CASE II. *When the product matrix on the right has rank* 1.

This case occurs when the factor matrices on the right have rank 1 and

$$[b_1 \, b_2 \dots b_m] \neq \pm\sqrt{-1}\, [a_1 \, a_2 \dots a_m].$$

Every solution then has rank 1, and we can put

$$\begin{bmatrix} x_1 & x_2 \dots x_m \\ y_1 & y_2 \dots y_m \end{bmatrix} = \begin{bmatrix} \xi \\ \eta \end{bmatrix} [z_1 \, z_2 \dots z_m], \quad \begin{bmatrix} a_1 & a_2 \dots a_m \\ b_1 & b_2 \dots b_m \end{bmatrix} = \begin{bmatrix} \alpha \\ \beta \end{bmatrix} [c_1 \, c_2 \dots c_m],$$

where all matrices on the right have rank 1, and $\alpha^2 + \beta^2 \neq 0$.

The equation (B) can then be replaced by

$$2\xi\eta \begin{bmatrix} z_1 \\ z_2 \\ \vdots \\ z_m \end{bmatrix} [z_1 \, z_2 \dots z_m] = (\alpha^2 + \beta^2) \begin{bmatrix} c_1 \\ c_2 \\ \vdots \\ c_m \end{bmatrix} [c_1 \, c_2 \dots c_m],$$

where $\alpha^2 + \beta^2 \neq 0, \quad 2\xi\eta \neq 0.$

Using Ex. i of § 160 we see that the general solution is given by

$$\begin{bmatrix} x_1 & x_2 \dots x_m \\ y_1 & y_2 \dots y_m \end{bmatrix} = \sqrt{\alpha^2 + \beta^2} \begin{bmatrix} \lambda \\ \mu \end{bmatrix} [c_1 \, c_2 \dots c_m], \text{ where } 2\lambda\mu = 1. \quad \text{.............}(B_2)$$

CASE III. *When the product matrix on the right has rank* 0.

This case occurs when $[b_1 \, b_2 \dots b_m] = \pm\sqrt{-1}\, [a_1 \, a_2 \dots a_m]$, and the equation (B) is satisfied when and only when

$$\text{either } [x_1 \, x_2 \dots x_m] = 0, \quad \text{or } [y_1 \, y_2 \dots y_m] = 0. \quad \text{..................}(B_3)$$

3. *The equation*

$$\begin{bmatrix} x_1 & y_1 \\ x_2 & y_2 \\ \dots\dots & \\ x_m & y_m \end{bmatrix} \begin{bmatrix} x_1 & x_2 \dots x_m \\ y_1 & y_2 \dots y_m \end{bmatrix} = \begin{bmatrix} a_1 & b_1 \\ a_2 & b_2 \\ \dots\dots & \\ a_m & b_m \end{bmatrix} \begin{bmatrix} b_1 & b_2 \dots b_m \\ a_1 & a_2 \dots a_m \end{bmatrix}. \quad \text{.......}(C)$$

CASE I. *When the product matrix on the right has rank* 2.

The factor matrices on both sides have rank 2, and the general solution is given by

$$\begin{bmatrix} x_1 & x_2 \dots x_m \\ y_1 & y_2 \dots y_m \end{bmatrix} = \begin{bmatrix} h_1 & k_1 \\ h_2 & k_2 \end{bmatrix} \begin{bmatrix} a_1 & a_2 \dots a_m \\ b_1 & b_2 \dots b_m \end{bmatrix}, \text{ where } \begin{bmatrix} h_1 & h_2 \\ k_1 & k_2 \end{bmatrix} \begin{bmatrix} h_1 & k_1 \\ h_2 & k_2 \end{bmatrix} = \begin{bmatrix} 0 & 1 \\ 1 & 0 \end{bmatrix}.$$

Using Ex. ii of § 162 we can express the general solution in the form

$$\begin{bmatrix} x_1 & x_2 \dots x_m \\ y_1 & y_2 \dots y_m \end{bmatrix} = \begin{bmatrix} \lambda, & \mu \\ \pm i\lambda, & \mp i\mu \end{bmatrix} \begin{bmatrix} a_1 & a_2 \dots a_m \\ b_1 & b_2 \dots b_m \end{bmatrix}, \text{ where } 2\lambda\mu = 1. \quad \text{............}(C_1)$$

CASE II. *When the product matrix on the right has rank* 1.

This case occurs when each factor matrix on the right has rank 1 and does not contain a long row of 0's. Every solution then has rank 1, and we can put

$$\begin{bmatrix} x_1 & x_2 \dots x_m \\ y_1 & y_2 \dots y_m \end{bmatrix} = \begin{bmatrix} \xi \\ \eta \end{bmatrix} [z_1 \, z_2 \dots z_m], \quad \begin{bmatrix} a_1 & a_2 \dots a_m \\ b_1 & b_2 \dots b_m \end{bmatrix} = \begin{bmatrix} \alpha \\ \beta \end{bmatrix} [c_1 \, c_2 \dots c_m],$$

where all matrices on the right have rank 1 and $\alpha\beta \neq 0$.

The equation (C) can then be replaced by

$$(\xi^2+\eta^2)\begin{bmatrix} z_1 \\ z_2 \\ \vdots \\ z_m \end{bmatrix}[z_1\, z_2 \dots z_m]=2\alpha\beta\begin{bmatrix} c_1 \\ c_2 \\ \vdots \\ c_m \end{bmatrix}[c_1\, c_2 \dots c_m],$$

where $2\alpha\beta \neq 0, \quad \xi^2+\eta^2\neq 0.$

Using Ex. i of § 160 we see that the general solution is given by

$$\begin{bmatrix} x_1 & x_2 \dots x_m \\ y_1 & y_2 \dots y_m \end{bmatrix}=\sqrt{2\alpha\beta}\begin{bmatrix} \lambda \\ \mu \end{bmatrix}[c_1\, c_2 \dots c_m], \text{ where } \lambda^2+\mu^2=1. \quad \dots\dots\dots\dots(C_2)$$

CASE III. *When the product matrix on the right has rank* 0.

This case occurs when a factor matrix on the right contains a long row of 0's, and the equation (C) is satisfied when and only when

$$[y_1\, y_2 \dots y_m]=\pm\sqrt{-1}\,[x_1\, x_2 \dots x_m], \text{ or } \begin{bmatrix} x_1 & x_2 \dots x_m \\ y_1 & y_2 \dots y_m \end{bmatrix}=\begin{bmatrix} l \\ m \end{bmatrix}[\xi_1\, \xi_2 \dots \xi_m] \quad \dots\dots(C_3)$$

where $l^2+m^2=0$, and $[\xi]_m$ is arbitrary.

4. *The equation*

$$\begin{bmatrix} x_1 & y_1 \\ x_2 & y_2 \\ \dots & \dots \\ x_m & y_m \end{bmatrix}\begin{bmatrix} y_1 & y_2 \dots y_m \\ x_1 & x_2 \dots x_m \end{bmatrix}=\begin{bmatrix} a_1 & b_1 \\ a_2 & b_2 \\ \dots & \dots \\ a_m & b_m \end{bmatrix}\begin{bmatrix} b_1 & b_2 \dots b_m \\ a_1 & a_2 \dots a_m \end{bmatrix}. \quad \dots\dots\dots\dots(D)$$

CASE I. *When the product matrix on the right has rank* 2.

The factor matrices on both sides have rank 2, and the general solution is given by

$$\begin{bmatrix} x_1 & x_2 \dots x_m \\ y_1 & y_2 \dots y_m \end{bmatrix}=\begin{bmatrix} h_1 & k_1 \\ h_2 & k_2 \end{bmatrix}\begin{bmatrix} a_1 & a_2 \dots a_m \\ b_1 & b_2 \dots b_m \end{bmatrix}, \text{ where } \begin{bmatrix} h_1 & h_2 \\ k_1 & k_2 \end{bmatrix}\begin{bmatrix} h_2 & k_2 \\ h_1 & k_1 \end{bmatrix}=\begin{bmatrix} 0 & 1 \\ 1 & 0 \end{bmatrix}.$$

Using Ex. v of § 162 we see that the general solution is given by the two formulae

$$\left.\begin{matrix}[x_1\, x_2 \dots x_m]=\lambda\,[a_1\, a_2 \dots a_m] \\ [y_1\, y_2 \dots y_m]=\mu\,[b_1\, b_2 \dots b_m]\end{matrix}\right\}, \quad \left.\begin{matrix}[x_1\, x_2 \dots x_m]=\lambda\,[b_1\, b_2 \dots b_m] \\ [y_1\, y_2 \dots y_m]=\mu\,[a_1\, a_2 \dots a_m]\end{matrix}\right\}, \quad \dots\dots\dots(D_1)$$

where $\lambda\mu=1.$

CASE II. *When the product matrix on the right has rank* 1.

This case occurs when each factor matrix on the right has rank 1 and does not contain a long row of 0's. Every solution then has rank 1, and we can put

$$\begin{bmatrix} x_1 & x_2 \dots x_m \\ y_1 & y_2 \dots y_m \end{bmatrix}=\begin{bmatrix} \xi \\ \eta \end{bmatrix}[z_1\, z_2 \dots z_m], \quad \begin{bmatrix} a_1 & a_2 \dots a_m \\ b_1 & b_2 \dots b_m \end{bmatrix}=\begin{bmatrix} \alpha \\ \beta \end{bmatrix}[c_1\, c_2 \dots c_m],$$

where all matrices on the right have rank 1, and $\alpha\beta\neq 0$.

The equation (D) can then be replaced by

$$\xi\eta\begin{bmatrix} z_1 \\ z_2 \\ \vdots \\ z_m \end{bmatrix}[z_1\, z_2 \dots z_m]=\alpha\beta\begin{bmatrix} c_1 \\ c_2 \\ \vdots \\ c_m \end{bmatrix}[c_1\, c_2 \dots c_m],$$

where $\alpha\beta\neq 0, \quad \xi\eta\neq 0.$

Using Ex. i of § 160 we obtain the general solution in the form

$$\begin{bmatrix} x_1 & x_2 \dots x_m \\ y_1 & y_2 \dots y_m \end{bmatrix} = \sqrt{\alpha\beta}\begin{bmatrix} \lambda \\ \mu \end{bmatrix}[c_1\, c_2 \dots c_m], \text{ where } \lambda\mu = 1. \quad \dots\dots\dots\dots (\mathrm{D}_2)$$

CASE III. *When the product matrix on the right has rank* 0.

This case occurs when a factor matrix on the right contains a long row of 0's, and the equation (D) is satisfied when and only when

$$\text{either } [x_1\, x_2 \dots x_m] = 0, \text{ or } [y_1\, y_2 \dots y_m] = 0. \quad \dots\dots\dots\dots (\mathrm{D}_3)$$

§ 164. Two special equations of the form $[x]_m^2 [y]_2^m = [a]_m^m$, where $[a]_m^m$ is a symmetric matrix whose rank does not exceed 2.

1. *The equation*
$$\begin{bmatrix} x_1 & y_1 \\ x_2 & y_2 \\ \dots & \dots \\ x_m & y_m \end{bmatrix}\begin{bmatrix} x_1 & x_2 & \dots & x_m \\ y_1 & y_2 & \dots & y_m \end{bmatrix} = [a]_m^m. \quad \dots\dots\dots\dots (\mathrm{A})$$

CASE I. *When* $[a]_m^m$ *has rank* 2.

If $\begin{bmatrix} x_1 & x_2 & \dots & x_m \\ y_1 & y_2 & \dots & y_m \end{bmatrix} = \begin{bmatrix} \alpha_1 & \alpha_2 & \dots & \alpha_m \\ \beta_1 & \beta_2 & \dots & \beta_m \end{bmatrix}$ is a particular solution, the general solution is

$$\begin{bmatrix} x_1 & x_2 & \dots & x_m \\ y_1 & y_2 & \dots & y_m \end{bmatrix} = \begin{bmatrix} l_1 & m_1 \\ l_2 & m_2 \end{bmatrix}\begin{bmatrix} \alpha_1 & \alpha_2 & \dots & \alpha_m \\ \beta_1 & \beta_2 & \dots & \beta_m \end{bmatrix}, \text{ where } \begin{bmatrix} l_1 & l_2 \\ m_1 & m_2 \end{bmatrix}\begin{bmatrix} l_1 & m_1 \\ l_2 & m_2 \end{bmatrix} = \begin{bmatrix} 1 & 0 \\ 0 & 1 \end{bmatrix}.$$

Sub-case 1. $a_{uu} \neq 0$.

If $s_{pq} = \begin{vmatrix} a_{uu} & a_{uq} \\ a_{pu} & a_{pq} \end{vmatrix}$, the matrix $[s]_m^m$ has rank 1, and we obtain a particular solution by writing

$$a_{uu}[a]_m^m = \begin{bmatrix} a_{u1} \\ a_{u2} \\ \vdots \\ a_{um} \end{bmatrix}[a_{u1}\, a_{u2} \dots a_{um}] + [s]_m^m, \quad [s]_m^m = \begin{bmatrix} \sqrt{s_{11}} \\ \sqrt{s_{22}} \\ \vdots \\ \sqrt{s_{mm}} \end{bmatrix}[\sqrt{s_{11}}\, \sqrt{s_{22}} \dots \sqrt{s_{mm}}].$$

The general solution is given by either of the formulae

$$\sqrt{a_{uu}}\begin{bmatrix} x_1 & x_2 & \dots & x_m \\ y_1 & y_2 & \dots & y_m \end{bmatrix} = \begin{bmatrix} \lambda, & \mu \\ \pm\mu, & \mp\lambda \end{bmatrix}\begin{bmatrix} a_{u1} & a_{u2} & \dots & a_{um} \\ \sqrt{s_{11}} & \sqrt{s_{22}} & \dots & \sqrt{s_{mm}} \end{bmatrix},$$

$$\sqrt{2a_{uu}}\begin{bmatrix} x_1 & x_2 & \dots & x_m \\ y_1 & y_2 & \dots & y_m \end{bmatrix} = \begin{bmatrix} \lambda, & \mu \\ \pm\mu, & \mp\lambda \end{bmatrix}\begin{bmatrix} a_{u1}+\sqrt{s_{11}}, & a_{u2}+\sqrt{s_{22}}, & \dots & a_{um}+\sqrt{s_{mm}} \\ a_{u1}-\sqrt{s_{11}}, & a_{u2}-\sqrt{s_{22}}, & \dots & a_{um}-\sqrt{s_{mm}} \end{bmatrix},$$

where
$$\lambda^2 + \mu^2 = 1. \quad \dots\dots\dots\dots (\mathrm{A}_1)$$

The signs inherent in the radicals must be suitably chosen, and we have $s_{uu} = 0$.

Sub-case 2. $a_{uu} = a_{vv} = 0$, $a_{uv} \neq 0$.

We obtain a particular solution by observing that

$$a_{uv}[a]_m^m = \begin{bmatrix} a_{u1} & a_{v1} \\ a_{u2} & a_{v2} \\ \dots & \dots \\ a_{um} & a_{vm} \end{bmatrix}\begin{bmatrix} 0 & 1 \\ 1 & 0 \end{bmatrix}\begin{bmatrix} a_{u1} & a_{u2} & \dots & a_{um} \\ a_{v1} & a_{v2} & \dots & a_{vm} \end{bmatrix}.$$

The general solution is given by

$$\sqrt{a_{uv}}\begin{bmatrix} x_1 & x_2 & \dots & x_m \\ y_1 & y_2 & \dots & y_m \end{bmatrix} = \begin{bmatrix} \lambda, & \mu \\ \pm i\lambda, & \mp i\mu \end{bmatrix}\begin{bmatrix} a_{u1} & a_{u2} & \dots & a_{um} \\ a_{v1} & a_{v2} & \dots & a_{vm} \end{bmatrix},$$

where $$i=\sqrt{-1}, \text{ and } 2\lambda\mu=1. \qquad \text{.......}(A_2)$$

CASE II. *When* $[a]_m^m$ *has rank* 1.

We can write $$[a]_m^m = \begin{bmatrix} \sqrt{a_{11}} \\ \sqrt{a_{22}} \\ \vdots \\ \sqrt{a_{mm}} \end{bmatrix} [\sqrt{a_{11}}\ \sqrt{a_{22}} \dots \sqrt{a_{mm}}],$$

and the general solution is then given by

$$\begin{bmatrix} x_1 & x_2 & \dots & x_m \\ y_1 & y_2 & \dots & y_m \end{bmatrix} = \begin{bmatrix} \lambda \\ \mu \end{bmatrix} [\sqrt{a_{11}}\ \sqrt{a_{22}} \dots \sqrt{a_{mm}}], \text{ where } \lambda^2+\mu^2=1. \qquad \text{.........}(A_3)$$

In this case $[s]_m^m \doteq 0$, and the formula (A_3) is included in the formulae (A_1).

CASE III. *When* $[a]_m^m$ *has rank* 0.

In this case the equation (A) is satisfied when and only when

$$[x_1\ x_2 \dots x_m] = \sqrt{-1}\,[y_1\ y_2 \dots y_m], \quad \text{or} \quad \begin{bmatrix} x_1 & x_2 & \dots & x_m \\ y_1 & y_2 & \dots & y_m \end{bmatrix} = \begin{bmatrix} \lambda \\ \mu \end{bmatrix} [\xi_1\ \xi_2 \dots \xi_m],$$

where $$\lambda^2+\mu^2=0, \text{ and } [\xi_1\ \xi_2 \dots \xi_m] \text{ is arbitrary.} \qquad \text{.......}(A_4)$$

Ex. 1. *The equation*

$$\begin{bmatrix} x_1 & x_2 \\ y_1 & y_2 \\ z_1 & z_2 \end{bmatrix}\begin{bmatrix} x_1 & y_1 & z_1 \\ x_2 & y_2 & z_2 \end{bmatrix} = \begin{bmatrix} a & h & g \\ h & b & f \\ g & f & c \end{bmatrix} = \phi$$

when the symmetric matrix ϕ *has rank* 2.

Using the notation of § 128 the radicals $\sqrt{A}, \sqrt{B}, \sqrt{C}$ (which do not all vanish) can be so chosen that

$$\begin{bmatrix} A & H & G \\ H & B & F \\ G & F & C \end{bmatrix} = \begin{bmatrix} \sqrt{A} \\ \sqrt{B} \\ \sqrt{C} \end{bmatrix} [\sqrt{A}\ \sqrt{B}\ \sqrt{C}].$$

Then particular solutions for the matrix $[xyz]_{12}$ in the cases $a \neq 0$, $b \neq 0$, $c \neq 0$ respectively are

$$\frac{1}{\sqrt{a}}\begin{bmatrix} a, & h, & g \\ 0, & \sqrt{C}, & -\sqrt{B} \end{bmatrix}, \quad \frac{1}{\sqrt{b}}\begin{bmatrix} h, & b, & f \\ -\sqrt{C}, & 0, & \sqrt{A} \end{bmatrix}, \quad \frac{1}{\sqrt{c}}\begin{bmatrix} g, & f, & c \\ \sqrt{B}, & -\sqrt{A}, & 0 \end{bmatrix}$$

and also $$\frac{1}{\sqrt{2a}}\begin{bmatrix} a, & h+\sqrt{C}, & g-\sqrt{B} \\ a, & h-\sqrt{C}, & g+\sqrt{B} \end{bmatrix}, \quad \frac{1}{\sqrt{2b}}\begin{bmatrix} h-\sqrt{C}, & b, & f+\sqrt{A} \\ h+\sqrt{C}, & b, & f-\sqrt{A} \end{bmatrix},$$

$$\frac{1}{\sqrt{2c}}\begin{bmatrix} g+\sqrt{B}, & f-\sqrt{A}, & c \\ g-\sqrt{B}, & f+\sqrt{A}, & c \end{bmatrix}.$$

Moreover particular solutions in the cases $b=0,\ c=0,\ f\neq 0$; $c=0,\ a=0,\ g\neq 0$; $a=0,\ b=0,\ h\neq 0$ respectively are

$$\frac{1}{\sqrt{2f}}\begin{bmatrix}1, & 1\\ i, & -i\end{bmatrix}\begin{bmatrix}h & 0 & f\\ g & f & 0\end{bmatrix},\quad \frac{1}{\sqrt{2g}}\begin{bmatrix}1, & 1\\ i, & -i\end{bmatrix}\begin{bmatrix}g & f & 0\\ 0 & h & g\end{bmatrix},\quad \frac{1}{\sqrt{2h}}\begin{bmatrix}1, & 1\\ i, & -i\end{bmatrix}\begin{bmatrix}0 & h & g\\ h & 0 & f\end{bmatrix}.$$

From these particular solutions the general solution can be obtained as in the text.

Ex. ii. *The equation* $\begin{bmatrix}x_1 & x_2\\ y_1 & y_2\\ z_1 & z_2\\ w_1 & w_2\end{bmatrix}\begin{bmatrix}x_1 & y_1 & z_1 & w_1\\ x_2 & y_2 & z_2 & w_2\end{bmatrix}=\begin{bmatrix}a & h & g & u\\ h & b & f & v\\ g & f & c & w\\ u & v & w & d\end{bmatrix}=\phi,$

when ϕ *has rank* 2 *and the rows of* ϕ *have a standard arrangement.*

Using the notation of § 129 the radicals $\sqrt{A_1}, \sqrt{B_1}, \ldots \sqrt{C_2}$ can be so chosen that

$$\begin{bmatrix}A_1 & H_1 & G_1 & U_0 & W_1 & V_2\\ H_1 & B_1 & F_1 & W_2 & V_0 & U_1\\ G_1 & F_1 & C_1 & V_1 & U_2 & W_0\\ U_0 & W_2 & V_1 & A_2 & H_2 & G_2\\ W_1 & V_0 & U_2 & H_2 & B_2 & F_2\\ V_2 & U_1 & W_0 & G_2 & F_2 & C_2\end{bmatrix}=\begin{bmatrix}\sqrt{A_1}\\ \sqrt{B_1}\\ \sqrt{C_1}\\ \sqrt{A_2}\\ \sqrt{B_2}\\ \sqrt{C_2}\end{bmatrix}[\sqrt{A_1}\ \sqrt{B_1}\ \sqrt{C_1}\ \sqrt{A_2}\ \sqrt{B_2}\ \sqrt{C_2}].$$

Then particular solutions in the two possible cases are :

Case I. $a\neq 0,\ C_1=ab-h^2\neq 0.$

$$\begin{bmatrix}x_1 & y_1 & z_1 & w_1\\ x_2 & y_2 & z_2 & w_2\end{bmatrix}=\frac{1}{\sqrt{a}}\begin{bmatrix}a, & h\ , & g\ , & u\\ 0, & \sqrt{C_1}, & -\sqrt{B_1}, & \sqrt{A_2}\end{bmatrix},$$

$$\begin{bmatrix}x_1 & y_1 & z_1 & w_1\\ x_2 & y_2 & z_2 & w_2\end{bmatrix}=\frac{1}{\sqrt{2a}}\begin{bmatrix}a, & h+\sqrt{C_1}, & g-\sqrt{B_1}, & u+\sqrt{A_2}\\ a, & h-\sqrt{C_1}, & g+\sqrt{B_1}, & u-\sqrt{A_2}\end{bmatrix}.$$

Case II. $a=b=c=d=0,\ C_1=ab-h^2\neq 0.$

$$\begin{bmatrix}x_1 & y_1 & z_1 & w_1\\ x_2 & y_2 & z_2 & w_2\end{bmatrix}=\frac{1}{\sqrt{2h}}\begin{bmatrix}1, & 1\\ i, & -i\end{bmatrix}\begin{bmatrix}0 & h & g & u\\ h & 0 & f & v\end{bmatrix}$$

From these particular solutions the general solution can be obtained as in the text.

2. *The equation* $\begin{bmatrix}x_1 & y_1\\ x_2 & y_2\\ \cdots & \cdots\\ x_m & y_m\end{bmatrix}\begin{bmatrix}y_1 & y_2 & \cdots & y_m\\ x_1 & x_2 & \cdots & x_m\end{bmatrix}=[a]_m^m$. ....(B)

Case I. *When* $[a]_m^m$ *has rank* 2.

If $\begin{bmatrix}x_1 & x_2 & \cdots & x_m\\ y_1 & y_2 & \cdots & y_m\end{bmatrix}=\begin{bmatrix}a_1 & a_2 & \cdots & a_m\\ \beta_1 & \beta_2 & \cdots & \beta_m\end{bmatrix}$ is a particular solution, the general solution (see § 163. 4) is given by the two formulae

$$\left.\begin{matrix}[x_1\ x_2\ \ldots\ x_m]=\lambda\,[a_1\ a_2\ \ldots\ a_m]\\ [y_1\ y_2\ \ldots\ y_m]=\mu\,[\beta_1\ \beta_2\ \ldots\ \beta_m]\end{matrix}\right\},\qquad \left.\begin{matrix}[x_1\ x_2\ \ldots\ x_m]=\lambda\,[\beta_1\ \beta_2\ \ldots\ \beta_m]\\ [y_1\ y_2\ \ldots\ y_m]=\mu\,[a_1\ a_2\ \ldots\ a_m]\end{matrix}\right\},$$

where $$\lambda\mu=1.$$

Sub-case 1. $a_{uu} \neq 0$.

Defining $[s]_m^m$ as in sub-article 1, we have, when the signs inherent in the radicals are suitably chosen,

$$\begin{bmatrix} \sqrt{-s_{11}} \\ \sqrt{-s_{22}} \\ \cdots\cdots \\ \sqrt{-s_{mm}} \end{bmatrix} [\sqrt{-s_{11}}\ \sqrt{-s_{22}} \ldots \sqrt{-s_{mm}}] = -[s]_m^m,$$

and the general solution is then given by

$$\sqrt{2a_{uu}}\,[x_1\, x_2 \ldots x_m] = \lambda\,[a_{u1} \pm \sqrt{-s_{11}},\ a_{u2} \pm \sqrt{-s_{22}},\ \ldots\ a_{um} \pm \sqrt{-s_{mm}}],$$

$$\sqrt{2a_{uu}}\,[y_1\, y_2 \ldots y_m] = \mu\,[a_{u1} \mp \sqrt{-s_{11}},\ a_{u2} \mp \sqrt{-s_{22}},\ \ldots\ a_{um} \mp \sqrt{-s_{mm}}],$$

where $\lambda\mu = 1$. ....(B_1)

Sub-case 2. $a_{uu} = a_{vv} = 0$, $a_{uv} \neq 0$.

The general solution is given by the two formulae

$$\sqrt{a_{uv}}\,[x_1\, x_2 \ldots x_m] = \lambda\,[a_{u1}\, a_{u2} \ldots a_{um}], \quad \sqrt{a_{uv}}\,[y_1\, y_2 \ldots y_m] = \mu\,[a_{v1}\, a_{v2} \ldots a_{vm}];$$

$$\sqrt{a_{uv}}\,[x_1\, x_2 \ldots x_m] = \lambda\,[a_{v1}\, a_{v2} \ldots a_{vm}], \quad \sqrt{a_{uv}}\,[y_1\, y_2 \ldots y_m] = \mu\,[a_{u1}\, a_{u2} \ldots a_{um}];$$

where in each case $\lambda\mu = 1$. ....(B_2)

Case II. *When* $[a]_m^m$ *has rank* 1.

The radicals being chosen as in Case II of sub-article 1 the general solution is given by

$$\sqrt{2}\begin{bmatrix} x_1 & x_2 & \ldots & x_m \\ y_1 & y_2 & \ldots & y_m \end{bmatrix} = \begin{bmatrix} \lambda \\ \mu \end{bmatrix} [\sqrt{a_{11}}\ \sqrt{a_{22}} \ldots \sqrt{a_{mm}}], \text{ where } \lambda\mu = 1. \quad \ldots\ldots\ldots(B_3)$$

Case III. *When* $[a]_m^m$ *has rank* 0.

In this case the equation (B) is satisfied when and only when

either $[x_1\, x_2 \ldots x_m] = 0$, or $[y_1\, y_2 \ldots y_m] = 0$. ....(B_4)

Note. *Deduction of the solutions of* (B) *from the solutions of* (A).

If $i = \sqrt{-1}$, the two matrices

$$\begin{bmatrix} x_1 & x_2 & \ldots & x_m \\ y_1 & y_2 & \ldots & y_m \end{bmatrix} = \sqrt{\frac{1}{2}}\begin{bmatrix} 1, & i \\ 1, & -i \end{bmatrix}\begin{bmatrix} \xi_1 & \xi_2 & \ldots & \xi_m \\ \eta_1 & \eta_2 & \ldots & \eta_m \end{bmatrix},$$

$$\begin{bmatrix} \xi_1 & \xi_2 & \ldots & \xi_m \\ \eta_1 & \eta_2 & \ldots & \eta_m \end{bmatrix} = \sqrt{\frac{1}{2}}\begin{bmatrix} 1, & 1 \\ -i, & i \end{bmatrix}\begin{bmatrix} x_1 & x_2 & \ldots & x_m \\ y_1 & y_2 & \ldots & y_m \end{bmatrix}$$

are so related that the first is a solution of the equation (B) when and only when the second is a solution of the equation (A).

Consequently we can derive the solutions of (B) from the solutions of (A) by prefixing the matrix $\sqrt{\frac{1}{2}}\begin{bmatrix} 1, & i \\ 1, & -i \end{bmatrix}$.

Similarly we can derive the solutions of (A) from the solutions of (B) by prefixing the matrix $\sqrt{\frac{1}{2}}\begin{bmatrix} 1, & 1 \\ -i, & i \end{bmatrix}$.

Ex. iii. *The equation* $\begin{bmatrix} x_1 & x_2 \\ y_1 & y_2 \\ z_1 & z_2 \end{bmatrix} \begin{bmatrix} x_2 & y_2 & z_2 \\ x_1 & y_1 & z_1 \end{bmatrix} = \begin{bmatrix} a & h & g \\ h & b & f \\ g & f & c \end{bmatrix} = \phi$, *when the symmetric matrix* ϕ *has rank* 2.

The radicals being chosen as in Ex. i particular solutions for the matrix $[x\,y\,z]_{12}$ in the cases $a \neq 0$, $b \neq 0$, $c \neq 0$ respectively are

$$\frac{1}{\sqrt{2a}} \begin{bmatrix} a, & h+\sqrt{-C}, & g-\sqrt{-B} \\ a, & h-\sqrt{-C}, & g+\sqrt{-B} \end{bmatrix}, \quad \frac{1}{\sqrt{2b}} \begin{bmatrix} h-\sqrt{-C}, & b, & f+\sqrt{-A} \\ h+\sqrt{-C}, & b, & f-\sqrt{-A} \end{bmatrix},$$

$$\frac{1}{\sqrt{2c}} \begin{bmatrix} g+\sqrt{-B}, & f-\sqrt{-A}, & c \\ g-\sqrt{-B}, & f+\sqrt{-A}, & c \end{bmatrix}.$$

Further in the cases $b=0$, $c=0$, $f \neq 0$; $c=0$, $a=0$, $g \neq 0$; $a=0$, $b=0$, $h \neq 0$ respectively particular solutions are

$$\frac{1}{\sqrt{f}} \begin{bmatrix} h & 0 & f \\ g & f & 0 \end{bmatrix}, \quad \frac{1}{\sqrt{g}} \begin{bmatrix} g & f & 0 \\ 0 & h & g \end{bmatrix}, \quad \frac{1}{\sqrt{h}} \begin{bmatrix} 0 & h & g \\ h & 0 & f \end{bmatrix}.$$

From these particular solutions the general solution can be obtained as in the text.

Ex. iv. *The equation* $\begin{bmatrix} x_1 & x_2 \\ y_1 & y_2 \\ z_1 & z_2 \\ w_1 & w_2 \end{bmatrix} \begin{bmatrix} x_2 & y_2 & z_2 & w_2 \\ x_1 & y_1 & z_1 & w_1 \end{bmatrix} = \begin{bmatrix} a & h & g & u \\ h & b & f & v \\ g & f & c & w \\ u & v & w & d \end{bmatrix} = \phi$,

when ϕ *has rank* 2 *and the rows of* ϕ *have a standard arrangement.*

The radicals being chosen as in Ex. ii particular solutions in the two possible cases are:

Case I. $a \neq 0$, $C_1 = ab - h^2 \neq 0$.

$$\begin{bmatrix} x_1 & y_1 & z_1 & w_1 \\ x_2 & y_2 & z_2 & w_2 \end{bmatrix} = \frac{1}{\sqrt{2a}} \begin{bmatrix} a, & h+\sqrt{-C_1}, & g-\sqrt{-B_1}, & u+\sqrt{-A_2} \\ a, & h-\sqrt{-C_1}, & g+\sqrt{-B_1}, & u-\sqrt{-A_2} \end{bmatrix}.$$

Case II. $a=b=c=d=0$, $C_1 = ab - h^2 \neq 0$.

$$\begin{bmatrix} x_1 & y_1 & z_1 & w_1 \\ x_2 & y_2 & z_2 & w_2 \end{bmatrix} = \frac{1}{\sqrt{h}} \begin{bmatrix} 0 & h & g & u \\ h & 0 & f & v \end{bmatrix}.$$

From these particular solutions the general solution can be obtained as in the text.

CHAPTER XVIII

THE EXTRAVAGANCES OF MATRICES AND OF SPACELETS IN HOMOGENEOUS SPACE

[In § 165 we define the extravagances of a matrix and obtain general formulae for a matrix whose orders, rank and extravagances are given. In §§ 166—7 we consider the properties of two mutually normal undegenerate matrices: we show that the corranged simple minor determinants of either one are proportional to the anti-correspondent affected simple minor determinants of the other; we obtain general formulae for two such matrices; and we show that they have the same extravagance. In §§ 168—9 we effect the reduction of any undegenerate matrix to an equivalent similar undegenerate matrix whose long rows are mutually orthogonal, and we define the cores of an undegenerate matrix. In §§ 170—1 we define the extravagance, core and plenum of a spacelet, and give general formulae for a spacelet whose rank and extravagance are known. The determination of all possible sets of unconnected mutually orthogonal solutions of any system of homogeneous linear algebraic equations, which was left unfinished in Chapter XI, is completed in § 172. The remaining three articles deal with the possible ranks and extravagances of spacelets satisfying certain conditions.]

§ 165. The degeneracy and the extravagances of a matrix.

1. *The degeneracy of any matrix.*

We will define the *degeneracy* of a matrix to be the amount by which its rank falls short of its efficiency, i.e. of the number of its long rows. Thus if $[a]_m^n$ is a matrix of rank r, its degeneracy is $m - r$ when $m \not> n$, and $n - r$ when $n \not> m$. A matrix is undegenerate when it has degeneracy 0. The degeneracy of a matrix may also be defined to be the greatest number of unconnected connections between its long rows.

2. *The extravagance of an undegenerate matrix.*

In § 72.2 an undegenerate matrix $[a]_r^n$ or $\underline{\overline{a}}{}^{r}_{n}$ of rank r has been called *extravagant* when the sum of the squares of its simple minor determinants vanishes, i.e. when $\det [a]_r^n \underline{\overline{a}}{}^{r}_{n} = 0$, or when the product $[a]_r^n \underline{\overline{a}}{}^{r}_{n}$ is degenerate.

We will now define the *extravagance* of an undegenerate matrix $[a]^n_r$ or $\underline{\overline{a}}{}^r_n$ of rank r to be the degeneracy of the product $\psi = [a]^n_r \underline{\overline{a}}{}^r_n$. Thus each of these undegenerate matrices has extravagance s when the product ψ has degeneracy s or rank $r - s$. An undegenerate matrix is *non-extravagant* when it has extravagance 0.

Ex. i. An undegenerate matrix and its conjugate have the same extravagance as well as the same rank.

Ex. ii. An undegenerate square matrix is necessarily non-extravagant.

Ex. iii. If $[a]^n_r$ is an undegenerate matrix of rank r, its extravagance is the number of unconnected solutions of the equation $[a]^n_r \underline{\overline{x}}{}_n = 0$ which are connected with the long rows of $[a]^n_r$.

This follows from Ex. iii of § 136.

Ex. iv. *Possible values of the extravagance of an undegenerate matrix.*

If r and n are given, and if $[a]^n_r$ or $\underline{\overline{a}}{}^r_n$ is an undegenerate matrix of rank r whose elements are arbitrary, the possible values of its extravagance s are those consistent with the conditions

$$s \nless 0, \quad s \ngtr r, \quad s \ngtr n - r. \qquad \text{.......(1)}$$

This follows from Theorem I *a* of § 136.

Ex. v. Every real undegenerate matrix is necessarily non-extravagant.

Ex. vi. *General formulae for an undegenerate matrix of given orders, given rank, and given extravagance.*

By Ex. xiv of § 160 and Theorem III of § 157 the undegenerate matrices $[a]^n_r$ of rank r which have extravagance s are given by any one of the following formulae in which $i = \sqrt{-1}$, $[\omega]^n_n$ is a derangement of the unit matrix $[1]^n_n$, and $[h]^r_r$ is an arbitrary undegenerate square matrix of order r:

(1)
$$[a]^n_r = [h]^r_r \begin{bmatrix} l, & 0 \\ \lambda, & i \end{bmatrix}^{n-s,\, s}_{r-s,\, s} [\omega]^n_n, \qquad \text{.......}(\mathrm{A}_1)$$

where $\begin{bmatrix} l \\ \lambda \end{bmatrix}^{n-s}_{r-s,\, s}$ is a semi-unit matrix of rank r, so that

$$\begin{bmatrix} l \\ \lambda \end{bmatrix}^{n-s}_{r-s,\, s} \underline{\overline{l,\ \lambda}}{}^{r-s,\, s}_{n-s} = [1]^r_r.$$

(2)
$$[a]^n_r = [h]^r_r [l]^n_r, \qquad \text{.......}(\mathrm{A}_2)$$

where $[l]^n_r$ is a solution of rank r of the equation $[l]^n_r \underline{\overline{l}}{}^r_n = \begin{bmatrix} 1, & 0 \\ 0, & 0 \end{bmatrix}^{r-s,\, s}_{r-s,\, s}$.

(3)
$$[a]^n_r = [h]^r_r \begin{bmatrix} 1, & 0, & 0 \\ 0, & 1, & i \end{bmatrix}^{r-s,\, s,\, s}_{r-s,\, s} \begin{bmatrix} l, & 0 \\ 0, & 1 \end{bmatrix}^{n-s,\, s}_{r,\, s} [\omega]^n_n, \qquad \text{.......}(\mathrm{A}_3)$$

where $[l]^{n-s}_r$ is a semi-unit matrix of rank r.

(4) $$[a]^n_r = [h]^r_r \begin{bmatrix}1, & 0, & 0\\ 0, & 1, & i\end{bmatrix}^{r-s,\,s,\,s}_{r-s,\,s} [l]^n_{r+s}, \qquad\text{.........................}(A_4)$$

where $[l]^n_{r+s}$ is a semi-unit matrix of rank $r+s$.

(5) $$[a]^n_r = [h]^r_r \begin{bmatrix}1, & 0, & 0, & 0\\ 0, & 1, & i, & 0\end{bmatrix}^{r-s,\,s,\,s,\,n-r-s}_{r-s,\,s} [l]^n_n, \qquad\text{....................}(A_5)$$

where $[l]^n_n$ is a square semi-unit matrix.

By equating the conjugates of both sides we obtain general formulae for all undegenerate matrices $\overline{\underline{a}}{}^{r}_{n}$ of rank r and extravagance s.

We will now prove the following theorem:

Theorem I. *Two similar undegenerate matrices $[a]^n_r$, $[b]^n_r$, both of rank r, have the same extravagance when and only when there exists a relation of the form*

$$[b]^n_r = [h]^r_r\, [a]^n_r\, [u]^n_n \qquad\text{.........................}(B)$$

where $[h]^r_r$ is undegenerate, and $[u]^n_n$ is a square semi-unit matrix.

Two similar undegenerate matrices $[a]^r_m$, $[b]^r_m$, both of rank r, have the same extravagance when and only when there exists a relation of the form

$$[b]^r_m = [v]^m_m\, [a]^r_m\, [k]^r_r, \qquad\text{......................}(B')$$

where $[k]^r_r$ is undegenerate, and $[v]^m_m$ is a square semi-unit matrix.

Let $[a]^n_r$ and $[b]^n_r$ both have rank r, and suppose in the first place that there exists a relation of the form (B).

Then $$[b]^n_r\, \overline{\underline{b}}{}^{r}_{n} = [h]^r_r\, [a]^n_r\, \overline{\underline{a}}{}^{r}_{n}\, \overline{\underline{h}}{}^{r}_{r}.$$

Therefore by Theorem II of § 131 the two products $[a]^n_r\, \overline{\underline{a}}{}^{r}_{n}$ and $[b]^n_r\, \overline{\underline{b}}{}^{r}_{n}$ have the same rank and the same degeneracy, i.e. $[a]^n_r$ and $[b]^n_r$ have the same extravagance.

Next suppose that $[a]^n_r$ and $[b]^n_r$ have the same extravagance s.

Then from formula (A_5) we deduce that

$$\begin{bmatrix}1, & 0, & 0, & 0\\ 0, & 1, & i, & 0\end{bmatrix}^{r-s,\,s,\,s,\,n-r-s}_{r-s,\,s} = [h]^r_r\, [a]^n_r\, [u]^n_n = [k]^r_r\, [b]^n_r\, [v]^n_n \qquad\text{......}(2)$$

where $[h]_r^r$ and $[k]_r^r$ are undegenerate, and $[u]_n^n$ and $[v]_n^n$ are square semi-unit matrices. From the last equality in (2) we obtain a relation of the form (B).

Thus the first part of the theorem is proved. The second part can be proved in a similar way, or deduced from the first part by equating the conjugates of both sides in (B).

Ex. vii. The simplest undegenerate matrices $[a]_r^n$, $[b]_m^r$ of rank r which have extravagances s and t respectively are

$$[a]_r^n = \begin{bmatrix} 1, & 0, & 0, & 0 \\ 0, & 1, & i, & 0 \end{bmatrix}_{r-s,\, s}^{r-s,\, s,\, s,\, n-r-s}, \qquad [b]_m^r = \begin{bmatrix} 1, & 0 \\ 0, & 1 \\ 0, & i \\ 0, & 0 \end{bmatrix}_{r-t,\, t,\, t,\, m-r-t}^{r-t,\, t}$$

From these all other such matrices can be derived by the formulae (B) and (B′) as is shown in formula (A_5).

Ex. viii. Any two equivalent similar undegenerate matrices have the same extravagance as well as the same rank.

NOTE 1. *Completely extravagant matrices.*

Let $[a]_r^n$ and $\overline{\underline{a}}_n^r$ be undegenerate matrices of rank r (so that $r \ngtr n$) whose elements are arbitrary. These matrices can have extravagance r when and only when $2r \ngtr n$, and in this case (see Ex. iv) r is the greatest possible value which the extravagance can have. Such matrices will be called completely extravagant matrices. They are identical with the undegenerate solutions of rank r of the equation $[a]_r^n \, \overline{\underline{a}}_n^r = 0$.

Thus an undegenerate matrix is *completely extravagant* when and only when its extravagance is equal to its rank; and it is possible for $[a]_r^n$ and $\overline{\underline{a}}_n^r$ to be completely extravagant matrices of rank r when and only when $r \nless 0$, $2r \ngtr n$.

The following are general formulae for all completely extravagant matrices $[a]_r^n$ of rank r when the necessary condition $2r \ngtr n$ is satisfied:

(1) $$[a]_r^n = [h]_r^r [l, i]_r^{n-r,r} [\omega]_n^n, \quad \text{.................................}(C_1)$$

where $[l]_r^{n-r}$ is a semi-unit matrix of rank r.

(2) $$[a]_r^n = [h]_r^r \, [1, i]_r^{r,r} \begin{bmatrix} l, & 0 \\ 0, & 1 \end{bmatrix}_{r,\, r}^{n-r,\, r} [\omega]_n^n, \quad \text{.........................}(C_2)$$

where $[l]_r^{n-r}$ is a semi-unit matrix of rank r.

(3) $$[a]_r^n = [h]_r^r [1, i]_r^{r,r} [l]_{2r}^n, \quad \text{.................................}(C_3)$$

where $[l]_{2r}^n$ is a real semi-unit matrix of rank $2r$.

(4) $$[a]^n_r = [h]^r_r [1, i, 0]^{r, r, n-2r}_r [l]^n_n, \quad \text{.........................(C}_4\text{)}$$

where $[l]^n_n$ is a real square semi-unit matrix.

(5) $$[a]^n_r = [h]^r_r [1, b]^{r, n-r}_r [\omega]^n_n, \quad \text{...............................(C}_5\text{)}$$

where $$[b]^{n-r}_r \lceil b \rfloor^{r}_{n-r} = -[1]^r_r.$$

In these formulae $i = \sqrt{-1}$, $[\omega]^n_n$ is any derangement of the unit matrix $[1]^n_n$, and $[h]^r_r$ is any undegenerate square matrix of order r.

We deduce the first four formulae from Theorem II of § 158. They are particular cases of the formulae of Ex. vi. It has been shown in § 159 that there is no loss of generality in restricting the semi-unit matrices $[l]^n_{2r}$ and $[l]^n_n$ to be real in (C_3) and (C_4).

The fifth formula follows from the first. We can also obtain it by observing that every undegenerate matrix $[a]^n_r$ of rank r can be expressed (see § 167.2) in the form

$$[a]^n_r = [h]^r_r [1, b]^{r, n-r}_r [\omega]^n_n,$$

where $[\omega]^n_n$ is a derangement of $[1]^n_n$ and $[h]^r_r$ is undegenerate. When this is done, the equation $[a]^n_r \lceil a \rfloor^{r}_{n} = 0$ is satisfied when and only when

$$[1, b]^{r, n-r}_r \lceil \begin{matrix} 1 \\ b \end{matrix} \rfloor^{r}_{r, n-r} = 0, \quad \text{or} \quad [b]^{n-r}_r \lceil b \rfloor^{r}_{n-r} + [1]^r_r = 0;$$

and this is possible when and only when $n - r \nless r$, or $2r \ngtr n$.

Ex. ix. If $\lceil a \rfloor^{r}_{n}$ is a completely extravagant matrix of rank r, there exist an undegenerate square matrix $[k]^r_r$ and two mutually orthogonal real semi-unit matrices $\lceil u \rfloor^{r}_{n}$, $\lceil v \rfloor^{r}_{n}$ such that

$$\lceil a \rfloor^{r}_{n} = \lceil w \rfloor^{r}_{n} [k]^r_r, \quad \lceil w \rfloor^{r}_{n} = \lceil u \rfloor^{r}_{n} + i \lceil v \rfloor^{r}_{n} = \lceil u, v \rfloor^{r, r}_{n} \begin{bmatrix} 1 \\ i \end{bmatrix}^{r}_{r, r}.$$

We then have $$\begin{bmatrix} u \\ v \end{bmatrix}^{n}_{r, r} \lceil u, v \rfloor^{r, r}_{n} = \begin{bmatrix} 1, & 0 \\ 0, & 1 \end{bmatrix}^{r, r}_{r, r} = [1]^{2r}_{2r}.$$

This result is equivalent to formula (C_3) and to the theorem of § 159.

Ex. x. If any completely extravagant matrix $\lceil a \rfloor^{r}_{n}$ of rank r is expressed in the form $\lceil a \rfloor^{r}_{n} = \lceil u \rfloor^{r}_{n} + i \lceil v \rfloor^{r}_{n}$, where $\lceil u \rfloor^{r}_{n}$ and $\lceil v \rfloor^{r}_{n}$ are real, we must have

$$[u]^n_r \lceil u \rfloor^{r}_{n} = [v]^n_r \lceil v \rfloor^{r}_{n}, \quad [u]^n_r \lceil v \rfloor^{r}_{n} = -[v]^n_r \lceil u \rfloor^{r}_{n}.$$

NOTE 2. *Plenarily extravagant matrices.*

Let $[a]^n_r$ and $\lceil a \rfloor^{r}_{n}$ be undegenerate matrices of rank r (so that $r \ngtr n$) whose elements are arbitrary. These matrices can have extravagance $n - r$ when and only when $2r \nless n$, and in this case (see Ex. iv) $n-r$ is the greatest possible value which the extravagance can have. Such matrices will be called plenarily extravagant matrices. They are the matrices of rank r which are such that $[a]^n_r \, \overline{\underline{a}}{}^r_n$ has rank $2r-n$.

Thus an undegenerate matrix is *plenarily extravagant* when and only when its extravagance is equal to the difference of its orders; and it is possible for $[a]^n_r$ and $\overline{\underline{a}}{}^r_n$ to be plenarily extravagant matrices of rank r when and only when $2r \nless n$, $r \ngtr n$.

The following are general formulae for all plenarily extravagant matrices $[a]^n_r$ of rank r when the necessary condition $2r \nless n$ is satisfied:

(1) $$[a]^n_r = [h]^r_r \begin{bmatrix} l, & 0 \\ \lambda, & i \end{bmatrix}^{r,\,n-r}_{2r-n,\,n-r} [\omega]^n_n, \quad \ldots\ldots\ldots\ldots(D_1)$$

where $\begin{bmatrix} l \\ \lambda \end{bmatrix}^{r}_{2r-n,\,n-r}$ is a semi-unit matrix of rank r.

(2) $$[a]^n_r = [h]^r_r \, [l]^n_r, \quad \ldots\ldots\ldots\ldots(D_2)$$

where $[l]^n_r$ is a solution of rank r of the equation $[l]^n_r \, \overline{\underline{l}}{}^r_n = \begin{bmatrix} 1, & 0 \\ 0, & 0 \end{bmatrix}^{2r-n,\,n-r}_{2r-n,\,n-r}$.

(3) $$[a]^n_r = [h]^r_r \begin{bmatrix} 1, & 0, & 0 \\ 0, & 1, & i \end{bmatrix}^{2r-n,\,n-r,\,n-r}_{2r-n,\,n-r} \begin{bmatrix} l, & 0 \\ 0, & 1 \end{bmatrix}^{r,\,n-r}_{r,\,n-r} [\omega]^n_n, \quad \ldots\ldots\ldots\ldots(D_3)$$

where $[l]^r_r$ is a square semi-unit matrix of rank r.

(4) $$[a]^n_r = [h]^r_r \begin{bmatrix} 1, & 0, & 0 \\ 0, & 1, & i \end{bmatrix}^{2r-n,\,n-r,\,n-r}_{2r-n,\,n-r} [l]^n_n, \quad \ldots\ldots\ldots\ldots(D_4)$$

where $[l]^n_n$ is a real square semi-unit matrix.

(5) $$[a]^n_r = [h]^r_r [1,\ b]^{r,\,n-r}_r [\omega]^n_n, \quad \ldots\ldots\ldots\ldots(D_5)$$

where $$\overline{\underline{b}}{}^r_{n-r} \, [b]^{n-r}_r = -[1]^{n-r}_{n-r}.$$

In these formulae $i = \sqrt{-1}$, $[\omega]^n_n$ is any derangement of the unit matrix $[1]^n_n$, and $[h]^r_r$ is any undegenerate square matrix of order r.

We deduce the first four formulae from Ex. vi of the present article and Ex. xiv of § 167. The fifth formula is that given in Ex. iv of § 94, and follows from Ex. xii of § 167. Proofs of it are given in Exs. xiv and xvii of § 167.

Ex. xi. *The matrix* $[1,\ b]^{r,\,n-r}_r$ *is*

(1) *completely extravagant when and only when* $[b]^{n-r}_r \, \overline{\underline{b}}{}^r_{n-r} = -[1]^r_r$;

(2) *plenarily extravagant when and only when* $\overline{\underline{b}}{}^r_{n-r} \, [b]^{n-r}_r = -[1]^{n-r}_{n-r}$.

The first of these results is obvious; and the second is proved in Ex. xvii of § 167, and can be deduced from Ex. iv of § 94 by means of Ex. xii of § 167.

Ex. xii. We can construct an undegenerate matrix $[a]_m^n$ of rank m in which the minor matrix $[a_{1q}]_m^{n-u}$ formed by striking out u short rows has rank ρ, when and only when ρ is an integer consistent with the conditions

$$\rho \nless 0, \quad \rho \nless m-u, \quad \rho \ngtr m, \quad \rho \ngtr n-u$$

which include $m \nless 0$, $m \ngtr n$, $u \nless 0$, $u \ngtr n$; and this is true even when the minor matrix $[a_{1q}]_m^{n-u}$ is given.

This result follows from Theorem Ia of § 104. We can use it to prove Ex. xiii.

Ex. xiii. **Theorem.** *If* $[a]_n^r$ *is a given undegenerate matrix of rank* r *and extravagance* ρ, *and if* $[b]_n^p$ *is any undegenerate matrix of rank* p *and extravagance* π *whose vertical rows are connected with the vertical rows of* $[a]_n^r$, *then the possible values of* p *and* π *are those consistent with the conditions*

$$p+\pi \ngtr r+\rho, \quad p-\pi \ngtr r-\rho, \quad \pi \nless 0, \quad \pi \ngtr p, \quad \text{.........................(3)}$$

which include $p \nless 0$, $p \ngtr r$.

From Ex. iv we see that the given numbers r and ρ can have any values consistent with the conditions

$$r \nless 0, \quad r \ngtr n; \quad \rho \nless 0, \quad \rho \ngtr r, \quad \rho \ngtr n-r, \quad \text{...........................(4)}$$

which, in accordance with (3), are equivalent to $r+\rho \ngtr n$, $\rho \nless 0$, $\rho \ngtr r$.

Using formula (A_2) we can write $[a]_n^r = [l]_n^r [k]_r^r$, where $[k]_r^r$ is a given undegenerate square matrix, and $[l]_n^r$ is a given solution of rank r of the equation

$$\underset{r}{\overline{\underline{l}}}^{\,n} [l]_n^r = \begin{bmatrix} 1, & 0 \\ 0, & 0 \end{bmatrix}_{r-\rho,\ \rho}^{r-\rho,\ \rho}.$$

Any matrix $[b]_n^p$ is connected with the vertical rows of $[a]_n^r$ when and only when it is connected with the vertical rows of $[l]_n^r$. Consequently the matrices $[b]_n^p$ of rank p whose vertical rows are connected with the vertical rows of $[a]_n^r$ are those given by the formula

$$[b]_n^p = [l]_n^r [t]_r^p,$$

where $[t]_r^p$ is an undegenerate matrix of rank p whose elements are arbitrary. We then have

$$\underset{p}{\overline{\underline{b}}}^{\,n} [b]_n^p = \underset{p}{\overline{\underline{t}}}^{\,r-\rho} [t]_{r-\rho}^p,$$

and $[b]_n^p$ has rank p and extravagance π when and only when $[t]_r^p$ has rank p and $\underset{p}{\overline{\underline{t}}}^{\,r-\rho} [t]_{r-\rho}^p$ has rank $p-\pi$.

Now by § 136 or § 157 we can determine a matrix $[t]_{r-\rho}^p$ of rank x such that the product $\underset{p}{\overline{\underline{t}}}^{\,r-\rho} [t]_{r-\rho}^p$ has rank $p-\pi$ when and only when

$$p-\pi \nless 0, \quad p-\pi \ngtr x, \quad (p-\pi)+(r-\rho) \nless 2x, \quad x \ngtr p; \quad \text{.................(5)}$$

and when $[t]_{r-\rho}^{p}$ has been thus determined it follows from Theorem I *b* of § 104 that we can form a matrix $[t]_r^p$ of rank p by adding final horizontal rows to $[t]_{r-\rho}^{p}$ when and only when

$$p \nless x, \quad p \ngtr x + \rho. \quad \text{......................................(6)}$$

When we eliminate x from (5) and (6), and omit conditions which are superfluous in virtue of (4), we obtain the inequalities (3). This shows that we can construct a matrix $[t]_r^p$ which has rank p and is such that the product $\underline{\overline{t}}_{p}^{\,r-\rho}\,[t]_{r-\rho}^{p}$ has rank $p-\pi$ when and only when the conditions (3) are satisfied.

When we eliminate π from the inequalities (3) we obtain $p \nless 0$, $p \ngtr r$.

Thus p can have any value consistent with these conditions, and when p is given, π can have any value consistent with the conditions (3).

3. *The horizontal and vertical extravagances of any matrix.*

We first notice the following property of any matrix:

If $[a]_m^n$ is any matrix of rank r, all undegenerate minor matrices formed with r unconnected horizontal rows of $[a]_m^n$ have the same extravagance, and all undegenerate minor matrices formed with r unconnected vertical rows of $[a]_m^n$ have the same extravagance.

For if $[a_{p1}]_r^n$, $[a_{u1}]_r^n$ are any two undegenerate horizontal minors of order r, there exists between them a relation of the form $[a_{u1}]_r^n = [h]_r^r\,[a_{p1}]_r^n$, where $[h]_r^r$ is undegenerate.

Therefore by Theorem I or Ex. viii the matrices $[a_{p1}]_r^n$, $[a_{u1}]_r^n$ have the same extravagance.

A similar proof can be given for the undegenerate vertical minors of reduced order r.

If s is the common extravagance of all undegenerate horizontal minors of reduced order r of a matrix $[a]_m^n$ whose rank is r, we will call s the *horizontal extravagance* of $[a]_m^n$. Similarly if t is the common extravagance of all undegenerate vertical minors of $[a]_m^n$ of reduced order r, we will call t the *vertical extravagance* of $[a]_m^n$.

The (unqualified) extravagance of an undegenerate matrix, as defined in sub-article 2, is its horizontal or vertical extravagance according as the long rows are horizontal or vertical.

From the above definitions we deduce the following theorem:

Theorem II. *If any matrix* $[a]_m^n$ *of rank r is expressed in the form*

$$[a]_m^n = [h]_m^r [k]_r^n, \quad \text{.............................(7)}$$

where $[h]_m^r$ *and* $[k]_r^n$ *are necessarily undegenerate matrices of rank r, then the horizontal extravagance of* $[a]_m^n$ *is equal to the (horizontal) extravagance of the undegenerate matrix* $[k]_r^n$, *and the vertical extravagance of* $[a]_m^n$ *is equal to the (vertical) extravagance of the undegenerate matrix* $[h]_m^r$.

Let $[a_{p1}]_r^n$ be any undegenerate horizontal minor of $[a]_m^n$ of reduced order r. Then from the equation (7) we see that

$$[a_{p1}]_r^n = [h_{p1}]_r^r [k]_r^n,$$

where by Theorem III of § 71 the square matrix $[h_{p1}]_r^r$ has rank r.

Therefore by Theorem I or Ex. viii, the undegenerate matrix $[a_{p1}]_r^n$ has the same extravagance as the undegenerate matrix $[k]_r^n$.

Similarly every undegenerate vertical minor $[a_{1q}]_m^r$ of $[a]_m^n$ of reduced order r has the same extravagance as the undegenerate matrix $[k]_r^n$.

Ex. xiv. Any two mutually conjugate matrices have the same rank, and the horizontal and vertical extravagances of one are respectively the vertical and horizontal extravagances of the other.

Ex. xv. If $[a]_m^n$ is a matrix of rank r, the (horizontal) extravagance of every undegenerate matrix $[k]_r^n$ of rank r whose horizontal rows are connected with the horizontal rows of $[a]_m^n$ is equal to the horizontal extravagance of $[a]_m^n$. Also the (vertical) extravagance of every undegenerate matrix $[h]_m^r$ whose vertical rows are connected with the vertical rows of $[a]_m^n$ is equal to the vertical extravagance of $[a]_m^n$.

Ex. xvi. *Any two horizontally equivalent matrices have the same horizontal extravagance as well as the same rank. Also any two vertically equivalent matrices have the same vertical extravagance as well as the same rank.*

Here we use the definitions of horizontal and vertical equivalence given in § 138.

Ex. xvii. *Possible values of the horizontal and vertical extravagances of any matrix* $[a]_m^n$.

When m and n are given and $[a]_m^n$ has the assigned rank r, r being any integer consistent with the conditions

$$r \nless 0, \quad r \ngtr m, \quad r \ngtr n, \quad \text{.................................(8)}$$

the horizontal and vertical extravagances s and t of $[a]^n_m$ can *independently of one another* have any values consistent with the conditions

$$s \nless 0, \quad s \ngtr r, \quad s \ngtr n-r \,; \quad \ldots\ldots(9)$$

$$t \nless 0, \quad t \ngtr r, \quad t \ngtr m-r. \quad \ldots\ldots(10)$$

This appears from Ex. iv when we express $[a]^n_m$ in the form (7). For when r is given we can choose $[h]^r_m$ to be an undegenerate matrix of rank r whose (vertical) extravagance is t, and $[k]^n_r$ to be an undegenerate matrix of rank r whose (horizontal) extravagance is s when and only when s and t are integers satisfying the conditions (9) and (10).

Since the elimination of s and t from (9) and (10) leads to (8), we see that:

When the value of r is not assigned, the possible values of r, s and t are those consistent with the conditions (9) *and* (10).

Again by eliminating r from (9) and (10) we see that:

When the value of r is not assigned, the possible values of s and t are those consistent with the conditions

$$s \nless 0, \;\; 2s \ngtr n \,; \quad t \nless 0, \;\; 2t \ngtr m \,; \quad s+t \ngtr m, \;\; s+t \ngtr n. \quad \ldots\ldots(11)$$

Ex. xviii. *Possible ranks of the product matrices* $\phi=[a]^n_m \,\overline{\underline{a}}{}^m_n$, $\psi=\overline{\underline{a}}{}^m_n\,[a]^n_m$.

When $[a]^n_m$ has any assigned rank r consistent with the conditions

$$r \nless 0, \quad r \ngtr m, \quad r \ngtr n, \quad \ldots\ldots(8)$$

and is expressed in the form (7), the ranks ρ and σ of ϕ and ψ are respectively equal to the ranks of $[k]^n_r \,\overline{\underline{k}}{}^r_n$ and $\overline{\underline{h}}{}^m_r\,[h]^r_m$. Therefore by § 136.1 ρ and σ can *independently of one another* have any values consistent with the conditions

$$\rho \nless 0, \quad \rho \ngtr r, \quad \rho+n \nless 2r \,; \quad \ldots\ldots(9')$$

$$\sigma \nless 0, \quad \sigma \ngtr r, \quad \sigma+m \nless 2r. \quad \ldots\ldots(10')$$

Each of the two sets of conditions (9), (10) and (9′), (10′) can be deduced from the other; for with the notation of Ex. xvii we have $s=r-\rho$, $t=r-\sigma$.

When the value of r is not assigned, the possible values of ρ and σ are those consistent with the conditions

$$\rho \nless 0, \;\; \rho \ngtr n \,; \quad \sigma \nless 0, \;\; \sigma \ngtr m \,; \quad 2\rho-\sigma \ngtr m, \;\; 2\sigma-\rho \ngtr n, \quad \ldots\ldots(11')$$

which include the conditions $\rho \ngtr m$, $\sigma \ngtr n$.

Ex. xix. *Possible degeneracies of the product matrices* $\phi=[a]^n_m \,\overline{\underline{a}}{}^m_n$, $\psi=\overline{\underline{a}}{}^m_n\,[a]^n_m$.

If we denote these by μ and ν respectively, then when $[a]^n_m$ has the assigned rank r, the possible values of μ and ν are (independently of one another) those consistent with the conditions

$$\mu \nless m-r, \quad \mu \ngtr m, \quad \mu \ngtr m+n-2r \,; \quad \ldots\ldots(9'')$$

$$\nu \nless n-r, \quad \nu \ngtr n, \quad \nu \ngtr m+n-2r. \quad \ldots\ldots(10'')$$

When the value of r is not assigned, the possible values of μ and ν are those consistent with the conditions

$$\mu \nless m-n,\quad \mu \ngtr m;\quad \nu \nless n-m,\quad \nu \ngtr n;\quad 2\mu-\nu \nless m-n;\quad 2\nu-\mu \nless n-m, \quad\text{...(11'')}$$

which include the conditions $\mu \nless 0$, $\nu \nless 0$.

Ex. xx. If $[a]_m^n$ has rank r and $[a_{p1}]_r^n$ is any undegenerate horizontal minor of reduced order r, then $[a]_m^n \overline{\underline{a}}{}_n^m$ has the same rank as $[a_{p1}]_r^n \overline{\underline{a_{p1}}}{}_n^r$. Also if $[a_{1q}]_m^r$ is any undegenerate vertical minor of reduced order r, then $\overline{\underline{a}}{}_n^m [a]_m^n$ has the same rank as $\overline{\underline{a_{1q}}}{}_r^m [a_{1q}]_m^r$.

Ex. xxi. If $[a]_m^n$ has rank r, then it has horizontal extravagance s when and only when $[a]_m^n \overline{\underline{a}}{}_n^m$ has rank $r-s$, and it has vertical extravagance t when and only when $\overline{\underline{a}}{}_n^m [a]_m^n$ has rank $r-t$.

Ex. xxii. If $[a]_m^n$ and $[b]_p^q$ have the same rank, then they have the same horizontal extravagance when and only when $[a]_m^n \overline{\underline{a}}{}_n^m$ and $[b]_p^q \overline{\underline{b}}{}_q^p$ have the same rank or are equigradent; and they have the same vertical extravagance when and only when $\overline{\underline{a}}{}_n^m [a]_m^n$ and $\overline{\underline{b}}{}_q^p [b]_p^q$ have the same rank or are equigradent.

This follows from the previous example.

Ex. xxiii. *General formulae for all matrices which have given orders, given rank, and given horizontal or vertical extravagance.*

A matrix $[a]_m^n$ has rank r and horizontal extravagance s when and only when $[a]_m^n = [h]_m^r [a]_r^n$, where $[h]_m^r$ has rank r and $[a]_r^n$ has any one of the forms given in Ex. vi.

Thus general formulae for all matrices $[a]_m^n$ having rank r and horizontal extravagance s are:

(1)
$$[a]_m^n = [h]_m^r \begin{bmatrix} l, & 0 \\ \lambda, & i \end{bmatrix}_{r-s,\,s}^{n-s,\,s} [\omega]_n^n, \quad\text{.....(E}_1\text{)}$$

where $\begin{bmatrix} l \\ \lambda \end{bmatrix}_{r-s,\,s}^{n-s}$ is a semi-unit matrix of rank r.

(2)
$$[a]_m^n = [h]_m^r [l]_r^n, \quad\text{.....(E}_2\text{)}$$

where $[l]_r^n$ is a solution of rank r of the equation $[l]_r^n \overline{\underline{l}}{}_n^r = \begin{bmatrix} 1, & 0 \\ 0, & 1 \end{bmatrix}_{r-s,\,s}^{r-s,\,s}$.

(3)
$$[a]_m^n = [h]_m^r \begin{bmatrix} 1, & 0, & 0 \\ 0, & 1, & i \end{bmatrix}_{r-s,\,s}^{r-s,\,s,\,s} \begin{bmatrix} l, & 0 \\ 0, & 1 \end{bmatrix}_{r,\,s}^{n-s,\,s} [\omega]_n^n, \quad\text{.....(E}_3\text{)}$$

where $[l]_r^{n-s}$ is a semi-unit matrix of rank r.

(4) $$[a]^n_m = [h]^r_m \begin{bmatrix}1, & 0, & 0\\ 0, & 1, & i\end{bmatrix}^{r-s,\, s,\, s}_{r-s,\, s} [l]^n_{r+s}, \quad \ldots\ldots\ldots\ldots (E_4)$$

where $[l]^n_{r+s}$ is a semi-unit matrix of rank $r+s$.

(5) $$[a]^n_m = [h]^m_m \begin{bmatrix}1, & 0, & 0, & 0\\ 0, & 1, & i, & 0\\ 0, & 0, & 0, & 0\end{bmatrix}^{r-s,\, s,\, s,\, n-r-s}_{r-s,\, s,\, m-r} [l]^n_n, \quad \ldots\ldots\ldots\ldots (E_5)$$

where $[l]^n_n$ is a square semi-unit matrix.

In these formulae $i=\sqrt{-1}$; $[\omega]^n_n$ is any derangement of the unit matrix $[1]^n_n$; and $[h]^r_m$, $[h]^m_m$ are arbitrary undegenerate matrices of ranks r, m.

By equating the conjugates of both sides we obtain general formulae for all matrices $\overline{\underline{a}}^m_n$ which have rank r and vertical extravagance s.

Ex. xxiv. *General formulae for all matrices $[a]^n_m$ which have rank r, horizontal extravagance s, and vertical extravagance t.*

Using Theorem II and Ex. vi, we see that general formulae for all such matrices are:

(1) $$[a]^n_m = \overline{\underline{u}}^{r+t}_m \; \overline{\underline{\begin{matrix}1, & 0\\ 0, & 1\\ 0, & i\end{matrix}}}^{r-t,\, t}_{r-t,\, t,\, t} \; [k]^r_r \begin{bmatrix}1, & 0, & 0\\ 0, & 1, & i\end{bmatrix}^{r-s,\, s,\, s}_{r-s,\, s} [v]^n_{r+s}, \quad \ldots\ldots\ldots\ldots (F_1)$$

where $\overline{\underline{u}}^{r+t}_m$, $[v]^n_{r+s}$ are semi-unit matrices of ranks $r+t$, $r+s$.

(2) $$[a]^n_m = \overline{\underline{u}}^{m}_m \; \overline{\underline{\begin{matrix}1, & 0\\ 0, & 1\\ 0, & i\\ 0, & 0\end{matrix}}}^{r-t,\, t}_{r-t,\, t,\, t,\, m-r-t} \; [k]^r_r \begin{bmatrix}1, & 0, & 0, & 0\\ 0, & 1, & i, & 0\end{bmatrix}^{r-s,\, s,\, s,\, n-r-s}_{r-s,\, s} [v]^n_n, \quad \ldots\ldots (F_2)$$

where $\overline{\underline{u}}^m_m$ and $[v]^n_n$ are square semi-unit matrices.

In these formulae $i=\sqrt{-1}$, and $[k]^r_r$ is any undegenerate square matrix of order r.

The matrix $[a]^n_m$ has rank r and neither horizontal nor vertical extravagance when and only when it can be expressed in the form

$$[a]^n_m = \overline{\underline{u}}^m_m \begin{bmatrix}k, & 0\\ 0, & 0\end{bmatrix}^{r,\, n-r}_{r,\, m-r} [v]^n_n,$$

where $\overline{\underline{u}}^m_m$ and $[v]^n_n$ are square semi-unit matrices, and $[k]^r_r$ is an undegenerate square matrix.

We will now prove the two theorems which follow:

Theorem III a. *Any two similar matrices* $[a]^n_m$, $[b]^n_m$ *have the same rank and the same horizontal extravagance when and only when there exists a relation of the form*

$$[b]^n_m = [h]^m_m [a]^n_m [u]^n_n, \qquad \text{.......................(G)}$$

where $[h]^m_m$ *is undegenerate and* $[u]^n_n$ *is a square semi-unit matrix.*

They have the same rank and the same vertical extravagance when and only when there exists a relation of the form

$$[b]^n_m = [v]^m_m [a]^n_m [k]^n_n, \qquad \text{.......................(G}')$$

where $[k]^n_n$ *is undegenerate and* $[v]^m_m$ *is a square semi-unit matrix.*

First suppose that there exists a relation of the form (G). Then by § 131 the matrices $[a]^n_m$ and $[b]^n_m$ *have the same rank.*

Further we have

$$[b]^n_m \overline{\underline{b}}{}^m_n = [h]^m_m . [a]^n_m \overline{\underline{a}}{}^m_n . \overline{\underline{h}}{}^m_m . \qquad \text{..................(12)}$$

Therefore $[a]^n_m \overline{\underline{a}}{}^m_n$ and $[b]^n_m \overline{\underline{b}}{}^m_n$ have the same rank; and it follows from Ex. xxii that $[a]^n_m$ and $[b]^n_m$ *have the same horizontal extravagance.*

The last result can also be deduced from (12) by writing

$$[a]^n_m = [\alpha]^r_m [a_{x1}]^n_r, \quad [b]^n_m = [\beta]^r_m [b_{y1}]^n_r,$$

where r is the common rank of $[a]^n_m$ and $[b]^n_m$, and $[a_{x1}]^n_r$, $[b_{y1}]^n_r$ are undegenerate horizontal minors of $[a]^n_m$, $[b]^n_m$.

Next suppose that $[a]^n_m$ and $[b]^n_m$ have the same rank r and the same horizontal extravagance s. Then by formula (E_5) we have

$$\begin{bmatrix} 1, & 0, & 0, & 0 \\ 0, & 1, & i, & 0 \\ 0, & 0, & 0, & 0 \end{bmatrix}^{r-s,\ s,\ s,\ n-r-s}_{r-s,\ s,\ m-r} = [h]^m_m [a]^n_m [u]^n_n = [k]^m_m [b]^n_m [v]^n_n, \quad \text{...(13)}$$

where $[h]^m_m$ and $[k]^n_n$ are undegenerate, and $[u]^n_n$ and $[v]^n_n$ are square semi-unit matrices. From the last equality in (13) we deduce a relation of the form (G).

Thus the first part of the theorem is proved; and the second part of the theorem can be proved in a similar way, or deduced from the first part by equating the conjugates of both sides in (G).

Theorem III b. *Any two matrices* $[a]^n_m$, $[b]^q_p$ *have the same rank and the same horizontal extravagance when and only when there exists a relation of the form*

$$[h]^m_\mu \, [a]^n_m \, [u]^\nu_n = [k]^p_\mu \, [b]^q_p \, [v]^\nu_q, \quad \text{......................(H)}$$

where $[h]^m_\mu$ *and* $[k]^p_\mu$ *have ranks* m *and* p, *and where* $[u]^\nu_n$ *and* $[v]^\nu_q$ *are semi-unit matrices of ranks* n *and* q.

They have the same rank and the same vertical extravagance when and only when there exists a relation of the form

$$[u]^m_\mu \, [a]^n_m \, [h]^\nu_n = [v]^p_\mu \, [b]^q_p \, [k]^\nu_q \quad \text{......................(H')}$$

where $[h]^\nu_n$ *and* $[k]^\nu_q$ *have ranks* n *and* q, *and where* $[u]^m_\mu$ *and* $[v]^p_\mu$ *are semi-unit matrices of ranks* m *and* p.

These results remain true when we restrict μ *to be the larger of the two numbers* m *and* p, *and* ν *to be the larger of the two numbers* n *and* q.

This theorem should be compared with that given in Ex. ii of § 146.

First suppose that there exists a relation of the form (H). Then $[a]^n_m$ and $[b]^q_p$ *have the same rank.* Further we have

$$[h]^m_\mu \cdot [a]^n_m \, \overline{\underline{a}}^{\,m}_{\,n} \cdot \overline{\underline{h}}^{\,\mu}_{\,m} = [k]^p_\mu \cdot [b]^q_p \, \overline{\underline{b}}^{\,p}_{\,q} \cdot \overline{\underline{k}}^{\,\mu}_{\,p}.$$

Therefore $[a]^n_m \, \overline{\underline{a}}^{\,m}_{\,n}$ and $[b]^q_p \, \overline{\underline{b}}^{\,p}_{\,q}$ have the same rank; and it follows from Ex. xxii that $[a]^n_m$ and $[b]^q_p$ *have the same horizontal extravagance.*

Next suppose that $[a]^n_m$ and $[b]^q_p$ have the same rank r and the same horizontal extravagance s, and let μ and ν be any integers such that

$$\mu \nless m, \quad \mu \nless p; \quad \nu \nless n, \quad \nu \nless q.$$

Then by formula (E_5) we have

$$\begin{bmatrix} 1, & 0, & 0, & 0 \\ 0, & 1, & i, & 0 \\ 0, & 0, & 0, & 0 \end{bmatrix}^{r-s,\, s,\, s,\, \nu-r-s}_{r-s,\, s,\, \mu-r} = \begin{bmatrix} h \\ 0 \end{bmatrix}^{m}_{m,\, \mu-m} [a]^n_m \, [u,\ 0]^{n,\, \nu-n}_n$$

$$= \begin{bmatrix} k \\ 0 \end{bmatrix}^{p}_{p,\, \mu-p} [b]^q_p \, [v,\ 0]^{q,\, \nu-q}_q, \quad \text{......(14)}$$

where $[h]_m^m$ and $[k]_p^p$ are undegenerate, and $[u]_n^n$ and $[v]_q^q$ are square semi-unit matrices; and the last equality in (14) is a relation of the form (H).

Thus the first part of the theorem is proved; and the second part of the theorem can be proved in a similar way.

Ex. xxv. *When $[u]_q^n$, $[v]_m^p$ are semi-unit matrices of ranks q, p, the equigradent transformations*

$$[b]_m^n=[h]_m^p\,[a]_p^q\,[u]_q^n,\quad [b]_m^n=[v]_m^p\,[a]_p^q\,[k]_q^n$$

convert $[a]_p^q$ respectively into a matrix $[b]_m^n$ having the same rank and the same horizontal extravagance as $[a]_p^q$, and into a matrix $[b]_m^n$ having the same rank and the same vertical extravagance as $[a]_p^q$.

Ex. xxvi. The simplest matrices $[a]_m^n$, $[b]_m^n$ which have respectively rank r and horizontal extravagance s, rank r and vertical extravagance t are

$$[a]_m^n=\begin{bmatrix}1, & 0, & 0, & 0\\ 0, & 1, & i, & 0\\ 0, & 0, & 0, & 0\end{bmatrix}_{r-s,\ s,\ m-r}^{r-s,\ s,\ s,\ n-r-s},\qquad [b]_m^n=\begin{matrix}1, & 0, & 0\\ 0, & 1, & 0\\ 0, & i, & 0\\ 0, & 0, & 0\end{matrix}\;{}_{r-t,\ t,\ t,\ m-r-t}^{r-t,\ t,\ n-r}$$

Ex. xxvii. The simplest matrix $[a]_m^n$ which has rank r, horizontal extravagance s, and vertical extravagance t is

$$[a]_m^n=\begin{matrix}1, & 0\\ 0, & 1\\ 0, & i\\ 0, & 0\end{matrix}\;{}_{r-t,\ t,\ t,\ m-r-t}^{r-t,\ t}\qquad \begin{bmatrix}1, & 0, & 0, & 0\\ 0, & 1, & i, & 0\end{bmatrix}_{r-s,\ s}^{r-s,\ s,\ s,\ n-r-s}$$

When $s \not< t$, this matrix is

$$[a]_m^n=\begin{matrix}1, & 0, & 0\\ 0, & 1, & 0\\ 0, & 0, & 1\\ 0, & 0, & i\\ 0, & 0, & 0\end{matrix}\;{}_{r-s,\ s-t,\ t,\ t,\ m-r-t}^{r-s,\ s-t,\ t}\qquad \begin{bmatrix}1, & 0, & 0, & 0, & 0, & 0\\ 0, & 1, & 0, & i, & 0, & 0\\ 0, & 0, & 1, & 0, & i, & 0\end{bmatrix}_{r-s,\ s-t,\ t}^{r-s,\ s-t,\ t,\ s-t,\ t,\ n-r-s}$$

$$=\begin{bmatrix}1, & 0, & 0, & 0, & 0, & 0\\ 0, & 1, & 0, & i, & 0, & 0\\ 0, & 0, & 1, & 0, & i, & 0\\ 0, & 0, & i, & 0, & -1, & 0\\ 0, & 0, & 0, & 0, & 0, & 0\end{bmatrix}_{r-s,\ s-t,\ t,\ t,\ m-r-t}^{r-s,\ s-t,\ t,\ s-t,\ t,\ n-r-s}$$

When $t \nless s$, it is

$$[a]^n_m = \begin{bmatrix} 1, & 0, & 0 \\ 0, & 1, & 0 \\ 0, & 0, & 1 \\ 0, & i, & 0 \\ 0, & 0, & i \\ 0, & 0, & 0 \end{bmatrix}^{r-t,\; t-s,\; s}_{r-t,\; t-s,\; s,\; t-s,\; s,\; m-r-t} \begin{bmatrix} 1, & 0, & 0, & 0, & 0 \\ 0, & 1, & 0, & 0, & 0 \\ 0, & 0, & 1, & i, & 0 \end{bmatrix}^{r-t,\; t-s,\; s,\; s,\; n-r-s}_{r-t,\; t-s,\; s}$$

$$= \begin{bmatrix} 1, & 0, & 0, & 0\;, & 0 \\ 0, & 1, & 0, & 0\;, & 0 \\ 0, & 0, & 1, & i\;, & 0 \\ 0, & i, & 0, & 0\;, & 0 \\ 0, & 0, & i, & -1, & 0 \\ 0, & 0, & 0, & 0\;, & 0 \end{bmatrix}^{r-t,\; t-s,\; s,\; s,\; n-r-s}_{r-t,\; t-s,\; s,\; t-s,\; s,\; m-r-t}$$

4. *The extravagance of a symmetric matrix.*

In the case of a symmetric matrix the horizontal and vertical extravagances are the same, and may be called simply the extravagance of the matrix. Thus *the extravagance of a symmetric matrix* of rank r is equal to the extravagance of any one of its undegenerate simple minors of reduced order r; and a symmetric matrix of rank r is non-extravagant when any one of its undegenerate simple minors of reduced order r is non-extravagant, in which case all such simple minors are non-extravagant.

Ex. xxviii. *Possible values of the extravagance of a symmetric matrix.*

If $[a]^m_m$ is a symmetric matrix of assigned rank r, then the possible values of the extravagance s of $[a]^m_m$ are those consistent with the conditions

$$s \nless 0, \quad s \ngtr r, \quad s \ngtr m-r.$$

This follows from Theorem II and Ex. iv when we express $[a]^m_m$ in the form

$$[a]^m_m = \overline{\underline{h}}^{\,r}_{\,m}\, [h]^m_r.$$

Ex. xxix. *General formulae for all symmetric matrices* $[a]^m_m$ *of rank* r *and extravagance* s.

If $i = \sqrt{-1}$, $[\omega]^m_m$ is any derangement of the unit matrix $[1]^m_m$, and $[k]^r_r$ is any undegenerate symmetric matrix of order r, it follows from Ex. vi that general formulae for all such matrices are:

$$(1) \qquad [a]^m_m = \overline{\underline{\omega}}^{\,m}_{\,m} \begin{bmatrix} l, & \lambda \\ 0, & i \end{bmatrix}^{r-s,\; s}_{m-s,\; s} [k]^r_r \begin{bmatrix} l, & 0 \\ \lambda, & i \end{bmatrix}^{m-s,\; s}_{r-s,\; s} [\omega]^m_m, \quad \ldots\ldots\ldots\ldots\ldots\ldots\ldots (\mathrm{I}_1)$$

where $\begin{bmatrix} l \\ \lambda \end{bmatrix}^{m-s}_{r-s,\; s}$ is a semi-unit matrix of rank r.

(2) $$[a]^m_m = \overline{\underline{l}}^{\,r}_{\,m}\, [k]^r_r\, [l]^m_r, \quad\text{.................................}(\mathrm{I}_2)$$

where $[l]^m_r$ is a solution of rank r of the equation $[l]^m_r\, \overline{\underline{l}}^{\,r}_{\,m} = \begin{bmatrix} 1, & 0 \\ 0, & 0 \end{bmatrix}^{r-s,\,s}_{r-s,\,s}$.

(3) $$[a]^m_m = \overline{\underline{l}}^{\,r+s}_{\,m}\, \overline{\underline{\begin{matrix} 1, & 0 \\ 0, & 1 \\ 0, & i \end{matrix}}}^{\,r-s,\,s}_{\,r-s,\,s,\,s}\, [k]^r_r \begin{bmatrix} 1, & 0, & 0 \\ 0, & 1, & i \end{bmatrix}^{r-s,\,s,\,s}_{r-s,\,s} [l]^n_{r+s}, \quad\text{...............}(\mathrm{I}_3)$$

where $[l]^n_{r+s}$ is a semi-unit matrix of rank $r+s$.

(4) $$[a]^m_m = \overline{\underline{l}}^{\,m}_{\,m}\, \overline{\underline{\begin{matrix} 1, & 0 \\ 0, & 1 \\ 0, & i \\ 0, & 0 \end{matrix}}}^{\,r-s,\,s}_{\,r-s,\,s,\,s,\,m-r-s}\, [k]^r_r \begin{bmatrix} 1, & 0, & 0, & 0 \\ 0, & 1, & i, & 0 \end{bmatrix}^{r-s,\,s,\,s,\,m-r-s}_{r-s,\,s} [l]^m_m, \quad\text{.....}(\mathrm{I}_4)$$

where $[l]^m_m$ is a square semi-unit matrix.

Ex. xxx. The simplest symmetric matrix $[a]^m_m$ of rank r and extravagance s is

$$[a]^m_m = \overline{\underline{\begin{matrix} 1, & 0 \\ 0, & 1 \\ 0, & i \\ 0, & 0 \end{matrix}}}^{\,r-s,\,s}_{\,r-s,\,s,\,s,\,m-r-s} \begin{bmatrix} 1, & 0, & 0, & 0 \\ 0, & 1, & i, & 0 \end{bmatrix}^{r-s,\,s,\,s,\,m-r-s}_{r-s,\,s}$$

$$= \begin{bmatrix} 1, & 0, & 0, & 0 \\ 0, & 1, & i, & 0 \\ 0, & i, & -1, & 0 \\ 0, & 0, & 0, & 0 \end{bmatrix}^{r-s,\,s,\,s,\,m-r-s}_{r-s,\,s,\,s,\,m-r-s}.$$

Ex. xxxi. *If a symmetric matrix $\phi = [a]^m_m$ has rank r, then it has extravagance s when and only when its square has rank $r-s$.*

This follows from Ex. xxi when we observe that $\phi^2 = [a]^m_m [a]^m_m = [a]^m_m\, \overline{\underline{a}}^{\,m}_{\,m}$.

Ex. xxxii. *If a symmetric matrix ϕ has rank r and is non-extravagant (in particular if it is real), then ϕ^2 and all the higher powers of ϕ have rank r.*

Let $\phi = [a]^m_m = [a]^r_m\, \overline{\underline{a}}^{\,m}_{\,r}$, $\psi = \overline{\underline{a}}^{\,m}_{\,r}\, [a]^r_m$, where $[a]^r_m$ has rank r.

By Ex. xxi the matrix $[a]^m_m\, \overline{\underline{a}}^{\,m}_{\,m} = [a]^r_m\, \psi\, \overline{\underline{a}}^{\,m}_{\,r}$ has rank r.

Therefore by Theorem II of § 131 or Theorem III of § 71 the matrix ψ has rank r. Thus ψ is an undegenerate square matrix of order r, and all the powers of ψ have rank r. Now using Theorem II of § 131 the equations

$$\phi^2 = [a]^r_m\, \psi\, \overline{\underline{a}}^{\,m}_{\,r}, \quad \phi^3 = [a]^r_m\, \psi^2\, \overline{\underline{a}}^{\,m}_{\,r}, \quad \dots \quad \phi^k = [a]^r_m\, \psi^{k-1}\, \overline{\underline{a}}^{\,m}_{\,r}, \quad \dots$$

show that $\phi^2, \phi^3, \dots \phi^k, \dots$ all have rank r.

Conversely if ϕ^2 has the same rank r as ϕ, then ϕ is non-extravagant, and all the powers of ϕ have rank r.

For in Ex. xxxi we have $s=0$, i.e. ϕ is non-extravagant.

§ 166. Minor determinants of the matrix $\phi=[1, a]_m^{m,n}$.

1. *Simple minor determinants.*

We will write $\phi=[1, a]_m^{m,n}=[a, a]_m^{m,n}$, where $[a]_m^m=[1]_m^m$.

Let u be any positive integer such that $u \not> m$, $u \not> n$.

Let $[p]_u$ and $[\lambda]_{m-u}$ be any two complementary corranged minors of the sequence $[1\ 2 \ldots m]$ of orders u and $m-u$;

and let $[q]_u$ be any corranged minor of order u of the sequence $[1\ 2 \ldots n]$.

Also let $\eta=\text{aff. } [a_{\lambda\lambda}]_{m-u}^{m-u} \text{ in } [a_{1\lambda}]_m^{m-u}=\text{aff. } [\lambda]_{m-u} \text{ in } [1\ 2 \ldots m]$,

$$\sigma=\text{aff. } [a_{1\lambda}, a_{1q}]_m^{m-u,u} \text{ in } [a, a]_m^{m,n},$$

$$\omega=\text{aff. } [p]_u \text{ in } [1\ 2 \ldots m], \quad \omega'=\text{aff. } [q]_u \text{ in } [1\ 2 \ldots n].$$

Then the general formula for a corranged simple minor determinant of ϕ is

$$\Delta=(a_{1\lambda}, a_{1q})_m^{m-u,u}=(-1)^\eta (a_{pq})_u^u=(-1)^{\omega+u+mu} (a_{pq})_u^u, \quad \ldots\ldots\ldots\ldots\ldots\text{(A)}$$

and the general formula for an affected simple minor determinant of ϕ is

$$\Delta'=(-1)^\sigma (a_{1\lambda}, a_{1q})_m^{m-u,u}=(-1)^{\eta+\sigma} (a_{pq})_u^u=(-1)^{\omega'+u} (a_{pq})_u^u. \quad \ldots\ldots\ldots\text{(A')}$$

We obtain the third term in (A) or (A′) from the second by expanding the determinant $(a_{1\lambda}, a_{1q})_m^{m-u,u}$ in terms of the simple minor determinants of the minor matrix $[a_{1\lambda}]_m^{m-u}$, observing that all these vanish except $(a_{\lambda\lambda})_{m-u}^{m-u}$ which has the value 1; and we obtain the fourth terms by observing (see Theorem VIII *a* and Theorem III of § 19) that

$$\omega+\eta=u(m-u);$$

$$\sigma=\text{aff. } [a_{1\lambda}]_m^{m-u} \text{ in } [a]_m^m+\text{aff. } [a_{1q}]_m^u \text{ in } [a_{1p}, a]_m^{u,n}$$

$$=\eta+u^2+\text{aff. } [a_{1q}]_m^u \text{ in } [a]_m^n$$

$$=\eta+u^2+\omega'=um-\omega+\omega';$$

and therefore $\quad \eta=um-u^2-\omega\equiv um+u+\omega \pmod{2};$

$$\eta+\sigma\equiv\omega'+u \pmod{2}.$$

Formulae (A) and (A′) give all the simple minor determinants of ϕ; and we see that:

If $(a_{pq})_u^u$ is any corranged minor determinant of $[a]_m^n$ of order u, where u does not exceed either m or n, there is one and only one simple minor determinant of ϕ which has, irrespective of sign, the identical value $(a_{pq})_u^u$, viz. that given by either of the formulae (A) *and* (A′).

In the special case when $u=0$, we have the simple minor determinant $(a)_m^m=1$.

Ex. i. If $\phi=[a, a]_m^{m,n}$, where $[a]_m^m=a[1]_m^m$, a being a scalar quantity, we must replace the third and fourth terms in (A) and (A′) by

$$\Delta=(-1)^\eta a^{m-u} (a_{pq})_u^u \quad =(-1)^{\omega+u+mu} a^{m-u} (a_{pq})_u^u, \quad \ldots\ldots\ldots\ldots\ldots\text{(B)}$$

$$\Delta'=(-1)^{\eta+\sigma} a^{m-u} (a_{pq})_u^u=(-1)^{\omega'+u} a^{m-u} (a_{pq})_u^u, \quad \ldots\ldots\ldots\ldots\ldots\ldots\text{(B')}$$

the second terms remaining as before.

Ex. ii. *Simple minor determinants of the matrix* $\psi=[a,\ 1]_m^{n,m}$.

Writing $\psi=[a,\ 1]_m^{n,m}=[a,\ a]_m^{n,m}$, where $[a]_m^m=[1]_m^m$, the general formulae for a corranged and an affected simple minor determinant are respectively

$$\Delta=(a_{1q},\ a_{1\lambda})_m^{u,m-u}=(-1)^{\eta+u(m-u)}(a_{pq})_u^u=(-1)^{\omega}(a_{pq})_u^u, \quad \text{..............(C)}$$

$$\Delta'=(-1)^{\tau}(a_{1q},\ a_{1\lambda})_m^{u,m-u}=(-1)^{\omega'+n(m-u)}(a_{pq})_u^u, \quad \text{........................(C')}$$

where u, $[p]_u$, $[q]_u$, $[\lambda]_{m-u}$, η, ω, ω' are defined as in the text, and where

$$\tau=\text{aff. }[a_{1q},\ a_{1\lambda}]_m^{u,m-u}\ \text{in}\ [a,\ a]_m^{n,m}=\omega'-\omega+n(m-u).$$

If $[a]_m^m=a[1]_m^m$, where a is a scalar quantity, we must supply the additional factor a^{m-u} in the third and fourth terms of formulae (C) and (C').

2. *Minor determinants of order s.*

We again write $\phi=[1,\ a]_m^{m,n}=[a,\ a]_m^{m,n}$, where $[a]_m^m=[1]_m^m$.

Let $[x]_s$ be a corranged minor of order s of the sequence $[1\ 2\ \dots\ m]$, so that

$$\phi_s=[a_{x1},\ a_{x1}]_s^{m,n}$$

is a corranged minor formed with s of the horizontal rows of ϕ.

Since all vertical rows of $[a_{x1}]_s^m$ except the x_1th, x_2th, ... x_sth are rows of 0's, the non-vanishing corranged minor determinants of order s of ϕ_s are the same as the corranged simple minor determinants of the corranged minor matrix

$$\phi_s'=[a_{xx},\ a_{x1}]_s^{s,n}=[1,\ a_{x1}]_s^{s,n}.$$

Let u be any positive integer such that $u \not> s$, $u \not> n$.

Let $[p]_u$ and $[\lambda]_{s-u}$ be any two complementary corranged minors of the sequence $[x]_s$ of orders u and $s-u$, and let $[q]_u$ be any corranged minor of order u of the sequence $[1\ 2\ \dots\ n]$.

Also let
$$\eta=\text{aff. }[a_{\lambda\lambda}]_{s-u}^{s-u}\ \text{in}\ [a_{x\lambda}]_s^{s-u}=\text{aff. }[\lambda]_{s-u}\ \text{in}\ [x]_s,$$

$$\sigma=\text{aff. }[a_{x\lambda},\ a_{xq}]_s^{s-u,u}\ \text{in}\ [a,\ a]_m^{m,n},$$

$$\omega=\text{aff. }[p]_u\ \text{in}\ [x]_s,\quad \omega'=\text{aff. }[q]_u\ \text{in}\ [1\ 2\ \dots\ n],$$

$$\epsilon=\text{aff. }[x]_s\ \text{in}\ [1\ 2\ \dots\ m].$$

Then every non-vanishing corranged minor determinant of order s of ϕ *is given by the formula*

$$\Delta=(a_{x\lambda},\ a_{xq})_s^{s-u,u}=(-1)^{\eta}(a_{pq})_u^u=(-1)^{\omega+u+su}(a_{pq})_u^u. \quad \text{...............(D)}$$

And every non-vanishing affected minor determinant of order s of ϕ *is given by the formula*

$$\Delta'=(-1)^{\sigma}(a_{x\lambda},\ a_{xq})_s^{s-u,u}=(-1)^{\sigma+\eta}(a_{pq})_u^u=(-1)^{\omega'+u}(a_{pq})_u^u. \quad \text{.........(D')}$$

These formulae, when $[x]_s$ is given, give the minor determinants of ϕ of order s which occur in ϕ_s'.

We can write down (D) at once from (A), or obtain it by observing that

$$\omega+\eta=u(s-u).$$

We deduce (D′) by observing that

$$\sigma = \text{aff. } [x]_s \text{ in } [1\ 2 \dots m] + \text{aff. } [a_{1\lambda}, a_{1q}]_m^{s-u, u} \text{ in } [a, \alpha]_m^{m,n}$$

$$= \epsilon + \text{aff. } [a_{1\lambda}]_m^{s-u} \text{ in } [a]_m^m + \text{aff. } [a_{1q}]_m^u \text{ in } [a_{1p}, \alpha]_m^{u,n}$$

$$\equiv \epsilon + (\epsilon + \eta) + u^2 + \text{aff. } [a_{1q}]_m^u \text{ in } [\alpha]_m^n$$

$$\equiv \eta + u^2 + \omega' \equiv \eta + u + \omega' \quad (\text{mod. } 2).$$

The remaining minor determinants of order s of ϕ are given by the formula

$$D = (a_{x\mu}, \alpha_{xq})_s^{s-u, u} = 0, \quad \dots\dots\dots\dots (D'')$$

where $[\mu]_{s-u}$ is a minor of $[1\ 2 \dots m]$ of order $s-u$ which is not also a minor of $[x]_s$.

When $[p]_u$ and $[q]_u$ are given, there are $\binom{m-u}{s-u}$ sequences $[x]_s$ which contain $[p]_u$ as a minor, and there are the same number of minor matrices ϕ_s each having one simple minor determinant whose identical value, irrespective of sign, is $(\alpha_{pq})_u^u$.

Thus if u is any integer not exceeding either s or n (where $s \not> m$), the matrix ϕ has exactly $\binom{m-u}{s-u}$ distinct minor determinants of order s which have one of the identical values $\pm(\alpha_{pq})_u^u$, and these are given by (D) *and* (D′) *when $[p]_u$ and $[q]_u$ are regarded as given.*

In particular the matrix ϕ has $\binom{m}{s}$ distinct minor determinants of order s which have the identical value 1, and these are given by the formula

$$\Delta = (a_{xx})_s^s = 1.$$

Ex. iii. When $\phi = [a, \alpha]_m^{m, n}$, where $[a]_m^m = a[1]_m^m$, and a is a scalar quantity, we must supply the additional factor a^{s-u} in the third and fourth terms of (D) and (D′).

Ex. iv. *Minor determinants of order s of the matrix $\psi = [\alpha, 1]_m^{n, m}$.*

We now write $\quad \psi = [\alpha, 1]_m^{n, m} = [\alpha, a]_m^{n, m}$, where $[a]_m^m = [1]_m^m$;

and $\quad \psi_s = [\alpha_{x1}, a_{x1}]_s^{n, m}, \quad \psi_s' = [\alpha_{x1}, a_{xx}]_s^{n, s}$.

Let u, $[p]_u$, $[q]_u$, $[\lambda]_{s-u}$, ω, ω', η be defined as in sub-article 2 of the text, and let σ be the affect of $(\alpha_{xq}, a_{x\lambda})_s^{u, s-u}$ in $(\alpha, a)_m^{n, m}$. Then general formulae for the corranged and affected non-vanishing minor determinants of order s of ψ are

$$\Delta = (\alpha_{xq}, a_{x\lambda})_s^{u, s-u} = (-1)^\omega (\alpha_{pq})_u^u, \quad \dots\dots\dots\dots (E)$$

$$\Delta' = (-1)^\sigma (\alpha_{xq}, a_{x\lambda})_s^{u, s-u} = (-1)^{\omega' + n(s-u)} (\alpha_{pq})_u^u. \quad \dots\dots\dots\dots (E')$$

There are $\binom{m-u}{s-u}$ distinct minor determinants of ϕ of order s which have one of the identical values $\pm(\alpha_{pq})_u^u$.

When $[a]_m^m = a[1]_m^m$, we must supply the additional factor a^{s-u} in the last terms of (E) and (E′).

§ 167. Properties of two mutually normal undegenerate matrices.

1. *Definition of mutually normal undegenerate matrices.*

As in Note 4 of § 114 two undegenerate matrices $[a]^n_r$, $[b]^n_s$ of ranks r, s both containing the same number of short rows, or two undegenerate matrices $\overline{\underline{a}}^{r}_{n}$, $\overline{\underline{b}}^{s}_{n}$ of ranks r, s both containing the same number of short rows, are said to be *mutually normal* when

$$[a]^n_r \overline{\underline{b}}^{s}_{n} = 0, \quad \text{and} \quad r+s=n, \quad \dots\dots\dots\dots\dots(1)$$

i.e. when they are mutually orthogonal and $r+s=n$.

The conditions (1) are satisfied when and only when the vertical rows of $\overline{\underline{b}}^{s}_{n}$ form a complete set of $n-r$ unconnected solutions of the equation $[a]^n_r \overline{\underline{x}}_{n} = 0$, and also when and only when the vertical rows of $\overline{\underline{a}}^{r}_{n}$ form a complete set of $n-s$ unconnected solutions of the equation $[b]^n_s \overline{\underline{x}}_{n} = 0$.

From this definition we see that two undegenerate matrices are mutually normal when and only when they have the forms

$$[a]^{m+n}_m, \; [b]^{m+n}_n; \quad \text{or} \quad \overline{\underline{a}}^{m}_{m+n}, \; \overline{\underline{b}}^{n}_{m+n};$$

where

$$[a]^{m+n}_m \overline{\underline{b}}^{n}_{m+n} = 0, \quad \text{or} \quad [b]^{m+n}_n \overline{\underline{a}}^{m}_{m+n} = 0.$$

NOTE 1. *Mutually normal and mutually orthogonal matrices in general.*

Two matrices $[a]^n_r$ and $[b]^n_s$ or $\overline{\underline{a}}^{r}_{n}$ and $\overline{\underline{b}}^{s}_{n}$ of ranks ρ and σ will often be called mutually normal when

$$[a]^n_r \overline{\underline{b}}^{s}_{n} = 0, \quad \text{and} \quad \rho+\sigma=n,$$

and mutually orthogonal when

$$[a]^n_r \overline{\underline{b}}^{s}_{n} = 0.$$

To avoid the ambiguity which may arise when both matrices are square, we may say more precisely that $[a]^n_r$ and $[b]^n_s$ are *horizontally* normal (or orthogonal), and that $\overline{\underline{a}}^{r}_{n}$ and $\overline{\underline{b}}^{s}_{n}$ are *vertically* normal (or orthogonal), when the above conditions are satisfied.

Ex. i. All undegenerate matrices normal to a given undegenerate matrix are mutually equivalent according to the definition of § 113.

Ex. ii. The matrices $[1, x]^{m,\,n}_m$, $[y, 1]^{m,\,n}_n$ are mutually normal when and only when $[y]^m_n = -\overline{\underline{x}}^{m}_{n}$; and they then have the forms

$$[1, x]^{m,\,n}_m, \quad \overline{\underline{-x, 1}}^{m,\,n}_{n}.$$

Ex. iii. If $[k]_m^m = k[1]_m^m$, $[k]_n^n = k[1]_n^n$, where k is a non-vanishing scalar quantity, the two matrices $[k, x]_m^{m, n}$, $[y, k]_n^{m, n}$ are mutually normal when and only when $[y]_n^m = -\boxed{x}_n^m$; and they then have the forms

$$[k, x]_m^{m, n}, \quad \boxed{-x, k}_n^{m, n} .$$

Ex. iv. If $\rho \not< 0$, $\rho \not> r$, $\rho \not> n-r$, then

$$[\alpha]_r^n = \begin{bmatrix} 1, & 0, & 0, & 0 \\ 0, & 1, & i, & 0 \end{bmatrix}_{r-\rho, \rho}^{r-\rho, \rho, \rho, n-r-\rho}, \quad [\beta]_{n-r}^n = \begin{bmatrix} 0, & 0, & 0, & 1 \\ 0, & 1, & i, & 0 \end{bmatrix}_{n-r-\rho, \rho}^{r-\rho, \rho, \rho, n-r-\rho}$$

are two mutually normal undegenerate matrices of ranks r and $n-r$. They both have extravagance ρ.

2. *General formulae for two mutually normal undegenerate matrices.*

Any two mutually normal undegenerate matrices can be reduced by equigradent transformations of a special character to the forms given in Ex. ii. In fact we have the following theorem:

Theorem I a. *All pairs of mutually normal undegenerate matrices* $[a]_m^{m+n}$, $[b]_n^{m+n}$ *are given by the formulae*

$$[a]_m^{m+n} = [h]_m^m [1, x]_m^{m, n} [\omega]_{m+n}^{m+n}, \quad [b]_n^{m+n} = [k]_n^n \boxed{-x, 1}_n^{m, n} [\omega]_{m+n}^{m+n}, \quad \ldots\text{(A)}$$

where $[h]_m^m$ *and* $[k]_n^n$ *are arbitrary undegenerate square matrices,* $[\omega]_{m+n}^{m+n}$ *is any derangement of the unit matrix* $[1]_{m+n}^{m+n}$*, and the elements of* $[x]_m^n$ *are arbitrary.*

It is clear that every pair of matrices given by the formulae (A) are undegenerate and mutually normal. We have to show further that every pair of undegenerate matrices $[a]_m^{m+n}$, $[b]_n^{m+n}$ which are mutually normal can be expressed in the forms (A).

Let $[h]_m^m = [a_{1p}]_m^m$ be any one of the undegenerate corranged vertical minors of $[a]_m^{m+n}$ of order m, and let $[c]_m^n = [a_{1q}]_m^n$ be the complementary corranged vertical minor of $[a]_m^{m+n}$ of order n, so that $[p]_m$ and $[q]_n$ are complementary corranged minors of the sequence $[1, 2, 3, \ldots (m+n)]$. Then by re-arranging the vertical rows of $[a]_m^{m+n}$ we can convert it into the matrix $[h, c]_m^{m, n}$, i.e. there exist relations of the forms

$$[h, c]_m^{m, n} = [a]_m^{m+n} \boxed{\omega}_{m+n}^{m+n}, \quad [a]_m^{m+n} = [h, c]_m^{m, n} [\omega]_{m+n}^{m+n},$$

where $[\omega]_{m+n}^{m+n}$ is a derangement of the unit matrix $[1]_{m+n}^{m+n}$.

Let $\overline{\underline{H}}{}^{m}_{m}$ be the inverse of $[h]^{m}_{m}$, so that

$$[h, c]^{m,n}_{m} = [h]^{m}_{m}\,\overline{\underline{H}}{}^{m}_{m}\,[h, c]^{m,n}_{m} = [h]^{m}_{m}\,[1, x]^{m,n}_{m},$$

where

$$[x]^{n}_{m} = \overline{\underline{H}}{}^{m}_{m}\,[c]^{n}_{m}.$$

Then

$$[a]^{m+n}_{m} = [h]^{m}_{m}\,[1, x]^{m,n}_{m}\,[\omega]^{m+n}_{m+n}. \quad \text{......................(2)}$$

The equation $[a]^{m+n}_{m}\,\overline{\underline{b}}{}^{n}_{m+n} = 0$ is now equivalent to

$$[1, x]^{m,n}_{m}\,[\omega]^{m+n}_{m+n}\,\overline{\underline{b}}{}^{n}_{m+n} = 0,$$

and by Exs. i and ii this equation is satisfied by an undegenerate matrix $\overline{\underline{b}}{}^{n}_{m+n}$ when and only when

$$[\omega]^{m+n}_{m+n}\,\overline{\underline{b}}{}^{n}_{m+n} = \begin{bmatrix} -x \\ 1 \end{bmatrix}^{n}_{m,n}\,\overline{\underline{k}}{}^{n}_{n},$$

or

$$[b]^{m+n}_{n} = [k]^{n}_{n}\,\overline{\underline{-x, 1}}{}^{m,n}_{n}\,[\omega]^{m+n}_{m+n}, \quad \text{......................(2')}$$

where $[k]^{n}_{n}$ is an undegenerate square matrix.

Thus when $[a]^{m+n}_{m}$ and $[b]^{m+n}_{n}$ are undegenerate and mutually normal, they can always be expressed in the forms (A).

Ex. v. The matrices $[a]^{m+n}_{m}$, $[b]^{m+n}_{n}$ of the theorem are formed from the matrices

$$[a]^{m+n}_{m} = [h]^{m}_{m}\,[1, x]^{m,n}_{m}, \quad [\beta]^{m+n}_{n} = [k]^{n}_{n}\,\overline{\underline{-x, 1}}{}^{m,n}_{n}$$

by corresponding derangements of vertical rows. Here $[a]^{m+n}_{m}$ is a matrix whose first m vertical rows are unconnected, and $[\beta]^{m+n}_{n}$ is a matrix whose last n vertical rows are unconnected.

Ex. vi. *Let* $[a]^{m+n}_{m}$, $[b]^{m+n}_{n}$ *be any two mutually normal undegenerate matrices, and let* $[p]_{m}$, $[q]_{n}$ *be complementary corranged minors of the sequence* $[1, 2, 3, \ldots (m+n)]$. *Then* $(a_{1p})^{m}_{m}$, $(b_{1q})^{n}_{n}$ *are either both zero or both not zero.*

Suppose that $(a_{1p})^{m}_{m} \neq 0$, and let the same notation be used as in the proof of Theorem I a and in Ex. v, so that $(a_{1p})^{m}_{m} = (h)^{m}_{m}$.

Then $[a_{1q}]^{n}_{m} = [c]^{n}_{m} = [h]^{m}_{m}[x]^{n}_{m}$, i.e. $[a_{1q}]^{n}_{m}$ is the corranged vertical minor of $[a]^{m+n}_{m}$ formed by its last n vertical rows.

Consequently $[b_{1q}]^{n}_{n}$ is the corranged vertical minor of $[\beta]^{m+n}_{n}$ formed by its last n vertical rows. Therefore we have $[b_{1q}]^{n}_{n} = [k]^{n}_{n}$, $(b_{1q})^{n}_{n} = (k)^{n}_{n} \neq 0$.

Similarly we can show that if $(b_{1q})^{n}_{n} \neq 0$, then $(a_{1p})^{m}_{m} \neq 0$.

Ex. vii. *If* $[a, a]^{m,n}_m$ *is an undegenerate matrix of rank* m *in which* $(a)^m_m \neq 0$, *and if* $\overline{\underline{A}}{}^m_m$ *is the inverse of* $[a]^m_m$, *we have*

$$[a, a]^{m,n}_m = [a]^m_m \overline{\underline{A}}{}^m_m [a, a]^{m,n}_m = [a]^m_m [1, x]^{m,n}_m,$$

where
$$[x]^n_m = \overline{\underline{A}}{}^m_m [a]^n_m.$$

Ex. viii. *If* $[\beta, b]^{m,n}_n$ *is an undegenerate matrix of rank* n *in which* $(b)^n_n \neq 0$, *and if* $\overline{\underline{B}}{}^n_n$ *is the inverse of* $[b]^n_n$, *we have*

$$[\beta, b]^{m,n}_n = [b]^n_n \overline{\underline{B}}{}^n_n [\beta, b]^{m,n}_n = [b]^n_n [y, 1]^{m,n}_n,$$

where
$$[y]^m_n = \overline{\underline{B}}{}^n_n [\beta]^m_n.$$

Ex. ix. *If* $[a, a]^{m,n}_m$, $[\beta, b]^{m,n}_n$ *are two mutually normal undegenerate matrices in which one and therefore (see Ex.* vi) *both of the inequalities* $(a)^m_m \neq 0$, $(b)^n_n \neq 0$ *are satisfied, we have*

$$[a, a]^{m,n}_m = [a]^m_m [1, x]^{m,n}_m, \quad [\beta, b]^{m,n}_n = [b]^n_n \overline{\underline{-x, 1}}{}^{m,n}_n,$$

where
$$[x]^n_m = \overline{\underline{A}}{}^m_m [a]^n_m = -\overline{\underline{\beta}}{}^n_m [B]^n_n.$$

For with the notation of Exs. vii and viii the two given undegenerate matrices are mutually normal when and only when $[1, x]^{m,n}_m$ and $[y, 1]^{m,n}_n$ are mutually normal, i.e. when and only when $\overline{\underline{y}}{}^n_m = -[x]^n_m$.

The next theorem shows that any two mutually normal undegenerate matrices can be reduced by equigradent transformations of a special character to the forms considered in Ex. iv.

Theorem I b. *All pairs of mutually normal undegenerate matrices* $[a]^n_r$, $[b]^n_{n-r}$ *of ranks* r, $n-r$ *are given by the formulae*

$$[a]^n_r = [h]^r_r [\alpha]^n_r [l]^n_n, \quad [b]^n_{n-r} = [k]^{n-r}_{n-r} [\beta]^n_{n-r} [l]^n_n, \quad \text{.........(B)}$$

where

$$[\alpha]^n_r = \begin{bmatrix} 1, & 0, & 0, & 0 \\ 0, & 1, & i, & 0 \end{bmatrix}^{r-\rho,\, \rho,\, \rho,\, n-r-\rho}_{r-\rho,\, \rho}, \quad [\beta]^n_{n-r} = \begin{bmatrix} 0, & 0, & 0, & 1 \\ 0, & 1, & i, & 0 \end{bmatrix}^{r-\rho,\, \rho,\, \rho,\, n-r-\rho}_{n-r-\rho,\, \rho};$$

$[h]^r_r$ *and* $[k]^{n-r}_{n-r}$ *are undegenerate square matrices;* $[l]^n_n$ *is a square semi-unit matrix; and* ρ *is an integer satisfying the conditions* $\rho \nless 0$, $\rho \ngtr r$, $\rho \ngtr n-r$.

It is clear that every pair of matrices given by the formulae (B) are undegenerate and mutually normal. We have to show further that every pair of undegenerate matrices $[a]^n_r$, $[b]^n_{n-r}$ which are mutually normal can be expressed in the forms (B).

Let $[a]^n_r$ be any given undegenerate matrix of rank r, and let it have extravagance ρ. Then by Ex. vi or Theorem I of § 165 we can express it in the form

$$[a]^n_r = [h]^r_r\,[\alpha]^n_r\,[l]^n_n, \qquad \text{.........................(3)}$$

where $[h]^r_r$ is undegenerate, and $[l]^n_n$ is a square semi-unit matrix.

Let $[b]^n_{n-r}$ be any undegenerate matrix (of rank $n-r$) normal to $[a]^n_r$, so that $\overline{\underline{b}}^{\,n-r}_{\,n}$ is any solution of rank $n-r$ of the equation

$$[a]^n_r\,\overline{\underline{b}}^{\,n-r}_{\,n} = 0, \quad \text{or} \quad [\alpha]^n_r\,[l]^n_n\,\overline{\underline{b}}^{\,n-r}_{\,n} = 0.$$

By Exs. i and iv this equation is satisfied by an undegenerate matrix $\overline{\underline{b}}^{\,n-r}_{\,n}$ of rank $n-r$ when and only when $[l]^n_n\,\overline{\underline{b}}^{\,n-r}_{\,n} = \overline{\underline{\beta}}^{\,n-r}_{\,n}\,\overline{\underline{k}}^{\,n-r}_{\,n-r}$, or

$$[b]^n_{n-r} = [k]^{n-r}_{n-r}\,[\beta]^n_{n-r}\,[l]^n_n, \qquad \text{....................(3′)}$$

where $[k]^{n-r}_{n-r}$ is an undegenerate square matrix.

Thus when $[a]^n_r$, $[b]^n_{n-r}$ are mutually normal undegenerate matrices of ranks r, $n-r$, they can always be expressed in the forms (B).

3. *Relations between the simple minor determinants of two mutually normal undegenerate matrices.*

Let the two mutually normal undegenerate matrices be $[a]^{m+n}_m$ and $[b]^{m+n}_n$, so that

$$[a]^{m+n}_m\,\overline{\underline{b}}^{\,n}_{\,m+n} = 0. \qquad \text{..........................(4)}$$

If $[p_1 p_2 \ldots p_m]$ and $[q_1 q_2 \ldots q_n]$ are any two complementary corranged minors of the sequence $[1, 2, 3, \ldots (m+n)]$, then the determinants

$$\Delta_1 = (a_{1p})^m_m, \quad \Delta_2 = (b_{1q})^n_n$$

will be called *anti-correspondent corranged simple minor determinants* of the two matrices $[a]^{m+n}_m$ and $[b]^{m+n}_n$.

Here Δ_1 is derived from the matrix $[a]^{m+n}_m$ by retaining its p_1th, p_2th, ... p_mth vertical rows and striking out its q_1th, q_2th, ... q_nth vertical rows;

and Δ_2 is derived from the matrix $[b]^{m+n}_n$ by striking out its p_1th, p_2th, ... p_mth vertical rows and retaining its q_1th, q_2th, ... q_nth vertical rows.

If Δ_1' is the determinant formed by providing Δ_1 with the sign determined by the affect of Δ_1 in $[a]^{m+n}_m$, and Δ_2' is the determinant formed by providing

Δ_2 with the sign determined by the affect of Δ_2 in $[b]_n^{m+n}$, then we will call Δ_2' *the affected simple minor determinant of* $[b]_n^{m+n}$ *anti-correspondent to* Δ_1, and we will call Δ_1' *the affected simple minor determinant of* $[a]_m^{m+n}$ *anti-correspondent to* Δ_2.

We will now consider two theorems, the first of which is included in the second.

Theorem II a. *If* $[a]_m^{m+n}$ *and* $[b]_n^{m+n}$ *are two mutually normal undegenerate matrices, then any two anti-correspondent simple minor determinants of* $[a]_m^{m+n}$ *and* $[b]_n^{m+n}$ *are either both zero or both not zero.*

Let $[p]_m$ and $[q]_n$ be two complementary minors of the sequence $[1, 2, 3, \ldots (m+n)]$, so that (4) can be replaced by

$$[a_{1p}]_m^m \, \overline{\underline{b_{1p}}}_m^n = -[a_{1q}]_m^n \, \overline{\underline{b_{1q}}}_n^n . \quad \ldots\ldots\ldots\ldots\ldots\ldots\ldots(4')$$

If $(b_{1q})_n^n \neq 0$, then by prefixing the inverse matrix of $\overline{\underline{b_{1q}}}_n^n$ on both sides of (4′), we see that the vertical rows of $[a_{1q}]_m^n$ are connected with the vertical rows of $[a_{1p}]_m^m$; and consequently the vertical rows of $[a_{1p}]_m^m$ are unconnected, as otherwise the rank of $[a]_m^{m+n}$ would be less than m. Thus if $(b_{1q})_n^n \neq 0$, then also $(a_{1p})_m^m \neq 0$. In a similar way we can show that if $(a_{1p})_m^m \neq 0$, then also $(b_{1q})_n^n \neq 0$.

Another proof of Theorem II a has been given in Ex. vi.

Theorem II b. *If* $[a]_m^{m+n}$ *and* $[b]_n^{m+n}$ *are two mutually normal undegenerate matrices, then the corranged simple minor determinants of either one of these matrices are proportional to the anti-correspondent affected simple minor determinants of the other matrix.*

Let $[p]_m$ and $[q]_n$ be any two complementary corranged minors of the sequence $[1, 2, 3, \ldots (m+n)]$, and let

$$\Delta_1 = (a_{1p})_m^m, \quad \Delta_2 = (b_{1q})_n^n; \quad \Delta_1' = (-1)^{\omega} (a_{1p})_m^m, \quad \Delta_2' = (-1)^{\omega'} (b_{1q})_n^n;$$

where

$$\omega = \text{aff.}\, [a_{1p}]_m^m \text{ in } [a]_m^{m+n} = \text{aff.}\, [p]_m \text{ in } [1, 2, 3, \ldots (m+n)],$$

and $$\omega' = \text{aff.}\, [b_{1q}]_n^n \text{ in } [b]_n^{m+n} = \text{aff.}\, [q]_n \text{ in } [1, 2, 3, \ldots (m+n)].$$

Then the theorem asserts that

$$\Delta_1 = k\Delta_2', \quad \Delta_2 = k'\Delta_1',$$

where k and k' are scalar constants which are finite and not zero, and are independent of the choice of $[p]_m$ and $[q]_n$.

It will appear that $$kk' = (-1)^{mn}.$$

FIRST SPECIAL CASE. *When* $[a]_m^{m+n} = [1, x]_m^{m,n}$, $[b]_n^{m+n} = \underline{\overline{-x, 1}}_n^{m,n}$.

In this case we can write

$$[a]_m^{m+n} = [1, x]_m^{m,n} = [\xi, x]_m^{m,n}, \quad [b]_n^{m+n} = [y, 1]_n^{m,n} = [y, \eta]_n^{m,n},$$

where
$$[y]_n^m = -\underline{\overline{x}}_n^m, \qquad \text{.......................................(5)}$$

and
$$[\xi]_m^m = [1]_m^m, \quad [\eta]_n^n = [1]_n^n.$$

The general formulae for Δ_1, Δ_2, Δ_1', Δ_2' are

$$\Delta_1 = (\xi_{1\lambda}, x_{1q})_m^{m-u,u}, \qquad \Delta_2 = (y_{1p}, \eta_{1\mu})_n^{u,n-u},$$

$$\Delta_1' = (-1)^{\sigma} (\xi_{1\lambda}, x_{1q})_m^{m-u,u}, \quad \Delta_2' = (-1)^{\tau} (y_{1p}, \eta_{1\mu})_n^{u,n-u},$$

where u is any positive integer which does not exceed either m or n, $[p]_u$ and $[\lambda]_{m-u}$ are complementary corranged minors of $[1\ 2 \ldots m]$, $[q]_u$ and $[\mu]_{n-u}$ are complementary corranged minors of $[1\ 2 \ldots n]$,

$$\sigma = \text{affect of } [\xi_{1\lambda}, x_{1q}]_m^{m-u,u} \text{ in } [\xi, x]_m^{m,n},$$

$$\tau = \text{affect of } [y_{1p}, \eta_{1\mu}]_n^{u,n-u} \text{ in } [y, \eta]_n^{m,n}.$$

Let ω = affect of $[p]_u$ in $[1\ 2 \ldots m]$, ω' = affect of $[q]_u$ in $[1\ 2 \ldots n]$.

Then by formulae (A), (C), (A′), (C′) of § 166 we see that

$$\Delta_1 = (-1)^{\omega+u+mu} (x_{pq})_u^u, \quad \Delta_2 = (-1)^{\omega'} (y_{pq})_u^u, \qquad \text{...........................(6)}$$

$$\Delta_1' = (-1)^{\omega'+u} (x_{pq})_u^u, \qquad \Delta_2' = (-1)^{\omega+m(n-u)} (y_{pq})_u^u. \qquad \text{...............(6′)}$$

Further from (5) we see that

$$(y_{pq})_u^u = (y_{qp})_u^u = (-1)^u (x_{pq})_u^u.$$

From (5), (6) and (6′) it follows that

$$\Delta_1 = (-1)^{mn} \Delta_2', \quad \Delta_2 = \Delta_1'. \qquad \text{..................................(7)}$$

Thus Theorem II a is true in this first special case.

SECOND SPECIAL CASE. *When the first m vertical rows of* $[a]_m^{m+n}$ *and therefore also the last n vertical rows of* $[b]_n^{m+n}$ *are unconnected.*

Using Ex. ix we can in this case write

$$[a]_m^{m+n} = [a]_m^m [1, x]_m^{m,n}, \quad [b]_n^{m+n} = [\beta]_n^n \underline{\overline{-x, 1}}_n^{m,n}, \qquad \text{...................(8)}$$

where $[a]_m^m$ and $[\beta]_n^n$ are undegenerate square matrices.

Let Δ_1, Δ_2 be any two mutually anti-correspondent corranged simple minor determinants of $[a]_m^{m+n}$, $[b]_n^{m+n}$; let Δ_1', Δ_2' be the affected simple minor determinants of $[a]_m^{m+n}$, $[b]_n^{m+n}$ corresponding to Δ_1, Δ_2; let D_1, D_2 be the two mutually anti-correspondent corranged simple minor determinants of $[1, x]_m^{m,n}$, $\overline{\underline{-x, 1}}_n^{m,n}$ formed from these matrices in the same way as Δ_1, Δ_2 are formed from $[a]_m^{m+n}$, $[b]_n^{m+n}$; and let D_1', D_2' be the affected simple minor determinants of $[1, x]_m^{m,n}$, $\overline{\underline{-x, 1}}_n^{m,n}$ corresponding to D_1, D_2.

Then by the first special case we have

$$D_1=(-1)^{mn}D_2', \quad D_2=D_1'. \qquad \text{.................................(9)}$$

Now if $A=(a)_m^m \neq 0$, $B=(\beta)_n^n \neq 0$, it follows from (8) that

$$\Delta_1=AD_1, \quad \Delta_1'=AD_1', \quad \Delta_2=BD_2, \quad \Delta_2'=BD_2'; \qquad \text{..................(10)}$$

and from (9) and (10) it follows that

$$\Delta_1=k\Delta_2', \quad \Delta_2=k'\Delta_1', \qquad \text{......................................(9')}$$

where $$k=(-1)^{mn}\frac{A}{B}, \quad k'=\frac{B}{A}, \quad kk'=(-1)^{mn}.$$

Thus Theorem II *b* is true in this second special case.

In particular when $\Delta_1=(a)_m^m=A$, we have $\Delta_2=B$, $\Delta_1'=A$, $\Delta_2'=(-1)^{mn}B$, and the equations (9') are satisfied.

General case. *When $[a]_m^{m+n}$ and $[b]_n^{m+n}$ are any two mutually normal undegenerate matrices.*

In this case we can determine two mutually normal undegenerate matrices $[a]_m^{m+n}$ and $[\beta]_n^{m+n}$ which are such that the first m vertical rows of $[a]_m^{m+n}$ are unconnected, the last n vertical rows of $[\beta]_n^{m+n}$ are unconnected, and

$$[a]_m^{m+n}=[a_{1\lambda}]_m^{m+n}, \quad [b]_n^{m+n}=[\beta_{1\lambda}]_n^{m+n},$$

where $[\lambda]_{m+n}$ is a fixed derangement of the sequence $[1, 2, 3, \ldots (m+n)]$.

For if $[a]_m^{m+n}$ and $[\beta]_n^{m+n}$ are two given matrices derived from $[a]_m^{m+n}$ and $[b]_n^{m+n}$ by corresponding derangements of their vertical rows, they are mutually normal undegenerate matrices, and if we choose $[a]_m^{m+n}$ so that its first m vertical rows are unconnected, then by Theorem II *a* the last n vertical rows of $[\beta]_n^{m+n}$ are unconnected.

Let $[p]_m$, $[q]_n$ be any two complementary corranged minors of the sequence $[1, 2, 3, \ldots (m+n)]$. Then $[a_{1p}]_m^m$, $[a_{1q}]_m^n$ are complementary corranged vertical minors of $[a]_m^{m+n}$, and are also complementary vertical minors (not corranged) of $[a]_m^{m+n}$; and $[b_{1p}]_n^m$, $[b_{1q}]_n^n$ are complementary corranged vertical minors of $[b]_n^{m+n}$, and are also complementary vertical minors (not corranged) of $[\beta]_n^{m+n}$.

Let $[a_{1u}]_m^m$, $[a_{1v}]_n^n$ be the complementary *corranged* vertical minors of $[a]_m^{m+n}$ formed with the same vertical rows as $[a_{1p}]_m^m$, $[a_{1q}]_m^n$; so that $[\beta_{1u}]_n^m$, $[\beta_{1v}]_n^n$ are the complementary *corranged* vertical minors of $[\beta]_n^{m+n}$ formed with the same vertical rows as $[b_{1p}]_n^m$, $[b_{1q}]_n^n$. Then $[u]_m$ and $[v]_n$ are complementary corranged minors of the sequence $[1, 2, 3, \ldots (m+n)]$.

Let ω, ω' be the affects of $[p]_m$, $[q]_n$ in $[1, 2, 3, \ldots (m+n)]$;

ϵ, ϵ' be the affects of $[u]_m$, $[v]_n$ in $[1, 2, 3, \ldots (m+n)]$;

η, η' be the affects of $[p]_m$, $[q]_n$ in $[u]_m$, $[v]_n$;

and σ be the affect of $[\lambda]_{m+n}$ in $[1, 2, 3, \ldots (m+n)]$,

which is the affect of $[a]_m^{m+n}$ in $[a]_m^{m+n}$, and of $[b]_n^{m+n}$ in $[\beta]_n^{m+n}$. Then by Theorem VIII *a* of § 19 we have

$$\omega+\omega'=\epsilon+\epsilon'=mn. \quad \text{.........(11)}$$

Further let

$$\Delta_1=(a_{1p})_m^m, \quad \Delta_2=(b_{1q})_n^n, \quad D_1=(a_{1u})_m^m, \quad D_2=(\beta_{1v})_n^n; \quad \text{.........(12)}$$

$$\Delta_1'=(-1)^{\omega}\Delta_1, \quad \Delta_2'=(-1)^{\omega'}\Delta_2, \quad D_1'=(-1)^{\epsilon}D_1, \quad D_2'=(-1)^{\epsilon'}D_2. \quad \text{.........(13)}$$

Then by the second special case we have

$$D_1=\kappa D_2', \quad D_2=\kappa' D_1', \quad \text{.........(14)}$$

where κ and κ' are invariable constants such that $\kappa\kappa'=(-1)^{mn}$; and to prove the theorem in the general case, we have to show that

$$\Delta_1=k\Delta_2', \quad \Delta_2=k'\Delta_1', \quad \text{.........(14')}$$

where k and k' are invariable constants such that $kk'=(-1)^{mn}$

Now $$\Delta_1=(-1)^{\eta}D_1, \quad \Delta_2=(-1)^{\eta'}D_2, \quad \text{.........(15)}$$

and from (13) and (15) it follows that

$$\Delta_1'=(-1)^{\omega+\epsilon+\eta}D_1', \quad \Delta_2'=(-1)^{\omega'+\epsilon'+\eta'}D_2'. \quad \text{.........(16)}$$

Further from (15), (14) and (16) it follows that

$$\Delta_1=(-1)^{\omega'+\epsilon'+\eta+\eta'}.\kappa\Delta_2', \quad \Delta_2=(-1)^{\omega+\epsilon+\eta+\eta'}.\kappa'\Delta_1'. \quad \text{.........(17)}$$

But ω horizontal moves change $[a]_m^{m+n}$ into $[a_{1p}, a_{1q}]_m^{m,n}$;

$\eta+\eta'$ horizontal moves change $[a_{1p}, a_{1q}]_m^{m,n}$ into $[a_{1u}, a_{1v}]_m^{m,n}$;

ϵ horizontal moves change $[a]_m^{m+n}$ into $[a_{1u}, a_{1v}]_m^{m,n}$;

ϵ horizontal moves change $[a_{1u}, a_{1v}]_m^{m,n}$ into $[a]_m^{m+n}$.

Therefore $\omega+\epsilon+\eta+\eta'$ horizontal moves change $[a]_m^{m+n}$ into $[a]_m^{m+n}$, and it follows from Theorem II *a* of § 19 or Theorem I *b* of § 25 that

$$\omega+\epsilon+\eta+\eta'\equiv\sigma \pmod{2}, \quad \omega'+\epsilon'+\eta+\eta'\equiv\sigma \pmod{2}.$$

Hence the equations (17) have the form (14') where

$$k=(-1)^{\sigma}\kappa, \quad k'=(-1)^{\sigma}\kappa', \quad kk'=\kappa\kappa'=(-1)^{mn}.$$

Since k and k' are invariable constants, this establishes the theorem.

4. *Extravagances of two mutually normal undegenerate matrices.*

When two mutually normal undegenerate matrices $[a]_r^n$, $[b]_{n-r}^n$ of ranks r and $n-r$ are expressed in the forms shown in Theorem I b, they have the same extravagances as $[\alpha]_r^n$, $[\beta]_{n-r}^n$, i.e. they both have the same extravagance ρ. Hence we have the following theorem:

Theorem III. *Any two mutually normal undegenerate matrices have equal extravagances.*

This theorem can also be stated as follows:

If $[a]_m^{m+n}$, $[b]_n^{m+n}$ *are any two mutually normal undegenerate matrices, then the first of the two product matrices*

$$\phi=[a]_m^{m+n}\,\overline{\underline{a}}^{\,m}_{\,m+n}, \quad \psi=[b]_n^{m+n}\,\overline{\underline{b}}^{\,n}_{\,m+n} \qquad \text{..............(18)}$$

has rank $m-s$ *when and only when the second has rank* $n-s$.

If ρ and σ are the ranks of ϕ and ψ, it follows from § 136.1 and the above result that the possible values of ρ and σ are those consistent with the conditions

$$\rho \nless 0,\ \ \rho \ngtr m,\ \ \rho \nless m-n;\quad \sigma \nless 0,\ \ \sigma \ngtr m,\ \ \sigma \nless n-m;\quad m-\rho=n-\sigma. \text{...(19)}$$

NOTE 2. *Generalisation of Theorem III.*

If $[a]_r^n$ *and* $[b]_s^n$ *are two (mutually normal) matrices of ranks* ρ *and* σ *such that*

$$[a]_r^n\,\overline{\underline{b}}^{\,s}_{\,n}=0, \quad \rho+\sigma=n,$$

then $[a]_r^n$ *and* $[b]_s^n$ *have equal horizontal extravagances. In other words if* $\phi=[a]_r^n\,\overline{\underline{a}}^{\,r}_{\,n}$ *and* $\psi=[b]_s^n\,\overline{\underline{b}}^{\,s}_{\,n}$, *then* ϕ *has rank* $\rho-t$ *when and only when* ψ *has rank* $\sigma-t$.

We deduce this result from Theorem III by writing

$$[a]_r^n=[h]_r^\rho\,[a]_\rho^n, \quad [b]_s^n=[k]_s^\sigma\,[\beta]_\sigma^n,$$

where $[h]_r^\rho$ and $[a]_\rho^n$ have rank ρ, and $[k]_s^\sigma$ and $[\beta]_\sigma^n$ have rank σ. Then $[a]_\rho^n$ and $[\beta]_\sigma^n$ are mutually normal *undegenerate* matrices of ranks ρ and σ; and the horizontal extravagances of $[a]_r^n$, $[b]_s^n$ are respectively the (horizontal) extravagances of $[a]_\rho^n$, $[\beta]_\sigma^n$, which are equal.

Ex. x. *Any two mutually normal undegenerate matrices are either both extravagant or both non-extravagant.*

In other words if $[a]_m^{m+n}$ *and* $[b]_n^{m+n}$ *are two mutually normal undegenerate matrices, then the two product matrices* (18) *are either both degenerate or both undegenerate.*

This particular case of Theorem III can be proved independently in the following way. By Theorem II b each corranged simple minor determinant of $[b]_n^{m+n}$ is k times the anticorrespondent affected simple minor determinant of $[a]_m^{m+n}$, where k is an invariable scalar constant which is finite and not zero. Hence since $\det\phi$ and $\det\psi$ are the sums of the squares of the distinct simple minor determinants of $[a]_m^{m+n}$ and $[b]_n^{m+n}$ respectively, we have

$$\det\psi = k^2 \det\phi.$$

Thus $\det\psi$ vanishes when and only when $\det\phi$ vanishes.

We can deduce the general theorem from this particular case.

Ex. xi. *The common extravagance s of two mutually normal undegenerate matrices $[a]_m^{m+n}$, $[b]_n^{m+n}$ of given orders can have any value consistent with the conditions*

$$s \not< 0, \quad s \not> m, \quad s \not> n.$$

For when s is any integer satisfying these conditions we can construct an undegenerate matrix $[a]_m^{m+n}$ of rank m having extravagance s. We can then determine an undegenerate matrix $[b]_n^{m+n}$ of rank n satisfying the equation $[a]_m^{m+n} \underline{\overline{b}}_{m+n}^{\,n} = 0$, and by Theorem III this matrix also has extravagance s.

Ex. xii. *One of two mutually normal undegenerate matrices is completely extravagant when and only when the other is plenarily extravagant.*

Let $[a]_r^n$ and $[b]_{n-r}^n$ be two mutually normal undegenerate matrices of ranks r and $n-r$.

If $[a]_r^n$ is completely extravagant (which is possible only when $2r \not> n$), it has extravagance r; therefore by Theorem III the matrix $[b]_{n-r}^n$ has extravagance r, i.e. $[b]_{n-r}^n$ is plenarily extravagant.

Again if $[a]_r^n$ is plenarily extravagant (which is possible only when $2r \not< n$), it has extravagance $n-r$; therefore by Theorem III the matrix $[b]_{n-r}^n$ has extravagance $n-r$, i.e. $[b]_{n-r}^n$ is completely extravagant.

Ex. xiii. *If $[a]_r^n$ and $[b]_{n-r}^n$ are two mutually normal undegenerate matrices of ranks r and $n-r$ of which $[a]_r^n$ is completely extravagant and $[b]_{n-r}^n$ is plenarily extravagant (which is possible only when $2r \not> n$), then we can write*

$$[a]_r^n = [h]_r^r\,[1,\ i,\ 0]_r^{r,\,r,\,n-2r}\,[l]_n^n, \quad [b]_{n-r}^n = [k]_{n-r}^{n-r}\begin{bmatrix}0, & 0, & 1\\ 1, & i, & 0\end{bmatrix}_{n-2r,\,r}^{r,\,r,\,n-2r}[l]_n^n,$$

where $i=\sqrt{-1}$, $[h]_r^r$ and $[k]_{n-r}^{n-r}$ are undegenerate square matrices, and $[l]_n^n$ is a real square semi-unit matrix.

That $[a]_r^n$ can be expressed in the form shown above follows from formula (C_4) of § 165 which is proved in § 159. That $[b]_{n-r}^n$ can then also be expressed in the form shown above follows from Theorem I *b* of the present article.

Ex. xiv. *General formula for all plenarily extravagant matrices $[a]_r^n$ of rank r.*

From Ex. xiii we see that if $[a]_r^n$ is a plenarily extravagant matrix of rank r (which is possible only when $2r \not< n$), then it can be expressed in the form

$$[a]_r^n = [h]_r^r\begin{bmatrix}0, & 0, & 1\\ 1, & i, & 0\end{bmatrix}_{2r-n,\,n-r}^{n-r,\,n-r,\,2r-n}[l]_n^n, \quad \text{.........................(20)}$$

where $[h]_r^r$ is an undegenerate square matrix, and $[l]_n^n$ is a *real* square semi-unit matrix.

Since every derangement of a square semi-unit matrix is also a square semi-unit matrix, an equivalent formula is

$$[a]^n_r = [h]^r_r \begin{bmatrix} 1, & 0, & 0 \\ 0, & 1, & i \end{bmatrix}^{2r-n,\, n-r,\, n-r}_{2r-n,\, n-r} [l]^n_n, \quad \text{.........................(21)}$$

where $[h]^r_r$ is an undegenerate square matrix, and $[l]^n_n$ is a *real* square semi-unit matrix.

Conversely every matrix $[a]^n_r$ given by either of the formulae (20) and (21) is a plenarily extravagant matrix of rank r whenever $[h]^r_r$ is undegenerate and $[l]^n_n$ is a square semi-unit matrix, real or not real.

Ex. xv. *Possible extravagances of two mutually orthogonal undegenerate matrices.*

Theorem. *If* $[a]^n_r$ *and* $[b]^n_s$ *are two mutually orthogonal undegenerate matrices of ranks* r *and* s, *so that*

$$[a]^n_r \, \underline{\overline{b}}^{\,s}_{n} = 0, \quad \text{...(22)}$$

and if ρ *and* σ *are the extravagances of* $[a]^n_r$ *and* $[b]^n_s$, *then when* n *is given, the possible values of* r *and* s *are those consistent with the conditions*

$$r \not< 0, \quad s \not< 0, \quad r + s \not> n, \quad \text{................................(23)}$$

and the possible values of r, s, ρ, σ *are those consistent with the conditions*

$$\rho \not< 0, \quad \rho \not> r; \quad \sigma \not< 0, \quad \sigma \not> s; \quad \rho - \sigma \not> n - r - s, \quad \sigma - \rho \not> n - r - s. \text{.........(24)}$$

For the sake of generality we consider that every matrix of rank 0, even when undegenerate, is replaceable as a factor by the scalar number 0.

The conditions (23) are clearly equivalent to

$$r \not< 0, \quad r \not> n, \quad r \not> n - s; \quad s \not< 0, \quad s \not> n, \quad s \not> n - r,$$

the second and fifth of the last conditions being superfluous.

That the possible values of r and s are those consistent with the conditions (23) follows from § 134. I.

When r and s have any given values consistent with the conditions (23) we can (see Ex. iv of § 165) determine an undegenerate matrix $[a]^n_r$ of rank r and extravagance ρ when and only when

$$\rho \not< 0, \quad \rho \not> r, \quad \rho \not> n - r. \quad \text{................................(25)}$$

Let $[a]^n_r$ be any given matrix determined in this way, and let $[\beta]^n_{n-r}$ be any given matrix (of rank $n-r$) normal to $[a]^n_r$, so that $[a]^n_r \, \underline{\overline{\beta}}^{\,n-r}_{n} = 0$. Then the possible values of $\underline{\overline{b}}^{\,s}_{n}$ are those of the form

$$\underline{\overline{b}}^{\,s}_{n} = \underline{\overline{\beta}}^{\,n-r}_{n} \, \underline{\overline{k}}^{\,s}_{n-r},$$

where $[k]^{n-r}_s$ is arbitrary subject to the condition that it is undegenerate and has rank s.

For the vertical rows of $\lceil b \rfloor^{s}_{n}$ must be connected with the vertical rows of $\lceil \beta \rfloor^{n-r}_{n}$, which are a complete set of $n-r$ unconnected solutions of the equation $[a]^{n}_{r} \lceil x \rfloor_{n} = 0$.

Since by Theorem III the given matrix $\lceil \beta \rfloor^{n-r}_{n}$ has rank $n-r$ and extravagance ρ, it follows from Ex. xiii of § 165 that the undegenerate matrix $\lceil b \rfloor^{s}_{n}$ whose rank is s can have extravagance σ when and only when

$$s+\sigma \not> (n-r)+\rho, \quad s-\sigma \not> (n-r)-\rho, \quad \sigma \not< 0, \quad \sigma \not> s. \quad \text{..............(26)}$$

Thus when r and s are given integers consistent with (23), the possible values of ρ and σ are those consistent with (25) and (26). Since the condition $\rho \not> n-r$ is superfluous, being obtainable from the others by the elimination of σ, the conditions (25) and (26) are equivalent to (24), and the possible values of ρ and σ are those consistent with the conditions (24).

The elimination of ρ and σ from (24) leads to the conditions (23). Consequently we can determine integers ρ and σ consistent with (24) when and only when r and s are integers consistent with (23); and the conditions (24) give the possible values of all four of the integers r, s, ρ, σ.

Ex. xvi. *Possible values of the ranks and extravagances of any two mutually orthogonal matrices.*

If $[a]^{n}_{p}$ and $[b]^{n}_{q}$ are any two (mutually orthogonal) matrices of ranks r and s satisfying the equation

$$[a]^{n}_{p} \lceil b \rfloor^{q}_{n} = 0,$$

and having horizontal extravagances ρ and σ, then when n, p, q are given, the possible values of r and s are those consistent with the conditions

$$r \not< 0, \quad r \not> p; \quad s \not< 0, \quad s \not> q; \quad r+s \not> n;$$

and when n, p, q, r, s are given, the possible values of ρ and σ are those consistent with the conditions (24).

We deduce the last of these results from Ex. xv by writing

$$[a]^{n}_{p} = [h]^{r}_{p}\,[a]^{n}_{r}, \qquad [b]^{n}_{q} = [k]^{s}_{q}\,[\beta]^{n}_{s}.$$

When p and q are arbitrary, the possible values of r, s, ρ, σ are the same as in Ex. xv.

Ex. xvii. *The matrix* $[1,\, b]^{r,n-r}_{r}$ *has extravagance* $n-r$ *or is plenarily extravagant when and only when* $\lceil b \rfloor^{r}_{n-r}\,[b]^{n-r}_{r} = -[1]^{n-r}_{n-r}$.

For the matrices $[1,\, b]^{r,n-r}_{r}$ and $\lceil -b,\, 1 \rfloor^{r,n-r}_{n-r}$ are undegenerate and mutually normal. Therefore the first of these matrices has extravagance $n-r$ when and only when the second has extravagance $n-r$, i.e. when and only when

$$\lceil -b,\, 1 \rfloor^{r,n-r}_{n-r} \begin{bmatrix} -b \\ 1 \end{bmatrix}^{n-r}_{r,n-r} = 0, \quad \text{or} \quad \lceil b \rfloor^{r}_{n-r}\,[b]^{n-r}_{r} + [1]^{n-r}_{n-r} = 0.$$

Ex. xviii. Let $[k]_r^r = k[1]_r^r$, $[k]_s^s = k[1]_s^s$, where k is a non-zero scalar quantity, and suppose that $s \not< r$. Then when $[x]_r^s$ is arbitrary, we have the following two equivalent results:

(1) *The two matrices*

$$[k, x]_r^{r,s} \overline{\underline{\begin{matrix} k \\ x \end{matrix}}}{}_{r,s}^{r} \quad \textit{or} \quad [x]_r^s \overline{\underline{x}}{}_s^r + k^2[1]_r^r$$

and

$$\overline{\underline{k, x}}{}_s^{s,r} \begin{bmatrix} k \\ x \end{bmatrix}_{s,r}^{s} \quad \textit{or} \quad \overline{\underline{x}}{}_s^r [x]_r^s + k^2[1]_s^s$$

have equal degeneracies, the possible values of their common degeneracy ρ being those consistent with the conditions $\rho \not< 0$, $\rho \not> r$.

(2) *The two matrices $[k, x]_r^{r,s}$ and $\overline{\underline{k, x}}{}_s^{s,r}$ have equal extravagances, the possible values of their common extravagance ρ being those consistent with the conditions $\rho \not< 0$, $\rho \not> r$.*

These results follow from Theorem III when we observe that the two matrices $[k, x]_r^{r,s}$ and $\overline{\underline{-x, k}}{}_s^{r,s}$ are undegenerate and mutually normal.

The following are particular cases:

(3) We have $[x]_r^s \overline{\underline{x}}{}_s^r = -[1]_r^r$ when and only when the matrix $\overline{\underline{1, x}}{}_s^{s,r} \begin{bmatrix} 1 \\ x \end{bmatrix}_{s,r}^{s}$ or $\overline{\underline{x}}{}_s^r [x]_r^s + [1]_s^s$ has rank $s-r$.

(4) We have $[x]_r^s \overline{\underline{x}}{}_s^r = [1]_r^r$ when and only when the matrix $\overline{\underline{i, x}}{}_s^{s,r} \begin{bmatrix} i \\ x \end{bmatrix}_{s,r}^{s}$ or $\overline{\underline{x}}{}_s^r [x]_r^s - [1]_s^s$ has rank $s-r$, where $i = \sqrt{-1}$.

We can deduce Ex. xvii from the particular case (3), which shows that

$$\overline{\underline{b}}{}_{n-r}^{r} [b]_r^{n-r} = -[1]_{n-r}^{n-r}$$

when and only when $[1, b]_r^{r,n-r}$ has extravagance $n-r$ or is plenarily extravagant.

Ex. xix. The corresponding results when $k=0$ are (see Exs. xviii and xix of § 165) as follows:

If $[x]_r^s$ is a matrix of assigned rank t (where $t \not< 0$, $t \not> r$, $t \not> s$), the degeneracies ρ and σ of $[x]_r^s \overline{\underline{x}}{}_s^r$ and $\overline{\underline{x}}{}_s^r [x]_r^s$ can have independently of one another all values consistent with the conditions

$$\rho \not< r-t, \quad \rho \not> r, \quad \rho \not> r+s-2t; \quad \sigma \not< s-t, \quad \sigma \not> s, \quad \sigma \not> r+s-2t.$$

When t is arbitrary, the possible values of ρ and σ are those consistent with the conditions

$$\rho \not< r-s, \quad \rho \not> r; \quad \sigma \not< s-r, \quad \sigma \not> s; \quad 2\rho - \sigma \not< r-s, \quad 2\sigma - \rho \not< s-r,$$

which include $\rho \not< 0$, $\sigma \not< 0$.

§ 168. Reduction of any undegenerate matrix to an equivalent similar undegenerate matrix whose long rows are mutually orthogonal.

We consider in this article undegenerate matrices whose long rows are horizontal. The corresponding results for matrices whose long rows are vertical will be evident. Theorems I and II can be deduced from formula (A_2) in Ex. vi of § 165. We give however independent proofs.

Theorem I. *If* $[a]_r^n$ *is an undegenerate matrix of rank* r, *we can determine equivalent similar matrices* $[b]_r^n$ *whose long rows are mutually orthogonal.*

If $[a]_r^n$ *has rank* r *and extravagance* ρ, *then every such matrix* $[b]_r^n$ *has* ρ *extravagant long rows and* $r-\rho$ *non-extravagant long rows.*

When $[a]_r^n$ has rank r and extravagance ρ, the symmetric matrix $[a]_r^n \; \overline{\underline{a}}\,{}_n^r$ has rank $r-\rho$, and as in § 147 we can determine in many ways an undegenerate square matrix $[h]_r^r$ and a quasi-scalar matrix of rank $r-\rho$ of the form

$$\begin{bmatrix} e, & 0 \\ 0, & 0 \end{bmatrix}_{r-\rho,\,\rho}^{r-\rho,\,\rho}, \quad \text{where } [e]_{r-\rho}^{r-\rho} = \begin{bmatrix} e_1 & 0 & \dots & 0 \\ 0 & e_2 & \dots & 0 \\ \dots & \dots & \dots & \dots \\ 0 & 0 & \dots & e_{r-\rho} \end{bmatrix}, \quad \dots\dots\dots\dots(1)$$

such that
$$[h]_r^r \cdot [a]_r^n \; \overline{\underline{a}}\,{}_n^r \cdot \overline{\underline{h}}\,{}_r^r = \begin{bmatrix} e, & 0 \\ 0, & 0 \end{bmatrix}_{r-\rho,\,\rho}^{r-\rho,\,\rho} = [e]_r^r.$$

Then $e_1, e_2, \dots e_{r-\rho}$ are all non-zero scalar quantities.

If $\quad [b]_r^n = [h]_r^r \, [a]_r^n$, we have $[b]_r^n \; \overline{\underline{b}}\,{}_n^r = \begin{bmatrix} e, & 0 \\ 0, & 0 \end{bmatrix}_{r-\rho,\,\rho}^{r-\rho,\,\rho}. \quad \dots\dots\dots(2)$

Every such matrix $[b]_r^n$ is equivalent to $[a]_r^n$ according to the definition of § 113, and its long rows are mutually orthogonal.

NOTE 1. *Domain of rationality of* $[a]_r^n$, $[h]_r^r$, $[b]_r^n$.

If the elements of $[a]_r^n$ all lie in some restricted domain of rationality Ω, then as shown in § 147 the matrices $[h]_r^r$, $[e]_r^r$, $[b]_r^n$ can be so chosen that their elements all lie in Ω. In particular if $[a]_r^n$ is real, then $[h]_r^r$, $[e]_r^r$, $[b]_r^n$ can be all real; but in this particular case we must have $\rho = 0$.

Now let $[b]_r^n$ be any undegenerate matrix which is similar to and equivalent to $[a]_r^n$, and whose long rows are mutually orthogonal. Then we have

$$[b]_r^n = [h]_r^r [a]_r^n, \quad [b]_r^n \; \overline{\underline{b}}\,{}_n^r = [k]_r^r,$$

where $[h]^r_r$ is an undegenerate square matrix, and $[k]^r_r$ is a quasi-scalar matrix of the form

$$[k]^r_r = \begin{bmatrix} k_1 & 0 & \dots & 0 \\ 0 & k_2 & \dots & 0 \\ \dots & \dots & \dots & \dots \\ 0 & 0 & \dots & k_r \end{bmatrix}.$$

Since $[k]^r_r = [h]^r_r \,.\, [a]^n_r \,\overline{\underline{a}}{}^r_n \,.\, \overline{\underline{h}}{}^r_r$, it follows from Theorem II of § 131 that the matrix $[k]^r_r$ has the same rank $r-\rho$ as $[a]^n_r \,\overline{\underline{a}}{}^r_n$. Therefore ρ of its diagonal elements $k_1, k_2, \dots k_r$ have the value 0, whilst the remaining $r-\rho$ elements have non-zero values. It follows that the matrix $[b]^n_r$ has ρ extravagant and $r-\rho$ non-extravagant horizontal or long rows.

Thus Theorem I is completely proved.

Ex. i. *If $[a]^n_r$ is an undegenerate matrix of rank r, we can determine equivalent similar undegenerate matrices $[b]^n_r$ such that*

$$[b]^n_r \,\overline{\underline{b}}{}^r_n = \begin{bmatrix} e, & 0 \\ 0, & 0 \end{bmatrix}^{r-\rho,\ \rho}_{r-\rho,\ \rho}, \qquad \dots\dots(3)$$

where $[e]^{r-\rho}_{r-\rho}$ is an undegenerate symmetric matrix, and also where $[e]^{r-\rho}_{r-\rho}$ is an undegenerate quasi-scalar matrix, when and only when $[a]^n_r$ has extravagance ρ.

For when any matrix $[b]^n_r$ equivalent to $[a]^n_r$ satisfies an equation of the form (3), both $[b]^n_r$ and $[a]^n_r$ must have extravagance ρ; and when $[a]^n_r$ has extravagance ρ, we can certainly determine an equivalent matrix $[b]^n_r$ satisfying an equation of the form (3).

Ex. ii. *If $[a]^n_m$ is any matrix having rank r and horizontal extravagance ρ, we can determine undegenerate matrices $[b]^n_r$ of rank r which are horizontally equivalent to $[a]^n_m$ and whose horizontal rows are mutually orthogonal.*

Every such matrix $[b]^n_r$ has ρ extravagant long rows and $r-\rho$ non-extravagant long rows.

We can write $[a]^n_m = [p]^r_m\,[a]^n_r$, where $[p]^r_m$ and $[a]^n_r$ are undegenerate matrices of rank r, and $[a]^n_r$ has (horizontal) extravagance ρ. Then an undegenerate matrix $[b]^n_r$ is horizontally equivalent to $[a]^n_m$ when and only when it is equivalent to the undegenerate matrix $[a]^n_r$.

Theorem II. *If $[a]^n_r$ is an undegenerate matrix of rank r, we can determine equivalent similar matrices $[b]^n_r$ such that*

$$[b]^n_r \,\overline{\underline{b}}{}^r_n = \begin{bmatrix} 1, & 0 \\ 0, & 0 \end{bmatrix}^{r-\rho,\ \rho}_{r-\rho,\ \rho} \qquad \dots\dots(4)$$

when and only when $[a]^n_r$ has extravagance ρ.

When $[a]^n_r$ has rank r and extravagance ρ, the symmetric matrix $[a]^n_r \underline{\overline{a}}{}^r_n$ has rank $r-\rho$, and as in § 147.2 we can determine an undegenerate square matrix $[h]^r_r$ such that

$$[h]^r_r . [a]^n_r \underline{\overline{a}}{}^r_n . \underline{\overline{h}}{}^r_r = \begin{bmatrix} 1, & 0 \\ 0, & 0 \end{bmatrix}^{r-\rho,\,\rho}_{r-\rho,\,\rho}$$

If $[b]^n_r = [h]^r_r [a]^n_r$, the equation (4) is satisfied.

Conversely if any undegenerate matrix $[b]^n_r$ which is equivalent to $[a]^n_r$ satisfies the equation (4), then both $[b]^n_r$ and $[a]^n_r$ have extravagance ρ.

NOTE 2. *Rows at unit intensity.*

Any non-extravagant row of a matrix will be said to be at unit intensity when the sum of the squares of its elements is equal to 1. It appears from Theorem II that when we reduce an undegenerate matrix to an equivalent similar undegenerate matrix whose long rows are mutually orthogonal, each of the non-extravagant long rows in the equivalent matrix can be so chosen as to be at unit intensity.

Ex. iii. *If $[a]^n_m$ is any matrix having rank r and horizontal extravagance ρ, we can determine undegenerate matrices $[b]^n_r$ of rank r which are horizontally equivalent to $[a]^n_m$ and are such that the equation* (4) *is satisfied.*

We express $[a]^n_m$ in the form $[a]^n_m = [p]^r_m [a]^n_r$. Then the undegenerate matrices $[b]^n_r$ of rank r which are horizontally equivalent to $[a]^n_m$ are those which are equivalent to $[a]^n_r$.

Ex. iv. If $[a]^n_r$ is an undegenerate matrix of rank r we can determine an equivalent undegenerate matrix $[b]^n_r = [h]^r_r [a]^n_r$ such that

$$[b]^n_r \underline{\overline{b}}{}^r_n = [e]^r_r,$$

where $[e]^r_r$ is an undegenerate square matrix or an undegenerate quasi-scalar matrix when and only when $[a]^n_r$ is non-extravagant.

Ex. v. *If $[a]^n_r$ is an undegenerate matrix of rank r, we can determine an equivalent semi-unit matrix $[l]^n_r$ of rank r when and only when $[a]^n_r$ is non-extravagant.*

Ex. vi. *If $[a]^n_r$ is a real undegenerate matrix of rank r, we can determine an equivalent real semi-unit matrix $[l]^n_r$ of rank r. We then have $[l]^n_r = [h]^r_r [a]^n_r$, where $[h]^r_r$ is a real undegenerate square matrix.*

In this case by Ex. v of § 149 the matrix $[a]^n_r \underline{\overline{a}}{}^r_n$ is a real and definite symmetric matrix of rank r whose signature is positive. Therefore by Theorem II of § 148 there exists a *real* undegenerate square matrix $[h]^r_r$ such that $[h]^r_r . [a]^n_r \underline{\overline{a}}{}^r_n . \underline{\overline{h}}{}^r_r = [1]^r_r$, and

$$[l]^n_r = [h]^r_r [a]^n_r \qquad \text{..(5)}$$

is then a *real* semi-unit matrix of rank r equivalent to $[a]^n_r$.

Again if $[l]_r^n$ is any real semi-unit matrix of rank r given by (5), we see by equating the purely imaginary portions of both sides that $[h]_r^r$ must be real.

Ex. vii. If $a_1, a_2, \ldots a_r$, where $a_i = [a_{i1}\, a_{i2} \ldots a_{in}]$, are r unconnected one-rowed matrices of order n, it is possible to determine r *unconnected mutually orthogonal* one-rowed matrices $\beta_1, \beta_2, \ldots \beta_r$ which are connected with them.

Further if $[a]_r^n$ has extravagance ρ, i.e. if $[a]_r^n \overline{\underline{a}}_n^r$ has degeneracy ρ or rank $r-\rho$, then ρ of the matrices $\beta_1, \beta_2, \ldots \beta_r$ are extravagant, and the remaining $r-\rho$ are non-extravagant. The non-extravagant matrices can be so chosen that any number of them or all of them have intensity 1.

These results are merely re-statements of Theorems I and II.

§ 169. The cores of an undegenerate matrix.

1. *Preliminary theorem.*

In this article we again consider undegenerate matrices whose long rows are horizontal. Similar results are clearly true for undegenerate matrices whose long rows are vertical. We commence by proving a theorem which leads to the definition of the cores of an undegenerate matrix.

Theorem I. *If $[a]_r^n$ is an undegenerate matrix of rank r and extravagance ρ, and if $\begin{bmatrix} b \\ \beta \end{bmatrix}_{r-\rho,\,\rho}^n$ is an equivalent similar matrix such that*

$$\begin{bmatrix} b \\ \beta \end{bmatrix}_{r-\rho,\,\rho}^n \overline{\underline{b,\ \beta}}_n^{r-\rho,\,\rho} = \begin{bmatrix} e, & 0 \\ 0, & 0 \end{bmatrix}_{r-\rho,\,\rho}^{r-\rho,\,\rho}, \quad \ldots\ldots\ldots\ldots\ldots\ldots(1)$$

where $[e]_{r-\rho}^{r-\rho}$ *is an undegenerate symmetric matrix, then*

(1) $[\beta]_\rho^n$ *is a completely extravagant matrix of rank* ρ.

(2) $[b]_{r-\rho}^n$ *is a non-extravagant matrix of rank* $r-\rho$.

(3) $[\beta]_\rho^n$ *and* $[b]_{r-\rho}^n$ *are mutually orthogonal.*

(4) $[\beta]_\rho^n$ *and* $[a]_r^n$ *are mutually orthogonal.*

(5) *Any one-rowed matrix $[x]_n$ which is connected with the horizontal rows of $[a]_r^n$ is orthogonal with $[a]_r^n$ (i.e. with all the horizontal rows of $[a]_r^n$) when and only when it is connected with the horizontal rows of $[\beta]_\rho^n$.*

NOTE. There exists a relation of the form (1) when $\begin{bmatrix} b \\ \beta \end{bmatrix}_{r-\rho,\,\rho}^n$ is an equivalent similar matrix whose long rows are mutually orthogonal, the last ρ being extravagant rows, and the first $r-\rho$ being non-extravagant rows. In this particular case $[e]_{r-\rho}^{r-\rho}$ is an undegenerate quasi-scalar matrix. If further every non-extravagant long row is so chosen as to be at unit intensity, then $[e]_{r-\rho}^{r-\rho} = [1]_{r-\rho}^{r-\rho}$.

We have $$[a]_r^n = [h]_r^r \begin{bmatrix} b \\ \beta \end{bmatrix}_{r-\rho,\rho}^{n}, \quad \text{..........................(2)}$$
where $[h]_r^r$ is a given undegenerate square matrix.

The first three properties of the theorem follow from the equations

$$[\beta]_\rho^n \overline{\underline{\beta}}_n^{\rho} = 0, \quad [b]_{r-\rho}^n \overline{\underline{b}}_n^{r-\rho} = [e]_{r-\rho}^{r-\rho}, \quad [b]_{r-\rho}^n \overline{\underline{\beta}}_n^{\rho} = 0.$$

The fourth property follows from the equation

$$[a]_r^n \overline{\underline{\beta}}_n^{\rho} = [h]_r^r \begin{bmatrix} b \\ \beta \end{bmatrix}_{r-\rho,\rho}^{n} \overline{\underline{\beta}}_n^{\rho} = 0.$$

To prove the fifth property we observe that $[x]_n$ is connected with the horizontal rows of $[a]_r^n$ when and only when it is connected with the horizontal rows of $\begin{bmatrix} b \\ \beta \end{bmatrix}_{r-\rho,\rho}^{n}$. Hence if $[x]_n$ is any one-rowed matrix connected with the horizontal rows of $[a]_r^n$, we can write

$$[x]_n = [l, \lambda]_{r-\rho,\rho} \begin{bmatrix} b \\ \beta \end{bmatrix}_{r-\rho,\rho}^{n};$$

and we then have $[a]_r^n \overline{\underline{x}}_n = 0$ if and only if

$$\begin{bmatrix} b \\ \beta \end{bmatrix}_{r-\rho,\rho}^{n} \overline{\underline{x}}_n = \begin{bmatrix} b \\ \beta \end{bmatrix}_{r-\rho,\rho}^{n} \overline{\underline{b, \beta}}_n^{r-\rho,\rho} \overline{\underline{\begin{matrix} l \\ \lambda \end{matrix}}}_{r-\rho,\rho} = \begin{bmatrix} e, & 0 \\ 0, & 0 \end{bmatrix}_{r-\rho,\rho}^{r-\rho,\rho} \overline{\underline{\begin{matrix} l \\ \lambda \end{matrix}}}_{r-\rho,\rho} = 0,$$

i.e. if and only if $$[e]_{r-\rho}^{r-\rho} \overline{\underline{l}}_{r-\rho} = 0.$$

Since $[e]_{r-\rho}^{r-\rho}$ is undegenerate, this is the case when and only when $\overline{\underline{l}}_{r-\rho} = 0$, or $[l]_{r-\rho} = 0$, i.e. when and only when $[x]_n$ has the form

$$[x]_n = [\lambda]_\rho [\beta]_\rho^n.$$

Ex. i. *Every one-rowed matrix $[\xi]_n$ connected with the horizontal rows of $[\beta]_\rho^n$ is orthogonal with every one-rowed matrix $[x]_n$ connected with the horizontal rows of $[a]_r^n$.*

For if $$[x]_n = [l]_r [a]_r^n \quad \text{and} \quad [\xi]_n = [\lambda]_\rho [\beta]_\rho^n,$$

then $$[x]_n \overline{\underline{\xi}}_n = [l]_r [a]_r^n \overline{\underline{\beta}}_n^{\rho} \overline{\underline{\lambda}}_\rho = 0.$$

Ex. ii. *If $[\gamma]_\rho^n = [h]_\rho^r [a]_r^n$ is any undegenerate matrix of rank ρ whose long rows are connected with the long rows of $[a]_r^n$ and are orthogonal with all the long rows of $[a]_r^n$, so that $[a]_r^n \overline{\underline{\gamma}}_n^{\rho} = 0$, then $[\gamma]_\rho^n$ is a completely extravagant matrix of rank ρ.*

For $$[\gamma]_\rho^n \overline{\underline{\gamma}}_n^{\rho} = [h]_\rho^r [a]_r^n \overline{\underline{\gamma}}_n^{\rho} = 0.$$

Ex. iii. *If* $[\gamma]_\rho^n$ *and* $[\gamma']_\rho^n$ *are any two (completely extravagant) undegenerate matrices of rank* ρ *whose long rows are connected with the long rows of* $[a]_r^n$ *and are orthogonal with all the long rows of* $[a]_r^n$, *then* $[\gamma]_\rho^n$ *and* $[\gamma']_\rho^n$ *are mutually equivalent.*

For by Theorem I the long rows of both these matrices are connected with the long rows of $[\beta]_\rho^n$. Therefore there exist relations of the forms

$$[\gamma]_\rho^n = [l]_\rho^\rho [\beta]_\rho^n, \quad [\gamma']_\rho^n = [l']_\rho^\rho [\beta]_\rho^n,$$

where the square matrices $[l]_\rho^\rho$ and $[l']_\rho^\rho$ are necessarily undegenerate; and it follows that there exists a relation of the form

$$[\gamma']_\rho^n = [k]_\rho^\rho [\gamma]_\rho^n,$$

where $[k]_\rho^\rho$ is an undegenerate square matrix. In fact the two matrices $[\gamma]_\rho^n$, $[\gamma']_\rho^n$ are both equivalent to $[\beta]_\rho^n$, and are therefore mutually equivalent.

2. *Definition of the cores of an undegenerate matrix.*

If $[a]_r^n$ is any undegenerate matrix of rank r and extravagance ρ whose long rows are horizontal, and if $[\gamma]_\rho^n$ is any (completely extravagant) undegenerate matrix of rank ρ whose long rows are connected with the long rows of $[a]_r^n$ and are orthogonal with all the long rows of $[a]_r^n$, so that $[a]_r^n \overline{\underline{\gamma}}{}_{n}^{\rho} = 0$, then $[\gamma]_\rho^n$ will be called a *core* of $[a]_r^n$.

Defining the cores of $\overline{\underline{a}}{}_{n}^{r}$ in a similar way we see that $\overline{\underline{\gamma}}{}_{n}^{\rho}$ is a core of $\overline{\underline{a}}{}_{n}^{r}$ when and only when $[\gamma]_\rho^n$ is a core of $[a]_r^n$.

From Ex. iii we obtain the following theorem:

Theorem II. *The cores of a given undegenerate matrix are mutually equivalent similar undegenerate matrices, and they are all known when any one of them is known.*

In Theorem I the matrix $[\beta]_\rho^n$ is a core of $[a]_r^n$.

Ex. iv. *In Theorem I a one-rowed matrix* $[x]_n$ *is both connected with the long rows of* $[a]_r^n$ *and orthogonal with* $[a]_r^n$ *when and only when it is connected with the long rows of every core of* $[a]_r^n$.

This follows from the fifth property proved in Theorem I.

Ex. v. *Two mutually equivalent undegenerate matrices have the same cores.*

Ex. vi. *A non-extravagant (in particular a real) undegenerate matrix has no core.*

Ex. vii. *A completely extravagant matrix is a core of itself.*

Ex. viii. *A plenarily extravagant matrix is normal to its cores.*

For if $[a]_r^n$ has rank r and extravagance $n-r$, and if $[b]_{n-r}^n$ is one of its cores, we have $[a]_r^n \overline{\underline{\,b\,}}{}_n^{n-r} = 0$, i.e. $[a]_r^n$ and $[b]_{n-r}^n$ are mutually normal.

Ex. ix. *Two mutually normal undegenerate matrices have the same cores.*

Let $[a]_r^n$ and $[b]_{n-r}^n$ be any two mutually normal undegenerate matrices whose long rows are horizontal, and whose common extravagance is ρ; and let any core of $[a]_r^n$ be

$$[\gamma]_\rho^n = [h]_\rho^r\, [a]_r^n.$$

Then
$$[a]_r^n \overline{\underline{\,b\,}}{}_n^{n-r} = 0, \quad [a]_r^n \overline{\underline{\,\gamma\,}}{}_n^{\rho} = 0, \quad [\gamma]_\rho^n \overline{\underline{\,\gamma\,}}{}_n^{\rho} = 0.$$

Since the vertical rows of $\overline{\underline{\,\gamma\,}}{}_n^{\rho}$ are solutions of the equation $[a]_r^n \overline{\underline{\,x\,}}{}_n = 0$, and the vertical rows of $\overline{\underline{\,b\,}}{}_n^{n-r}$ are a complete set of $n-r$ unconnected solutions of the same equation, there must exist a relation of the form $\overline{\underline{\,\gamma\,}}{}_n^{\rho} = \overline{\underline{\,b\,}}{}_n^{n-r} \overline{\underline{\,k\,}}{}_{n-r}^{\rho}$, or

$$[\gamma]_\rho^n = [k]_\rho^{n-r}\, [b]_{n-r}^n.$$

We also have
$$[b]_{n-r}^n \overline{\underline{\,\gamma\,}}{}_n^{\rho} = [b]_{n-r}^n \overline{\underline{\,a\,}}{}_n^{r} \overline{\underline{\,h\,}}{}_r^{\rho} = 0.$$

Thus $[\gamma]_\rho^n$ is a (completely extravagant) matrix of rank ρ whose long rows are connected with the long rows of $[b]_{n-r}^n$, and which is orthogonal with $[b]_{n-r}^n$, i.e. $[\gamma]_\rho^n$ is a core of $[b]_{n-r}^n$.

In a similar way we can show that if $[\gamma]_\rho^n = [k]_\rho^{n-r}\, [b]_{n-r}^n$ is any core of $[b]_{n-r}^n$, then it is also a core of $[a]_r^n$.

Ex. x. *The common cores of two mutually normal undegenerate matrices are identical with their complete intersections.*

Let $\phi = [a]_r^n$ and $\psi = [b]_{n-r}^n$ be any two mutually normal undegenerate matrices whose long rows are horizontal and whose common extravagance is ρ; and let $[\gamma]_\rho^n$ be any one of their common cores. Then the horizontal rows of $[\gamma]_\rho^n$ are connected both with the horizontal rows of ϕ and with the horizontal rows of ψ.

If $[x]_n = [h]_r\, [a]_r^n = [k]_{n-r}\, [b]_{n-r}^n$ is any one-rowed matrix connected with the horizontal rows of both ϕ and ψ, then $[x]_n$ must also be orthogonal with both ϕ and ψ; and therefore by Ex. iv it must be connected with the horizontal rows of $[\gamma]_\rho^n$. Consequently if $[x]_\sigma^n$ is an undegenerate matrix of rank σ whose horizontal rows are connected with the horizontal rows of both ϕ and ψ, the horizontal rows of $[x]_\sigma^n$ must be connected with the horizontal rows of $[\gamma]_\rho^n$, and σ cannot exceed ρ.

Thus $[\gamma]_\rho^n$ is an undegenerate matrix of the greatest possible rank whose horizontal rows are connected both with the horizontal rows of ϕ and with the horizontal rows of ψ, i.e. $[\gamma]_\rho^n$ is a complete (horizontal) intersection of ϕ and ψ.

The result stated now follows from the fact that all the complete (horizontal) intersections of ϕ and ψ are mutually equivalent.

Ex. xi. *An undegenerate matrix is completely extravagant when and only when it is contained in its normals; and it is plenarily extravagant when and only when it contains its normals.*

Let ϕ and ψ be any two mutually normal undegenerate matrices.

If ϕ is contained in ψ, i.e. if ψ contains ϕ, then ϕ is a complete intersection and therefore a common core of ϕ and ψ. Thus ϕ is a core of itself, and is completely extravagant. It follows (see Ex. xii of § 167) that ϕ is completely extravagant, and ψ is plenarily extravagant.

Conversely if ϕ is completely extravagant, i.e. if ψ is plenarily extravagant, then ϕ is a core of itself. Therefore ϕ is a common core and a complete intersection of ϕ and ψ. It follows that ϕ is contained in ψ, and ψ contains ϕ.

Ex. xii. *Let* $[a]_r^n$ *be an undegenerate matrix of rank* r *and extravagance* ρ *of which* $[\gamma]_\rho^n$ *is a core. Then* $\begin{bmatrix} c \\ \gamma \end{bmatrix}_{r-\rho,\,\rho}^n$ *is an undegenerate matrix similar to and equivalent to* $[a]_r^n$ *(see Ex.* iii *of* § 139) *when and only when* $[c]_{r-\rho}^n$ *is a non-extravagant matrix of rank* $r-\rho$ *whose long rows are connected with the long rows of* $[a]_r^n$.

First suppose that $\begin{bmatrix} c \\ \gamma \end{bmatrix}_{r-\rho,\,\rho}^n$ is undegenerate and equivalent to $[a]_r^n$. Then

$$\begin{bmatrix} c \\ \gamma \end{bmatrix}_{r-\rho,\,\rho}^n \overline{\underline{c,\ \gamma}}{}_n^{r-\rho,\,\rho}$$

has the form $\begin{bmatrix} k, & 0 \\ 0, & 0 \end{bmatrix}_{r-\rho,\,\rho}^{r-\rho,\,\rho}$, and has the same degeneracy ρ and the same rank $r-\rho$ as $[a]_r^n\ \overline{\underline{a}}{}_n^r$. Therefore the matrix $[k]_{r-\rho}^{r-\rho} = [c]_{r-\rho}^n\ \overline{\underline{c}}{}_n^{r-\rho}$ must have rank $r-\rho$ and must be undegenerate, i.e. $[c]_{r-\rho}^n$ is non-extravagant. Next suppose that $[c]_{r-\rho}^n$ is a non-extravagant undegenerate matrix of rank $r-\rho$ whose horizontal rows are connected with the horizontal rows of $[a]_r^n$. Then if $\begin{bmatrix} c \\ \gamma \end{bmatrix}_{r-\rho,\,\rho}^n$ were degenerate, there would exist a relation of the form

$$[l,\ \lambda]_{r-\rho,\,\rho} \begin{bmatrix} c \\ \gamma \end{bmatrix}_{r-\rho,\,\rho}^n = 0, \quad \text{or} \quad [l]_{r-\rho}\,[c]_{r-\rho}^n = -[\lambda]_\rho\,[\gamma]_\rho^n,$$

in which neither $[l]_{r-\rho}$ nor $[\lambda]_\rho$ vanishes, and by postfixing $\overline{\underline{c}}{}_n^{r-\rho}$ on both sides, we should obtain

$$[l]_{r-\rho}\cdot[c]_{r-\rho}^n\ \overline{\underline{c}}{}_n^{r-\rho} = 0.$$

But this is impossible since the postfactor on the left is undegenerate. Consequently $\begin{bmatrix} c \\ \gamma \end{bmatrix}_{r-\rho,\,\rho}^n$ is an undegenerate matrix of rank r whose long rows are connected with the long rows of $[a]_r^n$, i.e. it is an undegenerate matrix similar to and equivalent to $[a]_r^n$.

Ex. xiii. *If* $[a]_t^n$, *where* $t < \rho$, *is an undegenerate matrix of rank* t *whose long rows are connected with the long rows of* $[a]_r^n$ *in Ex.* xii *and are mutually orthogonal and extravagant, then by adding* $\rho - t$ *final horizontal rows to* $[a]_t^n$ *we can form a matrix* $[a]_\rho^n$ *which is a core of* $[a]_r^n$.

For the horizontal rows of $[a]_t^n$ are connected with the long rows of $[\gamma]_\rho^n$, and (as shown in Ex. iii of § 139) we can by adding final horizontal rows to $[a]_t^n$ form a matrix $[a]_\rho^n$ of rank ρ which is equivalent to $[\gamma]_\rho^n$. Then $[a]_\rho^n$ is a core of $[a]_r^n$.

Ex. xiv. *Let* $\overline{\underline{x}}{}^{r}_{n} = \overline{\underline{a,\ b,\ \dots\ c}}{}^{u,v,\dots w}_{n}$, *where* $u+v+\dots+w=r$, *be an undegenerate matrix of rank* r. *Then if the matrices* $\overline{\underline{a}}{}^{u}_{n}$, $\overline{\underline{b}}{}^{v}_{n}$, ... $\overline{\underline{c}}{}^{w}_{n}$ *are mutually orthogonal, the sum of their extravagances is the extravagance of* $\overline{\underline{x}}{}^{r}_{n}$.

We have

$$[x]_r^n\,\overline{\underline{x}}{}^{r}_{n} = \begin{bmatrix} \alpha, & 0, & \dots & 0 \\ 0, & \beta, & \dots & 0 \\ \dots & \dots & \dots & \dots \\ 0, & 0, & \dots & \gamma \end{bmatrix}_{u,\,v,\,\dots\,w}^{u,\,v,\,\dots\,w},$$

where

$$[\alpha]_u^u = [a]_u^n\,\overline{\underline{a}}{}^{u}_{n},\quad [\beta]_v^v = [b]_v^n\,\overline{\underline{b}}{}^{v}_{n},\quad \dots\quad [\gamma]_w^w = [c]_w^n\,\overline{\underline{c}}{}^{w}_{n}.$$

Let u', v', ... w', ρ be the extravagances of $\overline{\underline{a}}{}^{u}_{n}$, $\overline{\underline{b}}{}^{v}_{n}$, ... $\overline{\underline{c}}{}^{w}_{n}$, $\overline{\underline{x}}{}^{r}_{n}$.

Then since the rank of a compartite matrix is equal to the sum of the ranks of its parts, we have

$$r - \rho = (u - u') + (v - v') + \dots + (w - w'),$$

and therefore

$$\rho = u' + v' + \dots + w'.$$

In Ex. xii we have a special case of the corresponding theorem for an undegenerate matrix whose long rows are horizontal.

The second part of Theorem I of § 168 is that particular case in which $[a]_u^n$, $[b]_v^n$, ... $[c]_w^n$ are one-rowed matrices.

Ex. xv. *If two similar undegenerate matrices* $[a]_r^n$ *and* $[b]_r^n$ *of rank* r *having the same extravagance* ρ *are connected by a relation of the form*

$$[b]_r^n = [a]_r^n [l]_n^n, \quad or \quad [a]_r^n = [b]_r^n\,\overline{\underline{l}}{}^{n}_{n},$$

where $[l]_n^n$ *is a square semi-unit matrix, and if*

$$[\beta]_\rho^n = [a]_\rho^n\,[l]_n^n, \quad or \quad [a]_\rho^n = [\beta]_\rho^n\,\overline{\underline{l}}{}^{n}_{n},$$

then $[a]_\rho^n$ *is a core of* $[a]_r^n$ *when and only when* $[\beta]_\rho^n$ *is a core of* $[b]_r^n$.

Let $[a]_\rho^n = [\lambda]_\rho^r [a]_r^n$ be a core of $[a]_r^n$, so that $[\lambda]_\rho^r$ has rank ρ, and

$$[a]_r^n\,\overline{\underline{a}}{}^{\rho}_{n} = 0, \quad [a]_\rho^n\,\overline{\underline{a}}{}^{\rho}_{n} = 0.$$

Then if $[\beta]_\rho^n = [a]_\rho^n [l]_n^n = [\lambda]_\rho^r [b]_r^n$, the matrix $[\beta]_\rho^n$ has the same rank ρ and the same extravagance ρ as $[a]_\rho^n$, i.e. it is a completely extravagant matrix of rank ρ whose long rows are connected with the long rows of $[b]_r^n$. Moreover we have

$$[b]_r^n \underline{\overline{\beta}}{}_n^\rho = [a]_r^n [l]_n^n \underline{\overline{l}}{}_n^n \underline{\overline{a}}{}_n^\rho = [a]_r^n \underline{\overline{a}}{}_n^\rho = 0.$$

Therefore $[\beta]_\rho^n$ is a core of $[b]_r^n$.

In a similar way we can show that if $[\beta]_\rho^n$ is a core of $[b]_r^n$, then $[a]_\rho^n$ is a core of $[a]_r^n$.

Ex. xvi. *Let $[a]_r^n$ and $[b]_r^n$ be two similar undegenerate matrices of rank r having the same extravagance ρ, so that there exists a relation of the form*

$$[b]_r^n = [h]_r^r [a]_r^n [l]_n^n,$$

where $[h]_r^r$ is undegenerate, and $[l]_n^n$ is a square semi-unit matrix; also let

$$[\beta]_\rho^n = [a]_\rho^n [l]_n^n, \quad \text{or} \quad [a]_\rho^n = [\beta]_\rho^n \underline{\overline{l}}{}_n^n.$$

Then $[a]_\rho^n$ is a core of $[a]_r^n$ when and only when $[\beta]_\rho^n$ is a core of $[b]_r^n$.

This follows from Ex. xv when we observe that the undegenerate matrix $[a']_r^n = [h]_r^r [a]_r^n$ has the same cores as $[a]_r^n$.

3. *Determination of the cores of a given undegenerate matrix.*

When we determine any equivalent similar undegenerate matrix whose long rows are mutually orthogonal as in § 168, we at the same time determine a core of the given matrix, for the extravagant long rows of the equivalent matrix then form one of the cores. We can however determine the cores in an easier way.

Let $[a]_r^n$ be any given undegenerate matrix of rank r and extravagance ρ whose long rows are horizontal. Since the matrix $[a]_r^n \underline{\overline{a}}{}_n^r$ has rank $r-\rho$, we can determine an undegenerate matrix $[h]_\rho^r$ such that

$$[a]_r^n \underline{\overline{a}}{}_n^r \underline{\overline{h}}{}_r^\rho = 0,$$

the vertical rows of $\underline{\overline{h}}{}_r^\rho$ being a complete set of ρ unconnected solutions of the equation $[a]_r^n \underline{\overline{a}}{}_n^r \underline{\overline{x}}{}_r = 0.$

Let $$[\gamma]_\rho^n = [h]_\rho^r [a]_r^n.$$

Then by § 131 the matrix $[\gamma]_\rho^n$ has rank ρ; and we also have

$$[a]_r^n \underline{\overline{\gamma}}{}_n^\rho = 0, \quad [\gamma]_\rho^n \underline{\overline{\gamma}}{}_n^\rho = 0.$$

Thus $[\gamma]^{n}_{\rho}$ is a (completely extravagant) matrix of rank ρ whose long rows are connected with the long rows of $[a]^{n}_{r}$, and which is orthogonal with $[a]^{n}_{r}$, i.e. $[\gamma]^{n}_{\rho}$ is a core of $[a]^{n}_{r}$, and the cores of $[a]^{n}_{r}$ are the matrices similar to and equivalent to $[\gamma]^{n}_{\rho}$.

In the same way $\overline{\underline{\gamma}}{}^{\rho}_{n}$ is a core of the undegenerate matrix $\overline{\underline{a}}{}^{r}_{n}$ whose long rows are vertical.

§ 170. The extravagance, core and plenum of a spacelet in homogeneous space of $n-1$ dimensions or rank n.

1. *Preliminary remarks, chiefly recapitulatory.*

As indicated in §§ 113 and 138—40 any matrix $\overline{\underline{a}}{}^{p}_{n}$ of rank r can be regarded as representing a spacelet ω_r of $r-1$ dimensions or rank r in homogeneous space ω_n of $n-1$ dimensions or rank n. In particular any matrix $\overline{\underline{a}}{}^{p}_{n}$ of rank n, and any undegenerate square matrix $\overline{\underline{a}}{}^{n}_{n}$ of order and rank n, represents the complete space ω_n. Whenever we use the notations $\omega_r, \omega_r', \omega_r'', \ldots$ for a spacelet of ω_n, it will be understood that r is a positive integer not exceeding n, and that r is the rank and $r-1$ the number of dimensions of the spacelet. Two matrices $\overline{\underline{a}}{}^{p}_{n}$ and $\overline{\underline{b}}{}^{q}_{n}$ represent the same spacelet ω_r of ω_n when and only when they have the same rank r and have vertical equivalence, so that $\overline{\underline{a}}{}^{p}_{n} \equiv \overline{\underline{b}}{}^{q}_{n}$.

If $\omega_r \equiv \overline{\underline{a}}{}^{r}_{n}$ and $\omega_s \equiv \overline{\underline{b}}{}^{s}_{n}$ are any two spacelets in ω_n, then ω_r *contains* ω_s, i.e. ω_s *is contained in* or *lies in* ω_r when and only when there exists a relation of the form

$$\overline{\underline{b}}{}^{s}_{n} = \overline{\underline{a}}{}^{r}_{n}\,\overline{\underline{k}}{}^{s}_{r},$$

where $\overline{\underline{k}}{}^{s}_{r}$ necessarily has rank s; and this is possible only when $s \ngtr r$. More generally the spacelet $\omega_r \equiv \overline{\underline{a}}{}^{p}_{n}$ contains the spacelet $\omega_s \equiv \overline{\underline{b}}{}^{q}_{n}$ when and only when there exists a relation of the form $\overline{\underline{b}}{}^{q}_{n} = \overline{\underline{a}}{}^{p}_{n}\,\overline{\underline{k}}{}^{q}_{p}$, and this is only possible when $s \ngtr r$. The two spacelets ω_r and ω_s are *mutually incident* when one of them contains the other.

Two spacelets do or do not *intersect* according as they do or do not contain points in common; their *intersection* is the spacelet of greatest rank which is contained in both of them, and is the locus of all their common points; their *join* is the spacelet of smallest rank which contains both of them. The join of the two spacelets $\omega_r \equiv \overline{\underline{a}}{}^{p}_{n}$ and $\omega_s \equiv \overline{\underline{b}}{}^{q}_{n}$ is the spacelet represented by the matrix $\phi = \overline{\underline{a, b}}{}^{p,q}_{n}$, and these two spacelets are non-intersecting when and only when the rank of ϕ is equal to $r+s$. If $r+s>n$ the two spacelets necessarily intersect, and the rank of their intersection cannot be less than $r+s-n$.

If $\omega_\alpha \equiv \overline{\underline{a}}{}^{p}_{n}$, $\omega_\beta \equiv \overline{\underline{b}}{}^{q}_{n}$, ... $\omega_\gamma \equiv \overline{\underline{c}}{}^{r}_{n}$ are any number of spacelets in homogeneous space ω_n, their *intersection* is the spacelet of greatest rank which is contained in every one of them, and is the locus of all points common to all of them; and their *join* is the spacelet of smallest rank which contains all of them. Their *join* is the spacelet represented by the matrix

$$\phi = \overline{\underline{a, b, \ldots c}}{}^{p,q,\ldots r}_{n}$$

These spacelets are *unconnected* when and only when no one of them intersects the join of the others; and this is the case when and only when the rank of their join is equal to the sum of their ranks, i.e. when and only when ϕ has rank $\alpha+\beta+\ldots+\gamma$.

Any two spacelets ω_r and ω_s can always be represented in the forms

$$\omega_r \equiv \overline{\underline{c, a}}{}^{t,r-t}_{n}, \quad \omega_s \equiv \overline{\underline{c, b}}{}^{t,s-t}_{n},$$

where the matrix $\overline{\underline{c, a, b}}{}^{t,r-t,s-t}_{n}$ is undegenerate and has rank $T=r+s-t$; and their intersection and join are then respectively the spacelets

$$\omega_t \equiv \overline{\underline{c}}{}^{t}_{n}, \quad \omega_T \equiv \overline{\underline{c, a, b}}{}^{t,r-t,s-t}_{n}$$

Accordingly if ω_t and ω_T are respectively the intersection and the join of two spacelets ω_r and ω_s of ω_n, we always have

$$t+T=r+s.$$

Two spacelets ω_r and ω_s are *mutually complementary* when their join is ω_n, i.e. when $T=n$; and two spacelets ω_r and ω_s which both lie in a spacelet ω_p are *mutually complementary in* ω_p when their join is ω_p, i.e. when $T=p$.

Two spacelets $\omega_r \equiv \overline{\underline{a}}{}^{p}_{n}$ and $\omega_s \equiv \overline{\underline{b}}{}^{q}_{n}$ are *mutually orthogonal* when $[a]^{n}_{p}\, \overline{\underline{b}}{}^{q}_{n} = 0$; and they are *mutually normal* when $[a]^{n}_{p}\, \overline{\underline{b}}{}^{q}_{n} = 0$, and $r+s=n$.

They are mutually normal when and only when each of them is the locus of all points orthogonal with the other; and in this case each of them is called the *normal* to the other.

Ex. i. *If ω_u lies in ω_r, and ω_v does not intersect ω_r, then ω_u is the complete intersection of ω_r with the join of ω_u and ω_v, i.e. the join of ω_u and ω_v is a spacelet whose complete intersection with ω_r is ω_u.*

Because ω_v does not intersect ω_r, and ω_u lies in ω_r, therefore ω_v does not intersect ω_u. It follows that the join of ω_u and ω_v has rank $u+v$, and may be denoted by ω_{u+v}. The theorem asserts that the complete intersection of ω_r and ω_{u+v} is ω_u.

Since all spacelets which contain ω_r and ω_v also contain ω_u, they also contain ω_{u+v}. Therefore the join of ω_r and ω_v is the same as the join of ω_r and ω_{u+v}. But because ω_v does not intersect ω_r, the join of ω_r and ω_v has rank $r+v$. Therefore the join of ω_r and ω_{u+v} has rank $r+v$. It follows that the complete intersection of ω_r and ω_{u+v}, which contains ω_u, has rank u, and must be ω_u.

This proof depends on Theorem II of § 139. A direct proof is given in Ex. v of § 139. Another method of proof is indicated in Ex. iii.

Ex. ii. *If ω_u contains ω_r, and ω_v is complementary to ω_r, then ω_u is the join of ω_r with the complete intersection of ω_u and ω_v, i.e. the complete intersection of ω_u and ω_v is complementary to ω_r in ω_u.*

Because ω_v is complementary to ω_r, and ω_u contains ω_r, therefore ω_v is complementary to ω_u, i.e. the join of ω_u and ω_v has rank n. It follows that the intersection of ω_u and ω_v has rank $u+v-n$, and may be denoted by ω_{u+v-n}. The theorem asserts that the join of ω_r and ω_{u+v-n} is ω_u.

Since all spacelets which lie in ω_r and ω_v also lie in ω_u, they must also lie in ω_{u+v-n}. Therefore the intersection of ω_r and ω_v is the same as the intersection of ω_r and ω_{u+v-n}. But because ω_v is complementary to ω_r, the intersection of ω_r and ω_v has rank $r+v-n$. Therefore the intersection of ω_r and ω_{u+v-n} has rank $r+v-n$. It follows that the join of ω_r and ω_{u+v-n}, which lies in ω_u, has rank u, and must be ω_u.

Ex. iii. *If ω_{r+s} is the join of two given non-intersecting spacelets $\omega_r \equiv \left[a\right]^{r}_{n}$, $\omega_s \equiv \left[b\right]^{s}_{n}$, and if $\omega_\sigma \equiv \left[\beta\right]^{\sigma}_{n} \equiv \left[b\right]^{s}_{n}\left[k\right]^{\sigma}_{s}$ is a given spacelet lying in ω_s, then a general formula for any spacelet ω_p which lies in ω_{r+s} and has ω_σ for its complete intersection with ω_s is*

$$\omega_p \equiv \left[a,\ b\right]^{r,\,s}_{n} \begin{bmatrix} \lambda, & 0 \\ \mu, & k \end{bmatrix}^{p-\sigma,\,\sigma}_{r,\,s}, \quad \text{.................................(1)}$$

where $\left[\lambda\right]^{p-\sigma}_{r}$ is an arbitrary undegenerate matrix of rank $p-\sigma$ and $\left[\mu\right]^{p-\sigma}_{s}$ is arbitrary.

Since ω_p contains ω_σ, the condition $\sigma \not> p$ must be satisfied. A spacelet ω_p lies in ω_{r+s} and contains ω_σ when and only when it is the join of ω_σ and some spacelet

$$\omega_{p-\sigma} \equiv \left[\gamma\right]^{p-\sigma}_{n} \equiv \left[a,\ b\right]^{r,\,s}_{n} \begin{bmatrix} \lambda \\ \mu \end{bmatrix}^{p-\sigma}_{r,\,s} \quad \text{..............................(1')}$$

which lies in ω_{r+s} and does not intersect ω_σ, i.e. when and only when it is given by (1) where the postfactor on the right has rank p. The join of ω_p and ω_s is then the spacelet

$$\omega_T \equiv \left\lceil \begin{matrix} a, & b \end{matrix} \right\rfloor_n^{r,s} \left\lceil \begin{matrix} \lambda, & 0, & 0 \\ \mu, & k, & 1 \end{matrix} \right\rfloor_{r,s}^{p-\sigma,\sigma,s} \equiv \left\lceil \begin{matrix} a, & b \end{matrix} \right\rfloor_n^{r,s} \left\lceil \begin{matrix} \lambda, & 0 \\ \mu, & 1 \end{matrix} \right\rfloor_{r,s}^{p-\sigma,s} . \quad \ldots\ldots\ldots\ldots\ldots(2)$$

Every such spacelet ω_p has ω_σ for its complete intersection with ω_s when and only when the intersection of ω_p and ω_s has rank σ, i.e. when and only when the rank T of the join of ω_p and ω_s is $p+s-\sigma$, i.e. when and only when the last factor matrix on the right in (2) has rank $p+s-\sigma$, i.e. when and only when $\left\lceil \lambda \right\rfloor_r^{p-\sigma}$ is undegenerate and has rank $p-\sigma$. This establishes the theorem.

The join of ω_s with the spacelet $\omega_{p-\sigma}$ given by (1′) is

$$\omega_\tau \equiv \left\lceil \begin{matrix} \gamma, & b \end{matrix} \right\rfloor_n^{p-\sigma,\sigma} \equiv \left\lceil \begin{matrix} a, & b \end{matrix} \right\rfloor_n^{r,s} \left\lceil \begin{matrix} \lambda, & 0 \\ \mu, & 1 \end{matrix} \right\rfloor_{r,s}^{p-\sigma,s},$$

and the spacelets $\omega_{p-\sigma}$ and ω_s are non-intersecting when and only when the rank τ of their join is $p+s-\sigma$, i.e. when and only when $\left\lceil \lambda \right\rfloor_r^{p-\sigma}$ is undegenerate and has rank $p-\sigma$. Hence formula (1) is equivalent to the following result:

A spacelet lies in ω_{r+s} and has ω_σ for its complete intersection with ω_s when and only when it is the join of ω_σ with a spacelet which lies in ω_{r+s} and does not intersect ω_s.

This result is a generalisation of Ex. i above, and is equivalent to the theorem proved in Ex. xix of § 139. The particular case of Ex. i could be proved independently in a similar way by taking ω_r to be a spacelet complementary to ω_s.

2. *The extravagance of a spacelet.*

The *point* $\omega_1 \equiv \left\lceil x \right\rfloor_n$ is *extravagant* or *non-extravagant* according as

$$x_1^2 + x_2^2 + \ldots + x_n^2 = 0 \quad \text{or} \quad \neq 0.$$

Thus the point is extravagant when and only when $[x]_n \left\lceil x \right\rfloor_n = 0$, i.e. when and only when the point is orthogonal with itself.

The *spacelet* $\omega_r \equiv \left\lceil a \right\rfloor_n^r$ is *extravagant* or *non-extravagant* according as the product $\psi = [a]_r^n \left\lceil a \right\rfloor_n^r$ is degenerate or undegenerate; and the *extravagance* ρ of ω_r is defined to be the degeneracy of the product ψ, i.e. it is the extravagance of the undegenerate matrix $\left\lceil a \right\rfloor_n^r$.

From § 165.2 we see that the possible values of the extravagance ρ of a spacelet ω_r of given rank r are those consistent with the conditions

$$\rho \nless 0, \quad \rho \ngtr r, \quad \rho \ngtr n-r. \quad \ldots\ldots\ldots\ldots\ldots\ldots\ldots(3)$$

The spacelet ω_r is *completely extravagant* when it has extravagance r. There exist such spacelets when and only when $2r \not> n$; and when this condition is satisfied, r is the greatest value which the extravagance of ω_r can have.

The spacelet ω_r is *plenarily extravagant* when it has extravagance $n-r$. There exist such spacelets when and only when $2r \not< n$; and when this condition is satisfied, $n-r$ is the greatest value which the extravagance of ω_r can have.

Various properties of the extravagance of a spacelet can be deduced from the results obtained in § 165.

NOTE 1. *Real spacelets.*

We call ω_r a *real* spacelet when it can be represented in the form $\omega_r \equiv \overline{\underline{a}}{}^{r}_{n}$, where $\overline{\underline{a}}{}^{r}_{n}$ is a real matrix of rank r, i.e. when it contains r real unconnected points. A real spacelet is necessarily non-extravagant.

The complete space ω_n is real and non-extravagant.

Ex. iv. The extravagance ρ of the spacelet $\omega_r \equiv \overline{\underline{a}}{}^{p}_{n}$ is the vertical extravagance of $\overline{\underline{a}}{}^{p}_{n}$ or the horizontal extravagance of $[a]^{n}_{p}$, and the possible values of ρ are those consistent with the conditions (3).

In this case the product $[a]^{n}_{p}\, \overline{\underline{a}}{}^{p}_{n}$ can have rank $r-\rho$ when and only when the conditions (3) are satisfied, and ω_r has extravagance ρ when and only when this product has rank $r-\rho$.

Ex. v. *Two spacelets* $\omega_r \equiv \overline{\underline{a}}{}^{r}_{n}$, $\omega_r' \equiv \overline{\underline{b}}{}^{r}_{n}$ *of the same rank r have the same extravagance (see Theorem I of* § 165) *when and only when there exists a square semi-unit matrix* $[l]^{n}_{n}$ *such that*

$$\overline{\underline{b}}{}^{r}_{n} \equiv [l]^{n}_{n}\, \overline{\underline{a}}{}^{r}_{n}.$$

More generally two spacelets $\omega_r \equiv \overline{\underline{a}}{}^{p}_{n}$, $\omega_r' \equiv \overline{\underline{b}}{}^{q}_{n}$ have the same rank (denoted by r) and the same extravagance when and only when there exists a square semi-unit matrix $[l]^{n}_{n}$ such that

$$\overline{\underline{b}}{}^{q}_{n} \equiv [l]^{n}_{n}\, \overline{\underline{a}}{}^{p}_{n}.$$

Here the symbol $\equiv$ denotes vertical equivalence, and the two matrices connected by it are related as in Theorem II of § 138.

Ex. vi. *If* $\omega_u, \omega_v, \dots \omega_w$ *are a number of mutually orthogonal non-extravagant spacelets they must be unconnected.*

Let $\omega_u \equiv \overline{\underline{a}}{}^{u}_{n}$, $\omega_v \equiv \overline{\underline{b}}{}^{v}_{n}$, ... $\omega_w \equiv \overline{\underline{c}}{}^{w}_{n}$, $r = u+v+\dots+w$, and let

$$\phi = \overline{\underline{a,\, b,\, \dots\, c}}{}^{u,\, v,\, \dots\, w}_{n} = \overline{\underline{x}}{}^{r}_{n}.$$

Then the product $[x]^n_r \overline{\underline{x}}^{\,r}_{\,n}$ is a compartite matrix in standard form whose parts $[a]^n_u \overline{\underline{a}}^{\,u}_{\,n}$, $[b]^n_v \overline{\underline{b}}^{\,v}_{\,n}$, ... $[c]^n_w \overline{\underline{c}}^{\,w}_{\,n}$ have ranks $u, v, \ldots w$. Therefore $[x]^n_r \overline{\underline{x}}^{\,r}_{\,n}$ has rank r, and it follows that ϕ has rank r.

3. *The core of a spacelet.*

If $\omega_r \equiv \overline{\underline{a}}^{\,r}_{\,n}$ is any given spacelet of rank r of homogeneous space ω_n, we may define the *core* of ω_r to be the locus of all points of ω_r which are orthogonal with ω_r. The point $\overline{\underline{x}}_{\,n} \equiv \overline{\underline{a}}^{\,r}_{\,n} \overline{\underline{\xi}}_{\,r}$ of ω_r lies in the core of ω_r when and only when $[a]^n_r \overline{\underline{a}}^{\,r}_{\,n} \overline{\underline{\xi}}_{\,r} = 0$. Hence if ρ is the extravagance of ω_r, so that $[a]^n_r \overline{\underline{a}}^{\,r}_{\,n}$ has rank $r-\rho$, the core of ω_r is (as in § 169. 3) the completely extravagant spacelet ω_ρ of rank ρ given by

$$\omega_\rho \equiv \overline{\underline{a}}^{\,r}_{\,n} \overline{\underline{\xi}}^{\,\rho}_{\,r}, \quad \text{where } [a]^n_r \overline{\underline{a}}^{\,r}_{\,n} \,.\, \overline{\underline{\xi}}^{\,\rho}_{\,r} = 0,$$

and where $\overline{\underline{\xi}}^{\,\rho}_{\,r}$ has rank ρ. It is the uniquely determinate spacelet represented by the cores of the matrix $\overline{\underline{a}}^{\,r}_{\,n}$ and all equivalent similar undegenerate matrices; and it contains every (completely extravagant) spacelet which lies in ω_r and is orthogonal with ω_r. Conversely all spacelets which lie in the core of ω_r are completely extravagant, and are orthogonal with ω_r.

If ω_{n-r} is the normal to ω_r, i.e. the locus of all points orthogonal with ω_r, it follows from the above definition that the core of ω_r is the locus of all points common to ω_r and ω_{n-r}, i.e. the complete intersection of ω_r and ω_{n-r}, and is also the core of ω_{n-r}. Thus two mutually normal spacelets have the same core, and their common core is their complete intersection.

Ex. vii. *If ω_r is a spacelet (of extravagance ρ) whose core is $\omega_\rho \equiv \overline{\underline{a}}^{\,\rho}_{\,n}$, and if $\omega_{r-\rho}$ is any spacelet of rank $r-\rho$ which lies in ω_r and does not intersect ω_ρ, so that ω_r is the join of ω_ρ and $\omega_{r-\rho}$, we can write $\omega_{r-\rho} \equiv \overline{\underline{a}}^{\,r-\rho}_{\,n}$ and*

$$\omega_r \equiv \overline{\underline{a,\ a}}^{\,r-\rho,\ \rho}_{\,n}, \quad \textit{where } \begin{bmatrix} a \\ a \end{bmatrix}^n_{r-\rho,\ \rho} \overline{\underline{a,\ a}}^{\,r-\rho,\ \rho}_{\,n} = \begin{bmatrix} 1, & 0 \\ 0, & 0 \end{bmatrix}^{r-\rho,\ \rho}_{r-\rho,\ \rho}$$

For the spacelet $\omega_{r-\rho}$ is orthogonal with ω_ρ and is necessarily non-extravagant.

Ex. viii. *If ω_r is the join of a number of unconnected mutually orthogonal spacelets $\omega_u, \omega_v, \ldots \omega_w$, then the extravagance of ω_r is the sum of the extravagances of $\omega_u, \omega_v, \ldots \omega_w$, and the core of ω_r is the join of the cores of $\omega_u, \omega_v, \ldots \omega_w$.*

The first part of this theorem follows from Ex. xiv of § 169. Retaining the same notation, we obtain the second part of the theorem by observing that the join of $\omega_u, \omega_v, \ldots \omega_w$ is a completely extravagant spacelet of rank ρ which is contained in ω_r and is orthogonal with ω_r, and which must therefore be the core of ω_r.

Ex. ix. *If ω_r is a spacelet of rank r, and if $\omega_u \equiv \begin{bmatrix} a \end{bmatrix}_n^u$ is any non-extravagant spacelet of rank u lying in ω_r, we can always represent ω_r as the join of ω_u with a spacelet ω_{r-u} of rank $r-u$ which is orthogonal with ω_u. The spacelet ω_{r-u} is then the locus of all points of ω_r which are orthogonal with ω_u.*

Let ω_{n-u} be the normal to ω_u. If the intersection of ω_r and ω_{n-u} is ω_x, we have $r+(n-u)-x \not> n$, or $x \not< r-u$. Again since ω_u does not intersect ω_{n-u}, it does not intersect ω_x. Thus ω_u and ω_x are two non-intersecting spacelets lying in ω_r, and we have $u+x \not> r$, or $x \not> r-u$. Consequently we have $x=r-u$, and the complete intersection of ω_r and ω_{n-u}, i.e. the locus of all points of ω_r orthogonal with ω_u, is a spacelet ω_{r-u} of rank $r-u$. Since ω_u and ω_{r-u} are non-intersecting, their join is a spacelet of rank r which lies in ω_r and must be ω_r. This proves the first part of the theorem.

Again let ω_r be the join of ω_u with a spacelet ω'_{r-u} of rank $r-u$ which is orthogonal with ω_u and does not intersect ω_u, so that we can write

$$\omega_u \equiv \begin{bmatrix} a \end{bmatrix}_n^u, \quad \omega'_{r-u} \equiv \begin{bmatrix} b \end{bmatrix}_n^{r-u}, \quad \omega_r \equiv \begin{bmatrix} a, b \end{bmatrix}_n^{u,\, r-u};$$

and let $\omega_1 \equiv \begin{bmatrix} x \end{bmatrix}_n \equiv \begin{bmatrix} a, b \end{bmatrix}_n^{u,\, r-u} \begin{bmatrix} \lambda \\ \mu \end{bmatrix}_{u,\, r-u}$ be any point of ω_r. The point ω_1 is orthogonal with ω_u when and only when

$$[a]_u^n \begin{bmatrix} a, b \end{bmatrix}_n^{u,\, r-u} \begin{bmatrix} \lambda \\ \mu \end{bmatrix}_{u,\, r-u} = [a]_u^n \begin{bmatrix} a \end{bmatrix}_n^u \begin{bmatrix} \lambda \end{bmatrix}_u = 0;$$

and since $[a]_u^n \begin{bmatrix} a \end{bmatrix}_n^u$ is an undegenerate square matrix, this is the case when and only when $\begin{bmatrix} \lambda \end{bmatrix}_u = 0$, i.e. when and only when ω_1 lies in ω'_{r-u}. Thus ω'_{r-u} must be the locus of all points of ω_r orthogonal with ω_u, and is the spacelet ω_{r-u} defined above.

The spacelet ω_{r-u} has the same extravagance and the same core as ω_r.

This follows from Ex. viii.

Ex. x. *General formulae for a spacelet of rank r and extravagance ρ and its core.*

From Ex. vi of § 165 we see that general formulae for any spacelet ω_r of rank r and extravagance ρ and for its core ω_ρ are

$$\omega_r \equiv [l]_n^n \begin{bmatrix} 1, & 0 \\ 0, & 1 \\ 0, & i \\ 0, & 0 \end{bmatrix}_{r-\rho,\, \rho,\, \rho,\, n-r-\rho}^{r-\rho,\, \rho}, \qquad \omega_\rho \equiv [l]_n^n \begin{bmatrix} 0 \\ 1 \\ i \\ 0 \end{bmatrix}_{r-\rho,\, \rho,\, \rho,\, n-r-\rho}^{\rho}, \quad \text{.........(A)}$$

where $[l]_n^n$ is a square semi-unit matrix, and $i=\sqrt{-1}$.

Ex. xi. *If $\omega_r \equiv \begin{bmatrix} a \end{bmatrix}_n^r$, $\omega_r' \equiv \begin{bmatrix} b \end{bmatrix}_n^r$ are two equi-extravagant spacelets of rank r, so that there exists a relation of the form*

$$\begin{bmatrix} b \end{bmatrix}_n^r \equiv [l]_n^n \begin{bmatrix} a \end{bmatrix}_n^r,$$

where $[l]^n_n$ *is a square semi-unit matrix; and if*

$$\omega_\rho \equiv \overline{\underline{a}}^{\rho}_{n}, \quad \omega_\rho' \equiv \overline{\underline{\beta}}^{\rho}_{n} \equiv [l]^n_n \, \overline{\underline{a}}^{\rho}_{n};$$

then ω_ρ' *is the core of* ω_r' *when and only when* ω_ρ *is the core of* ω_r.

This follows from Ex. xvi of § 169.

Ex. xii. *If the spacelet* ω_p *lies in the spacelet* ω_r *and intersects the core of* ω_r *in the spacelet* ω_τ, *then the core of* ω_p *must contain* ω_τ.

Every point of ω_τ is orthogonal with all points of ω_r, and therefore with all points of ω_p. Thus all points of ω_τ lie in ω_p and are orthogonal with ω_p; and it follows that all points of ω_τ lie in the core of ω_p.

Ex. xiii. *Let* ω_r *be a given spacelet of rank* r *and extravagance* ρ *whose core is* ω_ρ; *let* ω_p *be any spacelet of rank* p *and extravagance* π *which lies in* ω_r *and whose core is* ω_π; *and let the complete intersection of* ω_p *with the core of* ω_r *be a spacelet* ω_τ *of rank* τ. *Then the possible values of* p, π *and* τ *are those consistent with the conditions*

$$\tau \nless 0, \quad \tau \ngtr \rho; \quad \pi \nless \tau, \quad \pi \ngtr p, \quad \pi \ngtr (r-\rho)-p+2\tau, \quad \text{.................(4)}$$

which include $p \nless 0$, $p \ngtr r$; *and this remains true when the spacelet* ω_τ *is given.*

In the figure the shaded areas represent completely extravagant spacelets. The spacelet ω_ρ lies in ω_r; the spacelet ω_π lies in ω_p; and ω_τ is the complete intersection of ω_p (and of ω_π) with ω_ρ.

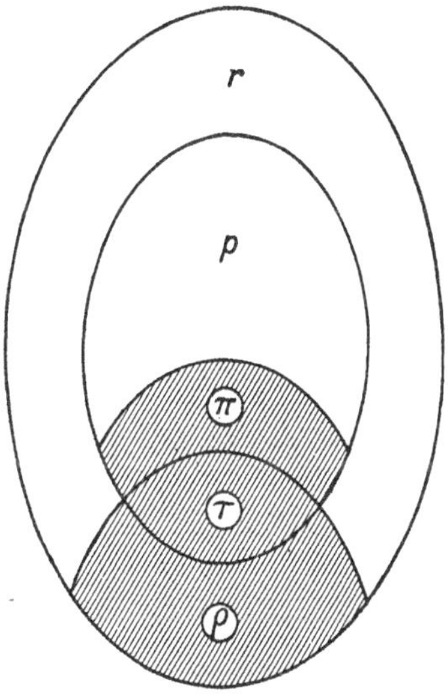

The given integers r and ρ may be any satisfying the necessary conditions

$$\rho \nless 0, \quad \rho \ngtr r, \quad r+\rho \ngtr n,$$

which include $r \nless 0$, $r \ngtr n$, $2\rho \ngtr n$; and we may assume that

$$\omega_r \equiv \overline{\underline{\alpha, a}}^{r-\rho,\,\rho}_{n},$$

where

$$\begin{bmatrix} \alpha \\ a \end{bmatrix}^n_{r-\rho,\,\rho} \overline{\underline{\alpha, a}}^{r-\rho,\,\rho}_{n} = \begin{bmatrix} 1, & 0 \\ 0, & 0 \end{bmatrix}^{r-\rho,\,\rho}_{r-\rho,\,\rho},$$

and where $\omega_\rho \equiv \overline{\underline{a}}^{\rho}_{n}$ is the core of ω_r.

When ω_τ is given, the possible values of τ are those consistent with the conditions

$$\tau \nless 0, \quad \tau \ngtr \rho; \quad \text{.................(5)}$$

and by Ex. iii a general formula for a spacelet ω_p which lies in ω_r and has ω_τ for its complete intersection with ω_ρ is

$$\omega_p \equiv \overline{\underline{b, \beta}}^{p-\tau,\,\tau}_{n} \equiv \overline{\underline{\alpha, a}}^{r-\rho,\,\rho}_{n} \overline{\underline{\begin{matrix} l, & 0 \\ \lambda, & k \end{matrix}}}^{p-\tau,\,\tau}_{r-\rho,\,\rho},$$

where $\lceil k \rfloor^{\tau}_{\rho}$ has rank τ, and $\lceil l \rfloor^{p-\tau}_{r-\rho}$ has rank $p-\tau$; therefore the possible values of p are those consistent with the conditions

$$p \nless \tau, \quad p-\tau \ngtr r-\rho, \quad \text{or} \quad p \nless \tau, \quad p+\rho-\tau \ngtr r, \quad \text{......(6)}$$

as can be deduced immediately from Ex. xvii of § 139.

If π is the extravagance of ω_p, then π is the degeneracy of the product

$$\begin{bmatrix} b \\ \beta \end{bmatrix}^{n}_{p-\tau,\,\tau} \lceil b, \beta \rfloor^{p-\tau,\,\tau}_{n} = \begin{bmatrix} e, & 0 \\ 0, & 0 \end{bmatrix}^{p-\tau,\,\tau}_{p-\tau,\,\tau}, \quad \text{where } [e]^{p-\tau}_{p-\tau} = [l]^{r-\rho}_{p-\tau} \lceil l \rfloor^{p-\tau}_{r-\rho}.$$

Let η be the rank of $[e]^{p-\tau}_{p-\tau}$. Then when p is given, it follows from § 136.1 that the possible values of η are given by

$$\eta \nless 0, \quad \eta \ngtr p-\tau, \quad \eta+r-\rho \nless 2(p-\tau).$$

Since $\pi = p-\eta$, it follows that when p is given, the possible values of π are those consistent with the conditions

$$\pi \nless \tau, \quad \pi \ngtr p, \quad \pi \ngtr (r-\rho)-p+2\tau. \quad \text{......(7)}$$

Thus the possible values of p, π and τ are those consistent with the conditions (5), (6) and (7). Since (6) is included in (7), the possible values of p, π and τ are those consistent with the conditions (4).

Ex. xiv. When τ is arbitrary in Ex. xiii, the possible values of p and π are those consistent with the conditions

$$p+\pi \ngtr r+\rho, \quad p-\pi \ngtr r-\rho, \quad \pi \nless 0, \quad \pi \ngtr p, \quad \text{......(8)}$$

which include $p \nless 0$, $p \ngtr r$; and when p and π have any assigned values consistent with (8), the possible values of τ are given by the conditions

$$\tau \nless 0, \quad \tau \ngtr p, \quad \tau \ngtr \pi, \quad 2\tau \nless (p+\pi)-(r-\rho), \quad \text{......(8')}$$

which include $\tau \nless p-r+\rho$.

Hence if ω_r is a given spacelet of rank r and extravagance ρ, and if ω_p is any spacelet of rank p and extravagance π which lies in ω_r, the possible values of p and π are those consistent with the conditions (8).

This has been proved before in Ex. xiii of § 165.

In particular if ω_π is a completely extravagant spacelet of rank π which lies in ω_r, the possible values of π are given by

$$\pi \nless 0, \quad 2\pi \ngtr r+\rho;$$

and π can be greater than ρ if $n \nless 2$; $r \nless 2$, $r \ngtr n$; and $\rho \nless 0$, $\rho \ngtr r-2$, $\rho \ngtr n-r$.

Ex. xv. If ω_p lies in ω_r and contains the core ω_ρ of ω_r, then the possible values of p and π are those consistent with the conditions

$$\pi \nless \rho, \quad \pi \ngtr p, \quad \pi \ngtr r+\rho-p, \quad \text{......(9)}$$

which include $p \nless \rho$, $p \ngtr r$.

If ω_p lies in ω_r and does not intersect the core of ω_r, then the possible values of p and π are those consistent with the conditions

$$\pi \nless 0, \quad \pi \ngtr p, \quad \pi \ngtr r-\rho-p, \quad \text{......(10)}$$

which include $p \nless 0$, $p \ngtr r-\rho$.

4. *The plenum of a spacelet.*

Let ω_r be any spacelet of rank r and extravagance ρ whose core is ω_ρ, and let ω_{n-r} be the normal to ω_r whose core is also ω_ρ. Further let $\omega_{r-\rho}$ be any (non-extravagant) spacelet of rank $r-\rho$ which lies in ω_r and does not intersect ω_ρ, and let $\omega_{n-r-\rho}$ be any (non-extravagant) spacelet of rank $n-r-\rho$ which lies in ω_{n-r} and does not intersect ω_ρ. Then since ω_r and ω_{n-r} are mutually orthogonal, we can write

$$\omega_\rho \equiv \boxed{c}^{\rho}_{n}, \quad \omega_{r-\rho} \equiv \boxed{a}^{r-\rho}_{n}, \quad \omega_{n-r-\rho} \equiv \boxed{b}^{n-r-\rho}_{n},$$

$$\omega_r \equiv \boxed{c,\ a}^{r-\rho,\ \rho}_{n}, \quad \omega_{n-r} \equiv \boxed{c,\ b}^{n-r-\rho,\ \rho}_{n},$$

where

$$\begin{bmatrix} c \\ a \\ b \end{bmatrix}^{n}_{\rho,\ r-\rho,\ n-r-\rho} \boxed{c,\ a,\ b}^{\rho,\ r-\rho,\ n-r-\rho}_{n} = \begin{bmatrix} 0, & 0, & 0 \\ 0, & 1, & 0 \\ 0, & 0, & 1 \end{bmatrix}^{\rho,\ r-\rho,\ n-r-\rho}_{\rho,\ r-\rho,\ n-r-\rho} = \begin{bmatrix} 0, & 0 \\ 0, & 1 \end{bmatrix}^{\rho,\ n-2\rho}_{\rho,\ n-2\rho}.$$

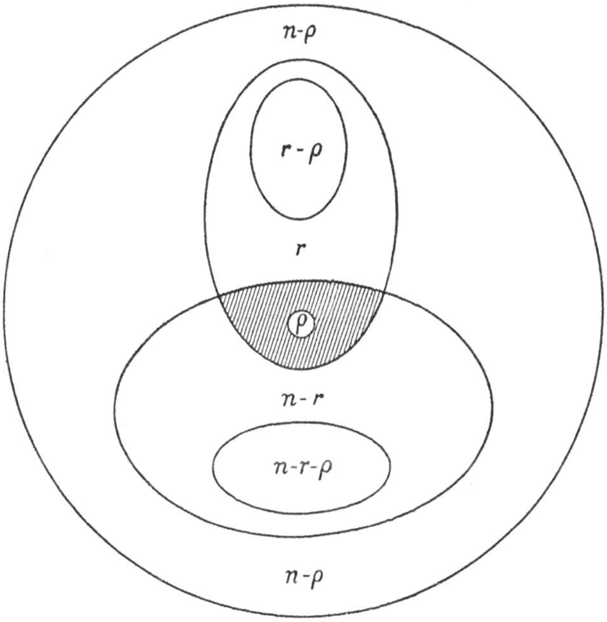

Since ω_ρ is the complete intersection of ω_r and ω_{n-r}, it follows that the join of ω_r and ω_{n-r} is the spacelet

$$\omega_{n-\rho} \equiv \boxed{c,\ a,\ b}^{\rho,\ r-\rho,\ n-r-\rho}_{n}$$

This is a plenarily extravagant spacelet of rank $n-\rho$ whose core is ω_ρ, and it is the normal to ω_ρ. We will call it the *plenum* of ω_r. Thus the core of ω_r is the complete intersection of ω_r and ω_{n-r}, the plenum of ω_r is the join of ω_r and ω_{n-r}, and the core and plenum of ω_r are mutually normal. The plenum of ω_r is the spacelet of smallest rank which contains all points of ω_r and also all points orthogonal with ω_r.

Ex. xvi. A real or non-extravagant spacelet has no core; and its plenum is the complete space ω_n.

Ex. xvii. A spacelet is plenarily extravagant when and only when it has its normal as core, or when and only when it has itself as plenum, or when and only when it contains its normal.

Ex. xviii. A spacelet is completely extravagant when and only when it has itself as core, or when and only when it has its normal as plenum, or when and only when it lies in its normal.

5. *Properties of the spacelets normal to given spacelets.*

One property deducible from Ex. ix of § 169 or from the preceding sub-articles is the following:

Two mutually normal spacelets have the same extravagance, the same core, and the same plenum. Their common core is their complete intersection, and their common plenum is their join. ....(11)

Another property can be expressed in the following two equivalent forms:

If the spacelet ω_u lies in the spacelet ω_v, then the normal to ω_u contains the normal to ω_v.

If the spacelet ω_u contains the spacelet ω_v, then the normal to ω_u lies in the normal to ω_v. ....(12)

It will be sufficient to prove the first statement. We suppose that ω_u lies in ω_v, and denote the normals to ω_u and ω_v by ω_{n-u} and ω_{n-v}. Then since all points normal to ω_v are necessarily normal to ω_u, it follows that ω_{n-v} lies in ω_{n-u}, so that ω_{n-u} contains ω_{n-v}.

The last property leads to the following theorem:

If ω_r and ω_s are any two spacelets whose complete intersection and join are respectively ω_t and ω_T, so that

$$t + T = r + s,$$

then the normals to ω_t and ω_T are respectively the join and the complete intersection of the normals to ω_r and ω_s. ....(13)

Let the normals to ω_r, ω_s, ω_t, ω_T be ω_{n-r}, ω_{n-s}, ω_{n-t}, ω_{n-T}.

Since ω_t lies in both ω_r and ω_s, therefore ω_{n-t} contains both ω_{n-r} and ω_{n-s}. Now let ω_u be any spacelet which contains both ω_{n-r} and ω_{n-s}, and let its normal be ω_{n-u}. Then since ω_u contains both ω_{n-r} and ω_{n-s}, therefore ω_{n-u} lies in both ω_r and ω_s, i.e. it lies in ω_t. It follows that ω_u contains ω_{n-t}, and must be either ω_{n-t} or a spacelet of larger dimensions than ω_{n-t}.

Thus ω_{n-t} is the spacelet of smallest dimensions which contains both ω_{n-r} and ω_{n-s}, and is the join of ω_{n-r} and ω_{n-s}.

Again since ω_T contains both ω_r and ω_s, therefore ω_{n-T} lies in both ω_{n-r} and ω_{n-s}. Now let ω_u be any spacelet which lies in both ω_{n-r} and ω_{n-s}, and let its normal be ω_{n-u}. Then since ω_u lies in both ω_{n-r} and ω_{n-s}, therefore ω_{n-u} contains both ω_r and ω_s, and by Ex. ii of § 139 we see that ω_{n-u} contains ω_T. It follows that ω_u lies in ω_{n-T}, and must be either ω_{n-T} or a spacelet of smaller dimensions than ω_{n-T}. Thus ω_{n-T} is the spacelet of greatest dimensions which lies in both ω_{n-r} and ω_{n-s}, and is the complete intersection of ω_{n-r} and ω_{n-s}.

We mention two other results which are immediately deducible from the last property.

If two spacelets are non-intersecting, their normals are mutually complementary.

If two spacelets are mutually complementary, their normals are non-intersecting. ....(14)

For if we use the same notation as before, the spacelets ω_r and ω_s are non-intersecting when and only when $t=0$, or $n-t=n$, i.e. when and only when ω_{n-t}, the join of ω_{n-r} and ω_{n-s}, is the complete space ω_n, i.e. when and only when ω_{n-r} and ω_{n-s} are mutually complementary.

NOTE 2. *Complementary theorems.*

It now appears that from any theorem concerning the joins, intersections, extravagances and cores of spacelets, we can deduce a complementary theorem by replacing each spacelet by its normal. We have two such mutually complementary theorems in Exs. i and ii.

Ex. xix. *General formulae for two mutually normal spacelets.*

From § 167.2 we see that general formulae for any two mutually normal spacelets ω_r and ω_{n-r} are

$$\omega_r \equiv [l]^n_n \begin{array}{c} \overbrace{\qquad}^{r-\rho,\,\rho} \\ 1,\ 0 \\ 0,\ 1 \\ 0,\ i \\ 0,\ 0 \\ \underbrace{\qquad}_{r-\rho,\,\rho,\,\rho,\,n-r-\rho} \end{array}, \qquad \omega_{n-r} \equiv [l]^n_n \begin{array}{c} \overbrace{\qquad}^{n-r-\rho,\,\rho} \\ 0,\ 0 \\ 0,\ 1 \\ 0,\ i \\ 1,\ 0 \\ \underbrace{\qquad}_{r-\rho,\,\rho,\,\rho,\,n-r-\rho} \end{array}, \quad \text{............(B)}$$

where $[l]^n_n$ is any square semi-unit matrix of rank n, and $i=\sqrt{-1}$.

Here ρ is the common extravagance; and the complete intersection or common core and the join or common plenum are the spacelets ω_ρ and $\omega_{n-\rho}$ given by

$$\omega_\rho \equiv [l]^n_n \begin{array}{c} \overbrace{\quad}^{\rho} \\ 0 \\ 1 \\ i \\ 0 \\ \underbrace{\quad}_{r-\rho,\,\rho,\,\rho,\,n-r-\rho} \end{array}, \qquad \omega_{n-\rho} \equiv [l]^n_n \begin{array}{c} \overbrace{\qquad\quad}^{r-\rho,\,n-r-\rho,\,\rho} \\ 1,\ 0,\ 0 \\ 0,\ 0,\ 1 \\ 0,\ 0,\ i \\ 0,\ 1,\ 0 \\ \underbrace{\qquad\quad}_{r-\rho,\,\rho,\,\rho,\,n-r-\rho} \end{array} \quad \text{............(B')}$$

Ex. xx. One of two mutually normal spacelets is completely extravagant when and only when the other is plenarily extravagant; and this is the case when and only when the two spacelets are mutually incident.

Ex. xxi. One of two mutually normal spacelets is non-extravagant when and only when the other is non-extravagant; and this is the case when and only when the two spacelets are non-intersecting, i.e. when and only when they are mutually complementary.

Ex. xxii. One of two mutually normal spacelets is real when and only when the other is real.

Ex. xxiii. *Let ω_r and ω_s be any two spacelets of ω_n, and let their normal spacelets be ω_{n-r} and ω_{n-s}. Then if x and y are the ranks of the complete intersections of ω_r, ω_s and of ω_{n-r}, ω_{n-s}, we have*

$$y+r+s=x+n.$$

This follows from (13) when we observe that $T=r+s-t$, $n-T=n-r-s+t$, i.e. $y=n-r-s+x$.

The complete intersection of ω_r and ω_s has rank x when and only when the complete intersection of ω_{n-r} and ω_{n-s} has rank $n-r-s+x$.

The complete intersection of ω_{n-r} and ω_{n-s} has rank y when and only when the complete intersection of ω_r and ω_s has rank $r+s-n+y$.

If ω_r and ω_s are non-intersecting, we must have $r+s \not> n$, and the complete intersection of ω_{n-r} and ω_{n-s} has rank $n-r-s$.

If ω_{n-r} and ω_{n-s} are non-intersecting, we must have $r+s \not< n$, and the complete intersection of ω_r and ω_s has rank $r+s-n$.

Ex. xxiv. *If X and Y are the ranks of the joins of ω_r, ω_s and of ω_{n-r}, ω_{n-s}, we have*

$$Y+r+s=X+n.$$

Ex. xxv. *The complete intersection and the join of the normals to any number of given spacelets are respectively the normals to the join and the complete intersection of the given spacelets themselves.*

We will denote the given spacelets by $a, b, \dots c$; their complete intersection and join by u and v; and the normals to $a, b, \dots c, u, v$ by $A, B, \dots C, U, V$.

Let the complete intersection of $A, B, \dots C$ be the spacelet x whose normal is X. Because x lies in each of the spacelets $A, B, \dots C$, therefore X contains each of the spacelets $a, b, \dots c$, and therefore also the spacelet v; consequently x lies in V. Because v contains each of the spacelets $a, b, \dots c$, therefore V lies in each of the spacelets $A, B, \dots C$; consequently V lies in x; it follows that x is the spacelet V.

Again let the join of $A, B, \dots C$ be the spacelet y whose normal is Y. Then by similar reasoning we can show that y contains U, and that U contains y. It follows that y is the spacelet U.

NOTE 3. *Interpretation of the terms 'extravagant', 'core', 'orthogonal' and 'normal'.*

The locus of those points $\omega_1 \equiv \overline{\underline{x}}_n$ of the homogeneous space ω_n which satisfy the equation

$$x_1^2+x_2^2+\dots+x_n^2=0$$

is called the *absolute quadric surface* of ω_n.

It will be shown in the chapters on Projective Space that the terms mentioned above indicate relations to the absolute quadric.

A point is *extravagant* when and only when it lies on the absolute quadric.

Two spacelets are *mutually orthogonal* when and only when they are *mutually conjugate* with respect to the absolute quadric.

Two spacelets are *mutually normal* when and only when they are *mutually polar* with respect to the absolute quadric.

The *core* of any spacelet ω_r is the locus of the double points on the curve of intersection of ω_r with the absolute quadric, and may be called the *spacelet of contact* of ω_r with the absolute quadric.

The *extravagance* of any spacelet ω_r is the greatest number of unconnected double points on the curve of intersection of ω_r with the absolute quadric, or the rank of the spacelet of contact of ω_r with the absolute quadric. It may be called the *rank of contact* of ω_r with the absolute quadric.

A spacelet ω_r is *extravagant* or *non-extravagant* according as its curve of intersection with the absolute quadric (which is a quadric curve) has or has not double points, i.e. according as it *touches* or *does not touch* the absolute quadric.

A *completely extravagant spacelet* is one every point of which lies on the absolute quadric, i.e. it is a *generating spacelet* of the absolute quadric.

§ 171. Standard representations of completely extravagant and plenarily extravagant spacelets.

Theorem I. *Every completely extravagant spacelet ω_r of rank r in homogeneous space ω_n can be represented in the standard form given by*

$$\omega_r \equiv \overline{\underline{w}}_{n}^{r}, \quad \overline{\underline{w}}_{n}^{r} = \overline{\underline{u}}_{n}^{r} + i\,\overline{\underline{v}}_{n}^{r} = \overline{\underline{u,\ v}}_{n}^{r,\ r} \begin{bmatrix} 1 \\ i \end{bmatrix}_{r,\ r}^{r}, \qquad \text{..................(A)}$$

where $i=\sqrt{-1}$, and $\overline{\underline{u}}_{n}^{r}$ and $\overline{\underline{v}}_{n}^{r}$ are two mutually orthogonal real semi-unit matrices of rank r, so that

$$\begin{bmatrix} u \\ v \end{bmatrix}_{r,\ r}^{n} \overline{\underline{u,\ v}}_{n}^{r,\ r} = \begin{bmatrix} 1, & 0 \\ 0, & 1 \end{bmatrix}_{r,\ r}^{r,\ r} = [1]_{2r}^{2r},$$

or

$$[u]_{r}^{n} \overline{\underline{u}}_{n}^{r} = [v]_{r}^{n} \overline{\underline{v}}_{n}^{r} = [1]_{r}^{r}, \quad [u]_{r}^{n} \overline{\underline{v}}_{n}^{r} = [v]_{r}^{n} \overline{\underline{u}}_{n}^{r} = 0.$$

One proof of this theorem has been given in § 159. We now give another proof, using the following lemma.

Lemma. *Every extravagant point ω_1 of homogeneous space ω_n can be represented in the form given by*

$$\omega_1 \equiv \overline{\underline{w}}_{n}, \quad \overline{\underline{w}}_{n} = \overline{\underline{u}}_{n} + i\,\overline{\underline{v}}_{n}$$

where $\overline{\underline{u}}_{n}$ and $\overline{\underline{v}}_{n}$ are real matrices of rank 1, *and*

$$\begin{bmatrix} u_1\, u_2 \dots u_n \\ v_1\, v_2 \dots v_n \end{bmatrix} \begin{bmatrix} u_1 & v_1 \\ u_2 & v_2 \\ \dots & \dots \\ u_n & v_n \end{bmatrix} = \begin{bmatrix} 1 & 0 \\ 0 & 1 \end{bmatrix},$$

or

$$u_1^2 + u_2^2 + \dots + u_n^2 = v_1^2 + v_2^2 + \dots + v_n^2 = 1, \quad u_1 v_1 + u_2 v_2 + \dots + u_n v_n = 0.$$

This is that special case of the above theorem which is proved first in § 159.

Since ω_r is completely extravagant, we have $2r \not> n$.

Let $\omega_{n-r} \equiv \ulcorner b \lrcorner_{n}^{n-r}$ be the spacelet normal to ω_r, so that $\ulcorner b \lrcorner_{n}^{n-r}$ is a plenarily extravagant matrix of rank $n-r$ whose core is ω_r.

Then we have $\omega_r \equiv \ulcorner w \lrcorner_{n}^{r}$, where $\ulcorner w \lrcorner_{n}^{r}$ is any solution of rank r of the equation $[b]_{n-r}^{n} \ulcorner w \lrcorner_{n}^{r} = 0$, so that the vertical rows of $\ulcorner w \lrcorner_{n}^{r}$ are any complete set of r unconnected solutions of the equation $[b]_{n-r}^{n} \ulcorner x \lrcorner_{n} = 0$.

Using for the values $1, 2, \dots r$ of s the notations

$$w_s = \begin{bmatrix} w_{s1} \\ w_{s2} \\ \vdots \\ w_{sn} \end{bmatrix} = \begin{bmatrix} u_{s1} \\ u_{s2} \\ \vdots \\ u_{sn} \end{bmatrix} + i \begin{bmatrix} v_{s1} \\ v_{s2} \\ \vdots \\ v_{sn} \end{bmatrix} = u_s + i v_s, \quad \psi_s = \begin{bmatrix} b \\ u \\ v \end{bmatrix}_{n-r,\, s,\, s}^{n},$$

where the u's and v's are real, and writing

$$\psi = \psi_0 = [b]_{n-r}^{n},$$

we can determine a matrix $\ulcorner w \lrcorner_{n}^{r}$ having the properties stated in the theorem by finding in succession non-zero solutions $w_1, w_2, \dots w_s, w_{s+1}, \dots w_r$ of the equations

$$\psi \ulcorner x \lrcorner_{n} = 0, \quad \psi_1 \ulcorner x \lrcorner_{n} = 0, \dots \psi_{s-1} \ulcorner x \lrcorner_{n} = 0, \quad \psi_s \ulcorner x \lrcorner_{n} = 0, \dots \psi_{r-1} \ulcorner x \lrcorner_{n} = 0.$$

To prove this suppose that $w_1, w_2, \dots w_s$ have been determined in this manner, and are such that $u_1, u_2, \dots u_s$ and $v_1, v_2, \dots v_s$ are real, and

$$\begin{bmatrix} u \\ v \end{bmatrix}_{s,\, s}^{n} \ulcorner u, v \lrcorner_{n}^{s,\, s} = [1]_{2s}^{2s}.$$

Then the matrix $\ulcorner w \lrcorner_{n}^{s} = \ulcorner u, v \lrcorner_{n}^{s,\, s} \begin{bmatrix} 1 \\ i \end{bmatrix}_{s,\, s}^{s}$ has rank s, and $[b]_{n-r}^{n} \ulcorner w \lrcorner_{n}^{s} = 0$.

Therefore the spacelet $\omega_s \equiv \ulcorner w \lrcorner_{n}^{s}$ is orthogonal with ω_{n-r} and lies in ω_r, i.e. in the core of ω_{n-r}. Consequently the horizontal rows of $[w]_{s}^{n}$ are connected with the horizontal rows of $[b]_{n-r}^{n}$, and there exists a relation of the form $[w]_{s}^{n} = [h]_{s}^{n-r} [b]_{n-r}^{n}$. It follows that the horizontal rows of $[v]_{s}^{n}$ are connected with the horizontal rows of $\begin{bmatrix} b \\ u \end{bmatrix}_{n-r,\, s}^{n}$, and that ψ_s has the same rank as $\begin{bmatrix} b \\ u \end{bmatrix}_{n-r,\, s}^{n}$.

Thus the rank of ψ_s cannot exceed $n-r+s$.

If $s<r$, or $n-r+s<n$, there is a connection between the vertical rows of ψ_s, and the equation $\psi_s \ulcorner x \lrcorner_{n} = 0$ admits of a non-zero solution. Let w_{s+1} be any non-zero solution of this equation. Then w_{s+1}, being a solution of the equation $[b]_{n-r}^{n} \ulcorner x \lrcorner_{n} = 0$, lies in the core of ω_{n-r} and is extravagant. Therefore by the lemma we can choose it to have the

form $w_{s+1}=u_{s+1}+iv_{s+1}$, where u_{s+1} and v_{s+1} are real mutually orthogonal semi-unit matrices of rank 1. Since $\begin{bmatrix} u \\ v \end{bmatrix}_{2s}^{n} w_{s+1}=0$, u_{s+1} and v_{s+1} are both orthogonal with all the matrices $u_1, u_2, \dots u_s, v_1, v_2, \dots v_s$, and we then have

$$\begin{bmatrix} u \\ v \end{bmatrix}_{s+1,\, s+1}^{n} \overline{\underline{u,\ v}}_{n}^{s+1,\, s+1} = [1]_{2s+2}^{2s+2}.$$

Also the matrix $\overline{\underline{w}}_{n}^{s+1} = \overline{\underline{u,\ v}}_{n}^{s+1,\, s+1} \begin{bmatrix} 1 \\ i \end{bmatrix}_{s+1,\, s+1}^{s+1}$ has rank $s+1$, and $[b]_{n-r}^{n} \overline{\underline{w}}_{n}^{s+1}=0$.

It follows that we can in succession determine $w_1, w_2, \dots w_r$ so that $\overline{\underline{w}}_{n}^{r}$ is a solution of rank r of the equation $[b]_{n-r}^{n} \overline{\underline{w}}_{n}^{r}=0$, and has the properties mentioned in the theorem, and we then have $\omega_r \equiv \overline{\underline{w}}_{n}^{r}$.

Ex. i. If ω_ρ is any completely extravagant spacelet of rank ρ in homogeneous space ω_n, so that $2\rho \not> n$, it can be represented in the forms

$$\omega_\rho \equiv [l]_{n}^{2\rho} \begin{bmatrix} 1 \\ i \end{bmatrix}_{\rho,\, \rho}^{\rho}, \quad \omega_\rho \equiv [l]_{n}^{n} \begin{bmatrix} 1 \\ i \\ 0 \end{bmatrix}_{\rho,\, \rho,\, n-2\rho}^{\rho}, \quad \text{..................(A}')$$

where $i=\sqrt{-1}$, and where $[l]_{n}^{2\rho}$ and $[l]_{n}^{n}$ are *real* semi-unit matrices of ranks 2ρ and n.

This follows from Theorem I when we write $\overline{\underline{u,\ v}}_{n}^{r,\, r} = [l]_{n}^{2r}$.

Ex. ii. *If ρ is any positive integer greater than* 1, *and if $i=\sqrt{-1}$, then*

$$\det \begin{bmatrix} 1, & i \\ i, & 1 \end{bmatrix}_{\rho,\, \rho}^{\rho,\, \rho} = 2^\rho \neq 0, \quad \det \begin{bmatrix} 1, & 1 \\ i, & -i \end{bmatrix}_{\rho,\, \rho}^{\rho,\, \rho} = (-2i)^\rho \neq 0.$$

Let

$$[e]_{2\rho}^{2\rho} = \begin{bmatrix} 1, & i \\ i, & 1 \end{bmatrix}_{\rho,\, \rho}^{\rho,\, \rho}, \quad \Delta_{2\rho}=(e)_{2\rho}^{2\rho}=\det \begin{bmatrix} 1, & i \\ i, & 1 \end{bmatrix}_{\rho,\, \rho}^{\rho,\, \rho}.$$

Then since $[e]_{2\rho+2}^{2\rho+2}$ is a symmetric derangement of $\begin{bmatrix} e, & 0 \\ 0, & e \end{bmatrix}_{2,\, 2\rho}^{2,\, 2\rho}$, we have

$$\Delta_{2\rho+2}=(e)_{2}^{2}\,(e)_{2\rho}^{2\rho}=\Delta_2\Delta_{2\rho}=2\Delta_{2\rho}.$$

It follows that $\Delta_{2\rho}=2^\rho$; and the second result can be deduced.

Ex. iii. *If $\overline{\underline{w}}_{n}^{\rho}$ is a completely extravagant matrix of rank ρ, so that $2\rho \not> n$, and if we express it in the form*

$$\overline{\underline{w}}_{n}^{\rho} = \overline{\underline{u}}_{n}^{\rho} + i\,\overline{\underline{v}}_{n}^{\rho} = \overline{\underline{u,\ v}}_{n}^{\rho,\, \rho} \begin{bmatrix} 1 \\ i \end{bmatrix}_{\rho,\, \rho}^{\rho}; \quad \text{..........................(1)}$$

where $\overline{\underline{u}}_{n}^{\rho}$ and $\overline{\underline{v}}_{n}^{\rho}$ are real, and $i=\sqrt{-1}$, then:

(1) *The real matrix $\overline{\underline{u,\ v}}_{n}^{\rho,\, \rho}$ is undegenerate and has rank 2ρ.*

(2) *There exists a real and definite undegenerate symmetric matrix* $[a]_{\rho}^{\rho}$ *and a real skew-symmetric matrix* $[\beta]_{\rho}^{\rho}$ *such that*

$$[u]_{\rho}^{n}\,\overline{\underline{u}}_{n}^{\rho}=[v]_{\rho}^{n}\,\overline{\underline{v}}_{n}^{\rho}=[a]_{\rho}^{\rho},\qquad [u]_{\rho}^{n}\,\overline{\underline{v}}_{n}^{\rho}=-[v]_{\rho}^{n}\,\overline{\underline{u}}_{n}^{\rho}=[\beta]_{\rho}^{\rho}.$$

If the real matrix $\overline{\underline{u,\,v}}_{n}^{\rho,\,\rho}$ were degenerate, there would exist a real non-vanishing matrix $[h,\,k]_{\rho,\,\rho}$ such that

$$\overline{\underline{u,\,v}}_{n}^{\rho,\,\rho}\,\overline{\underline{\begin{matrix}h\\k\end{matrix}}}_{\rho,\,\rho}=0,\quad\text{or}\quad \overline{\underline{u}}_{n}^{\rho}\,\overline{\underline{h}}_{\rho}+\overline{\underline{v}}_{n}^{\rho}\,\overline{\underline{k}}_{\rho}=0.$$

We should then have

$$\overline{\underline{w}}_{n}^{\rho}\left\{\overline{\underline{h}}_{\rho}-i\,\overline{\underline{k}}_{\rho}\right\}=i\left\{\overline{\underline{v}}_{n}^{\rho}\,\overline{\underline{h}}_{\rho}-\overline{\underline{u}}_{n}^{\rho}\,\overline{\underline{k}}_{\rho}\right\}.$$

This however is impossible; for the matrix on the left, being a non-vanishing one-rowed matrix connected with the vertical rows of $\overline{\underline{w}}_{n}^{\rho}$, must be extravagant, whereas the matrix on the right, being purely imaginary, cannot be extravagant. Accordingly the first part of the theorem is true.

We obtain the second part of the theorem by equating to zero the real and purely imaginary portions of the matrix on the left in the equation $[w]_{\rho}^{n}\,\overline{\underline{w}}_{n}^{\rho}=0$.

Further if

$$\overline{\underline{w'}}_{n}^{\rho}=\overline{\underline{u}}_{n}^{\rho}-i\,\overline{\underline{v}}_{n}^{\rho}=\overline{\underline{u,\,v}}_{n}^{\rho,\,\rho}\begin{bmatrix}1\\-i\end{bmatrix}_{\rho,\,\rho}, \qquad\text{..........................(2)}$$

then:

(1) $\overline{\underline{w'}}_{n}^{\rho}$ *is also a completely extravagant matrix of rank* ρ.

(2) $\overline{\underline{w,\,w'}}_{n}^{\rho,\,\rho}$ *and* $\overline{\underline{u,\,v}}_{n}^{\rho,\,\rho}$ *are two mutually equivalent undegenerate matrices of rank* 2ρ.

Since $\overline{\underline{u,\,v}}_{n}^{\rho,\,\rho}$ has rank 2ρ, the matrix $\overline{\underline{w'}}_{n}^{\rho}$ has rank ρ, and we see at once that $[w']_{\rho}^{n}\,\overline{\underline{w'}}_{n}^{\rho}=0$. The second result follows from Ex. ii and the equation

$$\overline{\underline{w,\,w'}}_{n}^{\rho,\,\rho}=\overline{\underline{u,\,v}}_{n}^{\rho,\,\rho}\begin{bmatrix}1, & 1\\ i, & -i\end{bmatrix}_{\rho,\,\rho}^{\rho,\,\rho}.$$

Lastly the join of the two completely extravagant spacelets $\omega_{\rho}\equiv\overline{\underline{w}}_{n}^{\rho}$, $\omega_{\rho}'\equiv\overline{\underline{w'}}_{n}^{\rho}$ *is the real spacelet* $\omega_{2\rho}$ *of rank* 2ρ *given by*

$$\omega_{2\rho}\equiv\overline{\underline{u,\,v}}_{n}^{\rho,\,\rho}\equiv\overline{\underline{w,\,w'}}_{n}^{\rho,\,\rho}$$

Ex. iv. If $2\rho \not> n$, every real spacelet $\omega_{2\rho}$ of rank 2ρ can be represented in the form $\omega_{2\rho}\equiv\overline{\underline{u,\,v}}_{n}^{\rho,\,\rho}$, where $\overline{\underline{u}}_{n}^{\rho}$ and $\overline{\underline{v}}_{n}^{\rho}$ are two mutually orthogonal real matrices of rank ρ. It is then the join of the two completely extravagant spacelets

$$\omega_{\rho}\equiv\overline{\underline{u}}_{n}^{\rho}+i\,\overline{\underline{v}}_{n}^{\rho},\qquad \omega_{\rho}'\equiv\overline{\underline{u}}_{n}^{\rho}-i\,\overline{\underline{v}}_{n}^{\rho}$$

Theorem II. *If ω_ρ and $\omega_{n-\rho}$ are any two mutually normal spacelets of homogeneous space ω_n of which the first is completely extravagant and the second plenarily extravagant, so that $2\rho \not> n$, then we can represent ω_ρ and $\omega_{n-\rho}$ in the standard forms*

$$\omega_\rho \equiv [l]_n^n \left\lceil \begin{matrix} 1 \\ i \\ 0 \end{matrix} \right\rfloor_{\rho,\rho,n-2\rho}^{\rho}, \quad \omega_{n-\rho} \equiv [l]_n^n \left\lceil \begin{matrix} 0, & 1 \\ 0, & i \\ 1, & 0 \end{matrix} \right\rfloor_{\rho,\rho,n-2\rho}^{n-2\rho,\rho}, \quad \text{.....................(B)}$$

where $[l]_n^n$ is a real square semi-unit matrix, and $i=\sqrt{-1}$.

This follows from Ex. i above and the general formulae for two mutually normal spacelets given in Ex. xix of § 170.

Theorem III. *Every plenarily extravagant spacelet ω_r of rank r in homogeneous space ω_n and its core ω_{n-r} can be represented in the standard forms*

$$\omega_r \equiv [l]_n^n \left\lceil \begin{matrix} 1, & 0 \\ 0, & 1 \\ 0, & i \end{matrix} \right\rfloor_{2r-n,\,n-r,\,n-r}^{2r-n,\,n-r}, \quad \omega_{n-r} \equiv [l]_n^n \left\lceil \begin{matrix} 0 \\ 1 \\ i \end{matrix} \right\rfloor_{2r-n,\,n-r,\,n-r}^{n-r}, \quad \text{...........(C)}$$

where $[l]_n^n$ is a square semi-unit matrix, and $i=\sqrt{-1}$.

This follows from Theorem II when we replace ρ by $n-r$. In this case we have $2r \not< n$.

Ex. v. *Every plenarily extravagant spacelet $\omega_{n-\rho}$ of rank $n-\rho$ can be represented in many ways as the join of its core $\omega_\rho \equiv \lceil w \rfloor_n^\rho$ and a real spacelet $\omega_{n-2\rho} \equiv \lceil c \rfloor_n^{n-2\rho}$, where $\lceil c \rfloor_n^{n-2\rho}$ is real and has rank $n-2\rho$, in the form*

$$\omega_{n-\rho} \equiv \lceil w, c \rfloor_n^{\rho,\,n-2\rho} \quad \text{...(C')}$$

If

$$\lceil w \rfloor_n^\rho = \lceil u \rfloor_n^\rho + i \lceil v \rfloor_n^\rho = \lceil u, v \rfloor_n^{\rho,\rho} \begin{bmatrix} 1 \\ i \end{bmatrix}_{\rho,\rho}^{\rho}, \quad \omega_\rho \equiv \lceil w \rfloor_n^\rho,$$

$$\lceil w' \rfloor_n^\rho = \lceil u \rfloor_n^\rho - i \lceil v \rfloor_n^\rho = \lceil u, v \rfloor_n^{\rho,\rho} \begin{bmatrix} 1 \\ -i \end{bmatrix}_{\rho,\rho}^{\rho}, \quad \omega_\rho' \equiv \lceil w' \rfloor_n^\rho,$$

then the real spacelet $\omega_{n-2\rho}$ is the normal to the real spacelet

$$\omega_{2\rho} \equiv \lceil w, w' \rfloor_n^{\rho,\rho} \equiv \lceil u, v \rfloor_n^{\rho,\rho}, \quad \text{...............................(3)}$$

which is the join of the two completely extravagant spacelets ω_ρ and ω_ρ'.

It has been shown in Ex. iii that the matrices on the right in (3) are two mutually equivalent undegenerate matrices of rank 2ρ, and that ω_ρ' is a completely extravagant spacelet of rank ρ.

By § 94 we can determine a real matrix $\lceil c \rfloor_n^{n-2\rho}$ of rank $n-2\rho$ satisfying the equation $\begin{bmatrix} u \\ v \end{bmatrix}_{\rho,\rho}^{n} \lceil c \rfloor_n^{n-2\rho} = 0$ and the equivalent equation $\begin{bmatrix} w \\ w' \end{bmatrix}_{\rho,\rho}^{n} \lceil c \rfloor_n^{n-2\rho} = 0$. Then by Ex. vi of § 170 the real matrix $\underline{\overline{u, v, c}}{}^{\rho, \rho, n-2\rho}_{n}$ and the equivalent matrix $\underline{\overline{w, w', c}}{}^{\rho, \rho, n-2\rho}_{n}$ are undegenerate and have rank n. Therefore $\underline{\overline{w, c}}{}^{\rho, n-2\rho}_{n}$ is a matrix of rank $n-\rho$ orthogonal with and normal to $\underline{\overline{w}}{}^{\rho}_{n}$, i.e. $\omega_{n-\rho}$ can be represented in the form (C′).

Conversely whenever $\omega_{n-\rho}$ is represented in the form (C′) and $\underline{\overline{c}}{}^{n-2\rho}_{n}$ is real, then the spacelet $\omega_{n-2\rho} \equiv \underline{\overline{c}}{}^{n-2\rho}_{n}$ is orthogonal with $\underline{\overline{w}}{}^{\rho}_{n}$ and therefore with both $\underline{\overline{u}}{}^{\rho}_{n}$ and $\underline{\overline{v}}{}^{\rho}_{n}$. It is therefore orthogonal with $\omega_{2\rho}$, and must be the normal to $\omega_{2\rho}$.

When the matrix $\underline{\overline{w}}{}^{\rho}_{n}$ is given, the completely extravagant spacelet ω_{ρ}' and the two mutually normal real spacelets $\omega_{2\rho}$ and $\omega_{n-2\rho}$ are known. Further since $\underline{\overline{w, w', c}}{}^{\rho, \rho, n-2\rho}_{n}$ has rank n, we see that:

The completely extravagant spacelet ω_{ρ}' does not intersect $\omega_{n-\rho}$, and is complementary to $\omega_{n-\rho}$.

NOTE. *The anti-cores of any spacelet.*

If ω_r is any spacelet of rank r and extravagance ρ whose core is ω_{ρ}, and if

$$\omega_{\rho} \equiv \underline{\overline{u}}{}^{\rho}_{n} + i\,\underline{\overline{v}}{}^{\rho}_{n}, \qquad \omega_{\rho}' \equiv \underline{\overline{u}}{}^{\rho}_{n} - i\,\underline{\overline{v}}{}^{\rho}_{n},$$

where $\underline{\overline{u}}{}^{\rho}_{n}$ and $\underline{\overline{v}}{}^{\rho}_{n}$ are real, and $i=\sqrt{-1}$, then we will call ω_{ρ}' an anti-core of ω_r. Since two mutually normal spacelets have the same core, they have the same anti-cores.

Let ω_r and ω_{n-r} be any two mutually normal spacelets of ranks r and $n-r$ whose common core is ω_{ρ}. Then their join is the plenarily extravagant spacelet $\omega_{n-\rho}$ normal to ω_{ρ}, and we see from the preceding examples that:

Every anti-core of ω_r and ω_{n-r} is a completely extravagant spacelet which does not intersect and is complementary to their join or common plenum $\omega_{n-\rho}$.

§ 172. Unconnected mutually orthogonal solutions of any system of homogeneous linear algebraic equations.

The results obtained in § 94 will now be generalised and completed. We will consider any system of r unconnected homogeneous linear algebraic equations in the n variables $x_1, x_2, \ldots x_n$, regarding these r equations as equivalent to the irreducible matrix equation

$$[a]^{n}_{r}\,\underline{\overline{x}}_{n} = 0, \qquad \text{(A)}$$

where $[a]^{n}_{r}$ is a given undegenerate matrix of rank r. The extravagance of $[a]^{n}_{r}$ will be denoted by ρ, ρ being some integer satisfying the conditions $\rho \nless 0$ $\rho \ngtr r$, $\rho \ngtr n-r$.

We will suppose that $\overline{\underline{z}}{}^{n-r}_{n}$ is any undegenerate matrix of rank $n-r$ whose successive vertical rows $z_1, z_2, \ldots z_{n-r}$ are any complete set of $n-r$ unconnected solutions of the equation (A); and that $\overline{\underline{x}}{}^{n-r}_{n}$ is any undegenerate matrix of rank $n-r$ whose successive vertical rows $x_1, x_2, \ldots x_{n-r}$ are any complete set of $n-r$ unconnected *mutually orthogonal* solutions of the equation (A).

Then $\overline{\underline{z}}{}^{n-r}_{n}$ is any undegenerate matrix of rank $n-r$ which has the indicated orders and satisfies the equation $[a]^n_r\, \overline{\underline{z}}{}^{n-r}_{n} = 0$, i.e. it is any undegenerate matrix normal to $[a]^n_r$, and it can be determined by the methods described in §§ 90 and 132. Since it is normal to $\overline{\underline{a}}{}^{r}_{n}$, it must always have the same extravagance ρ as $\overline{\underline{a}}{}^{r}_{n}$ or $[a]^n_r$.

When $\overline{\underline{z}}{}^{n-r}_{n}$ has been determined, $\overline{\underline{x}}{}^{n-r}_{n}$ is any undegenerate matrix similar and equivalent to $\overline{\underline{z}}{}^{n-r}_{n}$ whose long rows are mutually orthogonal, and we can determine it in the way shown in § 168. Since it is normal to $\overline{\underline{a}}{}^{r}_{n}$, it also must always have extravagance ρ, and the quasi-scalar matrix $[x]^n_{n-r}\, \overline{\underline{x}}{}^{n-r}_{n}$ must always have degeneracy ρ; i.e. ρ of the $n-r$ mutually orthogonal solutions $x_1, x_2, \ldots x_{n-r}$ must always be extravagant, and the remaining $n-r-\rho$ must always be non-extravagant.

Hence we have the following theorems:

Theorem I. *When $[a]^n_r$ is a given undegenerate matrix of rank r lying in any domain of rationality Ω we can always determine a complete set of $n-r$ unconnected mutually orthogonal solutions of the equation* (A) *all of which lie in Ω.*

Theorem II a. *If $[a]^n_r$ is non-extravagant, then all these $n-r$ solutions must be non-extravagant.*

Theorem II b. *If $[a]^n_r$ has extravagance ρ, then ρ of these $n-r$ solutions must be extravagant and the remaining $n-r-\rho$ of them must be non-extravagant.*

Theorem III. *In the general case when $[a]_r^n$ has extravagance ρ, the $n-r-\rho$ non-extravagant solutions can always be so chosen as to have unit intensities, i.e. so as to be unit solutions; but they then do not necessarily lie in Ω, except in the special case when Ω is the domain of all real numbers, in which case $\rho=0$.*

The special case in which Ω is the domain of all real numbers has been considered independently in § 94.

Ex. i. *All non-zero solutions of the equation* (A) *are extravagant when and only when $\rho=n-r$, i.e. when and only when $[a]_r^n$ is plenarily extravagant.*

First suppose that all non-zero solutions are extravagant. Then $\lceil z \rceil_n^{n-r}$ must be completely extravagant, and therefore $\rho=n-r$. Or again every non-vanishing solution $\lceil x \rceil_n = \lceil z \rceil_n^{n-r} \lceil k \rceil_{n-r}$ is extravagant; therefore the equation $[k]_{n-r}[z]_{n-r}^n \lceil z \rceil_n^{n-r} \lceil k \rceil_{n-r}=0$ is an identity in the elements of $[k]_{n-r}$; therefore by § 85 we must have $[z]_{n-r}^n \lceil z \rceil_n^{n-r}=0$; therefore $\lceil z \rceil_n^{n-r}$ is completely extravagant, i.e. $\rho=n-r$.

Next suppose that $\rho=n-r$. Then $\lceil z \rceil_n^{n-r}$ must be completely extravagant, and every non-vanishing solution $\lceil x \rceil_n = \lceil z \rceil_n^{n-r} \lceil k \rceil_{n-r}$ is extravagant.

Ex. ii. *If $y_1, y_2, \dots y_s$, the successive vertical rows of $\lceil y \rceil_n^s$, are any s unconnected mutually orthogonal non-extravagant solutions of the equation* (A), *then $\phi=\lceil a, y \rceil_n^{r,s}$ is an undegenerate matrix of rank $r+s$ and extravagance ρ which has the same cores as $\lceil a \rceil_n^r$.*

If there existed any connection between the vertical rows of ϕ such as

$$\lceil a, y \rceil_n^{r,s} \left\lceil \begin{matrix} h \\ k \end{matrix} \right\rceil_{r,s} = 0, \quad \text{or} \quad \lceil a \rceil_n^r \lceil h \rceil_r = -\lceil y \rceil_n^s \lceil k \rceil_s,$$

then $\lceil h \rceil_r$ and $\lceil k \rceil_s$ would both be non-zero matrices, and by prefixing $[y]_s^n$ on both sides we should obtain

$$[y]_s^n \lceil y \rceil_n^s \lceil k \rceil_s = 0.$$

But this is impossible since $[y]_s^n \lceil y \rceil_n^s$ is an undegenerate quasi-scalar matrix. Therefore ϕ has rank $r+s$. The rest of the theorem now follows from Ex. viii of § 170.

Hence all non-zero solutions of the equation $\begin{bmatrix} a \\ y \end{bmatrix}_{r,s}^n \lceil x \rceil_n = 0$ are extravagant when and only when $\rho=n-r-s$, or $s=n-r-\rho$.

This last result follows from Ex. i. The first result is a generalisation of Ex. iv of § 90.

Ex. iii. *If* $y_1, y_2, \ldots y_s$, *the successive vertical rows of* $\overline{\underline{y}}{}^{s}_{n}$, *are any* s *unconnected mutually orthogonal non-extravagant solutions of the equation* (A), *and if* $\overline{\underline{\eta}}{}^{t}_{n}$ *is any undegenerate matrix of rank* t *whose vertical rows* $\eta_1, \eta_2, \ldots \eta_t$ *are connected with the vertical rows of the cores of* $\overline{\underline{a}}{}^{r}_{n}$, *then the matrix*

$$\phi = \overline{\underline{y, \eta}}{}^{s,\, t}_{n}$$

is undegenerate and has rank $s+t$; *also* $y_1, y_2, \ldots y_s, \eta_1, \eta_2, \ldots \eta_t$ *are* $s+t$ *unconnected mutually orthogonal solutions of the equation* (A), *the last* t *of them being of course extravagant.*

Since the horizontal rows of $[\eta]^{n}_{t}$ are connected with the horizontal rows of the cores of $[a]^{n}_{r}$, we have

$$[a]^{n}_{r}\, \overline{\underline{\eta}}{}^{t}_{n} = 0, \quad [\eta]^{n}_{t}\, \overline{\underline{\eta}}{}^{t}_{n} = 0, \quad [\eta]^{n}_{t}\, \overline{\underline{y}}{}^{s}_{n} = 0.$$

Therefore $y_1, y_2, \ldots y_s, \eta_1, \eta_2, \ldots \eta_t$ are mutually orthogonal solutions of the equation (A).

Again if there existed any connection between the vertical rows of ϕ such as

$$\overline{\underline{y, \eta}}{}^{s,\, t}_{n} \; \overline{\underline{\begin{matrix} h \\ k \end{matrix}}}_{s,\, t} = 0, \quad \text{or} \quad \overline{\underline{y}}{}^{s}_{n}\, \overline{\underline{h}}_{s} = -\overline{\underline{\eta}}{}^{t}_{n}\, \overline{\underline{k}}_{t},$$

then $\overline{\underline{h}}_{s}$ and $\overline{\underline{k}}_{t}$ would both be non-zero matrices, and by prefixing $[y]^{n}_{s}$ on both sides we should obtain the equation

$$[y]^{n}_{s}\, \overline{\underline{y}}{}^{s}_{n} \,.\, \overline{\underline{h}}_{s} = 0.$$

This however is impossible because the prefactor on the left is an *undegenerate* quasi-scalar matrix.

Therefore ϕ has rank $s+t$; and it follows that $y_1, y_2, \ldots y_s, \eta_1, \eta_2, \ldots \eta_t$ are $s+t$ *unconnected* mutually orthogonal solutions of the equation (A).

Ex. iv. *If* $y_1, y_2, \ldots y_{n-r-\rho}$,—*the successive vertical rows of* $\overline{\underline{y}}{}^{n-r-\rho}_{n}$—, *are any* $n-r-\rho$ *unconnected mutually orthogonal non-extravagant solutions of the equation* (A), *then the matrix* $\phi = \overline{\underline{a, y}}{}^{r,\, n-r-\rho}_{n}$ *is a plenarily extravagant matrix of rank* $n-\rho$ *and extravagance* ρ, *and every solution* $\overline{\underline{x}}_{n}$ *of the equation*

$$\begin{bmatrix} a \\ y \end{bmatrix}^{n}_{r,\, n-r-\rho} \overline{\underline{x}}_{n} = 0 \quad \ldots\ldots\ldots\ldots\ldots\ldots\ldots\ldots\ldots\ldots \text{(B)}$$

is connected with the vertical rows of the cores of $\overline{\underline{a}}{}^{r}_{n}$, *and is extravagant when not zero.*

By Ex. ii the matrix $\phi = \overline{\underline{a, y}}{}^{r,\, n-r-\rho}_{n}$ is a plenarily extravagant matrix of rank $n-\rho$ and extravagance ρ whose cores are the cores of $\overline{\underline{a}}{}^{r}_{n}$. Let $\overline{\underline{\gamma}}{}^{\rho}_{n}$ be any matrix normal to ϕ

Then every solution $\lceil x\rfloor_n$ of the equation (B) is extravagant when not zero, and is connected with the vertical rows of $\lceil \gamma\rfloor^{\rho}_{n}$. But $\lceil \gamma\rfloor^{\rho}_{n}$ must be a core of ϕ and therefore a core of $\lceil a\rfloor^{r}_{n}$. Consequently every solution of the equation (B) is connected with the vertical rows of the cores of $\lceil a\rfloor^{r}_{n}$.

It follows that the equation (A) *cannot have more than* $n-r-\rho$ *unconnected mutually orthogonal non-extravagant solutions.*

Ex. v. *Let* $y_1, y_2, \dots y_{n-r-\rho}$ *be any given* $n-r-\rho$ *unconnected mutually orthogonal non-extravagant solutions of the equation* (A), *and let them be the successive vertical rows of the matrix* $\lceil y\rfloor^{n-r-\rho}_{n}$. *Then the successive vertical rows* $\eta_1, \eta_2, \dots \eta_t$ *of* $\lceil \eta\rfloor^{t}_{n}$ *form with* $y_1, y_2, \dots y_{n-r-\rho}$ *a set of* $n-r-\rho+t$ *unconnected mutually orthogonal solutions of the equation* (A) *when and only when* $\lceil \eta\rfloor^{t}_{n}$ *is an undegenerate matrix of rank* t *whose vertical rows are connected with the vertical rows of the cores of* $\lceil a\rfloor^{r}_{n}$.

First suppose that $y_1, y_2, \dots y_{n-r-\rho}, \eta_1, \eta_2, \dots \eta_t$ are $n-r-\rho+t$ unconnected mutually orthogonal solutions of the equation (A). Then $\lceil \eta\rfloor^{t}_{n}$ is an undegenerate matrix of rank t, and by Ex. iv its vertical rows are connected with the vertical rows of the cores of $\lceil a\rfloor^{r}_{n}$.

Next suppose that $\lceil \eta\rfloor^{t}_{n}$ is an undegenerate matrix of rank t whose vertical rows are connected with the vertical rows of the cores of $\lceil a\rfloor^{r}_{n}$. Then it follows from Ex. iii that $y_1, y_2, \dots y_{n-r-\rho}, \eta_1, \eta_2, \dots \eta_t$ are $n-r-\rho+t$ unconnected mutually orthogonal solutions of the equation (A).

The solutions $\eta_1, \eta_2, \dots \eta_t$ are necessarily extravagant.

As a particular case we see that:

The successive vertical rows $\eta_1, \eta_2, \dots \eta_\rho$ *of* $\lceil \eta\rfloor^{\rho}_{n}$ *form with* $y_1, y_2, \dots y_{n-r-\rho}$ *a complete set of* $n-r$ *unconnected mutually orthogonal solutions of the equation* (A) *when and only when* $\lceil \eta\rfloor^{\rho}_{n}$ *is a core of* $\lceil a\rfloor^{r}_{n}$.

We proceed to describe in greater detail two methods of determining the matrix $\lceil x\rfloor^{n-r}_{n}$, i.e. of determining $n-r$ unconnected mutually orthogonal solutions $x_1, x_2, \dots x_{n-r}$ of the equation (A).

FIRST METHOD OF SOLUTION.

This is the method indicated in the text above. We first determine any matrix $[z]^{n}_{n-r}$ which lies in Ω and is normal to $[a]^{n}_{r}$ either by the methods described in §§ 90 and 132 or by using the formulae of § 167. 2. Since $[z]^{n}_{n-r}$ has extravagance ρ, the symmetric product $[z]^{n}_{n-r}\lceil \acute{z}\rfloor^{n-r}_{n}$ has degeneracy ρ or rank $n-r-\rho$, and by Theorem II of § 147 we can determine an undegenerate square matrix $[h]_{n-r}^{n-r}$ whose elements all lie in Ω and non-vanishing scalar quantities $e_1, e_2, \dots e_{n-r-\rho}$ lying in Ω such that

$$[z]_{n-r}^{n} \, \overline{\underline{z}}_{n}^{n-r} = \overline{\underline{h}}_{n-r}^{n-r} \begin{bmatrix} e, & 0 \\ 0, & 0 \end{bmatrix}_{n-r-\rho,\ \rho}^{n-r-\rho,\ \rho} [h]_{n-r}^{n-r},$$

where $[e]_{n-r-\rho}^{n-r-\rho}$ is the undegenerate quasi-scalar matrix ${}^{1}[e]_{n-r-\rho}$. Then if $\overline{\underline{H}}_{n-r}^{n-r}$ is the inverse of $[h]_{n-r}^{n-r}$, and if $[x]_{n-r}^{n} = [H]_{n-r}^{n-r} [z]_{n-r}^{n}$, we have

$$[a]_{r}^{n} \, \overline{\underline{x}}_{n}^{n-r} = 0, \quad [x]_{n-r}^{n} \, \overline{\underline{x}}_{n}^{n-r} = \begin{bmatrix} e, & 0 \\ 0, & 0 \end{bmatrix}_{n-r-\rho,\ \rho}^{n-r-\rho,\ \rho}.$$

The successive vertical rows $x_1, x_2, \dots x_{n-r}$ of $\overline{\underline{x}}_{n}^{n-r}$ are then a complete set of $n-r$ unconnected mutually orthogonal solutions of (A) which all lie in Ω, the first $n-r-\rho$ being non-extravagant, and the last ρ being extravagant.

Dividing the 1st, 2nd, ... $(n-r-\rho)$th vertical rows of $\overline{\underline{x}}_{n}^{n-r}$ by $\sqrt{e_1}, \sqrt{e_2}, \dots \sqrt{e_{n-r-\rho}}$ respectively, we obtain a matrix $\overline{\underline{\xi}}_{n}^{n-r}$ such that

$$[a]_{r}^{n} \, \overline{\underline{\xi}}_{n}^{n-r} = 0, \quad [\xi]_{n-r}^{n} \, \overline{\underline{\xi}}_{n}^{n-r} = \begin{bmatrix} 1, & 0 \\ 0, & 0 \end{bmatrix}_{n-r-\rho,\ \rho}^{n-r-\rho,\ \rho};$$

and then $\xi_1, \xi_2, \dots \xi_{n-r}$ are a complete set of $n-r$ unconnected mutually orthogonal solutions of (A), the first $n-r-\rho$ being unit solutions, and the last ρ being extravagant solutions.

In the special case when Ω is the domain of all real numbers, we have $\rho=0$; also $e_1, e_2, \dots e_{n-r}$ are all non-zero real positive quantities, and $\xi_1, \xi_2, \dots \xi_{n-r}$ are all real unit solutions.

SECOND METHOD OF SOLUTION.

This is the method that was followed for a system of real equations in § 94.

We find in turn any non-vanishing and non-extravagant solutions $y_1, y_2, y_3, \dots y_s, \dots$ of the successive equations

$$[a]_{r}^{n} \, \overline{\underline{x}}_{n} = 0, \quad \begin{bmatrix} a \\ y \end{bmatrix}_{r,\ 1}^{n} \overline{\underline{x}}_{n} = 0, \quad \begin{bmatrix} a \\ y \end{bmatrix}_{r,\ 2}^{n} \overline{\underline{x}}_{n} = 0, \ \dots \ \begin{bmatrix} a \\ y \end{bmatrix}_{r,\ s-1}^{n} \overline{\underline{x}}_{n} = 0, \ \dots$$

as long as this is possible; $y_1, y_2, \dots y_s$ being the successive vertical rows of the matrix $\overline{\underline{y}}_{n}^{s}$.

By Ex. ii this is possible so long as $\rho \not> n-r-s+1$, or $s-1 \not> n-r-\rho$.

Thus we can determine $y_1, y_2, \dots y_{n-r-\rho}$, the vertical rows of $\overline{\underline{y}}_{n}^{n-r-\rho}$, in this way. These are $n-r-\rho$ unconnected mutually orthogonal non-extravagant solutions of the equation (A) which can be so chosen as to lie in Ω, and all such sets of $n-r-\rho$ solutions can be found in this way.

When any $n-r-\rho$ unconnected mutually orthogonal non-extravagant solutions $y_1, y_2, \dots y_{n-r-\rho}$ have been found, we see from Ex. v that $y_1, y_2, \dots y_{n-r-\rho}, \eta_1, \eta_2, \dots \eta_\rho$ will be a complete set of $n-r$ unconnected mutually orthogonal solutions if and only if

$\eta_1, \eta_2, \dots \eta_\rho$ are all extravagant and are the vertical rows of a matrix $\overline{\underline{\eta}}_n^{\,\rho}$ which is a core of $\overline{\underline{a}}_n^{\,r}$; and all such matrices $\overline{\underline{\eta}}_n^{\,\rho}$ can be found in the way described in § 169. 3.

In these ways all possible complete sets of $n-r$ unconnected mutually orthogonal solutions can be found.

Ex. vi. We can in general (viz. when $n \not< 2$; $r \not< 0$, $r \not> n-2$; and $\rho \not< 0$, $\rho \not> r$, $\rho \not> n-r-2$) find more than ρ unconnected mutually orthogonal extravagant solutions of the equation (A), but these cannot be all connected with the vertical rows of a core of $\overline{\underline{a}}_n^{\,r}$, and cannot all be members of a complete set of $n-r$ unconnected mutually orthogonal solutions of the equation (A).

This follows from Ex. xiv of § 170.

Ex. vii. If $y_1, y_2, \dots y_s, \eta_1, \eta_2, \dots \eta_t$ are any given $s+t$ unconnected mutually orthogonal solutions of the equation (A) of which the first s are non-extravagant and the last t are extravagant and connected with the vertical rows of the cores of $\overline{\underline{a}}_n^{\,r}$, so that $s \not> n-r-\rho$ and $t \not> \rho$, then we can determine $n-r-\rho-s$ other non-extravagant solutions $y_{s+1}, y_{s+2}, \dots y_{n-r-\rho}$ and $\rho-t$ other extravagant solutions $\eta_{t+1}, \eta_{t+2}, \dots \eta_\rho$ in such a manner that $y_1, y_2, \dots y_{n-r-\rho}, \eta_1, \eta_2, \dots \eta_\rho$ are a complete set of $n-r$ unconnected mutually orthogonal solutions.

We find $y_{s+1}, y_{s+2}, \dots y_{n-r-\rho}$ by determining non-vanishing and non-extravagant solutions of the successive equations

$$\begin{bmatrix} a \\ y \end{bmatrix}^n_{r,s} \overline{\underline{x}}_n = 0, \quad \begin{bmatrix} a \\ y \end{bmatrix}^n_{r,s+1} \overline{\underline{x}}_n = 0, \;\dots\; \begin{bmatrix} a \\ y \end{bmatrix}^n_{r,n-r-\rho-1} \overline{\underline{x}}_n = 0.$$

And we can find $\eta_{t+1}, \eta_{t+2}, \dots \eta_\rho$ by first determining any core $\overline{\underline{\gamma}}_n^{\,\rho}$ of $\overline{\underline{a}}_n^{\,r}$, and then forming an undegenerate matrix $\overline{\underline{\eta}}_n^{\,\rho}$ of rank ρ equivalent to $\overline{\underline{\gamma}}_n^{\,\rho}$ by adding final vertical rows to $\overline{\underline{\eta}}_n^{\,t}$ as in Ex. iii of § 139.

§ 173. Possible extravagances of two non-intersecting spacelets in homogeneous space of $n-1$ dimensions or rank n.

Theorem I. *If $\omega_r \equiv \overline{\underline{a}}_n^{\,r}$ and $\omega_s \equiv \overline{\underline{b}}_n^{\,s}$ are two non-intersecting spacelets of given ranks r and s in homogeneous space ω_n, and if ρ and σ are the extravagances of ω_r and ω_s, then ρ and σ can have independently of one another any values consistent with the necessary conditions*

$$\rho \not< 0,\ \ \rho \not> r,\ \ \rho \not> n-r; \quad \sigma \not< 0,\ \ \sigma \not> s,\ \ \sigma \not> n-s, \quad \text{.........(1)}$$

which include $r \not< 0$, $r \not> n$, $s \not< 0$, $s \not> n$. Accordingly the possible values of r, s, ρ, σ are those consistent with the conditions

$$r+s \not> n; \quad \rho \not< 0,\ \ \rho \not> r,\ \ \rho \not> n-r; \quad \sigma \not< 0,\ \ \sigma \not> s,\ \ \sigma \not> n-s. \quad \text{...(2)}$$

By Ex. x of § 139 the possible values of the given integers r and s are those consistent with the conditions

$$r \not< 0, \ \ s \not< 0, \ \ r+s \not> n; \quad \text{..........................(3)}$$

and by Ex. iv of § 165 the integers ρ and σ must satisfy the conditions (1). We have to show that when r and s are given integers satisfying (3), and ρ and σ are any integers satisfying (1), there exist non-intersecting spacelets ω_r and ω_s of ranks r and s whose extravagances are ρ and σ. We will consider four cases which include all possible cases.

Case I. $\rho \nless \sigma,\ n-r-s \nless \rho-\sigma;\ \ (n-r-s)-(\rho-\sigma)=\tau.$

This case can occur when r and s have any values whatever consistent with (3).

If $i=\sqrt{-1}$, the required conditions are satisfied when $\omega_r \equiv \left[\, a \,\right]^{r}_{n}$ and $\omega_s \equiv \left[\, b \,\right]^{s}_{n}$, where

$$\begin{bmatrix} a \\ b \end{bmatrix}^{n}_{r,s} = \left[\begin{array}{c} 1,\ 0,\ 0,\ 0,\ 0,\ 0,\ 0 \\ 0,\ 1,\ 0,\ i,\ 0,\ 0,\ 0 \\ 0,\ 0,\ 1,\ 0,\ 0,\ i,\ 0 \\ \hline 0,\ i,\ 0,\ 1,\ 0,\ 0,\ 0 \\ 0,\ 0,\ 0,\ 0,\ 1,\ 0,\ 0 \end{array}\right]^{r-\rho,\ \sigma,\ \rho-\sigma,\ \sigma,\ s-\sigma,\ \rho-\sigma,\ \tau}_{r-\rho,\ \sigma,\ \rho-\sigma,\ \sigma,\ s-\sigma}$$

Case II. $\rho \nless \sigma,\ n-r-s \ngtr \rho-\sigma,\ \ (\rho-\sigma)-(n-r-s)=\tau.$

This case can occur when and only when r and s satisfy the inequality $2r+s \nless n$ as well as the inequalities (3). The required conditions are satisfied when $\omega_r \equiv \left[\, a \,\right]^{r}_{n}$ and $\omega_s \equiv \left[\, b \,\right]^{s}_{n}$, where

$$\begin{bmatrix} a \\ b \end{bmatrix}^{n}_{r,s} = \left[\begin{array}{c} 1,\ 0,\ 0,\ 0,\ 0,\ 0,\ 0,\ 0 \\ 0,\ 1,\ 0,\ 0,\ i,\ 0,\ 0,\ 0 \\ 0,\ 0,\ 1,\ 0,\ 0,\ 0,\ 0,\ i \\ 0,\ 0,\ 0,\ 1,\ 0,\ 0,\ i,\ 0 \\ \hline 0,\ i,\ 0,\ 0,\ 1,\ 0,\ 0,\ 0 \\ 0,\ 0,\ 0,\ 0,\ 0,\ 1,\ 0,\ 0 \\ 0,\ 0,\ 0,\ 0,\ 0,\ 0,\ 1,\ 0 \end{array}\right]^{r-\rho,\ \sigma,\ n-r-s,\ \tau,\ \sigma,\ n-r-\rho,\ \tau,\ n-r-s}_{r-\rho,\ \sigma,\ n-r-s,\ \tau,\ \sigma,\ n-r-\rho,\ \tau}$$

Case III. $\sigma \nless \rho,\ n-r-s \nless \sigma-\rho;\ \ (n-r-s)-(\sigma-\rho)=\tau.$

This case can occur when r and s have any values whatever consistent with (3).

The required conditions are satisfied when $\omega_r \equiv \left[\, a \,\right]^{r}_{n}$ and $\omega_s \equiv \left[\, b \,\right]^{s}_{n}$, where

$$\begin{bmatrix} a \\ b \end{bmatrix}^{n}_{r,s} = \left[\begin{array}{c} 1,\ 0,\ 0,\ 0,\ 0,\ 0,\ 0 \\ 0,\ 1,\ 0,\ i,\ 0,\ 0,\ 0 \\ \hline 0,\ 0,\ 1,\ 0,\ 0,\ i,\ 0 \\ 0,\ i,\ 0,\ 1,\ 0,\ 0,\ 0 \\ 0,\ 0,\ 0,\ 0,\ 1,\ 0,\ 0 \end{array}\right]^{r-\rho,\ \rho,\ \sigma-\rho,\ \rho,\ s-\sigma,\ \sigma-\rho,\ \tau}_{r-\rho,\ \rho,\ \sigma-\rho,\ \rho,\ s-\sigma}$$

Case IV. $\sigma \nless \rho,\ n-r-s \ngtr \sigma-\rho;\ \ (\sigma-\rho)-(n-r-s)=\tau.$

This case can occur when and only when r and s satisfy the inequality $2s+r \nless n$ as well as the inequalities (3).

The required conditions are satisfied when $\omega_r \equiv \overline{a}^{r}_{n}$ and $\omega_s \equiv \overline{b}^{s}_{n}$, where

$$\begin{bmatrix} a \\ b \end{bmatrix}^{n}_{r,s} = \begin{bmatrix} 1, & 0, & 0, & 0, & 0, & 0, & 0, & 0 \\ 0, & 1, & 0, & 0, & 0, & 0, & 0, & 0 \\ 0, & 0, & 1, & 0, & 0, & i, & 0, & 0 \\ i, & 0, & 0, & 1, & 0, & 0, & 0, & 0 \\ 0, & 0, & 0, & 0, & 1, & 0, & 0, & i \\ 0, & 0, & i, & 0, & 0, & 1, & 0, & 0 \\ 0, & 0, & 0, & 0, & 0, & 0, & 1, & 0 \end{bmatrix}^{\tau,\, n-s-\sigma,\, \rho,\, \tau,\, n-r-s,\, \rho,\, s-\sigma,\, n-r-s}_{\tau,\, n-s-\sigma,\, \rho,\, \tau,\, n-r-s,\, \rho,\, s-\sigma}$$

In Cases I and II we have (see Ex. ii of § 171)

$$\det \begin{bmatrix} a \\ b \end{bmatrix}^{r+s}_{r,s} = \det \begin{bmatrix} 1, & i \\ i, & 1 \end{bmatrix}^{\sigma,\sigma}_{\sigma,\sigma} = 2^{\sigma} \neq 0;$$

and in Cases III and IV we have

$$\det \begin{bmatrix} a \\ b \end{bmatrix}^{r+s}_{r,s} = \det \begin{bmatrix} 1, & i \\ i, & 1 \end{bmatrix}^{\rho,\rho}_{\rho,\rho} = 2^{\rho} \neq 0.$$

Thus in all cases the matrix $\overline{a, b}^{r,s}_{n}$ has rank $r+s$, i.e. ω_r and ω_s are *non-intersecting* spacelets whose ranks are clearly r and s.

Again in all cases we have

$$[a]^{n}_{r} = \begin{bmatrix} 1, & 0, & 0, & 0 \\ 0, & 1, & i, & 0 \end{bmatrix}^{r-\rho,\, \rho,\, \rho,\, n-r-\rho}_{r-\rho,\, \rho} [u]^{n}_{n},$$

$$[b]^{n}_{s} = \begin{bmatrix} 0, & i, & 1, & 0 \\ 0, & 0, & 0, & 1 \end{bmatrix}^{n-s-\sigma,\, \sigma,\, \sigma,\, s-\sigma}_{\sigma,\, s-\sigma} [v]^{n}_{n},$$

where $[u]^{n}_{n}$ and $[v]^{n}_{n}$ are derangements of the unit matrix $[1]^{n}_{n}$. Therefore in all cases ω_r has extravagance ρ and ω_s has extravagance σ.

Ex. i. The possible values of r, s, ρ, σ are those consistent with the conditions (1) and (3), which are together equivalent to (2).

Ex. ii. The extravagances of the matrix $\overline{a, b}^{r,s}_{n}$ in Cases I, II, III, IV are respectively $\rho-\sigma$, $n-r-s$, $\sigma-\rho$, $n-r-s$.

Ex. iii. If ω_r is plenarily extravagant, only Case I is possible. If ω_r is completely extravagant, Case IV is impossible. If ω_r is non-extravagant only Cases III and IV are possible.

Theorem II. *Let r and s be any given positive integers satisfying the condition $r+s \not> n$. Then if ω_r is any given spacelet of rank r and extravagance ρ, we can always determine a spacelet ω_s of rank s which does not* *intersect* ω_r *and has any assigned extravagance* σ *consistent with the necessary conditions*

$$\sigma \not< 0, \quad \sigma \not> s, \quad \sigma \not> n-s, \quad \ldots\ldots\ldots\ldots\ldots\ldots\ldots(4)$$

i.e. with the necessary conditions (2).

By Theorem I we can determine matrices $\overline{\underline{\alpha}}{}^{r}_{n}$ and $\overline{\underline{\beta}}{}^{s}_{n}$ of ranks r and s whose extravagances are ρ and σ, and which are such that $\overline{\underline{\alpha, \beta}}{}^{r,s}_{n}$ has rank $r+s$. Let $\omega_r \equiv \overline{\underline{a}}{}^{r}_{n}$. Then since $\overline{\underline{a}}{}^{r}_{n}$ and $\overline{\underline{\alpha}}{}^{r}_{n}$ are similar and equi-extravagant undegenerate matrices, it follows from Theorem I of § 165 that there exists a relation of the form

$$\overline{\underline{a}}{}^{r}_{n} = [l]^{n}_{n} \, \overline{\underline{\alpha}}{}^{r}_{n} \, [k]^{r}_{r},$$

where $[k]^{r}_{r}$ is undegenerate, and $[l]^{n}_{n}$ is a square semi-unit matrix; i.e. we have $\omega_r \equiv [l]^{n}_{n} \, \overline{\underline{\alpha}}{}^{r}_{n}$.

Now let $\omega_s \equiv [l]^{n}_{n} \, \overline{\underline{\beta}}{}^{s}_{n}$. Then ω_s is a spacelet of rank s and extravagance σ, and the join of ω_r and ω_s is the spacelet $\omega_{r+s} \equiv [l]^{n}_{n} \, \overline{\underline{\alpha, \beta}}{}^{r,s}_{n}$ of rank $r+s$, i.e. ω_r and ω_s are non-intersecting.

Thus we can determine a spacelet ω_s of rank s which has the assigned extravagance σ and does not intersect ω_r.

Hence the condition that a spacelet is not to intersect a given spacelet imposes no restrictions on the possible values of its extravagance.

Ex. iv. *Possible extravagances of a number of unconnected spacelets of given ranks.*

If A, B, C, D, ... are a number of unconnected spacelets of given ranks $\alpha, \beta, \gamma, \delta, \ldots$, so that $\alpha+\beta+\gamma+\delta+\ldots \not> n$, they can independently of one another have any assigned extravagances $\alpha', \beta', \gamma', \delta', \ldots$ consistent with their ranks.

For we can determine two non-intersecting spacelets A and B of ranks α and β whose extravagances are α' and β'. We can then determine in succession a spacelet C of rank γ and extravagance γ' which does not intersect the join of A and B; a spacelet D of rank δ and extravagance δ' which does not intersect the join of A, B and C; and so on.

Ex. v. *Extravagances of spacelets lying in a given non-extravagant spacelet.*

If ω_p is a given non-extravagant spacelet of rank p of homogeneous space ω_n, we can write

$$\omega_p \equiv \overline{\underline{c}}{}^{p}_{n}, \quad \text{where} \quad [c]^{n}_{p} \, \overline{\underline{c}}{}^{p}_{n} = [1]^{p}_{p}.$$

Then if $\omega_r \equiv \overline{\underline{c}}{}^{p}_{n}\, \overline{\underline{h}}{}^{r}_{p}$ is any spacelet of rank r and extravagance ρ lying in ω_p, r and ρ are the rank and extravagance of the matrix $\overline{\underline{h}}{}^{r}_{p}$. Regarding $\Omega_r \equiv \overline{\underline{h}}{}^{r}_{p}$ as a spacelet of homogeneous space Ω_p of rank p, we see that there is a one-one correspondence between those spacelets ω_r of the complete space ω_n which lie in ω_p and the spacelets Ω_r of the complete space Ω_p, corresponding spacelets having the same rank and the same extravagance. Hence when ω_p is a given non-extravagant spacelet of ω_n, we have the following results :

If ω_r lies in ω_p and has extravagance ρ, the possible values of r and ρ are those consistent with the conditions

$$\rho \nless 0, \quad \rho \ngtr r, \quad r+\rho \ngtr p.$$

If ω_r and ω_s are two non-intersecting spacelets of given ranks r and s lying in ω_p, their extravagances ρ and σ can independently of one another have any values consistent with the necessary conditions

$$\rho \nless 0, \quad \rho \ngtr r, \quad r+\rho \ngtr p\,; \quad \sigma \nless 0, \quad \sigma \ngtr s, \quad s+\sigma \ngtr p\,;$$

and this is true even when one of the spacelets is given.

§ 174. Possible values of the rank and extravagance of a spacelet which lies in a given spacelet of homogeneous space ω_n.

Theorem. *If ω_p is a given spacelet of rank p and extravagance π, and if ω_r is any spacelet of rank r and extravagance ρ which lies in ω_p, then the possible values of r and ρ are those consistent with the conditions*

$$r+\rho \ngtr p+\pi, \quad r-\rho \ngtr p-\pi, \quad \rho \nless 0, \quad \rho \ngtr r, \quad \text{...........(A)}$$

which include the necessary conditions $r \nless 0$, $r \ngtr p$, $\rho \ngtr n-r$.

First proof of the theorem. By § 170. 2 the given integers p and π may have any values consistent with the necessary conditions

$$\pi \nless 0, \quad \pi \ngtr p, \quad p+\pi \ngtr n, \quad \text{.......................(1)}$$

and this result is in agreement with the theorem, since ω_p may be any spacelet of the complete space ω_n whose extravagance is 0.

We will write $\quad \omega_p \equiv \overline{\underline{a}}{}^{p}_{n}, \quad \omega_r \equiv \overline{\underline{b}}{}^{r}_{n}, \quad t = r-\rho.$

Then $\overline{\underline{a}}{}^{p}_{n}$ is a given undegenerate matrix of rank p and extravagance π; $\overline{\underline{b}}{}^{r}_{n}$ is any matrix given by the formula

$$\overline{\underline{b}}{}^{r}_{n} = \overline{\underline{a}}{}^{p}_{n}\, \overline{\underline{k}}{}^{r}_{p}, \quad \text{.............................(2)}$$

where $\overline{\underline{k}}{}^{r}_{p}$ is an arbitrary undegenerate matrix of rank r; and t is the rank of the product $[b]^{n}_{r}\, \overline{\underline{b}}{}^{r}_{n}$.

Let $$[a]_p^n \,\overline{\underline{a}}{}_n^p = [\alpha]_p^p, \quad [b]_r^n \,\overline{\underline{b}}{}_n^r = [\beta]_r^r,$$

so that $[\alpha]_p^p$ is a given symmetric matrix of rank $p-\pi$, and

$$[k]_r^p\,[\alpha]_p^p\,\overline{\underline{k}}{}_p^r = [\beta]_r^r. \quad \text{.........................(3)}$$

By § 136.2 it is possible to determine a matrix $[k]_r^p$ of rank r and a symmetric matrix $[\beta]_r^r$ of rank $r-\rho$ satisfying the equation (3) when and only when

$$r-\rho \nless 0, \quad r-\rho \ngtr r, \quad r-\rho \ngtr p-\pi; \quad r-\rho+2p \nless 2r+p-\pi, \quad r \ngtr p,$$

i.e. when and only when the conditions (A) are satisfied. Thus the rank r and the extravagance ρ of ω_r must satisfy the conditions (A).

Again whenever r and ρ satisfy the conditions (A), we can (as just shown) determine a matrix $[k]_r^p$ of rank r and a symmetric matrix $[\beta]_r^r$ of rank $r-\rho$ satisfying the equation (3); and whenever $[k]_r^p$ and $[\beta]_r^r$ have been so determined, the matrix $\overline{\underline{b}}{}_n^r$ given by (2) has rank r and extravagance ρ, and therefore $\omega_r \equiv \overline{\underline{b}}{}_n^r$ is a spacelet of rank r and extravagance ρ lying in ω_p.

Second proof of the theorem. Let ω_π be the core of ω_p, and let ω_r be any spacelet lying in ω_p whose complete intersection with ω_π is a spacelet ω_u of rank u. Then we can regard ω_r as the join of ω_u with a spacelet ω_x which lies in ω_p and does not intersect ω_π. Since we can always construct a (necessarily non-extravagant) spacelet of rank $p-\pi$ lying in ω_p which contains ω_x and does not intersect ω_π, it follows that when ω_r is arbitrary subject to the conditions stated above, we can regard ω_x as any spacelet lying in any (non-extravagant) spacelet $\omega_{p-\pi}$ of rank $p-\pi$ which lies in ω_p and does not intersect ω_π.

Let ρ and ξ be the extravagances of ω_r and ω_x. Then since ω_r is the join of the two unconnected mutually orthogonal spacelets ω_u and ω_x whose extravagances are u and ξ, we see from Ex. viii of § 170 that

$$r = u + x, \quad \rho = u + \xi. \quad \text{..........................(4)}$$

Further by Ex. v of § 173 the possible values of x and ξ are those consistent with the conditions

$$\xi \nless 0, \quad \xi \ngtr x, \quad x+\xi \ngtr p-\pi. \quad \text{....................(5)}$$

Eliminating x and ξ from (4) and (5) we obtain the following result:

When ω_r is arbitrary subject to the conditions that it lies in ω_p and that its complete intersection with ω_π is ω_u or has rank u, the possible values of r and ρ are those consistent with the conditions

$$\rho \nless u, \quad \rho \ngtr r, \quad r+\rho \ngtr (p-\pi)+2u. \quad \text{.................(6)}$$

When ω_u or u is arbitrary, the possible values of u are given by

$$u \not< 0, \quad u \not> \pi; \qquad \text{.......................(7)}$$

and when we eliminate u from (6) and (7), we obtain the result stated in the theorem.

Other proofs of the theorem have been given in Ex. xiii of § 165 and in Exs. xiii and xiv of § 170.

Ex. i. The conditions (A) which give the possible values of r and ρ are equivalent to

$$\rho \not< 0, \ \rho \not> r, \ \rho \not> n-r; \quad \rho-\pi \not> p-r, \ \pi-\rho \not> p-r.$$

Thus when r is given, the possible extravagances of ω_r are those consistent with its rank and the conditions $\rho-\pi \not> p-r$, $\pi-\rho \not> p-r$.

Ex. ii. Writing $r-\rho=t$ in (A) we see that the possible values of r and t are given by

$$2r-t \not> p+\pi, \ t \not> p-\pi, \ t \not< 0, \ t \not> r; \qquad \text{.......................(8)}$$

and that the possible values of ρ and t are given by

$$2\rho+t \not> p+\pi, \ t \not> p-\pi, \ t \not< 0, \ \rho \not< 0. \qquad \text{.......................(9)}$$

Ex. iii. The possible values of each of the integers r, ρ, t when the others are arbitrary are given respectively by

$$r \not< 0, \quad r \not> p; \qquad \text{.......................(10)}$$

$$\rho \not< 0, \quad 2\rho \not> p+\pi, \qquad \text{.......................(11)}$$

$$t \not< 0, \quad t \not> p-\pi. \qquad \text{.......................(12)}$$

We obtain these results by eliminating ρ from (A), r from (A), and r from (8) or ρ from (9).

Ex. iv. When r has any given value consistent with (10), the possible values of ρ and t are given respectively by

$$\rho \not< 0, \ \rho \not< \pi+r-p, \ \rho \not> \pi+p-r, \ \rho \not> r; \qquad \text{.......................(10')}$$

$$t \not> r, \ t \not> p-\pi, \quad t \not< 2r-p-\pi, \ t \not< 0. \qquad \text{.......................(10'')}$$

When ρ has any given value consistent with (11), the possible values of r and t are given respectively by

$$r \not< \rho, \ r \not> p+\pi-\rho, \quad r \not> p+\rho-\pi; \qquad \text{.......................(11')}$$

$$t \not< 0, \ t \not> p+\pi-2\rho, \ t \not> p-\pi. \qquad \text{.......................(11'')}$$

When t has any given value consistent with (12), the possible values of r and ρ are given respectively by

$$r \not< t, \quad 2r \not> p+\pi+t; \qquad \text{.......................(12')}$$

$$\rho \not< 0, \quad 2\rho \not> p+\pi-t. \qquad \text{.......................(12'')}$$

Ex. v. *Greatest values of r, ρ and t, where $t=r-\rho$.*

The greatest possible value of r is p, and r can have this value when and only when $\rho=\pi$, $t=p-\pi$.

The greatest possible value of ρ is the greatest integer which does not exceed $\frac{1}{2}(p+\pi)$. If $p+\pi=2k$, the greatest value of ρ is k, and ρ can have this value when and only when

$r=k$, $t=0$. If $p+\pi=2k+1$, the greatest value of ρ is k, and ρ can have this value when and only when either $r=k$, $t=0$, or $r=k+1$, $t=1$; both these sets of values being possible because in this case we must have $\pi < p$.

The greatest possible value of t is $p-\pi$, and t can have this value when and only when r has any value from $p-\pi$ to p, and therefore ρ has any value from 0 to π.

Ex. vi. *Non-extravagant spacelets lying in* ω_p.

There exists a non-extravagant spacelet ω_r of rank r lying in ω_p when and only when $r \nless 0$, $r \ngtr p-\pi$. For all such spacelets we have $\rho=0$, $t=r$.

Ex. vii. *Completely extravagant spacelets lying in* ω_p.

There exists a completely extravagant spacelet ω_r of rank r lying in ω_p when and only when $r \nless 0$, $r \ngtr \frac{1}{2}(p+\pi)$. For all such spacelets we have $\rho=r$, $t=0$.

Ex. viii. *Plenarily extravagant spacelets lying in* ω_p.

There exists a plenarily extravagant spacelet ω_r of rank r lying in ω_p when and only when $\pi=n-p$, $r \ngtr p$, $r \nless \frac{1}{2}n$.

These conditions can only be satisfied when ω_p is a plenarily extravagant spacelet, so that $\pi=n-p$, $p \nless \frac{1}{2}n$; and they are then satisfied when and only when $r \ngtr p$, $r \nless \frac{1}{2}n$. For all such spacelets we have $\rho=n-r$, $t=2r-n$.

We see that every spacelet which contains a plenarily extravagant spacelet must itself be plenarily extravagant.

The complete space ω_n is both non-extravagant and plenarily extravagant.

Ex. ix. If $p \nless 2$, the spacelet ω_p contains at least one extravagant point.

Ex. x. There exist spacelets which lie in ω_p and have extravagance greater than π when and only when $p-\pi \nless 2$; and this condition can be satisfied when and only when $p \nless 2$, $p \ngtr n$.

Ex. xi. *If* ω_r *lies in* ω_p *and does not intersect the core of* ω_p, *the possible values of* r *and* ρ *are given by*

$$\rho \nless 0, \quad \rho \ngtr r, \quad r+\rho \ngtr p-\pi.$$

This we see by putting $u=0$ in (6).

Hence we can determine a completely extravagant spacelet ω_ρ *of rank* ρ *whieh lies in* ω_p *and does not intersect the core of* ω_p *when and only when*

$$\rho \nless 0, \quad \rho \ngtr \tfrac{1}{2}(p-\pi).$$

The join of ω_π and ω_ρ is then a completely extravagant spacelet of rank $\pi+\rho$ lying in ω_p.

Ex. xii. *If* ω_r *lies in* ω_p *and contains its core* ω_π, *the possible values of* r *and* ρ *are those consistent with the conditions*

$$\rho \nless 0, \quad \rho \ngtr r, \quad r+\rho \ngtr p+\pi.$$

This we see by putting $u=\pi$ in (6).

Ex. xiii. If $\rho > \frac{1}{2}(p-\pi)$, every completely extravagant spacelet of rank ρ lying in ω_p must have an intersection with the core of ω_p whose rank is not less than $\rho-\frac{1}{2}(p-\pi)$.

Ex. xiv. *If ω_r and ω_s are two non-intersecting spacelets of ranks r and s and extravagances ρ and σ lying in ω_p whose join does not intersect ω_π, the possible values of r, s, ρ, σ are those consistent with the conditions*

$$r+s \not> p-\pi;\quad \rho \not< 0,\quad \rho \not> r,\quad r+\rho \not> p-\pi;\quad \sigma \not< 0,\quad \sigma \not> s,\quad s+\sigma \not> p-\pi.$$

This follows from Ex. v of § 173 when we regard ω_r and ω_s as two arbitrary non-intersecting spacelets of a (non-extravagant) spacelet of rank $p-\pi$ which lies in ω_p and does not intersect ω_π.

Ex. xv. *If ω_r and ω_s lie in ω_p and have ω_π for their complete intersection, the possible values of r, s, ρ, σ are those consistent with the conditions*

$$r+s \not> p+\pi;\quad \rho \not< \pi,\quad \rho \not> r,\quad r+\rho \not> p+\pi;\quad \sigma \not< \pi,\quad \sigma \not> s,\quad s+\sigma \not> p+\pi,$$

which include $r \not< \pi$, $r \not> p$, $s \not< \pi$, $s \not> p$.

We deduce this result from Ex. xiv by regarding ω_r and ω_s as the joins of ω_π with two non-intersecting spacelets ω_x and ω_y whose join does not intersect ω_π.

Ex. xvi. If ω_r and ω_s both lie in ω_p and both contain ω_π, the possible values of r, s, ρ, σ are given by

$$\rho \not< \pi,\quad \rho \not> r,\quad r+\rho \not> p+\pi;\quad \sigma \not< \pi,\quad \sigma \not> s,\quad s+\sigma \not> p+\pi.$$

Ex. xvii. *Graphical representation of the possible values of r and ρ in the theorem of the text.*

If we regard r and ρ as the Cartesian co-ordinates of a point R lying in a two-dimensional plane, then R can be any point with integral co-ordinates lying inside the area bounded by the straight lines

$$r-\rho=0,\quad r-\rho=p-\pi,\quad \rho=0,\quad r+\rho=p+\pi.$$

Note. *Tabular representations of the possible values of r, ρ and t in the theorem of the text.*

The following tables show the possible values of the remaining two of these integers when one of them has an assigned value. They can be constructed from the results given in Exs. iii and iv.

TABLE I.

Values of r and ρ when t is given.

Given value of t.	*Possible values of r.*	*Possible values of ρ.*
$t=0$	$r \not< 0, \quad r \not> \frac{1}{2}(p+\pi)$	$\rho \not< 0, \quad \rho \not> \frac{1}{2}(p+\pi)$
$t \not< 0, \quad t \not> p-\pi$	$r \not< t, \quad r \not> \frac{1}{2}(p+\pi-t)$	$\rho \not< 0, \quad \rho \not> \frac{1}{2}(p+\pi-t)$
$t=p-\pi$	$r \not< p-\pi, \quad r \not> p$	$\rho \not< 0, \quad \rho \not> \pi$

TABLE II.

Values of r and t when ρ is given.

Given value of ρ.	*Possible values of r.*	*Possible values of t.*
$\rho=0$	$r \nless 0,\ \ r \ngtr p-\pi$	$t \nless 0,\ \ t \ngtr p-\pi$
$\rho \nless 0,\ \ \rho \ngtr \pi$	$r \nless \rho,\ \ r \ngtr p-\pi+\rho$	$t \nless 0,\ \ t \ngtr p-\pi$
$\rho=\pi$	$r \nless \rho,\ \ r \ngtr p$	$t \nless 0,\ \ t \ngtr p-\pi$
$\rho \nless \pi,\ \ \rho \ngtr \frac{1}{2}(p+\pi)$	$r \nless \rho,\ \ r \ngtr p+\pi-\rho$	$t \nless 0,\ \ t \ngtr p+\pi-2\rho$
$\rho=\frac{1}{2}(p+\pi)-1=k$	$r=k,\ \ r=k+1$	$t=0,\ \ t=1$
$\rho=\frac{1}{2}(p+\pi)=k$	$r=k$	$t=0$

The cases shown in the last two lines are alternative according as $p+\pi$ is odd or even.

TABLE III*a*.

Values of ρ and t when r is given.

Case 1: $\pi \ngtr \frac{1}{3}p$, *or* $p-\pi \nless \frac{1}{2}(p+\pi)$.

Given value of r.	*Possible values of ρ.*	*Possible values of t.*
$r=0$	$\rho=0$	$t=0$
$r \nless 0,\ \ r \ngtr \frac{1}{2}(p+\pi)$	$\rho \nless 0,\ \ \rho \ngtr r$	$t \ngtr r,\ \ t \nless 0$
$r \nless \frac{1}{2}(p+\pi),\ \ r \ngtr p-\pi$	$\rho \nless 0,\ \ \rho \ngtr \pi+p-r$	$t \ngtr r,\ \ t \nless 2r-p-\pi$
$r=p-\pi$	$\rho \nless 0,\ \ \rho \ngtr 2\pi$	$t \ngtr p-\pi,\ \ t \nless p-3\pi$
$r \nless p-\pi,\ \ r \ngtr p$	$\rho \nless \pi+r-p,\ \ \rho \ngtr \pi+p-r$	$t \ngtr p-\pi,\ \ t \nless 2r-p-\pi$
$r=p$	$\rho=\pi$	$t=p-\pi$

TABLE III *b*.

Values of ρ *and* t *when* r *is given.*

Case 2: $\pi \nless \tfrac{1}{3}p$, *or* $p-\pi \ngtr \tfrac{1}{2}(p+\pi)$.

Given value of r.	*Possible values of* ρ.	*Possible values of* t.
$r=0$	$\rho=0$	$t=0$
$r \nless 0,\ \ r \ngtr p-\pi$	$\rho \nless 0,\ \ \rho \ngtr r$	$t \ngtr r,\ \ t \nless 0$
$r=p-\pi$	$\rho \nless 0,\ \ \rho \ngtr p-\pi$	$t \ngtr p-\pi,\ \ t \nless 0$
$r \nless p-\pi,\ \ r \ngtr \tfrac{1}{2}(p+\pi)$	$\rho \nless \pi+r-p,\ \ \rho \ngtr r$	$t \ngtr p-\pi,\ \ t \nless 0$
$r \nless \tfrac{1}{2}(p+\pi),\ \ r \ngtr p$	$\rho \nless \pi+r-p,\ \ \rho \ngtr \pi+p-r$	$t \ngtr p-\pi,\ \ t \nless 2r-p-\pi$
$r=p$	$\rho=\pi$	$t=p-\pi$

The case shown in Table III *b* can only occur when $4p \ngtr 3n$, or $p \ngtr 3(n-p)$.

§ 175. Possible values of the rank and extravagance of a spacelet of homogeneous space ω_n which contains a given spacelet.

Theorem. *If* ω_p *is a given spacelet of rank* p *and extravagance* π, *and if* ω_r *is any spacelet of rank* r *and extravagance* ρ *which contains* ω_p, *then the possible values of* r *and* ρ *are those consistent with the conditions*

$$r+\rho \nless p+\pi,\quad r-\rho \nless p-\pi,\quad \rho \nless 0,\quad \rho \ngtr n-r, \quad \text{..........(A)}$$

which include the necessary conditions $r \nless p$, $r \ngtr n$, $\rho \ngtr r$.

First proof of the theorem. By § 170. 2 or § 174 the possible values of the given integers p and π are those consistent with the conditions

$$\pi \nless 0,\quad \pi \ngtr p,\quad p+\pi \ngtr n. \quad \text{........................(1)}$$

We will write $\quad \omega_p \equiv \underline{\overline{a}}{}^{p}_{n},\quad \omega_r \equiv \underline{\overline{b}}{}^{r}_{n},\quad t=r-\rho.$

Then $\underline{\overline{a}}{}^{p}_{n}$ is a given undegenerate matrix of rank p and extravagance π; $\underline{\overline{b}}{}^{r}_{n}$ is any undegenerate matrix of rank r satisfying an equation of the form

$$\underline{\overline{a}}{}^{p}_{n} = \underline{\overline{b}}{}^{r}_{n}\ \underline{\overline{k}}{}^{p}_{r},$$

where $\overline{\underline{k}}{}^{p}_{r}$ is an arbitrary undegenerate matrix of rank r; and t is the rank of the product $[b]^{n}_{r}\,\overline{\underline{b}}{}^{r}_{n}$.

By § 174 or by Theorem I a of § 136 there exists a spacelet ω_r' of rank r and extravagance ρ when and only when r and ρ are integers satisfying the conditions

$$\rho \nless 0, \quad \rho \ngtr r, \quad r+\rho \ngtr n; \quad \ldots\ldots\ldots\ldots\ldots\ldots\ldots(2)$$

and when ω_r' is any such spacelet, it follows from § 174 that there exists a spacelet ω_p' which lies in ω_r' and has the given rank p and the given extravagance π when and only when the integers r and ρ are also consistent with the conditions

$$p+\pi \ngtr r+\rho, \quad p-\pi \ngtr r-\rho, \quad \pi \nless 0, \quad \pi \ngtr p. \quad \ldots\ldots\ldots(3)$$

In virtue of (1) the integers r and ρ satisfy the conditions (2) and (3) when and only when they satisfy the conditions (A). Consequently there exists a spacelet of rank r and extravagance ρ which contains *some* spacelet having the given rank p and the given extravagance π when and only when r and ρ satisfy the conditions (A). If then ω_r is a spacelet of rank r and extravagance ρ which contains ω_p, the conditions (A) must be satisfied. To complete the proof of the theorem it remains to show that whenever the conditions (A) are satisfied, we can determine a spacelet ω_r of rank r and extravagance ρ which contains the *given* spacelet ω_p.

Let r and ρ be any assigned integers satisfying the conditions (A); let $\omega_r' \equiv \overline{\underline{\beta}}{}^{r}_{n}$ be any spacelet of rank r and extravagance ρ; and (as has been shown to be possible) let $\omega_p' \equiv \overline{\underline{\alpha}}{}^{p}_{n} \equiv \overline{\underline{\beta}}{}^{r}_{n}\,\overline{\underline{k}}{}^{p}_{r}$ be any spacelet of rank p and extravagance π which lies in ω_r'. Then since ω_p and ω_p' are equi-extravagant spacelets of the same rank, therefore by Ex. v of § 170 or Theorem I of § 165 there exists a square semi-unit matrix $[l]^{n}_{n}$ such that

$$\overline{\underline{a}}{}^{p}_{n} \equiv [l]^{n}_{n}\,\overline{\underline{\alpha}}{}^{p}_{n},$$

i.e.

$$\overline{\underline{a}}{}^{p}_{n} \equiv \overline{\underline{b}}{}^{r}_{n}\,\overline{\underline{k}}{}^{p}_{r}, \quad \text{where} \quad \overline{\underline{b}}{}^{r}_{n} = [l]^{n}_{n}\,\overline{\underline{\beta}}{}^{r}_{n}.$$

Since $\overline{\underline{b}}{}^{r}_{n}$ has the same rank and the same extravagance as $\overline{\underline{\beta}}{}^{r}_{n}$, therefore

$$\omega_r \equiv \overline{\underline{b}}{}^{r}_{n} \equiv [l]^{n}_{n}\,\overline{\underline{\beta}}{}^{r}_{n}$$

is a spacelet of rank r and extravagance ρ which contains the given spacelet

$$\omega_p \equiv \overline{\underline{a}}{}^{p}_{n} \equiv [l]^{n}_{n}\,\overline{\underline{\alpha}}{}^{p}_{n}.$$

Second proof of the theorem. We can deduce it from the corresponding theorem of § 174 by the principle explained in Note 2 of § 170.

Let the normals to ω_p and ω_r be ω_{n-p} and ω_{n-r}. Then ω_r is a spacelet of rank r and extravagance ρ containing the given spacelet ω_p of rank p and extravagance π when and only when ω_{n-r} is a spacelet of rank $n-r$ and extravagance ρ lying in the given spacelet ω_{n-p} of rank $n-p$ and extravagance π. Hence by § 174 the possible values of r and ρ are those consistent with the conditions

$$(n-r)+\rho \ngtr (n-p)+\pi, \quad (n-r)-\rho \ngtr (n-p)-\pi, \quad \rho \nless 0, \quad \rho \ngtr n-r,$$

which are the conditions (A).

Ex. i. The conditions (A) which give the possible values of r and ρ are equivalent to

$$\rho \nless 0, \quad \rho \ngtr r, \quad \rho \ngtr n-r; \quad \rho-\pi \ngtr r-p, \quad \pi-\rho \ngtr r-p.$$

Thus when r is given, the possible extravagances of ω_r are those consistent with its rank and the conditions $\rho-\pi \ngtr r-p$, $\pi-\rho \ngtr r-p$.

Ex. ii. If the vertical rows of all the matrices which occur are unconnected, we have

$$[a]^{p}_{n} \equiv [b]^{r}_{n}\,[k]^{p}_{r} \quad \text{when and only when} \quad [b]^{r}_{n} \equiv [a,\ c]^{p,\ r-p}_{n}.$$

Ex. iii. *If* $[a]^{p}_{n}$ *is a given matrix of rank* p *and extravagance* π, *we can form a matrix* $[a,\ c]^{p,\ r-p}_{n}$ *of rank* r *and extravagance* ρ *by adding final vertical rows to* $[a]^{p}_{n}$ *when and only when* r *and* ρ *satisfy the conditions* (A).

By Ex. ii this is merely another form of the theorem of the text.

Ex. iv. *If* $[a]^{p}_{n}$ *is a given matrix of rank* p *and extravagance* π, *and if* $[c]^{r-p}_{n}$ *is so chosen that* $[a,\ c]^{p,\ r-p}_{n}$ *is undegenerate and has rank* r, *then the possible values of the rank* t *of the symmetric product matrix*

$$\begin{bmatrix} a \\ c \end{bmatrix}^{n}_{p,\ r-p} [a,\ c]^{p,\ r-p}_{n} = \begin{bmatrix} a, & \beta \\ \beta', & \gamma \end{bmatrix}^{p,\ r-p}_{p,\ r-p}$$

are those consistent with the conditions

$$t \nless p-\pi, \quad t \ngtr r, \quad t \ngtr 2r-p-\pi, \quad t \nless 2r-n. \qquad\qquad (4)$$

This is also another form of the theorem of the text. The first three conditions in (4) are the necessary conditions of Theorem II in § 106, and the last condition in (4) is the second necessary condition in the theorem of § 133.

Ex. v. Writing $t=r-\rho$ in (A) we see that the possible values of r and t are given by

$$2r-t \nless p+\pi, \quad t \nless p-\pi, \quad t \ngtr r, \quad 2r-t \ngtr n; \qquad\qquad (5)$$

and that the possible values of ρ and t are given by

$$2\rho+t \nless p+\pi, \quad t \nless p-\pi, \quad \rho \nless 0, \quad 2\rho+t \ngtr n. \qquad\qquad (6)$$

Ex. vi. The possible values of each of the integers r, ρ, t when the others are arbitrary are given respectively by

$$r \nless p, \qquad r \ngtr n; \quad \text{.......(7)}$$
$$\rho \nless 0, \qquad 2\rho \ngtr n-p+\pi; \quad \text{.......(8)}$$
$$t \nless p-\pi, \qquad t \ngtr n. \quad \text{.......(9)}$$

Ex. vii. When r has any given value consistent with (7), the possible values of ρ and t are given respectively by

$$\rho \nless 0, \quad \rho \ngtr \pi+p-r, \quad \rho \ngtr \pi+r-p, \quad \rho \ngtr n-r; \quad \text{.......(7')}$$
$$t \ngtr r, \quad t \ngtr 2r-p-\pi, \quad t \nless p-\pi, \quad t \ngtr 2r-n. \quad \text{.......(7'')}$$

When ρ has any given value consistent with (8), the possible values of r and t are given respectively by

$$r \nless p+\pi-\rho, \quad r \nless p-\pi+\rho, \quad r \ngtr n-\rho; \quad \text{.......(8')}$$
$$t \nless p+\pi-2\rho, \quad t \nless p-\pi, \quad t \ngtr n-2\rho. \quad \text{.......(8'')}$$

When t has any given value consistent with (9), the possible values of r and ρ are given respectively by

$$r \nless t, \quad 2r \nless p+\pi+t, \quad 2r \ngtr n+t; \quad \text{.......(9')}$$
$$\rho \nless 0, \quad 2\rho \nless p+\pi-t, \quad 2\rho \ngtr n-t. \quad \text{.......(9'')}$$

Ex. viii. *Greatest values of r, ρ and t, where $t=r-\rho$.*

The greatest possible value of r is n, and r can have this value when and only when $\rho=0$, $t=n$.

The greatest possible value of ρ is the greatest integer which does not exceed $\frac{1}{2}(n-p+\pi)$. If $n-p+\pi=2k$, the greatest value of ρ is k, and ρ can have this value when and only when $r=n-k$, $t=p-\pi=n-2k$. If $n-p+\pi=2k+1$, the greatest value of ρ is k, and ρ can have this value when and only when either $r=n-k$, $t=n-2k=p-\pi+1$, or $r=n-k-1$, $t=n-2k-1=p-\pi$; both these sets of values being possible because in this case we must have $p+\pi<n$.

The greatest possible value of t is n, and t can have this value when and only when $r=n$, $\rho=0$.

Ex. ix. *Least values of r, ρ and t.*

The least possible value of r is p, and r can have this value when and only when $\rho=\pi$, and therefore $t=p-\pi$.

The least possible value of ρ is 0, and this value of ρ can occur when and only when $r \nless p+\pi$, and therefore $t \nless p+\pi$.

The least possible value of t is $p-\pi$, and this value of t can occur when and only when $r \nless p-\pi$, $2r \ngtr n+p-\pi$, and therefore $\rho \nless 0$, $2\rho \ngtr n-p+\pi$.

Ex. x. *Non-extravagant spacelets containing ω_p.*

There exists a non-extravagant spacelet ω_r containing ω_p when and only when $r \nless p+\pi$, $r \ngtr n$. For all such spacelets we have $\rho=0$, $t=r$.

Ex. xi. *Completely extravagant spacelets containing ω_p.*

There exists a completely extravagant spacelet ω_r containing ω_p when and only when $\pi=p$, $r \nless p$, $r \ngtr \frac{1}{2}n$. These conditions can be satisfied only when ω_p is a completely extravagant spacelet, so that $p \ngtr \frac{1}{2}n$; and they are then satisfied when and only when $r \nless p$, $r \ngtr \frac{1}{2}n$. For all such spacelets we have $\rho=r$, $t=0$.

Ex. xii. *Plenarily extravagant spacelets containing ω_p.*

There exists a plenarily extravagant spacelet ω_r containing ω_p when and only when $2r \nless n+p-\pi$, $r \ngtr n$; and these conditions can always be satisfied.

Ex. xiii. *Graphical representation of the possible values of r and ρ in the theorem of the text.*

If we regard r and ρ as the Cartesian co-ordinates of a point R lying in a two-dimensional plane, then R can be any point with integral co-ordinates lying inside the area bounded by the straight lines

$$r+\rho=p+\pi, \quad r+\rho=n, \quad \rho=0, \quad r-\rho=p-\pi.$$

NOTE 1. *Tabular representations of the possible values of r, ρ and t in the theorem of the text.*

The following tables show the possible values of the remaining two of these integers when one of them has an assigned value. They can be constructed from the results given in Exs. vi and vii.

TABLE I.

Values of r and ρ when t is given.

Given value of t.	*Possible values of r.*	*Possible values of ρ.*
$t=p-\pi$	$r\nless p, \quad r \ngtr \frac{1}{2}(n+p-\pi)$	$\rho\nless \pi, \quad \rho\ngtr\frac{1}{2}(n-p+\pi)$
$t\nless p-\pi, \quad t\ngtr p+\pi$	$r\nless (p+\pi+t), \quad r\ngtr\frac{1}{2}(n+t)$	$\rho\nless\frac{1}{2}(p+\pi-t), \quad \rho\ngtr\frac{1}{2}(n-t)$
$t=p+\pi$	$r\nless p+\pi, \quad r\ngtr\frac{1}{2}(n+p+\pi)$	$\rho\nless \pi, \quad \rho\ngtr\frac{1}{2}(n-p+\pi)$
$t\nless p+\pi, \quad t\ngtr n$	$r\nless t, \quad r\ngtr\frac{1}{2}(n+t)$	$\rho\nless 0, \quad \rho\ngtr\frac{1}{2}(n-t)$
$t=n$	$r=n$	$\rho=0$

TABLE II.

Values of r and t when ρ is given.

Given value of ρ.	*Possible values of r.*	*Possible values of t.*
$\rho=0$	$r\nless p+\pi, \quad r\ngtr n$	$t\nless p+\pi, \quad t\ngtr n$
$\rho\nless 0, \quad \rho\ngtr\pi$	$r\nless p+\pi-\rho, \quad r\ngtr n-\rho$	$t\nless p+\pi-2\rho, \quad t\ngtr n-2\rho$
$\rho=\pi$	$r\nless p, \quad r\ngtr n-\pi$	$t\nless p-\pi, \quad t\ngtr n-2\pi$
$\rho\nless\pi, \quad \rho\ngtr\frac{1}{2}(n-p+\pi)$	$r\nless p-\pi+\rho, \quad r\ngtr n-\rho$	$t\nless p-\pi, \quad t\ngtr n-2\rho$
$\rho=\frac{1}{2}(n-p+\pi-1)=k$	$r=n-k-1, \quad r=n-k$	$t=n-2k-1, \quad t=n-2k$
$\rho=\frac{1}{2}(n-p+\pi)=k$	$r=n-k$	$t=n-2k$

The cases shown in the last two lines are alternative according as $n-p+\pi$ is odd or even.

TABLE III *a*.

Values of ρ *and* t *when* r *is given.*

Case 1: $\pi \ngtr \frac{1}{3}(n-p)$, *or* $p+\pi \ngtr \frac{1}{2}(n+p-\pi)$.

Given value of r.	*Possible values of* ρ.	*Possible values of* t.
$r=p$	$\rho=\pi$	$t=p-\pi$
$r \nless p$, $r \ngtr p+\pi$	$\rho \nless \pi+p-r$, $\rho \ngtr \pi+r-p$	$t \ngtr 2r-p-\pi$, $t \nless p-\pi$
$r=p+\pi$	$\rho \nless 0$, $\rho \ngtr 2\pi$	$t \ngtr p+\pi$, $t \nless p-\pi$
$r \nless p+\pi$, $r \ngtr \frac{1}{2}(n+p-\pi)$	$\rho \nless 0$, $\rho \ngtr \pi+r-p$	$t \ngtr r$, $t \nless p-\pi$
$r \nless \frac{1}{2}(n+p-\pi)$, $r \ngtr n$	$\rho \nless 0$, $\rho \ngtr n-r$	$t \ngtr r$, $t \nless 2r-n$
$r=n$	$\rho=0$	$t=n$

TABLE III *b*.

Values of ρ *and* t *when* r *is given.*

Case 2: $\pi \nless \frac{1}{3}(n-p)$, *or* $p+\pi \nless \frac{1}{2}(n+p-\pi)$.

Given value of r.	*Possible values of* ρ.	*Possible values of* t.
$r=p$	$\rho=\pi$	$t=p-\pi$
$r \nless p$, $r \ngtr \frac{1}{2}(n+p-\pi)$	$\rho \nless \pi+p-r$, $\rho \ngtr \pi+r-p$	$t \ngtr 2r-p-\pi$, $t \nless p-\pi$
$r \nless \frac{1}{2}(n+p-\pi)$, $r \ngtr p+\pi$	$\rho \nless \pi+p-r$, $\rho \ngtr n-r$	$t \ngtr 2r-p-\pi$, $t \nless 2r-n$
$r=p+\pi$	$\rho \nless 0$, $\rho \ngtr n-p-\pi$	$t \ngtr p+\pi$, $t \nless 2p+2\pi-n$
$r \nless p+\pi$, $r \ngtr n$	$\rho \nless 0$, $\rho \ngtr n-r$	$t \ngtr r$, $t \nless 2r-n$
$r=n$	$\rho=0$	$t=n$

The case shown in Table III *b* can only occur when $p \nless \frac{1}{4}n$.

NOTE 2. *Ranks and extravagances of spacelets which lie in one given spacelet and contain another given spacelet.*

Let ω_r be a given spacelet of rank r and extravagance ρ in homogeneous space ω_n, and let ω_s be another given spacelet of rank s and extravagance σ which lies in ω_r, so that r, ρ, s, σ are integers satisfying the conditions

$$r+\rho \not> n, \quad \rho \not< 0, \quad \rho \not> r; \quad s+\sigma \not> r+\rho, \quad s-\sigma \not> r-\rho, \quad \sigma \not< 0, \quad \sigma \not> s,$$

which include $$s+\sigma \not> n, \quad \sigma \not< 0, \quad \sigma \not> s, \quad s \not> r.$$

Then if ω_p is any spacelet of rank p and extravagance π which lies in ω_r and contains ω_s, the integers p and π must satisfy the conditions

$$p+\pi \not> r+\rho, \quad p-\pi \not> r-\rho; \quad p+\pi \not< s+\sigma, \quad p-\pi \not< s-\sigma; \quad \pi \not< 0, \quad \text{........(B)}$$

which include $p \not< s$, $p \not> r$ and $p+\pi \not> n$, $\pi \not< 0$, $\pi \not> p$.

For these are equivalent to the two sets of necessary conditions given by the theorems of this and the preceding article.

The conditions (B) include

$$2\pi \not> (r+\rho)-(s-\sigma) \not> \rho+\sigma+(r-s), \quad 2\pi \not< (s+\sigma)-(r-\rho) \not< \rho+\sigma-(r-s);$$

and if $\rho+\sigma \not< r-s$, the last of the conditions (B) is superfluous.

CHAPTER XIX

THE PARATOMY AND ORTHOTOMY OF TWO MATRICES AND OF TWO SPACELETS OF HOMOGENEOUS SPACE

[In § 176 we define the mutual paratomy and the mutual orthotomy of two spacelets of homogeneous space ω_n, and determine their possible values when only the ranks of the spacelets are given. In the following articles we determine the possible values of the mutual orthotomy of two spacelets which have given ranks and satisfy certain other conditions. The cases in which the two spacelets both lie in or both contain a given spacelet are considered in § 177 ; the cases in which the two spacelets do not intersect or are mutually complementary are considered in §§ 178 and 179 ; and the more important case in which the two spacelets have a given intersection is considered in §§ 180 and 181. In § 182 we determine the possible simultaneous values of the paratomy and orthotomy of two spacelets, or two real spacelets, whose ranks only are given ; and in § 183 we consider some general properties of mutually orthogonal spacelets.]

§ 176. The paratomy and orthotomy of any two spacelets in homogeneous space of $n-1$ dimensions or rank n.

Two matrices $\overline{\underline{a}}\,^{p}_{n}$ and $\overline{\underline{b}}\,^{q}_{n}$ of ranks r and s which both have the same number of vertical rows, or two matrices $[a]^{n}_{p}$ and $[b]^{n}_{q}$ of ranks r and s which both have the same number of horizontal rows, will be regarded as having the same mutual paratomy and the same mutual orthotomy as the homogeneous spacelets $\omega_r \equiv \overline{\underline{a}}\,^{p}_{n}$ and $\omega_s \equiv \overline{\underline{b}}\,^{q}_{n}$. We may therefore confine ourselves to a consideration of the paratomy and orthotomy of spacelets in homogeneous space.

1. *The mutual paratomy of two spacelets.*

The mutual paratomy p of any two spacelets ω_r and ω_s in homogeneous space ω_n will be defined to be the greatest number of unconnected points common to ω_r and ω_s. It is the rank of that spacelet which is the locus of all points common to ω_r and ω_s, i.e. the rank of the complete intersection of ω_r and ω_s.

Referring to Ex. ix of § 139 we see that:

If ω_r and ω_s are two arbitrary spacelets of ranks r and s, and if p is their mutual paratomy, the possible values of r, s and p are those consistent with the conditions

$$p \nless 0, \quad p \ngtr r, \quad p \ngtr s, \quad p \nless r+s-n, \quad \text{.................}(1)$$

which include the conditions giving the possible values of r and s, viz.

$$r \nless 0, \quad r \ngtr n, \quad s \nless 0, \quad s \ngtr n. \quad \text{......................}(2)$$

If the spacelets are non-intersecting, we have $r+s \ngtr n$ and $p=0$.

When r and s are given and $r \nless s$, the conditions (1) become

$$p \nless 0, \quad p \ngtr s, \quad p \nless r+s-n,$$

and the greatest possible value of p is s, this value of p occurring when and only when ω_s lies in ω_r.

When r and s are given and $r \ngtr s$, the conditions (1) become

$$p \nless 0, \quad p \ngtr r, \quad p \nless r+s-n,$$

and the greatest possible value of p is r, this value of p occurring when and only when ω_r lies in ω_s.

2. *The mutual orthotomy and the cross rank of two spacelets.*

The *mutual orthotomy* of any two spacelets $\omega_r \equiv \overline{\underline{a}}{}_n^{\,r}$ and $\omega_s \equiv \overline{\underline{b}}{}_n^{\,s}$ of homogeneous space ω_n will be defined to be the degeneracy of the product $\psi = [a]_r^n \, \overline{\underline{b}}{}_n^{\,s}$. It will be convenient to give also a name to the rank of the product ψ, and this we will call the *cross rank* of the spacelets ω_r and ω_s. Thus if t and τ are respectively the cross rank and the mutual orthotomy of ω_r and ω_s, then t and τ are respectively the rank and the degeneracy of the product ψ. They are integers independent of the particular matrices which represent ω_r and ω_s.

From Theorem I a of § 134 we obtain the following result:

If ω_r and ω_s are two arbitrary spacelets of ranks r and s, and if t is their cross rank, the possible values of r, s and t are those consistent with the conditions

$$t \nless 0, \quad t \ngtr r, \quad t \ngtr s, \quad t \nless r+s-n, \quad \text{.................}(3)$$

which include the conditions giving the possible values of r and s, viz.

$$r \nless 0, \quad r \ngtr n, \quad s \nless 0, \quad s \ngtr n. \quad \text{......................}(4)$$

Since $t = s-\tau$ or $r-\tau$ according as $r \nless s$ or $r \ngtr s$, we deduce the following theorem giving the possible values of τ:

If r and s have any assigned values consistent with the necessary conditions (4), *and if τ is the mutual orthotomy of ω_r and ω_s, then:*

(1) *When $r \nless s$, the possible values of τ are given by*

$$\tau \nless 0, \quad \tau \ngtr s, \quad \tau \ngtr n - r. \qquad \text{.........................(5)}$$

(2) *When $r \ngtr s$, the possible values of τ are given by*

$$\tau \nless 0, \quad \tau \ngtr r, \quad \tau \ngtr n - s. \qquad \text{.........................(6)}$$

(3) *In all cases the possible values of r, s and τ are given by*

$$\tau \nless 0, \quad \tau \ngtr r, \quad \tau \ngtr s, \quad \tau \ngtr n - r, \quad \tau \ngtr n - s. \qquad \text{...........(7)}$$

We obtain the last result by observing that the conditions (7) are equivalent to (5) when $r \nless s$, and to (6) when $r \ngtr s$, and that they include the conditions (4).

Let ω_{n-r} and ω_{n-s} be the normals to ω_r and ω_s. Then when $r \nless s$ we see from Ex. vi of § 134 or from Ex. iii below and Ex. xxiii of § 170 that:

(1) τ is the greatest number of unconnected points which lie in ω_s and are orthogonal with ω_r, i.e. it is the rank of the complete intersection of ω_s with ω_{n-r}.

(2) $r - s + \tau$ is the greatest number of unconnected points which lie in ω_r and are orthogonal with ω_s, i.e. it is the rank of the complete intersection of ω_r with ω_{n-s}.

The corresponding results when $r \ngtr s$ are obtained by interchanging r and s.

Ex. i. *If t is the cross rank of ω_r and ω_s, then:*

(1) *ω_r is the join of a spacelet of rank $r - t$ lying in ω_{n-s} and a spacelet of rank t which does not intersect ω_{n-s}.*

(2) *ω_s is the join of a spacelet of rank $s - t$ lying in ω_{n-r} and a spacelet of rank t which does not intersect ω_{n-r}.*

Thus t is the greatest possible rank of a spacelet which lies in ω_r and does not intersect the normal to ω_s;

and t is also the greatest possible rank of a spacelet which lies in ω_s and does not intersect the normal to ω_r.

Ex. ii. *If $\omega_r \equiv \overline{\underline{a}}^{\,r}_{\,n}$ is a spacelet of rank r whose normal is ω_{n-r}, and if $\omega_y \equiv \overline{\underline{k}}^{\,y}_{\,n}$ is a spacelet of rank y which does not intersect ω_{n-r}, then the product $[a]^n_r \, \overline{\underline{k}}^{\,y}_{\,n}$ is undegenerate and has rank y.*

For if there existed a connection of the form $[a]^n_r \, \overline{\underline{k}}^{\,y}_{\,n} \, \overline{\underline{\lambda}}_{\,y} = 0$, the point $\omega_1 \equiv \overline{\underline{k}}_{\,n} \, \overline{\underline{\lambda}}_{\,y}$ of ω_y would be orthogonal with ω_r and would lie in ω_{n-r}.

Ex. iii. With the same notation let $\omega_s \equiv \overline{\underline{b}}\,{}^{s}_{n}$ be a spacelet whose complete intersection with ω_{n-r} is $\omega_x \equiv \overline{\underline{h}}\,{}^{x}_{n}$. Then we can write $\omega_s \equiv \overline{\underline{h, k}}\,{}^{x,\,y}_{n}$, where $x+y=s$, and $\omega_y \equiv \overline{\underline{k}}\,{}^{y}_{n}$ is a spacelet which does not intersect ω_{n-r}; and because $[a]^{n}_{r}\,\overline{\underline{h}}\,{}^{x}_{n}=0$, therefore the product $\psi=[a]^{n}_{r}\,\overline{\underline{b}}\,{}^{s}_{n} \equiv [a]^{n}_{r}\,\overline{\underline{h, k}}\,{}^{x,\,y}_{n}$ has the same rank y as $[a]^{n}_{r}\,\overline{\underline{k}}\,{}^{y}_{n}$.

Thus if $r \not< s$, the degeneracy of ψ is equal to the rank x of the complete intersection of ω_s with ω_{n-r}.

Ex. iv. *The extravagance of any spacelet $\omega_r \equiv \overline{\underline{a}}\,{}^{r}_{n}$ is its orthotomy with itself.*

Ex. v. The mutual orthotomy of any two spacelets is the rank of the complete intersection of the smaller spacelet with the normal to the larger spacelet.

Ex. vi. *The normals to two spacelets have the same mutual orthotomy as the two spacelets themselves.*

For if $r \not< s$, so that $n-s \not< n-r$, the mutual orthotomy of ω_r and ω_s is the rank of the intersection of ω_s and ω_{n-r}, and the mutual orthotomy of ω_{n-r} and ω_{n-s} is the rank of the intersection ω_{n-r} and ω_s.

Ex. vii. *If t is the cross rank of ω_r and ω_s, and if t' is the cross rank of their normals ω_{n-r} and ω_{n-s}, then*

$$t'-t=n-r-s.$$

Let τ be the mutual orthotomy of both pairs of spacelets. Then if $r \not< s$, we have

$$t=s-\tau, \quad t'=n-r-\tau.$$

Ex. viii. *If ω_r and ω_s are two arbitrary spacelets of given ranks r and s whose normals are ω_{n-r} and ω_{n-s}, and if u and v are the ranks of the complete intersections of ω_r, ω_{n-s} and of ω_s, ω_{n-r}, then the possible values of u and v are given by either of the two equivalent sets of conditions*

$$u \not< 0,\quad u \not> r,\quad u \not> n-s,\quad u \not< r-s,\quad u-v=r-s\,; \qquad \text{(8)}$$

$$v \not< 0,\quad v \not> s,\quad v \not> n-r,\quad v \not< s-r,\quad u-v=r-s. \qquad \text{(9)}$$

For the possible ranks of the complete intersection of ω_r and ω_{n-s} are given by the first four inequalities in (8); the possible ranks of the complete intersection of ω_s and ω_{n-r} are given by the first four inequalities in (9); and by Ex. xxiii of § 170 we have

$$u-v=r-s. \qquad \text{(10)}$$

We can deduce the conditions (5) and (6) from (8) and (9); for if τ is the mutual orthotomy of ω_r and ω_s, we have

$$\tau=u, \text{ when } r \not> s; \quad \tau=v, \text{ when } r \not< s.$$

Ex. ix. *Conditions that each of two spacelets ω_r and ω_s shall be incident with the normal to the other.*

Let ω_{n-r} and ω_{n-s} be the normals to ω_r and ω_s. Then there are two possible cases of incidence.

Case I. If either one of the two spacelets ω_r and ω_s lies in the normal to the other, then each of them lies in the normal to the other, and the two spacelets are mutually orthogonal. This is possible only when $r+s \not> n$.

Case II. If either one of the two spacelets ω_r and ω_s contains the normal to the other, then each of them contains the normal to the other, and their two normals ω_{n-r} and ω_{n-s} are mutually orthogonal. This is possible only when $r+s \nless n$.

Special Case. When $r+s=n$, one of the spacelets is incident with the normal to the other when and only when the two spacelets are mutually normal.

Thus if either one of the two spacelets ω_r and ω_s is incident with the normal to the other, then each of them is incident with the normal to the other, and the incidence is necessarily of the first kind when $r+s \ngtr n$, and necessarily of the second kind when $r+s \nless n$.

NOTE 1. *Greatest possible value of the mutual orthotomy τ.*

Since the mutual orthotomy τ of ω_r and ω_s is the same as the mutual orthotomy of their normals ω_{n-r} and ω_{n-s}, it is immaterial which of these two pairs of spacelets we consider. There are two possible cases, in each of which we assume throughout that $r \nless s$.

Case I. When $r+s \ngtr n$. The greatest possible value of τ is s.

Each of the two spacelets ω_r and ω_s is incident with the normal to the other when and only when each of them lies in the normal to the other; and the mutual orthotomy τ has its greatest possible value s when and only when each or any one of the following equivalent conditions is satisfied:

(1) *The spacelets ω_r and ω_s are mutually orthogonal.*

(2) *Each of these two spacelets lies in the normal to the other.*

(3) *Each of these two spacelets is incident with the normal to the other.*

(4) *Every point of ω_r is orthogonal with every point of ω_s.*

These conditions can only be satisfied under the following circumstances:

(1) When ω_r and ω_s do not intersect.

(2) When the complete intersection of ω_r and ω_s is a completely extravagant (or self-orthogonal) spacelet.

From the conditions (7) we see that we can have $\tau=s$ only when

$$r \nless s, \quad r+s \ngtr n, \quad t=0.$$

Case II. When $r+s \nless n$. The greatest possible value of τ is $n-r$.

Each of the two spacelets ω_r and ω_s is incident with the normal to the other when and only when each of them contains the normal to the other; and the mutual orthotomy τ of ω_r and ω_s has its greatest possible value $n-r$ when and only when the mutual orthotomy of ω_{n-r} and ω_{n-s} is $n-r$, i.e. when and only when the cross rank of ω_{n-r} and ω_{n-s} is 0, i.e. when and only when each or any one of the following equivalent conditions is satisfied:

(1) *The normals ω_{n-r} and ω_{n-s} to ω_r and ω_s are mutually orthogonal.*

(2) *Each of the two spacelets ω_r and ω_s contains the normal to the other.*

(3) *Each of these two spacelets is incident with the normal to the other.*

(4) *Every point orthogonal with ω_r is orthogonal with every point orthogonal with ω_s.*

These conditions can only be satisfied under the following circumstances:

(1) When ω_r and ω_s are mutually complementary.

(2) When the join of ω_r and ω_s is a plenarily extravagant spacelet.

From the conditions (7) we see that we can have $\tau=n-r$ only when

$$r \nless s, \quad r+s \nless n, \quad t=r+s-n.$$

Both in Case I and in Case II the mutual orthotomy has its greatest possible value when and only when each of the two spacelets ω_r and ω_s is incident with the normal to the other.

In the particular case when $r+s=n$ the mutual orthotomy has its greatest possible value when and only when ω_r and ω_s are mutually normal.

Ex. x. *If* $t=0$ (*i.e. if* $r \nless s$, $\tau=s$ *or if* $r \ngtr s$, $\tau=r$), *and if the complete intersection of* ω_r *and* ω_s *is the (completely extravagant) spacelet* ω_π, *then* ω_r *and* ω_s *both lie in the normal* $\omega_{n-\pi}$ *to their complete intersection.*

Because ω_π lies in ω_s, therefore ω_{n-s} lies in $\omega_{n-\pi}$; and because ω_r lies in ω_{n-s}, therefore ω_r lies in $\omega_{n-\pi}$.

Ex. xi. *If* $t=r+s-n$ (*i.e. if* $r \nless s$, $\tau=n-r$ *or if* $r \ngtr s$, $\tau=n-s$), *and if the complete intersection of* ω_r *and* ω_s *is the spacelet* ω_p *whose core is* ω_π, *then the core of* ω_r *is the complete intersection of* ω_{n-r} *with* ω_π, *and the core of* ω_s *is the complete intersection of* ω_{n-s} *with* ω_π.

For the core of ω_r is the complete intersection of ω_r and ω_{n-r}. Because ω_{n-r} lies in ω_s, this is the complete intersection of ω_p and ω_{n-r}; and because ω_{n-r} lies in ω_{n-p}, this is the complete intersection of ω_π and ω_{n-r}.

NOTE 2. *Least possible value of the mutual orthotomy* τ.

The least possible value of τ is 0. Assuming that $r \nless s$, we have $\tau=0$ (or $t=s$) when and only when each or either of the following equivalent conditions is satisfied:

(1) *The spacelet* ω_s *does not intersect* ω_{n-r}.

(2) *The spacelet* ω_r *is one whose complete intersection with* ω_{n-s} *has rank* $r-s$.

From (7) we see that we can have $\tau=0$ whatever values r and s have.

§ 177. Mutual orthotomy of two spacelets which both lie in or both contain a given spacelet.

Theorem I. *If* ω_r *and* ω_s *are two spacelets of ranks* r *and* s *which both lie in a given spacelet* ω_p *of rank* p *and extravagance* π, *and if* t *is the cross rank of* ω_r *and* ω_s, *the possible values of* r, s *and* t *are those consistent with the conditions*

$$r \ngtr p, \quad s \ngtr p; \quad t \nless 0, \quad t \ngtr r, \quad t \ngtr s, \quad t \ngtr p-\pi, \quad t \nless r+s-p-\pi, \quad \ldots(1)$$

which include the conditions giving the possible values of r *and* s, *viz.*

$$r \nless 0, \quad r \ngtr p, \quad s \nless 0, \quad s \ngtr p. \quad \ldots\ldots(2)$$

If r and s have any given values consistent with the necessary conditions (2), and if τ is the mutual orthotomy of ω_r and ω_s, then when $r \nless s$ the possible values of τ are those consistent with the conditions

$$\tau \nless 0, \quad \tau \nless s-p+\pi, \quad \tau \ngtr s, \quad \tau \ngtr p+\pi-r. \quad \text{...........(3)}$$

Let
$$\omega_p \equiv \overline{\underline{c}}{}^{p}_{n}, \quad \omega_r \equiv \overline{\underline{a}}{}^{r}_{n}, \quad \omega_s \equiv \overline{\underline{b}}{}^{s}_{n}, \quad \text{..................(4)}$$

where
$$\overline{\underline{a}}{}^{r}_{n} = \overline{\underline{c}}{}^{p}_{n}\,\overline{\underline{h}}{}^{r}_{p}, \quad \overline{\underline{b}}{}^{s}_{n} = \overline{\underline{c}}{}^{p}_{n}\,\overline{\underline{k}}{}^{s}_{p}; \quad \text{..................(5)}$$

and let
$$[c]^{n}_{p}\,\overline{\underline{c}}{}^{p}_{n} = [\gamma]^{p}_{p}, \quad [a]^{n}_{r}\,\overline{\underline{b}}{}^{s}_{n} = [e]^{s}_{r},$$

so that $[\gamma]^{p}_{p}$ is a given symmetric matrix of rank $p-\pi$, and

$$[h]^{p}_{r}\,[\gamma]^{p}_{p}\,\overline{\underline{k}}{}^{s}_{p} = [e]^{s}_{r}. \quad \text{...........................(6)}$$

By sub-article 2 of § 134 we can determine matrices $\overline{\underline{h}}{}^{r}_{p}$, $\overline{\underline{k}}{}^{s}_{p}$ of ranks r, s and a matrix $[e]^{s}_{r}$ of rank t satisfying the equation (6) when and only when the conditions (1) are satisfied. And when they have been thus determined, the spacelets ω_r and ω_s defined by (4) and (5) lie in ω_p, have ranks r and s, and are such that their cross rank is t. Since the elimination of t from (1) leads to (2), the possible values of r and s are those consistent with (2).

We obtain the second part of the theorem by making the substitution $t = s - \tau$ in (1).

NOTE 1. *Spacelets lying in a given non-extravagant spacelet.*

Every theorem A concerning spacelets ω_r, ω_s, ... lying in a given non-extravagant spacelet ω_p of homogeneous space ω_n corresponds to and is deducible from a theorem B concerning corresponding spacelets Ω_r, Ω_s, ... of homogeneous space Ω_p. A rank, cross rank, paratomy, orthotomy or extravagance x occurs in the theorem A when and only when it occurs in exactly the same manner in the theorem B, in which n is replaced by p.

To see this let ω_p be represented in the form

$$\omega_p \equiv \overline{\underline{\lambda}}{}^{p}_{n}, \quad \text{where } [\lambda]^{n}_{p}\,\overline{\underline{\lambda}}{}^{p}_{n} = [1]^{p}_{p}.$$

Then if
$$\omega_r \equiv \overline{\underline{a}}{}^{r}_{n} \equiv \overline{\underline{\lambda}}{}^{p}_{n}\,\overline{\underline{\alpha}}{}^{r}_{p}, \quad \omega_s \equiv \overline{\underline{b}}{}^{s}_{n} \equiv \overline{\underline{\lambda}}{}^{p}_{n}\,\overline{\underline{\beta}}{}^{s}_{p}, \ \ldots$$

are spacelets of ranks r, s, ... lying in ω_p, we can regard

$$\Omega_r \equiv \overline{\underline{\alpha}}{}^{r}_{p}, \quad \Omega_s \equiv \overline{\underline{\beta}}{}^{s}_{p}, \ \ldots$$

as spacelets of ranks r, s, ... in homogeneous space Ω_p of rank p.

Thus there is a one-one correspondence between those spacelets ω_r, ω_s, ... of ω_n which lie in ω_p and the spacelets Ω_r, Ω_s, ... of Ω_p.

The spacelet ω_r lies in or contains ω_s when and only when Ω_r lies in or contains Ω_s, and the join and intersection of $\omega_r, \omega_s, \ldots$ correspond respectively to the join and intersection of $\Omega_r, \Omega_s, \ldots$.

Any two corresponding spacelets such as ω_r and Ω_r have the same rank and the same extravagance.

Also since $[a]_r^n \,\overline{\underline{b}}{}_n^s = [a]_r^p \,\overline{\underline{\beta}}{}_p^s$, any two pairs of corresponding spacelets such as ω_r, ω_s and Ω_r, Ω_s have the same cross rank and the same mutual orthotomy.

Let Ω_{p-r} be the normal to Ω_r. Then the locus of those points of ω_p which are orthogonal with ω_r is the spacelet ω_{p-r} which corresponds to Ω_{p-r}. If ω_{n-p} and ω_{n-r} are the normals to ω_p and ω_r, then ω_{p-r} is the complete intersection of ω_{n-r} with ω_p, and ω_{n-r} is the join of ω_{p-r} and ω_{n-p}. Consequently ω_{p-r} has the same extravagance and the same core as ω_{n-r}.

The core ω_ρ of ω_r is also the core of ω_{p-r} and the complete intersection of ω_r and ω_{p-r}, and it corresponds to the core Ω_ρ of Ω_r which is also the core of Ω_{p-r} and the complete intersection of Ω_r and Ω_{p-r}.

Thus the normal to Ω_r corresponds to the intersection of the normal to ω_r with ω_p; and the core Ω_ρ of Ω_r corresponds to the core ω_ρ of ω_r.

Illustrations of the principle of this note are given in Lemma B and Theorem II of § 178.

Ex. i. *If ρ and σ are the extravagances of two non-intersecting spacelets ω_r and ω_s which lie in a given non-extravagant spacelet ω_p, the possible values of r, s, ρ, σ are those consistent with the conditions*

$$r+s \ngtr p;\quad \rho \nless 0,\ \rho \ngtr r,\ \rho \ngtr p-r;\quad \sigma \nless 0,\ \sigma \ngtr s,\ \sigma \ngtr p-s,$$

which include the necessary conditions $r \nless 0$, $s \nless 0$, $r+s \ngtr p$; and this is true even when one of the spacelets is given.

Using the principle explained in Note 1 we deduce this result from Theorem I of § 173 by substituting p for n.

The possible extravagances of each spacelet are completely independent of the other spacelet.

Theorem II. *If ω_r and ω_s are two spacelets of ranks r and s which both contain a given spacelet ω_p of rank p and extravagance π, and if t is the cross rank of ω_r and ω_s, the possible values of r, s and t are those consistent with the conditions*

$$r \nless p,\ s \nless p;\ t \nless r+s-n,\ t \ngtr r,\ t \ngtr s,\ t \nless p-\pi,\ t \ngtr r+s-p-\pi, \quad \ldots(7)$$

which include the conditions giving the possible values of r and s, viz.

$$r \nless p,\ r \ngtr n,\ s \nless p,\ s \ngtr n. \quad \ldots\ldots(8)$$

If r and s have any given ranks consistent with the necessary conditions (8), *and if τ is the mutual orthotomy of ω_r and ω_s, then when $r \nless s$ the possible values of τ are those consistent with the conditions*

$$\tau \nless 0,\ \tau \nless p+\pi-r,\ \tau \ngtr s-p+\pi,\ \tau \ngtr n-r. \quad \ldots\ldots(9)$$

Let $\omega_{n-r}, \omega_{n-s}, \omega_{n-p}$ be the normals to $\omega_r, \omega_s, \omega_p$, so that ω_{n-p} is a given spacelet of rank $n-p$ and extravagance π; and let t' be the cross rank of

ω_{n-r} and ω_{n-s}. Then ω_{n-r} and ω_{n-s} are any two spacelets of ranks $n-r$ and $n-s$ which both lie in ω_{n-p}. The possible values of $n-r$, $n-s$ and t' are given by Theorem I when we replace r, s and t by $n-r$, $n-s$ and t' in (1). If then, using Ex. vii of § 176, we make the substitution $t'=n-r-s+t$, we see that the possible values of r, s and t are given by (7). Since the elimination of t from (7) leads to (8), the possible values of r and s are those consistent with (8).

We obtain the second part of the theorem by making the substitution $t=s-\tau$ in (7).

Ex. ii. *If* $[c]_n^p$ *is a given undegenerate matrix of rank* p *and extravagance* π, *and if* $[a]_n^{r-p}$ *and* $[b]_n^{s-p}$ *are matrices whose elements are arbitrary subject to the conditions that* $[c,\ a]_n^{p,\ r-p}$ *and* $[c,\ b]_n^{p,\ s-p}$ *are undegenerate and have ranks* r *and* s, *then the possible values of the rank* t *of the matrix*

$$\psi=\begin{bmatrix} c \\ a \end{bmatrix}_{p,\ r-p}^{n}[c,\ b]_n^{p,\ s-p}=\begin{bmatrix} \gamma, & \beta \\ a, & e \end{bmatrix}_{p,\ r-p}^{p,\ s-p}$$

are those consistent with the conditions

$$t \not< p-\pi,\quad t \not< r+s-n,\quad t \not> r,\quad t \not> s,\quad t \not> r+s-p-\pi. \quad\text{............(10)}$$

Here r and s are any given integers satisfying the conditions (8).

This result is clearly another form of stating Theorem II.

The conditions (10) are the necessary conditions given by Theorem II of § 104 and by the theorem of § 133.

Note 2. *Spacelets containing a given non-extravagant spacelet.*

Every theorem A concerning those spacelets ω_r, ω_s, ... of homogeneous space ω_n which contain a given non-extravagant spacelet ω_p corresponds to and is deducible from a theorem B concerning spacelets Ω_{r-p}, Ω_{s-p}, ... of homogeneous space Ω_{n-p}. A rank or cross rank x occurs in the theorem A when and only when a corresponding rank or cross rank $x-p$ occurs in exactly the same manner in the theorem B; and an orthotomy or extravagance y occurs in the theorem A when and only when the same orthotomy or extravagance y occurs in exactly the same manner in the theorem B, in which n is replaced by $n-p$.

To see this let ω_{n-p} be the normal to ω_p, and let ω_p and ω_{n-p} be represented in the forms

$$\omega_p \equiv [\lambda]_n^p,\quad \omega_{n-p}\equiv [\mu]_n^{n-p},\quad \text{where}\quad \begin{bmatrix} \lambda \\ \mu \end{bmatrix}_{p,\ n-p}^{n}[\lambda,\ \mu]_n^{p,\ n-p}=\begin{bmatrix} 1, & 0 \\ 0, & 1 \end{bmatrix}_{p,\ n-p}^{p,\ n-p}=[1]_n^n.$$

Then if ω_r, ω_s, ... are spacelets of ranks r, s, ... in ω_n which contain ω_p, it follows from Ex. ix of § 170 that we can represent them in the forms

$$\omega_r\equiv [a]_n^r \equiv [\lambda,\ \alpha']_n^{p,\ r-p},\quad \omega_s \equiv [b]_n^s \equiv [\lambda,\ \beta']_n^{p,\ s-p},\ \ldots,$$

where $[\alpha']_n^{r-p}$, $[\beta']_n^{s-p}$, ... are matrices of ranks $r-p$, $s-p$, ... orthogonal with $[\lambda]^p$. The spacelets

$$\omega'_{r-p}\equiv[\alpha']_n^{r-p},\quad \omega'_{s-p}\equiv[\beta']_n^{s-p},\ \ldots$$

are the complete intersections of $\omega_r, \omega_s, \ldots$ with ω_{n-p}; and $\omega_r, \omega_s, \ldots$ are the joins of ω_p with $\omega'_{r-p}, \omega'_{s-p}, \ldots$.

Thus there is a one-one correspondence between the spacelets $\omega_r, \omega_s, \ldots$ containing ω_p, and the spacelets $\omega'_{r-p}, \omega'_{s-p}, \ldots$ of the non-extravagant spacelet ω_{n-p}.

From Note 1 it follows that:

There is a one-one correspondence between the spacelets $\omega_r, \omega_s, \ldots$ of ω_n which contain ω_p and the spacelets $\Omega_{r-p}, \Omega_{s-p}, \ldots$ of homogeneous space Ω_{n-p}.

This last result can be obtained in a more precise form by writing

$$\overline{\underline{a'}}\,{}^{r-p}_{n} \equiv \overline{\underline{\mu}}\,{}^{n-p}_{n}\ \overline{\underline{a}}\,{}^{r-p}_{n-p}, \quad \overline{\underline{\beta'}}\,{}^{s-p}_{n} \equiv \overline{\underline{\mu}}\,{}^{n-p}_{n}\ \overline{\underline{\beta}}\,{}^{s-p}_{n-p}, \ldots$$

or

$$\omega_r \equiv \overline{\underline{a}}\,{}^{r}_{n} \equiv \overline{\underline{\lambda, \mu}}\,{}^{p,\,n-p}_{n} \begin{bmatrix} 1, & 0 \\ 0, & a \end{bmatrix}^{p,\,r-p}_{p,\,n-p}, \quad \omega_s \equiv \overline{\underline{b}}\,{}^{s}_{n} \equiv \overline{\underline{\lambda, \mu}}\,{}^{p,\,n-p}_{n} \begin{bmatrix} 1, & 0 \\ 0, & \beta \end{bmatrix}^{p,\,s-p}_{p,\,n-p}, \ldots.$$

Then

$$\Omega_{r-p} \equiv \overline{\underline{a}}\,{}^{r-p}_{n-p}, \quad \Omega_{s-p} \equiv \overline{\underline{\beta}}\,{}^{s-p}_{n-p}, \ldots$$

are the uniquely determinate spacelets of Ω_{n-p} which correspond to $\omega_r, \omega_s, \ldots$.

The spacelet ω_r lies in or contains ω_s when and only when Ω_{r-p} lies in or contains Ω_{s-p}; and the join and intersection of $\omega_r, \omega_s, \ldots$ correspond respectively to the join and intersection of $\Omega_{r-p}, \Omega_{s-p}, \ldots$.

If x and X are the ranks of two corresponding spacelets such as ω_r and Ω_{r-p}, we have $X = x - p$; and if y and Y are the extravagances of two such corresponding spacelets, we have $Y = y$.

Again if $[a]^{n-p}_{r-p}\ \overline{\underline{\beta}}\,{}^{s-p}_{n-p} = [u]^{s-p}_{r-p}$, we have $[a]^{n}_{r}\ \overline{\underline{b}}\,{}^{s}_{n} = \begin{bmatrix} 1, & 0 \\ 0, & u \end{bmatrix}^{p,\,s-p}_{p,\,r-p}$.

Hence if x and X are the cross ranks of the two pairs of corresponding spacelets such as (ω_r, ω_s) and $(\Omega_{r-p}, \Omega_{s-p})$, we have $X = x - p$; and if y and Y are the orthotomies of two such pairs of corresponding spacelets, we have $Y = y$.

The normal Ω_{n-r} to Ω_{r-p} corresponds to the join ω_{n-r+p} of ω_p with the normal to ω_r; and the core Ω_ρ of Ω_{n-r} corresponds to the join $\omega_{\rho+p}$ of ω_p with the core ω_ρ of ω_r.

Two spacelets ω_r and ω_s of ω_n whose complete intersection is ω_p correspond to two non-intersecting spacelets Ω_{r-p} and Ω_{s-p} of Ω_{n-p}.

Ex. iii. *If ρ and σ are the extravagances of two spacelets ω_r and ω_s of homogeneous space ω_n whose complete intersection is a given non-extravagant spacelet ω_p, the possible values of r, s, ρ, σ are given by*

$$r+s-p \not> n; \quad \rho \not< 0, \quad \rho \not> r-p, \quad \rho \not> n-r; \quad \sigma \not< 0, \quad \sigma \not> s-p, \quad \sigma \not> n-s;$$

and this is true even when one of the spacelets is given.

Using the principle explained in Note 2, we deduce this result from the theorems of § 173 by substituting $r-p, s-p, n-p$ for r, s, n and leaving ρ and σ unaltered.

Ex. iv. *If t and τ are the cross rank and mutual orthotomy of two spacelets ω_r and ω_s of homogeneous space ω_n which both contain a given non-extravagant spacelet ω_p, the possible values of r, s, t and the possible values of r, s, τ are given respectively by*

$$t \not< p, \quad t \not> r, \quad t \not> s, \quad t \not< r+s-n;$$

and

$$\tau \not< 0, \quad \tau \not> r-p, \quad \tau \not> s-p, \quad \tau \not> n-r, \quad \tau \not> n-s.$$

We deduce these results from the conditions (3) and (7) of § 176 by substituting $r-p$ $s-p, n-p, t-p$ for r, s, n, t and leaving τ unaltered.

NOTE 3. *Spacelets which lie in one given non-extravagant spacelet and contain another given non-extravagant spacelet.*

Every theorem A concerning those spacelets ω_r, ω_s, ... of homogeneous space ω_n which lie in one given non-extravagant spacelet ω_q and contain another given non-extravagant spacelet ω_p corresponds to and is deducible from a theorem B concerning spacelets Ω_{r-p}, Ω_{s-p}, ... of homogeneous space Ω_{q-p}. A rank or cross rank x occurs in the theorem A when and only when a corresponding rank or cross rank $x-p$ occurs in exactly the same manner in the theorem B; and an orthotomy or extravagance y occurs in the theorem A when and only when the same orthotomy or extravagance y occurs in exactly the same manner in the theorem B, in which n is replaced by $q-p$.

This result follows at once from Notes 1 and 2.

To give greater precision to the result let ω_{q-p} be the locus of those points of ω_q which are orthogonal with ω_p, i.e. the intersection of the normal to ω_q with ω_p, and let ω_q, ω_p, ω_{q-p} be represented in the forms

$$\omega_q \equiv \overline{\underline{h}}{}^{q}_{n}, \quad \omega_p \equiv \overline{\underline{h}}{}^{q}_{n}\, \overline{\underline{\lambda}}{}^{p}_{q}, \quad \omega_{q-p} \equiv \overline{\underline{h}}{}^{q}_{n}\, \overline{\underline{\mu}}{}^{q-p}_{q},$$

where
$$[h]^n_q\, \overline{\underline{h}}{}^{q}_{n} = [1]^q_q, \quad \begin{bmatrix} \lambda \\ \mu \end{bmatrix}^{q}_{p,\,q-p} \overline{\underline{\lambda,\ \mu}}{}^{p,\,q-p}_{q} = \begin{bmatrix} 1, & 0 \\ 0, & 1 \end{bmatrix}^{p,\,q-p}_{p,\,q-p} = [1]^q_q.$$

Any spacelet ω_r which lies in ω_q and contains ω_p can be regarded as the join of ω_p with the intersection ω'_{r-p} of ω_r and ω_{q-p}. Hence the spacelets ω_r, ω_s, ... can be represented in the forms

$$\omega_r \equiv \overline{\underline{a}}{}^{r}_{n} \equiv \overline{\underline{h}}{}^{q}_{n}\, \overline{\underline{\lambda,\ \mu}}{}^{p,\,q-p}_{q} \overline{\underline{\begin{matrix} 1, & 0 \\ 0, & a \end{matrix}}}{}^{p,\,r-p}_{p,\,q-p}, \quad \omega_s \equiv \overline{\underline{b}}{}^{s}_{n} \equiv \overline{\underline{h}}{}^{q}_{n}\, \overline{\underline{\lambda,\ \mu}}{}^{p,\,q-p}_{q} \overline{\underline{\begin{matrix} 1, & 0 \\ 0, & \beta \end{matrix}}}{}^{p,\,s-p}_{p,\,q-p}, \ \ldots,$$

where
$$\omega'_{r-p} \equiv \overline{\underline{h}}{}^{q}_{n}\, \overline{\underline{\mu}}{}^{q-p}_{q}\, \overline{\underline{a}}{}^{r-p}_{q-p}, \quad \omega'_{s-p} \equiv \overline{\underline{h}}{}^{q}_{n}\, \overline{\underline{\mu}}{}^{q-p}_{q}\, \overline{\underline{\beta}}{}^{s-p}_{q-p}, \ \ldots$$

are the intersections of ω_r, ω_s, ... with ω_{q-p}; and we can regard

$$\Omega_{r-p} \equiv \overline{\underline{a}}{}^{r-p}_{q-p}, \quad \Omega_{s-p} \equiv \overline{\underline{\beta}}{}^{s-p}_{q-p}, \ \ldots$$

as uniquely determinate spacelets of homogeneous space Ω_{q-p}.

Thus there is a one-one correspondence between those spacelets ω_r, ω_s, ... *of* ω_n *which lie in* ω_q *and contain* ω_p *and the spacelets* Ω_{r-p}, Ω_{s-p}, ... *of homogeneous space* Ω_{q-p}.

The spacelet ω_r lies in or contains ω_s when and only when Ω_{r-p} lies in or contains Ω_{s-p}; and the join and intersection of ω_r, ω_s, ... correspond respectively to the join and intersection of the corresponding spacelets Ω_{r-p}, Ω_{s-p}, ...

If x, X and y, Y are the ranks and extravagances of two corresponding spacelets such as ω_r, Ω_{r-p}, we have $X=x-p$, $Y=y$.

If x, X and y, Y are the cross ranks and orthotomies of two pairs of corresponding spacelets such as (ω_r, ω_s), $(\Omega_{r-p}, \Omega_{s-p})$, we have $X=x-p$, $Y=y$.

The normal to Ω_{r-p} is a spacelet Ω_{q-r} which corresponds to the join of ω_p with the intersection ω'_{q-r} of the normal to ω_r with ω_{q-p}, ω'_{q-r} being the locus of those points of ω_{q-p} which are orthogonal with ω'_{r-p}.

The core ω_ρ of ω_r, which is also the core of ω'_{r-p}, lies in ω_{q-p}, and the core Ω_ρ of Ω_{r-p} corresponds to the join of ω_p with ω_ρ.

Ex. v. *If ρ and σ are the extravagances of two spacelets ω_r and ω_s of homogeneous space ω_n which lie in one non-extravagant spacelet ω_q and have as their complete intersection another non-extravagant spacelet ω_p, the possible values of r, s, ρ, σ are those consistent with the conditions*

$$r+s-p \not> q;\quad \rho \not< 0,\quad \rho \not> r-p,\quad \rho \not> q-r;\quad \sigma \not< 0,\quad \sigma \not> s-p,\quad \sigma \not> q-s;$$

and this is true even when one of the spacelets is given.

We deduce this result from Theorem I of § 173 by substituting $r-p$, $s-p$, $q-p$ for r, s, n and leaving ρ and σ unaltered.

§ 178. Possible values of the mutual orthotomy of two non-intersecting spacelets of given ranks in homogeneous space ω_n.

Lemma A. *Let ω_r and ω_s be two non-intersecting spacelets in homogeneous space ω_n of ranks r and s whose cross rank is t. Then if ω_r is given and has extravagance ρ, the possible values of r, s, ρ and t are those consistent with the conditions*

$$r+s \not> n,\quad \rho \not< 0,\quad \rho \not> r,\quad t \not< 0,\quad t \not> r,\quad t \not> s,\quad t \not< r+s-n+\rho, \quad \text{...(1)}$$

which include the conditions giving the possible values of r, s and ρ, viz.

$$r+s \not> n,\quad s \not< 0,\quad \rho \not< 0,\quad \rho \not> r,\quad \rho \not> n-r; \quad \text{...........(2)}$$

and if ω_s is given and has extravagance σ, the possible values of r, s, σ and t are those consistent with the conditions

$$r+s \not> n,\quad \sigma \not< 0,\quad \sigma \not> s,\quad t \not< 0,\quad t \not> r,\quad t \not> s,\quad t \not< r+s-n+\sigma, \quad \text{...(1')}$$

which include the conditions giving the possible values of r, s and σ, viz.

$$r+s \not> n,\quad r \not< 0,\quad \sigma \not< 0,\quad \sigma \not> s,\quad \sigma \not> n-s. \quad \text{...........(2')}$$

Let ω_r be any spacelet whatever having rank r and extravagance ρ, so that r and ρ are any integers satisfying the necessary conditions

$$\rho \not< 0,\quad \rho \not> r,\quad r+\rho \not> n. \quad \text{.......................(3)}$$

Then ω_s will be a spacelet which does not intersect ω_r and which moreover is such that the cross rank of ω_r and ω_s is t when and only when it has a complete intersection of rank $r-t$ with ω_{n-r} and does not intersect ω_r. Regarding ω_s as the join of a spacelet ω_{s-t} lying in ω_{n-r} which does not intersect ω_r, i.e. does not intersect the core ω_ρ of ω_r, and a spacelet ω_t which does not intersect ω_{n-r} and does not intersect the join of ω_{s-t} and ω_r, we see as in Ex. xi of § 140 that when ω_r is given we can determine ω_s so as to have the required properties when and only when s and t satisfy the conditions

$$s-t \not< 0,\quad (s-t)+\rho \not> n-r,\quad t \not< 0,\quad t+(n-r) \not> n,\quad t+(r+s-t) \not> n,$$

or

$$r+s \not> n,\quad t \not< 0,\quad t \not> r,\quad t \not> s,\quad t \not< r+s-n+\rho. \quad \text{.........(4)}$$

Thus when ω_r is given, the possible values of r, s, ρ and t are those consistent with the conditions (3) and (4), which are together equivalent to the conditions (1); and this establishes the first part of the lemma. The conditions (1) clearly also give the possible values of r, s, ρ and t when both ω_r and ω_s are arbitrary subject to the condition that they do not intersect.

The second part of the lemma can be proved in a similar way.

Ex. i. Let τ be the mutual orthotomy of the two non-intersecting spacelets ω_r and ω_s, and suppose that ω_r is given and has extravagance ρ. Then when $r \nless s$, the possible values of r, s, ρ and τ are given by

$$r \nless s, \quad r+s \ngtr n, \quad \rho \nless 0, \quad \rho \ngtr r, \quad \tau \nless 0, \quad \tau \ngtr s, \quad \tau \ngtr n-r-\rho; \quad \text{.........(5)}$$

and when $r \ngtr s$, the possible values of r, s, ρ and τ are given by

$$r \ngtr s, \quad r+s \ngtr n, \quad \rho \nless 0, \quad \rho \ngtr r, \quad \tau \nless 0, \quad \tau \ngtr r, \quad \tau \ngtr n-s-\rho. \quad \text{..........(6)}$$

In both cases the possible values of r, s, ρ and τ are given by

$$r+s \ngtr n, \quad \rho \nless 0, \quad \rho \ngtr r, \quad \tau \nless 0, \quad \tau \ngtr r, \quad \tau \ngtr s, \quad \tau \ngtr n-r-\rho, \quad \tau \ngtr n-s-\rho. \quad \text{...(7)}$$

Ex. ii. Let τ be the mutual orthotomy of the two non-intersecting spacelets ω_r and ω_s, and suppose that ω_s is given and has extravagance σ. Then when $r \nless s$, the possible values of r, s, σ and τ are given by

$$r \nless s, \quad r+s \ngtr n, \quad \sigma \nless 0, \quad \sigma \ngtr s, \quad \tau \nless 0, \quad \tau \ngtr s, \quad \tau \ngtr n-r-\sigma; \quad \text{.........(5')}$$

and when $r \ngtr s$, the possible values of r, s, σ and τ are given by

$$r \ngtr s, \quad r+s \ngtr n, \quad \sigma \nless 0, \quad \sigma \ngtr s, \quad \tau \nless 0, \quad \tau \ngtr r, \quad \tau \ngtr n-s-\sigma. \quad \text{.........(6')}$$

In both cases the possible values of r, s, σ and τ are given by

$$r+s \ngtr n, \quad \sigma \nless 0, \quad \sigma \ngtr s, \quad \tau \nless 0, \quad \tau \ngtr r, \quad \tau \ngtr s, \quad \tau \ngtr n-r-\sigma, \quad \tau \ngtr n-s-\sigma. \quad \text{...(7')}$$

Ex. iii. When ω_r and ω_s do not intersect and have extravagances ρ and σ, the conditions

$$\tau \nless 0, \quad \tau \ngtr r, \quad \tau \ngtr s, \quad \tau \ngtr n-r-\rho, \quad \tau \ngtr n-r-\sigma, \quad \tau \ngtr n-s-\rho, \quad \tau \ngtr n-s-\sigma,$$

and the conditions

$$t \nless 0, \quad t \ngtr r, \quad t \ngtr s, \quad t \nless r+s-n+\rho, \quad t \nless r+s-n+\sigma$$

must always be satisfied.

Theorem I. *If ω_r and ω_s are two arbitrary non-intersecting spacelets of homogeneous space ω_n of ranks r and s whose cross rank is t, then the possible values of r, s and t are those consistent with the conditions*

$$r+s \ngtr n, \quad t \nless 0, \quad t \ngtr r, \quad t \ngtr s, \quad \text{....................(8)}$$

which include the conditions giving the possible values of r and s, viz.

$$r \nless 0, \quad s \nless 0, \quad r+s \ngtr n. \quad \text{........................(9)}$$

If r and s have any assigned values consistent with the necessary conditions (9), *and if τ is the mutual orthotomy of ω_r and ω_s, then when $r \nless s$ the possible values of τ are those consistent with the conditions*

$$\tau \nless 0, \quad \tau \ngtr s. \quad \text{.............................(10)}$$

We obtain (8) by eliminating ρ from the conditions (1); we obtain (9) by eliminating t from the conditions (8); and we obtain (10) by making the substitution $t = s - \tau$ in (8).

Ex. iv. When the restriction $r \nless s$ is removed, the possible values of r, s and τ are those consistent with the conditions

$$r+s \ngtr n, \quad \tau \nless 0, \quad \tau \ngtr r, \quad \tau \ngtr s. \qquad \text{.............................(11)}$$

Ex. v. *Alternative proof of the conditions* (10).

It follows from § 176 that τ must satisfy the conditions (10), and it remains to show that τ can have any value consistent with those conditions. This can be seen by taking ω_r to be a *non-extravagant* spacelet, so that ω_{n-r} does not intersect ω_r. Regarding ω_s as the join of a spacelet ω_τ which lies in ω_{n-r} and a spacelet $\omega_{s-\tau}$ which does not intersect ω_{n-r} and does not intersect the join of ω_τ and ω_r, we see as in Ex. xi of § 140 that the possible values of s and τ are those consistent with the conditions

$$\tau \nless 0, \quad \tau \ngtr n-r, \quad s-\tau \nless 0, \quad (s-\tau)+(n-r) \ngtr n, \quad (s-\tau)+(\tau+r) \ngtr n,$$

which in virtue of (9) and the condition $r \nless s$ are equivalent to the conditions (10).

Ex. vi. *Greatest possible value of the mutual orthotomy τ in Theorem I.*

Since $r+s \ngtr n$, each of the spacelets ω_r and ω_s is incident with the normal to the other when and only when each of them contains the normal to the other.

We assume throughout that $r \nless s$. Then the greatest possible value of τ is s, and τ has this value when and only when each or either of the following equivalent conditions is satisfied:

(1) *The spacelets ω_r and ω_s are mutually orthogonal.*

(2) *Each of these spacelets lies in (or is incident with) the normal to the other.*

If ω_r is given and has extravagance ρ, these conditions are satisfied when and only when ω_s lies in ω_{n-r} and does not intersect the core of ω_{n-r}; and this is possible when and only when s satisfies the conditions $s \nless 0$, $s+\rho \ngtr n-r$ or $s \nless 0$, $r+s \ngtr n-\rho$.

If ω_s is given and has extravagance σ, these conditions are satisfied when and only when ω_r lies in ω_{n-s} and does not intersect the core of ω_{n-s}; and this is possible when and only when r satisfies the conditions $r \nless 0$, $r+\sigma \ngtr n-s$ or $r \nless 0$, $r+s \ngtr n-\sigma$.

These results agree with Exs. i and ii.

Ex. vii. *Least possible value of the mutual orthotomy τ in Theorem I.*

We assume throughout that $r \nless s$. The least possible value of τ is 0, and τ has this value when and only when each or either of the following equivalent conditions is satisfied:

(1) *The spacelet ω_s does not intersect ω_{n-r}; and of course also does not intersect ω_r.*

(2) *The spacelet ω_r has a complete intersection of rank $r-s$ with ω_{n-s}; and of course does not intersect ω_s.*

If ω_r is given, we see from (1) that ω_s can be determined so that $\tau=0$ when and only when s satisfies the conditions $s \nless 0$, $s+(n-r) \ngtr n$, $s+r \ngtr n$ or $r \nless s$, $s \nless 0$, $r+s \ngtr n$; thus it can be so determined whenever s satisfies the necessary conditions (9) and is not greater than r.

If ω_s is given and has extravagance σ, we see from (2) that ω_r can be determined so that $\tau=0$ when and only when we can determine a spacelet ω_{r-s} of rank $r-s$ which lies in ω_{n-s} and does not intersect the core of ω_{n-s} and also a spacelet of rank s which does not intersect ω_{n-s} and does not intersect the join of ω_{r-s} and ω_s; and this is possible when and only when r is consistent with the conditions

$$r-s \nless 0, \quad (r-s)+\sigma \ngtr n-s, \quad s \nless 0, \quad s+(n-s) \ngtr n, \quad s+r \ngtr n,$$

i.e. when and only when r satisfies the conditions $r \nless s$, $r+s \ngtr n$, $r \ngtr n-\sigma$.

These results agree with Exs. i and ii.

Lemma B. *If ω_r and ω_s are two non-intersecting spacelets which both lie in a given non-extravagant spacelet ω_p of homogeneous space ω_n, if ω_r is given and has extravagance ρ, and if t is the cross rank of ω_r and ω_s, then the possible values of r, s, ρ and t are those consistent with the conditions*

$$r+s \not> p,\quad \rho \not< 0,\quad \rho \not> r,\quad t \not< 0,\quad t \not> r,\quad t \not> s,\quad t \not< r+s-p+\rho, \quad \ldots(12)$$

which include the conditions giving the possible values of r, s and ρ, viz.

$$r+s \not> p,\quad s \not< 0,\quad \rho \not< 0,\quad \rho \not> r,\quad \rho \not> p-r. \quad\ldots\ldots\ldots\ldots(13)$$

If ω_s is given and has extravagance σ, then the possible values of r, s, σ and t are those consistent with the conditions

$$r+s \not> p,\quad \sigma \not< 0,\quad \sigma \not> s,\quad t \not< 0,\quad t \not> r,\quad t \not> s,\quad t \not< r+s-p+\sigma, \quad \ldots(12')$$

which include the conditions giving the possible values of r, s and σ, viz.

$$r+s \not> p,\quad r \not< 0,\quad \sigma \not< 0,\quad \sigma \not> s,\quad \sigma \not> p-s. \quad\ldots\ldots\ldots(13')$$

Using the principle explained in Note 1 of § 177, we deduce Lemma B from Lemma A by substituting p for n.

Theorem II. *If ω_r and ω_s are two arbitrary non-intersecting spacelets which both lie in a given non-extravagant spacelet ω_p of homogeneous space ω_n, and if t is the cross rank of ω_r and ω_s, then the possible values of r, s and t are those consistent with the conditions*

$$r+s \not> p,\quad t \not< 0,\quad t \not> r,\quad t \not> s, \quad\ldots\ldots\ldots\ldots\ldots\ldots(14)$$

which include the conditions giving the possible values of r and s, viz.

$$r \not< 0,\quad s \not< 0,\quad r+s \not> p. \quad\ldots\ldots\ldots\ldots\ldots\ldots\ldots\ldots(15)$$

If r and s have any assigned values satisfying the necessary conditions (14), *and if τ is the mutual orthotomy of ω_r and ω_s, then when $r \not< s$ the possible values of τ are those consistent with the conditions*

$$\tau \not< 0,\quad \tau \not> s. \quad\ldots\ldots\ldots\ldots\ldots\ldots\ldots\ldots\ldots\ldots(16)$$

We can deduce Theorem II from Theorem I by substituting p for n, and we can deduce it from Lemma B by eliminating ρ from the conditions (12).

Ex. viii. When the restriction $r \not< s$ is removed, the possible values of r, s and τ in Theorem II are those consistent with the conditions

$$r+s \not> p,\quad \tau \not< 0,\quad \tau \not> r,\quad \tau \not> s.$$

Ex. ix. Let ω_r and ω_s lie in one non-extravagant spacelet ω_q and have as their complete intersection another non-extravagant spacelet ω_p. Then if ω_r is given and has extravagance ρ the possible values of r, s, ρ and t are given by

$$r+s-p \not> q,\quad \rho \not< 0,\quad \rho \not> r-p,\quad t \not< p,\quad t \not> r,\quad t \not> s,\quad t \not< r+s-q+\rho;$$

and when ω_s is given and has extravagance σ, the possible values of r, s, σ and t are given by

$$r+s-p \not> q,\quad \sigma \not< 0,\quad \sigma \not> s-p,\quad t \not< p,\quad t \not> r,\quad t \not> s,\quad t \not< r+s-q+\sigma.$$

We deduce this result from Lemma A by using Note 3 of § 177.

Ex. x. If ω_r and ω_s both lie in one non-extravagant spacelet ω_q and have as their complete intersection another non-extravagant spacelet ω_p, the possible values of r, s and t are given by

$$r+s-p \not> q, \quad t \not< p, \quad t \not> r, \quad t \not> s.$$

We deduce this result from Theorem I by using Note 3 of § 177.

NOTE. *Real non-intersecting spacelets.*

Theorem II remains true when ω_p is real and ω_r and ω_s are restricted to be real. In particular Theorem I remains true when we introduce the restrictions that ω_r and ω_s are to be real.

This will be evident from Ex. xvii of § 139.

§ 179. Possible values of the mutual orthotomy of two mutually complementary spacelets in homogeneous space ω_n.

Theorem I. *If ω_r and ω_s are two arbitrary mutually complementary spacelets of homogeneous space ω_n of ranks r and s whose cross rank is t, then the possible values of r, s and t are those consistent with the conditions*

$$r+s-n \not< 0, \quad t \not> r, \quad t \not> s, \quad t \not< r+s-n, \quad \text{..............(1)}$$

which include the conditions giving the possible values of r and s, viz.

$$r \not> n, \quad s \not> n, \quad r+s \not< n. \quad \text{........................(2)}$$

If r and s are any given integers satisfying the necessary conditions (2), *and if τ is the mutual orthotomy of ω_r and ω_s, then when $r \not< s$ the possible values of τ are those consistent with the conditions*

$$\tau \not< 0, \quad \tau \not> n-r. \quad \text{.............................(3)}$$

Let ω_{n-r} and ω_{n-s} be the normals to ω_r and ω_s, and let their cross rank and mutual orthotomy be t' and τ'. Then ω_{n-r} and ω_{n-s} are two arbitrary non-intersecting spacelets of ranks $n-r$ and $n-s$; and by Exs. vi and vii of § 176 we have $t'=t+n-r-s$, $\tau'=\tau$. The possible values of $n-r$, $n-s$, t', τ' are given by Theorem I of § 178, and substituting for t' and τ' the above values we obtain (1), (2) and (3).

Ex. i. *If ω_r and ω_s are mutually complementary, and if ω_r is given and has extravagance ρ, the possible values of r, s, ρ and t are those consistent with the conditions*

$$r+s \not< n, \quad \rho \not< 0, \quad \rho \not> n-r, \quad t \not> r, \quad t \not> s, \quad t \not< r+s-n, \quad t \not< \rho,$$

which include the conditions giving the possible values of r, s and ρ, viz.

$$r+s \not< n, \quad s \not> n, \quad \rho \not< 0, \quad \rho \not> r, \quad \rho \not> n-r.$$

Hence when $r \not< s$, the possible values of s and τ are given by

$$r \not< s, \quad r+s \not< n; \quad \tau \not< 0, \quad \tau \not> n-r, \quad \tau \not> s-\rho.$$

These results follow from Lemma A of § 178; and we can deduce Theorem I by eliminating t.

Ex. ii. *If ω_r and ω_s are mutually complementary, and if ω_s is given and has extravagance σ, the possible values of r, s, σ and t are those consistent with the conditions*

$$r+s \not< n, \quad \sigma \not< 0, \quad \sigma \not> n-s, \quad t \not> r, \quad t \not> s, \quad t \not< r+s-n, \quad t \not< \sigma,$$

which include the conditions giving the possible values of r, s and σ, viz.

$$r+s \not< n, \quad r \not> n, \quad \sigma \not< 0, \quad \sigma \not> s, \quad \sigma \not> n-s.$$

Hence when $r \not< s$, the possible values of r and τ are given by

$$r \not< s, \quad r+s \not< n; \quad \tau \not< 0, \quad \tau \not> n-r, \quad \tau \not> s-\sigma.$$

Ex. iii. When ω_r and ω_s are mutually complementary and have extravagances ρ and σ, the conditions

$$\tau \not< 0, \quad \tau \not> n-r, \quad \tau \not> n-s, \quad \tau \not> r-\rho, \quad \tau \not> r-\sigma, \quad \tau \not> s-\rho, \quad \tau \not> s-\sigma$$

and

$$t \not< r+s-n, \quad t \not> r, \quad t \not> s, \quad t \not< \rho, \quad t \not< \sigma$$

must always be satisfied.

Ex. iv. *Greatest possible value of the mutual orthotomy τ in Theorem I.*

We assume throughout that $r \not< s$. Then the greatest possible value of τ is $n-r$, and τ has this value when and only when each or either of the following equivalent conditions is satisfied, it being understood that ω_r and ω_s are mutually complementary.

(1) *The normals to ω_r and ω_s are mutually orthogonal.*

(2) *Each of the spacelets ω_r and ω_s contains (or is incident with) the normal to the other.*

If ω_r is given and has extravagance ρ, these conditions are satisfied when and only when ω_s is the normal to a spacelet ω_{n-s} which lies in ω_r and does not intersect the core of ω_r; and this is possible when and only when s satisfies the conditions $s \not> n$, $r+s \not< n+\rho$.

If ω_s is given and has extravagance σ, these conditions are satisfied when and only when ω_r is the normal to a spacelet ω_{n-r} which lies in ω_s and does not intersect the core of ω_s; and this is possible when and only when r satisfies the conditions $r \not> n$, $r+s \not< n+\sigma$.

These results agree with Exs. i and ii.

Ex. v. *Least possible value of the mutual orthotomy τ in Theorem I.*

We assume throughout that $r \not< s$. The least possible value of τ is 0, and τ has this value when and only when each or any one of the following equivalent conditions is satisfied:

(1) *The spacelet ω_r is the normal to a spacelet ω_{n-r} which intersects neither ω_s nor ω_{n-s}; i.e. ω_r is complementary to both ω_s and ω_{n-s}.*

(2) *The spacelet ω_s is the normal to a spacelet ω_{n-s} which has a complete intersection of rank $r-s$ with ω_r and does not intersect ω_{n-r}; i.e. ω_s is complementary to ω_r and does not intersect ω_{n-r}.*

If ω_s is given, we see from (1) that we can determine ω_{n-r} so that the mutual orthotomy of ω_r and ω_s is 0 when and only when r satisfies the conditions $r \not< s$, $r+s \not< n$, $r \not> n$; thus it can be so determined whenever r satisfies the necessary conditions (2) and is not less than s.

If ω_r is given and has extravagance ρ, we see from (2) as in Ex. vii of § 178 that we can determine ω_{n-s} so that the mutual orthotomy of ω_r and ω_s is 0 when and only when s satisfies the conditions $r \not< s$, $r+s \not< n$, $s \not< \rho$.

These results agree with Exs. i and ii.

Theorem II. *If ω_r and ω_s are two arbitrary spacelets of ranks r and s which both lie in a given non-extravagant spacelet ω_p of homogeneous space ω_n and are mutually complementary in ω_p, and if t is the cross rank of ω_r and ω_s, the possible values of r, s and t are those consistent with the conditions*

$$r+s-p \not< 0, \quad t \not> r, \quad t \not> s, \quad t \not< r+s-p, \qquad \text{..............(4)}$$

which include the conditions giving the possible values of r and s, viz.

$$r \not> p, \quad s \not> p, \quad r+s \not< p. \qquad \text{........................(5)}$$

If r and s are any given integers satisfying the necessary conditions (5), and if τ is the mutual orthotomy of ω_r and ω_s, then when $r \not< s$, the possible values of τ are those consistent with the conditions

$$\tau \not< 0, \quad \tau \not> p-r. \qquad \text{.............................(6)}$$

We can employ the same proof as that of Lemma B or Theorem II in § 178, observing that ω_r and ω_s are mutually complementary in ω_p when and only when the matrix $\left[\alpha, \beta\right]_p^{r,s}$ in Note 1 of § 177 has rank p, i.e. when and only when Ω_r and Ω_s are mutually complementary spacelets of the homogeneous space Ω_p. Thus we obtain Theorem II from Theorem I by simply substituting p for n.

NOTE 1. *Extensions of the results given in Exs.* i, ii *and* iii.

We can deduce from each of these results a corresponding result for spacelets ω_r and ω_s which lie in a given non-extravagant spacelet ω_p and are mutually complementary in ω_p by simply substituting p for n.

NOTE 2. *Real mutually complementary spacelets.*

Theorems I and II remain true when we introduce the restrictions that ω_r and ω_s are to be real, the given spacelet ω_p in Theorem II also being real.

§ 180. Possible values of the mutual orthotomy of two spacelets which have a given complete intersection and which both lie in the plenum of their intersection.

Lemma. *Let P be any spacelet of rank p and extravagance π in homogeneous space ω_n; let X and Y be two spacelets each of which is orthogonal with P and does not intersect P; let (P, X) denote the join of P and X, and (P, Y) denote the join of P and Y; and let*

$$t = \text{cross rank of } (P, X) \text{ and } (P, Y), \quad t' = \text{cross rank of } X \text{ and } Y,$$

$$\tau = \text{orthotomy of } (P, X) \text{ and } (P, Y), \quad \tau' = \text{orthotomy of } X \text{ and } Y.$$

Then
$$t = t' + (p-\pi), \quad \tau = \tau' + \pi.$$

PROOF OF THE LEMMA. If $\omega_\pi \equiv \left[\gamma\right]_n^{\pi}$ is the core of P, we can write

$$P \equiv \omega_p \equiv \left[\gamma, c\right]_n^{\pi,\, p-\pi}, \quad \text{where} \quad \begin{bmatrix}\gamma \\ c\end{bmatrix}_{\pi,\, p-\pi}^{n} \left[\gamma, c\right]_n^{\pi,\, p-\pi} = \begin{bmatrix}0, & 0 \\ 0, & 1\end{bmatrix}_{\pi,\, p-\pi}^{\pi,\, p-\pi}.$$

Let $X \equiv \omega_x \equiv \lceil a \rfloor^{x}_{n}$, $Y \equiv \omega_y \equiv \lceil b \rfloor^{y}_{n}$, so that

$$(P,\ X) \equiv \omega_{p+x} \equiv \lceil \gamma,\ c,\ a \rfloor^{\pi,\ p-\pi,\ x}_{n}, \quad (P,\ Y) \equiv \omega_{p+y} \equiv \lceil \gamma,\ c,\ b \rfloor^{\pi,\ p-\pi,\ y}_{n}.$$

Then if
$$\psi' = [a]^{n}_{x} \lceil b \rfloor^{y}_{n} = [d]^{y}_{x},$$

and
$$\psi = \begin{bmatrix} \gamma \\ c \\ a \end{bmatrix}^{n}_{\pi,\ p-\pi,\ x} \lceil \gamma,\ c,\ b \rfloor^{\pi,\ p-\pi,\ y}_{n} = \begin{bmatrix} 0, & 0, & 0 \\ 0, & 1, & 0 \\ 0, & 0, & d \end{bmatrix}^{\pi,\ p-\pi,\ y}_{\pi,\ p-\pi,\ x},$$

we have
$$t' = \text{rank of } \psi' = \text{rank of } [d]^{y}_{x},$$

and
$$t = \text{rank of } \psi = (p-\pi) + \text{rank of } [d]^{y}_{x} = (p-\pi) + t'.$$

In proving the second statement in the lemma we may suppose that $x \nless y$. Then

$$\tau' = \text{degeneracy of } \psi' = y - t', \quad \tau = \text{degeneracy of } \psi = p + y - t,$$

and therefore
$$\tau - \tau' = p - t + t' = \pi.$$

Ex. i. *The extravagance of* $(P,\ X) = \pi +$ *the extravagance of* X, *and the extravagance of* $(P,\ Y) = \pi +$ *the extravagance of* Y. *Also the core of* $(P,\ X)$ *is the join of the cores of* P *and* X, *and the core of* $(P,\ Y)$ *is the join of the cores of* P *and* Y.

The first two statements follow from the lemma, and all four statements follow from Ex. viii of § 170.

In the theorem which follows ω_p is a given spacelet of rank p and extravagance π, and ω_{n-p} is the normal to ω_p; ω_π is the common core and the complete intersection of ω_p and ω_{n-p}; $\omega_{n-\pi}$ is the plenarily extravagant spacelet normal to ω_π which is also the common plenum and the join of ω_p and ω_{n-p}; ω_r and ω_s are any two spacelets which have ω_p for their complete intersection and which both lie in the plenum $\omega_{n-\pi}$; ω_{n-r} and ω_{n-s} are the normals to ω_r and ω_s, and can be regarded as any two spacelets which lie in ω_{n-p}, are mutually complementary in ω_{n-p}, and both contain ω_π; ρ and σ are the extravagances of ω_r and ω_s; ω_ρ and ω_σ are respectively the common core of ω_r, ω_{n-r} and the common core of ω_s, ω_{n-s}, and are two spacelets lying in ω_{n-p} whose complete intersection is ω_π. Further $\omega_{p-\pi}$ is a non-extravagant spacelet of rank $p-\pi$ lying in ω_p, so that ω_p is the join of the two non-intersecting mutually orthogonal spacelets $\omega_{p-\pi}$ and ω_π; and $\omega_{n-p-\pi}$ is a non-extravagant spacelet of rank $n-p-\pi$ lying in ω_{n-p}, so that ω_{n-p} is the join of the two non-intersecting mutually orthogonal spacelets $\omega_{n-p-\pi}$ and ω_π. We can always write

$$\omega_\pi \equiv \lceil c \rfloor^{\pi}_{n}, \quad \omega_{p-\pi} \equiv \lceil a \rfloor^{p-\pi}_{n}, \quad \omega_{n-p-\pi} \equiv \lceil b \rfloor^{n-p-\pi}_{n},$$

$$\omega_p \equiv \lceil c,\ a \rfloor^{\pi,\ p-\pi}_{n}, \quad \omega_{n-p} \equiv \lceil c,\ b \rfloor^{\pi,\ n-p-\pi}_{n}, \quad \omega_{n-\pi} \equiv \lceil c,\ a,\ b \rfloor^{\pi,\ p-\pi,\ n-p-\pi}_{n},$$

where

$$\begin{bmatrix} c \\ a \\ b \end{bmatrix}_{\pi,\, p-\pi,\, n-p-\pi}^{n} \quad \begin{bmatrix} c, & a, & b \end{bmatrix}_{n}^{\pi,\, p-\pi,\, n-p-\pi} = \begin{bmatrix} 0, & 0, & 0 \\ 0, & 1, & 0 \\ 0, & 0, & 1 \end{bmatrix}_{\pi,\, p-\pi,\, n-p-\pi}^{\pi,\, p-\pi,\, n-p-\pi}.$$

In the figure the area inside any continuous curved line which is marked with a number x represents a spacelet ω_x; and the shaded area represents the completely extravagant spacelet ω_π which is the core of each of the spacelets ω_p, ω_{n-p}, $\omega_{n-\pi}$.

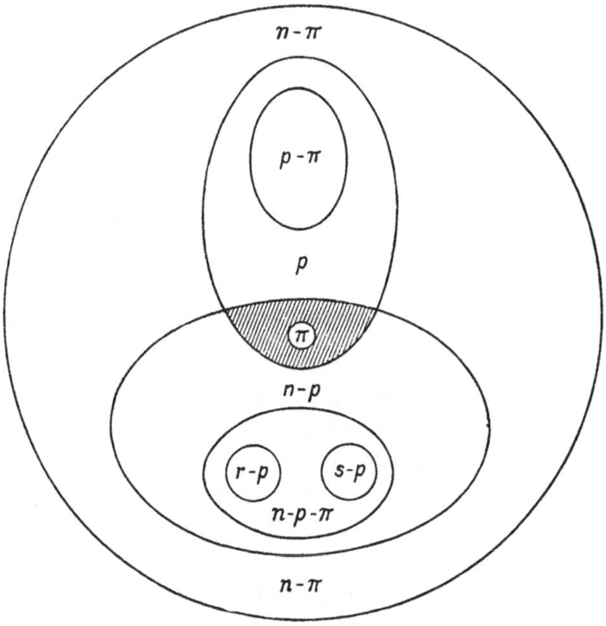

Theorem. *Let ω_r and ω_s be any two spacelets which have as their complete intersection the given spacelet ω_p whose core is ω_π, and which both lie in the plenum $\omega_{n-\pi}$ of ω_p. Then if t is the cross rank of ω_r and ω_s, the possible values of r, s and t are those consistent with the conditions*

$$r+s-p \ngtr n-\pi, \quad t \nless p-\pi, \quad t \ngtr r-\pi, \quad t \ngtr s-\pi, \quad \text{.........(1)}$$

which include the conditions giving the possible values of r and s, viz.

$$r \nless p, \quad s \nless p, \quad r+s-p \ngtr n-\pi. \quad \text{.....................(2)}$$

Further if r and s have any assigned values consistent with the necessary conditions (2), *and if τ is the mutual orthotomy of ω_r and ω_s, then when $r \nless s$ the possible values of τ are those consistent with the conditions*

$$\tau \nless \pi, \quad \tau \ngtr s-p+\pi. \quad \text{..........................(3)}$$

Here the given integers p and π can have any values consistent with the necessary conditions

$$\pi \nless 0, \quad \pi \ngtr p, \quad p + \pi \ngtr n. \qquad \text{(4)}$$

Before proving this theorem we will define certain other spacelets, the comprehension of which will be facilitated by the figure.

Definitions of $\omega_{r-p+\pi}$ *and* $\omega_{s-p+\pi}$. Since the join of ω_r and ω_{n-p} contains both ω_p and ω_{n-p} and also lies in $\omega_{n-\pi}$, it must be $\omega_{n-\pi}$; consequently the complete intersection of ω_r and ω_{n-p} has rank $r-p+\pi$; and similarly the complete intersection of ω_s and ω_{n-p} has rank $s-p+\pi$.

Thus the *complete intersection of* ω_r *with* ω_{n-p}, i.e. the locus of those points of ω_r which are orthogonal with ω_p, is a certain spacelet $\omega_{r-p+\pi}$ of rank $r-p+\pi$; and the *complete intersection of* ω_s *with* ω_{n-p}, i.e. the locus of those points of ω_s which are orthogonal with ω_p, is a certain spacelet $\omega_{s-p+\pi}$ of rank $s-p+\pi$. Since the points common to $\omega_{r-p+\pi}$ and $\omega_{s-p+\pi}$ are the points of ω_p which lie in ω_{n-p}, they are the points of ω_π.

Consequently $\omega_{r-p+\pi}$ and $\omega_{s-p+\pi}$ are two spacelets lying in ω_{n-p} whose complete intersection is ω_π.

Ex. ii. *If* $\omega_{p-\pi}$ *is any given non-extravagant spacelet of the greatest possible rank* $p-\pi$ *lying in* ω_p, *then* ω_r *is the join of the two non-intersecting mutually orthogonal spacelets* $\omega_{p-\pi}$ *and* $\omega_{r-p+\pi}$, *and* ω_s *is the join of the two non-intersecting mutually orthogonal spacelets* $\omega_{p-\pi}$ *and* $\omega_{s-p+\pi}$.

Since $\omega_{p-\pi}$ does not intersect ω_π, it does not intersect ω_{n-p}, and does not intersect $\omega_{r-p+\pi}$. Therefore the join of $\omega_{p-\pi}$ and $\omega_{r-p+\pi}$, which lies in ω_r, has rank r and must be ω_r.

Ex. iii. *The spacelet* $\omega_{r-p+\pi}$ *is the locus of those points of* ω_r *which are orthogonal with* $\omega_{p-\pi}$; *and the spacelet* $\omega_{s-p+\pi}$ *is the locus of those points of* ω_s *which are orthogonal with* $\omega_{p-\pi}$.

This follows from Ex. ii above and Ex. ix of § 170.

In fact since ω_r lies in $\omega_{n-\pi}$, all points of ω_r are orthogonal with ω_π. Hence the locus of those points of ω_r which are orthogonal with $\omega_{p-\pi}$ is the same as the locus of those points of ω_r which are orthogonal with ω_p, i.e. it is the complete intersection of ω_r and ω_{n-p} which has been defined to be $\omega_{r-p+\pi}$.

Ex. iv. *The complete intersection of* ω_p *and* $\omega_{r-p+\pi}$ *is* ω_π; *and the complete intersection of* ω_p *and* $\omega_{s-p+\pi}$ *is* ω_π.

For the points common to ω_p and $\omega_{r-p+\pi}$ must lie in ω_{n-p} and therefore in ω_π.

It follows that the joins of ω_p with $\omega_{r-p+\pi}$ and with $\omega_{s-p+\pi}$ have respectively ranks r and s.

Ex. v. *The spacelet* ω_r *is the join of the two mutually orthogonal spacelets* ω_p *and* $\omega_{r-p+\pi}$; *and the spacelet* ω_s *is the join of the two mutually orthogonal spacelets* ω_p *and* $\omega_{s-p+\pi}$.

For by Ex. iv the join of ω_p and $\omega_{r-p+\pi}$, which lies in ω_r, has rank r and must therefore be ω_r.

Ex. vi. *The spacelets $\omega_{r-p+\pi}$ and $\omega_{s-p+\pi}$ have the same mutual orthotomy as the spacelets ω_r and ω_s.*

In fact if t and t' are the cross ranks, and if τ and τ' are the mutual orthotomies of the pairs of spacelets (ω_r, ω_s) and $(\omega_{r-p+\pi}, \omega_{s-p+\pi})$, we see from Ex. ii above and the lemma that

$$t=t'+(p-\pi), \quad \tau=\tau'.$$

Ex. vii. *The spacelets $\omega_{r-p+\pi}$ and $\omega_{s-p+\pi}$ have the same extravagances and the same cores as the spacelets ω_r and ω_s.*

This follows from Ex. ii above and Ex. viii of § 170.

Definitions of ω_{r-p} and ω_{s-p}. Let ω_{r-p} be any spacelet of rank $r-p$ which lies in $\omega_{r-p+\pi}$ and does not intersect ω_π, i.e. does not intersect ω_p; and let ω_{s-p} be any spacelet of rank $s-p$ which lies in $\omega_{s-p+\pi}$ and does not intersect ω_π, i.e. does not intersect ω_p. Then $\omega_{r-p+\pi}$ is the join of the two non-intersecting mutually orthogonal spacelets ω_π, ω_{r-p}; and $\omega_{s-p+\pi}$ is the join of the two non-intersecting mutually orthogonal spacelets ω_π, ω_{s-p}. Since ω_p and ω_{r-p} do not intersect, their join, which lies in ω_r, has rank r and must be ω_r. Thus ω_r is the join of the two non-intersecting mutually orthogonal spacelets ω_p, ω_{r-p}; and similarly ω_s is the join of the two non-intersecting mutually orthogonal spacelets ω_p, ω_{s-p}. Again since ω_p is the complete intersection of ω_r and ω_s, it follows from the last result that the three spacelets ω_p, ω_{r-p}, ω_{s-p} are unconnected, i.e. ω_{r-p} and ω_{s-p} are two non-intersecting spacelets lying in ω_{n-p} whose ioin does not intersect ω_p, i.e. does not intersect ω_π.

Ex. viii. *We can always determine (non-extravagant) spacelets of rank $n-p-\pi$ lying in ω_{n-p} which contain both ω_{r-p} and ω_{s-p} and do not intersect ω_π, i.e. do not intersect ω_p.*

For the join of ω_{r-p} and ω_{s-p} is a spacelet of rank $r+s-2p$ which lies in ω_{n-p} and does not intersect ω_π, and we have $(r+s-2p)+\pi \not> n-p$, i.e. $r+s-2p \not> n-p-\pi$.

Ex. ix. *If $\omega_{n-p-\pi}$ is any given non-extravagant spacelet of the greatest possible rank $n-p-\pi$ lying in ω_{n-p}, we can always so choose the spacelets ω_{r-p} and ω_{s-p} that they both lie in $\omega_{n-p-\pi}$. They are then the complete intersections of ω_r and ω_s with $\omega_{n-p-\pi}$.*

Let the complete intersection of $\omega_{r-p+\pi}$ with $\omega_{n-p-\pi}$, which is also the complete intersection of ω_r with $\omega_{n-p-\pi}$, be ω_x. Then because the join of $\omega_{r-p+\pi}$ and $\omega_{n-p-\pi}$ must lie in ω_{n-p}, we have $(r-p+\pi)+(n-p-\pi)-x \not> n-p$, i.e. $x \not< r-p$. Again because $\omega_{n-p-\pi}$ does not intersect ω_π, it does not intersect ω_p, and therefore ω_x does not intersect ω_p; consequently ω_p and ω_x are two non-intersecting spacelets lying in ω_r, and we have $p+x \not> r$, i.e. $x \not> r-p$. Thus the complete intersection of $\omega_{r-p+\pi}$ or ω_r with $\omega_{n-p-\pi}$ is a spacelet ω'_{r-p} of rank $r-p$; and similarly the complete intersection of $\omega_{s-p+\pi}$ or ω_s with $\omega_{n-p-\pi}$ is a spacelet ω'_{s-p} of rank $s-p$.

The spacelets ω_{r-p} and ω_{s-p} lie in $\omega_{n-p-\pi}$ when and only when they are the spacelets ω'_{r-p} and ω'_{s-p}.

Ex. x. When ω_{r-p} and ω_{s-p} are any two non-intersecting spacelets of ranks $r-p$ and $s-p$ lying in ω_{n-p} whose join does not intersect ω_π, i.e. does not intersect ω_p, or when they are any two non-intersecting spacelets of ranks $r-p$ and $s-p$ lying in the given non-extravagant spacelet $\omega_{n-p-\pi}$, then the three spacelets ω_p, ω_{r-p}, ω_{s-p} are unconnected; therefore the join of the two spacelets ω_p, ω_{r-p} and the join of the two spacelets ω_p, ω_{s-p} are two spacelets of ranks r and s which lie in $\omega_{n-\pi}$ and have ω_p for their complete intersection.

Hence when ω_r and ω_s are arbitrary subject to the conditions that they have ω_p for their complete intersection and that they both lie in $\omega_{n-\pi}$, we can regard ω_{r-p} and ω_{s-p} as two arbitrary non-intersecting spacelets lying in ω_{n-p} whose join does not intersect ω_π, and we can also regard ω_{r-p} and ω_{s-p} as two arbitrary non-intersecting spacelets lying in the given non-extravagant spacelet $\omega_{n-p-\pi}$.

Ex. xi. If t, t', t'' are the cross ranks, and if τ, τ', τ'' are the mutual orthotomies of the three pairs of spacelets (ω_r, ω_s), $(\omega_{r-p+\pi}, \omega_{s-p+\pi})$, $(\omega_{r-p}, \omega_{s-p})$, we see from the lemma that

$$t'=t'', \quad t=t'+(p-\pi); \quad \tau=\tau'=\pi+\tau''.$$

Ex. xii. If ρ, ρ', ρ'' are the extravagances of ω_r, $\omega_{r-p+\pi}$, ω_{r-p}; and if σ, σ', σ'' are the extravagances of ω_s, $\omega_{s-p+\pi}$, ω_{s-p}; then

$$\rho=\rho'=\pi+\rho'', \quad \sigma=\sigma'=\pi+\sigma''.$$

Ex. xiii. The core of ω_r, which is also the core of $\omega_{r-p+\pi}$, is the join of ω_π with the core of ω_{r-p}; and the core of ω_s, which is also the core of $\omega_{s-p+\pi}$, is the join of ω_π with the core of ω_{s-p}.

Ex. xiv. If $\omega_{p-\pi}$ is any non-extravagant spacelet of rank $p-\pi$ which lies in ω_p (and does not intersect ω_π), then ω_r is the join of the three unconnected mutually orthogonal spacelets $\omega_{p-\pi}$, ω_π, ω_{r-p}; and ω_s is the join of the three unconnected mutually orthogonal spacelets $\omega_{p-\pi}$, ω_π, ω_{s-p}.

PROOF OF THE THEOREM. If ω_r and ω_s are two arbitrary spacelets of ranks r and s which lie in $\omega_{n-\pi}$ and have ω_p for their complete intersection, and if $\omega_{n-p-\pi}$ is any given non-extravagant spacelet of rank $n-p-\pi$ lying in ω_{n-p}, we have seen that ω_r and ω_s are the joins of ω_p with two arbitrary non-intersecting spacelets ω_{r-p} and ω_{s-p} of ranks $r-p$ and $s-p$ lying in $\omega_{n-p-\pi}$.

Let t and t'' be the cross ranks of the pairs of spacelets (ω_r, ω_s) and $(\omega_{r-p}, \omega_{s-p})$. Then by Theorem II of § 178 the possible values of $r-p$, $s-p$ and t'' are those consistent with the conditions

$$(r-p)+(s-p) \not> n-p-\pi, \quad t'' \not< 0, \quad t'' \not> r-p, \quad t'' \not> s-p. \quad \ldots(1')$$

Now by the lemma we have $t=t''+(p-\pi)$; and when we make the substitution $t''=t-p+\pi$ in (1′) we see that the possible values of r, s and t are those consistent with the conditions (1). When we eliminate t from (1), we obtain (2). Consequently, as can easily be seen directly, the possible values of r and s are those consistent with the conditions (2). If τ is the mutual orthotomy of ω_r and ω_s, then when $r \not< s$ we have $t=s-\tau$; and substituting this value for t in (1) we see that the possible values of τ are those consistent with the conditions (3).

We have thus proved the theorem.

Ex. xv. *If $\omega_{p-\pi}$ is any non-extravagant spacelet of rank $p-\pi$ lying in ω_p, then $\omega_{p-\pi}$ lies in both ω_r and ω_s but intersects neither ω_{n-r} nor ω_{n-s}.*

For $\omega_{p-\pi}$ does not intersect its normal, and ω_{n-r} and ω_{n-s} both lie in its normal. In fact ω_π is the complete intersection of ω_{n-r} with ω_p, and also the complete intersection of ω_{n-s} with ω_p.

Ex. xvi. *Verification of the limits to τ in the theorem of the text.*

Because ω_s contains a spacelet $\omega_{p-\pi}$ which does not intersect ω_{n-r}, therefore the complete intersection of ω_s with ω_{n-r} cannot have rank greater than $s-p+\pi$; and because all points of ω_π are common to ω_s and ω_{n-r}, therefore the complete intersection of ω_s with ω_{n-r} cannot have rank less than π.

Hence when $r \nless s$, we must have $r \ngtr s-p+\pi$ and $\tau \nless \pi$.

Ex. xvii. When we remove the restriction that $r \nless s$, the possible values of r, s and τ are those consistent with the conditions

$$r+s-p \ngtr n-\pi, \quad \tau \nless \pi, \quad \tau \ngtr r-p+\pi, \quad \tau \ngtr s-p+\pi. \quad \text{................(5)}$$

Ex. xviii. *The complete intersection of ω_{n-r} with $\omega_{r-p+\pi}$ is ω_ρ; and the complete intersection of ω_{n-s} with $\omega_{s-p+\pi}$ is ω_σ.*

For ω_ρ, which lies in ω_{n-r}, lies in ω_{n-p} and is the locus of those points of ω_{n-p} which are common to ω_r and ω_{n-r}, i.e. common to $\omega_{r-p+\pi}$ and ω_{n-r}.

Ex. xix. *The complete intersection of ω_{n-r} with ω_{r-p} is the core of ω_{r-p}; and the complete intersection of ω_{n-s} with ω_{s-p} is the core of ω_{s-p}.*

Since ω_π does not intersect ω_{r-p}, it follows from Exs. xiii and xviii that the core of ω_{r-p} is the locus of all points common to ω_ρ and ω_{r-p}, i.e. of all points common to ω_{n-r} and ω_{r-p}.

Ex. xx. *The complete intersections of the normals to $\omega_{r-p+\pi}$ and ω_{r-p} with ω_{n-p} are both ω_{n-r}; and the complete intersections of the normals to $\omega_{s-p+\pi}$ and ω_{s-p} with ω_{n-p} are both ω_{n-s}.*

For ω_r is the join of the two spacelets $\omega_{r-p+\pi}$, ω_π and also the join of the two spacelets ω_{r-p}, ω_p.

Ex. xxi. *Possible values of r, s, ρ, t in the theorem when ω_r is given or arbitrary.*

From Lemma B of § 178 we see that the possible values of r, s, ρ, t are those consistent with the conditions

$$r+s-p \ngtr n-\pi, \quad \rho \nless \pi, \quad \rho \ngtr r-p+\pi; \quad t \ngtr r-\pi, \quad t \ngtr s-\pi, \quad t \nless p-\pi, \quad t \nless r+s-n-\pi+\rho;$$

which include the conditions giving the possible values of r, s and ρ, viz.

$$r+s-p \ngtr n-\pi, \quad s \nless p; \quad \rho \nless \pi, \quad \rho \ngtr r-p+\pi, \quad \rho \ngtr n-r.$$

Ex. xxii. *Possible values of r, s, σ, t in the theorem when ω_s is given or arbitrary.*

From Lemma B of § 178 we see that the possible values of r, s, σ, t are those consistent with the conditions

$$r+s-p \ngtr n-\pi, \quad \sigma \nless \pi, \quad \sigma \ngtr s-p+\pi; \quad t \ngtr r-\pi, \quad t \ngtr s-\pi, \quad t \nless p-\pi, \quad t \nless r+s-n-\pi+\sigma;$$

which include the conditions giving the possible values of r, s and σ, viz.

$$r+s-p \ngtr n-\pi, \quad r \nless p; \quad \sigma \nless \pi, \quad \sigma \ngtr s-p+\pi, \quad \sigma \ngtr n-s.$$

Ex. xxiii. *Possible values of r, s, ρ, τ and of r, s, σ, τ in the theorem when $r \nless s$.*

When ω_r is either given or arbitrary the possible values of r, s, ρ, τ are those consistent with the conditions

$$r \nless s,\quad r+s-p \ngtr n-\pi,\quad \rho \nless \pi,\quad \rho \ngtr r-p+\pi;\quad \tau \nless \pi,\quad \tau \ngtr s-p+\pi,\quad \tau \ngtr n-r+\pi-\rho;$$

and when ω_s is either given or arbitrary the possible values of r, s, σ, τ are those consistent with the conditions

$$r \nless s,\quad r+s-p \ngtr n-\pi,\quad \sigma \nless \pi,\quad \sigma \ngtr s-p+\pi;\quad \tau \nless \pi,\quad \tau \ngtr s-p+\pi,\quad \tau \ngtr n-r+\pi-\sigma.$$

Ex. xxiv. *Possible values of r, s, ρ, σ in the theorem.*

If ρ'' and σ'' are the extravagances of ω_{r-p} and ω_{s-p}, we can obtain the possible values of $r-p$, $s-p$, ρ'', σ'' from Ex. i of § 177. Then making the substitutions $\rho''=\rho-\pi$, $\sigma''=\sigma-\pi$, we see that the possible values of r, s, ρ, σ are those consistent with the conditions

$$r+s-p \ngtr n-\pi;\quad \rho \nless \pi,\quad \rho \ngtr r-p+\pi,\quad \rho \ngtr n-r;\quad \sigma \nless \pi,\quad \sigma \ngtr s-p+\pi,\quad \sigma \ngtr n-s;$$

which include the conditions (2) giving the possible values of r and s.

We see that the possible extravagances of each of the spacelets ω_r and ω_s are entirely independent of the extravagance of the other spacelet, and are simply those due to the fact that it lies in $\omega_{n-\pi}$ and contains ω_p.

Ex. xxv. *Possible values of r and ρ, and possible values of s and σ in the theorem.*

The possible values of r and ρ are those consistent with the conditions

$$\rho \nless \pi,\quad \rho \ngtr r-p+\pi,\quad \rho \ngtr n-r,$$

which include

$$r \nless p,\quad r \ngtr n-\pi.$$

The possible values of s and σ are those consistent with the conditions

$$\sigma \nless \pi,\quad \sigma \ngtr s-p+\pi,\quad \sigma \ngtr n-s,$$

which include

$$s \nless p,\quad s \ngtr n-\pi.$$

NOTE 1. *Greatest possible value of the mutual orthotomy τ in the theorem.*

We may observe that in every one of the pairs of spacelets $(\omega_{r-p+\pi}, \omega_{s-p+\pi})$, $(\omega_{r-p}, \omega_{s-p})$, $(\omega_{r-p+\pi}, \omega_s)$, $(\omega_r, \omega_{s-p+\pi})$, (ω_{r-p}, ω_s), (ω_r, ω_{s-p}) each of the two spacelets is incident with the normal to the other when and only when each of them lies in the normal to the other, i.e. when and only when the two spacelets are mutually orthogonal. This follows from the necessary condition $r+s-p \ngtr n-\pi$.

We will now assume throughout that $r \nless s$. Then the greatest possible value of τ is $s-p+\pi$, and τ has this value when and only when the cross rank of $\omega_{r-p+\pi}$ and $\omega_{s-p+\pi}$ is 0, i.e. (see Ex. xi) when and only when the cross rank of ω_{r-p} and ω_{s-p} is 0; and this is the case when and only when each or any one of the following equivalent conditions is satisfied:

(1) *The spacelets $\omega_{r-p+\pi}$ and $\omega_{s-p+\pi}$ are mutually orthogonal, i.e. each of them lies in (or is incident with) the normal to the other.*

(2) *The spacelets ω_{r-p} and ω_{s-p} are mutually orthogonal, i.e. each of them lies in (or is incident with) the normal to the other.*

(3) *The spacelets $\omega_{r-p+\pi}$ and ω_s (or ω_r and $\omega_{s-p+\pi}$) are mutually orthogonal, i.e. each of them lies in (or is incident with) the normal to the other.*

(4) *The spacelets* ω_{r-p} *and* ω_s (*or* ω_r *and* ω_{s-p}) *are mutually orthogonal, i.e. each of them lies in* (*or is incident with*) *the normal to the other.*

(5) *Every point of* ω_r *which is orthogonal with* ω_p (*or with* $\omega_{p-\pi}$) *is orthogonal with every point of* ω_s *which is orthogonal with* ω_p (*or with* $\omega_{p-\pi}$).

(6) *Every point of* ω_r *which is orthogonal with* ω_p (*or with* $\omega_{p-\pi}$) *is also orthogonal with* ω_s.

(7) *Every point of* ω_s *which is orthogonal with* ω_p (*or with* $\omega_{p-\pi}$) *is also orthogonal with* ω_r.

The conditions (4) are equivalent to (2) because ω_r is the join of ω_p and ω_{r-p}, ω_s is the join of ω_p and ω_{s-p}, and ω_p is orthogonal with both ω_{r-p} and ω_{s-p}.

If ω_r is given and has extravagance ρ (r and ρ being consistent with the necessary conditions given in Ex. xxv), we see from (4) that these conditions are satisfied when and only when ω_{s-p} lies in ω_{n-r} and does not intersect $\omega_{r-p+\pi}$, i.e. does not intersect the core of ω_{n-r}; and this is possible when and only when $s-p \nless 0$, $(s-p)+\rho \ngtr n-r$; i.e. when and only when s satisfies both the necessary conditions (2) of the theorem and the condition $r+s-p \ngtr n-\rho$.

If ω_s is given and has extravagance σ (s and σ being consistent with the necessary conditions given in Ex. xxv), we see in the same way that these conditions are satisfied when and only when ω_{r-p} lies in ω_{n-s} and does not intersect the core of ω_{n-s}; and this is possible when and only when $r-p \nless 0$, $(r-p)+\sigma \ngtr n-s$; i.e. when and only when r satisfies both the necessary conditions (2) of the theorem and the condition $r+s-p \ngtr n-\sigma$.

These results agree with Ex. xxiii.

NOTE 2. *Least possible value of the mutual orthotomy* τ *in the theorem.*

We assume throughout that $r \nless s$. The least possible value of τ is π, and τ has this value when and only when the mutual orthotomy of ω_{r-p} and ω_{s-p} is 0, i.e. when and only when each or any one of the following conditions is satisfied:

(1) *The spacelet* ω_{s-p} *does not intersect the normal to* ω_{r-p}, *i.e. does not intersect* ω_{n-r}; *and of course also does not intersect* ω_r.

(2) *The spacelet* ω_{r-p} *has a complete intersection of the smallest possible rank* $r-s$ *with the normal to* ω_{s-p}, *i.e. with* ω_{n-s}; *and of course also does not intersect* ω_s.

(3) *The spacelet* ω_π *is the complete intersection of* $\omega_{s-p+\pi}$ *and* ω_{n-r}, *i.e. the complete intersection of* ω_s *and* ω_{n-r}.

If ω_r is given and has extravagance ρ (r and ρ being consistent with the necessary conditions given in Ex. xxv), we see from (1) that these conditions are satisfied when and only when ω_{s-p} lies in ω_{n-p} and intersects neither ω_{n-r} nor $\omega_{r-p+\pi}$; and this is possible when and only when $s-p \nless 0$, $(s-p)+(n-r) \ngtr n-p$, $(s-p)+(r-p+\pi) \ngtr n-p$; i.e. it is possible when and only when s is not greater than r and satisfies the necessary conditions (2) of the theorem.

If ω_s is given and has extravagance σ (s and σ being consistent with the necessary conditions given in Ex. xxv), we see from (2) that these conditions are satisfied when and only when ω_{r-p} is the join of a spacelet ω_{r-s} of rank $r-s$ lying in ω_{n-s} which does not intersect the core of ω_{n-s} and a spacelet ω'_{s-p} of rank $s-p$ lying in ω_{n-p} which does not intersect the join of ω_{r-s} and $\omega_{s-p+\pi}$; and this is possible when and only when $r-s \nless 0$, $(r-s)+\sigma \ngtr n-s$, $s-p \nless 0$, $(s-p)+(r-p+\pi) \ngtr n-p$; i.e. it is possible when and only when r is not less than s and satisfies both the necessary conditions (2) of the theorem and the condition $r \ngtr n-\sigma$.

These results agree with Ex. xxiii.

NOTE 3. *Possible values of the mutual orthotomy of two spacelets whose complete intersection is a given non-extravagant spacelet.*

The case in which ω_r and ω_s are two spacelets whose complete intersection is a given non-extravagant spacelet ω_p is that particular case of the theorem of the text in which $\pi = 0$ and $\omega_{n-\pi}$ is the complete space ω_n. The possible values of r and s are still given by (2); the possible values of r, s and t are given by

$$r+s-p \not> n, \quad t \not< p, \quad t \not> r, \quad t \not> s;$$

and when r and s are given and $r \not< s$, the possible values of τ are given by

$$\tau \not< 0, \quad \tau \not> s-p.$$

In this special case we can regard ω_r and ω_s as the joins of ω_p with two non-intersecting spacelets ω_{r-p} and ω_{s-p} which are orthogonal with ω_p, i.e. which lie in the spacelet ω_{n-p} which is now non-extravagant and does not intersect ω_p; and the mutual orthotomy of ω_r and ω_s is the same as the mutual orthotomy of ω_{r-p} and ω_{s-p}; moreover the extravagances and cores of ω_r and ω_s are the same as the extravagances and cores of ω_{r-p} and ω_{s-p}.

The mutual orthotomy τ has its greatest possible value $s-p$ when and only when each or any one of the following equivalent conditions is satisfied:

(1) *The spacelets ω_{r-p} and ω_{s-p} are mutually orthogonal, i.e. each of them lies in (or is incident with) the normal to the other.*

(2) *The spacelets ω_{r-p}, ω_s or the spacelets ω_{s-p}, ω_r are mutually orthogonal.*

(3) *Every point of ω_r orthogonal with ω_p is orthogonal with every point of ω_s orthogonal with ω_p.*

(4) *Every point of ω_r orthogonal with ω_p is orthogonal with ω_s; or every point of ω_s orthogonal with ω_p is orthogonal with ω_r.*

If ω_r is given, these conditions are satisfied when and only when ω_{s-p} lies in ω_{n-r} and does not intersect ω_{r-p}, i.e. does not intersect the core of ω_{n-r}. An exactly similar result applies when ω_s is given.

The mutual orthotomy τ has its least possible value 0 when and only when each or either of the following equivalent conditions is satisfied:

(1) *The spacelet ω_s does not intersect ω_{n-r}; or the complete intersection of ω_r with ω_{n-s} has rank $r-s$.*

(2) *The spacelet ω_{s-p} does not intersect the normal to ω_{r-p}, i.e. does not intersect ω_{n-r}; and of course also does not intersect ω_{r-p}.*

(3) *The spacelet ω_{r-p} has a complete intersection of rank $r-s$ with the normal to ω_{s-p}, i.e. with ω_{n-s}; and of course also does not intersect ω_{s-p}.*

We see from (2) how ω_s can be formed when ω_r is given, and from (3) how ω_r can be formed when ω_s is given.

NOTE 4. *Possible values of the mutual orthotomy of two real spacelets whose complete intersection is given.*

The possible values of r, s, t and τ when we introduce the restrictions that ω_r, ω_s and ω_p are all real are the same as in Note 3, and this is true even when one of the spacelets ω_r and ω_s is given. We use the facts (see Note 5 of § 181) that the normals to and the join and intersection of two real spacelets are all real. The proof is then similar to that of the theorem, all auxiliary spacelets being real.

§ 181. Possible values of the mutual orthotomy of any two spacelets whose complete intersection is given.

Lemma. *Let $\omega_{n-\pi}$ be a completely extravagant spacelet of homogeneous space ω_n whose core is ω_π; let ω_h and ω_k be two spacelets of ranks h and k which lie in $\omega_{n-\pi}$ and contain ω_π; let ω_u and ω_v be two spacelets of ranks u and v which lie completely outside or do not intersect $\omega_{n-\pi}$; let ω_r and ω_s be respectively the join of ω_h, ω_u and the join of ω_k, ω_v, so that $r=h+u$ and $s=k+v$; and let*

$$t = \text{cross rank of } \omega_r \text{ and } \omega_s, \quad t' = \text{cross rank of } \omega_h \text{ and } \omega_k.$$

Then
$$t = t' + u + v.$$

Proof of the lemma. We will write

$$\omega_\pi \equiv \lceil c \rfloor_n^{\pi}, \quad \omega_h \equiv \lceil c, a \rfloor_n^{\pi,\, h-\pi}, \quad \omega_k \equiv \lceil c, b \rfloor_n^{\pi,\, k-\pi}, \quad \omega_u \equiv \lceil a \rfloor_n^{u}, \quad \omega_v \equiv \lceil \beta \rfloor_n^{v},$$

so that
$$\omega_r \equiv \omega_{h+u} \equiv \lceil c, a, \alpha \rfloor_n^{\pi,\, h-\pi,\, u}, \quad \omega_s \equiv \omega_{k+v} \equiv \lceil c, b, \beta \rfloor_n^{\pi,\, k-\pi,\, v}.$$

Let
$$\psi' = \begin{bmatrix} c \\ a \end{bmatrix}_{\pi,\, h-\pi}^{n} \lceil c, b \rfloor_n^{\pi,\, k-\pi} = \begin{bmatrix} 0, & 0 \\ 0, & d \end{bmatrix}_{\pi,\, h-\pi}^{\pi,\, k-\pi},$$

and
$$\psi = \begin{bmatrix} c \\ a \\ \alpha \end{bmatrix}_{\pi,\, h-\pi,\, u}^{n} \lceil c, b, \beta \rfloor_n^{\pi,\, k-\pi,\, v} = \begin{bmatrix} 0, & 0, & x \\ 0, & d, & y \\ x', & y', & e \end{bmatrix}_{\pi,\, h-\pi,\, u}^{\pi,\, k-\pi,\, v}$$

Then
$$t' = \text{rank of } \psi' = \text{rank of } [d]_{h-\pi}^{k-\pi}, \quad t = \text{rank of } \psi.$$

Because $v \ngtr \pi$, and because ω_v does not intersect $\omega_{n-\pi}$, therefore the matrix

$$[x]_\pi^v = [c]_\pi^n \lceil \beta \rfloor_n^v$$

is undegenerate and has rank v, for no point of ω_v is orthogonal with ω_π. For similar reasons the matrix $[x']_u^\pi$ is undegenerate and has rank u. We may therefore assume without loss of generality that $(x)_v^v \neq 0$, $(x')_u^u \neq 0$; and ψ has then the same rank as the matrix

$$\chi = \begin{bmatrix} 0, & 0, & x \\ 0, & d, & y \\ x', & y', & e \end{bmatrix}_{v,\, h-\pi,\, u}^{u,\, k-\pi,\, v}$$

Now using Theorem IV of § 105 we see that

$$t = \text{rank of } \chi = u + \text{rank of } \begin{bmatrix} 0, & x \\ d, & y \end{bmatrix}_{v,\, h-\pi}^{k-\pi,\, v}$$
$$= u + v + \text{rank of } [d]_{h-\pi}^{k-\pi} = u + v + t';$$

and this proves the lemma.

Ex. i. *The extravagance of* ω_r = *the extravagance of* $\omega_h - u$; *and the extravagance of* ω_s = *the extravagance of* $\omega_k - v$.

These are particular cases of the lemma.

Ex. ii. *The join of* ω_u, $\omega_{n-\pi}$ *and the join of* ω_u, ω_π *both have extravagance* $\pi - u$; *and the join of* ω_v, $\omega_{n-\pi}$ *and the join of* ω_v, ω_π *both have extravagance* $\pi - v$.

These are particular cases of the lemma.

Ex. iii. *The join of* ω_u, $\omega_{n-\pi}$ *(which is also the join of* ω_r, $\omega_{n-\pi}$*) is a plenarily extravagant spacelet of rank* $n-\pi+u$ *whose core is a certain spacelet* $\omega_{\pi-u}$ *of rank* $\pi-u$ *lying in* ω_π; *and the join of* ω_v, $\omega_{n-\pi}$ *(which is also the join of* ω_s, $\omega_{n-\pi}$*) is a plenarily extravagant spacelet of rank* $n-\pi+v$ *whose core is a certain spacelet* $\omega_{\pi-v}$ *lying in* ω_π.

By Ex. ii the join of ω_u and $\omega_{n-\pi}$ has rank $n-\pi+u$ and extravagance $\pi - u$; and its core, which is orthogonal with $\omega_{n-\pi}$ (and also with ω_u and ω_r), must lie in ω_π.

If ω_{n-u}, ω_{n-v}, ω_{n-r}, ω_{n-s} are the normals to ω_u, ω_v, ω_r, ω_s, it follows that:

The complete intersection of ω_{n-r}, ω_π *(which is also the complete intersection of* ω_{n-u}, ω_π*) is the completely extravagant spacelet* $\omega_{\pi-u}$; *and the complete intersection of* ω_{n-s}, ω_π *(which is also the complete intersection of* ω_{n-v}, ω_π*) is the completely extravagant spacelet* $\omega_{\pi-v}$.

Again if ω_ρ and ω_σ are the cores of ω_r and ω_s, it follows that:

The spacelets $\omega_{\pi-u}$ *and* $\omega_{\pi-v}$ *are the complete intersections of* ω_ρ *and* ω_σ *with* ω_π.

For the intersection of ω_ρ with ω_π is the locus of all points common to ω_r, ω_{n-r} and ω_π, i.e. of all points common to ω_r and $\omega_{\pi-u}$; and since $\omega_{\pi-u}$ lies in ω_r, this locus must be $\omega_{\pi-u}$.

It will be observed that ω_{n-r}, ω_{n-s}, ω_ρ, ω_σ all lie in $\omega_{n-\pi}$.

Ex. iv. *The complete intersection of* $\omega_{\pi-u}$ *and* $\omega_{\pi-v}$ *is the normal to the join of* ω_u, ω_v, $\omega_{n-\pi}$, *i.e. the normal to the join of* ω_r, ω_s, $\omega_{n-\pi}$.

For the complete intersection of $\omega_{\pi-u}$ and $\omega_{\pi-v}$ is the complete intersection of ω_{n-u}, ω_{n-v} and ω_π, which is the normal to the join of ω_u, ω_v and $\omega_{n-\pi}$.

Hence the intersection of $\omega_{\pi-u}$ and $\omega_{\pi-v}$ has rank τ when and only when the join of ω_u, ω_v, $\omega_{n-\pi}$ has rank $n-\tau$. If ω_u and ω_v do not intersect, and if ω_{u+v} is their join, this is the case when and only when the intersection of ω_{u+v} with $\omega_{n-\pi}$ has rank $(u+v-\pi)+\tau$. If ω_u and ω_v have a complete intersection of rank κ, and if $\omega_{u+v-\kappa}$ is their join, this is the case when and only when the intersection of $\omega_{u+v-\kappa}$ with $\omega_{n-\pi}$ has rank $(u+v-\pi)-\kappa+\tau$.

Ex. v. *The join of* ω_u *and* ω_π *is a spacelet of rank* $\pi+u$ *whose core is* $\omega_{\pi-u}$; *and the join of* ω_v *and* ω_π *is a spacelet of rank* $\pi+v$ *whose core is* $\omega_{\pi-v}$.

By Ex. ii the join of ω_u and ω_π has extravagance $\pi - u$; and since $\omega_{\pi-u}$ is a completely extravagant spacelet of rank $\pi - u$ which lies in the join of ω_u and ω_π and is orthogonal with that join, it must be the core of that join.

Hence the intersections of ω_{n-u}, $\omega_{n-\pi}$ *and of* ω_{n-v}, $\omega_{n-\pi}$ *are spacelets of ranks* $n-\pi-u$ *and* $n-\pi-v$ *whose cores are* $\omega_{\pi-u}$ *and* $\omega_{\pi-v}$.

In the theorem which follows the spacelets ω_p, ω_{n-p}, ω_π, $\omega_{n-\pi}$, $\omega_{p-\pi}$, $\omega_{n-p-\pi}$ will be defined as in the theorem of § 180; ω_r and ω_s will now be any two spacelets which have ω_p for their complete intersection; ω_{n-r} and ω_{n-s} will still be the normals to ω_r and ω_s, and they can now be regarded as any two spacelets which lie in ω_{n-p} and are mutually complementary in ω_{n-p}; ω_ρ and ω_σ will still be respectively the common core of ω_r, ω_{n-r} and the common core of ω_s, ω_{n-s}, and they are two spacelets lying in ω_{n-p} which now do not in general contain ω_π; and ρ and σ are of course the extravagances of ω_r and ω_s.

In the figure ω_u and ω_v are two spacelets which do not intersect $\omega_{n-\pi}$, i.e. which lie completely outside $\omega_{n-\pi}$.

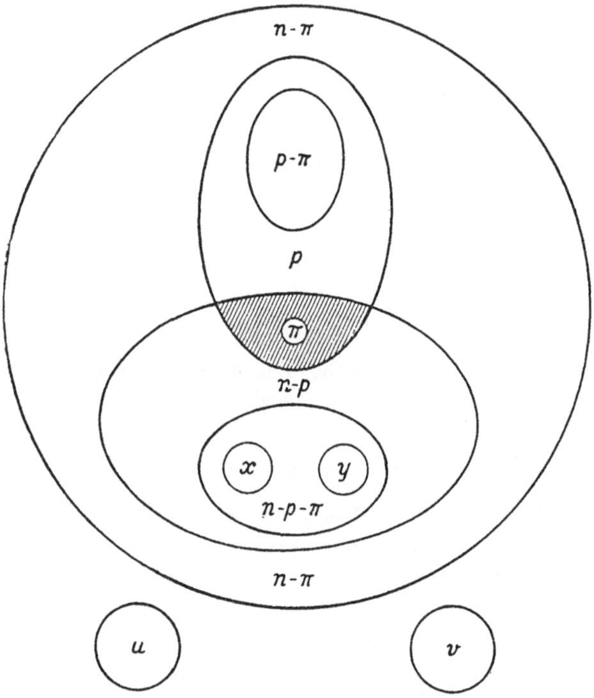

Theorem. *Let ω_r and ω_s be any two spacelets whose complete intersection is the given spacelet ω_p of rank p and extravagance π. Then if t is the cross rank of ω_r and ω_s, the possible values of r, s and t are those consistent with the conditions*

$$r \nless p, \quad s \nless p, \quad r+s-p \ngtr n, \quad \text{.....................(1)}$$

and $$t \nless p-\pi, \quad t \nless r+s-n, \quad t \ngtr r, \quad t \ngtr s, \quad t \ngtr r+s-p-\pi, \quad \text{......(2)}$$

the conditions (1) *being those which give the possible values of r and s.*

Further if r and s have any assigned values consistent with the necessary conditions (1), *and if τ is the mutual orthotomy of ω_r and ω_s, then when $r \nless s$ the possible values of τ are those consistent with the conditions*

$$\tau \nless 0, \quad \tau \nless p+\pi-r, \quad \tau \ngtr s-p+\pi, \quad \tau \ngtr n-r. \quad \text{.........(3)}$$

Here the given integers p and π can have any values consistent with the necessary conditions

$$\pi \not< 0, \quad \pi \not> p, \quad p + \pi \not> n. \qquad \text{.........................(4)}$$

Before proving this theorem we will define certain other spacelets, the comprehension of which will be facilitated by the figure.

Definitions of ω_{p+x} *and* ω_{p+y}. Let the complete intersections of ω_r and ω_s with $\omega_{n-\pi}$ (which both contain ω_p) be spacelets ω_{p+x} and ω_{p4y} of ranks $p+x$ and $p+y$, where x and y are positive integers each of which may be 0. Then ω_{p+x} and ω_{p+y} are two spacelets which lie in $\omega_{n-\pi}$ and have ω_p for their complete intersection; and we can write

$$r = p + x + u, \quad s = p + y + v, \qquad \text{.....................(5)}$$

where u and v also are positive integers each of which may be 0. The integers x, y, u, v are all known when ω_r and ω_s are known.

Ex. vi. The spacelet ω_r is the join of ω_{p+x} with a spacelet ω_u of rank u which does not intersect $\omega_{n-\pi}$; and the spacelet ω_s is the join of ω_{p+y} with a spacelet ω_v of rank v which does not intersect $\omega_{n-\pi}$. Since ω_p is the complete intersection of the join of ω_p, ω_u with the join of ω_p, ω_v, it follows that the three spacelets ω_p, ω_u, ω_v are unconnected.

Definitions of $\omega_{\pi+x}$ *and* $\omega_{\pi+y}$. The complete intersections of ω_{p+x} and ω_{p+y} with ω_{n-p} (which correspond to the spacelets $\omega_{r-p+\pi}$ and $\omega_{s-p+\pi}$ in § 180) are spacelets of ranks $\pi + x$ and $\pi + y$ which will be denoted by $\omega_{\pi+x}$ and $\omega_{\pi+y}$. These two spacelets lie in ω_{n-p} and have ω_π for their complete intersection. Clearly $\omega_{\pi+x}$ is the complete intersection of ω_r with ω_{n-p}, i.e. the locus of those points of ω_r which are orthogonal with ω_p; and $\omega_{\pi+y}$ is the complete intersection of ω_s with ω_{n-p}, i.e. the locus of those points of ω_s which are orthogonal with ω_p.

Ex. vii. The complete intersection of ω_p and $\omega_{\pi+x}$ is ω_π; and the complete intersection of ω_p and $\omega_{\pi+y}$ is ω_π.

Ex. viii. The spacelet ω_{p+x} is the join of the two mutually orthogonal spacelets ω_p and $\omega_{\pi+x}$; and the spacelet ω_{p+y} is the join of the two mutually orthogonal spacelets ω_p and $\omega_{\pi+y}$.

Definitions of ω_x *and* ω_y. Let ω_x be any spacelet of the greatest possible rank x which lies in $\omega_{\pi+x}$ and does not intersect ω_π (i.e. which lies in ω_r, is orthogonal with ω_p, and does not intersect ω_p); and let ω_y be any spacelet of the greatest possible rank y which lies in $\omega_{\pi+y}$ and does not intersect ω_π (i.e. which lies in ω_s, is orthogonal with ω_p, and does not intersect ω_p); so that ω_x and ω_y correspond to the spacelets ω_{r-p} and ω_{s-p} in § 180.

Then $\omega_{\pi+x}$ is the join of the two non-intersecting mutually orthogonal spacelets ω_π and ω_x; and $\omega_{\pi+y}$ is the join of the two non-intersecting mutually orthogonal spacelets ω_π and ω_y.

Since ω_π is the complete intersection of $\omega_{\pi+x}$ and $\omega_{\pi+y}$, it follows that the three spacelets ω_π, ω_x, ω_y are unconnected, i.e. ω_x and ω_y are two non-intersecting spacelets lying in ω_{n-p} whose join does not intersect ω_π, i.e. does not intersect ω_p.

Ex. ix. The spacelet ω_{p+x} is the join of the two unconnected mutually orthogonal spacelets ω_p and ω_x; and the spacelet ω_{p+y} is the join of the two unconnected mutually orthogonal spacelets ω_p and ω_y.

Ex. x. We can always determine (non-extravagant) spacelets of rank $n-p-\pi$ lying in ω_{n-p} which contain both ω_x and ω_y and do not intersect ω_π, i.e. do not intersect ω_p.

Ex. xi. If $\omega_{n-p-\pi}$ is any given non-extravagant spacelet of the greatest possible rank $n-p-\pi$ which lies in ω_{n-p} (and does not intersect ω_π), we can always so choose the spacelets ω_x and ω_y that they lie in $\omega_{n-p-\pi}$. They are then the complete intersections of ω_r and ω_s with $\omega_{n-p-\pi}$.

Definitions of ω_u and ω_v. Let ω_u be any spacelet of the greatest possible rank u which lies in ω_r and does not intersect $\omega_{n-\pi}$; and let ω_v be any spacelet of the greatest possible rank v which lies in ω_s and does not intersect $\omega_{n-\pi}$. Then ω_r is the join of the two non-intersecting spacelets ω_{p+x} and ω_u; and ω_s is the join of the two non-intersecting spacelets ω_{p+y} and ω_v. Therefore ω_r is the join of the three unconnected spacelets ω_p, ω_x, ω_u; and ω_s is the join of the three unconnected spacelets ω_p, ω_y, ω_v.

Since ω_p is the complete intersection of ω_r and ω_s, it follows that all the five spacelets ω_p, ω_x, ω_y, ω_u, ω_v are unconnected, i.e. ω_u and ω_v are two non-intersecting spacelets whose join ω_{u+v} does not intersect the join ω_{p+x+y} of ω_p, ω_x, ω_y; in particular the join of ω_u and ω_v does not intersect ω_p.

If $\omega_{p-\pi}$ is any given non-extravagant spacelet of rank $p-\pi$ lying in ω_p, it is shown in Ex. xxiii that ω_u and ω_v can always be so chosen that they are both orthogonal with $\omega_{p-\pi}$; and when ω_u and ω_v are given, it is shown in Ex. xiv that ω_x and ω_y can always be so chosen that ω_x is orthogonal with ω_u, and ω_y is orthogonal with ω_v.

Ex. xii. The joins of ω_r, $\omega_{n-\pi}$ and of ω_s, $\omega_{n-\pi}$, being the same as the joins of ω_u, $\omega_{n-\pi}$ and of ω_v, $\omega_{n-\pi}$, are (see Ex. iii) plenarily extravagant spacelets of ranks $n-\pi+u$ and $n-\pi+v$ whose cores lie in ω_π.

Ex. xiii. The complete intersections of ω_{n-r}, ω_π and of ω_{n-s}, ω_π, which are the same as the complete intersections of ω_{n-u}, ω_π and of ω_{n-v}, ω_π, are completely extravagant spacelets $\omega_{\pi-u}$ and $\omega_{\pi-v}$ of ranks $\pi-u$ and $\pi-v$.

Here ω_{n-u} and ω_{n-v} are the normals to ω_u and ω_v.

Ex. xiv. *When ω_u and ω_v are given, we can always choose the spacelets ω_x and ω_y so that ω_x is orthogonal with ω_u, and ω_y is orthogonal with ω_v.*

Because ω_u and ω_p do not intersect, therefore the intersection of ω_{n-u} and ω_{n-p} is a spacelet ω_{n-u-p} of rank $n-u-p$.

Because the complete intersection of ω_{n-u} with ω_π is a spacelet $\omega_{\pi-u}$ whose rank is $\pi-u$, therefore ω_{n-u-p} is the join of $\omega_{\pi-u}$ with a (non-extravagant) spacelet of rank $n-p-\pi$ which lies in ω_{n-p} and does not intersect ω_π, and $\omega_{\pi+x}$ has an intersection of rank x with this spacelet which can be taken to be ω_x. Then ω_x is orthogonal with ω_u.

Ex. xv. If t, t_1, t_2, t_3 are the cross ranks and if τ, τ_1, τ_2, τ_3 are the mutual orthotomies of the pairs of spacelets (ω_r, ω_s), $(\omega_{p+x}, \omega_{p+y})$, $(\omega_{\pi+x}, \omega_{\pi+y})$, (ω_x, ω_y), then from the lemmas of this article and of § 180 we see that

$$t_2=t_3, \quad t_1=t_3+(p-\pi), \quad t=t_1+u+v=t_3+u+v-\pi;$$

and

$$\tau_1=\tau_2=\pi+\tau_3.$$

Ex. xvi. If ρ, ρ_1, ρ_2, ξ are the extravagances of ω_r, ω_{p+x}, $\omega_{\pi+x}$, ω_x,

and if σ, σ_1, σ_2, η are the extravagances of ω_s, ω_{p+y}, $\omega_{\pi+y}$, ω_y,

then

$$\rho_2=\rho_1=\pi+\xi, \quad \rho=\rho_1-u=(\pi-u)+\xi;$$

and

$$\sigma_2=\sigma_1=\pi+\eta, \quad \sigma=\sigma_1-v=(\pi-v)+\eta.$$

Ex. xvii. The spacelets $\omega_{\pi+x}$ and ω_{p+x} have the same core, this being the join of ω_π with the core of ω_x; also the spacelets $\omega_{\pi+y}$ and ω_{p+y} have the same core, this being the join of ω_π with the core of ω_y.

Definitions of $\omega_{r-p+\pi}$ *and* $\omega_{s-p+\pi}$. Let $\omega_{p-\pi}$ be any non-extravagant spacelet of the greatest possible rank $p-\pi$ lying in ω_p, so that ω_p is the join of $\omega_{p-\pi}$ and ω_π. Then we can regard ω_r and ω_s as the joins of $\omega_{p-\pi}$ with spacelets $\omega_{r-p+\pi}$ and $\omega_{s-p+\pi}$ of ranks $r-p+\pi$ and $s-p+\pi$ which are orthogonal with and do not intersect $\omega_{p-\pi}$. The spacelet $\omega_{r-p+\pi}$ is the locus of all points of ω_r which are orthogonal with $\omega_{p-\pi}$; and the spacelet $\omega_{s-p+\pi}$ is the locus of all points of ω_s which are orthogonal with $\omega_{p-\pi}$.

Ex. xviii. The spacelets $\omega_{r-p+\pi}$ and $\omega_{s-p+\pi}$ have ω_π for their complete intersection.

Ex. xix. *The complete intersections of* $\omega_{r-p+\pi}$ *and* $\omega_{s-p+\pi}$ *with* $\omega_{n-\pi}$ *are* $\omega_{\pi+x}$ *and* $\omega_{\pi+y}$.

For $\omega_{\pi+x}$ is the locus of all those points of ω_r which are orthogonal both with $\omega_{p-\pi}$ and with ω_π, i.e. which are orthogonal with $\omega_{p-\pi}$ and lie in $\omega_{n-\pi}$.

Ex. xx. The complete intersection of $\omega_{r-p+\pi}$ and ω_p is ω_π, and the complete intersection of $\omega_{s-p+\pi}$ and ω_p is ω_π. Consequently the join of $\omega_{r-p+\pi}$ and ω_p is ω_r, and the join of $\omega_{s-p+\pi}$ and ω_p is ω_s.

It follows that the intersections of the normals to $\omega_{r-p+\pi}$ and $\omega_{s-p+\pi}$ with ω_{n-p} are respectively ω_{n-r} and ω_{n-s}.

Ex. xxi. *The spacelets* $\omega_{r-p+\pi}$ *and* $\omega_{s-p+\pi}$ *have the same mutual orthotomy as the spacelets* ω_r *and* ω_s.

In fact if t and t' are the cross ranks and if τ and τ' are the mutual orthotomies of the pairs of spacelets (ω_r, ω_s) and $(\omega_{r-p+\pi}, \omega_{s-p+\pi})$, we see from the lemma of § 180 that

$$t=t'+(p-\pi), \quad \tau=\tau'.$$

Ex. xxii. The spacelets $\omega_{r-p+\pi}$ and $\omega_{s-p+\pi}$ have the same extravagances ρ and σ, and the same cores ω_ρ and ω_σ, as the spacelets ω_r and ω_s.

Ex. xxiii. *We can choose the spacelets* ω_u *and* ω_v *so that they lie respectively in* $\omega_{r-p+\pi}$ *and* $\omega_{s-p+\pi}$, *i.e. so that they are both orthogonal with* $\omega_{p-\pi}$.

For since $u+(\pi+x)=r-p+\pi$, we can determine spacelets of rank u which lie in $\omega_{r-p+\pi}$ and do not intersect $\omega_{\pi+x}$, i.e. do not intersect $\omega_{n-\pi}$.

When ω_u *and* ω_v *are so chosen, then* $\omega_{r-p+\pi}$ *is the join of* $\omega_{\pi+x}$, ω_u, *i.e. the join of* ω_π, ω_x, ω_u; *and* $\omega_{s-p+\pi}$ *is the join of* $\omega_{\pi+y}$, ω_v, *i.e. the join of* ω_π, ω_y, ω_v.

Definitions of $\omega_{\pi-u}$, $\omega_{\pi-v}$, ω_ξ, ω_η. We will define $\omega_{\pi-u}$ to be the complete intersection of ω_π, ω_{n-r} or of ω_π, ω_{n-u}; $\omega_{\pi-v}$ to be the complete intersection of ω_π, ω_{n-s} or of ω_π, ω_{n-v}; and ω_ξ and ω_η to be the cores of ω_x and ω_y, so that ξ and η are the extravagances of ω_x and ω_y.

Ex. xxiv. The complete intersection of $\omega_{\pi+x}$, ω_{n-r} is ω_ρ; and the complete intersection of $\omega_{\pi+y}$, ω_{n-s} is ω_σ.

Ex. xxv. The spacelet $\omega_{\pi-u}$ is the complete intersection of ω_π, ω_ρ; and the spacelet $\omega_{\pi-v}$ is the complete intersection of ω_π, ω_σ.

Ex. xxvi. *If* ω_x *is orthogonal with* ω_u, *then* ω_ξ *lies in* ω_ρ; *and if* ω_y *is orthogonal with* ω_v, *then* ω_η *lies in* ω_σ.

For ω_ξ always lies in ω_r, and is always orthogonal with ω_x and ω_p; and when ω_x is orthogonal with ω_u, then ω_ξ is orthogonal with ω_u, ω_p, ω_x, i.e. with ω_r, and therefore lies in ω_{n-r} as well as in ω_r, i.e. it lies in ω_ρ.

Ex. xxvii. *The core of* $\omega_{\pi+x}$ (*which is the join of* ω_π, ω_ξ) *contains* ω_ρ; *and the core of* $\omega_{\pi+y}$ (*which is the join of* ω_π, ω_η) *contains* ω_σ.

For $\omega_{\pi+x}$ contains ω_ρ; and because $\omega_{\pi+x}$ lies in ω_r, therefore the normal to $\omega_{\pi+x}$ contains ω_{n-r}, and therefore contains ω_ρ.

Ex. xxviii. *If* ω_x *is so chosen as to be orthogonal with* ω_u, *then* ω_ρ *is the join of* $\omega_{\pi-u}$, ω_ξ; *and the complete intersection of* ω_{n-r}, ω_x *is* ω_ξ.

If ω_y *is so chosen as to be orthogonal with* ω_v, *then* ω_σ *is the join of* $\omega_{\pi-v}$, ω_η; *and the complete intersection of* ω_{n-r}, ω_y *is* ω_η.

For ω_ξ (which lies in $\omega_{n-\pi}$) is orthogonal with ω_π and therefore with $\omega_{\pi-u}$; and the join of $\omega_{\pi-u}$ and ω_ξ is a completely extravagant spacelet of rank $(\pi-u)+\xi$, i.e. of rank ρ. This spacelet will be identical with ω_ρ when and only when it lies in ω_ρ. Since $\omega_{\pi-u}$ lies in ω_ρ, this condition is satisfied when and only when ω_ξ lies in ω_ρ; and by Ex. xxvi it is satisfied when ω_x is orthogonal with ω_u.

Again since the complete intersections of ω_{n-r} with ω_π and $\omega_{\pi+x}$ have ranks $\pi-u$ and $(\pi-u)+\xi$, the complete intersection of ω_{n-r} with ω_x cannot have rank greater than ξ; and when ω_ξ lies in ω_{n-r} or in ω_ρ, it must be the complete intersection of ω_{n-r} with ω_x.

Ex. xxix. *The spacelets* $\omega_{\pi+x}$ *and* ω_{n-r} *are mutually complementary in* ω_{n-p} *when and only when* ω_x *is non-extravagant, i.e. when and only when* $\omega_{\pi-u}$ *is the core of* ω_r, *or* ω_π *is the core of* $\omega_{\pi+x}$.

Since ω_ρ is the complete intersection of $\omega_{\pi+x}$ and ω_{n-r}, it follows that ω_{n-p} is their join when and only when

$$(\pi+x)+(n-r)-\rho=n-p, \quad \text{i.e.} \quad \rho=\pi-(r-p-x)=\pi-u.$$

PROOF OF THE THEOREM. Replacing $p+x$ by h and $p+y$ by k, we have shown that when ω_r and ω_s are spacelets whose complete intersection is ω_p, we can always represent ω_r as the join of ω_h, ω_u, and ω_s as the join of ω_k, ω_v, where h, k, u, v are positive integers and

(*a*) $h+u=r$, $k+v=s$;

(*b*) ω_h and ω_k are two spacelets lying in $\omega_{n-\pi}$ whose complete intersection is ω_p;

(*c*) ω_u and ω_v are two non-intersecting spacelets lying entirely outside $\omega_{n-\pi}$ whose join ω_{u+v} does not intersect the join ω_{h+k-p} of ω_h and ω_k.

Whenever the conditions (a), (b), (c) are satisfied, the join of ω_h, ω_u and the join of ω_k, ω_v are spacelets of ranks r and s whose complete intersection is ω_p. Hence when ω_r and ω_s are arbitrary subject to the condition that their complete intersection is ω_p, then ω_h, ω_k, ω_u, ω_v are arbitrary subject to the conditions (b) and (c).

Let t' be the cross rank of ω_h and ω_k, and let t be the cross rank of ω_r and ω_s, so that by the lemma $t = t' + u + v$.

When ω_h and ω_k are arbitrary subject to the condition (b), we see from the theorem of § 180 that the possible values of h, k and t' are those consistent with the conditions

$$h + k - p \not> n - \pi, \quad t' \not< p - \pi, \quad t' \not> h - \pi, \quad t' \not> k - \pi. \quad \ldots\ldots(6)$$

And when ω_h and ω_k are given, and ω_u and ω_v are arbitrary subject to the condition (c), the possible values of u and v are those consistent with the conditions

$$u \not< 0, \quad u \not> \pi, \quad v \not< 0, \quad v \not> \pi, \quad u + v + h + k - p \not> n. \quad \ldots\ldots(7)$$

Also when ω_r is the join of ω_h and ω_u, ω_s is the join of ω_k and ω_v, and t is the cross rank of ω_r and ω_s, we have

$$h + u = r, \quad k + v = s, \quad t = t' + u + v. \quad \ldots\ldots\ldots\ldots\ldots\ldots(8)$$

Hence when ω_r and ω_s are arbitrary subject to the condition that their complete intersection is ω_p, the possible values of r, s and t are those consistent with the conditions (6), (7) and (8).

When we eliminate h, k and t' by making the substitutions $h = r - u$, $k = s - v$, $t' = t - u - v$ in (6) and (7), these conditions reduce by virtue of (4) to

$$\left.\begin{array}{c} r + s - p \not> n, \quad u \not< 0, \quad u \not> \pi, \quad v \not< 0, \quad v \not> \pi, \quad u + v + n - \pi \not< r + s - p, \\ t \not< u + v + p - \pi, \quad t \not> r + v - \pi, \quad t \not> s + u - \pi; \end{array}\right\}$$

and by further eliminating u and v, we see that the possible values of r, s and t are those consistent with the conditions (1) and (2) of the theorem.

Since the elimination of t from (1) and (2) leads to (1), the possible values of r and s are those consistent with (1); and we obtain (3) by making the substitution $t = s - \tau$ in (2). We have thus proved the theorem.

NOTE 1. *Alternative proof of the theorem.*

We can of course prove the theorem directly without using the theorem of § 180.

From the text we see that when ω_r and ω_s are spacelets whose complete intersection is ω_p, we can always represent ω_r as the join of ω_p, ω_x, ω_u and ω_s as the join of ω_p, ω_y, ω_v, where x, y, u, v are positive integers each of which may be 0, and where

(α) $p + x + u = r$, $p + y + v = s$;

(β) ω_x and ω_y lie in ω_{n-p};

(γ) ω_u and ω_v do not intersect $\omega_{n-\pi}$;

(δ) ω_p, ω_x, ω_y, ω_u, ω_v are unconnected.

Whenever the conditions (a), (β), (γ), (δ) are satisfied, the join of ω_p, ω_x, ω_u and the join of ω_p, ω_y, ω_v are spacelets of ranks r and s whose complete intersection is ω_p. Hence when ω_r and ω_s are arbitrary subject to the condition that their complete intersection is ω_p, then ω_x, ω_y, ω_u, ω_v are arbitrary subject to the conditions (β), (γ), (δ).

When $\omega_{n-p-\pi}$ is any given non-extravagant spacelet of rank $n-p-\pi$ which lies in ω_{n-p} (and does not intersect ω_π), the conditions (β), (γ), (δ) can be replaced by

(a') ω_x and ω_y lie in ω_{n-p} and are such that ω_x, ω_y, ω_p are unconnected, i.e. such that ω_x, ω_y, ω_π are unconnected; or ω_x and ω_y are two non-intersecting spacelets lying in ω_{n-p} whose join ω_{x+y} does not intersect ω_π;

or ω_x and ω_y are two non-intersecting spacelets lying in $\omega_{n-p-\pi}$;

(β') ω_u and ω_v do not intersect $\omega_{n-\pi}$, and are such that ω_u, ω_v, ω_p, ω_x, ω_y are unconnected.

We can first choose ω_x and ω_y arbitrarily subject to any one of the three alternative sets of conditions (a'), and afterwards choose ω_u and ω_v arbitrarily subject to the conditions (β').

Or again the conditions (β), (γ), (δ) can be replaced by

(a'') ω_u and ω_v do not intersect $\omega_{n-\pi}$, and are such that ω_u, ω_v, ω_p are unconnected;

or ω_u and ω_v are two non-intersecting spacelets lying entirely outside $\omega_{n-\pi}$ whose join ω_{u+v} does not intersect ω_p;

(β'') ω_x and ω_y lie in ω_{n-p}, and are such that ω_x, ω_y, ω_p, ω_u, ω_v are unconnected.

We can first choose ω_u and ω_v arbitrarily subject to the conditions (a''), and afterwards choose ω_x and ω_y arbitrarily subject to the conditiohs (β'').

Let t'', t', t be respectively the cross ranks of the three pairs of spacelets (ω_x, ω_y), $(\omega_{p+x}, \omega_{p+y})$, (ω_r, ω_s). Then since ω_{p+x} and ω_{p+y} are respectively the joins of the two pairs of mutually orthogonal spacelets (ω_p, ω_x) and (ω_p, ω_y), and ω_r and ω_s are respectively the joins of the two pairs of spacelets (ω_u, ω_{p+x}) and (ω_v, ω_{p+y}), we see from the lemmas of this article and § 180 that

$$t=t''+(p-\pi), \quad t=t'+u+v, \quad t=t''+(p-\pi)+u+v.$$

When ω_x and ω_y are arbitrary subject to the third set of conditions in (a'), it follows from Theorem II of § 178 that the possible values of x, y and t'' are those consistent with the conditions

$$x+y \not> n-p-\pi, \quad t'' \not< 0, \quad t'' \not> x, \quad t'' \not> y. \quad \text{.....................(9)}$$

And when ω_x and ω_y are given, and ω_u and ω_v are arbitrary subject to the conditions (β'), the possible values of u and v are those consistent with the conditions

$$u \not< 0, \quad u \not> \pi, \quad v \not< 0, \quad v \not> \pi, \quad u+v+x+y+p \not> n. \quad \text{............(10)}$$

Also when ω_r is the join of ω_p, ω_x, ω_u; ω_s is the join of ω_p, ω_y, ω_v; and t is the cross rank of ω_r and ω_s; we have

$$r=p+x+u, \quad s=p+y+v, \quad t=t''+(p-\pi)+u+v. \quad \text{..............(11)}$$

Hence when ω_r and ω_s are arbitrary subject to the condition that their complete intersection is ω_p, the possible values of r, s and t are those consistent with the conditions (9), (10) and (11).

Eliminating x, y and t'' by substituting from (11) in (9) and (10) we see that the possible values of u, v, r, s, t are given by

$$\left.\begin{array}{c} r+s-p \not> n, \quad u \not< 0, \quad u \not> \pi, \quad v \not< 0, \quad v \not> \pi, \quad u+v+(n-\pi) \not< r+s-p, \\ t+\pi \not< u+v+p, \quad t+\pi \not> r+v, \quad t+\pi \not> s+u. \end{array}\right\} \quad (12)$$

Or eliminating u, v and t'' by substituting from (11) in (9) and (10) we see that the possible values of x, y, r, s, t are given by

$$\left.\begin{array}{c} r+s-p \not> n, \quad x \not> r-p, \quad x \not< r-p-\pi, \quad y \not> s-p, \quad y \not< s-p-\pi, \quad x+y \not> n-p-\pi, \\ t+p+\pi \not< r+s-x-y, \quad t+p+\pi \not> r+s-x, \quad t+p+\pi \not> r+s-y. \end{array}\right\} \quad \ldots\ldots\ldots(13)$$

By eliminating u and v from (12) or by eliminating x and y from (13) we obtain the conditions (1) and (2) of the theorem; and we deduce the conditions (3) by making the substitution $t=s-\tau$ in (2).

Ex. xxx. The possible values of x, y, u, v, r, s and t in the theorem are given by

$$\left.\begin{array}{c} u \not< 0, \quad u \not> \pi, \quad v \not< 0, \quad v \not> \pi, \quad x+y+p+\pi \not> n, \quad x+y+u+v+p \not> n; \\ t+\pi \not< u+v+p, \quad t+\pi \not> u+v+p+x, \quad t+\pi \not> u+v+p+y; \\ r=p+x+u, \quad s=p+y+v. \end{array}\right\}$$

Ex. xxxi. The possible values of x, y, u, v, r, s are given by the conditions

$$\left.\begin{array}{c} u \not< 0, \quad u \not> \pi, \quad v \not< 0, \quad v \not> \pi, \quad x \not< 0, \quad y \not< 0, \quad x+y+p+\pi \not> n, \quad x+y+u+v+p \not> n; \\ r=p+x+u, \quad s=p+y+v; \end{array}\right\}$$

which include $\quad x+y+\pi \not> n-p, \quad u+v+p \not> n.$

The possible values of u, v, r, s are given by the conditions

$$\left.\begin{array}{c} u \not< 0, \quad u \not> \pi, \quad u \not> r-p; \quad v \not< 0, \quad v \not> \pi, \quad v \not> s-p; \\ r+s-p \not> n \quad r+s-p \not> u+v+n-\pi; \end{array}\right\}$$

which include $\quad u+v+p \not> n.$

The possible values of x, y, r, s are given by the conditions

$$\left.\begin{array}{c} x \not< 0, \quad x \not< r-p-\pi, \quad x \not> r-p; \quad y \not< 0, \quad y \not< s-p-\pi, \quad y \not> s-p; \\ r+s-p \not> n, \quad x+y \not> n-p-\pi. \end{array}\right\}$$

Ex. xxxii. The possible values of u and v, and the possible values of x and y are given respectively by the conditions

$$u \not< 0, \quad u \not> \pi, \quad v \not< 0, \quad v \not> \pi, \quad u+v+p \not> n;$$

and $\quad x \not< 0, \quad y \not< 0, \quad x+y \not> n-p-\pi.$

NOTE 2. *Possible extravagances of two spacelets ω_r and ω_s whose complete intersection is a given spacelet ω_p of rank p and extravagance π.*

When ω_r and ω_s are arbitrary subject to the condition that their complete intersection is ω_p, the conditions determining the possible values of x, y, ξ, η, u, v, r, s, ρ, σ are

$$\left.\begin{array}{c} x+y \not> n-p-\pi, \quad \xi \not< 0, \quad \xi \not> x, \quad x+\xi \not> n-p-\pi, \quad \eta \not< 0, \quad \eta \not> y, \quad y+\eta \not> n-p-\pi; \\ u \not< 0, \quad u \not> \pi, \quad v \not< 0, \quad v \not> \pi, \quad u+v+x+y+p \not> n; \\ r=p+x+u, \quad s=p+y+v, \quad \rho=\pi-u+\xi, \quad \sigma=\pi-v+\eta. \end{array}\right\} \ldots.(14)$$

Therefore the conditions determining the possible values of u, v, r, s, ρ, σ are

$$\left.\begin{array}{c} r+s-p \not> n; \quad u \not< 0, \quad u \not> \pi, \quad v \not< 0, \quad v \not> \pi, \quad u+v+n-\pi \not< r+s-p; \\ \rho \not< \pi-u, \quad \rho \not> n-r, \quad \rho+2u \not> r-p+\pi; \quad \sigma \not< \pi-v, \quad \sigma \not> n-s, \quad \sigma+2v \not> s-p+\pi. \end{array}\right\} \ldots(15)$$

Eliminating u and v we see that:

The possible values of r, s, ρ, σ are those consistent with the conditions

$$\left.\begin{array}{c} \rho \not< 0, \quad \rho \not> n-r, \quad r+\rho \not< p+\pi, \quad r-\rho \not< p-\pi; \\ \sigma \not< 0, \quad \sigma \not> n-s, \quad s+\sigma \not< p+\pi, \quad s-\sigma \not< p-\pi; \\ r+s-p \not> n. \end{array}\right\} \ldots\ldots\ldots\ldots\ldots\ldots(16)$$

Since by Ex. v of § 173 (or by a similar direct argument) the conditions (14) and (16) remain true when one of the two spacelets ω_r and ω_s is given, we see that:

When r and s are given, there is no necessary relation between the extravagances of ω_r and ω_s, even when one of these spacelets is given. The possible extravagances of each spacelet depend only on its rank and the condition that it contains ω_p.

When ω_r is an arbitrary spacelet containing ω_p, the possible values of x, ξ, u, r, ρ are given by

$$\xi \nless 0, \quad \xi \ngtr x, \quad x+\xi \ngtr n-p-\pi; \quad u \nless 0, \quad u \ngtr \pi; \quad r=p+x+u, \quad \rho=\pi-u+\xi; \quad \text{...(17)}$$

the possible values of u, r, ρ are given by

$$u \nless 0, \quad u \ngtr \pi; \quad \rho \nless \pi-u, \quad \rho \ngtr n-r, \quad \rho+2u \ngtr r-p+\pi, \quad \text{............(18)}$$

and the possible values of r and ρ by

$$r+\rho \nless p+\pi, \quad r-\rho \nless p-\pi, \quad \rho \nless 0, \quad \rho \ngtr n-r. \quad \text{....................(19)}$$

Ex. xxxiii. *Possible values of s, v and t in the theorem when ω_r is given.*

We will suppose that ω_r is given, so that r, ρ, u have any given values consistent with the necessary conditions (18). Then $\omega_x, \omega_u, x, \xi$ are also given, and we have $x=r-p-u$, $\xi=\rho-\pi+u$.

By using Lemma B of § 178 we see that the possible values of r, ρ, u, s, v and t are those consistent with (12) and with the additional conditions

$$\rho \nless \pi-u, \quad \rho+2u \ngtr r-p+\pi, \quad t \nless r+s-n-\pi+\rho+u,$$

the last condition being equivalent to $t \nless r+s-n+\xi$.

Therefore the possible values of s, v and t are those consistent with the conditions

$$\left.\begin{array}{c} r+s-p \ngtr n, \quad v \nless 0, \quad v \ngtr \pi, \quad u+v+n-\pi \nless r+s-p, \\ t+\pi \nless u+v+p, \quad t+\pi \ngtr r+v, \quad t+\pi \ngtr s+u, \quad t \nless r+s-n+\xi. \end{array}\right\} \quad \text{......(20)}$$

Eliminating v, we see that the possible values of s and t are given by

$$\left.\begin{array}{c} r+s-p \ngtr n, \quad t \nless r+s-n, \\ t+\pi \ngtr s+u, \quad t \ngtr r, \quad t+\pi \ngtr u+p, \quad t \nless r+s-n+\xi; \end{array}\right\} \quad \text{............(21)}$$

and further eliminating t we see that the possible values of s are given by

$$r+s-p \ngtr n, \quad s \nless p \quad \text{..................................(22)}$$

the condition $s \ngtr n-\xi$ being included in the condition $s \ngtr n-x$ or $s \ngtr n-r+p+u$, which is again included in $r+s-p \ngtr n$.

The set of conditions (20) is consistent with (18) and together with (18) includes the necessary conditions (1) and (2) of the theorem. From (22) we see that when ω_r is any given spacelet containing ω_p, we can always determine a spacelet ω_s which has ω_p for its complete intersection with ω_r, provided only that s is consistent with the necessary conditions (1) of the theorem.

The corresponding conditions determining the possible values of r, u and t when ω_s is given (and contains ω_p) can be written down from considerations of symmetry.

Ex. xxxiv. *Verification of the limits to the values of τ in the theorem.*

We suppose that $r \nless s$. Since τ is the mutual orthotomy both of ω_r, ω_s and of $\omega_{r-p+\pi}$, $\omega_{s-p+\pi}$, therefore by § 176.2 we must have $\tau \ngtr n-r$ and $\tau \ngtr s-p+\pi$. Again since the spacelet $\omega_{\pi-u}$ lies both in ω_s and in ω_{n-r}, we must have $\tau \nless \pi-u$, and because $r \nless p-u$, it follows that when u is arbitrary, we must have $\tau \nless p+\pi-r$.

NOTE 3. *Greatest possible values of the mutual orthotomy τ in the theorem.*

Since ω_s contains a spacelet $\omega_{p-\pi}$ which does not lie in ω_{n-r}, therefore ω_s cannot lie in the normal to ω_r; in fact the complete intersection of ω_s with the normal to ω_r cannot have rank greater than $s-p+\pi$, and similarly the complete intersection of ω_r with the normal to ω_s cannot have rank greater than $r-p+\pi$. Also the normal to ω_r cannot lie in ω_s unless the condition $(n-r)+(p-\pi) \not> s$, or $r+s-p \not< n-\pi$ is satisfied. These results are true both when $r \not< s$ and when $r \not> s$; and they show that it is impossible for each of the spacelets ω_r and ω_s to be incident with the normal to the other except when $r+s-p \not< n-\pi$.

Again it is impossible for the complete intersection of ω_s with the normal to ω_r to have rank $s-p+\pi$ (i.e. for the complete intersection of the normal to ω_s with ω_r to have rank $r-p+\pi$) unless the condition $s-p+\pi \not> n-r$, or $r+s-p \not> n-\pi$ is satisfied.

We will now assume throughout that $r \not< s$, and there will be two possible cases.

Case I. $r+s-p \not> n-\pi$. *The greatest possible value of* τ *is* $s-p+\pi$.

In this case it is possible for ω_r and ω_s to both lie in $\omega_{n-\pi}$. We have $\tau=s-p+\pi$ (or $t=p-\pi$) when and only when the cross rank of $\omega_{r-p+\pi}$ and $\omega_{s-p+\pi}$ is 0, i.e. when and only when the following condition is satisfied:

(1) *The spacelets $\omega_{r-p+\pi}$ and $\omega_{s-p+\pi}$ are mutually orthogonal; i.e. each of them lies in (or is incident with) the normal to the other; i.e. every point of ω_r orthogonal with $\omega_{p-\pi}$ is orthogonal with every point of ω_s orthogonal with $\omega_{p-\pi}$.*

Since the normals to $\omega_{r-p+\pi}$ and $\omega_{s-p+\pi}$ both lie in $\omega_{n-\pi}$, this is only possible when ω_r and ω_s both lie in $\omega_{n-\pi}$, i.e. when $u=v=0$. The spacelets $\omega_{r-p+\pi}$ and $\omega_{s-p+\pi}$ are then the same as $\omega_{\pi+x}$ and $\omega_{\pi+y}$, and the further conditions under which $\tau=s-p+\pi$ are the same as in Note 1 of § 180, the spacelets ω_x and ω_y being now the spacelets ω_{r-p} and ω_{s-p} of § 180.

If $\omega_{\pi+x}$ lies in ω_{n-s}, then ω_π must lie in ω_{n-s}; and since the complete intersection of ω_π and ω_{n-s} is $\omega_{\pi-v}$, this is only possible when $v=0$. In the same way we see that when $\omega_{\pi+y}$ lies in ω_{n-r}, we must have $u=0$. For similar reasons if $\omega_{r-p+\pi}$ lies in ω_{n-s}, we must have $v=0$; and if $\omega_{s-p+\pi}$ lies in ω_{n-r}, we must have $u=0$. Thus if all points of ω_r orthogonal with ω_p (or with $\omega_{p-\pi}$) are orthogonal with ω_s, we must have $v=0$, i.e. ω_s must lie in $\omega_{n-\pi}$; and if all points of ω_s orthogonal with ω_p (or with $\omega_{p-\pi}$) are orthogonal with ω_r, we must have $u=0$, i.e. ω_r must lie in $\omega_{n-\pi}$.

Consequently we see from Note 1 of § 180 that the conditions that the mutual orthotomy τ of ω_r and ω_s shall have its greatest possible value $s-p+\pi$ can also be expressed in any one of the following equivalent ways:

(2) *The spacelets ω_r and ω_s both lie in $\omega_{n-\pi}$, and moreover $\omega_{\pi+x}$ and $\omega_{\pi+y}$ (or ω_x and ω_y) are mutually orthogonal.*

(3) *The spacelets ω_r and ω_s both lie in $\omega_{n-\pi}$, and moreover every point of ω_r orthogonal with ω_p (or with $\omega_{p-\pi}$) is orthogonal with every point of ω_s orthogonal with ω_p (or with $\omega_{p-\pi}$).*

(4) *Every point of ω_r orthogonal with ω_p (or with $\omega_{p-\pi}$) is orthogonal with ω_s, and every point of ω_s orthogonal with ω_p (or with $\omega_{p-\pi}$) is orthogonal with ω_r.*

When the two conditions mentioned in (4) are satisfied, ω_r and ω_s necessarily both lie in $\omega_{n-\pi}$, and this condition need not be specified.

We may observe that in Case I the mutual orthotomy τ has its greatest possible value $s-p+\pi$ when and only when the cross rank t has its least possible value $p-\pi$, i.e. when and only when each of the spacelets ω_r and ω_s contains no spacelet of rank greater than $p-\pi$ which does not intersect the normal to the other.

Case II. $r+s-p>n-\pi$. *The greatest possible value of* τ *is* $n-r$.

In this case it is not possible for ω_r and ω_s to both lie in $\omega_{n-\pi}$.

It is of course possible for all points of ω_r orthogonal with ω_p to be orthogonal with all points of ω_s orthogonal with ω_p, this being the case whenever ω_x and ω_y are mutually orthogonal; but it is not possible for all points of ω_r orthogonal with ω_p (or with $\omega_{p-\pi}$) to be orthogonal with ω_s and at the same time for all points of ω_s orthogonal with ω_p (or with $\omega_{p-\pi}$) to be orthogonal with ω_r. In fact it is not possible for any one of the equivalent sets of conditions given in Case I to be satisfied.

In Case II we have $\tau=n-r$ (or $t=r+s-n$) when and only when the cross rank of the normals to ω_r and ω_s is 0, i.e. when and only when each or either of the following equivalent conditions is satisfied:

(1) *The normals to* ω_r *and* ω_s *are mutually orthogonal.*

(2) *Each of the two spacelets* ω_r *and* ω_s *contains* (*or is incident with*) *the normal to the other.*

These conditions cannot be satisfied in Case I, except when $r+s-p=n-\pi$. Whenever they are satisfied, we see from the equations (21) or from Ex. xi of § 176 that ξ and η must both be 0, i.e. ω_x and ω_y must be non-extravagant, or the cores of ω_r and ω_s must lie in ω_π, being the spacelets $\omega_{\pi-u}$ and $\omega_{\pi-v}$. Therefore the join of ω_r and ω_s, which contains the plenum of ω_r and ω_s, is a plenarily extravagant spacelet containing $\omega_{n-\pi}$.

When ω_r is given and is such that ω_x is non-extravagant, we have $\rho=\pi-u$, and (see Note 2) the possible values of r and u are given by $u\nless 0$, $u\ngtr\pi$, $u\ngtr r-p$, $\pi-u\ngtr n-r$. In this case ω_s will be such that $\tau=n-r$ when and only when ω_{n-s} is so chosen as to lie in $\omega_{\pi+x}$ and to be complementary to ω_{n-r} in ω_{n-p}. Since by Ex. xxix the spacelets $\omega_{\pi+x}$ and ω_{n-r} are mutually complementary in ω_{n-p}, this is possible when and only when s satisfies the conditions $r\nless s$, $r+s-p\ngtr n$, $s+x\nless n-\pi$; and when ω_s has been thus determined, it follows from Ex. xxxiii or from Ex. xi of § 176 that ω_y is necessarily non-extravagant.

When ω_s is given and is such that ω_y is non-extravagant, we have $\sigma=\pi-v$, and the possible values of s and v are given by $v\nless 0$, $v\ngtr\pi$, $v\ngtr s-p$, $\pi-v\ngtr n-s$. In this case ω_r will be such that $\tau=n-r$ when and only when ω_{n-r} is so chosen as to lie in $\omega_{\pi+y}$ and to be complementary to ω_{n-s} in ω_{n-p}; and this is possible when and only when r satisfies the conditions $r\nless s$, $r+s-p\ngtr n$, $r+y\nless n-\pi$.

These results agree with and can be deduced from Ex. xxxiii.

We may observe that in Case II we have $r+s-n>p-\pi$, and each of the two spacelets ω_r and ω_s necessarily contains a spacelet of rank $r+s-n$ which does not intersect the normal to the other. The mutual orthotomy τ has its greatest possible value $n-r$ when and only when the cross rank t has its least possible value $r+s-n$, i.e. when and only when each of the spacelets ω_r and ω_s contains no spacelet of rank greater than $r+s-n$ which does not intersect the normal to the other.

The construction of ω_r and ω_s so as to make τ have its maximum value $n-r$ in Case II is more easily effected through that of their normals ω_{n-r} and ω_{n-s}. We have to choose these to be two mutually orthogonal spacelets whose join is ω_{n-p}. Referring to Ex. ii of § 183 we see that this is possible if and only if the given positive integers r and s satisfy the conditions

$$n-r\ngtr n-p \quad n-s\ngtr n-p, \quad (n-r)+(n-s)-(n-p)\nless 0, \quad (n-r)+(n-s)-(n-p)\ngtr\pi,$$

or $$r\nless p, \quad s\nless p, \quad n-r-s+p\nless 0, \quad n-r-s+p\ngtr\pi;$$

and these conditions are satisfied in Case II but not in Case I. If we write $n-r-s+p=\kappa$, then to make $\tau=n-r$ in Case II we select any sub-space ω_κ of ω_π of rank κ, and take ω_{n-r} and ω_{n-s} to be the joins of ω_κ with two mutually orthogonal non-intersecting spacelets $\omega_{n-r-\kappa}$ and $\omega_{n-s-\kappa}$ (or ω_{s-p} and ω_{r-p}) lying in ω_{n-p} whose join is complementary to ω_κ in ω_{n-p}. Then ω_{n-r} and ω_{n-s} have ω_κ for their complete intersection; their cores ω_ρ and ω_σ are their complete intersections with ω_π; and ω_π is the join of ω_ρ and ω_σ; these results following from Theorem I of § 183. Moreover $\omega_{n-r-\kappa}$, $\omega_{n-s-\kappa}$ have cores $\omega_{\rho-\kappa}$, $\omega_{\sigma-\kappa}$ which lie in ω_π; and ω_ρ and ω_σ are respectively the joins of ω_κ, $\omega_{\rho-\kappa}$ and ω_κ, $\omega_{\sigma-\kappa}$. It is clear that ω_ρ and ω_σ are the spacelets $\omega_{\pi-u}$ and $\omega_{\pi-v}$ of the text; and since the normal to $\omega_{\pi+x}$ (being the join of ω_{n-r} and ω_p, or the join of the two mutually orthogonal non-intersecting spacelets $\omega_{n-r-\kappa}$ and ω_p) has ω_π as its core, we see again that ω_x (and similarly ω_y) is non-extravagant in Case II. The join of ω_u and ω_v intersects $\omega_{n-\pi}$ in a spacelet ω_t whose rank t is given by $(u+v-t)+(n-\pi)=r+s-p$, as we see from Ex. iv or from the fact that the join of ω_r and ω_s contains $\omega_{n-\pi}$; and the join of ω_p, ω_x, ω_y, ω_t is $\omega_{n-\pi}$.

Summarising the constructions obtained in both cases, and writing $\kappa=n-(r+s-p)$, so that $\kappa \nless \pi$, $\kappa \ngtr n-p$ in Case I, and $\kappa \nless 0$, $\kappa \ngtr \pi$ in Case II, we see that:

To make τ a maximum in Case I we select a sub-space ω_κ of ω_{n-p} which contains ω_π; represent ω_{n-p} as the join of two non-intersecting spacelets ω_κ, $\omega_{n-p-\kappa}$; represent $\omega_{n-p-\kappa}$ as the join of two mutually orthogonal non-intersecting spacelets ω_{r-p}, ω_{s-p} (these being any two mutually orthogonal non-intersecting spacelets of ranks $r-p$, $s-p$ lying in ω_{n-p} whose join does not intersect ω_π); and take ω_r and ω_s to be respectively the joins of the two pairs of mutually orthogonal non-intersecting spacelets ω_p, ω_{r-p} and ω_p, ω_{s-p}.

To make τ a maximum in Case II we select a sub-space ω_κ of ω_π; represent ω_{n-p} as the join of two (mutually orthogonal) non-intersecting spacelets ω_κ, $\omega_{n-p-\kappa}$; represent $\omega_{n-p-\kappa}$ as the join of two mutually orthogonal non-intersecting spacelets $\omega_{n-r-\kappa}$, $\omega_{n-s-\kappa}$ (so that ω_κ, $\omega_{n-r-\kappa}$, $\omega_{n-s-\kappa}$ are three mutually orthogonal unconnected spacelets whose join is ω_{n-p}); and take ω_{n-r} and ω_{n-s} to be respectively the joins of the two pairs of mutually orthogonal non-intersecting spacelets ω_κ, $\omega_{n-r-\kappa}$ and ω_κ, $\omega_{n-s-\kappa}$.

NOTE 4. *Least possible values of the mutual orthotomy τ in the theorem.*

We assume throughout that $r \nless s$, so that τ is the rank of the complete intersection of ω_s and ω_{n-r}. Since $\omega_{\pi-u}$ is common to ω_s and ω_{n-r}, we must always have $\tau \nless \pi-u$.

Case I. $r \nless p+\pi$, $r \nless s$. *The least possible value of τ is* 0.

In this case it is possible for u to have the value π, the corresponding value of x being $r-p-\pi$. Since $\tau \nless \pi-u$, we cannot have $\tau=0$ except when $u=\pi$.

When ω_r is a given spacelet containing ω_p for which $u=\pi$ and therefore $x=r-p-\pi$, $\rho=\xi$, the possible values of r and ξ are by (17) those consistent with the conditions $\xi \nless 0$, $\xi \ngtr r-p-\pi$, $\xi \ngtr n-r$. In this case ω_s will be such that $\tau=0$ (or $t=s$) when and only when it is the join of ω_p with a spacelet ω_{s-p} which does not intersect ω_r and does not intersect the join of ω_p and ω_{n-r}. Since ω_p does not intersect ω_{n-r}, we can determine such a spacelet ω_{s-p} when and only when s satisfies the conditions $r \nless s$, $s-p \nless 0$, $(s-p)+\pi \ngtr n$, $(s-p)+(n-r+p) \ngtr n$, or $r \nless s$, $s \nless p$, $r+s-p \ngtr n$. Thus s can have any value which is not greater than r and satisfies the necessary conditions (1) of the theorem.

When ω_s is a given spacelet containing ω_p, the possible values of s, y and η are given by

$$y \ngtr s-p, \quad y+\pi \nless s-p, \quad \eta \nless 0, \quad \eta \ngtr y, \quad y+\eta \ngtr n-p-\pi;$$

and from Ex. xxxiii we see that we can determine a spacelet ω_r which has ω_p for its complete intersection with ω_s and is such that $\tau=0$ (or $t=s$) when and only when $u=\pi$

and r satisfies the conditions $r+s-p \not> n$, $r-s \not< \pi-v$. In this case y is the greatest possible rank of a spacelet which lies in ω_x and does not intersect ω_{n-s}, and ω_x is the join of a spacelet of rank y which does not intersect ω_{n-s} and a spacelet of rank $(r-s)-(\pi-v)$ lying in ω_{n-s}. The last result shows how ω_r can be constructed.

Case II. $r \not> p+\pi$, $r \not< s$. *The least possible value of* τ *is* $p+\pi-r$.

Since the spacelet $\omega_{\pi-u}$ lies in the intersection of ω_s, ω_{n-r}, and the spacelet $\omega_{\pi-v}$ lies in the intersection of ω_r, ω_{n-s}, we must have $\tau \not< \pi-u$ and $\tau+r-s \not< \pi-v$. Hence when $\tau=p+\pi-r$, we must have $u \not< r-p$ and $v \not< s-p$, i.e. we must have $u=r-p$ and $v=s-p$, or $x=0$ and $y=0$. Conversely whenever $x=0$ and $y=0$, i.e. whenever $u=r-p$ and $v=s-p$, we have $t=(p-\pi)+u+v=r+s-p-\pi$, and therefore $\tau=p+\pi-r$, it being assumed that $r \not< s$.

Thus when $r \not< s$, we have $\tau=p+\pi-r$ when and only when $x=0$ and $y=0$; and this is of course only possible when $r \not> p+\pi$.

If we remove the restriction $r \not< s$, we have $t=r+s-p-\pi$ when and only when $x=0$ and $y=0$; and this result follows immediately from the conditions (12) or (13) in Note 1.

When ω_r is a given spacelet containing ω_p for which $x=0$, the possible values of r, ρ and u are given by $r \not< p$, $r \not> p+\pi$, $u=r-p$, $\rho=\pi-u=p+\pi-r$. In this case ω_{n-r} is a spacelet lying in ω_{n-p} whose complete intersection with ω_π or ω_p is a spacelet $\omega_{p+\pi-r}$ of rank $p+\pi-r$ which is also the core of ω_r; and the join of ω_p and ω_{n-r} is $\omega_{n-\pi}$. If ω_s contains ω_p and is the join of ω_{s-p} and ω_p, the complete intersection of ω_s and ω_{n-r} will have rank $p+\pi-r$ (being then $\omega_{p+\pi-r}$) when and only when the join of ω_s and ω_{n-r} has rank $(s-p)+(n-\pi)$, i.e. when and only when the join of ω_{s-p}, ω_p, ω_{n-r} or the join of ω_{s-p}, $\omega_{n-\pi}$ has rank $(s-p)+(n-\pi)$, i.e. when and only when ω_{s-p} does not intersect $\omega_{n-\pi}$. Hence if $r \not< s$, the spacelet ω_s will have ω_p for its complete intersection with ω_r and will be such that $\tau=p+\pi-r$ when and only when it is the join of ω_p with a spacelet ω_{s-p} which does not intersect ω_r and does not intersect the join $\omega_{n-\pi}$ of ω_p and ω_{n-r}. This is possible when and only when $y=0$, and s satisfies the conditions $r \not< s$, $s-p \not< 0$, $(s-p)+r \not> n$, $(s-p)+(n-\pi) \not> n$, which are equivalent to $r \not< s$, $s \not< p$, $r+s-p \not> n$.

When ω_s is a given spacelet containing ω_p for which $y=0$, the possible values of s, σ and v are given by $s \not< p$, $s \not> p+\pi$, $v=s-p$, $\sigma=\pi-v=p+\pi-s$. In this case ω_{n-s} is a spacelet lying in ω_{n-p} whose complete intersection with ω_π or ω_p is a spacelet $\omega_{p+\pi-s}$ of rank $p+\pi-s$ which is also the core of ω_s; and the join of ω_p and ω_{n-s} is $\omega_{n-\pi}$. If ω_r contains ω_p and is the join of ω_{r-p} and ω_p, the complete intersection of ω_r and ω_{n-s} will have rank $p+\pi-s$ (being then $\omega_{p+\pi-s}$) when and only when the join of ω_r and ω_{n-s} has rank $(r-p)+(n-\pi)$, i.e. when and only when the join of ω_{r-p}, ω_p, ω_{n-s} or the join of ω_{r-p}, $\omega_{n-\pi}$ has rank $(r-p)+(n-\pi)$, i.e. when and only when ω_{r-p} does not intersect $\omega_{n-\pi}$. Hence if $r \not< s$, the spacelet ω_r will have ω_p for its complete intersection with ω_s and will be such that $\tau=p+\pi-r$ when and only when it is the join of ω_p with a spacelet ω_{r-p} which does not intersect ω_s and does not intersect the join $\omega_{n-\pi}$ of ω_p and ω_{n-s}. This is possible when and only when $x=0$, and r satisfies the conditions $r \not< s$, $r-p \not< 0$, $(r-p)+s \not> n$, $(r-p)+(n-\pi) \not> n$, which are equivalent to $r \not< s$, $r \not< p$, $r+s-p \not> n$, $r \not> p+\pi$.

Thus if r and s are any integers satisfying the necessary conditions (1) of the theorem and the conditions $r \not< s$, $r \not> p+\pi$, and if ω_r in the theorem is a given spacelet for which $x=0$, we can always determine ω_s so that $\tau=p+\pi-r$; and if ω_s in the theorem is a given spacelet for which $y=0$, we can always determine ω_r so that $\tau=p+\pi-r$.

These results agree with and can be deduced from Ex. xxxiii.

NOTE 5. *The joins, intersections and normals of real spacelets are real.*

It is obvious that the join of any number of real spacelets is real. That the normal to a real spacelet is real follows immediately from the methods of solving a matrix equation of the first degree, and has been proved in the theorem of § 94. That the complete intersection of any number of real spacelets is real follows from the fact that it is the normal to the (real) join of the normal spacelets. These results have been used in Note 4 of § 180 to obtain the possible values of the mutual orthotomy of two real spacelets which have a given complete intersection.

It should be observed however that the joins and intersections of non-extravagant spacelets are not in general non-extravagant (except in the particular case when the spacelets are real), as can at once be seen from particular examples. For instance, if $i=\sqrt{-1}$, the first of the matrices

$$\begin{bmatrix} 1 & 1 \\ 0 & 1 \\ 0 & i \end{bmatrix}, \quad \begin{bmatrix} 1 & 1 \\ i & 0 \\ 0 & 1 \end{bmatrix}, \quad \begin{bmatrix} 1 & 0 \\ i & 1 \\ 0 & 1 \end{bmatrix}$$

represents the join of two non-extravagant points in homogeneous space of rank 3, and this join is an extravagant spacelet; the second and third matrices represent two non-extravagant spacelets whose complete intersection is an extravagant point.

Ex. xxxv. *Orthotomy and extravagances of two spacelets which have a given join.*

Let ω_p be a given spacelet of extravagance π; let ω_r and ω_s be any two spacelets (of extravagances ρ and σ) which have ω_p for their join, so that the possible values of r and s are given by

$$r \not> p, \quad s \not> p, \quad r+s-p \not< 0;$$

and let t and τ be respectively the cross rank and the mutual orthotomy of ω_r and ω_s.

Then when r and s have any assigned values, the possible values of t are given by

$$t \not< 0, \quad t \not< r+s-p-\pi, \quad t \not> r, \quad t \not> s, \quad t \not> p-\pi;$$

and when r and s have assigned values such that $r \not< s$, the possible values of τ are given by

$$\tau \not< 0, \quad \tau \not< s-p+\pi, \quad \tau \not> s, \quad \tau \not> p+\pi-r;$$

moreover the possible values of r, s, ρ, σ are given by

$$\left.\begin{array}{l} \rho \not< 0, \quad \rho \not> r, \quad r+\rho \not> p+\pi, \quad r-\rho \not> p-\pi; \\ \sigma \not< 0, \quad \sigma \not> s, \quad s+\sigma \not> p+\pi, \quad s-\sigma \not> p-\pi; \\ \qquad r+s-p \not< 0. \end{array}\right\}$$

These results are the theorems complementary to the theorem of the text and that given in Note 2. We obtain them by the principle explained in Note 2 of § 170, using Exs. vi and vii of § 176.

When r and s are given, there is no necessary relation between the extravagances of ω_r and ω_s, even when one of these spacelets is given. The possible extravagances of each spacelet depend only on its rank and the condition that it lies in ω_p.

Ex. xxxvi. *Orthotomy and extravagances of two spacelets which have a given join and contain its core.*

In the last example let ω_r and ω_s both contain the core of ω_p, so that the possible values of r and s are given by

$$r \not> p, \quad s \not> p, \quad r+s-p \not< \pi.$$

Then the possible values of r, s and t are given by

$$r+s-p \nless \pi, \quad t \nless r+s-p-\pi, \quad t \ngtr r-\pi, \quad t \ngtr s-\pi;$$

and when r and s have assigned values such that $r \nless s$, the possible values of τ are given by

$$\tau \nless \pi, \quad \tau \ngtr p+\pi-r;$$

moreover the possible values of r, s, ρ, σ are given by

$$r+s-p \nless \pi; \quad \rho \nless \pi, \quad \rho \ngtr r, \quad r+\rho \ngtr p+\pi; \quad \sigma \nless \pi, \quad \sigma \ngtr s, \quad s+\sigma \ngtr p+\pi.$$

These are the results complementary to the theorem of § 180 and to Ex. xxiv of § 180.

When r and s are given, the possible extravagances of each of the spacelets ω_r and ω_s depend only on its rank and the conditions (see Ex. xv of § 170) that it lies in ω_p and contains the core of ω_p; and this is true even when the other spacelet is given.

Ex. xxxvii. *Orthotomy of two real spacelets which have a given join.*

If ω_r and ω_s are two real spacelets which have a given join ω_p, the possible values of r, s and t are given by

$$r+s-p \nless 0, \quad t \nless r+s-p, \quad t \ngtr r, \quad t \ngtr s;$$

and when r and s have assigned values such that $r \nless s$, the possible values of τ are given by

$$\tau \nless 0, \quad \tau \ngtr p-r.$$

This is the result complementary to Note 4 of § 180.

§ 182. Possible simultaneous values of the paratomy and orthotomy of two arbitrary spacelets whose ranks only are given.

By eliminating π from the conditions (1), (2) and (4) or the conditions (1), (3) and (4) of the theorem of § 181 we obtain the following theorem in which each of the two parts is deducible from the other.

Theorem I. *If ω_r and ω_s are two spacelets of homogeneous space ω_n which have the assigned ranks r and s but are otherwise arbitrary, then:*

(1) *The mutual paratomy p and the cross rank t of ω_r and ω_s can have independently of one another any values consistent with the necessary conditions*

$$p \nless 0, \quad p \ngtr r, \quad p \ngtr s, \quad p \nless r+s-n; \quad t \nless 0, \quad t \ngtr r, \quad t \ngtr s, \quad t \nless r+s-n. \quad \text{...(A)}$$

(2) *The mutual paratomy p and the mutual orthotomy τ of ω_r and ω_s can have independently of one another any values consistent with the necessary conditions*

$$p \nless 0, \; p \ngtr r, \; p \ngtr s, \; p \nless r+s-n; \; \tau \nless 0, \; \tau \ngtr r, \; \tau \ngtr s, \; \tau \ngtr n-r, \; \tau \ngtr n-s. \quad \text{...(B)}$$

Both (A) and (B) include the conditions which r and s must satisfy, viz.

$$r \nless 0, \quad r \ngtr n; \quad s \nless 0, \quad s \ngtr n. \quad \text{.................................(C)}$$

To simplify the conditions (B) we consider two cases separately, and further make the legitimate assumption that $r \nless s$.

CASE I. $r+s-n \ngtr 0$, $r \nless s$. *The possible values of p and τ are given by*

$$p \nless 0, \quad p \ngtr s; \quad \tau \nless 0, \quad \tau \ngtr s. \quad \text{.................................}(B_1)$$

In this case, denoting the normals to ω_r and ω_s by ω_{n-r} and ω_{n-s}, and the extravagances of ω_r and ω_s by ρ and σ, we see that:

(a) The *least possible value of* p is 0, this value occurring when and only when ω_r and ω_s are non-intersecting.

(b) The *greatest possible value of* p is s, this value occurring when and only when ω_r and ω_s are mutually incident (i.e. when ω_s lies in ω_r).

(c) The *least possible value of* τ is 0, this value occurring when and only when no point of the smaller spacelet ω_s is orthogonal with the larger spacelet ω_r (i.e. when ω_s does not intersect ω_{n-r}).

(d) The *greatest possible value of* τ is s, this value occurring when and only when each of the two spacelets ω_r and ω_s is incident with the normal to the other (i.e. when ω_s lies in ω_{n-r}).

To obtain (a) and (c) together, we choose ω_s so that it intersects neither ω_r nor ω_{n-r}, this being possible because $s+r \not> n$, and $s+(n-r) \not> n$.

To obtain (a) and (d) together, we choose ω_s so that it lies in ω_{n-r} and does not intersect ω_r. This is possible if $s \not> n-r$, and $s+\rho \not> n-r$, i.e. when ω_r is so chosen that $\rho \not> n-r-s$.

To obtain (b) and (c) together, we choose ω_s so that it lies in ω_r and does not intersect ω_{n-r}. This is possible if $s \not> r$, and $s+\rho \not> r$, i.e. when ω_r is so chosen that $\rho \not> r-s$.

To obtain (b) and (d) together, we choose ω_s so that it lies both in ω_r and in ω_{n-r}. This is possible when ω_r is so chosen that its extravagance ρ satisfies the condition $\rho \not< s$ as well as the conditions $\rho \not< 0$, $\rho \not> r$, $r+\rho \not> n$; and in the present case these four conditions are compatible with one another.

We cannot have (b) and (d) together when both the spacelets ω_r and ω_s are real or non-extravagant.

CASE II. $r+s-n \not< 0$, $r \not< s$. *The possible values of* p *and* τ *are given by*

$$p \not< r+s-n, \quad p \not> s; \quad \tau \not< 0, \quad \tau \not> n-r. \qquad \text{......................(B}_2\text{)}$$

In this case, denoting the normals to ω_r and ω_s by ω_{n-r} and ω_{n-s}, and the extravagances of ω_r and ω_s by ρ and σ, we see that:

(a) The *least possible value of* p is $r+s-n$, this value occurring when and only when ω_r and ω_s are mutually complementary (i.e. when ω_{n-r} and ω_{n-s} are non-intersecting).

(b) The *greatest possible value of* p is s, this value occurring when and only when ω_r and ω_s are mutually incident (i.e. when ω_s lies in ω_r, or ω_{n-r} lies in ω_{n-s}).

(c) The *least possible value of* τ is 0, this value occurring when and only when no point of the smaller spacelet ω_s is orthogonal with the larger spacelet ω_r (i.e. when ω_s does not intersect ω_{n-r}).

(d) The *greatest possible value of* τ is $n-r$, this value occurring when and only when each of the two spacelets ω_r and ω_s is incident with the normal to the other (i.e. when ω_{n-r} lies in ω_s).

To obtain (a) and (c) together, we choose ω_r to be the normal to a spacelet ω_{n-r} which intersects neither ω_s nor ω_{n-s}, this being possible because

$$(n-r)+s \not> n, \quad \text{and} \quad (n-r)+(n-s) \not> n.$$

To obtain (*a*) and (*d*) together, we choose ω_r to be the normal to a spacelet ω_{n-r} which lies in ω_s and does not intersect ω_{n-s}. This is possible if $n-r \not> s$, and $(n-r)+\sigma \not> s$, i.e. when ω_s is so chosen that $\sigma \not> r+s-n$.

To obtain (*b*) and (*c*) together, we choose ω_r to be the normal to a spacelet ω_{n-r} which lies in ω_{n-s} and does not intersect ω_s. This is possible if $n-r \not> n-s$, and $n-r+\sigma \not> n-s$, i.e. when ω_s is so chosen that $\sigma \not> r-s$.

To obtain (*b*) and (*d*) together, we choose ω_r to be the normal to a spacelet ω_{n-r} which lies both in ω_s and in ω_{n-s}. This is possible when ω_s is so chosen that its extravagance σ satisfies the condition $\sigma \not< n-r$ as well as the conditions $\sigma \not< 0$, $\sigma \not> s$, $s+\sigma \not> n$; and in the present case these four conditions are compatible with one another.

We cannot have (*b*) and (*d*) together when both the spacelets ω_r and ω_s are real or non-extravagant.

Ex. i. If p, t, τ are the mutual paratomy, the cross rank, the mutual orthotomy of two arbitrary spacelets of ω_n, the possible simultaneous values of p, t and τ; of p and t; of p and τ; and of t and τ are given respectively by

$$p \not< 0,\quad t \not< 0,\quad \tau \not< 0,\quad t+\tau \not< p,\quad t+2\tau \not> n,\quad 2(t+\tau) \not> n+p; \quad \text{.........(1)}$$

$$p \not< 0,\quad t \not< 0,\quad 2p \not> n+t,\quad 2t \not> n+p; \quad \text{.........(2)}$$

$$p \not< 0,\quad \tau \not< 0,\quad p+\tau \not> n,\quad 2\tau \not> n; \quad \text{.........(3)}$$

$$t \not< 0,\quad \tau \not< 0,\quad t+2\tau \not> n. \quad \text{.........(4)}$$

We can prove the first result by making the assumption that $r \not< s$, so that $s=t+\tau$, and then eliminating r and s from these two conditions and the conditions (A) or (B). The other results can be deduced by eliminating τ, t or p.

Ex. ii. If ω_r and ω_s are arbitrary subject to the condition that they both lie in the plenum of their intersection, it follows from § 180 that we have the same results as in Theorem I.

When the two spacelets ω_r and ω_s are restricted to be real, the possible values of their mutual paratomy and mutual orthotomy are not independent of one another. In fact from Notes 3 and 4 of § 180 we obtain the theorem which follows.

Theorem II. *If ω_r and ω_s are two real spacelets of homogeneous space ω_n which have the assigned ranks r and s but are otherwise arbitrary, and if p, t and τ are the mutual paratomy, the cross rank, and the mutual orthotomy of ω_r and ω_s, then:*

(1) *The possible simultaneous values of p and t are those consistent with the conditions*

$$p \not< 0,\quad p \not< r+s-n,\quad t \not< p,\quad t \not> r,\quad t \not> s. \quad \text{.........(A')}$$

(2) *The possible simultaneous values of p and τ are those consistent with the conditions*

$$p \not< 0,\quad p \not< r+s-n,\quad \tau \not< 0,\quad \tau \not> r-p,\quad \tau \not> s-p. \quad \text{.........(B')}$$

Both (A′) and (B′) include the conditions which r and s must satisfy, viz.

$$r \not< 0,\quad r \not> n;\quad s \not< 0,\quad s \not> n. \quad \text{.........(C')}$$

Simplifying the conditions (B′) by the assumption that $r \not< s$, we have two separate cases.

CASE I. $r+s-n \not> 0$, $r \not< s$. The conditions giving the possible values of p and τ can be expressed in either of the forms

$$p \not< 0, \quad p \not> s; \quad \tau \not< 0, \quad \tau \not> s-p; \quad \text{.............................(5)}$$

$$\tau \not< 0, \quad \tau \not> s; \quad p \not< 0, \quad p \not> s-\tau; \quad \text{.............................(6)}$$

where in (5) we regard τ as dependent on p, and in (6) we regard p as dependent on τ.

In this case the *greatest possible value of* p is s, and this value can only occur when $\tau=0$; the *greatest possible value of* τ is s, and this value can only occur when $p=0$, i.e. when the two spacelets are non-intersecting.

CASE II. $r+s-n \not< 0$, $r \not< s$. The conditions giving the possible values of p and τ can be expressed in either of the forms

$$p \not< r+s-n, \quad p \not> s; \qquad \tau \not< 0, \qquad \tau \not> s-p; \quad \text{..................(7)}$$

$$\tau \not< 0, \qquad \tau \not> n-r; \quad p \not< r+s-n, \quad p \not> s-\tau; \quad \text{..................(8)}$$

where in (7) we regard τ as dependent on p, and in (8) we regard p as dependent on τ.

In this case the *greatest possible value of* p is s, and this value can only occur when $\tau=0$; the *greatest possible value of* τ is $n-r$, and this value can only occur when $p=r+s-n$, i.e. when the two spacelets are mutually complementary.

Ex. iii. If p, t, τ are the mutual paratomy, the cross rank, the mutual orthotomy of two arbitrary *real* spacelets of ω_n, the possible simultaneous values of p, t and τ; of p and t; of p and τ; and of t and τ are given respectively by

$$p \not< 0, \quad t \not< p, \quad \tau \not< 0, \quad 2(t+\tau) \not> n+p; \quad \text{........................(1')}$$

$$p \not< 0, \quad \tau \not< p, \quad 2t \not> n+p; \quad \text{................................(2')}$$

$$p \not< 0, \quad \tau \not< 0, \quad p+2\tau \not> n; \quad \text{..............................(3')}$$

$$t \not< 0, \quad \tau \not< 0, \quad t+2\tau \not> n. \quad \text{................................(4')}$$

We can prove the first result by making the assumption that $r \not< s$, so that $s=t+\tau$, and then eliminating r and s from these two conditions and the conditions (A') or (B'). The other results can be deduced by eliminating τ, t or p.

The conditions (1') and (2') are equivalent respectively to the conditions (1) and (2) of Ex. i together with the additional condition $t \not< p$; and the conditions (3') are equivalent to the conditions (3) of Ex. i together with the additional condition $p+2\tau \not> n$.

Ex. iv. If ω_r and ω_s are arbitrary subject to the condition that their complete intersection is non-extravagant, we have the same results as in Theorem II.

§ 183. Properties of mutually orthogonal spacelets.

If two spacelets are mutually orthogonal, their intersection, being orthogonal with itself, must be a completely extravagant spacelet; consequently two mutually orthogonal real spacelets must be non-intersecting. It follows that if the normals to two spacelets are mutually orthogonal, the join of the two spacelets must be plenarily extravagant; consequently two real spacelets whose normals are mutually orthogonal must be mutually complementary. Again if any number of spacelets are mutually orthogonal, their intersections with one another, the intersections of their joins, and their complete intersection must all be completely extravagant spacelets; consequently mutually orthogonal real spacelets must be unconnected. The properties of mutually orthogonal spacelets are given more precisely in the two theorems which follow.

Theorem I. *If ω_r and ω_s are any two mutually orthogonal spacelets in homogeneous space ω_n whose cores are ω_ρ and ω_σ, and if the join of ω_r and ω_s is a spacelet ω_p whose core is ω_π, then:*

(1) *The complete intersection of ω_r and ω_s is a completely extravagant spacelet lying in ω_π which may be denoted by ω_κ, where $\kappa = r + s - p$.*

(2) *The cores ω_ρ and ω_σ are respectively the complete intersections of ω_r and ω_s with ω_π.*

(3) *The join and complete intersection of ω_ρ and ω_σ are respectively the spacelets ω_π and ω_κ.*

The complete intersection of ω_r and ω_s necessarily has rank κ and may be denoted by ω_κ. Since ω_κ is orthogonal with ω_r and ω_s, it is a spacelet lying in ω_p which is orthogonal with ω_p, and must therefore be a sub-space of ω_π and be completely extravagant. Again since all points of ω_ρ lie in ω_π (being orthogonal with ω_r and ω_s, and therefore with ω_p), and since all points common to ω_r and ω_π lie in ω_ρ (being orthogonal with ω_p, and therefore with ω_r), therefore ω_ρ is the complete intersection of ω_r with ω_π; and similarly ω_σ is the complete intersection of ω_s with ω_π. Hence ω_κ (which lies in ω_π and is the complete intersection of ω_r and ω_s) must contain and be the complete intersection of ω_ρ and ω_σ. Lastly if we represent the spacelet ω_r as the join of ω_ρ, $\omega_{r-\rho}$, the spacelet ω_s as the join of ω_σ, $\omega_{s-\sigma}$, and denote the join of ω_ρ and ω_σ by ω_λ, then ω_p is the join of the three mutually orthogonal unconnected spacelets $\omega_{r-\rho}$, $\omega_{s-\sigma}$, ω_λ of which the first two are non-extravagant and the third is completely extravagant. Consequently by Ex. viii of § 170 the core ω_π of ω_p is the spacelet ω_λ, i.e. the join of ω_ρ and ω_σ is ω_π.

Ex. i. By the properties of a join and intersection we have $\kappa + \pi = \rho + \sigma$, or

$$r + s - p = \rho + \sigma - \pi = \kappa. \qquad \text{(1)}$$

Ex. ii. When ω_p is given, it is clear that r and s must satisfy the conditions

$$r \not> p, \quad s \not> p, \quad r + s - p \not< 0, \quad r + s - p \not> \pi. \qquad \text{(2)}$$

Conversely let r and s be any two (positive) integers satisfying these conditions; let $r + s - p = \kappa$, so that $\kappa \not< 0$, $\kappa \not> \pi$; and let ω_κ be any sub-space of ω_π of rank κ. Then we can represent ω_p as the join of ω_κ with another spacelet $\omega_{p-\kappa}$ (orthogonal with ω_κ), and $\omega_{p-\kappa}$ as the join of two mutually orthogonal non-intersecting spacelets $\omega_{r-\kappa}$ and $\omega_{s-\kappa}$. When this is done, the joins of ω_κ, $\omega_{r-\kappa}$ and ω_κ, $\omega_{s-\kappa}$ are two mutually orthogonal spacelets having ω_p as their join.

Thus we can construct two mutually orthogonal spacelets ω_r and ω_s of ranks r and s which have as their join a given spacelet ω_p of extravagance π when and only when r and s are integers satisfying the conditions (2).

In the next examples it is to be understood that ω_p, ω_r, ω_s, ω_t are spacelets of ranks p, r, s, t and extravagances π, ρ, σ, τ whose cores are ω_π, ω_ρ, ω_σ, ω_τ.

Ex. iii. *Ranks and extravagances of two mutually orthogonal spacelets having a given join* ω_p.

If ω_r and ω_s are any two mutually orthogonal non-intersecting spacelets which have ω_p for their join, the possible values of r, s, ρ, σ are given by

$$\rho \nless 0, \quad \sigma \nless 0, \quad \rho \ngtr r, \quad \sigma \ngtr s, \quad r+s=p, \quad \rho+\sigma=\pi; \quad \text{.................(3)}$$

and therefore the possible values of r, s and of ρ, σ are given respectively by

$$r \nless 0, \quad s \nless 0, \quad r+s=p, \quad \textit{and} \quad \rho \nless 0, \quad \sigma \nless 0, \quad \rho+\sigma=\pi.$$

For two such spacelets can always be constructed by representing ω_p as the join of p unconnected mutually orthogonal points, of which π must be extravagant (forming the core of ω_p) and $p-\pi$ must be non-extravagant; dividing the $p-\pi$ non-extravagant points into two groups of x and y points forming non-extravagant spacelets ω_x and ω_y; dividing the π extravagant points into two groups of ξ and η points forming completely extravagant spacelets ω_ξ and ω_η; and taking ω_r and ω_s to be respectively the joins of the two pairs of mutually orthogonal non-intersecting spacelets ω_x, ω_ξ and ω_y, ω_η. The possible values of x, y, ξ, η are then given by

$$x \nless 0, \quad y \nless 0, \quad x+y=p-\pi; \quad \xi \nless 0, \quad \eta \nless 0, \quad \xi+\eta=\pi;$$

and we have

$$r=x+\xi, \quad \rho=\xi, \quad s=y+\eta, \quad \sigma=\eta.$$

If ω_r and ω_s are any two mutually orthogonal spacelets which have ω_p for their join and have a given complete intersection ω_κ (which must be a sub-space of ω_π, so that $\kappa \nless 0$, $\kappa \ngtr \pi$), the possible values of r, s, ρ, σ are given by

$$\rho \nless \kappa, \quad \sigma \nless \kappa, \quad \rho \ngtr r, \quad \sigma \ngtr s, \quad r+s-p=\rho+\sigma-\pi=\kappa; \quad \text{...............(3')}$$

and therefore the possible values of r, s and ρ, σ are given respectively by

$$r \nless \kappa, \quad s \nless \kappa, \quad r+s=p+\kappa, \quad \textit{and} \quad \rho \nless \kappa, \quad \sigma \nless \kappa, \quad \rho+\sigma=\pi+\kappa.$$

For we can regard ω_p as the join of ω_κ and a spacelet $\omega_{p-\kappa}$ of extravagance $\pi-\kappa$, and take ω_r and ω_s to be respectively the joins of ω_κ, ω_x and ω_κ, ω_y, where ω_x and ω_y are any two mutually orthogonal *non-intersecting* spacelets (of extravagances ξ and η) which have $\omega_{p-\kappa}$ as their join.

If ω_r and ω_s are any two mutually orthogonal spacelets which have ω_p for their join, the possible values of r, s, ρ, σ are given by

$$\rho \ngtr \pi, \quad \sigma \ngtr \pi, \quad \rho+\sigma \nless \pi, \quad r \nless \rho, \quad s \nless \sigma, \quad r+s-p=\rho+\sigma-\pi; \quad \text{.........(3'')}$$

and therefore the possible values of r, s and ρ, σ are given respectively by

$$r \ngtr p, \quad s \ngtr p, \quad r+s \nless p, \quad r+s \ngtr p+\pi, \quad \textit{and} \quad \rho \ngtr \pi, \quad \sigma \ngtr \pi, \quad \rho+\sigma \nless \pi.$$

We see this by eliminating κ from the conditions (3') and the additional condition $\kappa \nless 0$; the condition $\kappa \ngtr \pi$ being superfluous.

The results (3), (3'), (3'') clearly remain true when one of the two spacelets ω_r and ω_s is given.

Ex. iv. *Ranks and extravagances of two mutually orthogonal spacelets lying in a given spacelet ω_p.*

If ω_r and ω_s are any two mutually orthogonal non-intersecting spacelets which lie in the given spacelet ω_p, the possible values of r, s, ρ, σ are given by

$$\rho \not< 0, \quad \sigma \not< 0, \quad \rho \not> r, \quad \sigma \not> s, \quad (r-\rho)+(s-\sigma) \not> p-\pi, \quad (r+\rho)+(s+\sigma) \not> p+\pi; \; \ldots(4)$$

and therefore the possible values of r, s and ρ, σ are given respectively by

$$r \not< 0, \quad s \not< 0, \quad r+s \not> p, \quad \textit{and} \quad \rho \not< 0, \quad \sigma \not< 0, \quad 2(\rho+\sigma) \not> p+\pi.$$

For if the join of ω_r and ω_s is a spacelet ω_t of extravagance τ, the possible values of t, τ, r, s, ρ, σ are given by

$$\tau \not< 0, \quad \tau \not> t, \quad t+\tau \not> p+\pi, \quad t-\tau \not> p-\pi;$$

$$\rho \not< 0, \quad \sigma \not< 0, \quad \rho \not> r, \quad \sigma \not> s, \quad r+s=t, \quad \rho+\sigma=\tau.$$

If ω_r and ω_s are any two mutually orthogonal spacelets which lie in the given spacelet ω_p and have a given complete intersection ω_κ (which must be completely extravagant, so that $\kappa \not< 0$, $2\kappa \not> p+\pi$), the possible values of r, s, ρ, σ are given by

$$\rho \not< \kappa, \quad \sigma \not< \kappa, \quad \rho \not> r, \quad \sigma \not> s, \quad (r-\rho)+(s-\sigma) \not> p-\pi, \quad (r+\rho)+(s+\sigma) \not> p+\pi+2\kappa; \; \ldots(4')$$

and therefore the possible values of r, s and ρ, σ are given respectively by

$$r \not< \kappa, \quad s \not< \kappa, \quad r+s \not> p+\kappa, \quad r+s \not> p+\pi, \quad \textit{and} \quad \rho \not< \kappa, \quad \sigma \not< \kappa, \quad 2(\rho+\sigma) \not> p+\pi+2\kappa.$$

For if the join of ω_r and ω_s is a spacelet ω_t of extravagance τ (so that ω_κ is a sub-space of ω_τ), the possible values of t, τ, κ, r, s, ρ, σ are given by

$$\tau \not< 0, \quad \tau \not> t, \quad t+\tau \not> p+\pi, \quad t-\tau \not> p-\pi; \quad \kappa \not< 0, \quad \kappa \not> \tau;$$

$$\rho \not< \kappa, \quad \sigma \not< \kappa, \quad \rho \not> r, \quad \sigma \not> s, \quad r+s-t=\rho+\sigma-\tau=\kappa.$$

If ω_r and ω_s are any two mutually orthogonal spacelets which lie in the given spacelet ω_p, the possible values of r, s, ρ, σ are given by

$$\rho \not< 0, \quad \sigma \not< 0, \quad \rho \not> r, \quad \sigma \not> s, \quad (r-\rho)+(s-\sigma) \not> p-\pi,$$

$$\rho-\sigma \not> p+\pi-r-s, \quad \sigma-\rho \not> p+\pi-r-s; \; \ldots\ldots\ldots\ldots(4'')$$

and therefore the possible values of r, s and ρ, σ are given respectively by

$$r \not< 0, \quad s \not< 0, \quad r \not> p, \quad s \not> p, \quad r+s \not> p+\pi,$$

and

$$\rho \not< 0, \quad \sigma \not< 0, \quad 2\rho \not> p+\pi, \quad 2\sigma \not> p+\pi.$$

We see this by eliminating κ from the conditions (4′) and the additional condition $\kappa \not< 0$; the condition $2\kappa \not> p+\pi$ being superfluous.

Ex. v. *Ranks and extravagances of any two mutually orthogonal spacelets in homogeneous space ω_n.*

Taking ω_p in Ex. iv to be the complete space ω_n, we obtain the following particular results:

If ω_r and ω_s are any two mutually orthogonal non-intersecting spacelets in homogeneous space ω_n, the possible values of r, s, ρ, σ are given by

$$\rho \not< 0, \quad \sigma \not< 0, \quad \rho \not> r, \quad \sigma \not> s, \quad (r+\rho)+(s+\sigma) \not> n; \; \ldots\ldots\ldots\ldots(5)$$

and the possible values of r, s and ρ, σ are given respectively by

$$r \not< 0, \quad s \not< 0, \quad r+s \not> n, \quad \textit{and} \quad \rho \not< 0, \quad \sigma \not< 0, \quad 2(\rho+\sigma) \not> n.$$

If ω_r and ω_s are any two mutually orthogonal spacelets of homogeneous space ω_n which have a given complete intersection ω_κ (which must be completely extravagant, so that $\kappa \not< 0$, $2\kappa \not> n$), the possible values of r, s, ρ, σ are given by

$$\rho \not< \kappa, \quad \sigma \not< \kappa, \quad \rho \not> r, \quad \sigma \not> s, \quad (r+\rho)+(s+\sigma) \not> n+2\kappa\,; \quad \text{..............(5')}$$

and the possible values of r, s and ρ, σ are given respectively by

$$r \not< \kappa, \quad s \not< \kappa, \quad r+s \not> n, \quad \text{and} \quad \rho \not< \kappa, \quad \sigma \not< \kappa, \quad 2(\rho+\sigma) \not> n+2\kappa.$$

If ω_r and ω_s are any two mutually orthogonal spacelets of homogeneous space ω_n, the possible values of r, s, ρ, σ are given by

$$\rho \not< 0, \quad \sigma \not< 0, \quad \rho \not> r, \quad \sigma \not> s, \quad \rho-\sigma \not> n-r-s, \quad \sigma-\rho \not> n-r-s\,; \quad \text{.........(5'')}$$

and the possible values of r, s and ρ, σ are given respectively by

$$r \not< 0, \quad s \not< 0, \quad r+s \not> n, \quad \text{and} \quad \rho \not< 0, \quad \sigma \not< 0, \quad 2\rho \not> n, \quad 2\sigma \not> n.$$

These results clearly remain true when one of the two spacelets ω_r and ω_s is given. They can be obtained directly by observing that ω_s is orthogonal with a given spacelet ω_r when and only when it lies in the normal to ω_r, and using Ex. xiii of § 170. The last of the three results was obtained in a different way in Ex. xv of § 167.

Theorem II. *If ω_r, ω_s, ω_t, ... are any number of mutually orthogonal spacelets, and if their join is the spacelet ω_p whose core is ω_π, then:*

(1) *The complete intersection of ω_r, ω_s, ω_t, ... is a completely extravagant spacelet ω_κ lying in ω_π.*

(2) *The cores of the spacelets ω_r, ω_s, ω_t, ... are their respective complete intersections with ω_π.*

(3) *The join and complete intersection of the cores of ω_r, ω_s, ω_t, ... are respectively the spacelets ω_π and ω_κ.*

These results follow by induction from Theorem I, and they are generalisations of Ex. viii of § 170. The following is an independent proof of Theorem II.

Proof of Theorem II. Let the cores of the given mutually orthogonal spacelets ω_r, ω_s, ω_t, ... be ω_ρ, ω_σ, ω_τ, ...; let the complete intersection of these cores be

$$\omega_\kappa = [\,w\,]^{\kappa}_{n},$$

this being necessarily a completely extravagant spacelet; let

$$\omega_\rho \equiv [\,a, w\,]^{\rho-\kappa,\,\kappa}_{n}, \quad \omega_\sigma \equiv [\,\beta, w\,]^{\sigma-\kappa,\,\kappa}_{n}, \quad \omega_\tau \equiv [\,\gamma, w\,]^{\tau-\kappa,\,\kappa}_{n}, \ \ldots,$$

$$\omega_r \equiv [\,a, a, w\,]^{r-\rho,\,\rho-\kappa,\,\kappa}_{n}, \quad \omega_s \equiv [\,b, \beta, w\,]^{s-\sigma,\,\sigma-\kappa,\,\kappa}_{n}, \ \ldots;$$

and let ω_π and $\omega_{p-\pi}$ be the spacelets defined by

$$\omega_\pi \equiv [\,a, \beta, \gamma, \ldots w\,]^{\rho-\kappa,\,\sigma-\kappa,\,\tau-\kappa,\,\ldots\,\kappa}_{n}, \quad \omega_{p-\pi} \equiv [\,a, b, c, \ldots\,]^{r-\rho,\,s-\sigma,\,t-\tau,\,\ldots}_{n}.$$

The spacelet ω_π is the join of $\omega_\rho, \omega_\sigma, \omega_\tau, \ldots$; and it is completely extravagant because the constituents of the matrix which represents it are mutually orthogonal completely extravagant matrices.

Because the constituents of the matrix which represents $\omega_{p-\pi}$ are mutually orthogonal non-extravagant undegenerate matrices, it follows from Exs. vi and viii of § 170 that they are vertically unconnected, that $\omega_{p-\pi}$ is non-extravagant, and that

$$p-\pi=(r-\rho)+(s-\sigma)+(t-\tau)+\ldots \qquad (6)$$

The join of $\omega_r, \omega_s, \omega_t, \ldots$ is the join of the two mutually orthogonal spacelets $\omega_{p-\pi}$ and ω_π; and because $\omega_{p-\pi}$ is non-extravagant and ω_π is completely extravagant, it follows as in Ex. iii of § 172 that this join has rank p; moreover by Ex. viii of § 170 its core is ω_π. Thus the join of $\omega_r, \omega_s, \omega_t, \ldots$ is the spacelet

$$\omega_p = \boxed{a, b, \ldots \alpha, \beta, \ldots w}_{\,n}^{\,r-\rho,\, s-\sigma,\, \ldots\, \rho-\kappa,\, \sigma-\kappa,\, \ldots\, \kappa}$$

whose core is ω_π.

The results stated in the theorem now all follow immediately. In the first place because the spacelets $\omega_{r-\rho} \equiv \boxed{a}_{\,n}^{\,r-\rho}$, $\omega_{s-\sigma} \equiv \boxed{b}_{\,n}^{\,s-\sigma}$, ... are unconnected, the complete intersection of $\omega_r, \omega_s, \omega_t, \ldots$ is the same as that of $\omega_\rho, \omega_\sigma, \omega_\tau, \ldots$, i.e. it is the spacelet ω_κ, which clearly lies in ω_π. In the second place because $\omega_{r-\rho}$ does not intersect ω_π, and ω_ρ lies in ω_π, therefore ω_ρ is the complete intersection of ω_r with ω_π; and similarly $\omega_\sigma, \omega_\tau, \ldots$ are the complete intersections of $\omega_s, \omega_t, \ldots$ with ω_π. In the third place the join ω_π of $\omega_\rho, \omega_\sigma, \omega_\tau, \ldots$ has been shown to be the core of ω_p; and because $\omega_{r-\rho}, \omega_{s-\sigma}, \ldots$ and ω_π are unconnected, the complete intersection of $\omega_r, \omega_s, \omega_t, \ldots$ is that of $\omega_\rho, \omega_\sigma, \omega_\tau, \ldots$, which is ω_κ.

Ex. vi. The equation (6) can be written in the form

$$r+s+t+\ldots-p=\rho+\sigma+\tau+\ldots-\pi, \qquad (7)$$

and we can give interpretations to the expressions occurring on the two sides of (7).

If any number of spacelets $\omega_r, \omega_s, \omega_t, \ldots$ have as their join a spacelet ω_p of rank p, we may call $r+s+t+\ldots-p$ the *number of connections* between them; and it is easily seen that with this definition the number of connections between them is the rank of the complete intersection of any one of the spacelets with the join of the others increased by the number of connections between the others. When the spacelets are mutually orthogonal, the intersection of any one of the spacelets with the join of the others is the intersection of the core of that one with the join of the cores of the others; and from this fact the equation (7) follows by induction.

Ex. vii. In the theorem the core of the join of any number of the spacelets $\omega_r, \omega_s, \omega_t, \ldots$ is the intersection of that join with ω_π.

APPENDIX A

ADDITION TO CHAPTER XIII

§ 117 a. Utility of the relations obtained in Chapter XIII.

1. *Regular minor determinants.*

We will define a *regular* minor determinant of a matrix *whose elements are constants* to be one which does not vanish. According to this definition no minor determinant of order greater than the rank of the matrix can be regular. From Ex. i of § 116 we see that:

Every regular minor determinant whose order is not less than 2 *has a primary subdeterminant which is regular; and every regular minor determinant whose order is less than the rank of the matrix has a primary superdeterminant which is regular.*

NOTE 1. If we were to consider that every minor determinant whose order is greater than the rank of the matrix is regular although it necessarily vanishes, and if we at the same time adopt the usual convention that a determinant of order 0 has the value 1, we could say more concisely that:

Every regular minor determinant has a primary subdeterminant which is regular, and also a primary superdeterminant which is regular.

The corresponding properties of regular diagonal minor determinants of symmetric and skew-symmetric matrices are given in §§ 126 and 127.

2. *The fundamental identity of* § 116.

The importance of this identity is due to the following facts which can be deduced from it, r being any positive integer which is less than both m and n.

The fundamental identity (A) *of* § 116 *enables us to express every element of the matrix* $A = [a]_m^n$ *as a rational function of any one regular minor determinant* Δ *of order* r *and the elements, primaries and primary superdeterminants of* Δ. ....(a)

Arbitrary values can be ascribed to the elements, the primaries and the primary superdeterminants of any regular minor determinant Δ of order r; and the values of all elements of the matrix A are then completely and uniquely determined. ....(b)

If the values ascribed to the primary superdeterminants of Δ are such that their matrix has rank x, then the rank ρ of the matrix A is given by

$$\rho = x + r. \qquad \text{(c)}$$

When arbitrary values are ascribed to the elements of a *regular* minor determinant Δ, it is to be understood that they are arbitrary subject to the condition that $\Delta \neq 0$.

In proving the results (a), (b), (c) we may suppose without loss of generality that Δ is the leading minor determinant of A of order r, or that $\Delta = (a)_r^r$, so that the identity (A) of § 116 is

$$\Delta\,[a]_m^n = [a]_m^r \,\overline{\underline{A}}_r^r\, [a]_r^n + \begin{bmatrix} 0, & 0 \\ 0, & Q \end{bmatrix}_{r,\,m-r}^{r,\,n-r}; \qquad \text{(A')}$$

and $[Q]_{m-r}^{n-r}$ is what we call the matrix of the primary superdeterminants of Δ. The equation (A) clearly expresses every element of A as a rational function of Δ, the elements of Δ, the remaining elements of the vertical and horizontal minors $H = [a]_m^r$ and $K = [a]_r^n$ of A which contain Δ, and the elements of $[Q]_{m-r}^{n-r}$. Since by § 108 all elements of H and K can be expressed in terms of the elements and the horizontal and vertical primaries of Δ, we see that (a) is true. To obtain direct expressions for the elements of A of the nature stated, we multiply both sides of (A′) by Δ, and replace it as in Note 5 of § 116 by the identity

$$\Delta^2 [a]_m^n = \begin{bmatrix} \Delta \\ p \end{bmatrix}_{r,\,m-r}^{r} [a]_r^r\,[\Delta,\, q]_r^{r,\,n-r} + \Delta \begin{bmatrix} 0, & 0 \\ 0, & Q \end{bmatrix}_{r,\,m-r}^{r,\,n-r}, \qquad \text{(B')}$$

or

$$\Delta^2 [a]_m^n = \begin{bmatrix} \Delta, & 0 \\ p, & 1 \end{bmatrix}_{r,\,m-r}^{r,\,m-r} \begin{bmatrix} a, & 0 \\ 0, & \Delta Q \end{bmatrix}_{r,\,m-r}^{r,\,n-r} \begin{bmatrix} \Delta, & q \\ 0, & 1 \end{bmatrix}_{r,\,n-r}^{r,\,n-r}. \qquad \text{(C')}$$

The equation (B′) expresses every element of A as a rational function of the elements of $[a]_r^r$ or Δ, the p's (which are the horizontal primaries of Δ), the q's (which are the vertical primaries of Δ), and the Q's (which are the primary superdeterminants of Δ).

If in (B′) or (C′) we ascribe arbitrary values to the elements of Δ such that $\Delta \neq 0$, and arbitrary values to the p's, q's and Q's, we obtain a matrix $[a]_m^n$ every element of which is known. Therefore (b) is true.

Lastly, regarding (C′) as an equigradent transformation, we see from § 131 that the matrix $A=[a]_m^n$ has the same rank as the bipartite matrix which is the middle factor on the right; and if ρ and x are the ranks of A and $[Q]_{m-r}^{n-r}$, it follows from § 100 that $\rho = x + r$. Thus (c) is also true.

NOTE 2. We can of course prove the results (a), (b), (c) without regarding Δ as a *leading* minor determinant. If $\Delta=(a_{pq})_r^r$, the identity (A) of § 116 is

$$\Delta\,[a]_m^n = [a_{1q}]_m^r \,\boxed{A_{pq}}_r^r\, [a_{p1}]_r^n + [P]_m^n\,; \quad\text{.........................(A)}$$

and if we multiply both sides of (A) by Δ, we obtain as in Note 5 of § 116 the identity

$$\Delta^2\,[a]_m^n = [a_{1p}]_m^r\,[a_{pq}]_r^r\,[\beta_{q1}]_r^n + \Delta\,[P]_m^n, \quad\text{..........................(B)}$$

or

$$\Delta^2\,[a]_m^n = [a]_m^m\,[a']_m^n\,[\beta]_n^n, \quad\text{...(C)}$$

where $[a_{pp}]_r^r = \Delta\,[1]_r^r,\quad [a_{\mu\mu}]_{m-r}^{m-r} = [1]_{m-r}^{m-r},\quad [a_{p\mu}]_r^{m-r} = 0,\quad (a)_m^m = \pm\Delta^r,$

$$[\beta_{qq}]_r^r = \Delta\,[1]_r^r,\quad [\beta_{\nu\nu}]_{n-r}^{n-r} = [1]_{n-r}^{n-r},\quad [\beta_{\nu q}]_{n-r}^r = 0,\quad (\beta)_n^n = \pm\Delta^r,$$

$$[P_{pq}]_r^r = 0,\qquad [P_{\mu q}]_{m-r}^r = 0,\qquad [P_{p\nu}]_r^{n-r} = 0,$$

$[a_{\mu p}]_{m-r}^r$ is the matrix of the horizontal primaries of Δ,

$[\beta_{q\nu}]_r^{n-r}$ is the matrix of the vertical primaries of Δ,

$[P_{\mu\nu}]_{m-r}^{n-r}$ is the matrix of the primary superdeterminants of Δ.

The equation (B) expresses every element of A as a rational function of Δ and the elements, primaries and primary superdeterminants of Δ; by assigning arbitrary values to the elements of $[a_{pq}]_r^r$ consistent with the condition $\Delta \neq 0$, and arbitrary values to the elements of $[a_{\mu p}]_{m-r}^r$, $[\beta_{q\nu}]_r^{n-r}$, $[P_{\mu\nu}]_{m-r}^{n-r}$ in (B) or (C) we obtain a matrix $A=[a]_m^n$ every element of which is known; and since (C′) is an equigradent transformation, and $[a']_m^n$ is a bipartite matrix whose two parts are $[a_{pq}]_r^r$ and $\Delta\,[P_{\mu\nu}]_{m-r}^{n-r}$, we see that when the latter matrix has rank x, the rank ρ of A has the value given in (c).

Ex. i. The result (c) is a generalisation of Note 4 of § 116.

Ex. ii. We can give greater precision to Ex. x of § 116. If the determinant $\Delta=(a_{pq})_r^r=0$, then by § 124.6 or § 73 the matrix $\boxed{A_{pq}}_r^r$ has rank 1 or 0 according as the rank of $[a_{pq}]_r^r$ is equal to or less than $r-1$. Accordingly the identity (A) shows that:

If $[a_{pq}]_r^r$ has rank less than $r-1$, then $[P]_m^n$ has rank 0.

If $[a_{pq}]_r^r$ has rank less than $r-1$, the rank of $[P]_m^n$ is either 0 or 1.

Ex. iii. If in (C′) we write $[a]_m^n = \begin{bmatrix} a, & c \\ b, & d \end{bmatrix}_{r,\, m-r}^{r,\, n-r}$, we have

$$[b]_{m-r}^{r} = \Delta\,[p]_{m-r}^{r}\,[a]_r^r, \quad [c]_r^{n-r} = \Delta\,[a]_r^r\,[q]_r^{n-r},$$

$$[d]_{m-r}^{n-r} = [p,\ 1]_{m-r}^{r,\, m-r} \begin{bmatrix} a, & 0 \\ 0, & \Delta Q \end{bmatrix}_{r,\, m-r}^{r,\, n-r} \begin{bmatrix} q \\ 1 \end{bmatrix}_{r,\, n-r}^{n-r}$$

$$= [p]_{m-r}^{r}\,[a]_r^r\,[q]_r^{n-r} + \Delta\,[Q]_{m-r}^{n-r}.$$

3. *The identities of* §§ 108 *and* 109.

The results which correspond to (a) and (b) when r is equal to m or n are contained in § 108, being deduced from the formulae (A) and (C) and the formulae (A′) and (C′) of that article. Adding to them the results obtained in § 109 (referring in particular to Ex. xi of § 109), we see that:

The identities (A) *and* (A′) *of* § 108 *express every element of the matrix* $A = [a]_m^n$ *as a rational function of degree* 1 *of any one regular simple minor determinant* Δ *and the elements and primaries of* Δ. ....(a′)

Arbitrary values can be ascribed to the elements and primaries of any one regular simple minor determinant Δ; *and the values of all elements of the matrix* A *are then completely and uniquely determined.*......(b′)

The identities of § 109 *serve to express every simple minor determinant of the matrix* $A = [a]_m^n$ *as a rational function of degree* 1 *of any one regular simple minor determinant* Δ *and its primaries.* ....(c′)

Arbitrary values can be ascribed to any one regular simple minor determinant Δ *and its primaries; and the values of all simple minor determinants are then completely and uniquely determined.* ....(d′)

4. *The equations of* §§ 115 *and* 117.

By putting $[P]_m^n = 0$, $[Q]_{m-r}^{n-r} = 0$ in the equations (A), (B), (C), (A′), (B′), (C′) of sub-article 2 we obtain the corresponding equations of § 115, which however are not identities in the elements of A; and from these we deduce two results corresponding to (a) and (b). Adding to these the results obtained in § 117, we see that:

The equation (A) *of* § 115 *enables us to express every element of a matrix* $A=[a]_m^n$ *whose rank is r as a rational function of degree* 1 *of any one regular minor determinant* Δ *of order r and the elements and primaries of* Δ. ....(a″)

In a matrix $A=[a]_m^n$ *of rank r arbitrary values can be ascribed to the elements and primaries of any one regular minor determinant* Δ *of order r; and the values of all other elements of A are then completely and uniquely determined.* ....(b″)

The equations of § 117 *serve to express every minor determinant of order r of a matrix* $A=[a]_m^n$ *whose rank is r as a rational function of degree* 1 *of any one regular minor determinant* Δ *of order r and its primaries.* ....(c″)

In a matrix $A=[a]_m^n$ *of rank r arbitrary values can be ascribed to any one regular minor determinant* Δ *of order r; and the values of all other minor determinants of order r are then completely and uniquely determined.* ....(d″)

Ex. iv. We can regard (a′), (b′), (c′), (d′) as particular cases of (a″), (b″), (c″), (d″).

5. *The standard relations between the minor determinants of a matrix.*

The *standard identities* of §§ 109—111 are all expressible in the form

$$\Delta^{s-1}D=(\delta)_s^s. \quad\text{..........(D)}$$

In § 109, if m is the number of long rows in a matrix A, we have identities of this form in which Δ and D are any two simple minor determinants of A having exactly $m-s$ of the short rows of A in common, and the elements of the determinant on the right are the primaries of Δ formed with the rows of D. These identities serve to express every simple minor determinant D as a rational function (of degree 1) of any one regular simple minor determinant Δ and its primaries; and they also give the values of determinants of the primaries of a simple minor determinant.

In § 110 we have identities of this form in which Δ is a minor determinant of order r contained in a determinant D of order $r+s$, and the elements of the determinant on the right are the primary superdeterminants of Δ which lie in D. They serve to express every superdeterminant D of a regular minor determinant Δ as a rational function (of degree 1) of Δ and the *primary* superdeterminants of Δ; and they also give the values of determinants of primary superdeterminants.

In § 111 we have identities of this form in which Δ is a determinant of order $r+s$ which contains a minor determinant D of order r, and the elements of the determinant on the right are the primary subdeterminants of Δ which contain D. They serve to express every subdeterminant D of a regular determinant Δ as a rational function (of degrec 1) of Δ and the *primary* subdeterminants of Δ; and they also give the values of determinants of primary subdeterminants.

NOTE 3. The identities of § 112 give the values of sums of terms formed from a product ΔD of two simple minor determinants of a matrix by replacing according to a fixed law a certain number s of the short rows of Δ by rows of D, and the s rows taken from D by rows of Δ.

The *standard equations* of § 117 all have the form

$$\Delta^{\rho+\sigma-1} D = (\alpha)_\rho^\rho \, (\beta)_\sigma^\sigma. \quad \text{..........................(E)}$$

If Δ and D are any two minor determinants of order r of a matrix $A=[a]_m^n$ of rank r, and if there are exactly $r-\rho$ horizontal rows and exactly $s-\sigma$ vertical rows of A which intersect both Δ and D, we have an equation of this form in which the elements of the first and second determinants on the right are respectively the horizontal and vertical primaries of Δ formed with the rows of A which intersect D. These equations serve to express every minor determinant D of order r of a matrix A whose rank is r as a rational function (of degree 1) of any one regular minor determinant Δ of order r and its primaries.

APPENDIX B

ADDITION TO CHAPTER XIV

§ 130 a. The Pfaffian of a skew-symmetric matrix of even order.

1. *Definition of a Pfaffian.*

If $A = [a]_m^m$ is a skew-symmetric matrix with arbitrary elements of even order m, m being greater than 2, its reciprocal $[A]_m^m$ is also skew-symmetric, and if $\Delta_m = (a)_m^m$, $\Delta_{m-2} = \begin{pmatrix} 34\ldots m \\ a \\ 34\ldots m \end{pmatrix}$, we have by § 124.4 or § 111

$$(A)_2^2 = A_{12}^2 = \Delta_m \Delta_{m-2}.$$

Since the determinants Δ_m and Δ_{m-2} are functions of their elements which do not vanish identically, it follows that if Δ_{m-2}, which is a skew-symmetric determinant of even order $m-2$, is a perfect square, then Δ_m is a perfect square. Since moreover every skew-symmetric matrix of order 2 is certainly a perfect square, it follows by induction that:

The determinant of every skew-symmetric matrix A of even order is the square of a rational integral function of the elements of A.

It is shown in Theorem II *a* that this function can be identified with the Pfaffian of A defined in the two theorems which follow.

Theorem I a. *If $A = [a]_m^m$ is a skew-symmetric matrix of even order $m = 2r$; if the elements of the sequence $[1\,2\ldots m]$ are divided in all possible ways into r pairs $(u_1, v_1), (u_2, v_2), \ldots (u_r, v_r)$, the order of arrangement of the elements in each pair being immaterial, and the order of arrangement of the r pairs being immaterial; and if ω is the affect of the sequence $[u_1 v_1 u_2 v_2 \ldots u_r v_r]$ in $[1\,2\ldots m]$; then the equation*

$$P_{12\ldots m} = \Sigma(-1)^{\omega} a_{u_1 v_1} a_{u_2 v_2} \ldots a_{u_r v_r} \quad \ldots\ldots\ldots\ldots\ldots\ldots\text{(A)}$$

defines a uniquely determinate rational integral function of the elements of A which is called the Pfaffian of A.

The number of terms in the sum on the right in (A) is the odd number

$$\frac{m!}{2^r . r!} = 1 . 3 . 5 . \ldots . (m-1).$$

To prove the theorem we have to show that if P is the sum (A), and if

$$P' = \Sigma (-1)^{\omega'} a_{p_1 q_1} a_{p_2 q_2} \dots a_{p_r q_r} \quad \dots\dots\dots\dots\dots\dots(A')$$

is any other sum formed in the same way, ω' being the affect of

$$[p_1 q_1 p_2 q_2 \dots p_r q_r] \text{ in } [1\, 2 \dots m],$$

then P' is identical with P.

The number of terms in P' is the same as in P, and to each term

$$T = (-1)^{\omega} a_{u_1 v_1} a_{u_2 v_2} \dots a_{u_r v_r} = (-1)^{\omega} t$$

of P there corresponds a term

$$T' = (-1)^{\omega'} a_{p_1 q_1} a_{p_2 q_2} \dots a_{p_r q_r} = (-1)^{\omega} t'$$

of P', where the product $t' = a_{p_1 q_1} a_{p_2 q_2} \dots a_{p_r q_r}$ is formed from the product $t = a_{u_1 v_1} a_{u_2 v_2} \dots a_{u_r v_r}$ by interchanging the two suffixes in some of the factors of t, and by altering the order of arrangement of the factors.

If t' is formed from t by merely replacing $a_{u_i v_i}$ by $a_{v_i u_i}$, then since $a_{v_i u_i} = -a_{u_i v_i}$, we have $t' = -t$. In this case it follows from § 21 that ω' differs from ω by an odd integer, and therefore $T' = T$.

If t' is formed from t by merely interchanging two factors $a_{u_i v_i}$ and $a_{u_j v_j}$, we have $t' = t$. In this case the sequence $[p_1 q_1 p_2 q_2 \dots p_r q_r]$ is formed from the sequence $[u_1 v_1 u_2 v_2 \dots u_r v_r]$ by two interchanges of two elements; therefore ω' differs from ω by an even integer, and we again have $T' = T$.

Since t' is always formed from t by a number of operations of these two kinds, we always have $T' = T$; and it follows that we always have $P' = P$.

Ex. i. For the matrix $A = \begin{bmatrix} a_{11} & a_{12} \\ a_{21} & a_{22} \end{bmatrix} = \begin{bmatrix} 0, & a_{12} \\ -a_{12}, & 0 \end{bmatrix}$ we have

$$P_{12} = a_{12}.$$

Ex. ii. For the matrix

$$A = \begin{bmatrix} a_{11} & a_{12} & a_{13} & a_{14} \\ a_{21} & a_{22} & a_{23} & a_{24} \\ a_{31} & a_{32} & a_{33} & a_{34} \\ a_{41} & a_{42} & a_{43} & a_{44} \end{bmatrix} = \begin{bmatrix} 0, & a_{12}, & a_{13}, & a_{14} \\ -a_{12}, & 0, & a_{23}, & a_{24} \\ -a_{13}, & -a_{23}, & 0, & a_{34} \\ -a_{14}, & -a_{24}, & -a_{34}, & 0 \end{bmatrix}$$

we have $$P_{1234} = a_{12} a_{34} - a_{13} a_{24} + a_{14} a_{23}.$$

Ex. iii. If in (A) we introduce the restrictions that $u_1 < v_1$, $u_2 < v_2$, ... $u_r < v_r$, we can regard $P_{12\dots m}$ as a rational integral function of those elements of A which lie above the leading diagonal. Taking this point of view, Sir Thomas Muir uses the notation

$$P_{12\dots m} = \left| \begin{matrix} a_{12} & a_{13} & a_{14} & \dots & a_{1m} \\ & a_{23} & a_{24} & \dots & a_{2m} \\ & & a_{34} & \dots & a_{3m} \\ & & & \dots\dots\dots & \\ & & & & a_{m-1,m} \end{matrix} \right|$$

Ex. iv. The Pfaffian of A is isobaric in the suffixes of the a's, the sum of the suffixes in each term being $1+2+\ldots+m=\frac{1}{2}m(m+1)$.

Ex. v. For the special skew-symmetric matrix

$$\begin{bmatrix} \omega, & 0, & \ldots & 0 \\ 0, & \omega, & \ldots & 0 \\ \ldots & \ldots & \ldots & \ldots \\ 0, & 0, & \ldots & \omega \end{bmatrix}_{2,2,\ldots 2}^{2,2,\ldots 2}, \quad \text{where } [\omega]_2^2 = \begin{bmatrix} 0, & 1 \\ -1, & 0 \end{bmatrix},$$

the Pfaffian is 1.

Theorem I b. *If $B=[a_{hh}]_\mu^\mu$ is any diagonal minor of even order $\mu=2\rho$ of a skew-symmetric matrix $A=[a]_m^m$, being therefore any skew-symmetric matrix of order μ; if the elements of the sequence $[h_1 h_2 \ldots h_\mu]$ are divided in all possible ways into ρ pairs $(u_1, v_1), (u_2, v_2), \ldots (u_\rho, v_\rho)$, the order of arrangement of the two elements in each pair being immaterial, and the order of arrangement of the ρ pairs being immaterial; and if ω is the affect of the sequence $[u_1 v_1 u_2 v_2 \ldots u_\rho v_\rho]$ in $[h_1 h_2 \ldots h_\mu]$; then the equation*

$$P_{h_1 h_2 \ldots h_\rho} = \Sigma (-1)^\omega a_{u_1 v_1} a_{u_2 v_2} \ldots a_{u_\rho v_\rho}$$

defines a uniquely determinate rational integral function of the elements of B which is the Pfaffian of B.

This follows from Theorem I *a* when we write $[a_{hh}]_\mu^\mu=[b]_\mu^\mu$. In the special case when m is even and $\mu=m$, B may be any symmetric derangement of A.

Ex. vi. *If $[k]_\mu$ is any derangement of $[h]_\mu$ in Theorem I b, then*

$$P_{k_1 k_2 \ldots k_\rho} = (-1)^\eta P_{h_1 h_2 \ldots h_\rho},$$

where η is the affect of $[k]_\mu$ in $[h]_\mu$.

For if $P=P_{h_1 h_2 \ldots h_\rho}$ and $P'=P_{k_1 k_2 \ldots k_\rho}$, we have

$$P' = \Sigma (-1)^{\omega'} a_{u_1 v_1} a_{u_2 v_2} \ldots a_{u_\rho v_\rho},$$

where ω' is the affect of the sequence $[u_1 v_1 u_2 v_2 \ldots u_\rho v_\rho]$ in $[k]_\mu$. But by Theorems V *b* and VII *a* of § 19 we have

$$\omega' \equiv \omega + \eta \pmod{2},$$

and it follows that

$$P' = (-1)^\eta P.$$

Ex. vii. For the matrix

$$\begin{bmatrix} a_{44} & a_{43} & a_{41} & a_{42} \\ a_{34} & a_{33} & a_{31} & a_{32} \\ a_{14} & a_{13} & a_{11} & a_{12} \\ a_{24} & a_{23} & a_{21} & a_{22} \end{bmatrix} = \begin{bmatrix} 0, & a_{43}, & a_{41}, & a_{42} \\ -a_{43}, & 0, & a_{31}, & a_{32} \\ -a_{41}, & -a_{31}, & 0, & a_{12} \\ -a_{42}, & -a_{32}, & -a_{12}, & 0 \end{bmatrix}$$

we have

$$\begin{aligned} P_{4312} &= a_{43}a_{12} - a_{41}a_{32} + a_{42}a_{31} \\ &= -(a_{12}a_{34} - a_{13}a_{24} + a_{14}a_{23}) = -P_{1234}. \end{aligned}$$

2. *Expansion of the Pfaffian of a skew-symmetric matrix A of even order in terms of the elements of any horizontal or vertical row of A.*

Let the matrix be $A=[a]_m^m$, where $m=2r$, and let its Pfaffian be $P=P_{12\ldots m}$. On account of the relations $a_{ji}=-a_{ij}$ it will be clear that P can be expressed as a homogeneous linear function of the elements of any given horizontal or vertical row of A.

First let a_{ij} be any non-diagonal element of A, and let ϖ_{ij} be the coefficient of a_{ij} in P when a_{ji} has everywhere been replaced by $-a_{ij}$. Let $[\lambda_1\lambda_2\ldots\lambda_{m-2}]$ be the corranged complement of the minor sequence $[ij]$ in $[1\,2\ldots m]$, and let

$$\eta = \text{affect of } [ij] \text{ in } [1\,2\ldots m].$$

To determine ϖ_{ij} we observe that $a_{ij}\varpi_{ij}$ is the sum of all those terms of P in which a_{ij} occurs as a factor. Hence if we divide the elements of $[\lambda]_{m-2}$ in all possible ways into $r-1$ pairs (u_1, v_1), (u_2, v_2), ... (u_{r-1}, v_{r-1}) as in Theorem I*b*, we have

$$a_{ij}\varpi_{ij} = a_{ij}\,.\,\Sigma(-1)^{\omega}\,a_{u_1v_1}a_{u_2v_2}\ldots a_{u_{r-1}v_{r-1}},$$

where

$$\begin{aligned}\omega &= \text{aff.}\,[ij\,u_1v_1u_2v_2\ldots u_{r-1}v_{r-1}] \text{ in } [1\,2\ldots m]\\ &= \text{aff.}\,[ij] \text{ in } [1\,2\ldots m] + \text{aff.}\,[u_1v_1u_2v_2\ldots u_{r-1}v_{r-1}] \text{ in } [\lambda]_{m-2}\\ &= \eta+\omega';\end{aligned}$$

and therefore

$$\varpi_{ij}=(-1)^{\eta}.\,\Sigma(-1)^{\omega'}a_{u_1v_1}a_{u_2v_2}\ldots a_{u_{r-1}v_{r-1}}=(-1)^{\eta}P_{\lambda_1\lambda_2\ldots\lambda_{m-2}}.\;\ldots(1)$$

If $i<j$, then η is the least number of moves by which the element a_{ij} can be brought to the position of a_{12} in A, and we have

$$\varpi_{ji}=-\varpi_{ij}.$$

We may call $P_{\lambda_1\lambda_2\ldots\lambda_{m-2}}$ the *corranged Pfaffian complementary to* a_{ij} in A or P, it being the Pfaffian of the matrix formed by striking out the ith and jth horizontal and vertical rows of A; and we will call ϖ_{ij} the *Pfaffian co-factor* of a_{ij} in A or P.

Next let a_{ii} be any diagonal element of A. Since $a_{ii}=0$, we can write

$$\varpi_{ii}=0 \;\ldots\ldots\ldots\ldots\ldots\ldots\ldots\ldots\ldots\ldots(1')$$

and regard ϖ_{ii} as the co-factor of a_{ii} in P.

It will now be clear that the expansion of P in terms of the elements of the ith horizontal row of A can be expressed in the form

$$P_{12\ldots m}=\Sigma(-1)^{\eta}a_{ij}P_{\lambda_1\lambda_2\ldots\lambda_{m-2}},$$

where in the summation j receives all the values 1, 2, ... m except i, and also in the form

$$P_{12\ldots m} = a_{i1}\varpi_{i1} + a_{i2}\varpi_{i2} + \ldots + a_{im}\varpi_{im}, \quad\ldots\ldots(2)$$

where the ϖ's are given by (1) and (1′). In fact P is the sum of the products obtained by multiplying each element of the ith horizontal row of A by its Pfaffian co-factor.

Similarly the expansion of P in terms of the elements of the jth vertical row of A is

$$P_{12\ldots m} = a_{1j}\varpi_{1j} + a_{2j}\varpi_{2j} + \ldots + a_{mj}\varpi_{mj}. \quad\ldots\ldots(2')$$

Ex. viii. For the skew-symmetric matrix $A = [a]_6^6$ we have

$$P_{123456} = a_{12}P_{3456} - a_{13}P_{2456} + a_{14}P_{2356} - a_{15}P_{2346} + a_{16}P_{2345}$$
$$= a_{12}(a_{34}a_{56} - a_{35}a_{46} + a_{45}a_{36}) - a_{13}(a_{24}a_{56} - a_{25}a_{46} + a_{45}a_{26})$$
$$+ a_{14}(a_{23}a_{56} - a_{25}a_{36} + a_{35}a_{26}) - a_{15}(a_{23}a_{46} - a_{24}a_{36} + a_{34}a_{26})$$
$$+ a_{16}(a_{23}a_{45} - a_{24}a_{35} + a_{34}a_{25}).$$

3. *Determinants formed by bordering a given skew-symmetric determinant.*

We consider now the value of any determinant Δ' formed by bordering any given skew-symmetric determinant Δ with one additional horizontal row and one additional vertical row. We can always regard Δ' as a minor determinant of some skew-symmetric matrix $A = [a]_m^m$ which has Δ as a diagonal minor determinant, and then use the notation of Theorem I *b*.

Theorem II a. *If* $\Delta = (a)_{2r-1}^{2r-1}$ *is a skew-symmetric determinant of odd order* $2r-1$ *with arbitrary elements, and if* $\Delta' = \begin{pmatrix} v, 1, 2, \ldots 2r-1 \\ a \\ u, 1, 2, \ldots 2r-1 \end{pmatrix}$ *is a determinant of order* $2r$ *formed by bordering* Δ *with one additional horizontal row and one additional vertical row, the elements of the bordering rows being arbitrary subject to the conditions*

$$a_{uj} = -a_{ju}, \quad a_{iv} = -a_{vi}, \quad\ldots\ldots(3)$$

then

$$\Delta' = P_{u,1,2,\ldots 2r-1} \cdot P_{v,1,2,\ldots 2r-1}. \quad\ldots\ldots(B)$$

In particular if $\Delta' = (a)_{2r}^{2r}$ *is a skew-symmetric determinant of even order* $2r$, *then*

$$\Delta' = P^2_{1,2,\ldots 2r}. \quad\ldots\ldots(B')$$

Theorem II b. *If* $\Delta = (a)_{2r}^{2r}$ *is a skew-symmetric matrix of even order* $2r$ *with arbitrary elements, and if* $\Delta' = \begin{pmatrix} v, 1, 2, \ldots 2r \\ a \\ u, 1, 2, \ldots 2r \end{pmatrix}$ *is a determinant of order* $2r+1$ *formed by bordering* Δ *with one additional horizontal row and one additional vertical row, the elements of the bordering rows being arbitrary subject to the conditions* (3), *then*

$$\Delta' = P_{1,2,\ldots 2r} \cdot P_{u,v,1,2,\ldots 2r}. \quad\ldots\ldots(C)$$

In particular if $\Delta' = (a)_{2r+1}^{2r+1}$ *is a skew-symmetric determinant of odd order* $2r+1$, *then* Δ' *vanishes identically.*

In both theorems the conditions (3) to be satisfied by the elements of the bordering rows are introduced in order that we may be able to regard Δ as a leading diagonal minor determinant of some skew-symmetric matrix $A=[a]_m^m$, and Δ' as a minor determinant of A containing Δ. They impose no restrictions on the possible values of the bordering elements so long as we regard u and v as different integers, neither of which is a suffix of any element of Δ. Of course in the special case when $v=u$, we must have $a_{uv}=0$.

We will prove these two theorems on the hypothesis that they are true when r is replaced by any smaller positive integer. Since they are obviously true when $r=1$, this will establish the general truth of the theorems by induction.

Proof of Theorem II*a*. We denote the co-factor of any element a_{ij} of Δ' by A_{ij}, and expand Δ' in terms of the elements of its first horizontal row. From the hypothesis that Theorem II*b* is true when r is replaced by $r-1$ and from Ex. vi it follows that

$$A_{uv}=(a)_{2r-1}^{2r-1}=0,$$

and that when $i \neq v$,

$$A_{ui}=(-1)^i\begin{pmatrix} v,\,1,\,2,\,\ldots\,i-1,\,i+1,\,\ldots\,2r-1 \\ a \\ 1,\,2,\,\ldots\,i-1,\,i,\,i+1,\,\ldots\,2r-1 \end{pmatrix}=(-1)^{2i-1}\begin{pmatrix} v,\,1,\,2,\,\ldots\,i-1,\,i+1,\,\ldots\,2r-1 \\ a \\ i,\,1,\,2,\,\ldots\,i-1,\,i+1,\,\ldots\,2r-1 \end{pmatrix}$$

$$=(-1)^{2i-1}P_{1,2,\ldots i-1,i+1,\ldots 2r-1}\cdot P_{i,v,1,2,\ldots i-1,i+1,\ldots 2r-1}$$

$$=(-1)^{i-1}P_{1,2,\ldots i-1,i+1\ldots 2r-1}\cdot P_{v,1,2,\ldots i-1,i,i+1,\ldots 2r-1}.$$

Therefore $$\Delta'=P_{v,1,2,\ldots 2r-1}\cdot\Sigma(-1)^{i-1}a_{ui}P_{1,2,\ldots i-1,i+1,\ldots 2r-1},$$

where in the summation i receives the values $1, 2, \ldots 2r-1$; and because $i-1$ is the affect of the sequence $[u\,i]$ in $[u, 1, 2, \ldots 2r-1]$, it follows from the expansion (2) in sub-article 2 that the identity (B) is true.

We obtain the second part of Theorem II*a* by considering the special case in which $v=u$.

Proof of Theorem II*b*. We again denote the co-factor of any element a_{ij} of Δ' by A_{ij}, and expand Δ' in terms of the elements of its first horizontal row. From the hypothesis that Theorem II*a* is true when r is replaced by $r-1$ and from Ex. vi it follows that

$$A_{uv}=(a)_{2r}^{2r}=P^2_{1,2,\ldots 2r},$$

and that when $i \neq v$,

$$A_{ui}=(-1)^i\begin{pmatrix} v,\,1,\,2,\,\ldots\,i-1,\,i+1,\,\ldots\,2r \\ a \\ 1,\,2,\,\ldots\,i-1,\,i,\,i+1,\,\ldots\,2r \end{pmatrix}=(-1)^{2i-1}\begin{pmatrix} v,\,1,\,2,\,\ldots\,i-1,\,i+1,\,\ldots\,2r \\ a \\ i,\,1,\,2,\,\ldots\,i-1,\,i+1,\,\ldots\,2r \end{pmatrix}$$

$$=(-1)^{2i-1}P_{i,1,2,\ldots i-1,i+1,\ldots 2r}\cdot P_{v,1,2,\ldots i-1,i+1,\ldots 2r}$$

$$=(-1)^i\,.\,P_{1,2,\ldots i-1,i,i+1,\ldots 2r}\cdot P_{v,1,2,\ldots i-1,i+1,\ldots 2r}.$$

Therefore $$\Delta'=P_{1,2,\ldots 2r}\cdot\{a_{uv}P_{1,2,\ldots 2r}+\Sigma(-1)^i a_{ui}P_{v,1,2,\ldots i-1,i+1,\ldots 2r}\},$$

where in the summation i receives the values 1, 2, ... $2r$; and because i is the affect of the sequence $[u\,i]$ in $[u, v, 1, 2, \dots 2r]$, it follows from the expansion (2) in sub-article 2 that the identity (C) is true.

The second part of Theorem II b has been shown to be true in § 127; but it can also be obtained from the first part of the theorem (see Ex. ix) by considering the special case in which $v=u$.

The following are more general forms of the last two theorems.

Theorem III a. *If* $\Delta=(a_{hh})_{2r-1}^{2r-1}$ *is a diagonal minor determinant of odd order* $2r-1$ *of a skew-symmetric matrix* $A=[a]_m^m$ *whose elements are arbitrary, and if* $\Delta'=\begin{pmatrix} v\,h_1\,h_2\dots h_{2r-1} \\ a \\ u\,h_1\,h_2\dots h_{2r-1}\end{pmatrix}$ *is any minor determinant of* A *of order* $2r$ *which contains* Δ, *then*

$$\Delta'=P_{u\,h_1\,h_2\dots h_{2r-1}}\,.\,P_{v\,h_1\,h_2\dots h_{2r-1}}\,. \quad \dots\dots\dots\dots\dots\dots\text{(D)}$$

In particular if $\Delta'=(a_{\lambda\lambda})_{2r}^{2r}$ *is any diagonal minor determinant of* A *of even order* $2r$, *i.e. any skew-symmetric matrix of even order* $2r$, *then*

$$\Delta'=P^2_{\lambda_1\lambda_2\dots\lambda_{2r}}\,. \quad \dots\dots\dots\dots\dots\dots\dots\dots\text{(D')}$$

Theorem III b. *If* $\Delta=(a_{hh})_{2r}^{2r}$ *is a diagonal minor determinant of even order* $2r$ *of a skew-symmetric matrix* $A=[a]_m^m$ *whose elements are arbitrary, and if* $\Delta'=\begin{pmatrix} v\,h_1\,h_2\dots h_{2r} \\ a \\ u\,h_1\,h_2\dots h_{2r}\end{pmatrix}$ *is any minor determinant of* A *of order* $2r+1$ *which contains* Δ, *then*

$$\Delta'=P_{h_1\,h_2\dots h_{2r}}\,.\,P_{uv\,h_1\,h_2\dots h_{2r}}\,. \quad \dots\dots\dots\dots\dots\dots\text{(E)}$$

These two theorems can be proved in the same ways as Theorems II a and II b, or they can be regarded as other ways of expressing those two theorems.

Ex. ix. *The Pfaffian* $P_{\lambda_1\lambda_2\dots\lambda_{2r}}$ *vanishes identically when two of the suffixes* $\lambda_1, \lambda_2, \dots \lambda_{2r}$ *are equal.*

For by (D′) its square is then identically equal to a determinant in which two of the horizontal rows (and two of the vertical rows) are the same.

NOTE 1. *The reciprocal of a skew-symmetric matrix of odd order.*

Let $A=[a]_m^m$ be a skew-symmetric matrix of odd order m; let a_{ij} be any element of A; let $[u]_{m-1}$ and $[v]_{m-1}$ be the sequences formed from $[1\,2\dots m]$ by striking out the elements i and j respectively; let

$$\Pi_i=(-1)^{i-1}P_{u_1u_2\dots u_{m-1}},\quad \Pi_j=(-1)^{j-1}P_{v_1v_2\dots v_{m-1}}$$

be the affected Pfaffians of the skew-symmetric matrices formed respectively by striking out the ith horizontal and vertical rows of A, and by striking out the jth horizontal and vertical rows of A, the signs prefixed to them being those determined by the affects of i and j in $[1\ 2 \dots m]$; and let

$$\omega=(i-1)+(j-1)$$

be the affect of the element a_{ij} in A.

Then if $[A]_m^m$ *is the reciprocal of* $[a]_m^m$, *we have*

$$A_{ij}=(-1)^{\omega}\, P_{u_1 u_2 \dots u_{m-1}} \,.\, P_{v_1 v_2 \dots v_{m-1}}=\Pi_i \Pi_j, \quad \text{.....................(F)}$$

and therefore

$$[A]_m^m = \begin{bmatrix} \Pi_1 \\ \Pi_2 \\ \Pi_3 \\ \vdots \\ \Pi_m \end{bmatrix} [\Pi_1,\ \Pi_2,\ \Pi_3,\ \dots\ \Pi_m]. \quad \text{........................(F')}$$

To prove this we first observe that (F) is true when $j=i$, for we have

$$A_{ii}=\begin{pmatrix} 1, 2, \dots i-1, i+1, \dots m \\ a \\ 1, 2, \dots i-1, i+1, \dots m \end{pmatrix}=P^2_{1, 2, \dots i-1, i+1, \dots m}=\Pi_i^2.$$

When $j \neq i$, let $[\lambda]_{m-2}$ be the sequence formed from $[1\ 2 \dots m]$ by striking out both the elements i and j, and let

$$\epsilon=\text{affect of } i \text{ in } [v]_{m-1}, \quad \epsilon'=\text{affect of } j \text{ in } [u]_{m-1}.$$

Then $$A_{ij}=(-1)^{\omega}\begin{pmatrix} v_1\, v_2 \dots v_{m-1} \\ a \\ u_1\, u_2 \dots u_{m-1} \end{pmatrix}=(-1)^{\omega+\epsilon+\epsilon'}\begin{pmatrix} i\, \lambda_1\, \lambda_2 \dots \lambda_{m-2} \\ a \\ j\, \lambda_1\, \lambda_2 \dots \lambda_{m-2} \end{pmatrix};$$

and therefore by Theorem IIIa and Ex. vi we have

$$A_{ij}=(-1)^{\omega+\epsilon+\epsilon'} P_{j \lambda_1 \lambda_2 \dots \lambda_{m-2}} \,.\, P_{i \lambda_1 \lambda_2 \dots \lambda_{m-2}}=(-1)^{\omega} P_{u_1 u_2 \dots u_{m-1}} \,.\, P_{v_1 v_2 \dots v_{m-1}}.$$

Thus (F) is true in this case also; and from (F) we deduce (F').

Ex. x. When $j \neq i$, the difference between ϵ and ϵ' is equal to the number of integers lying between i and j; therefore $\omega+\epsilon+\epsilon'$ is an odd integer, and we have

$$A_{ij}=-P_{i\lambda_1 \lambda_2 \dots \lambda_{m-2}} \,.\, P_{j \lambda_1 \lambda_2 \dots \lambda_{m-2}}.$$

Ex. xi. The reciprocal matrix $[A]_m^m$ is symmetric, and by § 124 its rank cannot exceed 1, because $(a)_m^m=0$. Therefore by § 125 we can so choose the radicals that

$$[A]_m^m = \begin{bmatrix} \sqrt{A_{11}} \\ \sqrt{A_{22}} \\ \vdots \\ \sqrt{A_{mm}} \end{bmatrix} [\sqrt{A_{11}},\ \sqrt{A_{22}},\ \dots\ \sqrt{A_{mm}}].$$

We can now deduce (F') by means of the equations

$$A_{11}=\Pi_1^2, \quad A_{22}=\Pi_2^2, \quad \dots \quad A_{mm}=\Pi_m^2,$$

determining the signs from a consideration of special skew-symmetric matrices whose non-zero elements all have one of the values 1 and -1.

Ex. xii. If $[a]_3^3 = \begin{bmatrix} 0, & a_{12}, & a_{13} \\ -a_{12}, & 0, & a_{23} \\ -a_{13}, & -a_{23}, & 0 \end{bmatrix}$, then $[A]_3^3 = \begin{bmatrix} a_{23} \\ -a_{13} \\ a_{12} \end{bmatrix} [a_{23}, -a_{13}, a_{12}]$.

NOTE 2. *The reciprocal of a skew-symmetric matrix of even order.*

Let $A = [a]_m^m$ be a skew-symmetric matrix of even order m; let a_{ij} be any non-diagonal element of A, so that $j \neq i$; let $[\lambda]_{m-2}$ be the sequence formed from the sequence $[1\,2 \ldots m]$ by striking out the two elements i and j; let η be the affect of $[ij]$ in $[1\,2 \ldots m]$; and let

$$P = P_{12\ldots m}, \quad \varpi_{ij} = (-1)^\eta P_{\lambda_1 \lambda_2 \ldots \lambda_{m-2}}$$

be respectively the Pfaffian of A and the Pfaffian co-factor of a_{ij} in A or P, so that $\varpi_{ji} = -\varpi_{ij}$.

Then if $[A]_m^m$ *is the reciprocal of* $[a]_m^m$, *we have*

$$A_{ij} = (-1)^\eta P_{\lambda_1 \lambda_2 \ldots \lambda_{m-2}} \cdot P_{12\ldots m} = P \cdot \varpi_{ij}, \textit{ when } j \neq i; \quad \ldots\ldots\ldots\ldots\ldots\text{(G)}$$

and we can therefore write

$$[A]_m^m = P_{12\ldots m} \cdot [\varpi]_m^m = P \cdot \begin{bmatrix} 0, & \varpi_{12}, & \varpi_{13}, & \ldots & \varpi_{1m} \\ -\varpi_{12}, & 0, & \varpi_{23}, & \ldots & \varpi_{2m} \\ -\varpi_{13}, & -\varpi_{23}, & 0, & \ldots & \varpi_{3m} \\ \cdots & \cdots & \cdots & \cdots & \cdots \\ -\varpi_{1m}, & -\varpi_{2m}, & -\varpi_{3m}, & \ldots & 0 \end{bmatrix} \quad \ldots\ldots\ldots\text{(G')}$$

Using the same notation as in Note 1 and Ex. x, it follows from Theorem III b and Ex. vi that, when $j \neq i$,

$$A_{ij} = -\begin{pmatrix} i\,\lambda_1\,\lambda_2 \ldots \lambda_{m-2} \\ a \\ j\,\lambda_1\,\lambda_2 \ldots \lambda_{m-2} \end{pmatrix} = -P_{\lambda_1 \lambda_2 \ldots \lambda_{m-2}} \cdot P_{ji\lambda_1 \lambda_2 \ldots \lambda_{m-2}}$$

$$= (-1)^\eta P_{\lambda_1 \lambda_2 \ldots \lambda_{m-2}} \cdot P_{12\ldots m},$$

i.e. (G) is true. Again when $j = i$, we have $A_{ii} = 0$, because A_{ii} is a skew-symmetric matrix of odd order; and by definition we have $\varpi_{ii} = 0$. Therefore the identity (G′) is also true.

We can verify that the matrix $[A]_m^m$ given in (G′) is the reciprocal of $[a]_m^m$ by evaluating the product $[a]_m^m \overline{\underline{A}}_m^m$, making use of the expansions of sub-article 2.

Ex. xiii. If $[a]_4^4 = \begin{bmatrix} 0, & a_{12}, & a_{13}, & a_{14} \\ -a_{12}, & 0, & a_{23}, & a_{24} \\ -a_{13}, & -a_{23}, & 0, & a_{34} \\ -a_{14}, & -a_{24}, & -a_{34}, & 0 \end{bmatrix}$,

then $[A]_4^4 = \begin{bmatrix} 0, & a_{34}, & -a_{24}, & a_{23} \\ -a_{34}, & 0, & a_{14}, & -a_{13} \\ a_{24}, & -a_{14}, & 0, & a_{12} \\ -a_{23}, & a_{13}, & -a_{12}, & 0 \end{bmatrix}$.

NOTE 3. *Determinants formed by bordering a skew-symmetric determinant symmetrically.*

Let $A = [a]_m^m$ be any skew-symmetric matrix of order m, and let its reciprocal be $[A]_m^m$. We first observe that in all cases the equation

$$[x]_m [a]_m^m \overline{\underline{x}}_m = 0 \quad \ldots\ldots\ldots\ldots\ldots\ldots\ldots\ldots\ldots\ldots\text{(4)}$$

is an identity in $x_1, x_2, \ldots x_m$; and we then proceed to evaluate the determinant

$$\Delta' = \begin{bmatrix} 0 & x_1 & x_2 & \ldots & x_m \\ x_1 & a_{11} & a_{12} & \ldots & a_{1m} \\ x_2 & a_{21} & a_{22} & \ldots & a_{2m} \\ \cdots & \cdots & \cdots & \cdots & \cdots \\ x_m & a_{m1} & a_{m2} & \ldots & a_{mm} \end{bmatrix},$$

there being two cases to be considered according as m is even or odd. By Ex. xi of § 62 we have in both cases

$$\Delta' = -\det [x]_m \, \overline{\underline{A}}{}_m^m \, \overline{\underline{x}}{}_m . \qquad \text{(5)}$$

Case I. When m is even.

Using Note 2 we see from (G′) and the identity (4) that in this case

$$\Delta' = 0. \qquad \text{(6)}$$

This result also follows from Theorem IIb, which shows that Δ' is the product of the Pfaffian of $[a]_m^m$ and the Pfaffian of the skew-symmetric matrix

$$\begin{bmatrix} 0, & 0, & x_1, & x_2, & \ldots & x_m \\ 0, & 0, & -x_1, & -x_2, & \ldots & -x_m \\ -x_1, & x_1, & a_{11}, & a_{12}, & \ldots & a_{1m} \\ -x_2, & x_2, & a_{21}, & a_{22}, & \ldots & a_{2m} \\ \cdots & \cdots & \cdots & \cdots & \cdots & \cdots \\ -x_m, & x_m, & a_{m1}, & a_{m2}, & \ldots & a_{mm} \end{bmatrix},$$

which is clearly 0.

If the leading element of Δ' is x_0 instead of 0, we have

$$\Delta' = x_0 (a)_m^m = x_0 P^2_{1, 2, \ldots m} . \qquad \text{(6')}$$

Case II. When m is odd.

Using the same notation as in Note 1, we see from (5) and (F′) that in this case

$$\Delta' = -(\Pi_1 x_1 + \Pi_2 x_2 + \ldots + \Pi_m x_m)^2. \qquad \text{(7)}$$

This result also follows from Theorem Ia, which shows that Δ' is the product of the Pfaffians of the two skew-symmetric matrices

$$\begin{bmatrix} 0, & x_1, & x_2, & \ldots & x_m \\ -x_1, & a_{11}, & a_{12}, & \ldots & a_{1m} \\ -x_2, & a_{21}, & a_{22}, & \ldots & a_{2m} \\ \cdots & \cdots & \cdots & \cdots & \cdots \\ -x_m, & a_{m1}, & a_{m2}, & \ldots & a_{mm} \end{bmatrix} \quad \text{and} \quad \begin{bmatrix} 0, & -x_1, & -x_2, & \ldots & -x_m \\ x_1, & a_{11}, & a_{12}, & \ldots & a_{1m} \\ x_2, & a_{21}, & a_{22}, & \ldots & a_{2m} \\ \cdots & \cdots & \cdots & \cdots & \cdots \\ x_m, & a_{m1}, & a_{m2}, & \ldots & a_{mm} \end{bmatrix}.$$

The equation (7) is still true when the leading element of Δ' is x_0 instead of 0.

APPENDIX C

ADDITION TO CHAPTER XVIII

§ 175 a. Equigradent transformations in which one of the transforming factors is a semi-unit matrix.

1. *Unilaterally semi-unit equigradent transformations of matrices.*

Let $$[h]_m^r [a]_r^s [k]_s^n = [b]_m^n, \quad \text{or} \quad HAK = B \quad \text{.................(1)}$$

be an equigradent transformation converting the matrix $A = [a]_r^s$ into the matrix $B = [b]_m^n$, so that $H = [h]_m^r$ and $K = [k]_s^n$ are matrices of ranks r and s with constant elements.

When one of the transforming factors H and K is a semi-unit matrix, we may call (1) a *unilaterally semi-unit equigradent transformation.* Every such transformation converts a matrix A whose elements are constants into another matrix B which has the same rank as A and which also has the same horizontal or vertical extravagance as A according as K or H is a semi-unit matrix. Equigradent transformations of these special kinds occur throughout Chapter XVIII, and play a very important part in the theory of extravagant matrices. From formulae (E_4) and (E_5) of § 165 we see that a matrix whose elements are constants can always be reduced to (or derived from) a known standard form by such a transformation when its rank and one of its extravagances are given.

2. *Unilateral equigradent transformations of matrices.*

When one of the transforming factors H and K is a unit matrix, which can be struck out, we may call (1) a *unilateral equigradent transformation,* such transformations being particular cases of those just considered. Every such transformation converts a matrix A whose elements are constants into a matrix B horizontally or vertically equivalent to A according as K or H is a unit matrix. The reduction of a matrix whose elements are constants by unilateral equigradent transformations is given in § 168.

3. *Semi-unit transformations of matrices.*

When both the transforming factors H and K are semi-unit matrices, we may call (1) a *bilaterally semi-unit equigradent transformation,* but we generally call it more simply a *semi-unit transformation.* Every such transformation converts a matrix A whose elements are constants into another matrix B which has the same rank and also the same two extravagances as A. These transformations have not been studied in this volume, but from formulae (F_1) and (F_2) of § 165 we see that a matrix whose elements are constants can always be reduced to (or derived from) a standard form by a semi-unit transformation when its rank and its two extravagances are known. The standard form is not however a completely determinate matrix.

4. *Equigradent transformations of spacelets.*

Let $A = \overline{\underline{a}}{}^{p}_{n}$ and $B = \overline{\underline{b}}{}^{q}_{n}$ be matrices representing two spacelets in homogeneous space ω_n. An equigradent transformation $HAK = B$ converting A into B, as well as an equigradent transformation $HBK = A$ converting B into A, can, so far as the spacelets are concerned, be replaced by a relation of the form

$$[h]^{n}_{n}\,\overline{\underline{a}}{}^{p}_{n} \equiv \overline{\underline{b}}{}^{q}_{n}, \quad \text{.............................(2)}$$

where $[h]^{n}_{n}$ is an undegenerate square matrix, and the symbol $\equiv$ denotes vertical equivalence. On this account we may define (2) to be an equigradent transformation converting the *spacelet* represented by A into the *spacelet* represented by B, and, if $\overline{\underline{H}}{}^{n}_{n}$ is the inverse matrix of $[h]^{n}_{n}$, we can associate with it the inverse relation

$$\overline{\underline{H}}{}^{n}_{n}\,\overline{\underline{b}}{}^{q}_{n} \equiv \overline{\underline{a}}{}^{p}_{n}, \quad \text{.............................(2')}$$

which is an equigradent transformation converting the *spacelet* represented by B into the *spacelet* represented by A. Since moreover the existence of a relation of the form (2) is the necessary and sufficient condition that *one* of the matrices A and B shall be convertible into the other by an equigradent transformation, i.e. that the two matrices shall be equigradent or have the same rank, we see that:

We perform an equigradent transformation on a spacelet of homogeneous space by prefixing an undegenerate square matrix whose elements are constants; and each of two spacelets is convertible into the other by an equigradent transformation when and only when the two spacelets have the same rank.

Accordingly general formulae for any spacelet ω_r and any spacelet ω_r' which can be derived from ω_r by an equigradent transformation are

$$\omega_r \equiv \overline{\underline{a}}{}^{p}_{n}, \quad \omega_r' \equiv [h]^{n}_{n}\,\overline{\underline{a}}{}^{p}_{n}, \quad \text{.......................(3)}$$

where $[h]_n^n$ is an arbitrary undegenerate square matrix; and when spacelets are represented by undegenerate matrices, we can replace (3) by

$$\omega_r \equiv \overline{\underline{a}}{}_n^r, \quad \omega_r' \equiv [h]_n^n \overline{\underline{a}}{}_n^r; \quad \text{......................(3')}$$

the matrix $\overline{\underline{a}}{}_n^p$ in (3) and the matrix $\overline{\underline{a}}{}_n^r$ in (3′) having of course rank r.

An equigradent transformation converting a *point* $\overline{\underline{x}}{}_n$ into a *point* $\overline{\underline{y}}{}_n$ is a relation of the form

$$[h]_n^n \overline{\underline{x}}{}_n \equiv \overline{\underline{y}}{}_n,$$

where $[h]_n^n$ is an undegenerate square matrix. When the transformation (4) is applied to every point of ω_n, the prefactor $[h]_n^n$ being the same for all points, we call it an *equigradent transformation of the points of space.* It then converts any spacelet $\omega_r \equiv \overline{\underline{a}}{}_n^p$ into the spacelet ω_r' given in (3); and we regard it as given when the matrix $[h]_n^n$ is given. We can in this case replace (4) by

$$[h]_n^n \overline{\underline{x}}{}_n = \lambda \overline{\underline{y}}{}_n, \quad \text{..........................(4')}$$

where λ is for each separate point of space some unspecified finite non-zero scalar quantity. Accordingly it is the same thing as a *projective transformation* of the points of space.

5. *Semi-unit transformations of spacelets.*

The transformations of spacelets in sub-article 4 are called semi-unit transformations when $[h]_n^n$ is a square semi-unit matrix. Accordingly a semi-unit transformation converting a *spacelet* represented by the matrix $A = \overline{\underline{a}}{}_n^p$ into a *spacelet* represented by the matrix $B = \overline{\underline{b}}{}_n^q$ is a relation of the form

$$[l]_n^n \overline{\underline{a}}{}_n^p \equiv \overline{\underline{b}}{}_n^q, \quad \text{............................(5)}$$

where $[l]_n^n$ is a square semi-unit matrix, and the symbol $\equiv$ denotes vertical equivalence; and with (5) we can associate the inverse relation

$$\overline{\underline{l}}{}_n^n \overline{\underline{b}}{}_n^q \equiv \overline{\underline{a}}{}_n^p, \quad \text{............................(5')}$$

which is a semi-unit transformation converting the *spacelet* represented by B into the *spacelet* represented by A. When we observe that the existence of a relation of the form (5) is a necessary and sufficient condition that *one* of the matrices A and B shall be convertible into the other by a unilaterally semi-unit equigradent transformation, or when we refer to Ex. v of § 170, we see that:

We perform a semi-unit transformation on a spacelet of homogeneous space by prefixing a square semi-unit matrix; and each of two spacelets is convertible into the other by a semi-unit transformation when and only when the two spacelets have the same rank and the same extravagance.

General formulae for any spacelet ω_r and any spacelet ω_r' which can be derived from ω_r by a semi-unit transformation are

$$\omega_r \equiv \underline{\overline{a}}{}^{p}_{n}, \quad \omega_r' \equiv [l]^{n}_{n}\, \underline{\overline{a}}{}^{p}_{n}, \quad \ldots\ldots\ldots\ldots\ldots\ldots\ldots\ldots(6)$$

where $[l]^{n}_{n}$ is an arbitrary square semi-unit matrix of order n.

A semi-unit transformation converting a *point* $\underline{\overline{x}}{}_{n}$ into a *point* $\underline{\overline{y}}{}_{n}$ is a relation of the form

$$[l]^{n}_{n}\, \underline{\overline{x}}{}_{n} \equiv \underline{\overline{y}}{}_{n}, \quad \ldots\ldots\ldots\ldots\ldots\ldots\ldots\ldots\ldots(7)$$

where $[l]^{n}_{n}$ is a square semi-unit matrix. When the transformation (7) is applied to every point of ω_n, the prefactor $[l]^{n}_{n}$ being the same for all points, we call it a *semi-unit transformation of the points of space*, regarding the transformation as given when $[l]^{n}_{n}$ is given. It then converts any spacelet $\omega_r \equiv \underline{\overline{a}}{}^{p}_{n}$ into the spacelet ω_r' given in (6). It is frequently called an *orthogonal transformation* of the points of space.

INDEX

The references are to pages. The following is a list of the abbreviations used.

alg.	=algebraic,	equiv.	=equivalent,	parat.	=paratomy,
arb.	=arbitrary,	extrav.	=extravagant,	rat.	=rational,
compl.	=complementary,	homog.	=homogeneous,	recip.	=reciprocal,
conj.	=conjugate,	horiz.	=horizontal,	reg.	=regular,
corresp.	=correspondent,	inc.	=incident,	sim.	=similar,
corresp.	=corresponding,	int.	=intersecting,	sp.	=spacelet,
degen.	=degenerate,	intersec.	=intersection,	sq.	=square,
detant	=determinant,	inv.	=inverse,	sym.	=symmetric,
detoid	=determinoid,	non-int.	=non-intersecting,	transfn.	=transformation,
diag.	=diagonal,	mut.	=mutually,	vert.	=vertical,
el.	=element,	orthog.	=orthogonal,	uncon.	=unconnected,
equigr.	=equigradent,	orthot.	=orthotomy,	undeg.	=undegenerate.

Absolute quadric, 434–5.
Affected:
minor detants of $[1, a]_m^{m, n}$ and $[a, 1]_m^{n, m}$, 395–7;
simple minor detants of two mut. normal undeg. matrices, 403–6;
Pfaffian compl. to any el. of a skew-sym. matrix of even order, 524;
compl. to any diag. el. of a skew-sym. matrix of odd order, 527–8.
Affects: formulae involving, 43, 57, 97, 101, 131.
Algebraic equations: Equiv. systems of linear alg. equations, 80–4, 207–8.
Uncon. mut. orthog. solutions of any system of homog. linear alg. equations, 440–6.
Algebraic expressions or forms: Linear transfns. of bilinear and quadratic forms, 234–5.
Real roots of a real and definite quadratic form; positive and negative real quadratic forms, 300–1.
Reduction of a quadratic form to a sum of squares of linear expressions, 290.
Anti-cores of a spacelet: defined, 440;
join of the core to an anti-core is real, 437–8.
Anti-correspondent matrices of the minor detants:
of two co-joint complete matrices of the minor detants of a sq. matrix, defined, 115; relation between, 116–7;
of a sq. matrix and its recip., 117–8, 129.
Anti-correspondent minor detants:
of two co-joint complete matrices of the minor detants of a sq. matrix, defined, 115; relation between, 118–9;
of a sq. matrix and its recip., 119, 128.
Anti-correspondent simple minor detants of two mut. normal undeg. matrices, defined, 402; relations between, 403.
Arbitrary spacelets: *see* Extravagance, Orthotomy.
Arbitrary values:
of any reg. minor detant of a matrix and its els., primaries and primary superdetants, 516;
of any reg. simple minor detant and its els. and primaries, 518;
of any reg. minor detant of order r of a matrix of rank r and its els. and primaries, 519.

Bilinear forms: Linear transfns. of, 234.
Bordered determinants: formed by
bordering any sq. matrix, 123–7;
bordering a skew-sym. matrix, 525;
symmetrically, 529.

Co-factors:
of minor detants, 97, 108–9, 111, 157;
see Co-joint complete matrices;
Pfaffian, of the els. of a skew-sym. matrix of even order, 524.
Co-incident spacelets, 219.
Co-joint complete matrices of the minor detants of a sq. matrix:
defined, 108; fundamental property, 109;
determinants, 112–3; ranks, 122–3;
reciprocals, 113–4.
A sq. matrix and its recip. are two co-joint matrices of the minor detants of the sq. matrix, 111.
Definitions of corresponding and anti-cor-

Co-joint complete matrices (*cont.*):
respondent minor detants and matrices of minor detants, 115.
Relations between two:
anti-corresp. minor detants, 108–9;
matrices of minor detants, 117;
corresp. complete matrices of minor detants, 120.
Co-joint matrix of a complete matrix of the minor detants of a rectangular matrix, 111–2.
Common core and plenum of two mut. normal spacelets, 432–3.
Compartite matrix: defined, parts, standard form, successive parts, 6–7;
with quasi-scalar parts, 8;
reciprocals and inverses, 11.
Rank, is sum of ranks of parts, 7–8.
Signants of one which is sym. and real, are sums of signants of parts, 299.
Compartite matrix:
Equigr. transfns. of one and its parts, 233–4.
Reduction of a matrix in Ω by equigr. transfns. in Ω to a compartite matrix whose non-zero parts are undeg. sq. matrices:
any matrix with constant els., 252–9, 262–3;
sym. matrix, 259, 263; 269–71;
real, 298;
skew-sym. matrix, 260, 264; 269–71; 307–8.
Complementary spacelets: defined, 219, 423;
extravagances of two, are independent, 505;
normals to two, are non-int., 433;
orthotomy (or cross rank) of two, 478–80;
when real, 480.
Two real sps. which necessarily intersect and have the greatest orthot. consistent with their ranks are compl., 509.
Spacelets which lie in a given sp. and are mut. compl. in it, 480, 505.
Complementary theorems concerning spacelets, 433.
Complete intersection:
of two sps., 210–19, 422–3;
of any number of sps., 219–20, 423;
is the normal to join of the normal sps., 434;
is real when the sps. are real, 505;
of two mut. normal sps., is their common core, 432;
of mut. orthog. sps., 509–14;
see Intersection.
Complete matrix of the minor detants:
of a rectangular matrix, rank, 122–3;
of a square matrix, determinant, 112;
of a definite matrix, 303;
of a skew-sym. matrix, 141–2;
see Co-joint complete matrices.
Completely extravagant matrices: defined, 381;
general formulae for, 381–3, 349;
normal matrices, 408, 419.
Cores of an undeg. matrix, 417.
Completely extravagant spacelet:
defined, 426, 381; interpreted, 435;
is core of itself, is contained in its normal which is its plenum, 432, 434.
Every contained sp. is completely extrav., 459.
Completely extravagant spacelet (*cont.*):
Standard representation, 435–9.
(*See* Core.)
Completely extravagant spacelets:
containing a given sp., 459;
lying in a given sp., 453.
Completion of two mutually inverse matrices, 235.
Composition of equigradent transformations, 231;
of derangements, of unitary and quasi-scalar transfns., 233.
Compound matrices:
defined, constituent matrices, 1–5;
quasi-scalar, 4;
multiplication, 5–6;
notation for scalar constituents, 4.
Compound sq. semi-unit matrix with four real or purely imaginary constituents, 327, 362.
Conditions for the equivalence of
two sim. undeg. matrices, 76;
two systems of linear alg. equations, 81, 207–8;
for the horiz. or vert. equivalence of any two matrices, 205.
Conditions that a matrix may have rank r:
any matrix, 93, 516;
sym. matrix, 136;
skew-sym. matrix, 142.
Conditions that two matrices may have same rank:
any two matrices, 272–3, 229;
two sym. matrices, 294;
two skew-sym. matrices, 308;
that two real sym. matrices may be equisignant, 298.
Conditions that two matrices may have the same rank and the same (horiz. or vert.) extravagance:
any two matrices, 391;
two sim. matrices, 390;
two sim. undeg. matrices, 380;
that two spacelets may have the same rank and the same extravagance, 426.
Conditions that two spacelets or matrices may be:
mut. compl., 219, 423;
mut. inc., 219, 422;
mut. orthog., 84, 398, 423;
mut. normal, 85, 398, 423–4;
non-int., 210–11, 423;
that any number of them may be uncon., 220, 224, 225–6, 423.
Conditions that two spacelets may have the greatest possible mutual orthotomy:
two arb. sps. of given ranks, 467–8, 507, (either their intersec. is completely extrav., or their join is plenarily extrav.);
two arb. real sps. of given ranks, 509, (they are either non-int. or compl.);
two sps. of given ranks having a given complete intersec. ω_p, 501–3, (either they both lie in the plenum of ω_p, or their join contains the plenum of ω_p);
two real sps. of given ranks having a given complete intersec. ω_p, 489,

Conditions that two spacelets (*cont.*):
(they are the joins of ω_p with two mut. orthog. sps. orthog. with ω_p).
Conjugate matrices: Expression of a sym. (or skew-sym.) matrix as a sum (or difference) of two mut. conj. matrices, 141.
Ranks of the products formed with two mut. conj. matrices, 194, 387–8, 407, 411.
(*See* Extravagance.)
Conjugate of a semi-unit matrix: is its inv., 321.
Conjugate reciprocals of:
a matrix with one el., 89;
a product of sq. matrices, 107;
some special matrices, 8–13, 240–1;
two co-joint matrices, 112;
see Inverse, Reciprocal.
Connections between matrices: defined, 223–5.
Condition that matrices may be uncon., 225.
Distinction between connected rows and connected one-rowed matrices, 224.
Equivalent matrices, 75, 205.
Intersecs. and joins of matrices, 222–3.
Matrices connected with a given matrix, their ranks and extravagances, 384.
Uncon. mut. orthog. one-rowed matrices connected with a given matrix, 412–5, 440–6.
Connections between the short rows of an undeg. matrix, 41–6;
between the rows of a degen. matrix, 46, 86.
Connections between spacelets: defined, 225–7.
Condition that sps. may be uncon., 220, 225–6, 423.
Incidences of sps., 79, 219, 422, 466.
Intersecs. and joins, 209–11, 219–20, 422–3.
Zero sps., have no connections, 211, 225.
(*See* Spacelets and sub-heads.)
Constituents of a compound matrix:
are the matrices by the juxtaposition of which it is formed, xxiv, 1–6;
of a compound sq. semi-unit matrix when each is real or purely imaginary, 327, 362.
Conversion of any equigr. transfn. and its inv. into two mut. inv. equigr. transfns. of sim. matrices, 236.
Conversion of any matrix into a standard form by derangements of its rows, 94, 140, 145;
compartite matrix, 7;
of any sym. matrix into a standard form by sym. derangements and additions of rows, 291.
Conversion of a matrix with constant elements:
into a bipartite or compartite matrix by equigr. transfns., 241–71;
into a quasi-scalar matrix or a standard form by equigr. transfns., 272–308;
into a standard form by a semi-unit and a general equigr. transfn., 378–94, 401;
into a standard form by two semi-unit transfns., 389, 394;
into an equiv. matrix whose long rows are mut. orthog., 412–5;
see Reduction.
Core of a spacelet:
of any sp. (of extravagance ρ), is a sp. (of rank ρ) which is the locus of all points
Core of a spacelet (*cont.*):
which lie in the sp. and are orthog. with it, 427;
of completely and plenarily extrav. sps., 432;
of the join of mut. orthog. sps., 437, 509–14;
of two mut. normal sps., 432.
General formulae for:
a sp. and its core, 427, 428;
a sp. having a given core, 427;
two mut. normal sps., their common core and common plenum, 431, 433.
Real or non extrav. sp. has no core, 432.
Semi-unit transfns. of a sp. and its core, 428.
(*See* Anti-core.)
Core of a spacelet which:
contains a given sp. and lies in its plenum, 485;
contains a given plenum, 491;
contains a given sp., 496.
Cores of an undeg. matrix (of extravagance ρ): 417–22;
are the completely extrav. matrices of rank ρ connected with the given matrix, are all mut. equiv., 417.
Determination of the cores, 421.
Two mut. normal matrices have the same extravagance and the same cores, 407, 418;
their common cores are their complete intersecs., 418.
Correction to Ex. vi of § 83, 24.
Correspondences between all sps. of a certain complete homog. space and all those sps. of a given homog. space which:
contain a given non-extrav. sp., 471;
lie in a given non-extrav. sp., 469, 217;
lie in one and contain another given non-extrav. sp., 473.
Corresponding matrices of the minor detants:
of two co-joint complete matrices of the minor detants of a sq. matrix;
defined, 115; relation between them, 120–2;
of a sq. matrix and its recip., 121, 129.
Criteria for the equivalence of two systems of linear alg. equations, 80–2, 207–8.
Cross rank of two matrices:
is that of the sps. which they represent;
of two special matrices, 471.
Cross rank of two spacelets:
defined, relation to orthotomy, 464;
interpreted, 465;
of the normals to two sps., 466;
of special sps., 480, 490;
of two sps. having a given intersec., 485, 495.
Cross rank, possible values for two spacelets of given ranks which:
are arb., 464;
contain a given sp., 470;
lie in a given sp., 468;
are non-intersecting, 475, 477;
one being given, 474, 477;
both being real, 478;
are mut. complementary, 478, 480;
one being given, 478;
both being real, 480;
lie in or contain a given non-extrav. sp., or lie in one and contain another given non-extrav. sp., 469, 471, 473; 477, 480.

Cross rank, possible values for two sps. of given ranks which:
have a given complete intersec., 492, (one sp. being given, 500);
and lie in its plenum, 482, (one sp. being given, 486);
have a given non-extrav. intersec., 489;
are real and have a given intersec., 489;
have a given join, 505;
are real and have a given join, 506.
Cross rank and paratomy: Possible simultaneous values of these for
two arb. sps. of given ranks, 506;
two arb. real sps. of given ranks, 508.
Cross rank, paratomy and orthotomy: Possible simultaneous values of these or any two of them for
two entirely arb. sps., 508;
two entirely arb. real sps., 509.

Definite (real sym.) matrix: 300–3;
general formula, 302; semi-definite, 303.
Diag. detants of order *s* (and diag. els.) all have same sign, 303.
Sym. matrices of minor detants are definite, 303.
Definite real quadratic forms:
are essentially positive or negative, 301;
real roots, 300.
Degeneracy of a matrix: defined, 378;
possible degeneracies of special products, 387, 407, 411.
Extravagance is a degeneracy, 379, 385, 425.
Orthotomy is a degeneracy, 464.
(*See* Extravagance, Orthotomy, Rank.)
Degree: Matrix equations of
first degree, 168–77;
second degree, 309–77; 179–85, 194–6;
third degree, 185–9, 196–9;
any degree, 189–94, 199–205.
Derangements (transfns. of a matrix):
defined, resultant of, 232–3.
Formulae involving derangements, 314, 339, 341, 346, 360, 379, 381–3, 388, 399–400.
Inverse of a derangement, 233.
Reduction of a matrix to a standard form by derangements:
any matrix, 94;
compartite matrix, 7;
sym. matrix, 140;
skew-sym. matrix, 145.
Solutions of the sym. equation $X'X=A'A$ derivable from one another by derangements, 336.
Derangements (transformed matrices):
of semi-unit and unit matrices, are all semi-unit matrices, 321;
inverse of a derangement of a unit matrix, is its conjugate, 233.
Inverses of the derangements of special sq. matrices, 240–1.
Recip. of any derangement of a sq. matrix, 131.
Determinant:
of a complete matrix of the minor detants of a sq. matrix, 112
of the primaries of one simple minor detant Δ formed with the rows of another D, 48, 51, 55; 57, 59, 61; 63, 65; 519;

Determinant (*cont.*):
of the primary subdetants of a detant Δ which contain a given minor detant D, 70, 71, 519–20;
of the primary superdetants of a minor detant Δ which lie in a given detant D, 66, 67, 519–20;
of a skew-sym. matrix, 142, 521, 525–6;
of a special sq. matrix, 437;
of a sq. compound matrix, notation, 1–3.
Sq. matrix whose detant is ± 1, 280.
(*See* Standard identities.)
Determinant, minor (or simple minor):
primaries (defined, notation), 13–17;
primary subdetants (defined), 18;
primary superdetants (defined), 17.
Arb. values of the els., primaries and primary superdetants of a reg. minor detant, 516.
(*See* sub-heads; *see also* Relations.)
Determinants, anti-correspondent minor:
of two co-joint matrices of the minor detants of a sq. matrix,
defined, 115; relative between, 118–9;
of a sq. matrix and its recip., 119, 128–9, 70–2.
(*See* Anti-correspondent matrices.)
Determinants, anti-corresp. simple minor:
of two mut. normal undeg. matrices, 403.
Determinants, bordered: formed by
bordering any sq. matrix, 123–7;
bordering a skew-sym. matrix, 525;
symmetrically, 529.
Determinants, diagonal:
see Diagonal minor determinants.
Determinants: Equigr. transfns. between two sim. undeg. quasi-scalar or sq. matrices whose detants are equal, 278–9.
Determinants, minor:
of $[1, a]_m^{m,n}$ and $[a, 1]_m^{n,m}$, 395–7;
simple, 395–6; of order *s*, 396–7;
of a sym. matrix of order 3, 145–6;
of order 4, 147–56.
Relations between the minor detants of order *r* of a matrix of rank *r*, 98–106, 519–20;
see Relations, Standard equations.
Determinants, non-vanishing or regular:
see Regular minor determinants.
Determinants, simple minor:
of two equiv. undeg. matrices, 76–8;
of two mut. normal undeg. matrices, 403.
Relations between the simple minor detants of any matrix, 46–66, 73–5, 518–20;
see Relations (identical), Standard identities.
Determinoid:
of a compound matrix (notation), 1–3;
of the primaries of a simple minor detant Δ formed with the rows of a superior simple minor detoid D, 65;
of the primary superdetants of a minor detant Δ which lies in a minor detoid D, 69.
Diagonal elements:
of a definite matrix, have same sign, 302;
of a sym. matrix, 133, 137, 139;
of rank 1, 133;

Diagonal elements (*cont.*):
of a skew-sym. matrix, are all 0's, 241;
their compl. Pfaffians, 524, 527-8.
Diagonal minor determinants:
defined, 107-8;
of a definite matrix, 302-3;
all of order *s* have same sign; 303;
of a sym. matrix, 135-41;
every sym. matrix of order *r* has a non-vanishing diag. detant of order *r*, 135;
of a skew-sym. matrix, 142-5;
every diag. detant of odd order vanishes, 142;
every skew-sym. matrix of order *r* has a non-vanishing diag. detant of order *r*, 142.
Successive reg. leading diag. detants:
in an arranged matrix, 94, 273-7;
sym., 138, 280-92;
skew-sym., 143, 303-7;
in an unarranged matrix, 252-64;
sym. or skew-sym., 269-71.
Diagonal minor matrices: defined, 29, 107-8;
possible ranks of a sym. matrix containing
a given diag. minor, 29-32;
a given zero diag. minor, 32-6;
possible ranks of a diag. minor of a sym. matrix whose rank is given, 31;
compl. to a given zero diag. minor, 36.
Dimensions of a complete homogeneous space or a spacelet: are less by 1 than the rank, 78-80, 422;
see Rank.
Distinct solutions of the sym. equations $X'X=A'A$, $X'X=0$: 336, 340, 346, 348.
Domain of rationality: 229, 241-2, 441-2.

Element: Recip. of a matrix having only one el., 89.
Elements: Fundamental identities satisfied by the els. of any matrix, 90-3, 515, 157;
special cases, 38-9, 41, 92-3, 518;
derived relations, 46-66, 518, 519.
Relations between the els. of a matrix of rank *r*, 85, 518-20;
derived relations, 98-106, 519-20.
(*See* Relations.)
Elements: Pfaffian co-factors of the els. of a skew-sym. matrix of even order or its Pfaffian, 524.
Pfaffians compl. to the diag. els. of a skew-sym. matrix of odd order, 527-8.
Elements, diagonal, of a:
definite matrix (all have same sign), 302;
sym. matrix, 133, 137, 139;
of rank 1 (do not all vanish), 133;
skew-sym. matrix, 141;
their compl. Pfaffians, 524, 527-8.
Elimination of a variable from a system of inequalities, 18.
Equal: Matrices having equal ranks, 229, 272-3, 294, 308.
Matrices having equal ranks and equal extravagances, 380, 390, 391.
Spacelets having equal ranks and equal extravagances, 426.
Undeg. sq. matrices whose detants are equal, 278-9.
(*See* Conditions, Equi-.)

Equations: satisfied by the els. of a matrix whose rank does not exceed *r*, 85-9, 98-106;
see Relations, Standard equations.
Equations, algebraic: Equiv. systems of linear alg. equations, 80-3, 207-8.
Uncon. mut. orthog. solutions of any system of homog. linear alg. equations, 442-6.
Equations, identical: satisfied by the els. and minor detants of any matrix, 37-75, 90-8;
see Relations (identical); *see also* Co-joint complete matrices.
Equations, matrix:
of the first degree, 168-77;
of the second degree, 309-77, 179-85, 194-6;
of the third degree, 185-9, 196-9;
of any degree, 189-94, 199-205, 309;
see Matrix equations.
Equi-extravagant matrices: 380, 390, 391;
(two mut. normal undeg. matrices), 407, 411.
Two horizontally (or vertically) equiv. or normal matrices have the same horiz. (or vert.) extravagance, 386, 407.
Equi-extravagant spacelets: 426, 432.
Equigradent matrices: defined, 229;
are matrices of equal rank, 272-3;
symmetrically equigr. matrices, 230; 295, 308; 299.
Equigradent transformations: defined, 228;
between two sim. matrices, 230-1;
in a domain Ω, 229; 241-2, 441-2;
of a compartite matrix and its parts, 233-4;
composition, resultant, 231, 232-3;
elementary, 231-2, 258;
inverses of, 229, 230, 232-3;
properties, 237-40;
symmetric, 229-30, 231.
Conversion of an equigr. transfn. and its inv. into two mut. inv. equigr. transfns. of sim. matrices, 236.
Correspondences of equigr. transfns. of matrices whose els. are constants with linear transfns. of bilinear and quadratic alg. forms, 234-5.
Equigradent transformations (special classes):
derangements, 232-3;
quasi-scalar and scalar transfns., 233;
unitary transfns., 232-3, 278-80.
Equigradent transformations of special form, 240;
(occurrence) 257; 242, 278; 281, 286, 305.
Equigradent transformations in Ω which convert any matrix in Ω with constant els.
into a bipartite matrix one of whose parts is a given undeg. sq. minor or non-vanishing el.:
any matrix, 242-52; 93-4, 248-9, 516-7;
special cases, 247-8, 251-2;
sym. or skew-sym. matrix, 264-9;
special cases, 265-7, 268-9;
into a compartite matrix whose non-zero parts are all undeg. sq. matrices:
any matrix,
unitary transfns., 252-60;
non-unitary, 260-4;
sym. or skew-sym. matrix,
sym. unitary transfns., 269-70
non-unitary, 271;

Equigradent transformations (*cont.*):
specially arranged matrix,
unitary transfns., 273–5;
non-unitary, 276–7;
specially arranged sym. matrix,
sym. unitary transfns., 280–2, 283–8;
non-unitary, 282–3, 289;
specially arranged skew-sym. matrix,
sym. unitary transfns., 303–5;
non-unitary, 306–7.
Equigradent transformations: Reduction of a matrix in Ω with constant els. to a quasi-scalar matrix or a standard form by equigr. transfns.:
any matrix by equigradent transfns. in Ω, 272, 278;
by derangements and unitary transfns. in Ω, 277;
sym. matrix by sym. equigr. transfns. (not in Ω), 293;
by sym. derangements and unitary transfns. in Ω, 292;
real sym. matrix by real sym. equigr. transfns., 298;
by sym. derangements and real sym. unitary transfns., 296;
skew-sym. matrix by sym. equigr. transfns. in Ω, 308;
by sym. derangements and unitary transfns. in Ω, 307.
Derivation of a matrix from a unit matrix by equigr. transfns., 272, 294.
Equigradent transformations: Reduction of a matrix with constant els. to a standard form
by two equigr. transfns. of which one is semi-unit:
any matrix, 388–9;
undeg. matrix, 379–80;
completely extrav., 381–2, 349, 435;
plenarily extrav., 383, 439;
two mut. normal undeg. matrices, 401;
by two semi-unit transformations:
any matrix, 389;
sym. matrix, 393–4.
Reduction to an equiv. matrix whose long rows are mut. orthog., 412–5.
Equigradent transformations between two matrices with constant els.
which have the same rank:
any two matrices, 272;
two sim. matrices, 278;
two sym. matrices, 294;
real and equisignant, 298;
two skew-sym. matrices, 308;
two equiv. matrices, 75–6, 205;
which have the same rank and the same (horiz. or vert.) extravagance, 380, 390, 391, 531.
Equigradent transformations:
of a sq. matrix whose detant is ± 1, 280;
between two sim. undeg. sq. or quasi-scalar matrices whose detants are equal, 278–9.
Equigradent transformations: *see* Invariants.
Equisignant (real sym.) matrices, 297, 298–9.
Equivalence of two sim. undeg. matrices;
horiz. and vert. equivalence of two matrices:
defined, 75–6; 205;
Equivalence of two sim. undeg. matrices (*cont.*):
conditions for equivalence, 75–6, 76–8; 205;
sign of equivalence, 78, 208.
Equivalences of two sim. sq. matrices, 206.
Equivalence of two systems of linear alg. equations, 80–3; 207–8.
Equivalent matrices: defined, 75–6, 205.
Cores of an undeg. matrix are mut. equiv., 417.
Mut. equiv. matrices define a spacelet, 78–9, 208.
Normals to a given matrix are mut. equiv., 85, 398.
Equivalent matrix:
Reduction of a matrix (of extravagance ρ)
to an equiv. undeg. matrix whose long rows are mut. orthog., 412–5;
exactly ρ long rows are extrav., 412;
to one which is the join of a core and a semi-unit matrix, 413–4.
Reduction of a non-extrav. (or real) matrix
to an equiv. semi-unit (or real semi-unit) matrix, 414.
Expansions:
of certain bordered detants in terms of the simple minor detants of the bordering rows, 123–7;
of the Pfaffian of a skew-sym. matrix of even order, 524;
of a symmetrically bordered skew-sym. detant of odd order, 529.
Extravagance (defined):
of an undeg. matrix, 378;
complete, 381; plenary, 382–3; zero, 379;
of any matrix (horiz. or vert.), 385;
of a sym. (or skew-sym.) matrix, 392;
of a spacelet, 425; interpreted, 435;
is its orthot. with itself, 466.
Extravagance: Matrices and spacelets having the same rank and the same extravagance:
matrices, 380, 390, 391;
equiv. matrices, 381, 386;
spacelets, 426;
having the same extravagance, mut. normal, 407, 411, 432;
for formulae see Extravagant.
Extravagance of:
the join of uncon. mut. orthog. sps., 427;
of any mut. orthog. sps., 513;
a sp. containing a given sp., 495, 485, 428;
a sp. lying in a given non-extrav. sp., 469;
containing a given non-extrav. sp., 471;
lying in one and containing another given non-extrav. sp., 473.
Extravagances, horizontal and vertical:
of any matrix, 385;
of a matrix of rank r expressed as a product of passivity r, 386.
Extravagances, possible, of a:
matrix of given orders, 386–7;
undeg. matrix, 379;
sym. matrix, 393;
matrix connected with a given matrix, 384;
spacelet of given rank, 425;
containing a given sp., 456;
and lying in its plenum, 487;
lying in a given sp., 450; 430, 384;

Extravagances, possible, of a (*cont.*):
and having a given intersec. with its core, 429;
lying in one and containing another given sp., (462), 474;
spacelet which does not intersect a given sp., 448–9;
which has a given complete intersec. with a given sp., 499–500.
Extravagances, possible, of:
two mut. normal matrices, 379, 401, 408;
two mut. orthog. matrices, 409, 410;
two mut. normal sps., 425, 401; 194, 279, 408;
two mut. orthog. sps., 409;
which satisfy certain conditions, 511–3;
two non-int. sps., 446, 511;
two sps. which have a given intersec., 499–500;
and lie in its plenum, 487;
two sps. which have a given join, 505–6;
and contain its core, 505;
uncon. sps., 449.
Extravagant undeg. matrix: defined, 378;
completely extrav., 381;
cores of a matrix, 415–22;
plenarily extrav., 382;
non-extrav., 379;
see sub-heads.
Extravagant matrices: Formulae for
undeg. matrix of given extravagance, 379–80, 381;
completely extrav. matrix, 381–2; 349, 351, 408, 437, 439;
plenarily extrav. matrix, 383, 408–9, 439;
two mut. normal undeg. matrices, 401;
matrix of given rank of which
one extravagance is given, 388; 392;
both extravagances are given, 389; 392–3;
sym. matrix, 393–4;
both extravagances are zero, 389.
Extravagant and non-extravagant points: defined, 425; interpreted, 435.
Extravagant rows: 412, 415, 440–6.
Extravagant solutions of any system of homog. linear alg. equations, 441–6.
Extravagant spacelet:
defined, 425; interpreted, 435;
completely extrav. sp., 426, 435;
core of a sp., 427, 435;
plenarily extrav. sp., 426;
plenum of a sp., 431–2;
non-extrav. sp., 426, 435;
see sub-heads.
Extravagant spacelets: Formulae for
a sp. of given rank and extravagance, 428;
completely extrav. sp., 435, 437;
plenarily extrav. sp., 439;
two mut. normal sps., 433, 431.

Factorisation of a matrix: 93–4, 248–9, 516–7. (*See* Matrix equations.)
Factors: Possible ranks of the matrix factors in any matrix product, 193;
any sym. matrix product, 205.
Restrictions on the ranks of the matrix factors in any matrix product, 177.
(*See* Ranks, possible.)
Factors: of $\begin{bmatrix}1&0\\0&1\end{bmatrix}$ and $\begin{bmatrix}0&1\\1&0\end{bmatrix}$, 367;
of any matrix of rank 1, 89, 95, 316;
of any sym. matrix of order 2, 365–9;
of rank 2, 373–7;
of rank 1, 134; 352–3;
of a zero matrix, 344–52, 409–10.
First degree: *see* Matrix equations.
Flat loci, 78.
Formulae (general) for matrices:
matrix of given orders and rank, 88;
sym., 294, 355; real sym., 298, 361;
real and definite, 302; skew-sym., 308;
matrix equiv. to a given undeg. matrix, 329;
matrix of given orders, given rank, and given extravagance or extravagances, 379–80, 388–9, (sym.) 393–4.
compartite matrix to which any matrix can be reduced
by unitary equigr. transfns., 253;
by non-unitary equigr. transfns., 263;
completely extrav. undeg. matrix, 381–2;
plenarily extrav. undeg. matrix, 383;
sq. semi-unit matrix of order 2, 328;
of order 3, 329;
two mut. normal undeg. matrices, 399, 401.
Formulae (general) for spacelets:
sp. containing a given sp., 212, 216, 471;
lying in a given sp., 216, 450, 469;
lying in one and containing another given sp., 216, 473;
sp. which lies in a given sp. and has a given complete intersec. with one of its sub-spaces, 424;
sp. which lies in a given sp. and does not intersect two of its sub-spaces, 221;
two sps., their intersec. and join, 214–5, 216–7, 423;
join of any number of sps., 219, 423;
sp. of rank r and extravagance ρ, 428;
sp. having a given core, 427;
completely extrav. sp., 435, 437;
plenarily extrav. sp., 439;
two mut. normal undeg. matrices (or a sp., its core and plenum), 431, 433.
Formulae (general) for the solutions of a matrix equation; *see* Matrix equations.
Functional matrices: Rank of a matrix product in which an extreme factor matrix has rank equal to its passivity, 167.
Fundamental identity satisfied by the els. of any matrix, 90;
utility of it, 515.

General solutions of matrix equations: *see* Matrix equations.
Greatest possible orthotomy of two spacelets whose ranks are given:
two arb. sps., 467, 507,
(either their intersec. is completely extrav., or their join is plenarily extrav.);
two arb. real sps., 467, 509,
(they must be either non-int. or compl.);
two non-int. sps., 476;
two mut. compl. sps., 479;
two sps. which have a given intersec. and lie in its plenum, 487–8;

Greatest possible orthotomy (*cont.*):
two sps. having a given intersec. ω_p, 501–2, (either they both lie in the plenum of ω_p, or their join is a plenarily extrav. sp. which contains the plenum of ω_p);
two sps. having a given non-extrav. intersec., 489;
two real sps. having a given intersec., 489.
Greatest possible paratomy of two spacelets whose ranks are given:
two arb. sps., 464, 507;
two arb. real sps., 464, 509.

Homogeneous linear equations: Uncon. mut. orthog. solutions of any system, 440–6.
Homogeneous space: dimensions and rank, 78–80, 422.
Horizontal extravagance of a matrix: 385–93; *see* Extravagance, Extravagant.
Horizontal primaries: defined, 13; *see* Primaries.
Horizontally:
equivalent matrices, 205–7;
have the same horiz. extravagance, 386;
normal matrices, 398;
have the same horiz. extravagance, 407;
all matrices horizontally normal to a given matrix are horizontally equiv., 398;
orthogonal matrices, 398;
see sub-heads.

Identities in the elements of a matrix:
see Relations (identical), Standard identities;
see also Co-joint matrices.
Identities to which special names are given:
Kronecker's Identity, 98;
Sylvester's Identities, 66–70.
Identities which represent equigr. transfns. converting a matrix into a bipartite matrix, 93–4, 248–9, 516–7.
Incident spacelets: defined, 219, 422;
normals to two inc. sps. are inc., 432.
Incidence of each of two sps. with the normal to the other, 466–7.
Two arb. sps. of given ranks have their greatest orthot. when each is inc. with the normal to the other, 467–8.
(*See* Greatest possible orthotomy.)
Indefinite: real sym. matrices, 300;
real quadratic forms, 301.
Infinite: Equivalence of two systems of linear alg. equations which have only infinite solutions, 82.
Intensity: Non-extrav. rows at unit intensity, 414–5.
Interpretation of the terms 'extravagance,' 'core,' 'orthogonal' and 'normal,' 435.
Intersecting matrices, intersections of matrices, 222–3.
Intersecting spacelets: defined, condition that two sps. may be int. or non-int., 210–11, 422–3;
sp. of rank 0 has no intersecs., 211;
two sps. are non-int. when and only when their normals are compl., 433.
(*See also* Incident, Non-intersecting.)

Intersection, complete, of two spacelets:
defined, 210, 422–3;
is the normal to join of the normal sps., 432;
is therefore real when the two sps. are real, 489;
of two mut. normal sps., is their common core, 427, 432;
of two mut. orthog. sps., 509–10.
Formulae for two sps., their intersec. and join; relation between the ranks of the intersec. and join, 215, 216–7, 423.
Possible ranks of the intersec. of two sps. of given ranks, 215;
lying in or containing a given sp., or lying in one and containing another given sp., 217–8.
Ranks of the intersec. and join of the normals to two given sps., 434.
Intersection, complete, of any number of spacelets: defined, 219–20, 423;
is the normal to join of the normal sps., 434;
is therefore real when the sps. are real, 489;
of mut. orthog. sps., 509–14.
Intersection: Properties of two spacelets having a given complete intersection, 480, 506;
their extravagances, are independent, 487, 499–500; 485, 495;
their orthotomy, 485, 495; 482, 492; 487–9, 501–4; 511–2;
see Greatest possible orthotomy.
Intersections of spacelets: 209–23, 422–35, 463–5;
of sps. and their normals, 432–4.
Formulae for a sp. lying in or containing a given sp., or lying in one and containing another given sp., 216–7;
for a sp. which lies in a given sp. and does not intersect two of its sub-spaces, 221;
for a sp. which lies in the join of two sps. and has a given intersec. with one of them, 424.
Paratomy and orthotomy of two sps., 463–5.
Possible ranks of a sp. which lies in a given sp. and:
does not intersect a given sub-space of it, 218;
and has a given intersec. with another, 226–7;
does not intersect two given sub-spaces of it, 220, 222;
all sps. being real, 221;
has a given intersec. with a given sub-space of it, 218–9, 429–30.
Invariants of a matrix (with constant els.):
rank of any matrix, is invariant in all equigr. transfns., 165–7, 228–30;
extravagances of any matrix, are invariant in all semi-unit transfns., 391, 531–2;
horiz. extravagance in all equigr. transfns. of horiz. rows, 386, 391;
vert. extravagance in all equigr. transfns. of vert. rows, 386, 391;
extravagance of an undeg. matrix in all equigr. transfns. of long rows, 380–1, 386;

Invariants of a matrix (*cont.*):
signants and signature of a real sym. matrix are invariant in all real sym. equigr. transfns., 396.
Invariants of a spacelet in a given transfn. of the points of space:
rank of a sp. is invariant in every equigr. transfn., vii, 165–7, 532;
extravagance of a sp. is invariant in every semi-unit transfn., vii, 426, 533.
A semi-unit transfn. which converts one sp. into another also converts the core of the first into the core of the second, 428–9.
Invariants of two spacelets:
paratomy is invariant in every equigr. transfn., vii, 463, 532;
orthotomy or cross rank is invariant in every semi-unit transfn., vii, 464, 533.
An equigr. transfn. which converts two sps. into two other sps. also converts the intersec. and join of the first two into the intersec. and join of the second two, 214–5, 532.
A semi-unit transfn. which converts one sp. into another also converts the normal to the first sp. into the normal to the second, 433, 533.
Inverse matrices:
inv. of a semi-unit matrix is its conj., 321;
inverses of special matrices, 8–13, 240.
Completion of two mut. inv. matrices, 235.
Inverse transformations:
inv. of an equigr. transfn., 229;
(sym) 230; (between sim. matrices) 231; (derangement, unitary or quasi-scalar transfn.) 232–3.
Conversion of an equigr. transfn. and its inv. into two mut. inv. equigr. transfns. of sim. matrices, 236.

Join of two or more matrices, 222–3.
Join of two spacelets: 209–19, 422–35, 509–11; defined, 209, 423;
is real when the two sps. are real, 505;
is the normal to intersec. of the normal sps., 432;
of two compl. sps., 219, 423;
two sps. are compl. when and only when their normals are non-int., 433;
of two mut. normal sps., is their common plenum, 432;
of two mut. orthog. sps., 427, 509–11;
of two non-extrav. sps., need not be non-extrav., 505;
of two non-int. sps., 211;
of two sps. of which one is inside and the other outside a given sp., 211, 424; 218, 425; 429;
of a core and anti-core, 437–8, 440.
Formulae for two sps. and their join and intersec.; relation between the ranks of their join and intersec., 215, 216–7, 423.
If two sps. lie in a given sp., every point of their join lies in that sp., 210.
Possible ranks of the join of two sps. of given ranks, 215;

Join of two spacelets (*cont.*):
lying in or containing a given sp., or lying in one and containing another given sp., 216–7.
Ranks of the join and intersec. of the normals to two given sps., 434.
Representation of any sp. as the join of its core and a non-extrav. sp., 427, 428, 413;
as the join of two mut. orthog. non-int. sps., 428, 511;
of a plenarily extrav. sp. as the join of its core and a real sp., 439.
Join of any number of given spacelets: 219–27, 422–35, 509–14; defined, 219, 423;
is real when the given sps. are real, 505;
is the normal to intersec. of the normal sps., 434;
of mut. orthog. sps., 427–8, 509–14;
of sps. lying in or containing a given sp., or lying in one and containing another given sp., 216–7.
Representation of any sp. as the join of mut. orthog. uncon. points, 412–5.
Join of a given spacelet and
any non-int. sp., 494–5;
lying in its plenum, 484–5;
lying outside its plenum, 490–1, 495;
of a given plenum and a sp. lying outside it, 490–1.

Kronecker's Identity, 98.

Least possible orthotomy of two spacelets whose ranks are given:
two arb. sps., 468, 507;
two arb. real sps., 468, 509;
two non-int. sps., 476;
two mut. compl. sps., 479;
two sps. which have a given intersec., 502–4;
and lie in its plenum, 488;
have a given non-extrav. intersec., 489;
two real sps. which have a given intersec., 489.
Least possible paratomy of two spacelets whose ranks are given:
two arb. sps., 464, 507;
two arb. real sps., 464, 509.
Line: *see* Straight line.
Linear equations: *see* Equations, algebraic.
Linear transformations:
Correspondences between equigr. transfns. of a matrix with constant els. and linear transfns. of bilinear and quadratic forms, 234–5.
Reduction of a quadratic form to a sum of squares by a linear transfn., 290.
Long rows:
of an undeg. matrix represent points, 78.
Equivalence of sim. undeg. matrices, 75.
Extravagance of an undeg. matrix, 378, 380.
Matrices with mut. orthog. long rows, 412–5, 442–6.
Mut. orthog. undeg. matrices, 84.
Relations between the simple minor detants of a matrix having m long rows, 63–6, 73–5.
Semi-real undeg. matrices, 360–1.

Matrices or matrix (qualifying terms):
Compartite; compound; definite; degenerate; extravagant (horizontally, vertically, completely, plenarily); indefinite; non-extravagant; one-rowed; quasi-scalar; real; scalar; semi-definite; semi-real; semi-unit; skew-symmetric; square; symmetric; undegenerate; unit; zero.
See sub-heads.
Matrices or matrix (qualifying terms indicating relations between matrices):
Co-joint (of minor detants); conjugate; conjugate reciprocal; connected (horizontally, vertically); equi-extravagant; equigradent; equisignant; equivalent (horizontally, vertically); incident (lying in, containing); inverse; normal or orthogonal (horizontally, vertically); reciprocal; skew-conjugate; unconnected.
See sub-heads.
Matrices or matrix (properties of a single matrix):
Cores; degeneracy; determinant (of a sq. matrix); determinoid; derangements; elements (diagonal, non-diagonal); extravagances or extravagance; invariants; minor determinants (diagonal, non-diagonal, regular, simple); minor determinoids; minor matrices (diagonal, simple, square); orders; Pfaffian (of a skew-symmetric matrix of even order); rank; rows (long, short, horizontal, vertical, extravagant, non-extravagant, orthogonal, connections between); signants and signature; standard forms.
See sub-heads.
Matrices (properties of two or more):
Connections; cross rank; equivalences; incidences; intersections; invariants; joins; orthotomy; paratomy.
See sub-heads.
Matrices or matrix (operations on or with):
Conversions; derangements; multiplication; reductions; transformations (equigradent, quasi-scalar, scalar, semi-unit, unitary, non-unitary).
See sub-heads.
Matrices, anti-correspondent, of the minor detants of:
a sq. matrix and its recip., 117–8, 129;
two co-joint matrices of the minor detants of a sq. matrix, 115, 116–7.
Matrices, complete, of the minor detants of:
any matrix (their ranks), 122–3;
any sq. matrix (their detants), 112–3;
a definite matrix, 303;
a skew-sym. matrix, 141;
a sym. matrix of order 3 or 4, 145–56.
(*See* Co-joint complete matrices.)
Matrices, corresponding, of the minor detants of:
a sq. matrix and its recip., 121, 129;
two co-joint complete matrices of the minor detants of a sq. matrix, 115, 120–2.
Matrices: Formulae for; *see* Formulae.
Matrices, special:
having a given minor, 19–36;
Matrices, special (*cont.*):
having a simple minor formed with an undeg. sq. matrix and zero els., 26;
having a zero order, 5;
having only one el., 89;
compound square, 8–13, 240–1, 437;
semi-unit, 327;
sq. semi-unit of order 2, 328;
of order 3, 329;
sq. whose detants are ± 1, 11–13, 257, 280;
whose detants are equal, 278–9;
of rank 1, 89, 95, 316;
of rank r, 85–9, 319;
sym. of order 2, 365–9;
of order 3, 145–6, 374, 377;
of order 4, 147–56, 375, 377;
of rank 1, 133–5, 352, 373–7;
of rank 2, 373–7;
of rank r, 135, 355;
whose orders differ by 1, 43;
$[1, a]_m^{m,n}$ and $[a, 1]_m^{n,m}$, 395–7.
Matrix of primary superdeterminants in any matrix: identity for, 90, 516;
rank, 96–7, 516–8;
properties, 97–8;
in a sq. matrix: detant of, 66;
properties, 157–8.
Matrix equations of the form $X_1X_2\ldots X_n=C$: 307; 179–94; (sym.) 194–205.
Possible ranks of the solutions, 193;
$(n=2)$ 183, $(n=3)$ 187;
for sym. equations of this form, 205;
$(n=2)$ 196, $(n=3)$ 198.
Matrix equations of the first degree:
General solutions and possible ranks of the solutions of the equations
$AX=C$, 168–70;
$XB=C$, 170–2;
$AXB=C$, 173–5;
of the sym. equation
$A'XA=C$, 176–7.
Matrix equations of the second degree (not sym.):
$XY=AB$, 310–5;
formulae, 310, 312, 314;
$XY=C$, 316–20;
formulae, 316, 317.
Matrix equations of the second degree (sym.):
$X'X=I$, where I is a unit matrix, 320–31;
determination of all real solutions, 322–3;
of all solutions, 325–6;
general solutions in special cases, 328–9.
Equations of a more general form, 343.
(*See* semi-unit matrices.)
Matrix equations of the second degree (sym.):
$X'X=A'A$, 333–44;
distinct solutions, 336, 340, 343;
formulae, 341–2; 333, 335, 338, 339;
$X'X=0$, 344–52;
distinct solutions, 346, 348;
formulae, 346–7; 345, 349, 351–2;
undegenerate solutions, 349–51;
$X'X=C$, 352–63;
formulae, 353, 356;
particular solutions, 356–8;
semi-real solutions when C is real, 360–1;
$X'X=A'BA$, 359;
$X'BX=A'A$, 359;
$X'BX=A'CA$, 359–60;

Matrix equations of the second degree (sym.) (*cont.*):
$X'AX=C$, 363–5.
Matrix equations of the second degree: Special equations of passivity 2 of the form $XY=C$, 365–77;
factorisation of a sym. matrix
of order 2, 365–9;
of rank 2, 373–7.
Matrix factors (possible ranks): 165–205; *see* Ranks, possible.
Matrix factors: Expression of a matrix of rank 1 as a product of two one-rowed matrices:
any matrix, 89, 95, 316;
sym. matrix, 134, 352–3;
of a matrix of rank r as a product of two factors of passivity r:
any matrix, 88; 85–6, 277, 317;
sym. matrix, 292, 294, 355–9;
real, 360;
real and definite, 302;
of a sym. matrix of rank 2 or 1 as a product of two factors of passivity 2, 373–7;
sym. matrix of order 2, 365–9.
(*See* Factors, Matrix equations.)
Matrix products (possible ranks): 165–205; *see* Ranks, possible.
Matrix products (special):
product in which an extreme factor matrix has rank equal to its passivity, 165–6;
products of two mut. conj. matrices, 194, 387, 407, 411;
products of two matrices, 471, 481, 490;
sym. product of three matrices in which the middle factor is undeg., 199.
Extravagances of a product of two matrices which both have rank and passivity r, 386.
Minor determinant: primaries of, 13–17;
primary subdetants and superdetants, 17–18;
see Primaries, Primary, Relations.
Minor determinant, regular or non-vanishing:
of any matrix, 94; defined, 515;
diagonal, of a sym. matrix, 135, 138;
of a skew-sym. matrix, 142–4;
of a definite matrix, 303.
(*See* Regular, Relations.)
Minor determinants: Relations between the minor detants of order r of a matrix of rank r, 98–106, 519–20;
see Relations, Standard equations.
Minor determinants of a square matrix:
determinant of, 112:
matrices of, 108–23, 127–33, 145–56.
Relations between two anti-correspondent minor detants of:
a sq. matrix and its recip., 119, 128, 70–1;
two co-joint matrices of minor detants, 118–9.
(*See* Determinants, Matrices of minor detants, Co-joint matrices.)
Minor determinants, diagonal: defined, 107–8;
of a definite (real sym.) matrix, 302–3;
of a sym. matrix, 135–41;
of a skew-sym. matrix, 142–5;
see Diagonal minor determinants.
Minor determinants, simple:
of two equiv. undeg. matrices, 76–7;
of two mut. normal undeg. matrices, 302–3.
Identical relations between the simple minor detants of any matrix, 46–66, 73–5, 518–20.
Primaries of a simple minor detant, 13–17, 518–9.
(*See* Primaries, Relations, Standard identities.)
Minor determinants of special matrices:
of $[1, a]_m^{m,n}$ and $[a, 1]_m^{n,m}$, 395–7;
of a sym. matrix of order 3, 145–6;
of order 4, 147–56.
Minor determinoids, 65, 69.
Minor matrices: Recip. of a minor compl. to a sq. minor in any matrix, 97.
Recip. of a sq. minor of the recip. of a sq. matrix, 157.
Possible ranks of a matrix containing:
a given minor, 19–22;
a given zero minor, 22–9.
Possible ranks of a minor of a matrix of given rank, 22
compl. to a given zero minor, 28.
Minor matrices, diagonal: defined, 29, 107;
of sym. matrices, 135–41;
real and definite, 303;
of skew-sym. matrices, 141–5.
Possible ranks of a sym. matrix containing:
a given diag. minor, 29–32;
a given zero diag. minor, 32–6.
Possible ranks of a diag. minor of a sym. matrix of given rank, 31;
compl. to a given zero diag. minor, 36.
Successive undeg. leading diag. minors of any matrix, 252, 260, 269;
in an arranged matrix, 94, 138, 143.
(*See* Diagonal minor determinants.)
Muir's notation for Pfaffians, 522.
Multiplication of compound matrices, 5.
Mutual orthotomy of two spacelets: *see* Orthotomy;
paratomy of two spacelets: *see* Paratomy.
Mutually conjugate matrices: *see* Conjugate;
equivalent matrices: *see* Equivalent.
Mutually incident matrices and spacelets: 219, 422, 432, 466–7; *see* Incident.
Mutually inverse:
equigradent transformations, 231, 232–3;
conversion of an equigr. transfn. and its inv. into two mut. inv. equigr. transfns. of sim. matrices, 236;
matrices, 8–13, 240, 321;
completion of two mut. inv. matrices, 235.
Mutually normal:
matrices, defined, 85, 398;
properties, 398, 411;
spacelets, defined, 84, 423–4;
properties, 427, 432–4, 466–7;
see Normal.
Mutually orthogonal:
matrices, defined, 84, 398;
properties, 409–10, 416–7, 420–2;

Mutually orthogonal (*cont.*):
points, defined, 84; interpreted, 435;
spacelet represented as join of, 412–5;
rows, 412–5;
solutions, 440–6;
spacelets, defined, 84, 423;
properties, 409, 426–8, 431–2, 466–7, 476, 489, 509–14;
see Orthogonal.

Negative real quadratic forms, 301.
Negative signant of a real sym. matrix, 297.
Non-extravagant matrices:
undegenerate, defined, 426;
mut. orthog., are uncon., 426;
normals to, are non-extrav., 407;
sym., powers of, 394.
Matrices whose extravagances are both 0, 389.
Non-extravagant point, interpreted, 435.
Non-extravagant rows, at unit intensity, 414.
Non-extravagant solutions, 440–6.
Non-extravagant spacelet:
defined, 425; interpreted, 435;
has no core, has complete space as plenum, 432;
has a non-extrav. normal which does not intersect it and is compl. to it, 434.
Representation of any sp. as the join of two mut. orthog. sps. of which one is its core, 427;
a given non-extrav. sub-space, 428.
Spacelets lying in or containing a given non-extrav. sp., or lying in one and containing another given non-extrav. sp.; their properties deduced from those of corresponding sps. of a complete space, 469–73.
Non-extravagant spacelets:
containing a given sp., 459;
lying in a given sp., 453;
mut. orthog., are uncon., 426;
intersections and joins need not be non-extrav., 505.
Non-incident spacelets, 219.
Non-intersecting spacelets: defined, 210, 422;
condition for non-intersection, 211, 423;
normals to two, are compl., 434.
Extravagances of two non-int. sps., 446–50;
lying in a given non-extrav. sp., 450, 470.
Joins of two non-int. sps., 211, 424; 215, 423; 218, 425; 490, 491, 493.
Joins of two mut. orthog. non-int. sps., 427, 428, 431, 439, 484.
Representations of any sp. as a join of two mut. orthog. non-int. sps., 511.
Orthotomy of two non-int. sps. 474–8;
when one is given, 477;
lying in a given non-extrav. sp., 477;
when one is given, 477.
Spacelet of rank 0 has no intersecs., 211.
Spacelets which do not intersect either of two given sps., 220–2.
(*See* Intersections.)
Normal matrices: defined, 85, 398;
horizontally or vertically normal, 398;
properties, 398–411, 418–9.

Normal matrices (*cont.*):
All matrices horizontally (or vertically) normal to a given matrix are horizontally (or vertically) equiv., 398.
Normals to completely and plenarily extrav. matrices, 408, 418–9.
Two mut. normal undeg. matrices:
have the same extravagance, 407;
have the same cores, these being their complete intersecs., 418;
formulae for, 399, 401;
possible values of their ranks and extravagance, 401; 379, 408;
simple minor detants of, 403;
special pairs, 398–9.
Normal spacelets (two mutually normal):
defined, 84, 423–4; interpreted, 435.
Two mut. normal sps. have the same extravagance, the same core, and the same plenum; their common core is their complete intersec. and the core of their plenum; their common plenum is their join and the plenum of their core; their core and plenum are mut. normal, 431–2.
Formulae for two mut. normal sps., their core and their plenum, 431, 433.
Possible values of the ranks and extravagances of two mut. normal sps., 425, 401; 194, 279, 408.
Normal to a spacelet: defined, is locus of all points orthog. with it, 424;
to a completely extrav. sp., is its plenum, 432, 434;
to a plenarily extrav. sp., is its core, 432, 434;
to a non-extrav. sp., is non-extrav., 434;
to a real sp., is real, 434, 505.
Normals to given spacelets: properties, 432–4;
to mut. inc. sps., are mut. inc., 432;
to mut. compl. sps., are non-int., 433;
to non-int. sps., are mut. compl., 433.
The complete intersec. and join of the normal sps. are the normals to the join and complete intersec. of the given sps., 434.
Incidence of two sps. each with the normal to the other, 466–7.
Ranks of the intersec. and join of the normals to two given sps., 434.

One-rowed matrices: notations, 4;
mut. orthog., 415, 440–6; zero, 224;
see Points, Rows.
Orders: Matrices whose orders differ by 1, 43;
which have a zero order, 5.
Relations between the minor detants of order r of a matrix of rank r, 98–106.
See Matrices (special), Relations.
Orthogonal matrices: defined, 84, 398;
horizontally or vertically orthog., 398;
joins of mut. orthog. uncon. matrices, 420, 427;
mut. orthog. non-extrav. matrices are uncon., 426;
possible ranks and extravagances of two mut. orthog. matrices, 409, 410.
Cores of an (undeg.) matrix are the

Orthogonal matrices (*cont.*):
matrices of greatest rank connected with it and orthog. with it; their common rank is the extravagance of the matrix, 416–22.
See Orthogonal spacelets.
Orthogonal points: defined, 84; interpreted, 435;
see Orthogonal spacelets.
Orthogonal rows: undeg. matrices whose long rows are mut. orthog., 412–5, 440–6;
see Reduction.
Orthogonal solutions: uncon. mut. orthog. solutions of any system of linear alg. equations, 440–6.
Orthogonal spacelets:
defined, 84, 423; interpreted, 435;
intersec. join and cores of mut. orthog. sps.; their intersec. is the intersec. of their cores, and is completely extrav.; their cores are their intersecs. with the core of their join; the join of their cores is the core of their join, 509–14;
joins of mut. orthog. uncon. sps., 427;
mut. orthog. non-extrav. sps. are uncon., 426.
Core of a sp. is the locus of all points which lie in it and are orthog. with it; extravagance of a sp. is the number of uncon. points which lie in it and are orthog. with it, 427.
Orthotomy of two sps. is the number of uncon. points of the smaller sp. which are orthog. with the larger sp., 465.
Cases in which two sps. which are mut. orthog. have the greatest orthot. consistent with their ranks, 467, 476, 487 ($p=\pi$).
Cases in which two sps. whose normals are mut. orthog. have the greatest orthot. consistent with their ranks, 467–8, 479, 502.
Plenum of a sp. is the locus of all points orthog. with its core, 431–2.
Possible ranks and extravagances of two mut. orthog. sps., 409;
which satisfy certain conditions, 511–3.
Representation of a sp. as a join:
of mut. orthog. uncon. points, 412–5;
of two mut. orthog. non-int. sps. of which
one is its core, 427;
one is non-extrav., 428;
one is given, 484;
of two mut. orthog. non-int. sps., 511;
of two mut. orthog. sps., 511.
Orthogonal transformations:
of a spacelet and its core, 428–9.
See Semi-unit transformations.
Orthotomy of two matrices: is that of the spacelets which they represent, 463.
Orthotomy of two spacelets: 464–509;
defined, relation to cross rank, 464;
is intersec. of smaller sp. with normal to larger sp., 465;
of two special sps., 480;
of two sps. having a given intersec., 495, 485.
Extravagance of a sp. is its orthot. with itself, 466.
Orthotomy of two spacelets (*cont.*):
Two sps. and their two normals have the same orthot., 466.
(*See* Invariants.)
Orthotomy: Possible values for two sps. of given ranks which
are arb., 465;
lie in a given sp., 468;
contain a given sp., 470;
are non-intersecting, 475, 477,
(one being given), 474, 477;
are mut. complementary, 478, 480,
(one being given), 478,
(both being real), 480;
lie in or contain a given non-extrav. sp., or lie in one and contain another given non-extrav. sp., 469, 471, 473; 477, 480.
(*See* Greatest, Least.)
Orthotomy: Possible values for two sps. of given ranks which
have a given complete intersec., 492,
(one sp. being given), 500;
and lie in its plenum, 482,
(one sp. being given), 486;
have a given non-extrav. intersec., 489;
are real and have a given intersec., 489.
(*See* Greatest, Least.)
Orthotomy and paratomy: Possible simultaneous values of these for
two arb. sps. of given ranks, 506;
two arb. real sps. of given ranks, 508.
Orthotomy, paratomy and cross rank: Possible simultaneous values of these or any two of them for
two entirely arb. sps., 508;
two entirely arb. real sps., 509.

Paratomy of two matrices: is that of the spacelets which they represent, 463.
Paratomy of two spacelets:
defined, is rank of their intersec., 463.
possible values for two sps. of given ranks which
are arb., 464, 215;
lie in or contain a given sp., or lie in one and contain another given sp., 217–8.
(*See* Intersection, Invariants, Greatest, Least.)
Paratomy and Orthotomy:
see Orthotomy.
Paratomy, orthotomy and cross rank:
see Orthotomy.
Parts of a compartite matrix, successive parts of one in standard form, 6–7;
see Compartite matrix.
Pfaffian: defined, 521, 523;
co-factors and complements, 524, 527–8;
expansions of a Pfaffian, 524.
Detant of a skew-sym. matrix, 142, 525–6;
bordered, 525;
bordered symmetrically, 529–30.
Recip. of a skew-sym. matrix, 527–9.
Plane in homogeneous space of rank n:
is a sp. of rank $n-1$, 78–80.
Plenarily extravagant matrices:
defined, 382–3;
general formulae for, 383, 408, 410;
normal matrices, 418–9.

Plenarily extravagant spacelet:
defined, 426, 382–3;
is plenum of itself, contains its core which is its normal, 432, 434;
is the join of its core and a real sp., 439.
Every sp. which contains it is plenarily extravagant, 453.
Standard representation, 439.
(*See* Plenum.)
Plenarily extravagant spacelets:
containing a given sp., 459;
lying in a given sp., 453.
Plenum of a spacelet:
is the normal to its core, is the join of the sp. and its normal, 431–2;
of a completely or plenarily extrav. sp., 432;
of two mut. normal sps., 432;
real or non-extrav. sp. has the complete space as plenum, 432.
Formulae for a sp. and its normal, core and plenum, 431, 433.
Join of a given sp. with one lying outside its plenum, 490–500; 494, 495.
Sps. which contain a given sp. and lie in its plenum, 480–7; 484, 485.
Points of homogeneous space: defined, 79–80;
extrav. and non-extrav., 425;
mut. orthog., 84; interpretations, 435;
lying in neither of two given sps., 221–2.
Sp. of rank r is join of r uncon. points, 78.
Representation of any sp. as a join of uncon. mut. orthog. points, 412–5, 440–6.
Positive real quadratic forms, 301.
Positive signant of a real sym. matrix, 297.
Possible ranks of:
a matrix containing a given minor, 19–29, 29–36.
a minor of a matrix of given rank, 22, 28, 31, 36;
matrix factors and products, 177–205;
the intersec. and join of two sps., 215–8, 463–4.
See Ranks, possible.
Possible ranks and degeneracies of special matrix products, 165–6, 199, 387–8, 407, 411, 471, 481, 490.
Possible values of
cross rank, or ranks and cross rank;
extravagances, or ranks and extravagances;
orthotomy, or ranks and orthotomy;
paratomy, or ranks and paratomy;
see Cross rank, Extravagances, Orthotomy, Paratomy.
Primaries of a minor determinant Δ:
defined, number of primaries, 13;
notation, formulae involving primaries, 14–17;
properties, 85–9, 98–106, 515–20.
Arb. values of the els., primaries and primary superdetants of Δ when Δ is reg., 516;
of the els. and primaries of Δ when Δ is a reg. minor detant of order r of a matrix whose rank is r, 519.
Relations between the els. of any matrix of rank r, serving to express every el. as a rat. function of Δ and the els. and primaries of Δ when Δ is a reg. minor detant of order r, 85–9, 519.
Relations between the minor detants of order r of any matrix of rank r, serving to express all of them as rat. functions of Δ and its primaries when Δ is any one reg. minor detant of order r, 98–106, 519, 520.
See Relations, Standard equations.
Primaries of a simple minor determinant Δ of a matrix A:
defined, number of primaries, 13–14;
properties, 37–46, 46–66, 518–9.
Arb. values of the els. and primaries of Δ when Δ is regular, 62–3, 518.
Determinants of the primaries of Δ formed with the rows of another simple minor detant D, 63, 65; 48, 51, 55; 57, 59, 61; 519.
Identical relations between the els. of A, serving to express every el. as a rat. function of Δ and the els. and primaries of Δ when Δ is reg., 37–46, 518.
Relations (identical in the els.) between the simple minor detants of Δ, serving to express all of them as rat. functions of Δ and its primaries when Δ is reg., 46–66, 518, 519.
See Relations (identical), Standard identities.
Primary subdeterminants of a determinant Δ:
defined, 18; properties, 70–3, 519–20.
Determinants of primary subdetants;
identities which serve to express every subdetant of Δ as a rat. function of Δ and the primary subdetants of Δ when Δ is reg., 70, 519–20.
See Relations (identical), Standard identities.
Primary superdeterminants of a minor determinant Δ:
defined, 17; properties, 66–70, 90–8, 157–64, 515–8, 520.
Arb. values of the els., primaries and primary superdetants of Δ when Δ is reg., 516.
Determinants of primary superdetants;
identities which serve to express every superdetant of Δ as a rat. function of Δ and the primary superdetants of Δ when Δ is reg., 66, 519–20.
Matrix of primary superdetants:
identities for, 90–8, 157–64; 93–4, 516;
rank, 96–7, 516.
See Relations (identical), Standard identities.
Product:
of square matrices, 107–8.
Possible ranks of the product-matrix
in any matrix product, 193;
in any sym. matrix product, 205.
Restrictions on the rank of any matrix product, 177–9.
See Ranks, possible.
Products, special: *see* Matrix products.

Quadratic forms: linear transfns., 234–5;
reduction of one to a sum of squares, 290.

Quadratic forms, real and definite:
essentially positive or negative, 301;
real roots, 300.
Quadric surface, absolute, 434–5.
Quasi-scalar equigradent transformations:
defined, composition, 233; real, 294.
Quasi-scalar matrices: notations, 4–5;
equigr. transfns. between two which are sim. and undeg., 278–9.
Reduction of a matrix in Ω to a quasi-scalar matrix by equigr. transfns. in Ω:
any matrix, 277;
sym. matrix by sym. transfns., 292;
real sym. matrix, 298.

Rank of a matrix: not altered by equigr. transfns., 165–7, 228–9.
Conditions that:
a matrix may have rank r, 93, 516;
sym. matrix, 136;
skew-sym. matrix, 142;
two matrices may have the same rank, 272–3;
two sym. matrices, 294;
two skew-sym. matrices, 308.
See Conditions.
Rank of a:
compartite matrix, is the sum of the ranks of its parts, 7;
matrix having a simple minor formed with an undeg. sq. matrix and zero els., 26;
matrix product in which an extreme factor matrix has rank equal to its passivity, 165–7;
matrix of minor detants, 122–3;
matrix of primary superdetants, 96–7, 516;
sym. matrix, 135–7;
skew-sym. matrix, 142–3.
Rank of a spacelet:
is the number of uncon. points lying in it, is greater by 1 than the number of its dimensions, 78–9, 208–9, 422.
Ranks of:
two mut. normal sps., 85, 398;
the intersec. and join of
two sps., 215, 423;
the normals to two given sps., 434.
Ranks, possible, of a:
matrix containing a given minor, 19–22;
a given zero minor, 22–9;
sym. matrix containing a given diag. minor, 29, 32;
a given zero diag. minor, 32–6;
minor of a matrix of given rank, 22;
compl. to a given zero minor, 28;
diag. minor of a sym. matrix of given rank, 31;
compl. to a given zero diag. minor, 36.
Ranks, possible, of the product matrix and the factor matrices in:
any matrix product, 293;
with two factors, 183; 179–80;
with three factors, 187; 185;
any sym. matrix product, 205;
with two factors, 196; 194;
with three factors, 198; 196.
Restrictions on the rank of any matrix product and the ranks of its factors, 177–9.
Ranks, possible, of the two products of two mut. conj. matrices, 387.
Ranks, possible, of special matrix products:
products of two mut. conj. matrices, 194, 387, 407, 411;
products of two matrices containing two given undeg. mut. conj. minors, 471, 481, 490;
sym. product of three matrices in which the middle factor is undeg., 199.
Ranks, possible, of the solutions of matrix equations of the first degree:
$AX=C$, 168;
$XB=C$, 171;
$AXB=C$, 173;
of sym. solutions of the sym. equation
$A'XA=C$, 176.
Ranks, possible, of the intersec. and join of two sps. of given ranks, 215, 464;
lying in or containing a given sp., or lying in one and containing another given sp., 217–8.
(*See also* Paratomy.)
Ranks, possible, of a spacelet which lies in a given sp. and:
has a given intersec. with a given sub-space of it, 218–9;
has a given intersec. with one given sub-space and does not intersect another, 226–7;
does not intersect a given sub-space, 218;
does not intersect two given sub-spaces, 220, 222;
all sps. being real, 221.
Rational integral functional matrices:
Equigr. transfns., 228–41.
Rank of a product in which an extreme factor matrix has rank equal to its passivity, 167.
Real matrices:
are non-extrav., have no cores, 379, 417;
intersecs., joins and normals are real, 505;
semi-unit, 322–5, 329, 330–2;
sym., 133, 295–303;
of rank 1, 133; definite, 300–3.
Join of two algebraically conjugate completely extrav. undeg. matrices, 437–40.
Reduction of a real undeg. matrix to an equiv. real semi-unit matrix, 414.
Real solutions of real sym. equations:
$X'X=A'A$, 337, 339;
$X'X=\ 0\ $, 345;
semi-real solutions of the real sym. equation
$X'X=A$, 361–2;
see also Real semi-unit matrices.
Real spacelets:
are non-extrav., have no cores, 432;
intersecs., joins and normals are real, 505;
mut. normal, are non-int. and compl., 432;
mut. orthog., are non-int., 509–10;
which do not intersect two given real sps., 221.
Join of a core with an anti-core, 437–40.
Orthotomy (or cross rank) of two real sps. which:
are non-int., 478;
are mut. compl., 480;
have a given intersec., 489;
have a given join, 506.
Paratomy and orthotomy (or cross rank), possible simultaneous values for
two arb. real sps. of given ranks, 508;

Real spacelets (*cont.*):
two entirely arb. real sps., 509.
Plenarily extrav. sp. is the join of its core and a real sp., 439.
Reciprocal of a square matrix: properties, 127–33;
detant of the recip., 114, 128;
rank of the recip., 130;
reciprocal of the recip., 128.
Reciprocal of a product of sq. matrices, 107.
Square matrix and its recip., relations between
anti-corresp. minor detants, 119, 128–9;
anti-corresp. matrices of minor detants, 117–8, 129;
corresp. matrices of minor detants, 121, 129–30.
Reciprocals of:
a minor compl. to a sq. minor in any matrix, 97;
a sq. minor of the recip. of a sq. matrix, 157;
sq. minors of a sym. matrix of order 4 and its recip., 152–6;
two co-joint complete matrices of the minor detants of a sq. matrix, 113–4.
Reduction of a matrix to a standard form by derangements of its rows:
any matrix, 94; compartite, 7;
sym. matrix, 140;
skew-sym. matrix, 145;
of a sym. matrix to a standard form by sym. derangements and additions of rows, 291.
Reduction of a matrix in Ω with constant els. by equigr. transfns. in Ω:
to a bipartite matrix one of whose parts is a given undeg. sq. minor or non-vanishing el.
by unitary transfns., 242–52, 264–9;
to a compartite matrix whose non-zero parts are all undeg. sq. matrices
by unitary transfns., 252–60, 269–70;
by corresp. non-unitary transfns., 260–4, 271.
Reduction of a matrix in Ω with constant els. to a quasi-scalar matrix or a standard form by equigr. transfns.:
by derangements and unitary transfns. in Ω:
any matrix, 277;
sym. matrix, 292;
skew-sym. matrix, 307;
by unrestricted equigr. transfns. in Ω:
any matrix, 272, 278;
real sym. matrix, 298;
skew-sym. matrix, 308;
by unrestricted equigr. transfns. (not in Ω):
sym. matrix, 293.
Reduction of a matrix with constant els. to a standard form by a semi-unit and a general equigr. transfn.:
any matrix, 389;
undeg. matrix, 380;
completely extrav., 382;
plenarily extrav., 383;
by two semi-unit transfns.:
any matrix, 389;
sym. matrix, 394.
Reduction:
of an undeg. matrix to an equiv. undeg. matrix whose long rows are mut. orthog., 412–5;
to join of a core and a semi-unit matrix, 413–4;
of a non-extrav. (or real) undeg. matrix to an equiv. semi-unit (or real semi-unit) matrix, 414;
of two mut. normal undeg. matrices to two respectively equiv. matrices of standard forms, 401;
of a completely extrav. undeg. matrix to an equiv. undeg. matrix of standard form, 349–52, 435–7;
of a plenarily extrav. undeg. matrix to an equiv. undeg. matrix of standard form, 439.
Reduction of a quadratic form to a sum of squares, 290.
Regular minor determinant of a matrix with constant elements: defined, 515;
has a primary subdetant which is reg., and a primary superdetant which is reg., 94, 515;
diagonal, of a sym. matrix, 135–9;
of a skew-sym. matrix, 142–4.
Properties of any reg. minor detant and:
its primaries and primary superdetants, 90–8, 157–64, 515–6;
its primary subdetants, 70–3, 519–20;
its primary superdetants, 66–70, 96, 516–7, 519–20;
of a reg. simple minor detant and:
its primaries, 37–66, 76–7, 518, 519;
of a reg. minor detant of maximum rank and:
its primaries, 85–9, 98–106, 519, 520.
Successive regular leading diag. minor detants, 252, 260, 269;
in an arranged matrix, 94, 138, 143.
Relations (identical in the els.) between the els. and minor detants of any matrix:
Fundamental identity, which serves to express every el. as a rat. function of any one reg. minor detant Δ and the els., primaries and primary superdetants of Δ, 90–4, 157;
utility, 515–7.
Identities which give all connections between the short rows of the matrix when it is undeg., and which serve to express every el. as a rat. function of degree 1 of any one reg. simple minor detant Δ and the els. and primaries of Δ, 37–46;
utility, 62, 518.
Relations between the simple minor detants which serve to express every simple minor detant D as a rat. function of degree 1 of any one reg. simple minor detant Δ and the primaries of Δ, and which give the values of determinants of the primaries of a simple minor detant, 46–66;
utility, 62–3, 518–9.
Relations between the simple minor detants which give the sums of terms

Relations (*cont.*):
formed from a product ΔD of two simple minor detants by replacing s short rows of Δ by rows of D, and the s rows taken from D by rows of Δ, 73–5.
Identities which serve to express every superdetant D of a reg. minor detant Δ as a rat. function of degree 1 of Δ and the primary superdetants of Δ contained in D, and which give the values of determinants of primary superdetants, 66–70;
utility, 519–20.
Identities which serve to express every subdetant D of a reg. detant Δ as a rat. function of degree 1 of Δ and the primary subdetants of Δ containing D, and which give the values of determinants of primary subdetants, 70–3;
utility, 519–20.
See Standard identities.
Relations (not identical in the els.) between the els. and minor detants of order r of a matrix whose rank is r:
Relations which serve to express every el. as a rat. function of degree 1 of any one reg. minor detant Δ of order r and the els. and primaries of Δ, 85–9;
utility, 518–9.
Relations between the minor detants of order r which serve to express every minor detant D of order r as a rat. function of degree 1 of any one reg. minor detant Δ of order r and the primaries of Δ, 98–106;
utility, 518–9, 520.
See Standard equations.
Representation by matrices of:
bilinear and quadratic forms, 234–5, 290, 301;
rotations of a rigid body, 330;
spacelets of homog. space, 78–9, 208–9, 422;
two sps., their intersec. and join, 214–5, 423;
mut. normal sps., 431, 433;
standard representations of completely and plenarily extrav. sps., 435–9.
(*See* Join.)
Restrictions on the rank of a matrix product and the ranks of its factor matrices, 177.
Resultant of successive equigr. transfns., 231;
derangements, unitary and quasi-scalar transfns., 232–3;
equigr. transfns. of the parts of a compartite matrix, 234;
unitary transfns. of a special form, 240.
Rotations of a rigid body: represented by sq. semi-unit matrices of order 3, 330;
about O, resolved into three component rotations about rectangular axes through O, 331;
about fixed axes through O, equiv. to rotations about axes moving with the body, 331–2.
Rows: extrav., non-extrav., orthog.; *see* Matrices (one-rowed).
Rows: connected, connections between, 37–45, 46, 224;
non-extrav., at unit intensity, 414;
standard arrangements, 7, 94, 140, 145.
Rows, horiz. and vert.: Circumstances under which two matrices are horizontally or vertically
equiv., 205;
normal or orthog., 398.
Horiz. and vert. extravagance, 385.
Rows, short: Connections between the short rows of an undeg. matrix, 37–46.
Rows, long: represent points, 85.
Circumstances under which two undeg. matrices are
equiv., 75;
normal or orthog., 84, 398.
Extravagance of an undeg. matrix, 379.
Identical relations between the simple minor detants of a matrix with m long rows, 63, 73.
Semi-real undeg. matrices, 360.
Undeg. matrices whose long rows are mut. orthog., 412–5, 440–6.

Scalar matrix: occurring as a constituent of a compound matrix, 4.
Scalar transformations, 233.
Second degree: *see* Matrix equations.
Self-conjugate: *see* Symmetric.
Self-conjugate factors of a sym. matrix of order 2, 366–7, 368–9.
Semi-definite (real sym.) matrix: is a definite (real sym.) matrix which is degen., 303.
Semi-real undeg. matrix: is one in which every long row is either real or purely imaginary; semi-real solutions of the real sym. equation $X'X=A$, 360–2.
Semi-unit matrices: 320–32; defined, 320;
determination of all, 325–7; of all real, 322–4;
include all derangements of unit matrices, 321.
Semi-unit matrix:
derangements and long minors are all semi-unit matrices, 321;
enlargement by the addition of long rows, 326–7;
(real), 324;
inverse of, is the conj. matrix, 321;
number of arbitrary parameters in, 323, 326.
Reduction of any undeg. matrix to an equiv. sim. undeg. matrix which is the join of a core and a semi-unit matrix, 412–5;
the semi-unit matrix being real when the given matrix is plenarily extrav., 439, 408.
Reduction of a non-extrav. (or real) undeg. matrix to an equiv. semi-unit (or real semi-unit) matrix, 414.
Semi-unit matrix, square:
compound, with four real and purely imaginary constituents, 327;
most general of order 2, 328;
of order 3, 329;
reciprocal of, 328.
Rotations represented by one of order 3, 330–2.
Semi-unit transformations:
between two sps. having the same rank and the same extravagance, 426, 533;

Semi-unit transformations (*cont.*):
of a sp. and its core, 428–9;
of two mut. normal sps., 483.
Equigr. transfns. between two matrices with constant els. which have the same rank and the same (horiz. or vert.) extravagance, 380, 390, 391, 531.
Reduction of a matrix to a standard form by two equigr. transfns. of which one is semi-unit:
any matrix, 388–9;
undeg. matrix, 379–80;
completely extrav., 381–2, 349, 435;
plenarily extrav., 383, 439;
two mut. normal undeg. matrices, 401;
by two semi-unit transfns.:
any matrix, 389;
sym. matrix, 393–4.
(*See* Invariants.)
Short rows: Connections between the short rows of an undeg. matrix, 37–46.
Sign of equivalence, 78–9, 208–9.
Signants and signature of a real sym. matrix, 295–303.
If one signant is 0, the matrix is definite, 300.
Similar matrices:
having the same rank and the same (horiz. or vert.) extravagance, 390.
Equigr. transfns. between two, 230–1;
derangements, unitary and quasi-scalar transfns., 232–3.
Equivalences of two sim. sq. matrices, 206.
Similar undegenerate matrices:
equiv., 75–8; equi-extrav., 380;
sq. or quasi-scalar, whose detants are equal in value, 278–9.
Reduction of an undeg. matrix to an equiv. undeg. matrix whose long rows are mut. orthog., 412–5;
to the join of a core and a semi-unit matrix, 413–4;
of a non-extrav. (or real) undeg. matrix to an equiv. semi-unit (or real semi-unit) matrix, 414.
Simple minor determinants:
of two equiv. undeg. matrices, 76–8;
of two mut. normal undeg. matrices, 403;
of $[1, a]_m^{m,n}$ and $[a, 1]_m^{n,m}$, 395–6;
primaries of, 13–14, 38, 41, 63, 518–9;
relations between, 46–66, 73–5, 518–20.
See Primaries, Relations (identical).
Simple minor matrices, sq. and undeg., 399–400, 401.
See Regular simple minor determinants.
Simultaneous values of the extravagances of:
two non-int. sps., 446–50;
two sps. having a given intersec., 499;
two sps. having a given join, 505;
any number of uncon. sps., 449.
See Extravagance.
Simultaneous values of the paratomy and orthotomy (or cross rank) of:
two sps. of given ranks, 506;
two arb. sps., 508;
two real sps. of given ranks, 508;
two arb. real sps., 509.
Simultaneous values of the paratomy, orthotomy and cross rank of:
two arb. sps., 508;
two arb. real sps., 509.
Skew-conjugate matrices, 141.
Skew-symmetric determinant:
of odd order, vanishes, 142, 526;
of even order, is sq. of its Pfaffian, 525;
bordered, is a product of two Pfaffians, 525;
symmetrically bordered, 529.
Skew-symmetric matrices: 141–5, 252, 260, 264–71, 303–8, 521–9;
equigr. transfns., 264–71, 303–8;
general properties, 141–5.
Condition that two may have same rank, 308.
Skew-symmetric matrix:
complete matrix of its minor detants, 141;
diag. els., are all 0's, 141;
diag. minor detants, 142–5;
those of odd order vanish; matrix of rank r has a reg. diag. detant of order r, 142;
is difference of two mut. conj. matrices, 141;
Pfaffian of one of even order, 521–9;
rank (is always even), 142–3;
reciprocal, 527–9;
reduction by sym. derangements, 145.
Reduction of a skew-sym. matrix in Ω by sym. equigr. transfns. in Ω:
to a bipartite matrix, 264–9, 242–52;
to a compartite matrix whose non-zero parts are all undeg. skew-sym. matrices
by unitary transfns., 269–70, 260;
non-unitary, 271, 264;
to a standard form
by equigr. transfns., 308;
by derangements and unitary transfns., 307.
See Skew-symmetric determinant.
Solutions of matrix equations:
of the first degree, 168–77;
of the second degree, 309–77.
Methods of determining all solutions of $X_1X_2 \ldots X_n = C$, 168–205, 309.
See Matrix equations.
Spacelet or sub-space of homogeneous space:
rank, is number of uncon. points in it, is greater by 1 than the number of its dimensions, 78–9, 208–9, 422;
represented by any one of a system of mut. equiv. matrices, 78–80, 208–9, 422;
by an undeg. matrix with mut. orthog. long rows, or as the join of uncon. mut. orthog. points, 412–5;
as the join of a core and a semi-unit matrix, 413–4, 427;
by a semi-unit matrix, 414;
as the join of two mut. orthog. non-int. sps., 427, 428, 484, 511;
as the join of two mut. orthog. sps., 511.
Spacelets:
complementary, 219; 423, 433, 478–80, 509;
completely extrav., 426–7; 432, 434–40, 453, 459;
connected, connections of, 225–7;
containing a given sp., 79, 422; 210, 216–7; 456–62, 470–2, 480, 490;
cores of, 427–30; 432–3, 485, 491, 496, 513;
cross rank of, 464; 465–509;

Spacelets (*cont.*):
extravagant, extravagances of, 425–7; 428–40, 446–62, 466, 470, 472, 474, 487, 499, 505;
having a given intersection, 480, 490;
having a given join, 480, 505;
incident, incidences of, 219, 422; 432, 466–7, 479, 501–2, 507;
invariants of, 165–7, 426, 463–4;
joins of, 209, 219, 423; 209–19, 422–35, 484–5, 489, 491, 495, 509–14;
lying in a given sp., 79, 422; 216–7, 220, 226, 429, 449, 450–6, 468–70, 477, 480–9, 512;
non-extrav., 425; 414, 426, 428, 432, 434–5, 449–50, 453, 459, 469–74, 477–8, 480, 489, 505;
non-incident, 219;
non-intersecting, 210, 422; 211, 218, 474–8, 446–50, 474–8, 509, 511;
normal, 85, 423–4; 427, 431–5, 465;
orthogonal, 84, 423; 426–8, 432, 467, 476, 487–9, 501–2, 509–14;
orthotomy of, 464; 465–509;
ranks of, 78–9, 208–9, 422; 215–22, 226–7;
paratomy of, 463; 217–8, 463–509;
plenarily extrav., 426–7; 431–2, 434–5, 439, 453, 459;
plenums of, 431–2; 480–7, 490–500;
real, 426; 221, 432, 437–40, 478, 480, 489, 505, 508–10;
semi-unit transfns. of, 426, 428–9;
unconnected, 220, 225–6, 423; 426–7, 484, 494.
(*See* sub-heads.)
Square matrices: general, 107–33, 157–8;
bordered, 123–7;
co-joint, 108–23;
quasi-scalar, 4–5, 277–9, 292;
reciprocals of, 127–33.
See sub-heads;
see also Symmetric, Skew-symmetric;
see also Determinants.
Square matrices:
Determinant of a complete matrix of the minor detants of a sq. matrix, 112.
Equigr. transfns. between two sim. undeg. sq. matrices whose detants are equal, 279.
Equivalences of two sim. sq. matrices, 206–7.
Matrix of primary superdetants in a sq. matrix, 157–8.
Products of sq. matrices, 107–8.
Reciprocals of derangements of a sq. matrix, 131.
Special sq. matrices, 8–13, 240–1, 437.
Undeg. sq. matrix whose detant is ± 1, 280.
Square minors: Recip. of a sq. minor of the recip. of a sq. matrix, 157.
Reciprocals of sq. minors of a sym. matrix of order 4 and its recip. 152–4.
See Diagonal minor, Minor determinant.
Square semi-unit matrices:
compound, with four real and purely imaginary constituents, 327;
general, of orders 2 and 3, 328–9;
include all derangements of unit matrices, 321;
Square semi-unit matrices (*cont.*):
particular, of orders 3 and 4, 324–7;
reciprocals, 328.
Rotations represented by one of order 3, 330–2.
Standard arrangements of rows:
in a compartite matrix, 7;
in any matrix, 94;
in a sym. matrix, 140;
in a skew-sym. matrix, 145.
Standard equations, giving the relations between the minor detants of order r of a matrix of rank r, and having the form

$$\Delta^{\rho+\sigma-1}D = (\alpha)^{\rho}_{\rho}\,(\beta)^{\sigma}_{\sigma},$$

where Δ and D are any two minor detants of order r, and the α's and β's are horiz. and vert. primaries of Δ, 101, 102, 103, 520.
Standard form of a compartite matrix, 7.
Standard form to which any sym. matrix can be reduced by sym. derangements and additions of rows, 291–2.
Standard forms to which any matrix can be reduced by derangements, 7, 94, 140, 145.
See Standard arrangements.
Standard forms to which a matrix with constant els. can be reduced:
by equigr. transfns. 272, 280, 298, 303;
by a semi-unit and an equigr. transfn., 379–80; 388–9;
by two semi-unit transfns., 389, 394.
See Reduction.
Standard identities, giving relations between the minor detants of any matrix, and having the form

$$\Delta^{s-1}D = (\delta)^{s}_{s}:$$

Relations between simple minor detants, Δ and D being two simple minor detants, and the δ's being the primaries of Δ formed with the rows of D, 48, 51, 55; 57, 59, 61; 63, 65; 519–20.
Relations between a minor detant Δ and its superdetants, D being any superdetant of Δ, and the δ's being the primary superdetants of Δ which lie in D, 66, 67; 519–20.
Relations between a minor detant Δ and its subdetants, D being any subdetant of Δ, and the δ's being the primary subdetants of Δ which contain D, 70, 71; 519–20.
Standard representations of:
a completely extrav. sp., 435, 437;
a plenarily extrav. sp., 439;
two mut. normal matrices, 401;
two mut. normal sps., 433.
Straight line in homogeneous space: is the join of two uncon. points, is a sp. of rank 2, 78–80.
Subdeterminants, primary: defined, 18;
determinants of, 70–3;
regular, 94, 138, 143;
see Primary, Standard identities.
Sub-spaces, 78–9: *see* Spacelets.

Successive:
parts of a compound matrix in standard form, 7;
reg. leading diag. minor detants
in an arranged matrix, 94, 273–7;
sym., 138, 280–92;
skew-sym., 143, 303–7;
in an unarranged matrix, 252–64;
sym. or skew-sym., 269–71.
Superdeterminants, primary: defined, 17;
determinants of, 66–70;
matrices of, 90–8;
regular, 94, 138, 144.
See Primary, Standard identities.
Sylvester's Identities, 66–70.
Symmetric matrices: 29–36, 133–41, 145–56, 158–64, 194–208, 264–71, 280–95, 320–65, 393–5.
Condition that two shall have the same rank, 294;
that two real shall be equisignant, 298.
(*See also* Quasi-scalar matrices.)
Symmetric matrix:
defined, 133; general properties, 135–41;
diag. minor detants, 135–41;
matrix of rank r has a reg. diag. detant of order r, 135;
is sum of two mut. conj. matrices, 141;
rank, 136–7;
reduction by sym. derangements, 140–1;
and additions of rows, 291–2.
Symmetric matrix in Ω: Sym. equigr. transfns. in Ω, 264–71, 280–93.
Reduction to a bipartite matrix one of whose parts is a given reg. diag. minor, 264, 267;
special cases, 265–7, 268–9.
Reduction to a compartite matrix whose non-zero parts are undeg. sym. matrices:
any matrix by
unitary transfns., 269;
non-unitary, 271;
specially arranged matrix by
unitary transfns., 280, 283;
non-unitary, 282, 289.
Reduction to a standard form by derangements and unitary transfns., 292.
Symmetric matrix in Ω: Sym. equigr. transfns. not in Ω, 293–5, 393–4.
Reduction to a standard form
by equigr. transfns., 293;
by semi-unit transfns., 393–4.
Symmetric matrix, real: 295–300, 360–2.
Signants and signature, 297–300.
Reduction to a standard form by real sym. equigr. transfns., 298.
Resolution into semi-real factors, 360–2.
Symmetric matrix, real and definite: 300–3;
diag. detants of order s all have same sign, 303;
standard form, product of two real matrices, 302;
sym. matrix of minor detants is definite, 303.
Symmetric matrix of rank 1: 133–5, 352–5, 366, 368, 374, 376;
diag. els., do not all vanish, all have the same sign when the matrix is real, 133;
Symmetric matrix of rank 1 (*cont.*):
expressed as a sym. product of two one-rowed matrices, 134, 352–5.
Symmetric matrix (extravagance): 393–5.
Formulae for matrix of given extravagance, 393–4.
Possible values of the extravagance, 393.
Powers of a non-extrav. sym. matrix, 394–5.
Symmetric matrix (factors):
of rank 2, 373–7; of rank r, 294, 355–9;
of order 2, 365–9; (self-conj. factors), 366, 368–9.
Symmetric matrix (minors):
of order 3, 145–6, 100–1, 374, 377;
of order 4, 147–56, 161–4, 375, 377.
Symmetric matrix:
Possible ranks of one which contains
a given diag. minor, 29–32;
a given zero diag. minor, 32–6.
Possible ranks of a diag. minor of one which has a given rank, 31;
compl. to a given zero diag. minor, 36.
Symmetric matrix equations:
of the first degree, 176–7;
of the second degree, 320–65;
of any degree, 194–205.
See Matrix equations.
Symmetric matrix products: Possible ranks of the product matrix and the factor matrices, 194–205;
any number of factors, 205;
two factors, 196; 194;
three factors, 198; 196.
(*See also* Matrix products.)
Systems of linear alg. equations:
Conditions for the equivalence of two, 80–3, 207–8.
General solutions, 168–70.
Uncon. mut. orthog. solutions of any system of homog. linear equations, 440–6.

Transformations (equigr.) of matrices: 228–308, 531–2;
unitary, quasi-scalar, scalar, 233;
unilateral, 75–6, 205, 531;
unilaterally semi-unit, 380, 390, 391; 379–80, 381–2, 383, 388–9, 401, 530;
semi-unit, 389, 531.
See Equigradent transformations, Reduction.
Transformations (semi-unit) of spacelets, 426, 428–9, 433, 435, 437, 439, 533.

Unconnected matrices: defined, 224;
join of mut. orthog., 420, 427;
mut. orthog. one-rowed, 412–5, 440–6.
Mut. orthog. non-extrav. matrices are uncon., 426.
Unconnected points:
A sp. of rank r is join of r uncon. points, 78;
is join of r mut. orthog. uncon. points, 412–5.
Unconnected solutions, mut. orthog., of any system of homog. linear alg. equations, 440–6.

Unconnected spacelets: defined, 220, 225–6, 423;
join of mut. orthog. uncon. sps., 427;
possible extravagances of uncon. sps., 449;
set of 5 uncon. sps., 494;
sets of 3 uncon. sps., 215, 423; 431; 484;
two uncon. sps. are non-int., 210–11, 226, 423.
Unilateral equigr. transfns., 530.
Unilaterally semi-unit equigr. transfns., 530.
Unit intensity of a non-extrav. row or point, 414.
Unit matrix: mut. conj. factors of, 320.
Unitary transformations, 233, 280, 241–60, 264–70.
See Equigradent transformations, Reduction.
Unrestricted equigr. transfns., reducing a matrix to a standard form, 272; sym., 293, 308.

Vertical extravagance of a matrix, 385–93; *see* Extravagance, Extravagant.
Vertical primaries: defined, 13; *see* Primaries.
Vertically:
equivalent matrices, 205–7;
have the same vert. extravagance, 386;
normal matrices, 398;
have the same vert. extravagance, 407;
all matrices vertically normal to a given matrix are vertically equiv., 398;
orthogonal matrices, 398.
See sub-heads.

Way: $(r-1)$-way spacelet has $r-1$ dimensions and rank r, 78.

Zero extravagance, 379, 389;
see Non-extravagant.
Zero matrices: notation, 4; connections, 224; factors, 344–52, 409–10.
Zero order of a matrix, 5, 363.
Zero spacelets, sps. of rank 0: defined, 79;
have no intersecs. and no connections, 211, 225.